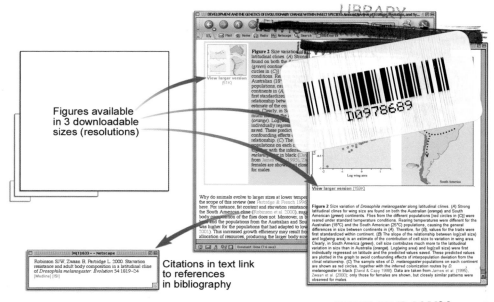

Figures available in 3 downloadable sizes (resolutions)

Robinson SJW, Zwaan B, Partridge L. 2000. Starvation resistance and adult body composition in a latitudinal cline of *Drosophila melanogaster*. *Evolution* 54:1819–24 [Medline] [ISI]

Citations in text link to references in bibliography

References in Annual Reviews chapter bibliography link out to sources of cited articles online

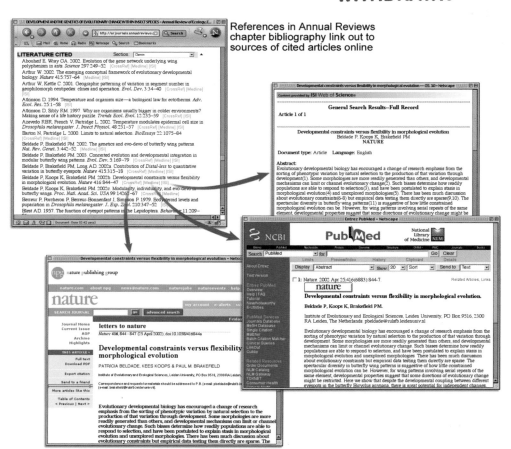

ANNUAL REVIEW OF
ECOLOGY, EVOLUTION, AND SYSTEMATICS

ANNUAL REVIEW OF ECOLOGY, EVOLUTION, AND SYSTEMATICS

VOLUME 36, 2005

DOUGLAS J. FUTUYMA, *Editor*
State University of New York, Stony Brook

H. BRADLEY SHAFFER, *Associate Editor*
University of California, Davis

DANIEL SIMBERLOFF, *Associate Editor*
University of Tennessee

www.annualreviews.org   science@annualreviews.org   650-493-4400

ANNUAL REVIEWS
4139 El Camino Way • P.O. Box 10139 • Palo Alto, California 94303-0139

ANNUAL REVIEWS
Palo Alto, California, USA

International Standard Serial Number: 1543-592X
International Standard Book Number: 0-8243-1436-0
Library of Congress Catalog Card Number: 71-135616

All Annual Reviews and publication titles are registered trademarks of Annual Reviews.

♾ The paper used in this publication meets the minimum requirements of American National
Standards for Information Sciences—Permanence of Paper for Printed Library Materials.
ANSI Z39.48-1992.

Annual Reviews and the Editors of its publications assume no responsibility for the
statements expressed by the contributors to this *Annual Review*.

TYPESET BY TECHBOOKS, FAIRFAX, VA
PRINTED AND BOUND BY MALLOY INCORPORATED, ANN ARBOR, MI

Annual Review of Ecology, Evolution, and Systematics
Volume 36, 2005

CONTENTS

ERRATA
An online log of corrections to *Annual Review of Ecology,
Evolution, and Systematics* chapters may be found at
http://ecolsys.annualreviews.org/errata.shtml

Related Articles

From the *Annual Review of Earth and Planetary Sciences*, Volume 33 (2005)

Feathered Dinosaurs, Mark A. Norell and Xing Xu

Molecular Approaches to Marine Microbial Ecology and the Marine Nitrogen Cycle, Bess B. Ward

Evolution of Fish-Shaped Reptiles (Reptilia: Ichthyopterygia) in Their Physical Environments and Constraints, Ryosuke Motani

The Ediacara Biota: Neoproterozoic Origin of Animals and Their Ecosystems, Guy M. Narbonne

From the *Annual Review of Entomology*, Volume 50 (2005)

Pheromone-Mediated Aggregation in Nonsocial Arthropods: An Evolutionary Ecological Perspective, Bregje Wertheim, Erik-Jan A. van Baalen, Marcel Dicke, and Louise E.M. Vet

Egg Dumping in Insects, Douglas W. Tallamy

The Evolution of Male Traits in Social Insects, Jacobus J. Boomsma, Boris Baer, and Jürgen Heinze

Evolutionary and Mechanistic Theories of Aging, Kimberly A. Hughes and Rose M. Reynolds

Ecology of Interactions Between Weeds and Arthropods, Robert F. Norris and Marcos Kogan

Evolutionary Ecology of Insect Immune Defenses, Paul Schmid-Hempel

From the *Annual Review of Environment and Resources*, Volume 30 (2005)

Archaeology and Global Change: The Holocene Record, Patrick V. Kirch

Feedback in the Plant-Soil System, Joan G. Ehrenfeld, Beth Ravit, and Kenneth Elgersma

The Role of Protected Areas in Conserving Biodiversity and Sustaining Local Livelihoods, Lisa Naughton-Treves, Margaret Buck Holland, and Katrina Brandon

From the *Annual Review of Microbiology*, Volume 59 (2005)

Yeast Evolution and Comparative Genomics, Gianni Liti and Edward J. Louis

Biology of Bacteriocyte-Associated Endosymbionts of Plant Sap-Sucking Insects, Paul Baumann

ANNUAL REVIEWS is a nonprofit scientific publisher established to promote the advancement of the sciences. Beginning in 1932 with the *Annual Review of Biochemistry*, the Company has pursued as its principal function the publication of high-quality, reasonably priced *Annual Review* volumes. The volumes are organized by Editors and Editorial Committees who invite qualified authors to contribute critical articles reviewing significant developments within each major discipline. The Editor-in-Chief invites those interested in serving as future Editorial Committee members to communicate directly with him. Annual Reviews is administered by a Board of Directors, whose members serve without compensation.

Annu. Rev. Ecol. Evol. Syst. 2005. 36:1–21
doi: 10.1146/annurev.ecolsys.36.102003.152640
First published online as a Review in Advance on July 18, 2005

THE GENETICS AND EVOLUTION OF FLUCTUATING ASYMMETRY

Larry J. Leamy[1] and Christian Peter Klingenberg[2]

[1]*Department of Biology, University of North Carolina, Charlotte, North Carolina 28223; email: ljleamy@email.uncc.edu*
[2]*Faculty of Life Sciences, The University of Manchester, Manchester M13 9PT, United Kingdom; email: cpk@manchester.ac.uk*

Key Words developmental instability, dominance, epistasis, heritability

■ **Abstract** Variation in the subtle differences between right and left sides of bilateral characters, or fluctuating asymmetry (FA), has long been considered to be primarily environmental in origin, and this has promoted its use as a measure of developmental instability (DI) in populations. There is little evidence for specific genes that govern FA per se. Numerous studies show that FA levels in various characters are influenced by dominance and especially epistatic interactions among genes. An epistatic genetic basis for FA may complicate its primary use in comparisons of DI levels in outbred or wild populations subjected or not subjected to various environmental stressors. Although the heritability of FA typically is very low or zero, epistasis can generate additive genetic variation for FA that may allow it to evolve especially in populations subjected to bottlenecks, hybridizations, or periods of rapid environmental changes caused by various stresses.

INTRODUCTION

What role does genetics play in producing variation in the subtle differences between right and left sides of bilateral characters known as fluctuating asymmetry (FA)? It might be supposed that the answer to this question should be clear, because there have been a number of quantitative genetic studies of FA in various characters since the classical artificial selection experiment was conducted by Mather (1953) more than 50 years ago. Most show the heritability of FA to be quite low, but significant heritabilities have been found and therefore the extent of additive genetic variation for FA across various characters is uncertain (Fuller & Houle 2003, Leamy 1997, Markow & Clarke 1997, Palmer 2000, Palmer & Strobeck 1997). Few studies have attempted to assess the importance of nonadditive (dominance and epistatic) genetic effects in producing FA, and of those conducted, results have been mixed (reviewed in Leamy 2003). As a consequence, FA remains an elusive character whose genetic architecture is still largely unknown.

FA generally is not of interest for its own sake, but rather for what it is thought to assess: developmental instability (DI) in populations. DI results from internal or external stressors that disturb the development of structures along their normal developmental pathway in a given environment and produce developmental "noise" (Palmer 1994, Waddington 1957, Zakharov 1992). Thus, DI can be thought of as variation around the expected (target) phenotype that should be produced by a specific genotype in a specific environment. Developmental noise affects left and right sides of bilateral characters separately and therefore produces FA, but the eventual level of FA in a character depends on how successful developmental stability (DS) processes are in reducing this noise (Zakharov 1992). Although developmental noise theoretically is purely environmental in origin, DS has long been assumed to be at least partly under genetic control (Mather 1943, 1953; Palmer 1994; Waddington 1940, 1942, 1957; Zakharov 1992).

It is unfortunate that we do not yet fully understand the nature and extent of the genetic basis of FA. This kind of knowledge is essential if we are to use FA properly as a measure of DI and know, for example, whether sufficient genetic variance in DI exists so that it can respond to selection. FA in certain characters also must have genetic variability if it is used as a cue for female choice of males in "good genes" models of sexual selection (Møller & Pomiankowski 1993). Knowledge of the genetic basis of FA would help us understand whether there are organism-wide developmental mechanisms that act to ensure symmetry of multiple bilateral characters or whether DI is character-specific (Gangestad & Thornhill 1999, Leamy 1993, Polak et al. 2003). More generally, a better understanding of the genetic architecture of FA should provide a much-needed perspective for sorting out the sometimes unexpected or contradictory patterns of differences in FA levels between populations subjected or not subjected to various genetic or environmental stressors or among individuals differing in fitness.

In this review, we first explain how FA and DI are assessed, and provide a brief, genetical perspective of the relationship of FA to stress, fitness, and integration in populations. We then describe the theoretical origin of FA and its implications for the genetic basis of FA. This is followed by an assessment of genetic studies of FA. Based on inferences from the assumed origin of DI and evidence especially from recent genetic studies, we conclude that it is likely there are no genes that govern FA per se, and develop the hypothesis that FA levels in various characters are influenced by dominance and especially epistatic interactions among a number of genes affecting these or other characters. We use this hypothesized genetic architecture as a basis to discuss the implications of the use of FA in ecological and evolutionary studies and offer some final thoughts about the kinds of studies that would be appropriate to test this hypothesis.

FLUCTUATING ASYMMETRY

FA is perhaps easiest to visualize at the population level as variation of differences in left-minus-right sides (L − R) of a bilateral character. Individual values of FA are most often assessed by unsigned left-minus-right-side differences,

|L − R|, although a number of other indices of FA are in common use (Palmer & Strobeck 2003). Besides FA of various size measures, methods recently have been developed to generate shape FA measures from the results of the Procrustes procedure applied to digitized landmark points on bilateral structures such as wings or mandibles (Klingenberg et al. 2002, Klingenberg & McIntyre 1998). The precision of measurement of subtle asymmetries has improved in recent years with better technology and especially with an increasing realization of the importance of minimizing measurement error (Palmer 1994). Most investigators now repeatedly measure both left and right sides of each character in order to obtain estimates of the extent of measurement error and to test for the significance of FA.

Beyond FA, two other kinds of asymmetries are sometimes found in bilateral characters. Directional asymmetry (DA) is fairly common and occurs when one side consistently differs from the other side, whereas antisymmetry (AS) is rarer and is characterized by a mixture of left- and right-biased individuals (Palmer & Strobeck 1986, 2003). It has been debated whether these asymmetries may reflect increased DI (Graham et al. 1993, Klingenberg 2003b), and in fact some empirical studies have shown changes from FA to either DA (Leamy et al. 1999a) or AS (McKenzie & Clarke 1988) in stressed populations. However, FA is the most commonly accepted measure of DI and we restrict our review to this type of asymmetry. Appropriate statistical procedures for adjusting for DA, AS and/or size effects on FA, and for testing for significant differences in FA among two or more groups, are discussed in some detail by Palmer & Strobeck (2003).

Because FA is calculated from only two sides (one degree of freedom) of a bilateral character in a given individual, it is a rather imprecise estimator of DI variability. Thus, sampling variation should result in many FA values at or near zero (the expected mean for FA) even in individuals with very high levels of DI. The correspondence between FA and DI has been parameterized by the hypothetical repeatability, $\Re$, which expresses the degree to which differences in FA reflect differences in DI. Theoretical formulas for calculating $\Re$ that are based on variation in FA have been derived by Whitlock (1996, 1998) and Van Dongen (1998), and estimates of $\Re$ from a number of FA studies have averaged about 0.08 (Gangestad & Thornhill 2003). This very low value serves as an important reminder of the difficulty of obtaining good estimates of DI in populations from measurement of FA in various characters.

Fluctuating Asymmetry and Stress

FA has been widely used to compare presumed DI levels in populations subjected or not subjected to a number of environmental stressors such as temperature, nutrition, radiation, chemicals, population density, noise, parasites, light conditions, predation risk, and habitat structure (reviewed in Hoffmann & Woods 2003, Møller & Swaddle 1997). The working hypothesis in such comparisons is that DI, and therefore its most easily observable outcome, FA, will be higher in the more stressed populations compared to the control or unstressed populations. In fact, this result often has been found (e.g., Graham et al. 2000, Pankakoski et al. 1992).

But for all of these studies suggesting that stress can increase FA, there are others where no or even opposite effects of stress on FA have been observed (Leamy et al. 1999a, Markow 1995, Woods et al. 1999). A number of factors such as uncertainty about the degree of stress imposed and the choice of characters and FA indices have been invoked as potential reasons for this inconsistent relationship between stress and FA (Hoffmann & Woods 2003).

In these studies it is often assumed that FA has no genetic basis and that differences in FA levels between stressed and unstressed populations are purely environmental in origin (Palmer 1994). Yet unless genetically identical (isogenic) populations have been used, FA levels may also partly reflect genotypic differences in response to stress (i.e., DS differences). In fact, environmental stresses in genetically variable populations may result in selection of certain genes that ameliorate the immediate stress but that themselves perturb developmental processes and result in increased asymmetry. An outstanding example of this effect is seen in populations of the Australian blowfly, *Lucilia cuprina,* that have developed resistance to exposure to various insecticides such as dieldrin, diazinon, or malathion but that also show increased levels of FA in bristle numbers compared to susceptible (nonexposed) populations (McKenzie 1997, 2003).

Fluctuating Asymmetry and Fitness

If FA truly measures DI, it is natural to imagine that individuals with the greatest levels of DI would be the least fit, and thus that there should be a negative relationship of FA with fitness or individual condition. Such a relationship has been reported in some studies (for example, Badyaev et al. 2000), and several meta-analyses have found a significant negative relationship, albeit a weak one, between FA and various fitness components (Møller & Thornhill 1998, Thornhill et al. 1999, Tracy et al. 2003). Some findings in the early 1990s suggesting that FA levels may provide a cue for female choice in sexual selection (Møller 1990, 1992; Møller & Hoglund 1991) also generated much interest in searching for a possible connection between FA and fitness. But this connection has not always been found in subsequent studies (reviewed in Simmons et al. 1999), and thus the precise role of FA in the sexual selection process is controversial (Tomkins & Simmons 2003). Many other studies also have not discovered a link between FA and various fitness components, suggesting extreme caution in the use of FA as an overall indicator of fitness (Clarke 1995a,b, 1998a, 2003; Leung & Forbes 1996).

Fluctuating Asymmetry and Integration

Many investigators have measured FA for multiple traits, and the question therefore arises whether the asymmetries are integrated or independent among traits. There are two different contexts in which the FA of multiple traits has been studied. The majority of studies have used FA of multiple traits to quantify organism-wide DI as a measure of individual quality or exposure to stress, whereas others use FA as a tool to investigate developmental integration among traits.

If FA is to be used as a measure of organism-wide DI, then clearly there should be a correspondence between the amounts of FA of different traits. Individuals with higher DI should be generally more prone to developmental noise in all their traits, and therefore, it should be possible to use a combined index for FA in multiple traits to increase the precision of the estimate of organism-wide DI (Lens et al. 2002, Leung et al. 2000). This reasoning assumes that there is a positive correlation between the amounts of FA in different traits, that is, of the unsigned asymmetries (absolute values of the trait asymmetries). This type of correlation is traditionally called the individual asymmetry parameter (IAP) (Clarke 1998b, Leamy 1993, Polak et al. 2003). In order to provide independent evidence on DI, the traits should be developmentally unrelated, which means they should be chosen from different parts of the organism. The correlations of FAs found in empirical studies are generally weak but tend to be greater than zero (Polak et al. 2003). Such a correlation, indicating organism-wide DI, is of critical importance for the use of FA as an indicator of exposure to stress, individual quality, or fitness in good genes models of selection (Lens et al. 2002, Møller & Swaddle 1997, Tomkins & Simmons 2003).

A different use of correlated FA is inferring whether there are developmental interactions between traits (Klingenberg 2003a, 2004b). Because FA is due to random perturbations in development, the signed asymmetries (signed values of the trait differences between left and right sides) of two traits will be correlated only if there is a developmental interaction that can transmit the effects of perturbations to both traits jointly. If the precursors of the traits develop separately, without signaling or other interactions among them, random perturbations cannot be transmitted between them and there will be no statistical association of the resulting asymmetries. Therefore, a correlation of the signed asymmetries between traits indicates interactions of the respective developmental pathways (Klingenberg 2003a, 2004b). Such correlations have been shown in a number of studies, and, as is to be expected, they have been limited to sets of traits that are anatomically and developmentally related (Klingenberg & Zaklan 2000; Klingenberg et al. 2001a, 2003; Leamy 1993).

This second use of correlated FA differs fundamentally from the preceding one. In studies of organism-wide DI, this second approach can be used as a preliminary step to assess whether the traits are developmentally independent, in which case the correlations between signed asymmetries should be zero.

ORIGIN OF FLUCTUATING ASYMMETRY AND BUFFERING

FA originates from small perturbations that produce a component of random noise in developmental processes (Klingenberg 2003b). At the molecular level, most cellular processes are stochastic—the way in which individual molecules bind to each other, take part in metabolic reactions, or are transported from one location to another are all in part chance events (McAdams & Arkin 1999). If a kind of molecule exists in many copies in a cell, then the random variability will not be apparent in the behavior of the cell as a whole, because the random variability will be averaged out over the large number of molecules present. If developmentally important molecules are present only in small numbers, however, random variation will be apparent at the cell, tissue, and morphological levels.

In particular, DNA is a type of molecule that is present in cells only in small numbers. In diploid organisms, single-copy genes (i.e., those that are not duplicated) are present in only two alleles per nucleus. Therefore, understanding the stochastic component in the dynamics of gene expression is particularly relevant to the study of developmental noise (Blake et al. 2003, Cook et al. 1998, Ozbudak et al. 2002, Raser & O'Shea 2004). Because regulation of gene activity is substantially through switching between "on" and "off" states by the formation and decay of macromolecular complexes, a component of random noise is generated as part of the normal cell function (Fiering et al. 2000). Cook et al. (1998) used a simplified mathematical model of gene switching to study the effects of gene dosage on the levels of gene products and found that the loss of a copy of the gene not only decreased the average level of the gene product in the cell, but above all, that it substantially increased the variability. Experiments with yeast showed that differences in promoter sequences, upstream activating sequences, and deletions of components of chromatin-remodeling complexes produce changes in the level of stochastic variation of gene expression that are not necessarily tied to the absolute level of gene expression (Raser & O'Shea 2004). Accordingly, there can be genetic regulation of variability in gene expression, and this variability itself can therefore evolve.

Nonlinear Developmental Mapping and Developmental Instability

The molecular origin of developmental noise is just one of the components required for a full understanding of DI. A different question is how this variability is translated into morphological outcomes like FA or, in other words, how noise is transmitted through the developmental system. The link to the genetics of FA is to ask how genetic variation in the system mediates this transmission of variability.

One approach is to use models of simple developmental processes such as diffusion-threshold models in which the parameters are each controlled by a separate gene (Nijhout & Paulsen 1997) in combination with a small component of random noise that is independent of the genotype (Klingenberg & Nijhout 1999). FA was simulated by running the model twice for each genotype, with separate values for the noise component, and calculating the difference between the two resulting trait values. The differences in the behavior of the system caused by the genetically controlled variation of model parameters can amplify or dampen the effects of the random noise. Therefore, the model generates genetic variation of FA through genetically mediated expression of perturbations that are themselves strictly nongenetic (Klingenberg & Nijhout 1999).

These conclusions do not rely on the specific nature of the particular model, but are valid for a broad range of nonlinear models. Because most developmental models are nonlinear, complex genetic behaviors like dominance and epistasis of the trait invariably emerge. These nonadditive effects are particularly important for the genetic architecture of FA in these models and can produce consequences such as an asymmetric response to selection on FA, with a greater response to selection for increased than for decreased FA (Klingenberg & Nijhout 1999).

A more general way to think of the developmental origin and genetic control of FA is in terms of a mathematical mapping from the genetic and environmental factors that influence a trait to the resulting phenotypic value, which can be called "developmental mapping" (Klingenberg 2003b, 2004a; Klingenberg & Nijhout 1999). This is related to the idea of a "genotype–phenotype map" (Wagner & Altenberg 1996) but explicitly includes nongenetic factors as well as the genotype (Rice 2002). These mapping functions will normally be nonlinear; that is, plots of a phenotypic trait against the genetic and environmental factors will be curved (if multiple factors are considered, the mappings will be curved surfaces). As a result, the slope of the mapping depends on the genetic and environmental factors. These slopes determine the change of the phenotypic value in response to a small perturbation of the factors, that is, the sensitivity to developmental noise and therefore DI. As a consequence of nonlinear developmental mapping, different genotypes can therefore have different DI and FA.

Because nonlinear developmental mapping is associated with nonadditive genetic effects, dominance and epistasis are also expected to be the prevalent features of the genetic architecture of FA (Klingenberg 2003b, 2004a; Nijhout & Klingenberg 1999).

Origins of Developmental Buffering

FA is the observable outcome of developmental noise as it has been "filtered" by the developmental system. It is possible that a substantial proportion of variability is not apparent because it has been absorbed by the action of developmental buffering.

Simulation and experimental studies have found that developmental systems are remarkably robust against perturbations (e.g., Bergman & Siegal 2003, Houchmandzadeh et al. 2002, von Dassow et al. 2000). This means that there is a considerable buffering capacity built into the networks of developmental interactions that set up morphological structures, which may be a major determinant of DS.

Moreover, from the perspective of developmental mapping, buffering and DS result from mapping functions that are relatively flat, for instance, a curve that reaches an asymptote or a surface with a plateau (Klingenberg 2003b, 2004a). Because flat mapping functions are associated with the absence of change in the mean phenotype in response to the genetic or other inputs, this perspective on buffering assigns a common origin to DS and canalization (e.g., Meiklejohn & Hartl 2002), a view that is not shared universally (e.g., Debat & David 2001). Canalization is the ability to develop the same phenotype despite genetic or environmental variability

(Klingenberg 2003b). Viewed from this perspective, genetic variation of buffering capacity is linked to genetic variation in the steepness of the developmental mapping function, which is again linked to dominance and epistasis in the genetic architecture of the phenotype under study (Klingenberg 2004a).

There have also been studies of buffering that have focused on particular mechanisms rather than on global properties of the developmental system as a whole. A specific gene that has received particular attention in this context is the heat-shock protein 90 (*Hsp90*; Queitsch et al. 2002, Rutherford 2000, Rutherford & Lindquist 1998, Sangster et al. 2004). Inhibition of *Hsp90* activity led to the regular occurrence of various morphological anomalies in various qualitative traits in *Drosophila* and *Arabidopsis* (Rutherford & Lindquist 1998, Queitsch et al. 2002). The specific anomalies produced in these experiments depended on the genetic background, and their prevalence could be increased by selection (Rutherford & Lindquist 1998), indicating an important role for genetic interactions. Further experiments yielded evidence for both genetic and epigenetic interactions of *Hsp90* with other genes (Sollars et al. 2003). *Hsp90* is an attractive candidate for a specific "buffering gene" because it encodes a molecular chaperone protein that takes part in stabilizing a variety of signal transduction and other cellular processes (Rutherford 2000). However, an empirical study of bristle counts and wing size traits in *Drosophila* found that inhibiting *Hsp90* activity pharmacologically or by mutation did not lead to increased FA or variation among individuals (Milton et al. 2003).

Further experimental studies are needed to test the generality of *Hsp90* as a potential buffering mechanism and to explore whether there are other mechanisms of this kind. In general, the relationship between the mechanisms that generate or buffer against variation within and between individuals is still largely unclear. A result emerging consistently from the available studies, however, is the central role of gene interaction in these processes.

GENETICS OF DEVELOPMENTAL STABILITY

Heritability of Fluctuating Asymmetry

Most empirical studies estimating the heritability of FA have yielded very low, nonsignificant values (for example, Leamy 1999, Pelabon et al. 2004), but occasional significant heritabilities have been found as well (Polak & Starmer 2001, Santos 2002, Scheiner et al. 1991, Thornhill & Sauer 1992). In an effort to test for a potential global heritability for FA, Møller & Thornhill (1997) performed a meta-analysis of 34 studies that yielded a significant heritability of FA that averaged 0.19 over a number of characters and taxa. But the estimates of heritabilities from the studies sampled often were rather poor because of various measurement, statistical, and/or experimental difficulties, and the meta-analysis itself suffered from several errors and misinterpretations (Fuller & Houle 2003, Leamy 1997, Markow & Clarke 1997, Palmer 2000, Palmer & Strobeck 1997). Other analyses show lower FA heritability values (Gangestad & Thornhill 1999, Whitlock &

Fowler 1997); the most recent by Fuller & Houle (2003) estimated the average heritability of FA to be just 0.026. Most investigators now appear to agree that there is very little additive genetic variation for FA in most characters, although in some cases this variation may be statistically significant.

It is important to discover whether even very low heritabilities exist for FA in various characters, because they may translate into moderate to high heritabilities of DI (Gangestad & Thornhill 2003, Houle 2000). Because FA is an imprecise estimator of DI, an unbiased heritability of DI must be obtained by dividing the heritability of FA in a given character by its repeatability, $\Re$ (Van Dongen 1998; Whitlock 1996, 1998). Unfortunately, heritabilities of DI estimated in this manner have not been very informative because they have varied erratically from less than 0 to well over 1.0 (Houle 1997). Fuller & Houle (2003) have suggested that the inherent variability in these estimates either is too large for precise estimates, or that the assumptions on which $\Re$ has been formulated are wrong. It is likely that developmental errors are not additive and independent and therefore do not produce a normal distribution of left-minus-right-side differences as has been assumed in the standard model relating FA with DI (Klingenberg 2003b). Given that the precise relationship between FA and DI remains speculative, our present state of knowledge of the extent of additive genetic variation for DI is even less than that for FA.

Single Gene Effects

Although rare, there are a few unambiguous cases of single genes that significantly affect FA in one or more characters. The best-known examples come from the extensive studies by McKenzie and colleagues on mutants in Australian blowflies and in *Drosophila* that cause increased levels of FA in bristle number (review in McKenzie 2003). *Rop-1* and *Rdl* confer resistance to various insecticides in blowflies and act with partial or complete dominance specifically on bristle FA and not on FA levels in various wing characters (Clarke et al. 2000). The action of *Rop-1* is totally ameliorated by a modifier gene, *Scl* (scalloped wings). *Scl* is homologous to the *Notch* gene in *Drosophila* that is involved in bristle formation. Various *Notch* mutants affecting specific bristle types typically cause elevated asymmetry as well, but only in the specific bristle type that they affect (Indrasamy et al. 2000). *Notch* is not an "asymmetry" gene per se, but rather a gene that is character-specific in its action on specific bristles and that also affects the asymmetry level in these bristles. These sorts of genes represent precisely the kind of genetic basis of asymmetry predicted by the perspective of developmental mapping (Klingenberg & Nijhout 1999).

Some genes may act more generally by affecting FA levels in a suite of characters. For example, alleles at the lactose dehydrogenase (LDH) locus in killifish affect DI in a number of scale and fin ray characters (Mitton 1993). Atchley et al. (1984) found that mice with one dose of the muscular dysgenesis gene $(+/mdg)$ generally produce higher FA than wild genotypes $(+/+)$ in several different regions of the mandible, all of which are associated with skeletal muscle attachment. In addition, Leamy et al. (2001) discovered that mice carrying the *t-locus* haplotype exhibit more DI than wild mice when assessed by a composite FA index calculated over four different skeletal characters. The *t-locus* is a complex one consisting of inversions that suppress recombination and that are thought to harbor various mutations (Silver 1985), some of which may well affect the pathways leading to bone formation.

Dominance and Heterozygosity

Historically, developmental buffering (including both canalization and stability) in organisms has been thought to depend on the presence of specific gene complexes that become coadapted over time by natural selection (Mather 1943; Waddington 1940, 1942, 1957). Early investigators such as Dobzhansky (1950) and Lerner (1954) believed that selection primarily favored heterozygotes at many loci (see Woolf & Markow 2003), and this belief stimulated a number of experimental studies that focused on testing whether FA levels were less (and thus DS greater) in heterozygotes compared with homozygotes. This comparison was especially facilitated by the development of molecular markers such as allozymes, and in fact several allozyme studies did find that reduced FA tended to be associated with heterozygosity (for example, Leary et al. 1984, Soulé 1979, Vrijenhoek & Lerman 1982). But the typically few allozyme loci used in such studies were recognized to be poor indicators of overall genomic heterozygosity (Chakraborty 1981, Mitton & Grant 1984, Mitton & Pierce 1980), suggesting instead that the specific loci themselves might be important in maintaining DS. Mitton (1993) reviewed a number of investigations showing an association between heterozygosity at certain allozyme (or protein) loci with reduced FA levels, and he has argued that these loci may be exerting their effects by regulating overall metabolism and/or other physiological processes.

In recent years, the presumed link between allozyme/protein heterozygosity and FA has been increasingly questioned (see Clarke 1993). Partly this has been because most studies purporting to show such a link suffer from the same difficulties as those testing for FA/fitness associations. Perhaps worse, a number of studies that have been conducted with various allozyme or protein loci have not shown a significant FA/heterozygosity association (see Clarke 1993, Vollestad et al. 1997). L. Leamy, E.J. Routman & J.M. Cheverud (unpublished data) also did not find a significant relationship between FA of molar size and shape and the percentage of heterozygous microsatellite markers in a population of mice. In addition, in studies with various haplo-diploid insects (Clarke 1997, Clarke et al. 1992), the loss of heterozygosity from inbreeding did not appear to affect FA levels in various characters. Thus, though there may well be a potential role for dominance of alleles at loci influencing FA (see below), a universal heterozygosity/FA association does not appear to be supported.

Epistasis

One factor that complicates virtually every test for the effects of dominance or heterozygosity on FA in genetically variable populations is epistasis. A given locus A, for example, may interact with another locus B so that the effects of A (the differences among individuals with genotypes AA, Aa, and aa) depend on whether the genotype at locus B is BB, Bb, or bb. This is clearly seen in Australian sheep blowflies homozygous or heterozygous for the mutant allele *Rop-1* that confers resistance to the insecticide diazinon. This mutation also increases FA in bristle number over that of wild types $(+/+)$, but only if a dominant modifier allele at another locus is absent. Over time and with continuous exposure to diazinon, this modifier allele typically increases in frequency and renders FA levels for all three genotypes at the *Rop-1* locus virtually identical (McKenzie 2003). Thus, the *Rop-1* allele acts as a complete dominant in its effect on FA in one genetic environment but has no effect at all in another genetic environment. Unambiguous tests for dominance effects at single loci can be done only in co-isogenic populations where all loci other than the one being tested are homozygous (see Leamy 1981), although in these cases the results are valid only for the genetic background used.

Beyond allozyme and protein studies, epistasis is a particularly important confounding factor in experiments intended to assess the effects of heterozygosity on FA in inbred parents versus their hybrid offspring (Alibert & Auffray 2003, Clarke 1993). Generally, we expect that hybrids produced from crossing inbred lines should show positive heterosis and have lower FA levels than the mean of their inbred parents because the inbreeding process tends to fix deleterious recessive alleles whose harmful effects are masked by dominant alleles in the hybrids. But even in those cases where the parents are homozygous at all loci (as in fully inbred laboratory strains) and only the hybrid progeny contain some heterozygous loci, differences in FA between inbreds and hybrids may be the result not of heterozygosity but rather of the epistatic interaction of one or more of these newly formed heterozygous loci with various other loci. Lower FA levels in hybrids than in either parent (for example, Leamy 1984, Mather 1953) are especially likely to be due to epistatic effects unless we assume that underdominance of individual loci for FA is pervasive.

Hybrids generated from two different species or subspecies, taxa more distinct than inbred laboratory strains, often show outbreeding depression characterized by FA levels that are greater than those in their parents (reviews in Alibert & Auffray 2003, Graham 1992). Given that the hybrids are expected to be more heterozygous than their parents, heterozygosity is not acting to reduce FA in these crosses (Clarke 1993). Instead, the conventional explanation for such results is that selection has produced coadapted gene complexes (epistatic interactions) unique to each taxon that are broken down once the hybrids are formed (Clarke 1993). If so, the greater the differences in these gene complexes between the parental taxa, the greater should be the probability that DI (as measured by FA levels) will be higher rather than lower in the hybrids compared with that in the parents. This concept appears to be supported by the results of a survey done by Alibert & Auffray (2003, table 8.3) in which the number of studies showing hybrids with increased or decreased DI compared to their parents was 17/1 from crosses of different genera or species, but 8/10 from crosses of subspecies, races, or lines ($\chi_1^2 = 10.6$; $P < 0.01$). Such results provide indirect evidence that epistatic effects can be important in controlling FA levels.

Quantitative Trait Loci

In the past few years, the discovery of molecular markers such as microsatellites has made it possible to perform whole genome searches for genes (quantitative trait loci = QTLs) affecting quantitative characters (Erickson et al. 2004, Lynch & Walsh 1998). The approach typically starts with a cross of two inbred lines pheno-typically divergent for some character of interest, such as body weight, and geno-typically divergent for a number of polymorphic molecular markers. The F_1 hybrid individuals produced from the intercross of the inbred lines are themselves crossed to produce the F_2 individuals, which are phenotyped (for body weight, weighed) and genotyped for sufficient numbers of molecular markers to ensure uniform coverage of all chromosomes throughout the genome. This breeding scheme creates maximum linkage disequilibrium among all loci on each chromosome (Lynch & Walsh 1998). Because of this, significant differences in the mean of the character among the three genotypes for any marker on a given chromosome suggest that a QTL for the character exists at or near the position of that marker. Various statistical techniques have been developed to test appropriately the strength of phenotypic-genotypic associations along the length of each chromosome (including in intervals between markers) and, where a QTL is found, to estimate its additive and dominance effects on the character (Haley & Knott 1992, Lynch & Walsh 1998).

Using this approach, Leamy and colleagues searched for QTLs for FA in various morphometric characters in F_2 mice generated from an original cross of the Large (LG/J) and Small (SM/J) inbred strains. In the first such analysis, Leamy et al. (1997) discovered 11 QTLs affecting FA in 10 mandible dimensions, but 9.5 QTLs were expected by chance alone so it is difficult to say if any of these QTLs are real. Interestingly, 9 of the 11 putative QTLs exhibited significant dominance effects (Leamy et al. 1997), whereas QTLs for most characters generally show predominantly additive effects (for example, Workman et al. 2002). In a follow-up study, Klingenberg et al. (2001b) calculated mandible size and shape measures in these same mice but found only 1 QTL for FA of size and 1 for shape, neither of which were well-supported statistically. On the other hand, Leamy et al. (1998) found 13 QTLs for FA in 6 discrete skeletal characters in this same population of mice, this number being greater than the approximately 6 QTLs expected by chance alone. Again, dominance effects predominated in these QTLs (Leamy et al. 1998). Finally, no QTLs were found for FA in 15 mandible characters in a backcross mouse population created from a cross of the F_1 between a wild strain (CAST/Ei) and a strain (M16i) selected for rapid growth rate (Leamy et al. 2000). In general, therefore, these results suggest that there is little evidence for individual QTLs affecting FA in the morphometric characters used in these two populations of mice.

QTL analyses also allow tests for significant interactions among pairs of QTLs (Cheverud 2000a, Cheverud & Routman 1995, Leamy 2003, Routman & Cheverud

1995). Two QTL studies tested for epistasis for FA, both again making use of the F_2 population of mice generated from an original intercross of the Large and Small inbred strains previously described. In a full-genome scan, Leamy et al. (2002) found an abundance of epistasis for FA in mandible size, despite the fact that only one QTL of marginal significance was identified for this character. More recently, Leamy et al. (2005) conducted a follow-up study with these mice to test if epistasis might be present for FA in molar size and shape as well, and if so, whether the loci involved would be common for FA in the mandible and molar characters. Results of the single-gene analysis showed no individual QTLs for FA of molar size and just two significant QTLs for FA of molar shape, both of marginal significance and jointly contributing less than 5% of the total variance. However, numerous pair-wise combinations of QTLs again exhibited significant epistasis for FA in both molar size and shape. The (unknown) QTLs involved in these interactions differed for FA in the two molar characters (and from those for FA in mandible size), but their contribution to the total variance was nearly the same (about 20%) for FA in both molar characters. These results suggest that the genetic basis of FA in the molar characters consists almost entirely of character-specific epistatic effects whose contribution to the total variance is considerable.

A HYPOTHESIS

From all these results, it seems reasonable to hypothesize that FA has a predominantly nonadditive genetic basis with substantial dominance and especially epistasis. The genes participating in these epistatic interactions influencing FA in a given bilateral character most likely will be character-specific, and perhaps involved in some way with the formation of the bilateral character itself rather than being genes for FA per se. Occasional genes may be found in some populations whose single-locus effects on FA may generate small amounts of additive genetic variation, and where this is the case, these genes might be expected to exhibit dominance.

IMPLICATIONS OF THE HYPOTHESIS

An important implication of an epistatic genetic basis for FA is that it may complicate the primary use of FA in comparisons of DI levels in outbred or wild populations subjected or not subjected to various environmental stressors. It has been a tacit assumption that there is little if any additive genetic variation to confound such comparisons, but a potential confounding role for epistasis has generally not been considered. The situation with the *Rop-1* gene in blowflies previously described provides an example. If FA levels are measured in populations of blowflies exposed (stressed) and not exposed to diazinon (control), elevated asymmetry levels in bristle number would faithfully reflect the expected increased DI in populations first exposed to diazinon, but not some generations later after the *Scl* locus has exerted its modifying effects (McKenzie 2003). This suggests that the failure to detect FA differences between stressed versus nonstressed populations may sometimes be a consequence of epistatic adaptation that has occurred in the stressed population.

This example also suggests a way in which DI may evolve in spite of the low to zero additive genetic variation typically found for FA. Mutations such as *Rop-1* that are selected because of rapid environmental change clearly can alter DI levels in populations, and in this process, generate additive genetic variation (Cheverud & Routman 1995). Cheverud & Routman (1996) have shown how additive-by-additive (*aa*), additive-by dominance (*ad*), and dominance-by-dominance (*dd*) epistatic types all can contribute to additive genetic variation, especially as populations pass through bottlenecks. The *dd* epistatic type that exerted the greatest effect on FA in molar size (Leamy et al. 2005) produces the maximum additive genetic variance when an allele is fixed at one locus and the frequency of alleles at the other epistatic locus is either 0.15 or 0.85 (Cheverud & Routman 1996). However, the consistently low levels of heritability of FA (Fuller & Houle 2003) suggest any additive genetic variation generated by epistasis is continually eroded, perhaps by selection that favors fixation of alleles at most individual loci affecting FA. Except for unusual population events such as rapid environmental changes caused by various stresses, bottlenecks, and/or hybridizations, epistasis may generally act to suppress additive genetic variation in FA (Leamy et al. 2002).

Although DI may be capable of evolving, another implication of our hypothesis is that this evolution is character-specific, and may be detectable by change in FA levels only in certain characters. Because many genes appear to control only certain characters or character complexes (Cheverud 2000b, Leamy et al. 1999b), it would not be surprising to find that epistatic interactions affecting FA levels also are unique to specific characters. In support of this hypothesis, Leamy et al. (2005) discovered that entirely different QTLs were involved in the epistatic combinations that significantly affected FA levels in mandible size, molar size, and molar shape in mice. Depending on the type, magnitude, and sign of the epistatic effects, FA levels in these characters would be expected to respond differently (increase, decrease, or show no change) to various stressors such as inbreeding or selection (Leamy et al. 2005). Clearly the choice of characters in FA analyses is critical (Indrasamy et al. 2000, Woods et al. 1999) and may account for many ambiguous results in FA studies, including those cases in which no differences in FA levels have been detected between various parental races or subspecies and their hybrid offspring (Alibert & Auffray 2003, Schneider et al. 2003).

A slightly heritable, predominantly epistatic genetic architecture of FA parallels that of fitness components such as litter size and maternal performance (Peripato et al. 2002, 2004) rather than characters such as body size and skeletal size and shape that are affected by many single-locus QTLs each with generally small effects (Cheverud et al. 1996, Klingenberg et al. 2001b, Leamy et al. 1999b, Workman et al. 2002). Further, Gangestad & Thornhill (2003) analyzed the repeatabilities for FA in a number of studies and found them to be consistent with a coefficient of variation of DI that is within the range for typical fitness characters (Houle 1992). Although neither of these results is definitive, they do suggest that FA may have some kind of role as a fitness indicator. But we need better estimates than typically have been made of potential associations between FA levels, fitness parameters, and epistatic effects within populations. Until we obtain such improved estimates, we should be extremely cautious in the use of FA as an indicator of fitness.

Final Thoughts

Our hypothesis of an epistatic genetical basis of FA amounts to a modern version of some of the basic ideas first promoted by Waddington (1957), Lerner (1954), and other early researchers in this area. The major implication of this finding is that we cannot view FA as a purely environmentally determined character because in many cases epistatic interactions of genes may be important as well. This may well explain why we do not always obtain consistent increases of FA in various characters in populations subjected to various stressors. Perhaps we need to consider performing more experiments using isogenic strains to avoid these sorts of effects. This is practical with laboratory organisms such as mice and *Drosophila*, and may even be possible in wild populations (Kristensen et al. 2004).

All implications of our hypothesis apply only if the hypothesis holds up with additional testing. We need more, and better, tests of the genetical basis of FA. For tests of epistasis, QTL analyses should be especially useful because they are more powerful statistically than line crosses, diallel crosses, or other comparable approaches (see Leamy 2003). QTL analyses also allow us to scan the entire genome of an organism to search for individual genes affecting FA in a character, and if these are found, to test for additive and dominance effects for such genes. Traditional studies estimating the heritability of FA by parent–offspring regression, for example, cannot provide this kind of information and, in fact, are poorly suited even to detect significant additive genetic variation in FA (Fuller & Houle 2003). However, QTL studies are inherently difficult and require substantial amounts of effort and resources, particularly those searching for dominance and epistasis (Carlborg & Haley 2004, Erickson et al. 2004, Lynch & Walsh 1998). Moreover, it is a further substantial challenge to go from a QTL to specific candidate genes (e.g., Flint & Mott 2001). But only when we know the nature of the genes participating in single-locus or two-locus epistatic effects on FA will we be able to test whether they are involved in the development of the bilateral character itself and learn precisely how they affect the overall level of FA.

ACKNOWLEDGMENTS

It is a pleasure to thank Jim Cheverud, Geoff Clarke, Vincent Debat, Nelly Gidaszewski, and David Houle for constructive suggestions for revisions on earlier drafts of this review. This work was supported in part by funds made available by the University of North Carolina at Charlotte, the Wellcome Trust, and the Biotechnology and Biological Sciences Research Council (UK).

The *Annual Review of Ecology, Evolution, and Systematics* is online at
http://ecolsys.annualreviews.org

LITERATURE CITED

Alibert P, Auffray J-C. 2003. Genomic coadaptation, outbreeding depression, and developmental instability. See Polak 2003, pp. 116–34

Atchley WR, Herring SW, Riska B, Plummer AA. 1984. Effects of the muscular dysgenesis gene on developmental stability in the mouse mandible. *J. Craniofac. Genet. Dev. Biol.* 4:179–89

Badyaev AV, Foresman KR, Fernandes MV. 2000. Stress and developmental stability: vegetation removal causes increased fluctuating asymmetry in shrews. *Ecology* 81:336–45

Bergman A, Siegal ML. 2003. Evolutionary capacitance as a general feature of complex gene networks. *Nature* 424:549–52

Blake WJ, Kaern M, Cantor CR, Collins JJ. 2003. Noise in eukaryotic gene expression. *Nature* 422:633–37

Carlborg Ö, Haley CS. 2004. Epistasis: too often neglected in complex trait studies? *Nat. Rev. Genet.* 5:618–25

Chakraborty R. 1981. The distribution of the number of heterozygous loci in an individual in natural populations. *Genetics* 98:461–66

Cheverud JM. 2000a. Detecting epistasis among quantitative trait loci. In *Epistasis and the Evolutionary Process*, ed. B Wolf, EDI Brodie, JM Wade, pp. 58–81. New York: Oxford Univ. Press

Cheverud JM. 2000b. The genetic architecture of pleiotropic relations and differential epistasis. In *The Character Concept in Evolutionary Biology*, ed. GP Wagner, pp. 411–33. San Diego: Academic

Cheverud JM, Routman EJ. 1995. Epistasis and its contribution to genetic variance components. *Genetics* 139:1455–61

Cheverud JM, Routman EJ. 1996. Epistasis as a source of increased additive genetic variance at population bottlenecks. *Evolution* 50:1042–51

Cheverud JM, Routman EJ, Duarte FAM, Swinderen BV, Cothran K, Perel C. 1996. Quantitative trait loci for murine growth. *Genetics* 142:1305–19

Clarke GM. 1993. The genetical basis of developmental stability. I. Relationship between stability, heterozygosity and genomic coadaptation. *Genetica* 89:15–23

Clarke GM. 1995a. Relationships between developmental stability and fitness: application for conservation biology. *Conserv. Biol.* 9:18–24

Clarke GM. 1995b. Relationships between fluctuating asymmetry and fitness: how good is the evidence? *Pac. Conserv. Biol.* 2:146–49

Clarke GM. 1997. The genetic basis of developmental stability. III. Haplo-diploidy: Are males more unstable than females? *Evolution* 51:2021–28

Clarke GM. 1998a. Developmental stability and fitness: the evidence is not quite so clear. *Am. Nat.* 152:762–66

Clarke GM. 1998b. The genetic basis of developmental stability. IV. Individual and population asymmetry parameters. *Heredity* 80:553–61

Clarke GM. 2003. Developmental stability-fitness relationships in animals: some theoretical considerations. See Polak 2003, pp. 187–95

Clarke GM, Yen JL, McKenzie JA. 2000. Wings and bristles: character specificity of the asymmetry phenotype in insecticide resistant strains of *Lucilia cuprina. Proc. R. Soc. London Ser. B* 267:1815–18

Clarke GM, Oldroyd BP, Hunt P. 1992. The genetic basis of developmental stability in *Apis mellifera*: heterozygosity versus genic balance. *Evolution* 46:753–62

Cook DL, Gerber AN, Tapscott SJ. 1998.

Modeling stochastic gene expression: implications for haploinsufficiency. *Proc. Natl. Acad. Sci. USA* 95:15641–46

Debat V, David P. 2001. Mapping phenotypes: canalization, plasticity and developmental stability. *Trends Ecol. Evol.* 16:555–61

Dobzhansky Th. 1950. Genetics of natural populations. XIX. Origin of heterosis through natural selection in populations of *Drosophila pseudoobscura*. *Genetics* 35:288–302

Erickson DL, Fenster CB, Stenøien HK, Price D. 2004. Quantitative trait locus analyses and the study of evolutionary process. *Mol. Ecol.* 13:2505–22

Fiering S, Whitelaw E, Martin DIK. 2000. To be or not to be active: the stochastic nature of enhancer action. *BioEssays* 22:381–87

Flint J, Mott R. 2001. Finding the molecular basis of quantitative traits: successes and pitfalls. *Nat. Rev. Genet.* 2:437–45

Fuller RC, Houle D. 2003. Inheritance of developmental instability. See Polak 2003, pp. 157–86

Gangestad W, Thornhill R. 1999. Individual differences in developmental precision and fluctuating asymmetry: A model and its implications. *J. Evol. Biol.* 12:402–16

Gangestad W, Thornhill R. 2003. Fluctuating asymmetry, developmental instability, and fitness: toward model-based interpretation. See Polak 2003, pp. 62–80

Graham JH. 1992. Genomic coadaptation and developmental stability in hybrid zones. *Acta Zool. Fenn.* 191:121–31

Graham JH, Fletcher D, Tigue J, McDonald M. 2000. Growth and developmental stability of *Drosophila melanogaster* in low-frequency magnetic fields. *Bioelectromagnetics* 21:465–72

Graham JH, Freeman DC, Emlen JM. 1993. Antisymmetry, directional asymmetry, and dynamic morphogenesis. *Genetica* 89:121–37

Haley CS, Knott SA. 1992. A simple regression method for mapping quantitative trait loci in line crosses using flanking markers. *Heredity* 69:315–24

Hoffmann AA, Woods RE. 2003. Associating environmental stress with developmental stability: problems and patterns. See Polak 2003, pp. 387–401

Houchmandzadeh B, Wieschaus E, Leibler S. 2002. Establishment of developmental precision and proportions in the early *Drosophila* embryo. *Nature* 415:798–802

Houle D. 1992. Comparing evolvability and variability of traits. *Genetics* 130:195–204

Houle D. 1997. Comment on "A meta-analysis of the heritability of developmental stability" by Møller and Thornhill. *J. Evol. Biol.* 10:17–20

Houle D. 2000. A simple model of the relationship between asymmetry and developmental stability. *J. Evol. Biol.* 13:720–30

Indrasamy H, Woods RE, McKenzie JA, Batterham P. 2000. Fluctuating asymmetry of specific bristle characters in Notch mutants of *Drosophila melanogaster*. *Genetica* 109:151–59

Klingenberg CP. 2003a. Developmental instability as a research tool: using patterns of fluctuating asymmetry to infer the developmental origins of morphological integration. See Polak 2003, pp. 427–42

Klingenberg CP. 2003b. A developmental perspective on developmental instability: theory, models, and mechanisms. See Polak 2003, pp. 14–34

Klingenberg CP. 2004a. Dominance, nonlinear developmental mapping and developmental stability. In *The Biology of Genetic Dominance*, ed. RA Veitia, pp. 37–51. Austin, TX: Landes Biosci.

Klingenberg CP. 2004b. Integration, modules and development: molecules to morphology to evolution. In *Phenotypic Integration: Studying the Ecology and Evolution of Complex Phenotypes*, ed. M Pigliucci, K Preston, pp. 213–30. New York: Oxford Univ. Press

Klingenberg CP, Badyaev AV, Sowry SM, Beckwith NJ. 2001a. Inferring developmental modularity from morphological integration: analysis of individual variation and asymmetry in bumblebee wings. *Am. Nat.* 157:11–23

Klingenberg CP, Barluenga M, Meyer A.

2002. Shape analysis of symmetric structures: quantifying variation among individuals and asymmetry. *Evolution* 56:1909–20

Klingenberg CP, Leamy LJ, Routman EJ, Cheverud JM. 2001b. Genetic architecture of mandible shape in mice: effects of quantitative trait loci analyzed by geometric morphometrics. *Genetics* 157:785–802

Klingenberg CP, McIntyre GS. 1998. Geometric morphometrics of developmental instability: analyzing patterns of fluctuating asymmetry with Procrustes methods. *Evolution* 52:1363–75

Klingenberg CP, McIntyre GS, Zaklan SD. 1998. Left-right asymmetry of fly wings and the evolution of body axes. *Proc. R. Soc. London Ser. B* 265:1255–59

Klingenberg CP, Mebus K, Auffray J-C. 2003. Developmental integration in a complex morphological structure: how distinct are the modules in the mouse mandible? *Evol. Dev.* 5:522–31

Klingenberg CP, Nijhout HF. 1999. Genetics of fluctuating asymmetry: a developmental model of developmental instability. *Evolution* 53:358–75

Klingenberg CP, Zaklan SD. 2000. Morphological integration between developmental compartments in the Drosophila wing. *Evolution* 54:1273–85

Kristensen TN, Pertoldi C, Pedersen LD, Andersen DH, Bach LA, Loeschcke V. 2004. The increase of fluctuating asymmetry in a monoclonal strain of collembolans after chemical exposure—discussing a new method for estimating the environmental variance. *Ecol. Indic.* 4:73–81

Leamy L. 1981. Effects of alleles at the albino locus on odontometric traits in the 129/J strain of house mice. *J. Hered.* 72:199–204

Leamy L. 1984. Morphometric studies in inbred and hybrid house mice. V. Directional and fluctuating asymmetry. *Am. Nat.* 123:579–93

Leamy L. 1993. Morphological integration of fluctuating asymmetry in the house mouse mandible. *Genetica* 89:139–53

Leamy L. 1997. Is developmental stability heritable? *J. Evol. Biol.* 10:21–29

Leamy L. 1999. Heritability of directional and fluctuating asymmetry for mandibular characters in random-bred mice. *J. Evol. Biol.* 12:146–55

Leamy L. 2003. Dominance, epistasis, and fluctuating asymmetry. See Polak 2003, pp. 142–56

Leamy LJ, Doster MJ, Huet-Hudson YM. 1999a. Effects of methoxychlor on directional and fluctuating asymmetry of mandible characters in mice. *Ecotoxicology* 8:63–71

Leamy LJ, Meagher S, Taylor S, Carroll L, Potts WK. 2001. Size and fluctuating asymmetry of morphometric characters in mice: their associations with inbreeding and *t*-haplotype. *Evolution* 55:2333–41

Leamy LJ, Pomp D, Eisen EJ, Cheverud JM. 2000. Quantitative trait loci for directional but not fluctuating asymmetry of mandible characters in mice. *Genet. Res.* 76:27–40

Leamy LJ, Routman EJ, Cheverud JM. 1997. A search for quantitative trait loci affecting asymmetry of mandibular characters in mice. *Evolution* 51:957–69

Leamy LJ, Routman EJ, Cheverud JM. 1998. Quantitative trait loci for fluctuating asymmetry of discrete skeletal characters in mice. *Heredity* 80:509–18

Leamy LJ, Routman EJ, Cheverud JM. 1999b. Quantitative trait loci for early and late developing skull characters in mice: a test of the genetic independence model of morphological integration. *Am. Nat.* 153:201–14

Leamy LJ, Routman EJ, Cheverud JM. 2002. An epistatic genetic basis for fluctuating asymmetry of mandible size in mice. *Evolution* 56:642–53

Leamy LJ, Workman MS, Routman EJ, Cheverud JM. 2005. An epistatic genetic basis for fluctuating asymmetry of tooth size and shape in mice. *Heredity* 94:316–25

Leary RF, Allendorf FW, Knudsen LK. 1984. Superior developmental stability of heterozygotes at enzyme loci in salmonid fishes. *Am. Nat.* 12:540–51

Lens L, Van Dongen S, Kark S, Matthysen E. 2002. Fluctuating asymmetry as an indicator of fitness: can we bridge the gap between studies? *Biol. Rev.* 77:27–38

Lerner IM. 1954. *Genetic Homeostasis.* New York: Wiley

Leung B, Forbes MR. 1996. Fluctuating asymmetry in relationship to stress and fitness: effects of trait type as revealed by meta-analysis. *Ecoscience* 3:400–13

Leung B, Forbes MR, Houle D. 2000. Fluctuating asymmetry as a bioindicator of stress: comparing efficacy of analyses involving multiple traits. *Am. Nat.* 155:101–15

Lynch M, Walsh B. 1998. *Genetics and Analysis of Quantitative Traits.* Sunderland, MA: Sinauer

Markow TA. 1995. Evolutionary ecology and developmental instability. *Annu. Rev. Entomol.* 40:105–20

Markow TA, Clarke GM. 1997. Meta-analysis of the heritability of developmental stability: a giant step backward. *J. Evol. Biol.* 10:31–37

Mather K. 1943. Polygenic balance in the canalization of development. *Nature* 151:68–71

Mather K. 1953. Genetical control of stability in development. *Heredity* 7:297–336

McAdams HH, Arkin A. 1999. It's a noisy business! Genetic regulation at the nanomolecular scale. *Trends Genet.* 15:65–69

McKenzie JA. 1997. Stress and asymmetry during arrested development of the Australian sheep blowfly. *Proc. R. Soc. London Ser. B* 264:1749–56

McKenzie JA. 2003. The analysis of the asymmetry phenotype: single genes and the environment. See Polak 2003, pp. 135–41

McKenzie JA, Clarke GM. 1988. Diazinon resistance, fluctuating asymmetry and fitness in the Australian sheep blowfly, *Lucilia cuprina. Genetics* 120:213–20

Meiklejohn CD, Hartl DL. 2002. A single mode of canalization. *Trends Ecol. Evol.* 17:468–73

Milton CC, Huynh B, Batterham P, Rutherford SL, Hoffmann AA. 2003. Quantitative trait symmetry independent of Hsp90 buffering: distinct modes of genetic canalization and developmental stability. *Proc. Natl. Acad. Sci. USA* 100:13396–401

Mitton JB. 1993. Enzyme heterozygosity, metabolism, and developmental stability. *Genetica* 89:47–66

Mitton JB, Grant MC. 1984. Associations among protein heterozygosity, growth rate, and developmental homeostasis. *Annu. Rev. Ecol. Syst.* 15:479–99

Mitton JB, Pierce BA. 1980. The distribution of individual heterozygosity in natural populations. *Genetics* 95:1043–54

Møller AP. 1990. Fluctuating asymmetry in male sexual ornaments may reliably reveal male quality. *Anim. Behav.* 40:1187–95

Møller AP. 1992. Female swallow preference for symmetrical male sexual ornaments. *Nature* 357:238–40

Møller AP, Hoglund J. 1991. Patterns of fluctuating asymmetry in avian feather ornaments: implications for models of sexual selection. *Proc. R. Soc. London Ser. B* 245:1–5

Møller AP, Pomiankowski A. 1993. Fluctuating asymmetry and sexual selection. *Genetica* 89:267–79

Møller AP, Swaddle JP. 1997. *Asymmetry, Developmental Stability and Evolution.* Oxford: Oxford Univ. Press

Møller AP, Thornhill R. 1997. A meta-analysis of the heritability of developmental stability. *J. Evol. Biol.* 10:1–16

Møller AP, Thornhill R. 1998. Bilateral symmetry and sexual selection: a meta-analysis. *Am. Nat.* 151:174–92

Nijhout HF, Paulsen SM. 1997. Developmental models and polygenic characters. *Am. Nat.* 149:394–405

Ozbudak EM, Thattai M, Kurtser I, Grossman AD, van Oudenaarden A. 2002. Regulation of noise in the expression of a single gene. *Nat. Genet.* 31:69–73

Palmer AR. 1994. Fluctuating asymmetry analyses: a primer. In *Developmental Instability: Its Origins and Evolutionary Implications,* ed. TA Markow, pp. 335–64. Dordrecht: Kluwer

Palmer AR. 2000. Waltzing with asymmetry: is

fluctuating asymmetry a powerful new tool for biologists or just an alluring new dance step? *BioScience* 46:518–32

Palmer AR, Strobeck C. 1986. Fluctuating asymmetry: measurement, analysis, patterns. *Annu. Rev. Ecol. Syst.* 17:391–421

Palmer AR, Strobeck C. 1997. Fluctuating asymmetry and developmental stability: heritability of observable variation vs. heritability of inferred cause. *J. Evol. Biol.* 10:39–49

Palmer AR, Strobeck D. 2003. Fluctuating asymmetry analyses revisited. See Polak 2003, pp. 279–319

Pankakoski E, Koivisto I, Hyvarinen H. 1992. Reduced developmental stability as an indicator of heavy metal pollution in the common shrew, *Sorex araneus. Acta Zool. Fenn.* 191:137–44

Pelabon C, Hansen TF, Carlson ML, Armbruster WS. 2004. Variational and genetic properties of developmental stability in *Dalechampia scandens. Evolution* 58:504–14

Peripato AC, de Brito RA, Matioli SR, Pletscher LS, Vaughn TT, Cheverud JM. 2004. Epistasis affecting litter size in mice. *J. Evol. Biol.* 17:593–602

Peripato AC, de Brito RA, Vaughn TT, Pletscher LS, Matioli SR, Cheverud JM. 2002. Quantitative trait loci for maternal performance for offspring survival in mice. *Genetics* 162:1341–53

Polak M, ed. 2003. *Developmental Instability. Causes and Consequences.* New York: Oxford Univ. Press

Polak M, Møller AP, Gangestad S, Kroeger DE, Manning J, Thornhill R. 2003. Does an individual asymmetry parameter exist? A meta-analysis. See Polak 2003, pp. 81–98

Polak M, Starmer WT. 2001. The quantitative genetics of fluctuating asymmetry. *Evolution* 55:498–511

Queitsch C, Sangster TA, Lindquist S. 2002. Hsp90 as a capacitor of phenotypic variation. *Nature* 417:618–24

Raser JM, O'Shea EK. 2004. Control of stochasticity in eukaryotic gene expression. *Science* 304:1811–14

Rice SH. 2002. A general population genetic theory for the evolution of developmental interactions. *Proc. Natl. Acad. Sci. USA* 99:15518–23

Routman EJ, Cheverud JM. 1997. Gene effects on a quantitative trait: two-locus epistatic effects measured at microsatellite markers and at estimated QTL. *Evolution* 51:1654–62

Rutherford SL. 2000. From genotype to phenotype: buffering mechanisms and the storage of genetic information. *BioEssays* 22:1095–105

Rutherford SL, Lindquist S. 1998. Hsp90 as a capacitor for morphological evolution. *Nature* 396:336–42

Sangster TA, Lindquist S, Queitsch C. 2004. Under cover: causes, effects and implications of Hsp90-mediated genetic capacitance. *BioEssays* 26:348–62

Santos M. 2002. Genetics of wing size asymmetry in *Drosophila buzzatii. J. Evol. Biol.* 15:720–34

Scheiner SM, Caplan RL, Lyman RF. 1991. The genetics of phenotypic plasticity. III. Genetic correlations and fluctuating asymmetries. *J. Evol. Biol.* 4:51–68

Schneider SS, Leamy LJ, Lewis LA, DeGrandi-Hoffman G. 2003. The influence of hybridization between African and European honeybees, *Apis mellifera,* on asymmetries in wing size and shape. *Evolution* 57:2350–64

Silver LM. 1985. Mouse *t* haplotypes. *Annu. Rev. Genet.* 19:179–208

Simmons LW, Tomkins JL, Kotiaho JS, Hunt J. 1999. Fluctuating paradigm. *Proc. R. Soc. London Ser. B* 266:593–95

Sollars V, Lu X, Xiao L, Wang X, Garfinkel MD, Ruden DM. 2003. Evidence for an epigenetic mechanism by which Hsp90 acts as a capacitor for morphological evolution. *Nat. Genet.* 33:70–74

Soulé ME. 1979. Heterozygosity and developmental stability: another look. *Evolution* 331:396–401

Thornhill R, Møller AP, Gangestad SW. 1999. The biological significance of fluctuating asymmetry and sexual selection: a reply to Palmer. *Am. Nat.* 154:234–41

Thornhill R, Sauer P. 1992. Genetic sire effects on the fighting ability of sons and daughters and mating success of sons in a scorpionfly. *Anim. Behav.* 43:255–64

Tomkins JL, Simmons LW. 2003. Fluctuating asymmetry and sexual selection: paradigm shifts, publication bias, and observer expectation. See Polak 2003, pp. 231–61

Tracy M, Freeman DC, Duda JL, Miglia KJ, Graham JH, Hough RA. 2003. Developmental instability: an appropriate indicator of plant fitness components? See Polak 2003, pp. 196–212

Van Dongen S. 1998. How repeatable is the estimation of developmental stability by fluctuating asymmetry? *Proc. R. Soc. London Ser. B* 265:1423–27

Vollestad LA, Hindar K, Møller AP. 1997. A meta-analysis of fluctuating asymmetry in relation to heterozygosity. *Heredity* 83:206–18

von Dassow G, Meir E, Munro EM, Odell GM. 2000. The segment polarity network is a robust developmental module. *Nature* 406:188–92

Vrijenhoek RC, Lerman S. 1982. Heterozygosity and developmental stability under sexual and asexual breeding systems. *Evolution* 36:768–76

Waddington CH. 1940. *Organizers and Genes.* Cambridge, UK: Cambridge Univ. Press

Waddington CH. 1942. Canalization of development and the inheritance of acquired characters. *Nature* 150:563–65

Waddington CH. 1957. *The Strategy of the Genes.* New York: Macmillan

Wagner GP, Altenberg L. 1996. Complex adaptations and the evolution of evolvability. *Evolution* 50:967–76

Whitlock MC. 1996. The heritability of fluctuating asymmetry and the genetic control of developmental stability. *Proc. R. Soc. London Ser. B* 263:849–54

Whitlock MC. 1998. The repeatability of fluctuating asymmetry: a revision and extension. *Proc. R. Soc. London Ser. B.* 265:1429–31

Whitlock MC, Fowler K. 1997. The instability of studies of instability. *J. Evol. Biol.* 10:63–67

Woolf CM, Markow TA. 2003. Genetic models for developmental homeostasis: historical perspectives. See Polak 2003, pp. 99–115

Woods RE, Sgro CM, Hercus MJ, Hoffmann AA. 1999. The association between fluctuating asymmetry, trait variability, trait heritability, and stress: A multiply replicated experiment on combined stresses in *Drosophila melanogaster. Evolution* 53:493–505

Workman MS, Leamy LJ, Routman EJ, Cheverud JM. 2002. Analysis of QTL effects on the size and shape of mandibular molars in mice. *Genetics* 160:1573–86

Zakharov VM. 1992. Population phenogenetics: analysis of developmental stability in natural populations. *Acta Zool. Fenn.* 191:7–30

Annu. Rev. Ecol. Evol. Syst. 2005. 36:23–46
doi: 10.1146/annurev.ecolsys.36.102003.152631
Copyright © 2005 by Annual Reviews. All rights reserved
First published online as a Review in Advance on July 25, 2005

LIFE-HISTORY EVOLUTION IN REPTILES

Richard Shine

Biological Sciences, University of Sydney, NSW 2006, Australia;
email: rics@bio.usyd.edu.au

Key Words   behavioral thermoregulation, ecothermy, plasticity, viviparity

■ **Abstract**   Two consequences of terrestrial ectothermy (low energy needs and behavioral control of body temperatures) have had major consequences for the evolution of reptile life-history traits. For example, reproducing females can manipulate incubation temperatures and thus phenotypic traits of their offspring by retaining developing eggs in utero. This ability has resulted in multiple evolutionary transitions from oviparity to viviparity in cool-climate reptile populations. The spatial and temporal heterogeneity of operative temperatures in terrestrial habitats also has favored careful nest-site selection and a matching of embryonic reaction norms to thermal regimes during incubation (e.g., via temperature-dependent sex determination). Many of the life-history features in which reptiles differ from endothermic vertebrates—such as their small offspring sizes, large litter sizes, and infrequent reproduction—are direct consequences of ectothermy, reflecting freedom from heat-conserving constraints on body size and energy storage. Ectothermy confers immense flexibility, enabling a dynamic matching of life-history traits to local circumstances. This flexibility has generated massive spatial and temporal variation in life-history traits via phenotypic plasticity as well as adaptation. The diversity of life histories in reptiles can best be interpreted within a conceptual framework that views reptiles as low-energy, variable-temperature systems.

INTRODUCTION

Reptiles display immense diversity in their rates and routes of reproductive output, and the life-history traits that generate that variation display corresponding diversity. For example, even closely related reptiles living in the same area may differ in their modes of sex determination and modes of reproduction, as well as in their ages at first reproduction, in the numbers and sizes of offspring that they produce, and when and how often they produce them. Seasonally inundated floodplains in tropical Australia provide a good example of the kinds of life-history diversity that can occur. For example, two of the most abundant floodplain snakes are keelbacks (*Tropidonophis mairii*) and filesnakes (*Acrochordus arafurae*). Keelbacks grow rapidly and can mature at a few months of age; females produce several large clutches of small eggs each year (Brown & Shine 2002). In contrast, female filesnakes grow slowly, mature at about four years, and produce a single small litter of relatively large live-born offspring about once every four years thereafter (Madsen & Shine 2001). Female pythons (*Morelia spilota*) exhibit complex parental care, not only remaining with their eggs until hatching, but generating metabolic heat to keep the eggs warm (Harlow & Grigg 1984). Saltwater crocodiles (*Crocodylus porosus*) build large nests, defend them, and protect the offspring after hatching (Webb & Manolis 2002). In the same billabongs, longneck turtles (*Chelodina rugosa*) lay their eggs underwater, so that the beginning of embryonic development will be synchronized with the seasonal drying-out of the floodplain (Kennett et al. 1993). What factors have stimulated (and constrained) the evolution of such diversity in life-history traits?

The strong causal link between life-history traits and individual reproductive success has encouraged many researchers to look for an adaptive basis to variation in life-history traits. That search has revealed immense diversity among taxa, with some traits exhibiting strong phylogenetic conservatism among major lineages (e.g., the predominance of oviparity in birds versus viviparity in mammals), whereas other traits display remarkable convergence and parallelism (e.g., placental nutrition of embryos in some lizards, as well as in mammals; Flemming & Blackburn 2003). Any review of a topic as diverse as life-history traits in reptiles must make arbitrary choices about what to leave out, as well as what to include; thus, I will ignore some of the themes emphasised in previous syntheses (e.g., reproductive physiology, life-history trade-offs) and focus instead on a framework that is centered upon two of the most distinctive ecological attributes of reptiles—the ability to behaviorally regulate their own body temperature, and the freedom from heat-conserving and energy-flow constraints allowed by ectothermy. My aim is to explore the degree to which these characteristics may have shaped evolutionary divergence in life-history traits.

Necessarily, such an analysis must pose questions at a variety of taxonomic levels and incorporate an understanding of the phylogenetic relationships among groups. For example, living "reptiles" are not monophyletic, and the four main lineages usually described as reptiles (turtles, crocodilians, tuatara, squamates) are actually very different kinds of animals and have pursued independent evolutionary trajectories since the Triassic (Pough et al. 1998). Similarly, living birds and mammals result from separate, phylogenetically distinct transitions from ectothermy to endothermy (Pough et al. 1998). Despite this complexity, a useful first step is to pose simple questions about similarities and differences in life-history traits. For example, why do reptiles tend to produce larger clutches of smaller offspring than do mammal and bird species of similar mean adult body mass (Figure 1)? Why do we see major life-history similarities—as well as divergences—among the four distantly related types of reptiles? Are such similarities due to the shared retention of ectothermy and behavioral thermoregulation? If so, what causes the differences? Last, what factors have generated life-history diversity within each major reptilian lineage, especially the highly speciose squamates (lizards and snakes)?

WHAT IS SPECIAL ABOUT REPTILES?

In ecological terms, the primary defining characteristics of living reptiles center on their mechanism of temperature regulation. Unlike the endothermic mammals and birds that maintain relatively high and constant internal temperatures via their own metabolic processes, reptiles are ectotherms. By exploiting ambient thermal heterogeneity to control its internal temperature, a reptile may be able to achieve a body temperature as high as that of a sympatric endotherm at much less energy cost—for example, simply by basking in a patch of sunlight rather than generating heat through the breakdown of previously ingested food (Pough 1980). Most vertebrates probably exploit thermal heterogeneity by behavioral means, but behavior plays a far greater role in this respect for reptiles than for the other vertebrate groups. In most mammals and birds, behavioral thermoregulation simply acts to keep the animal within a set of ambient conditions at which its internal machinery is capable of maintaining precise thermal control. In fishes, the high thermal conductivity of water reduces spatial thermal heterogeneity, thus limiting an animal's ability to behaviorally modify its own temperature by moving to a cooler or warmer location. The moist external covering of amphibians renders them vulnerable to desiccation if exposed to direct sunlight, and thus limits (though it does not eliminate) their capacity to maintain high internal temperatures in exposed terrestrial sites (Huey 1982, Pough et al. 1998). Thus, terrestrial reptiles stand out as the organisms best able to exert behavioral control over their own internal temperatures over a broad range.

Clearly, there will be many exceptions to these gross generalizations. Notably, aquatic reptiles will be exposed to the same constraints on behavioral regulation of body temperature as are other aquatic organisms, although semiaquatic taxa (such as many turtles and crocodilians) may haul out to bask on convenient logs and shoreline in a way that most fishes and frogs do not. Also, spatial heterogeneity in operative temperatures (and thus, opportunities for thermoregulation) may be minimal for fossorial and nocturnal reptiles, or those living in heavily shaded habitats (Huey 1974, 1982). Thus, behavioral thermoregulation will be unimportant for some reptiles. Equally, reptiles are not the only terrestrial ectotherms; especially, there are many parallels between reptiles and terrestrial invertebrates. In both groups, ectothermy, relatively impervious external coverings (scales, exoskeleton), and high spatial and temporal thermal heterogeneity allow precise behavioral thermoregulation. Nonetheless, the small body sizes of invertebrates mean that they heat and cool very rapidly with little hysteresis (Grigg et al. 1979), so that maintaining high stable temperatures will be more difficult than for (larger) reptiles. Thus, as a general rule, terrestrial reptiles behaviorally control their own body temperatures over a wider range than do other organisms.

Below, I explore the hypothesis that ectothermy and behavioral thermoregulation are causally related to patterns of life-history evolution in reptiles. To do so, I frame the discussion around specific attributes of reptile biology that might have influenced patterns of life-history evolution.

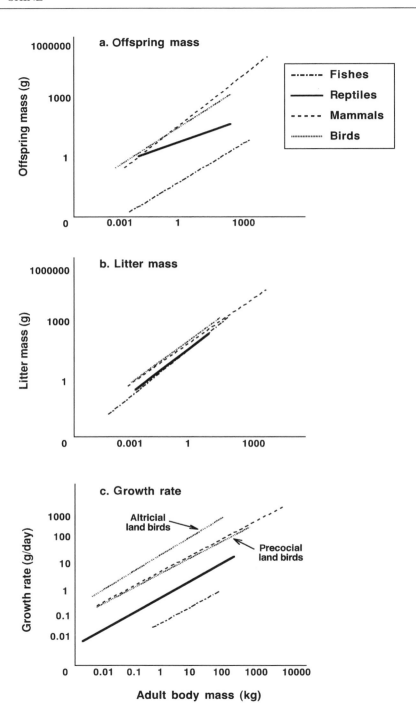

REPRODUCING FEMALE REPTILES CAN CONTROL THEIR BODY TEMPERATURES OVER A WIDE RANGE

First and most obviously, diurnal reptiles not only can adjust their own body temperatures by behavioral means, but do so precisely and over a considerable thermal range. In many diurnal heliothermic (sun-basking) reptiles, a single individual may have access to adjacent sites offering operative temperatures over a 30°C range (Peterson 1987). Temperature influences the rate and sometimes the trajectory of biochemical reactions, and thus affects a multitude of vital parameters at the level of the whole organism. For example, higher temperatures result in higher rates of metabolic expenditure, locomotion, growth, and digestion (Huey & Slatkin 1976). Especially important for life-history evolution, temperature also affects reproductive traits such as rates of gonadal recrudescence and embryogenesis (Huey 1991, Huey & Slatkin 1976). The ability of a reproducing female to modify not only her own body temperature, but also that of her developing offspring, has been a major influence on life-history evolution in reptiles.

A central role for behavioral thermoregulation in the evolution of reptile life-history traits is not a new idea. More than 60 years ago, three workers publishing at about the same time in three different languages (Mell in German, Weekes in English, Sergeev in Russian) all suggested essentially the same hypothesis: that maternal selection of high body temperatures might provide the major selective advantage for the evolution of viviparity (production of fully-formed live young) from oviparity (egg-laying) in reptiles. The basic idea was that in a cool climate, eggs laid in the nest will develop only slowly or not at all, whereas eggs retained within a sun-basking female's uterus will be kept warmer, and thus can complete development within the short summer season (Mell 1929, Sergeev 1940, Weekes 1933). All three of these researchers recognized that the evolutionary transition from oviparity to viviparity had occurred repeatedly in reptiles, and that present-day live-bearing species were generally found in colder areas than were egg-laying species. Hence, thermal factors were likely involved in the transition. Remarkably, this "cold-climate hypothesis" for the evolution of reptilian viviparity has stood the test of time and retains widespread support many decades later (Blackburn 2000; Shine 1985, 2004b).

Figure 1   Allometry of reproductive output and offspring growth rates in reptiles compared to other vertebrate groups. (*a*) Relative to the mass of their mother, hatchling reptiles are larger than the offspring of fishes but smaller than those of birds or mammals. However, the relative offspring size is much greater for species of reptiles with small absolute adult mass, whereas this trend is weaker in the other taxa. (*b*) Litter mass relative to maternal mass is relatively similar among groups. (*c*) Hatchling reptiles grow more rapidly than fishes, but not as quickly as endothermic vertebrates (birds and mammals). Data for (*a*) and (*b*) from Blueweiss et al. 1978; (*c*) Case 1978.

By superimposing reproductive-mode data onto a phylogenetic framework, it is clear that the evolutionary shift from oviparity to viviparity has taken place independently in at least 100 lineages of lizards and snakes, but never in turtles or crocodilians (Lee & Shine 1998, Shine 1985). Other cases continue to accumulate (e.g., Bauer & Sadlier 2000, Ota et al. 1991, Schulte et al. 2000, Staub & Emberton 2002) so that the total number of transitions may well be much higher. Surprisingly, the process appears to be one-way, with no clear examples of any reversion from viviparity back to oviparity (Lee & Shine 1998). The high frequency of phylogenetic transitions in reproductive mode within reptiles (sometimes, occurring multiple times within a single species: Fairbairn et al. 1998, Heulin et al. 2002; see Figure 2) bears striking contrast to the conservatism of this trait in other vertebrates. Viviparity has arisen from oviparity only about 12 times in bony fishes (Goodwin et al. 2002), 10 times in elasmobranch fishes (Dulvy & Reynolds 1997, Reynolds et al. 2002), about 6 times in amphibians (Duellman & Trueb 1986, Wake & Dickie 1998), 1 time in mammals, and not at all in birds (Reynolds et al. 2002). Thus, the control over body (and thus incubation) temperatures afforded by terrestrial ectothermy appears to have generated an immense difference between the vertebrate Classes in phylogenetic lability of this major life-history trait.

How strong is the evidence that thermal differentials between the nest and the uterus were the key to this evolutionary transition, and if so, what kinds of thermal differentials were involved, and how did these translate into selective forces? The evidence initially used to support this hypothesis (the current distribution of live-bearers in cooler-than-average climates) has now been substantially boosted. First, phylogenetically based tests have shown that wherever we can make a clear comparison between closely related oviparous and viviparous taxa (e.g., within the same genus, or even the same species), the viviparous forms tend to inhabit cooler climates (Hodges 2004, Mendez-de la Cruz et al. 1998, Shine 1985). Thus, viviparity not only is more common in cooler areas, but appears to have evolved there. Second, we now have detailed field studies that support many of the assumptions critical to the cold-climate hypothesis, such as different thermal regimes inside females than in nests and beneficial effects of maternal incubation temperatures on developmental rates of embryos (Andrews 2000; Shine 1983b, 2004b). Third, a recent study recreated transitional forms between "normal" oviparity and viviparity by incubating eggs partway through development at maternal body temperatures, then depositing them in field nests. Hatching success was increased by prolonged "uterine" (maternal-temperature) retention of eggs at high elevations in the field, above (but close to) the upper elevational limit for successful oviparous reproduction (Shine 2002b; Figure 3).

These recent studies also have clarified the kinds of thermal modifications that might enhance offspring viability. Although earlier researchers focused on very straightforward aspects of incubation regimes (mean temperatures only) and hatchling fitness (egg survival and developmental rates only), recent work reveals a more complex picture whereby subtle changes in the temperatures experienced by an egg also influence the developmental trajectories of the embryo inside it. Thus,

Figure 2   Viviparity has arisen at least 100 times within squamate reptiles (lizards and snakes), and there are many cases where both oviparous (egg-laying) and viviparous (live-bearing) species occur within a single genus. For example, within the Australian venomous (elapid) snakes of the genus *Pseudechis*, Collett's Snakes (*Pseudechis colletti*) are egg-layers (*upper panel*), whereas Common Blacksnakes (*P. porphyriacus*) are live-bearers (*lower panel*). As in most such comparisons, the oviparous taxon inhabits warmer climates than the closely related viviparous species (Shine 1985).

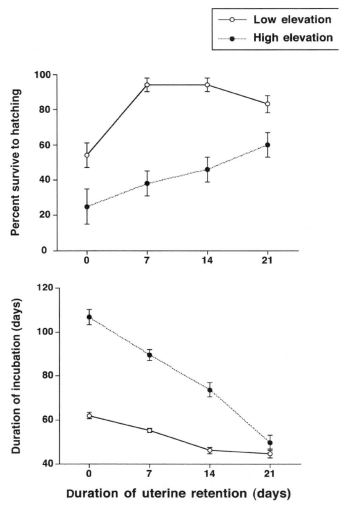

Figure 3   Direct evidence on the adaptive significance of prolonged uterine retention of eggs in cold-climate reptiles comes from an experimental study on montane scincid lizards (*Bassiana duperreyi*) in southeastern Australia (Shine 2002b). Newly laid eggs were translocated to the laboratory, where they were kept under thermal regimes mimicking maternal body temperatures for periods of 0, 7, 14 or 21 days; then the eggs were returned to the field site and placed in artificial nests at a range of elevations. The graphs show data for low elevation nests (1615 m above sea level, the actual upper limit for natural oviposition sites in this area) and higher up the same mountain (high elevation, 1710 m). Increasing durations of retention at maternal temperatures (the intermediate stages toward viviparity) substantially increased hatching success of eggs in field nests (*upper graph*) and reduced incubation periods (*lower graph*).

a change in thermal regimes during incubation not only can enhance hatching success and accelerate hatching, but also can modify fitness-relevant traits of the hatchling such as its body size, shape, and locomotor speed (Deeming 2004a, Ji & Brana 1999, Van Damme et al. 1992). Hence, the offspring from maternally retained eggs may be more viable because of their phenotypic traits as well as (or rather than) the timing or success of hatching (Shine 1995). Importantly, this idea is more general than the original cold-climate hypothesis, because it suggests a reason for viviparity to enhance offspring (and thus maternal) fitness in any kind of environment where uterine retention of eggs allows the mother to provide optimal incubation conditions for her offspring. For example, even in a hot climate a pregnant female can select less variable temperatures than would be available in any external nest.

Not only do thermal regimes affect multiple fitness-relevant traits, but it has also become clear that several facets of "incubation temperature" play a role. An early focus on mean temperatures has been replaced by a growing understanding that hatching success, developmental rates, and offspring phenotypes also can be affected by the ways in which incubation temperatures vary around any given mean value. Compared to a constant-temperature nest at the same mathematical mean temperature, the addition of daily thermal variance (as is typical in natural nests, at least of shallow-nesting temperate-zone reptiles) can significantly modify hatchling traits (Shine & Harlow 1996). Even a short period of cooler-than-usual or warmer-than-usual conditions, especially if it occurs early in development, can affect the offspring's phenotype (Shine 2002a). So can gradual heating or cooling over the incubation period, as often occurs in nests but presumably not inside thermoregulating females (Shine 2004c). Finally, the phenotypic traits of neonates from simulated maternal regimes (high stable temperature by day, cold at night) differed from those of their siblings incubated under simulated nest regimes (smoothly sinusoidal curves of heating and cooling on a diel cycle, but with the same mean value), suggesting that the evolution of viviparity might enhance offspring fitness because of retention at stable (not high) temperatures (Shine 2004b).

Although we are still learning about the ways that specific attributes of thermal incubation regimes influence offspring phenotypes, it seems increasingly likely that thermal consequences of uterine retention played a causal role in the selective advantage of extended uterine retention of eggs that eventually has led to viviparity in so many reptile lineages. The development of miniature temperature-sensitive data-loggers has revolutionized the field, allowing us to make direct measures of thermal regimes inside a nest or inside a female reptile. One intriguing result has been to confirm that behavioral thermoregulation allows a female to provide incubation regimes unavailable at any potential nest site, notably by breaking the mathematical links that must exist between minimum, mean, and maximum temperatures at any fixed location. As one moves to higher (colder) elevations, the inevitable decrease in overnight thermal minima in a nest means that to attain a high-enough mean temperature for embryonic development, the maximum must be higher also. This soon reaches the point that eggs would need to be in very exposed locations where they would experience lethally high daytime temperatures. Similarly, the hottest nests may often be the driest, whereas uterine retention can keep the eggs moist as well as warm. Thus, oviparity may be precluded in cold climates not because there are no potential nest sites with high-enough mean incubation temperatures but because eggs in such a nest would desiccate or (ironically) overheat (Andrews 2000, Shine et al. 2003).

Given that viviparity enables mothers to enhance the viability of their offspring, why is this reproductive mode relatively common only in cool climates? The answer presumably lies in associated costs: for example, viviparous females are heavily burdened with the large mass and volume of the clutch, perhaps increasing their vulnerability to predators (Servan et al. 1989). Similarly, viviparity constrains the female to a single litter per season, whereas oviparous females can cycle through multiple clutches within the same period. Thus, the relative fitness of oviparous and viviparous strategies depends upon the balance between costs and benefits. The evolution of reptilian viviparity also provides examples of the kinds of pathways by which physiological constraints and lineage-specific factors may limit life-history diversity. Crocodilians and turtles have not evolved viviparity, and one reason may be that their embryos do not continue development for very long even if retained in utero (Shine 1983a). Thus, prolonging the period of retention would not take these animals any closer to viviparity. In contrast, most oviparous squamates retain developing embryos for at least one quarter of the total developmental period, and this situation may have facilitated evolutionary shifts to even longer retention and ultimately, to viviparity. The scarcity of intermediate stages between normal oviparity and full viviparity, despite the many transitions from one state to the other in reptile phylogeny, suggests that these intermediate stages may be disadvantageous relative to either end of the oviparity–viviparity continuum (Blackburn 1995, Qualls et al. 1997, Shine 1983a). Physiological constraints to prolonged retention are documented in some squamates, and even differ between closely related taxa. Some species may be unable to retain developing embryos without reducing developmental rates or hatchling viability (Andrews 2004, Andrews & Mathies 2000); in other taxa, adaptations to low-temperature incubation may be incompatible with maternal retention at high body temperatures (Shine 1999).

REPRODUCING FEMALES SELECT EXTERNAL INCUBATION REGIMES FOR THEIR OFFSPRING

Even in oviparous taxa, a reproducing female can influence the incubation conditions experienced by her offspring. So can females of other vertebrates, but the range of potential nest conditions available to the female is reduced by embryonic inability to tolerate (or develop over) a broad range of temperatures (endotherms) or hydric conditions (amphibians), or by limited thermal heterogeneity in the local environment (fishes). Reflecting embryonic sensitivities, many kinds of animals regulate the incubation environment of their offspring; for example, most birds maintain their developing offspring at close-to-parental temperatures by brooding. Viviparity in reptiles (and mammals) is an extreme development of this kind of parental thermal buffering. In contrast to birds, the lack of parental brooding in most oviparous reptiles exposes their embryos to substantial ambient fluctuations in moisture and temperature. Given the sensitivity of hatching success and hatchling phenotypic traits to incubation conditions (above; Deeming 2004b), we expect strong selection in reptiles for maternal manipulation of incubation conditions in ways that enhance offspring (and thus, maternal) fitness (Packard et al. 1993, Qualls & Andrews 1999a, Shine 2004a, Shine & Harlow 1996).

The most obvious way in which a reproducing female reptile in a thermally heterogeneous environment can modify the conditions experienced by her eggs is simply by judicious selection of an appropriate nest site. There is growing evidence of such maternal selectivity, and of its consequences for offspring fitness. Even within a single small area, potential nest-sites may differ in ways that will significantly affect not only hatching success, but also developmental rate (and thus, date of hatching) as well as phenotypic traits of the offspring (Packard & Packard 1988, Warner & Andrews 2002). One such trait is sex. Some species determine sex by genetic factors (sometimes involving heteromorphic sex chromosomes), whereas others tie sex determination to environmental stimuli experienced by the organism—often, early in development (Bull & Charnov 1989). Genetic sex determination (GSD) is almost universal in birds and mammals (but see Göth & Booth 2004), whereas sex is determined by incubation temperature (temperature-dependent sex determination, or TSD) in all crocodilians, many turtles, sphenodontians, and a phylogenetically diverse array of lizards (Bull & Charnov 1989, Rhen & Lang 1999). There have been many independent evolutionary shifts between GSD and TSD within reptiles, and the underlying physiological mechanisms for these two sex-determining modes may be very similar (Sarre et al. 2004). Remarkably, TSD and GSD co-occur in at least one lizard species (Shine et al. 2002a) and one turtle species (Servan et al. 1989), and TSD has evolved in viviparous as well as oviparous lizards (Robert & Thompson 2001, Wapstra et al. 2004). Thus, the variable-temperature incubation regimes of reptiles have allowed the evolution of a sex-determining mechanism (TSD) that would be impossible under constant-temperature incubation. Unsurprisingly, GSD and TSD co-occur also in the other ectothermic groups, the amphibians and fishes (Sarre et al. 2004).

The selective forces responsible for shifts between GSD and TSD have attracted considerable research. The most widely supported adaptationist hypothesis invokes sex differences in norms of reaction in response to incubation temperatures, such that TSD allows a reproducing female to bias the sex ratio of her clutch in a way that maximizes the fitness of her offspring. For example, if low-temperature incubation produces viable sons but low-quality daughters, TSD may enhance overall clutch fitness by allowing all the eggs within a cold nest to develop as sons. Exactly this situation is reported in alpine scincid lizards (Shine et al. 2002a) and could occur even within viviparous species, if the conditions selected by pregnant females differentially affected fitness of their sons versus daughters (Robert & Thompson

2001, Wapstra et al. 2004). Leakage of steroid hormones between adjacent embryos in utero, thus modifying development of sexually dimorphic features (Uller & Olsson 2003), suggests another possible selective force for environmental sex determination in viviparous species: litters consisting of a single sex might be fitter than mixed-sex litters. Another plausible mechanism in oviparous taxa is that seasonal shifts in soil temperatures may be consistent enough that incubation temperature serves as a reliable predictor of seasonality. In such a system, TSD may allow seasonal shifts in offspring sex ratio such that each sex is produced at the time of year when its fitness would be higher (on average) than that of the other sex (Harlow & Taylor 2000).

Maternal control over temperature may have facilitated the evolution not only of nest-site selection and embryonic norms of reaction, but also of parental care. Parental care of eggs has evolved multiple times within squamate reptiles, and one major selective pressure for this behavior may involve maternal control over incubation conditions (Shine 1988, Aubret et al. 2003). As in the evolution of viviparity from oviparity, maternal attendance at the nest may provide the eggs with a combination of incubation conditions (e.g., warm, moist) unavailable in nonattended nests (Somma 1990, 2003). Hydric as well as thermal regimes are likely to have been important in this respect, but nest-attending females can warm the eggs by intermittent sun-basking followed by retreat to the nest, or in one lineage (pythonid snakes) by shivering thermogenesis (Aubret et al. 2003). Indeed, facultative thermogenesis to enhance offspring viability may have been the initial selective force for the evolution of endothermy (Farmer 2003).

ECTOTHERMY PERMITS A WIDE RANGE OF ECOLOGICALLY VIABLE BODY SIZES

Maintaining a thermal differential between the body and the ambient environment (as is essential for endotherms) places major constraints on the relationship between an organism's body volume (and thus, its heat-generating capacity) and its surface area (and thus, the rate at which it loses or gains heat from the environment). As a result, endotherms are characterized by large body mass and rounded shape (Pough 1980). Ectothermy significantly reduces these constraints, allowing organisms to be virtually any size. That freedom has had an immense impact on the life-history traits of ectotherms, for the following reasons:

1. *Offspring size* is limited by heat-transfer rates in endotherms; by the time they leave their parents, endothermic offspring must be large enough to control their own body temperatures. Ectothermy removes this constraint, with the result that neonatal reptiles often are an order of magnitude smaller than their parents at hatching or birth and receive no parental nutrient input posthatching (Somma 2003; see Figure 1). Offspring size relative to parental size is maximized in species of small absolute body size as adults, perhaps reflecting

minimum viable offspring sizes (Shine 1978); interestingly, reptiles seem to differ in this respect from other vertebrate groups (Figure 1). Especially in large species, hatchling reptiles (and fishes and amphibians) are so much smaller than their parents that they depend upon different prey types, live in different habitats, and are vulnerable to a different array of predators (Fitch 1960, Shine et al. 1998b). The range of ecologically viable body sizes is thus much greater in a population of ectotherms than in one of endotherms. For the same reason, sexual size dimorphism can take more extreme values in ectotherms than endotherms; and the degree of sexual divergence in body size shows strong but lineage-specific allometry within reptiles (e.g., Shine 1994, Shine et al. 1998a).

2. *Fecundity* is inversely related to offspring size; a given allocation of energy to reproduction can be divided either into a few large offspring or many small ones. Experimental manipulation of clutch sizes by follicle ablation or hormonal stimulation shows a direct mechanistic tradeoff in this respect: A decrease in clutch size translates directly into an increased offspring size, and vice versa (Sinervo & Licht 1991, Sinervo et al. 1992). Accordingly, one consequence of the small offspring sizes permitted by ectothermy is that clutch sizes of reptiles generally are much higher than those of endotherms: Total litter mass relative to maternal size is similar among groups, but reptiles have much smaller offspring and thus, more of them (see Figure 1).

Fecundity is not only higher on average in reptiles than in endotherms, but also is more variable. This variability is seen at several levels—from ontogenetic shifts within the lifetime of a single adult female through to among-female comparisons within a single population, to among-population and among-species comparisons. One underlying cause for such variation is maternal body size: Larger females produce larger clutches in most taxa that have been studied (Fitch 1970). Thus, clutch size increases with maternal body size within a female's lifetime (sometimes changing by an order of magnitude over that period: Fitch 1970, Shine et al. 1998a), and populations or species with larger mean adult female body sizes tend to have larger clutch sizes (Seigel & Ford 1987). The major exceptions to this generalization involve small lizards that produce a small fixed (constant) clutch size of either one or two eggs (e.g., anoles, geckos); but in almost all other taxa across all the major reptile lineages, maternal body size and clutch size are strongly linked (Greer 1989, Shine & Greer 1991). The reasons for this linkage may involve simple physical constraint, with females delaying reproduction until they have accumulated enough energy to produce a clutch as large as they can physically accommodate with their body cavity (Vitt & Congdon 1978, Vitt & Price 1982). In keeping with this hypothesis, interspecific variations in body shape (and thus, abdominal volume) predict interspecific patterns in clutch mass within squamate reptiles (Shine 1992) with females apparently filling their body cavities to some optimal degree (Qualls & Andrews 1999b, Qualls & Shine 1995).

The trend for clutch sizes to increase with maternal body size in females presumably has important consequences for patterns of selection; for example, if the costs of reproduction are independent of fecundity, selection may favor females to delay reproduction until they have accumulated enough energy for a very large clutch (Bull & Shine 1979). Such a selective pressure might influence traits such as ages and body sizes at female maturation (and thus, patterns of sexual size dimorphism); early maturation might reduce lifetime fitness for females under such a selective regime.

More generally, the wide range of body sizes of adult females within a single population of reptiles (compared to endotherms) will tend to generate an equally broad range in clutch sizes. The same factor is likely to cause variation in fecundity on spatial and temporal scales—for example, a short-term decrease in survival or growth rates, and thus a decrease in mean adult female body size (perhaps due to local weather conditions) will cause a corresponding decrease in mean clutch sizes. Likewise, geographic variation in such traits, or in food resources, may substantially modify reproductive output per female as an indirect consequence of effects on body-size distributions (Seigel & Fitch 1985). In total, the resultant scenario involves higher levels of variation in reproductive traits in reptile populations than might be expected in endotherms. Indeed, for many reptile populations in highly stochastic environments, mean values may be a "meaningless" abstraction, and variance may be more important. In extreme cases, successive annual cohorts of offspring within the same population may encounter very different circumstances and exhibit very different reproductive histories; this may be true for males (Madsen & Shine 1992) as well as for females (Madsen & Shine 2000b).

3. *Parental care* is almost ubiquitous in endotherms, but rare in reptiles (Farmer 2003, Somma 2003). Reflecting the minimum body-size constraints of endothermy, parents in most mammalian and avian species provide their progeny with nutrients until they attain close-to-adult body sizes (Farmer 2003). Indeed, lactation is a defining feature of mammals. In the minority of reptile species in which parents care for their young after hatching or birth, however, the benefits involve protection rather than feeding (Webb & Manolis 2002, O'Connor & Shine 2004). One parent can probably defend the offspring as well as two, so that uniparental rather than biparental care is the norm. The energy demands of juvenile endotherms, in contrast, frequently require biparental care (Farmer 2003).

ECTOTHERMY PERMITS LONG-TERM ENERGY STORAGE WITHOUT COMPROMISING THERMAL EXCHANGE

Endotherms are finely tuned machines that face high costs for energy storage; large fat bodies are energetically expensive to carry around and impede heat flow (Bonnet et al. 1998). Thus, for example, migrating birds lay down fat reserves only immediately before they embark on their migration, and most species must stop and refuel along the way (Dingle 1980). In contrast, ectotherms can lay down large energy stores without paying such high costs (Bonnet et al. 1998, Jonsson 1997). The consequent ability to decouple the time of feeding (energy acquisition) from reproduction (energy expenditure) allows ectotherms to fuel reproduction from stored capital instead of (or as well as) immediate income. The more rapid energy flow through endotherms precludes this tactic, forcing them instead toward income breeding (Bonnet et al. 1998, 2001).

Reflecting their low metabolic rates, infrequent activity, and often low rates of feeding, most ectotherms probably rely to a large degree on capital breeding, at least for the first clutch each season. In environments where resources are scarce, or available only sporadically in response to stochastic climatic events, female reptiles may postpone reproduction for many years. Indeed, in many snake populations, the majority of adult-sized females may be nonreproductive in most years (Seigel & Ford 1987). Bursts of prey resources may be followed by bursts of reproductive output, creating enormous temporal heterogeneity in age structure within such a population (Madsen & Shine 2000b, Shine & Madsen 1997); a stable age distribution may be a mathematical fallacy in most reptile populations. The exact proportion of adult-sized animals that reproduce in response to a given level of prey abundance may be regulated in complex ways by prior history. In viperid snakes from Europe, the body-condition (energy-storage) threshold needed to initiate breeding in female snakes is relatively constant from year to year, although variation in prey abundance generates corresponding variation in the proportion of adult-size female snakes that will reproduce in any given year (Bonnet et al. 2002). In the more highly stochastic habitats of tropical Australia, however, where annual variation in wet-season rainfall generates correspondingly dramatic fluctuations in food supply, acrochordid and pythonid snakes show a flexible adjustment of the body-condition threshold to prior resource levels. Females exposed to several "bad" years in succession will initiate reproduction even when they are in relatively poor condition, whereas females exposed to several "good" years delay reproduction until they have accumulated very high energy stores (Madsen & Shine 1999, 2000b).

More generally, history as well as present conditions can influence demographic traits in reptile populations. In water pythons, not only do growth rates vary in direct response to prey availability, but the conditions that a hatchling snake experiences in its first few months of life determine the nature of that relationship thereafter. Young pythons fortunate to hatch in a year when rats are common thereafter grow more rapidly, at any given level of rat abundance, than do conspecifics hatched out in "poor" seasons (Madsen & Shine 2000a). Similarly, viviparous scincid lizards modify their allocation of energy to offspring based on the resource levels they experienced in the preceding year (Doughty & Shine 1998). More generally, reproductive output in many infrequently reproducing "capital breeders" can be influenced by food availability over a period of several years prior to the reproductive episode (Brown 1993).

Because ectothermy breaks the temporal link between energy acquisition and expenditure imposed by endothermy, many ectotherms can store enough fat reserves to withstand months or years of starvation (Pough 1980). Accordingly, many ectotherms become anorexic while reproducing, presumably because of a trade-off between feeding and reproductive activities. For example, feeding may be incompatible with effective mate-searching for males (Aldridge & Brown 1995), or maintenance of high, stable body temperatures by females (Shine 1980). Also, feeding may be too risky for pregnant females whose locomotory abilities are severely impaired by the mass or volume of the developing offspring (Seigel et al. 1987b, Shine 1980). By foregoing feeding, such an animal may be able to translate a potential survival cost of reproduction (vulnerability to predators) into an energy cost (lost feeding opportunities), and thereby enhance overall fitness (Brodie 1989).

The ability of reproducing ectotherms to stop feeding may facilitate shifts in behavior and/or microhabitat selection associated with reproduction. For example, pregnant females of several reptile taxa aggregate in microhabitats that provide thermally optimal conditions but no access to feeding opportunities (Graves & Duvall 1995). By minimizing the conflict between feeding and reproduction (i.e., focusing on one at a time rather than doing both simultaneously), a female may be able to "afford" a greater degree of locomotor impairment (and thus, a greater clutch mass) than would be compatible with continued feeding during pregnancy. However, pregnancy can be prolonged; in extreme cases, viviparous female skinks and geckos carry their fullterm oviductal embryos over the winter hibernation period, giving birth the following summer almost 12 months after ovulation (Olsson & Shine 1998, Rock & Cree 2003).

Foraging biology and reproductive investment interact in another way also. Squamate reptiles that rely principally upon ambushing their prey tend to have heavyset bodies, whereas active searchers are more elongate and slender-bodied (Vitt & Congdon 1978). Because abdominal volume constrains clutch volume, this difference in body shape translates into a difference in relative clutch mass; ambush foragers typically produce heavier clutches relative to maternal mass than do active searchers (Vitt & Congdon 1978). This divergence may be enhanced by the selective disadvantages of impaired maternal mobility for an active hunter, which relies upon speed for foraging and for escape from predators. In contrast, an ambush forager that relies upon crypsis, and moves only rarely, may be less burdened by a large clutch or litter (Vitt 1981, Vitt & Congdon 1978).

Ectothermy also may facilitate dietary specialization. An organism that needs to feed only infrequently can afford to specialize on specific prey types that are available only occasionally, whereas reliance on such prey may be impossible for an endotherm that needs to feed on a more frequent basis (Pough 1980). Accordingly, many ectotherms, including reptiles, rely upon only a few prey types (Greene 1997). In extreme cases, feeding may be restricted to brief periods each year, with long intervening periods of starvation (Schwaner & Sarre 1990, Shine et al. 2002b). In combination with gape-limitation, the ability to specialize favors intrapopulation divergence in niche dimensions. For example, juvenile snakes frequently feed on entirely different kinds of prey than are eaten by conspecific adults (Greene 1997). This kind of niche divergence may decouple fluctuations in prey resources among different age or sex groups within a single population. Thus, for example, a climatically induced reduction in numbers of large prey might decrease feeding rates and reproductive output of large females while not affecting smaller females within the same population. Similarly, dependence of juvenile water pythons (*Liasis fuscus*) on juvenile (ingestible-size) rodents causes stochastic year-to-year variation in juvenile survival rates depending on the seasonal timing of rat breeding relative to python hatching, even when adult rodent numbers remain high enough to sustain survival and reproduction of adult snakes (Shine & Madsen 1997). The presence of ecologically diverse subunits within a single reptile population can thus further destabilize age structures in such systems, and prevent any approach to steady-state conditions (see above).

ECTOTHERMY ENABLES VARIABLE-RATE LIFE HISTORIES

To maintain their high rates of metabolism and growth (Figure 1), endotherms need to·maintain high and relatively constant rates of food intake. In contrast, ectotherms have low metabolic rates and, thus, can afford more flexible, opportunistic lifestyles in which they synchronize rates of growth and reproduction with spatial and temporal variation in resource availability. Thus, variable-rate life histories are typical of ectotherms, especially in environments with episodic or stochastic variation in food supply, predator abundance, or the availability of appropriate thermal or hydric conditions. For example, the rates of survival, growth, and reproduction of filesnakes and pythons in tropical Australia display immense annual variation, driven by annual fluctuations in the abundance of prey (fishes and rodents, respectively), which in turn are driven by annual variation in rainfall (Madsen & Shine 2000b, Shine & Madsen 1997). Low energy costs for maintenance mean that ectotherms can survive long periods of resource (e.g., food, water) scarcity and hence are well suited to habitats with stochastic fluctuations in resource levels. In such places, resources sufficient to fuel reproduction may be available only sporadically, in turn generating variation through time and space in reproductive activity. This situation will impose selection for flexibility, and thus for variable-rate life histories. Many avian and mammalian populations also exhibit demographic changes in response to environmental fluctuations (e.g., Hau et al. 2004) but reptiles may provide more dramatic examples. For example, individual Galapagos marine iguanas (*Amblyrhynchus cristatus*) actually shrunk significantly (by up to 20%!) in body size when an El Niño event reduced their algal food supply (Wikelski & Thom 2000). In another spectacular example of animals retaining flexibility long into adulthood, adult tuatara (*Sphenodon guntheri*) that had been stable in body size during seven years of prior monitoring, all rapidly recommenced growing when translocated to an island offering more abundant food (Nelson et al. 2002).

The ability of ectotherms to link their life-history traits to environmental fluctuations means that spatial and temporal heterogeneity in thermal regimes will generate corresponding (albeit buffered) variation in body temperatures, and thus in the rates of feeding, growth, and reproduction. We might thus predict not only that reptiles will display more dramatic spatiotemporal heterogeneity in such life-history traits than will endothermic vertebrates, but that we will see consistent links between average temperature and the life history. For example, a reptile living in warm climates is likely to grow rapidly, mature early, and reproduce frequently (because it is constantly active at high temperatures), whereas a genetically identical reptile living in a cold climate will grow more slowly, mature later, and reproduce less frequently (Adolph & Porter 1993).

Body temperature is not the only variable that influences an ectotherm's metabolic rate. Even at identical temperatures, species vary considerably in their metabolic rates and, thus, energy needs. For example, snakes tend to have lower metabolic rates than lizards, and slow-moving relatively inactive lizards have lower metabolic rates than fast-moving active species (Andrews & Pough 1985). This range of interspecific variation in rates of energy metabolism in ectotherms is further amplified by variation in mean body temperatures. Thus, we might expect to see "slow" life histories (slow growth, delayed maturation, infrequent reproduction) in low-metabolic-rate taxa as well as cold-climate taxa, and "fast" life histories (rapid growth, early maturation, frequent reproduction) in high-metabolic-rate taxa as well as hot-climate taxa. Comparative analyses support this proposition, at least for the small subset of reptile species that have been studied intensively enough for inclusion (Andrews & Pough 1985, Parker & Plummer 1987). However, we need much more extensive analyses before we can claim to understand the nature of any such link between the rates of physiological and life-history processes, let alone the nature of the causal link, if any, between the two.

CONCLUSIONS

Many of the distinctive life-history traits of reptiles—the ways in which they differ from other vertebrates, as well as the diversity apparent within the Reptilia—can usefully be interpreted in a conceptual framework built around the ecological consequences of ectothermy. The common feature of ectothermy may explain why the four main lineages of reptiles, although phylogenetically distinct from each other since the Triassic, display many similarities in life-history traits. For example, all produce large clutches of relatively small offspring, and the thermal conditions of maternally selected nest sites strongly modify important fitness-relevant traits (including sex, in some taxa within all four lineages) of the hatchlings that emerge from those nests. Similarities are apparent also between reptiles and the other ectothermic vertebrates (fishes, amphibians), whereas many endothermic vertebrates exhibit higher and less variable rates of most life-history processes, and accordingly have less flexibility to adjust life-history traits to local environments.

What was the phylogenetic history of the emergence of reptilian reproduction? That is, how and why did the life-history features of early reptiles diverge from those of their amphibian-like progenitors? Female reptiles display synchronous production of smaller clutches of larger, more energy-rich eggs than do amphibians. Studies on living salamanders suggest that large yolky eggs are typical of terrestrial-breeding taxa with parental care (Salthe 1969), suggesting that maternal attendance and control over incubation conditions may have been a critical feature during this transition. Especially in species without parental care, females of many amphibian species are capable of ovipositing their available eggs immediately in response to unpredictable rainfall events that provide larval habitat (Duellman & Trueb 1986, Salthe 1969), whereas female reptiles show synchronized vitellogenesis of an entire clutch, often timed such that oviposition occurs during the season when suitable nesting conditions (in cooler climates, sufficiently high temperatures) are available (Fitch 1970). Thus, linking reproductive output to thermal rather than hydric cues, and a central role for maternal control over incubation environments, may have been major themes of life-history adaptation in reptiles throughout their evolutionary history.

ACKNOWLEDGMENTS

I thank the Australian Research Council for funding, and my colleagues (especially Xavier Bonnet, Mats Olsson, Mike Wall, and Raju Radder) for comments on this review.

The *Annual Review of Ecology, Evolution, and Systematics* is online at
http://ecolsys.annualreviews.org

LITERATURE CITED

Adolph SC, Porter WP. 1993. Temperature, activity, and lizard life-histories. *Am. Nat.* 142:273–95

Aldridge RD, Brown WS. 1995. Male reproductive cycle, age at maturity, and cost of reproduction in the timber rattlesnake (*Crotalus horridus*). *J. Herpetol.* 29:399–407

Andrews RM. 2000. Evolution of viviparity in squamate reptiles (*Sceloporus* spp.): a variant of the cold-climate model. *J. Zool.* 250:243–53

Andrews RM. 2004. Patterns of embryonic development. See Deeming 2004b, pp. 75–102

Andrews RM, Mathies T. 2000. Natural history of reptilian development: constraints on the evolution of viviparity. *BioScience* 50:227–38

Andrews RM, Pough FH. 1985. Metabolism of squamate reptiles—allometric and ecological relationships. *Physiol. Zool.* 58:214–31

Aubret F, Bonnet X, Shine R, Maumelat S. 2003. Clutch size manipulation, hatching success and offspring phenotype in the ball python (*Python regius*, Pythonidae). *Biol. J. Linn. Soc.* 78:263–72

Bauer AM, Sadlier RA. 2000. *The Herpetofauna of New Caledonia.* Ithaca NY: Soc. Study Amphib. Reptiles. 310 pp.

Blackburn DG. 1995. Saltationist and punctuated equilibrium models for the evolution of viviparity and placentation. *J. Theor. Biol.* 174:199–216

Blackburn DG. 2000. Reptilian viviparity: past research, future directions, and appropriate models. *Comp. Biochem. Physiol. A* 127: 391–409

Blueweiss L, Fox H, Kudzma V, Nakashima D, Peters RH, Sams S. 1978. Relationships between body size and some life history parameters. *Oecologia* 37:257–72

Bonnet X, Bradshaw D, Shine R. 1998. Capital versus income breeding: an ectothermic perspective. *Oikos* 83:333–42

Bonnet X, Lourdais O, Shine R, Naulleau G. 2002. Reproduction in a typical capital breeder: costs, currencies, and complications in the aspic viper. *Ecology* 83:2124–35

Bonnet X, Naulleau G, Shine R, Lourdais O. 2001. Short-term versus long-term effects of food intake on reproductive output in a viviparous snake, *Vipera aspis. Oikos* 92: 297–308

Brodie EDI. 1989. Behavioral modification as a means of reducing the cost of reproduction. *Am. Nat.* 134:225–38

Brown GP, Shine R. 2002. Reproductive ecology of a tropical natricine snake, *Tropidonophis mairii* (Colubridae). *J. Zool.* 258: 63–72

Brown WS. 1993. Biology, status and management of the timber rattlesnake (*Crotalus horridus*): a guide for conservation. *Herpetol. Circ.* 22:1–78

Bull JJ, Charnov EL. 1989. Enigmatic reptilian sex ratios. *Evolution* 43:1561–66

Bull JJ, Shine R. 1979. Iteroparous animals that skip opportunities for reproduction. *Am. Nat.* 114:296–316

Case TJ. 1978. On the evolution and adaptive significance of post-natal growth in terrestrial vertebrates. *Q. Rev. Biol.* 53:243–82

Deeming DC. 2004a. Post-hatching phenotypic effects of incubation in reptiles. See Deeming 2004b, pp. 229–51

Deeming DC, ed. 2004b. *Reptilian Incubation. Environment Evolution and Behaviour.* Nottingham: Nottingham Univ. Press. 362 pp.

Dingle H. 1980. Ecology and evolution of migration. In *Animal Migration Orientation and Navigation*, ed. SAJ Gauthreaux, pp. 1–103. New York: Academic. 387 pp.

Doughty PD, Shine R. 1998. Energy allocation to reproduction in a viviparous lizard species (*Eulamprus tympanum*): the role of long-term energy stores. *Ecology* 79:1073–83

Duellman WE, Trueb L. 1986. *Biology of Amphibians.* New York: McGraw-Hill. 670 pp.

Dulvy NK, Reynolds JD. 1997. Evolutionary transitions among egg-laying, live-bearing and maternal inputs in sharks and rays. *Proc. R. Soc. London Ser. B* 264:1309–15

Fairbairn J, Shine R, Moritz C, Frommer M. 1998. Phylogenetic relationships between oviparous and viviparous populations of an Australian lizard (*Lerista bougainvillii*, Scincidae). *Mol. Phylogenet. Evol.* 10:95–103

Farmer CG. 2003. Reproduction: the adaptive significance of endothermy. *Am. Nat.* 162: 826–40

Fitch HS. 1960. Autecology of the copperhead. *Univ. Kansas Publ. Mus. Nat. Hist.* 13:85–288

Fitch HS. 1970. Reproductive cycles in lizards and snakes. *Univ. Kansas Mus. Nat. Hist. Misc. Publ.* 52:1–247

Flemming AF, Blackburn DG. 2003. Evolution of placental specializations in viviparous African and South American lizards. *J. Exp. Zool. A* 299:33–47

Goodwin NB, Dulvy NK, Reynolds JD. 2002. Life-history correlates of the evolution of live bearing in fishes. *Philos. Trans. R. Soc. London Ser. B* 357:259–67

Göth A, Booth DT. 2004. Temperature-dependent sex ratio in a bird. *Biol. Lett.* 1:31–33

Graves BM, Duvall D. 1995. Aggregation of squamate reptiles associated with gestation, oviposition, and parturition. *Herpetol. Monogr.* 9:102–19

Greene HW. 1997. *Snakes. The Evolution of Mystery in Nature.* Berkeley: Univ. Calif. Press. 351 pp.

Greer AE. 1989. *The Biology and Evolution of Australian Lizards.* Chipping Norton, NSW: Surrey Beatty & Sons. 264 pp.

Grigg GC, Drane CR, Courtice GP. 1979. Time constants of heating and cooling in the eastern water dragon, *Physignathus lesueurii*, and some generalisations about heating and cooling in reptiles. *J. Therm. Biol.* 4:95–103

Harlow PS, Grigg G. 1984. Shivering thermogenesis in a brooding diamond python, *Python spilotes spilotes. Copeia* 1984:959–65

Harlow PS, Taylor JE. 2000. Reproductive ecology of the jacky dragon (*Amphibolurus muricatus*): an agamid lizard with temperature-dependent sex determination. *Aust. Ecol.* 25:640–52

Hau M, Wikelski M, Gwinner H, Gwinner E. 2004. Timing of reproduction in a Darwin's finch: temporal opportunism under spatial constraints. *Oikos* 106:489–500

Heulin B, Ghielmi S, Vogrin N, Surget-Groba Y, Guillaume CP. 2002. Variation in eggshell characteristics and in intrauterine egg retention between two oviparous clades of the lizard *Lacerta vivipara*: insight into the oviparity-viviparity continuum in squamates. *J. Morphol.* 252:255–62

Hodges WL. 2004. Evolution of viviparity in horned lizards (*Phrynosoma*): testing the cold-climate hypothesis. *J. Evol. Biol.* 17:1230–37

Huey RB. 1974. Behavioral thermoregulation in lizards: importance of associated costs. *Science* 184:1001–3

Huey RB. 1982. Temperature, physiology, and the ecology of reptiles. In *Biology of the Reptilia*, ed. C Gans, FH Pough, 12:25–91. London: Academic. 536 pp.

Huey RB. 1991. Physiological consequences of habitat selection. *Am. Nat.* 137:S91–115

Huey R, Slatkin M. 1976. Costs and benefits of lizard thermoregulation. *Q. Rev. Biol.* 51:363–84

Ji X, Brana F. 1999. The influence of thermal and hydric environments on embryonic use of energy and nutrients, and hatchling traits, in the wall lizards (*Podarcis muralis*). *Comp. Biochem. Physiol. A* 124:205–13

Jonsson KI. 1997. Capital and income breeding as alternative tactics of resource use in reproduction. *Oikos* 78:57–66

Kennett R, Georges A, Palmer-Allen M. 1993. Early developmental arrest during immersion of eggs of a tropical freshwater turtle, *Chelodina rugosa* (Testudinata: Chelidae), from northern Australia. *Aust. J. Zool.* 41:37–45

Lee MSY, Shine R. 1998. Reptilian viviparity and Dollo's Law. *Evolution* 52:1441–50

Madsen T, Shine R. 1992. Temporal variability in sexual selection acting on reproductive tactics and body size in male snakes. *Am. Nat.* 141:167–71

Madsen T, Shine R. 1999. The adjustment of reproductive threshold to prey abundance in a capital breeder. *J. Anim. Ecol.* 68:571–80

Madsen T, Shine R. 2000a. Silver spoons and snake sizes: prey availability early in life influences long-term growth rates of free-ranging pythons. *J. Anim. Ecol.* 69:952–58

Madsen T, Shine R. 2000b. Rain, fish and snakes: climatically-driven population dynamics of Arafura filesnakes in tropical Australia. *Oecologia* 124:208–15

Madsen T, Shine R. 2001. Conflicting conclusions from long-term versus short-term studies on growth and reproduction of a tropical snake. *Herpetologica* 57:147–56

Mell R. 1929. *Beiträge zur Fauna Sinica. IV. Grundzüge einer Ökologie der Chinesischen Reptilien und einer Herpetologischen Tiergeographie Chinas.* Berlin: Walter de Gruyter. 282 pp.

Mendez-de la Cruz FR, Villagran-Santa Cruz M, Andrews RM. 1998. Evolution of viviparity in the lizard genus *Sceloporus. Herpetologica* 54:521–32

Nelson NJ, Keall SN, Brown D, Daugherty CH. 2002. Establishing a new wild population of tuatara (*Sphenodon guntheri*). *Conserv. Biol.* 16:887–94

O'Connor D, Shine R. 2004. Parental care protects against infanticide in the lizard *Egernia saxatilis* (Scincidae). *Anim. Behav.* 68:1361–69

Olsson MM, Shine R. 1998. Timing of parturition as a maternal care tactic in an alpine lizard species. *Evolution* 52:1861–64

Ota H, Iwanaga S, Itoman K, Nishimura M,

Mori A. 1991. Reproductive mode of a natricine snake, *Amphiesma pryeri* (Colubridae: Squamata), from the Ryukyu Archipelago, with special reference to the viviparity of *A. p. ishigakiensis. Biol. Mag. Okinawa* 29:37–43

Packard GC, Miller K, Packard MJ. 1993. Environmentally induced variation in body size of turtles hatching in natural nests. *Oecologia* 93:445–48

Packard GC, Packard MJ. 1988. The physiological ecology of reptilian eggs and embryos. In *Biology of the Reptilia*, ed. C Gans, RB Huey, 16:523–605. New York: Liss. 659 pp.

Parker WS, Plummer MV. 1987. Population ecology. See Seigel et al. 1987a, pp. 253–301

Peterson CW. 1987. Daily variation in the body temperatures of free-ranging garter snakes. *Ecology* 68:160–69

Pough FH. 1980. The advantages of ectothermy for tetrapods. *Am. Nat.* 115:92–112

Pough FH, Andrews RM, Cadle JE, Crump ML, Savitsky AH, Wells KD. 1998. *Herpetology*. Upper Saddle River NJ: Prentice Hall. 577 pp.

Qualls CP, Andrews RM. 1999a. Cold climates and the evolution of viviparity in reptiles: cold incubation temperatures produce poor-quality offspring in the lizard, *Sceloporus virgatus. Biol. J. Linn. Soc.* 67:353–76

Qualls CP, Andrews RM. 1999b. Maternal body volume constrains water uptake by lizard eggs in utero. *Funct. Ecol.* 13:845–51

Qualls CP, Andrews RM, Mathies T. 1997. The evolution of viviparity and placentation revisited. *J. Theor. Biol.* 185:129–35

Qualls CP, Shine R. 1995. Maternal body-volume as a constraint on reproductive output in lizards: evidence from the evolution of viviparity. *Oecologia* 103:73–78

Reynolds JD, Goodwin NB, Freckleton RP. 2002. Evolutionary transitions in parental care and live bearing in vertebrates. *Philos. Trans. R. Soc. London Ser. B* 357:269–81

Rhen T, Lang JW. 1999. Temperature during embryonic and juvenile development influ-ences growth in hatchling snapping turtles, *Chelydra serpentina. J. Therm. Biol.* 24:33–41

Robert KA, Thompson MB. 2001. Viviparous lizard selects sex of embryos. *Nature* 412:698–99

Rock J, Cree A. 2003. Intraspecific variation in the effect of temperature on pregnancy in the viviparous gecko *Hoplodactylus maculatus. Herpetologica* 59:8–22

Salthe SN. 1969. Reproductive modes and the number and sizes of urodeles. *Am. Midl. Nat.* 81:467–90

Sarre SD, Georges A, Quinn A. 2004. The ends of a continuum: genetic and temperature-dependent sex determination in reptiles. *BioEssays* 26:639–45

Schulte JAI, Macey JR, Espinoza RE, Larson A. 2000. Phylogenetic relationships in the iguanid lizard genus *Liolaemus*: multiple origins of viviparous reproduction and evidence for recurring Andean vicariance and dispersal. *Biol. J. Linn. Soc.* 69:75–102

Schwaner TD, Sarre SD. 1990. Body size and sexual dimorphism in mainland and island Tiger Snakes. *J. Herpetol.* 24:320–22

Seigel RA, Collins JT, Novak SS, eds. 1987a. *Snakes: Ecology and Evolutionary Biology*. New York: Macmillan. 414 pp.

Seigel RA, Fitch HS. 1985. Annual variation in reproduction in snakes in a fluctuating environment. *J. Anim. Ecol.* 54:497–505

Seigel RA, Ford NB. 1987. Reproductive ecology. See Seigel et al. 1987a, pp. 210–52

Seigel RA, Huggins MM, Ford NB. 1987b. Reduction in locomotor ability as a cost of reproduction in snakes. *Oecologia* 73:481–85

Sergeev AM. 1940. Researches in the viviparity of reptiles. *Mosc. Soc. Nat.* (Jubilee Issue):1–34

Servan J, Zaborski P, Dorizzi M, Pieau C. 1989. Female-biased sex ratio in adults of the turtle *Emys obicularis* at the northern limit of its distribution in France: a probable consequence of interaction of temperature and genotypic sex determination. *Can. J. Zool.* 67:1279–84

Shine R. 1978. Growth rates and sexual maturation in six species of Australian elapid snakes. *Herpetologica* 34:73–79

Shine R. 1980. "Costs" of reproduction in reptiles. *Oecologia* 46:92–100

Shine R. 1983a. Reptilian reproductive modes: the oviparity-viviparity continuum. *Herpetologica* 39:1–8

Shine R. 1983b. Reptilian viviparity in cold climates: testing the assumptions of an evolutionary hypothesis. *Oecologia* 57:397–405

Shine R. 1985. The evolution of viviparity in reptiles: an ecological analysis. In *Biology of the Reptilia*, ed. C Gans, F Billett, 15:605–94. New York: Wiley. 731 pp.

Shine R. 1988. Parental care in reptiles. In *Biology of the Reptilia*, ed. C Gans, RB Huey, 16:275–330. New York: Liss. 659 pp.

Shine R. 1992. Relative clutch mass and body shape in lizards and snakes: is reproductive investment constrained or optimized? *Evolution* 46:828–33

Shine R. 1994. Sexual size dimorphism in snakes revisited. *Copeia* 1994:326–46

Shine R. 1995. A new hypothesis for the evolution of viviparity in reptiles. *Am. Nat.* 145:809–23

Shine R. 1999. Egg-laying reptiles in cold climates: determinants and consequences of nest temperatures in montane lizards. *J. Evol. Biol.* 12:918–26

Shine R. 2002a. Eggs in autumn: responses to declining incubation temperatures by the eggs of montane lizards. *Biol. J. Linn. Soc.* 76:71–77

Shine R. 2002b. Reconstructing an adaptationist scenario: what selective forces favor the evolution of viviparity in montane reptiles? *Am. Nat.* 160:582–93

Shine R. 2004a. Adaptive consequences of developmental plasticity. See Deeming 2004b, pp. 187–210

Shine R. 2004b. Does viviparity evolve in cold climate reptiles because pregnant females maintain stable (not high) body temperatures? *Evolution* 58:1809–18

Shine R. 2004c. Seasonal shifts in nest temperature can modify the phenotypes of hatchling lizards, regardless of overall mean incubation temperature. *Funct. Ecol.* 18:43–49

Shine R, Elphick M, Barrott EG. 2003. Sunny side up: lethally high, not low, temperatures may prevent oviparous reptiles from reproducing at high elevations. *Biol. J. Linn. Soc.* 78:325–34

Shine R, Elphick M, Donnellan S. 2002a. Co-occurrence of multiple, supposedly incompatible modes of sex determination in a lizard population. *Ecol. Lett.* 5:486–89

Shine R, Greer AE. 1991. Why are clutch sizes more variable in some species than in others? *Evolution* 45:1696–706

Shine R, Harlow PS. 1996. Maternal manipulation of offspring phenotypes via nest-site selection in an oviparous reptile. *Ecology* 77:1808–17

Shine R, Harlow PS, Keogh JS, Boeadi. 1998a. The allometry of life-history traits: insights from a study of giant snakes (*Python reticulatus*). *J. Zool.* 244:405–14

Shine R, Harlow PS, Keogh JS, Boeadi. 1998b. The influence of sex and body size on food habits of a giant tropical snake, *Python reticulatus*. *Funct. Ecol.* 12:248–58

Shine R, Madsen T. 1997. Prey abundance and predator reproduction: rats and pythons on a tropical Australian floodplain. *Ecology* 78:1078–86

Shine R, Sun L, Zhao E, Bonnet X. 2002b. A review of 30 years of ecological research on the Shedao pit-viper. *Herpetol. Nat. Hist.* 9:1–14

Sinervo B, Doughty P, Huey RB, Zamudio K. 1992. Allometric engineering: a causal analysis of natural selection on offspring size. *Science* 285:1927–30

Sinervo B, Licht P. 1991. Proximate constraints on the evolution of egg size, number, and total clutch mass in lizards. *Science* 252:1300–2

Somma LA. 1990. A categorization and bibliographic survey of parental behavior in lepidosaurian reptiles. *Smithson. Herpetol. Info. Serv.* 81:1–53

Somma LA. 2003. *Parental Behavior in Lepidosaurian and Testudinian Reptiles*. Malabar FL: Krieger. 174 pp.

Staub RE, Emberton J. 2002. *Eryx jayakari* (Arabian Sand Boa). Reproduction. *Herpetol. Rev.* 33:214

Uller T, Olsson M. 2003. Prenatal sex ratios influence sexual dimorphism in a reptile. *J. Exp. Zool. A* 295:183–87

Van Damme R, Bauwens D, Brana F, Verheyen RF. 1992. Incubation temperature differentially affects hatching time, egg survival and sprint speed in the lizard *Podarcis muralis*. *Herpetologica* 48:220–28

Vitt LJ. 1981. Lizard reproduction: habitat specificity and constraints on relative clutch mass. *Am. Nat.* 117:506–14

Vitt LJ, Congdon JD. 1978. Body shape, reproductive effort, and relative clutch mass in lizards: resolution of a paradox. *Am. Nat.* 112:595–608

Vitt LJ, Price HJ. 1982. Ecological and evolutionary determinants of relative clutch mass in lizards. *Herpetologica* 38:237–55

Wake MH, Dickie R. 1998. Oviduct structure and function and reproductive modes in amphibians. *J. Exp. Zool.* 282:477–506

Wapstra E, Olsson M, Shine R, Edwards A, Swain R, Joss JMP. 2004. Maternal basking behaviour determines offspring sex in a viviparous reptile. *Proc. R. Soc. London Ser. B* 271:S230–32

Warner DA, Andrews RM. 2002. Nest-site selection in relation to temperature and moisture by the lizard *Sceloporus undulatus*. *Herpetologica* 58:399–407

Webb G, Manolis C. 2002. *Australian Crocodiles: A Natural History*. Sydney: Reed New Holland. 160 pp.

Weekes HC. 1933. On the distribution, habitat and reproductive habits of certain European and Australian snakes and lizards, with particular regard to their adoption of viviparity. *Proc. Linn. Soc. NSW* 58:270–74

Wikelski M, Thom C. 2000. Marine iguanas shrink to survive El Niño. *Nature* 403:37–38

Annu. Rev. Ecol. Evol. Syst. 2005. 36:47–79
doi: 10.1146/annurev.ecolsys.36.091704.175539
Copyright © 2005 by Annual Reviews. All rights reserved
First published online as a Review in Advance on July 25, 2005

THE EVOLUTIONARY ENIGMA OF MIXED MATING SYSTEMS IN PLANTS: Occurrence, Theoretical Explanations, and Empirical Evidence

Carol Goodwillie,[1] Susan Kalisz,[2] and Christopher G. Eckert[3]

[1]*Department of Biology, East Carolina University, Greenville, North Carolina 27858;*
email: goodwilliec@mail.ecu.edu
[2]*Department of Biological Sciences, University of Pittsburgh, Pittsburgh,*
Pennsylvania 15260; email: kalisz+@pitt.edu
[3]*Department of Biology, Queen's University, Kingston, Ontario K7L 3N6 Canada;*
email: eckertc@biology.queensu.ca

Key Words inbreeding, intermediate outcrossing, pollination, reproductive
strategy, selfing

■ **Abstract** Mixed mating, in which hermaphrodite plant species reproduce by both
self- and cross-fertilization, presents a challenging problem for evolutionary biologists.
Theory suggests that inbreeding depression, the main selective factor opposing the evolution of selfing, can be purged with self-fertilization, a process that is expected to yield
pure strategies of either outcrossing or selfing. Here we present updated evidence suggesting that mixed mating systems are frequent in seed plants. We outline the floral and
pollination mechanisms that can lead to intermediate outcrossing, review the theoretical models that address the stability of intermediate outcrossing, and examine relevant
empirical evidence. A comparative analysis of estimated inbreeding coefficients and
outcrossing rates suggests that mixed mating often evolves despite strong inbreeding
depression. The adaptive significance of mixed mating has yet to be fully explained for
any species. Recent theoretical and empirical work suggests that future progress will
come from a better integration of studies of floral mechanisms, genetics, and ecology,
and recognition of how selective pressures vary in space and time.

INTRODUCTION

Evolutionary biologists have long been interested in mixed strategies, in which a
species uses more than one tactic, each with distinct fitness consequences (Dawkins
1980, Maynard Smith 1982). For example, organisms exhibit mixed strategies for
defense against enemies (Fornoni et al. 2004), timing of reproduction (Satake et al.
2001), dormancy (Spencer et al. 2001), and dispersal of offspring (Venable 1985).
The maintenance of alternative tactics within populations is presumed to involve

frequency-dependent selection or condition-dependence where the tactic an individual uses is influenced by its physiological state or environmental circumstance (Maynard Smith 1982). However, the evolutionary processes that maintain mixed strategies have remained controversial (Flaxman 2000, Plaistow et al. 2004).

This review focuses on the evolution of mixed mating strategies in hermaphroditic plants, where reproduction occurs by both self-fertilization (selfing) and mating with other individuals (outcrossing). The evolutionary transition from outcrossing to predominant selfing has occurred in many plant groups, earning it the reputation as a pathway that "has probably been followed by more different lines of evolution in flowering plants than has any other" (Stebbins 1974). The rate of selfing can vary widely among closely related species and even among populations within species, suggesting that it can respond rapidly to natural selection (Jain 1976). Moreover, the evolutionary shift toward self-fertilization is associated with changes in floral biology, life history, and ecology (Barrett et al. 1996, Ornduff 1969), and has manifold consequences for population genetics and evolution (Charlesworth 1992, Grant 1981, Stebbins 1974, Takebayashi & Morrell 2001). Accordingly, questions surrounding the evolution of self-fertilization and the maintenance of mixed mating systems have remained at the forefront of evolutionary biology.

Whether mixed mating systems of plants are evolutionarily stable has been especially controversial. Here we (*a*) provide a historical and theoretical background to this controversy, (*b*) present new comparative data on the occurrence of mixed mating systems, (*c*) outline the functional mechanisms of mixed mating in terms of floral biology and pollination, (*d*) review theoretical explanations for evolutionarily stable mixed mating systems and the progress with empirically testing them, and (*e*) identify worthwhile opportunities for future research.

HISTORICAL AND THEORETICAL BACKGROUND

Inbreeding Depression Versus Automatic Selection

Like many enduring problems in evolutionary biology, interest in the evolution of mating systems is rooted in the work of Darwin and Fisher (reviewed in Holsinger 1996). Darwin's (1876) experimental demonstrations of inbreeding depression led him to question the adaptive value of self-fertilization. In contrast, Fisher (1941) showed that an allele for self-fertilization should spread in an outcrossing population because it increases its own transmission. Outcrossing individuals are mother to their own seed, whereas selfers are both mother and father to their own seed. All else being equal, an allele for selfing will spread if selfed progeny are at least half as fit as outcrossed progeny (see also Lloyd 1979, Nagylaki 1976).

Twenty years ago, Lande & Schemske (1985) showed that inbreeding depression, the main genetic cost of selfing, is not a constant but should evolve jointly with the level of selfing. As a population inbreeds, deleterious recessive or partially recessive alleles that cause inbreeding depression are expressed in homozygous form and can be purged by selection. This positive feedback facilitates the spread of an allele that increases selfing in a population that is already partially selfing. Consequently, Lande & Schemske (1985) predicted only two stable endpoints of mating system evolution: predominant outcrossing with strong inbreeding depression ($\delta = 1 -$ [fitness of selfed progeny/fitness of outcrossed progeny] > 0.5) and predominant selfing with weak inbreeding depression ($\delta < 0.5$). To test this hypothesis, Schemske & Lande (1985) collected the available estimates of the proportion of seeds produced through outcrossing (t) for plant species. The generally bimodal frequency distribution of those data supported their prediction and led to the suggestion that mixed mating in many species might represent transitional states or incidental by-products of other adaptive mechanisms (Schemske & Lande 1985). Building on the groundbreaking work of Lloyd (1979), Allard (1975), Jain (1976) and others, this provocative combination of theory and comparative data set the stage for a renaissance of work on plant mating system evolution.

The Distribution of Mating Systems Among Seed Plants

The distribution of outcrossing rates in seed plants is central and contentious evidence in the debate on the stability of mixed mating systems (Aide 1986, Barrett & Eckert 1990, Schemske & Lande 1987, Waller 1986, Vogler & Kalisz 2001), with controversy surrounding both the interpretation and adequacy of data. As a starting point for our review, we updated Schemske & Lande's (1985) analysis of 55 species to include estimates now available for 345 species in 78 families (data available from the authors on request). The distributions from the two surveys differ significantly (likelihood ratio $\chi^2 = 21.1$, df $= 4$, $P = 0.0003$; see Figure 1). Compared to the 1985 analysis, the expanded survey shows a much lower representation of predominantly selfing taxa ($t \leq 0.2$, $\chi^2 = 23.7$, df $= 1$, $P < 0.0001$) such that the marked bimodality of the earlier distribution is no longer apparent. The frequency of species exhibiting what has broadly been classified as mixed mating systems ($0.2 < t \leq 0.8$) has increased from 31% in Schemske & Lande's (1985) analysis to 42% in ours, although this difference is not quite significant ($\chi^2 = 2.5$, df $= 1$, $P = 0.11$).

As with all previous comparative analyses of plant mating system variation, our results should be viewed with caution. Although the updated survey is extensive, the 345 species included are far from a random sample of seed plants. For instance, 44% of the species belong to five families: Fabaceae, Pinaceae, Myrtaceae (mostly *Eucalyptus*), Asteraceae, and Poaceae. Thus, our sample may not be representative of the overall distribution of mating systems, and lack of phylogenetic independence may bias inferences regarding associations between mating systems and ecological or life-history traits (but see Barrett et al. 1996). Importantly, there is likely a bias against estimating t for taxa that are expected to be either completely outcrossing (e.g., strongly self-incompatible) or wholly selfing, and lack of polymorphic markers can preclude estimation of t for highly selfing species. It seems unlikely, however, that the addition of these species would yield a markedly

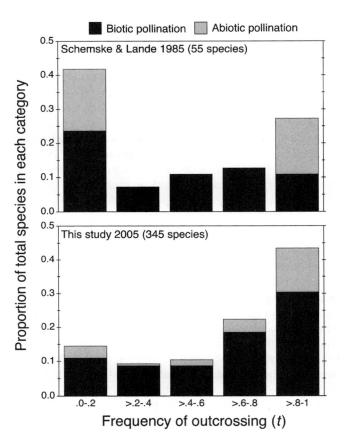

Figure 1   The distribution of estimated outcrossing rates in species of seed plants. The *top panel* shows the distribution of 55 species presented in Schemske & Lande (1985). The *bottom panel* shows the updated distribution of 345 species. The species in each of the five outcrossing categories are classified as biotically pollinated (usually insects, birds or bats) or abiotically pollinated (wind or water). All estimates of *t* are based on at least five maternal plants in natural populations or taken from natural populations and pollinated in a common garden and were derived from the assay of open-pollinated progeny arrays using genetic markers (usually allozyme polymorphisms). When estimates of *t* were available for more than one population of a given species, we averaged across populations. When individual populations were studied for multiple years, we averaged *t* across years before averaging across populations (following Schemske & Lande 1985, Barrett & Eckert 1990).

bimodal distribution, given the relative rarity of strongly selfing taxa (Takebayashi & Morrell 2001). "Apparent selfing" due to biparental inbreeding (sensu Ritland 1984), which is functionally and evolutionarily distinct from true selfing, may produce estimates of $t < 1$ even in fully self-incompatible (SI) or outcrossing species. Finally, estimates of t may differ from the primary, or initial, rate of outcrossing when selection against selfed zygotes occurs between fertilization and sampling of progeny arrays. These issues notwithstanding, our results indicate that mixed mating systems occur frequently in a wide variety of plants, motivating continued exploration of the factors that promote their evolutionary stability.

Emerging Importance of Pollination Ecology

As in previous analyses (Aide 1986, Barrett & Eckert 1990), we find that the distribution of t differs significantly between biotically and abiotically pollinated species (Figure 1; $\chi^2 = 13.1$, df $= 4$, $P = 0.011$), with animal-pollinated taxa almost twice as likely to exhibit a mixed mating system (46.4% of 267 species) than those pollinated by wind or water (26.9% of 78 species, $\chi^2 = 9.8$, df $= 1$, $P = 0.0017$). A higher resolution graphical analysis (Figure 2, following Vogler & Kalisz 2001) indicates that the distribution of t for animal-pollinated species

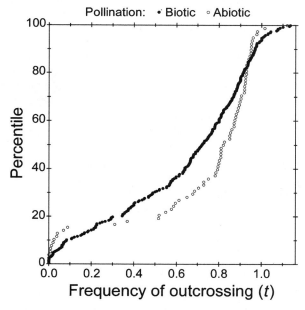

Figure 2   High-resolution distribution of estimated outcrossing (t) for 267 species with biotic pollination (*closed circles*) and 78 species with abiotic pollination (*open circles*). For each pollination mode, species were ranked by estimated t, and ranks were converted to percentiles (y-axis).

is almost continuous, whereas estimates of t for abiotically pollinated taxa are concentrated at either endpoint. This may indicate that variability in biotic pollination limits the rate of evolutionary response to selection on the mating system (Schemske & Lande 1986). An alternative explanation is that factors that lend evolutionary stability to mixed mating systems are more prevalent in animal-than in wind-pollinated taxa (Aide 1986, Barrett & Eckert 1990, Barrett et al. 1996, Vogler & Kalisz 2001). Thus, the effects of pollination ecology demand consideration.

Beginning with the landmark work of Lloyd (1979), theoretical studies of the past few decades clarify how selective pressures related to pollination ecology can influence the evolution of mating systems (reviewed in Holsinger 1996). First, selfing can be selected as a mechanism of reproductive assurance, allowing plants to produce offspring when pollinators and/or potential mates are scarce (Lloyd 1979, 1992). Second, selfing can incur the costs of gamete discounting, where self-fertilization reduces production of outcrossed seed (seed discounting, Lloyd 1992), siring of outcrossed seed on other individuals (pollen discounting, Harder & Wilson 1998, Schoen et al. 1996), or both. Third, the ecological costs and benefits of selfing depend on how and when self-fertilization occurs (Lloyd & Schoen 1992). As discussed below, some models incorporating these costs and benefits predict stable mixed mating, addressing the disparity between previous theory and the occurrence of these mating systems in natural populations.

THE MECHANICS OF MIXED MATING

Mixed mating, in which the population outcrossing rate departs significantly from both zero and one, can result from three types of reproductive systems (Cruden & Lyon 1989). First, a genetically based selfing rate polymorphism can exist, as for instance in the relatively rare case where populations contain both self-compatible and self-incompatible individuals (Stone 2002). Second, species can exhibit heteromorphic flower systems (Masuda et al. 2004, Schoen & Lloyd 1984), such as cleistogamous (purely selfing) and chasmogamous flowers (both outcrossing and selfing possible). In the third and by far most common system, individual plants produce a single flower type, and fruits may contain selfed, outcrossed, or a mixture of progeny types (Schoen & Brown 1991). Here, the proportions of selfed versus outcrossed progeny are determined by the timing and relative amount of self- and outcross-pollination and postpollination processes (Holsinger 1991, Kalisz et al. 2004, Lloyd & Schoen 1992).

Modes of Selfing and Their Selective Effects

Self-pollination contributing to mixed mating can occur within (autogamy) or among (geitonogamy) flowers on an individual plant, and within-flower selfing can be either vector-mediated (facilitated) or autonomous. Three modes of autonomous selfing can be distinguished with respect to the period of potential outcross pollen receipt (Lloyd 1979, Lloyd & Schoen 1992). On one end of the continuum is prior selfing, which occurs before the receipt of outcross pollen. Prior selfing is conferred by early spatial and developmental overlap of male and female functions within a flower (e.g., Fishman & Wyatt 1999), and self-pollination may take place before the flower opens. Competing selfing occurs concurrently with outcross pollen receipt (e.g., Leclerc-Potvin & Ritland 1994), and delayed selfing occurs after the opportunity for outcross pollen receipt has passed. Delayed selfing is achieved through a variety of processes, including developmental changes in stigma-anther position and floral age-dependent breakdown in SI (Lloyd & Schoen 1992, reviewed in Kalisz et al. 1999). With outcross pollen limitation, all modes of autonomous selfing can confer reproductive assurance with little or no pollen and seed discounting (Lloyd 1992, Schoen & Brown 1991). When pollen is not limited, prior and competing selfing can incur costs of gamete discounting, but delayed selfing avoids such costs, because it occurs only after opportunities for pollen export and deposition of outcross pollen. On this basis, delayed selfing is expected to be nearly always advantageous.

In reality, autonomous selfing does not fall into three strict classes nor is floral lifespan invariant. Rather, the timing of autonomous self-pollination is continuously distributed, can be related to floral age or developmental stage (Kalisz & Vogler 2003), and can reflect plasticity in floral longevity (Arathi et al. 2002). A further complication is that the distribution of floral resources may not reflect the timing of pollination. For example, selfed ovules might compete with outcrossed ovules for resources even with delayed selfing.

Facilitated selfing occurs when pollinators transfer pollen from anther to stigma within a flower. This mode of selfing could provide reproductive assurance when mates are scarce, but generally not when pollinators are limiting (but see Anderson et al. 2003). Facilitated selfing is expected to extract high selective costs because of gamete discounting (Lloyd 1992), which has been shown in natural populations (Eckert 2000, Ushimaru & Kikuzawa 1999). Geitonogamy occurs when a biotic or abiotic vector moves self-pollen among flowers on the same plant, the rate of which may be a function of the number of simultaneously open flowers and the pattern of pollinator movement (Eckert 2000, Harder & Barrett 1995, Karron et al. 2004). Like facilitated selfing, geitonogamy offers no reproductive assurance when pollinators are scarce and causes severe or total gamete discounting. Thus, it is considered an unavoidable by-product of selection for outcrossing success if multiple flowers are required to attract a pollinator (reviewed by de Jong et al. 1993, Harder & Barrett 1995) and may impose particularly high costs in pollen discounting because it occurs after pollen has been successfully placed for potential outcross export on the pollinator's body (Lloyd 1992). Although different modes of selfing can be distinguished, some aspects of floral variation are likely to affect more than one mode of selfing (Schoen et al. 1996). For instance, reduced herkogamy (spatial separation of stigma and anthers) or dichogamy (temporal separation of male and female functions) that promotes delayed selfing may also increase the potential for competing or facilitated selfing and enhance the probability of geitonogamy (see Leclerc-Potvin & Ritland 1994). In this case, selection for reproductive assurance through delayed selfing could result in indirect selection for competing autogamy and geitonogamy (Schoen et al. 1996).

Postpollination Processes

The measured outcrossing rate is frequently higher than that expected from the relative amounts of self versus cross pollen deposited on the stigma (Husband & Schemske 1996). This selective filtering of pollen or zygotes can occur in a variety of forms, including the rejection of self pollen or selfed ovules with SI (de Nettancourt 1997, Dickinson 1994, Lipow & Wyatt 2000, Seavey & Bawa 1986), cryptic SI caused by differential pollen tube growth (Cruzan & Barrett 1993, Eckert & Allen 1997, Jones 1994, Weller & Ornduff 1977), the differential provisioning of outcrossed versus selfed embryos or fruits (Korbecka et al. 2002, Marshall & Ellstrand 1986, Rigney 1995, Stephenson 1981), and the death of selfed embryos expressing lethal recessive alleles (Husband & Schemske 1996, Lande et al. 1994). It is often difficult to identify the actual mechanism of postpollination selection. For example, some forms of late-acting self-incompatibility are difficult to distinguish from early-inbreeding depression (Seavey & Bawa 1986). Many postpollination mechanisms can be environmentally and developmentally plastic (Becerra & Lloyd 1992, Goodwillie et al. 2004, Marshall & Diggle 2001, Marshall & Ellstrand 1986, Vogler et al. 1998). In addition, pre- and postpollination processes can jointly influence the timing of self-fertilization. For instance, in a species of *Agalinus*, self-pollination occurs in the floral bud, but the germination of this self-pollen is delayed (Stewart et al. 1996). In summary, the mechanics of mixed mating are varied and interacting and have complex consequences for selection and the potential for evolution.

THE EVOLUTIONARY STABILITY OF INTERMEDIATE OUTCROSSING: A REVIEW OF THE THEORY AND EMPIRICAL EVIDENCE

The evolution of mixed mating poses a general theoretical problem: What prevents any pure strategy from going to fixation? The stabilizing effects of a wide range of ecological, genetic, and demographic factors have been explored in theoretical models (Table 1). In some, mixed mating is maintained by frequency dependence or some form of negative feedback, whereby increased selfing weakens a factor that selects for selfing. In others, stochastic variation in a selective factor promotes stability of mixed mating through the general principle of nonlinear averaging, or Jensen's inequality. Framed in an evolutionary context, this principle states that the fitness effect in the average environment will differ systematically from the fitness averaged across all environments (Cheptou & Schoen 2002, Morgan & Wilson 2005).

With the striking exception of the "mass action" model of Holsinger (1991), which predicts that selfing can evolve and mixed mating can be maintained

TABLE 1 Selected models that predict stability of mixed mating systems in plants. PD = pollen discounting, RA = reproductive assurance (represented in most cases by seed discounting parameter), δ = inbreeding depression. Models with "fixed" inbreeding depression do not consider its evolutionary dynamics, although expression may be variable. Monomorphic stability describes the evolution and maintenance of a single phenotype with an intermediate rate of outcrossing. With polymorphic stability, variation for outcrossing rate is maintained within populations

Model citation	Phenotypic or genetic	Ecological/ pollination parameters	Inbreeding depression (δ)	Conditions required for stability	Monomorphic/ polymorphic stability
Inbreeding depression					
Holsinger 1986	Genetic	PD, seed dispersal	Fixed, relative fitness of selfed progeny differ depending on dispersal	PD < 1, selfed progeny less successful as migrants, high migration rate	Polymorphic
Holsinger 1988	Genetic	PD	Evolves, mating system and fitness loci not independent	δ overdominant	Monomorphic
Charlesworth & Charlesworth 1990	Genetic		Evolves, mating system and fitness loci not independent	δ overdominant	Both
Uyenoyama & Waller 1991b	Genetic	PD	Evolves, mating system and fitness loci not independent	δ overdominant, low PD	Monomorphic
Latta & Ritland 1993	Genetic	PD	Fixed, modeled in consecutive generations of selfing	Fitness declines with generations of selfing, selfing rate polygenic	Both

(Continued)

TABLE 1 (*Continued*)

Model citation	Phenotypic or genetic	Ecological/ pollination parameters	Inbreeding depression (δ)	Conditions required for stability	Monomorphic/ polymorphic stability
Latta & Ritland 1994	Genetic		Evolves, modeled in consecutive generations of selfing	Fitness declines with generations of selfing, δ caused by weakly deleterious alleles	Monomorphic
Rausher & Chang 1999	Genetic		Fixed, differs in male and female function	δ moderate, lower in male than in female expression	Both
Cheptou & Mathias 2001	Phenotypic		Fixed, spatial/temporal variation	δ temporally variable	Both
Cheptou & Schoen 2002	Phenotypic		Fixed, variation among years/generations	δ varies among years/generations	Monomorphic
Cheptou & Dieckmann 2002	Phenotypic	Density regulation, competition	Fixed, expression density-dependent	Wide range, depends on competitive interactions between inbred and outbred progeny	Monomorphic
Biparental inbreeding					
Uyenoyama 1986	Genetic	Biparental inbreeding	Fixed	Biparental inbreeding occurs	Monomorphic
Ronfort & Couvet 1995	Genetic	Population dynamics, limited pollen and seed dispersal	Evolves, purging depends on population structure	Density-dependent recruitment, limited pollen and seed dispersal	Monomorphic

Pollen discounting					
Holsinger 1991	Genetic	PD	None	Whenever selfing can evolve	Polymorphic
Johnston 1998	Phenotypic	PD, RA	Evolves, partial dominance	PD increases with selfing rate, or selfing reduces seed set	Monomorphic
Porcher & Lande 2005c	Both	PD, RA	Evolves, recessive lethals and nearly additive mildly deleterious alleles	Without pollen limitation, substantial PD required. With pollen limitation, $0.9 <$ selfing rate < 1 for all levels of δ and PD	Monomorphic
Pollen limitation/variability					
Schoen & Brown 1991	Phenotypic	RA, variable pollination	Fixed	Pollination environment variable, $\delta < 1$	Monomorphic
Sakai & Ishii 1999	Phenotypic	RA, variable pollination, seed size/number trade-off	Fixed	Optimal seed size small, pollen delivery low and unpredictable	Monomorphic
Vallejo-Marin & Uyenoyama 2004	Genetic	RA, SI	Fixed	Pollen limitation, low S-allele diversity	Monomorphic

(Continued)

TABLE 1 (*Continued*)

Model citation	Phenotypic or genetic	Ecological/ pollination parameters	Inbreeding depression (δ)	Conditions required for stability	Monomorphic/ polymorphic stability
Morgan & Wilson 2005	Phenotypic	RA, PD, variable pollination, population dynamics	Fixed	Delayed selfing always selected, variable pollination allows intermediate prior selfing	Monomorphic
Resource allocation					
Iwasa 1990	Phenotypic	RA	Fixed	High resource availability, low pollen availability	Monomorphic
Sakai 1995	Phenotypic	RA, PD, pollinator visitation a function of flower size/number	Fixed	Nonlinear constraints on flower size and number with delayed and competing selfing	Monomorphic

regardless of the relative fitness of selfed and outcrossed progeny (see below), most plant mating system theory includes inbreeding depression, which is the key factor that opposes the advantages of selfing. Where the stability of mixed mating is explained theoretically by variation or frequency dependence in other factors, most models find that its maintenance can occur only with certain values of inbreeding depression, although these vary widely among models. In light of Lande & Schemske's (1985) result, the added challenge is to understand the maintenance of some outcrossing given that selective purging of inbreeding depression can generate positive feedback for the evolution of selfing. Most theory on mixed mating assumes a constant value for inbreeding depression (Table 1), and therefore does not account for this potentially destabilizing force. The magnitude of inbreeding depression in species with mixed mating systems and the extent to which purging occurs, then, is relevant to our understanding of mating system dynamics. We turn first to theory and empirical work that explores the evolution of inbreeding depression with the mating system.

The Evolutionary Dynamics of Inbreeding Depression

THEORY   In Lande & Schemske's (1985) model, partial selfing was found to reduce equilibrium levels of inbreeding depression, and purging occurred more easily for nearly recessive lethals than for more nearly additive mildly deleterious mutations. The existence of both types of mutation (Charlesworth & Charlesworth 1999, Husband & Schemske 1996) suggests that their combined effect on the dynamics of inbreeding depression are complex. Inbreeding depression may be difficult to purge if the number of lethal recessive alleles maintained with outcrossing is sufficiently high. In this case, very few selfed progeny survive, thus opportunities for purging are diminished (Lande et al. 1994). Other models of partial dominance relax the assumption that the loci influencing inbreeding depression and selfing are independent (Charlesworth et al. 1990, Holsinger 1988, Uyenoyama & Waller 1991a). Holsinger (1988) showed that this allows inbreeding depression to be selectively purged to a greater extent in more selfing lineages. As a result, selfing can evolve at well above the standard threshold of $\delta \leq 0.5$. However, with inbreeding depression caused by partial dominance, interactions between fitness and mating system loci have not been shown to stabilize mixed mating.

In contrast, inbreeding depression caused by overdominance (heterozygote advantage) is expected to increase with selfing under some conditions (Charlesworth & Charlesworth 1987), providing a negative feedback that can maintain mixed mating. This prediction also holds for models that consider the evolution of multilocus genotypes (Charlesworth & Charlesworth 1990, Holsinger 1988, Uyenoyama & Waller 1991b). Several models build on the expectation that inbreeding depression will increase over generations of selfing. Maynard Smith (1977) first suggested that this could stabilize intermediate outcrossing if the selective consequences of selfing differ for inbred and outcrossed plants (see also Damgaard et al. 1992). Using an explicit genetic model, Latta & Ritland (1993) found that this process is most likely to yield an intermediate t when the selfing rate is controlled by many loci. When purging occurs concurrently, intermediate t can be stable if inbreeding depression is caused by partially recessive, weakly deleterious alleles (Latta & Ritland 1994). In other words, the short-term negative feedback of consecutive inbreeding on the fitness effects of an allele for increased selfing can more than compensate for the long-term positive feedback of purging, when selection on fitness loci is very mild. Recent theory also suggests that the extent and consequences of purging can be affected by reproductive compensation, a process by which aborted zygotes are replaced by viable ones when resources limit seed production. Reproductive compensation is expected to modify the relations between selfing, inbreeding depression, and genetic load and, under restricted conditions, may stabilize mixed mating (Porcher & Lande 2005b).

EMPIRICAL INVESTIGATION   Do the dynamics of inbreeding depression play a central role in either promoting or preventing the stability of intermediate outcrossing? Theory tells us that the answer hinges on its genetic basis. Experiments in which various fitness components are measured through consecutive generations of inbreeding generally detect purging (reviewed in Crnokrak & Barrett 2002), which is inconsistent with the overdominance hypothesis and suggests that partial dominance is the basis of inbreeding depression. Biometric investigations of the genetic basis of inbreeding depression also largely reject overdominance (reviewed in Carr & Dudash 2003, Charlesworth & Charlesworth 1999). Quantitative trait locus (QTL) studies have detected apparently overdominant fitness loci, but these are difficult to distinguish from the effect of deleterious recessive alleles linked in repulsion (reviewed in Carr & Dudash 2003). Experimental evidence for rapid purging is also inconsistent with the hypothesis that intermediate outcrossing is maintained by short-term declines in inbred fitness with recurrent selfing (Damgaard et al. 1992; Latta & Ritland 1993, 1994; Maynard Smith 1977). Declining inbred fitness has been demonstrated in cultivated plant species (Hallauer & Sears 1973), but generally not in natural populations (Dudash 1990, Schoen 1983). In summary, there is relatively little evidence that changes in inbreeding depression associated with the mating system play a positive role in maintaining mixed mating systems.

What is the evidence that purging of inbreeding depression represents a strong destabilizing force? A comparative study of 54 species found significantly lower inbreeding depression in primarily selfing species (mean $\delta = 0.23$) than in primarily outcrossing species (mean $\delta = 0.53$), suggesting that purging is common (Husband & Schemske 1996). In contrast, a meta-analysis of studies comparing inbreeding depression in related species that differ in inbreeding history found only equivocal support for purging (Byers & Waller 1999). The incongruence between the results of these two surveys could reflect differences in timescale; that is, purging might be more consistently detectable in the long term than in the divergence of closely related species or populations. Taken together, experimental and comparative studies indicate that purging occurs but is inconsistent across plant taxa.

Recent theoretical results suggest that a range of factors such as population size and structure can influence purging and standing levels of inbreeding depression (Glemin 2003, Whitlock et al. 2000) and could account for some of this variation.

AN ALTERNATIVE EMPIRICAL ASSESSMENT OF INBREEDING DEPRESSION AND MATING SYSTEMS   The strength of inbreeding depression can vary with environmental conditions and is often expressed more strongly under field than greenhouse conditions (Roff 1997). Thus, the evolutionary consequences of inbreeding depression may be hard to predict from studies conducted under benign experimental conditions. An alternative approach is to estimate inbreeding depression from the inbreeding coefficient, F, of mature plants relative to the expected F of progeny based on the selfing rate ($s = 1 - t$; Ritland 1990). In the absence of inbreeding depression, the expected equilibrium value of F for mature plants is $F_e = s/(2 - s)$. Inbreeding depression (δ) reduces F to $F_e = sw/(2 - 2s + sw)$, where w is the fitness of selfed offspring compared to outcrossed offspring (i.e., $w = 1 - \delta$). Inbreeding depression can, therefore, be estimated as:

$$\delta = 1 - 2\left[\frac{(1 - s)F}{s(1 - F)}\right].$$

This estimator of δ assumes that populations are at inbreeding equilibrium, selfing is the only form of inbreeding, and the marker polymorphisms are neutral and not physically linked to polymorphic loci affecting fitness (Ritland 1990, Charlesworth 1991; for additional discussion of the assumptions of this method see Eckert & Barrett 1994, Routley et al. 1999). It integrates episodes of mortality in natural populations from zygote formation, seed maturation, and dispersal, through survival to reproductive maturity over several years and has generally revealed very strong inbreeding depression (Eckert & Barrett 1994, Herlihy & Eckert 2002, Kohn & Biardi 1995, Routley et al. 1999).

We plotted the estimated F of mature plants against $s = 1 - t$ for 150 species where both parameters were estimated (Figure 3). Overall, mature plants were generally much less inbred than expected in the absence of inbreeding depression. For 88% of the 64 species that exhibited broadly defined mixed mating systems ($0.2 \leq s < 0.8$), F was lower than that expected with no inbreeding depression [i.e., $F < s/(2 - s)$]. Using Ritland's (1990) estimator, mean $\delta \pm$ SD $= 0.81 \pm 0.70$, and inferred $\delta \geq 0.5$ for 72% of species. There are probably large standard errors around any given point (Ritland 1990), and it is possible that the assumptions of this method are violated for some of these species (though most of these violations lead to δ being underestimated). However, it is difficult to envision that some ubiquitous ecological or genetic factor is consistently biasing the estimated F downward and the estimated s upward for such a large fraction of these species. Hence, we view this result as consistent with the possibility that substantial self-fertilization has evolved in many species despite strong inbreeding depression and, conversely, that strong inbreeding depression has been maintained in the face

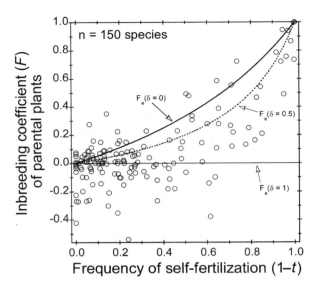

Figure 3   The relation between estimated levels of self-fertilization ($s = 1 - t$) and the inbreeding coefficient (F) of reproductively mature individuals in natural populations of 150 seed plant species. Estimates of t were collected from the literature as described in Figure 1. Estimates of F were from the same populations as t and were usually based on direct assays of mature plants or maternal genotypes inferred from segregation patterns in progeny arrays (usually ≥ 20 maternal genotypes per population). The *heavy solid line* shows the expected relation between s and F in populations at equilibrium (F_e) with no inbreeding depression ($\delta = 1 -$ [fitness of selfed progeny/fitness of outcrossed progeny] $= 0$). The *broken line* shows the equilibrium F if selfed progeny are only half as fit as outcrossed progeny (i.e., $\delta = 0.5$) and the *thin solid line* at $F = 0$ would occur if selfed offspring never survive to reproductive maturity (i.e., $\delta = 1$).

of substantial selfing. What maintains high selfing when selfed offspring are so infrequently recruited into the adult population? This analysis also suggests that some species exhibit substantial outcrossing with weak inbreeding depression. Are those species with low inbreeding depression in transition toward higher selfing rates, or can mixed mating be stabilized by other factors in the face of low δ? We turn our attention to these factors, beginning with theory that explores variation in inbreeding depression.

Variability in Inbreeding Depression

Theory shows that differential expression of inbreeding depression in male and female functions can promote the fixation of an allele conferring intermediate selfing or a stable polymorphism of selfing and outcrossing (Rausher & Chang 1999). In the few taxa for which this has been studied, inbreeding depression is shown to differ in gender expression, but the specific values of inbreeding depression necessary for stability have generally not been found (Carr & Dudash 1995, Chang & Rausher 1999, del Castillo 1998, Robertson et al. 1994). Although further empirical work is warranted, the limited data available do not suggest a general explanation for mixed mating. Other models indicate that temporal variation in the expression of inbreeding depression can lead to stable mixed mating through the principle of nonlinear averaging (Cheptou & Mathias 2001, Cheptou & Schoen 2002). With temporal variation, the time-averaged fitness of selfed progeny is reduced (Cheptou & Schoen 2002), which provides negative feedback from increased selfing that can stabilize mixed mating. Stochasticity in population density can also lead to stable intermediate outcrossing if competitive interactions cause variation in the expression of inbreeding depression (Cheptou & Dieckmann 2002). Evidence from a few field and greenhouse experiments indicate that competition and stress can influence the expression of inbreeding depression (Cheptou et al. 2000a,b; Eckert & Barrett 1994; Hauser & Loeschcke 1996; Johnston 1992; Wolfe 1993). Nevertheless, the extent of temporal variation of inbreeding depression in the field is largely unexplored. Thus, although these results are theoretically intriguing, their biological relevance has yet to be tested. Finally, a model by Holsinger (1986) explored a form of variation in inbreeding depression in which the relative fitness of selfed and outcrossed progeny differ depending upon whether progeny disperse to a new habitat, which can yield stable intermediate outcrossing. The applications of this model are fairly limited; however it provides one of the first considerations of how a mating system could be affected by ecological and spatial parameters.

Biparental Inbreeding

The potential effects of biparental inbreeding (mating between related individuals) on mating system evolution are complex. Biparental inbreeding reduces the genetic cost of outcrossing because it increases the relatedness of parents to their outcrossed progeny. Its rate is expected to increase with the population selfing rate because, with selfing, individuals in the population come to share more genes that are identical by descent. For this reason, biparental inbreeding might promote frequency-dependent selection that stabilizes intermediate outcrossing (Uyenoyama 1986). Alternatively, biparental inbreeding could have a destabilizing effect in mating system evolution if, as it increases with the selfing rate, it reduces the fitness of outcrossed progeny, thereby reducing realized inbreeding depression (Lloyd 1979, Uyenoyama & Antonovics 1987), a factor not included in Uyenoyama's (1986) original formulation. Mating among relatives can also influence the evolution of inbreeding depression, but the rate of purging depends on parameters of population size and structure (Waller 1993). In a stochastic model, mixed mating was found to be stable when population structure developed through limited pollen and seed dispersal. In this case, biparental inbreeding conferred stability because population structure promoted an increase in inbreeding depression with increased selfing above some threshold selfing rate (Ronfort & Couvet 1995). The ultimate effect of biparental inbreeding on mating system dynamics, therefore, is difficult to predict and may depend on current levels of inbreeding depression (Uyenoyama & Antonovics 1987, Yahara 1992).

Relatively little empirical work addresses the magnitude and effect of biparental inbreeding on mating system evolution. Pollen and seed dispersal is often limited (Levin & Kerster 1974) but plant populations vary widely in their degree of genetic structure (Heywood 1991, Loveless & Hamrick 1984). Marker-based estimates of selfing rate can include some amount of close biparental inbreeding (Brown 1990, Ennos & Clegg 1982). The amount decreases as the number of marker loci used increases so that biparental inbreeding can be inferred from the difference between the mean single locus and multilocus estimates (Ritland & Jain 1981, Shaw et al. 1981). Inferential estimates of biparental inbreeding vary widely among plant species but tend to be higher in species with mixed mating systems (Brown 1990). Several experimental techniques have been used to distinguish between biparental inbreeding and selfing (Herlihy & Eckert 2004, Kelly & Willis 2002, Lu 2000). For example, Griffin & Eckert (2003) used a transplant experiment to estimate that about 30% of matings involved close relatives in natural populations of *Aquilegia canadensis*. Biparental inbreeding depression, measured either directly through experimental populations (Heywood 1993) or indirectly, by comparing progeny fitness from crosses at different distances (Waser & Price 1994), appears to be substantial in some species. Clearly, the varied effects of population structure on the dynamics of mating systems warrant further attention. Given the wide variation across plant taxa in the extent of biparental inbreeding, comparative approaches might be used to ask whether this factor has a consistent and important effect on mating system dynamics.

Pollen Discounting

Modifications to floral morphology that increase self-pollination can reduce the export of pollen to flowers on other plants (Holsinger 1996). Can pollen discounting provide the negative feedback required to stabilize mixed selfing and outcrossing? In a model that included reproductive assurance, pollen discounting, and purging of inbreeding depression, Johnston (1998) showed that intermediate outcrossing can be evolutionarily stable if pollen discounting increases with the selfing rate as, for example, if evolutionary changes in floral traits that increase selfing incur increasingly higher pollen discounting costs.

Holsinger (1991) found that stable mixed mating could result from frequency-dependent pollination processes. In his mass action model, the selfing rate of an individual is a function of the proportion of self versus outcross pollen grains deposited on its stigma, which in turn is determined by its rate of pollen export and that of other plants in the population. In contrast to genetic models of mating system evolution that balance inbreeding depression against the cost of outcrossing and assume no pollen discounting, Holsinger's (1991) model includes no inbreeding depression and treats pollen discounting as the trait that is selected, yielding an inherent positive relation between pollen discounting and selfing. The model predicts that, in a highly selfing population, a rare variant that exports pollen will always have a transmission advantage; thus, whenever selfing can evolve, the stable outcome is mixed mating. This process can also maintain mixed mating when inbreeding depression and its evolutionary dynamics are included (Porcher & Lande 2005c). The incorporation of realistic levels of pollen limitation, a factor not considered in Holsinger's (1991) model, yields the expectation of selfing rates that are high but less then one under a wide range of parameters (Porcher & Lande 2005c).

The mass action perspective on mating system evolution yields unexpected and valuable insights, yet the generality of its results may be limited. This approach adds realism to plant mating system theory in that it models the frequency dependence of the selfing rate. However, it may be difficult to reconcile with the floral biology of mating system transitions, in that, for instance, variation in stigma-anther separation could have dramatic effects on the selfing rate with little effect on pollen export. In other words, because it focuses explicitly on pollen discounting, it tells us little about the cases where this parameter is negligible. Moreover, the models assume that self and outcross pollen arrives simultaneously on stigmas. Although strict prior selfing, such as bud pollination, is probably rare, a continuum from prior to competing selfing undoubtedly exists, and autonomous self-pollination may often occur before outcrossing.

Harder & Wilson (1998) extended the mass action approach to a broader array of floral scenarios by considering a more complex partitioning of pollen fates, including a distinction between discounting and nondiscounting sources of self-pollination. Nondiscounting pollen is unavailable for export from the flower, and thus its participation in self-fertilization does not decrease outcross siring. This approach explores how changes in pollen fate affect complex stepwise selection on floral traits that may occur during the evolution of mating systems. It examines the conditions that allow evolutionary shifts between different modes of selfing, a largely unexplored area of theory that is likely to provide insights into the stability of mixed mating. A similar consideration of stepwise changes in ovule fate with floral evolution may also be fruitful.

Despite the potential evolutionary importance of pollen discounting, this parameter is quantified only rarely. The male outcrossing success of plants that vary in selfing rate can be compared using genetic markers and experimental arrays (Chang & Rausher 1998, Fishman 2000, Kohn & Barrett 1994, Rausher et al. 1993). Such studies yield variable estimates and suggest that pollen discounting may sometimes depend on the frequency of selfing variants. For example, Chang & Rausher (1998) show that negative frequency dependence of outcross success can contribute to the mixed mating system of *Ipomoea purpurea*. In a different approach, pollen discounting was estimated in *Erythronium grandiflorum* using a pollen color polymorphism (Holsinger & Thomson 1994). Although measurement of pollen discounting poses formidable challenges, accumulating theoretical results suggest that knowledge of this parameter may be critical to our understanding of mixed mating.

Reproductive Assurance

The often cited "best of both worlds" hypothesis (Becerra & Lloyd 1992, Cruden & Lyon 1989) holds that certain forms of mixed mating evolve because they promote outcrossing but provide reproductive assurance when pollinators or mates are scarce, combining the advantages of both reproductive strategies. This view is supported by the higher frequency of mixed mating systems among taxa pollinated by animals, which may exhibit marked spatiotemporal variation in pollinator service (Eckert 2002, Wolfe & Barrett 1988). Despite its intuitive appeal and frequent invocation, this hypothesis and the role of reproductive assurance have only recently received extensive theoretical attention. Indeed, many mating system models do not include a parameter that allows selfing to increase total seed set, a measure of reproductive assurance (Table 1). Lloyd (1979, 1992) showed that delayed selfing provides reproductive assurance, has no gamete discounting costs, and is therefore nearly always favored, yielding a mixed mating system when outcross pollen is limited. However, the selective value of other forms of selfing can be eroded by seed discounting (Lloyd 1992). Moreover, reproductive assurance selfing in perennials could reduce future survival and discount later production of outcrossed seeds within or between reproductive seasons (Morgan et al. 1997). Thus, selection for reproductive assurance is more complex than might be expected.

Because cross-incompatibility among plants with the same SI genotype can limit mate availability and intensify pollen limitation (Byers & Meagher 1992), reproductive assurance is expected to play a part in the evolutionary breakdown of SI (Charlesworth & Charlesworth 1979, Porcher & Lande 2005a). Theory suggests that, with pollen limitation, partial SI (yielding mixed mating) can be maintained because of the complex interaction of direct selective effects (inbreeding depression, cost of outcrossing, reproductive assurance) and genetic associations between the self-incompatibility (S) locus and alleles modifying the strength of SI (Vallejo-Marin & Uyenoyama 2004). The conditions for stability are restrictive in this model, however, requiring a narrow range of inbreeding depression and low S-allele numbers, which suggests either that many partially SI species are in transition toward self-compatibility or that other factors can also promote stability.

The models described above consider the evolution of selfing, and mixed mating a consequence when outcross pollen is chronically limited. However, both pollen limitation and selection for reproductive assurance are likely to be variable, which can, in principle, lead to the evolution of stable mixed mating. A simple model that includes variation in pollinator service (Schoen & Brown 1991) gives the result that selfing induced only in the absence of cross-fertilization is selected whenever pollination is variable and $\delta < 1$. In two other models, stability of mixed mating derives from variable pollination and the principle of nonlinear averaging. Morgan & Wilson (2005) use a population dynamic approach and find that the effect of variance in pollen delivery depends on the mode of autonomous selfing. Delayed selfing goes to fixation regardless of variance in pollination, such that the population outcrossing rate is determined proximately by the level of outcross pollination. In contrast, selection on prior selfing is influenced by the degree of variance in pollination; stable mixed mating is expected for a wide range of parameters. Sakai & Ishii (1999) showed that variable pollination may or may not select for a mixed strategy when the trade-off between seed size and number is considered. In other words, when selfed progeny have lower fitness, it may be better to forego reproductive assurance selfing and make fewer, larger outcross seeds. Accordingly, mixed mating is more likely to be stable when the optimal seed size is small. Thus, the model provides a scenario in which a best of both worlds mating system may not, in fact, be best.

The selective value of reproductive assurance may also change during the course of mating system evolution because autonomous selfing allows populations to grow under conditions where insufficient outcross pollination would otherwise limit reproduction. Cheptou (2004) showed that this, in turn, alleviates outcross pollen limitation, thereby reducing the selective advantage of selfing. Moreover, increased population density may also intensify the expression of inbreeding depression, which could lead to reduced population density. These demographic feedback loops have a variety of counterintuitive consequences, but, interestingly, do not appear to explain stable mixed mating.

A clear theoretical understanding of the role of reproductive assurance in the evolution of mixed mating has not yet emerged, and empirical work lags behind theory. Indirect evidence for reproductive assurance includes autonomous (especially delayed) selfing (e.g., Cheptou et al. 2002, Dole 1992, Rathcke & Real 1993; see also Lloyd & Schoen 1992), reduced pollen limitation in selfing relative to outcrossing populations or species (Goodwillie 2001, Kasagi & Kudo 2003, Larson & Barrett 2000) and low rates of cross-pollination (Burd 1994, Fausto et al. 2001, Ramsey & Vaughton 1996). However, direct tests of reproductive assurance that compare seed set by emasculated flowers versus intact control flowers subjected to open pollination (Schoen & Lloyd 1992) have been carried out for relatively few species with variable results (reviewed in Cruden & Lyon 1989, Elle & Carney 2003, Herlihy & Eckert 2004, Holsinger 1996, Schoen et al. 1996). Moreover, demonstrating that selfing confers reproductive assurance in the face of variable pollination requires substantial temporal and spatial replication of the emasculation experiment (Eckert & Schaefer 1998), which has been attempted only rarely. Kalisz et al. (2004) showed that autonomous delayed selfing increases seed production during years of pollinator failure in three populations of *Collinsia verna*, yielding intermediate outcrossing and reproductive assurance when outcross-pollination is limiting. In contrast, Herlihy & Eckert (2002) found that autonomous selfing always increased seed production in 12 populations of *Aquilegia canadensis*; however the extent of selfing at the population level was unrelated to the degree of pollinator failure. To date, there has been no theoretical investigation of the longer-term, between-season trade-offs of reproductive assurance and seed discounting (Morgan et al. 1997) or empirical studies of how reproductive assurance varies with population demography.

Resource Allocation Models

Reduced allocation to attractive structures (e.g., reduced corolla size) has long been observed in highly selfing plant taxa (Ornduff 1969). Although changes in allocation are often considered to occur in response to evolution of the mating system, they can also be a direct determinant of outcrossing rates, as they affect deposition of outcross pollen and pollen export. The evolution of selfing rate has been modeled as a function of resource allocation to floral structures. In a model of delayed selfing, Iwasa (1990) showed that trade-offs between allocation to pollen capture promoting outcrossing (large flowers) and progeny fitness (inbreeding depression) can result in evolutionarily stable intermediate selfing rates. Sakai (1995) found that when the relationship between flower size and number is considered, mixed mating with delayed or competing selfing can be stable only when that function is nonlinear and flower number is strongly constrained. Empirical testing of the effects of shifts in resource allocation on the evolution of the mating systems has lagged far behind theory, probably because changes in allocation are often interpreted as a secondary adaptation to optimize fitness in response to changes in the selfing rate (reviewed in Ornduff 1969, Takebayashi & Morrell 2001). However, a recent QTL study of the genetics of floral differences between a selfing and a predominantly outcrossing species of *Mimulus* (Fishman et al. 2002) found evidence of linkage and/or pleiotropy between some of the QTLs for traits that directly affect selfing (i.e., herkogamy) and traits involving allocation to attraction (i.e., corolla size). Although it is impossible to determine the order in which various gene (or QTL) substitutions occurred in the transition from selfing to outcrossing, this approach may be valuable for evaluating models of mating system evolution (see discussion in Fishman et al. 2002).

MIXED MATING SYSTEMS: A BROADER PERSPECTIVE

As reviewed here, the challenge to explain the occurrence of mixed mating has inspired remarkably diverse theoretical approaches and hypotheses. Most are directly applicable to only a subset of the mechanisms or modes of selfing that produce mixed mating (Table 1). For example, models of reproductive assurance are most relevant for autonomously selfing species. General models have limited application to partial SI because its stability may be affected by the diversity (Vallejo-Marin & Uyenoyama 2004) and dynamics of S-alleles. Few models address the evolution of geitonogamy. Finally, theory that predicts a balanced polymorphism may not be applicable to the evolution of a monomorphic basis for mixed mating. To assess the potential importance of different hypotheses, then, we would like to know the broad distribution of selfing modes and floral mechanisms that produce intermediate outcrossing rates.

To that end, we compiled both anecdotal and experimental information on autonomous selfing, geitonogamy, and postpollination phenomena for the angiosperm species included in our survey (Figure 4, see legend for methods). The

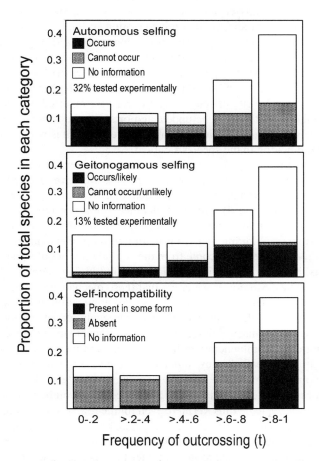

Figure 4   Mechanisms that contribute to the distribution of outcrossing rates in angiosperm species in our survey. Mechanisms that were reported to be only partially effective (e.g., partial SI, partial seed set due to autonomous selfing) were recorded as positive responses. We recorded information from the t estimate studies and then searched more broadly for journal articles with additional information on each species. We distinguished between anecdotal evidence (e.g., a verbal hypothesis that geitonogamy is likely or unlikely based upon inflorescence size and pollinator behavior) and experimental results (e.g., bagging experiments to test for autonomous selfing ability) and present the proportion of all species in the data set for which experimental data are available. Data available from the authors on request.

available data are remarkably limited, but support a few preliminary observations. A substantial proportion of the species with intermediate outcrossing ($0.2 < t < 0.8$) are at least partially capable of autonomous selfing, indicating a possible role for reproductive assurance in their mixed mating systems. Although experimental investigations are few, the potential for geitonogamous selfing is relatively common in species with high intermediate outcrossing rates, suggesting that a proportion of selfed progeny in these species may be an indirect result of selection for outcrossing success (e.g., Eckert 2000). SI is relatively common for highly outcrossing species, whereas the majority of species with lower t are reportedly fully self-compatible. Although partial SI may be undetected or attributed to inbreeding depression in some cases, the information available suggests that it is not a primary contributor to mixed mating in the species sampled.

Most strikingly, the proximate mechanisms contributing to the estimated outcrossing rates are unknown for most species. Moreover, for only a small fraction of species surveyed do we have any quantitative information on the mode of self-fertilization (Figure 4). Although detailed studies of reproductive biology have been undertaken for many plant species, these are often not linked to outcrossing rate estimates, limiting their contribution to our understanding of mating system evolution. Progress will come from the compilation of outcrossing estimates with at least basic functional information on floral biology across a broader range of plant species. Moreover, we argue that these mechanisms, when known, must be taken into greater consideration in the valid application of mating system theory.

A more complete understanding of the evolution of mating systems in any species requires that we dissect the rate of selfing into its components parts, identify the floral traits influencing each component, and then experimentally alter these traits in some realistic way to determine the consequences for fitness and hence the evolutionary stability of the resident mating system (Herlihy & Eckert 2004). However, to date, the adaptive significance and evolutionary stability of mixed mating has yet to be fully verified, even for the few species in which the functional aspects and many of the relevant costs benefits have been studied (Barrett 2003).

The literature on plant mating system evolution focuses primarily on what maintains mixed mating systems, rather than assessing whether, or which, existing mixed mating systems are evolutionarily stable. The substantial frequency of species with intermediate t offers evidence that mixed mating systems can be stable, but we have no quantitative expectation as to how frequent mixed mating should be under either hypothesis. A danger inherent in this debate is the creation of a false dichotomy—that mixed mating as a whole is either fundamentally stable or unstable. Phylogenetic studies show that transitions between predominantly selfing and outcrossing states are common in some plant taxa (Goodwillie 1999, Kohn et al. 1996, Wyatt 1988), and we can reasonably assume that some of the existing mixed mating systems are transitional. A more rigorous examination of the frequency of mixed mating systems within a phylogenetic perspective is warranted. The serious biases inherent in surveys of published outcrossing rates could be remedied by systematic efforts, in which species are sampled randomly and intensively from individual plant families or lineages. Although phylogenetic approaches have been used in the study of breeding system stability (Dorken et al. 2002, Husband & Barrett 1993, Weller et al. 1995), they have contributed surprisingly little to the problem of mixed mating. Most of the phylogenetically based studies of mating or breeding systems use qualitative character states (Barrett et al. 1996, Bena et al. 1998, Goodwillie 1999, Igic et al. 2004, Kohn et al. 1996, Wyatt 1988) and have not considered intermediate outcrossing. Transitional mixed mating should occur only on the tips of branches, whereas the occurrence of deeper phylogenetic lineages with intermediate t would indicate stability. Combining phylogenetic approaches with estimates of inbreeding depression can allow us to ask whether purging is a key determinant in the stability of mixed mating. For example, in *Leptosiphon* (Polemoniaceae), transitions from SI to predominant selfing have occurred multiple times (Goodwillie 1999), and partial SI with mixed mating has been observed in only one species of the genus, *L. jepsonii* (Goodwillie et al. 2004). The phylogenetic data and the finding of low δ in some populations of *L. jepsonii* (Goodwillie 2000) suggest that, in this case, partial SI may represent a transitional state in the breakdown of SI. With the increasing availability of genetic data for both mating system and phylogenetic analyses, such approaches could yield insights in other taxa.

Lande & Schemske's (1985) provocative model has inspired a wealth of studies of inbreeding depression and set in motion the developing theory of mixed mating systems. Although large empirical gaps remain, a broad understanding of the genetic basis of inbreeding depression has emerged (Charlesworth & Charlesworth 1999). In contrast, empirical estimates of the pollination parameters required to test ecological mating system models are still extremely limited. Further theoretical exploration is needed as well. To date no compelling evidence suggests that we have identified factors or processes that can account for mixed mating across a broad range of plant taxa. The conditions for stable mixed mating are restrictive in many models and others have failed to find empirical support. Moreover, existing models do not yet provide an explanation for observed differences in the frequency of mixed mating with biotic versus abiotic pollination. In addition to the recent expansion of theory that emphasizes pollination biology (e.g., Harder & Wilson 1998, Morgan & Wilson 2005, Porcher & Lande 2005c), several promising, largely untested, theoretical directions are emerging that may provide further insights on mixed mating. A common feature is that the evolution of the mating system is considered in a broader context of, for instance, life-history strategies (Morgan et al. 1997, Tsitrone et al. 2003), population dynamics (Cheptou 2004), metapopulation dynamics (Pannell & Barrett 1998, 2001), plant and insect communities (Fausto et al. 2001, Moeller 2005), and coevolutionary interactions between plants and their parasites (Agrawal & Lively 2001, Busch et al. 2004). Indeed, although the search for general explanations for mixed mating is compelling, distinct factors may well be important for its maintenance in different plant taxa. Whether or not unifying explanations are found, the enduring enigma of mixed mating promises to motivate yet more fruitful exploration into the evolution of plant mating systems.

ACKNOWLEDGMENTS

We thank D.W. Schemske for helpful discussion and encouragement, S.C.H. Barrett and J. Busch for sharing data collected for previous comparative analyses of plant mating systems, K. Neville for help with the updated database of mating system parameters, the National Science Foundation for funding to C.G. and S.K., and the Natural Sciences and Engineering Research Council of Canada for funding to C.G.E. We thank R. Lande, E. Porcher, and an anonymous reviewer for insightful comments that substantially improved the manuscript.

The *Annual Review of Ecology, Evolution, and Systematics* is online at
http://ecolsys.annualreviews.org

LITERATURE CITED

Agrawal AF, Lively CM. 2001. Parasites and the evolution of self-fertilization. *Evolution* 55:869–79

Aide TM. 1986. The influence of wind and animal pollination on variation in outcrossing rates. *Evolution* 40:434–35

Allard RW. 1975. The mating system and microevolution. *Genetics* 79:115–26

Anderson B, Midgley JJ, Stewart BA. 2003. Facilitated selfing offers reproductive assurance: a mutualism between a hemipteran and carnivorous plant. *Am. J. Bot.* 90:1009–15

Arathi HS, Rasch A, Cox C, Kelly JK. 2002. Autogamy and floral longevity in *Mimulus guttatus*. *Int. J. Plant. Sci.* 163:567–73

Barrett SCH. 2003. Mating strategies in flowering plants: the outcrossing-selfing paradigm and beyond. *Philos. Trans. R. Soc. London Ser. B* 358:991–1004

Barrett SCH, Eckert CG. 1990. Variation and evolution of mating systems in seed plants. In *Biological Approaches and Evolutionary Trends in Plants*, ed. S Kawano, 14:229–54. London/San Diego: Academic. 417 pp.

Barrett SCH, Harder LD, Worley AC. 1996. The comparative biology of pollination and mating in flowering plants. *Philos. Trans. R. Soc. London Ser. B* 351:1271–80

Becerra JX, Lloyd DG. 1992. Competition-dependent abscission of self-pollinated flowers of *Phormium tenax* (Agavaceae): a sec-

ond action of self-incompatibility at the whole flower level. *Evolution* 46:458–69

Bena G, Lejeune B, Prosperi JM, Olivieri I. 1998. Molecular phylogenetic approach for studying life-history evolution: the ambiguous example of the genus *Medicago* L. *Proc. R. Soc. London Ser. B* 265:1141–51

Brown AHD. 1990. Genetic characterization of plant mating systems. In *Plant Population Genetics, Breeding and Genetic Resources*, ed. AHD Brown, MT Clegg, A Kahler, B Weir, pp. 145–62. Sunderland, MA: Sinauer

Burd M. 1994. Bateman's principle and plant reproduction: the role of pollen limitation in fruit and seed set. *Bot. Rev.* 60:83–139

Busch JW, Neiman M, Koslow JM. 2004. Evidence for maintenance of sex by pathogens in plants. *Evolution* 58:2584–90

Byers DL, Meagher TR. 1992. Mate availability in small populations of plant species with homomorphic sporophytic self-incompatibility. *Heredity* 68:353–59

Byers DL, Waller DM. 1999. Do plant populations purge their genetic load? Effects of population size and mating history on inbreeding depression. *Annu. Rev. Ecol. Syst.* 30:479–513

Carr DE, Dudash MR. 1995. Inbreeding depression under a competitive regime in *Mimulus guttatus*: consequences for potential male and female function. *Heredity* 75:437–45

Carr DE, Dudash MR. 2003. Recent approaches

into the genetic basis of inbreeding depression in plants. *Philos. Trans. R. Soc. London Ser. B* 358:1071–84

Chang SM, Rausher MD. 1998. Frequency-dependent pollen discounting contributes to maintenance of a mixed mating system in the common morning glory *Ipomoea purpurea*. *Am. Nat.* 152:671–83

Chang SM, Rausher MD. 1999. The role of inbreeding depression in maintaining the mixed mating system of the common morning glory, *Ipomoea purpurea*. *Evolution* 53:1366–76

Charlesworth B. 1992. Evolutionary rates in partially self-fertilizing species. *Am. Nat.* 140:126–48

Charlesworth B, Charlesworth D. 1999. The genetic basis of inbreeding depression. *Genet. Res.* 74:329–40

Charlesworth D. 1991. The apparent selection on neutral marker loci in partially inbreeding populations. *Genet. Res.* 57:159–75

Charlesworth D, Charlesworth B. 1979. The evolution and breakdown of *S*-allele systems. *Heredity* 43:41–55

Charlesworth D, Charlesworth B. 1987. Inbreeding depression and its evolutionary consequences. *Annu. Rev. Ecol. Syst.* 18:237–68

Charlesworth D, Charlesworth B. 1990. Inbreeding depression with heterozygote advantages and its effect on selection for modifiers changing the outcrossing rate. *Evolution* 44:870–88

Charlesworth D, Morgan MT, Charlesworth B. 1990. Inbreeding depression, genetic load, and the evolution of outcrossing rates in a multilocus system with no linkage. *Evolution* 44:1469–89

Cheptou PO. 2004. Allee effect and self-fertilization in hermaphrodites: reproductive assurance in demographically stable populations. *Evolution* 58:2613–21

Cheptou PO, Berger A, Blanchard A, Collin C, Escarre J. 2000a. The effect of drought stress on inbreeding depression in four populations of the Mediterranean outcrossing plant *Crepis sancta* (Asteraceae). *Heredity* 85:294–302

Cheptou PO, Dieckmann U. 2002. The evolution of self-fertilization in density-regulated populations. *Proc. R. Soc. London Ser. B* 269:1177–86

Cheptou PO, Imbert E, Lepart J, Escarre J. 2000b. Effects of competition on lifetime estimates of inbreeding depression in the outcrossing plant *Crepis sancta* (Asteraceae). *J. Evol. Biol.* 13:522–31

Cheptou PO, Lepart J, Escarre J. 2002. Mating system variation along a successional gradient in the allogamous and colonizing plant *Crepis sancta* (Asteraceae). *J. Evol. Biol.* 15:753–62

Cheptou PO, Mathias A. 2001. Can varying inbreeding depression select for intermediary selfing rates? *Am. Nat.* 157:361–73

Cheptou PO, Schoen DJ. 2002. The cost of fluctuating inbreeding depression. *Evolution* 56:1059–62

Crnokrak P, Barrett SCH. 2002. Perspective: purging the genetic load: a review of the experimental evidence. *Evolution* 56:2347–58

Cruden RW, Lyon DL. 1989. Facultative xenogamy: examination of a mixed mating system. In *The Evolutionary Ecology of Plants*, ed. JH Bock, YB Linhart, pp. 171–207. Boulder, CO:Westview

Cruzan MB, Barrett SCH. 1993. Contribution of cryptic incompatibility to the mating system of *Eichhornia paniculata* (Pontederiaceae). *Evolution* 47:925–34

Damgaard C, Couvet D, Loeschcke V. 1992. Partial selfing as an optimal mating strategy. *Heredity* 69:289–95

Darwin CR. 1876. *The Effects of Cross and Self-fertilization in the Vegetable Kingdom*. London: Murray

Dawkins R. 1980. Good strategy or evolutionarily stable strategy? In *Sociobiology: Beyond Nature/Nurture*, ed. GWBJ Silverberg, pp. 331–57. Boulder, CO: Westview

de Jong TJ, Waser NM, Klinkhamer PGL. 1993. Geitonogamy: the neglected side of selfing. *Trends Ecol. Evol.* 8:321–25

del Castillo RF. 1998. Fitness consequences of maternal and nonmaternal components of inbreeding in the gynodioecious *Phacelia dubia*. *Evolution* 52:44–60

de Nettancourt D. 1997. Self-incompatability in angiosperms. *Sex. Plant Reprod.* 10:185–99

Dickinson HG. 1994. Self-pollination: simply a social disease. *Nature* 367:517–18

Dole JA. 1992. Role of corolla abscission in delayed self-pollination of *Mimulus guttatus* (Scrophulariaceae). *Am. J. Bot.* 77:1505–7

Dorken ME, Friedman J, Barrett SCH. 2002. The evolution and maintenance of monoecy and dioecy in *Sagittaria latifolia* (Alismataceae). *Evolution* 56:31–41

Dudash MR. 1990. Relative fitness of selfed and outcrossed progeny in a self-compatible, protandrous species, *Sabatia angularis* L. (Gentianaceae): a comparison in three environments. *Evolution* 44:1129–39

Eckert CG. 2000. Contributions of autogamy and geitonogamy to self-fertilization in a mass-flowering, clonal plant. *Ecology* 81:532–42

Eckert CG. 2002. Effect of geographical variation in pollinator fauna on the mating system of *Decodon verticillatus* (Lythraceae). *Int. J. Plant Sci.* 163:123–32

Eckert CG, Allen M. 1997. Cryptic self-incompatibility in tristylous *Decodon verticillatus* (Lythraceae). *Am. J. Bot.* 84:1391–97

Eckert CG, Barrett SCH. 1994. Inbreeding depression in partially self-fertilizing *Decodon verticillatus* (Lythraceae): population genetic and experimental analyses. *Evolution* 48:952–64

Eckert CG, Schaefer A. 1998. Does self-pollination provide reproductive assurance in *Aquilegia canadensis* (Ranunculaceae)? *Am. J. Bot.* 85:919–24

Elle E, Carney R. 2003. Reproductive assurance varies with flower size in *Collinsia parviflora* (Scrophulariaceae). *Am. J. Bot.* 90:888–96

Ennos RA, Clegg MT. 1982. Effect of population substructuring on estimates of outcrossing rate in plant populations. *Heredity* 48:283–92

Fausto JA, Eckhart VM, Geber MA. 2001. Reproductive assurance and the evolutionary ecology of self-pollination in *Clarkia xantiana* (Onagraceae). *Am. J. Bot.* 88:1794–800

Fisher RA. 1941. Average excess and average effect of a gene substitution. *Ann. Eugen.* 11:53–63

Fishman L. 2000. Pollen discounting and the evolution of selfing in *Arenaria uniflora* (Caryophyllaceae). *Evolution* 54:1558–65

Fishman L, Kelly AJ, Willis JH. 2002. Minor quantitative trait loci underlie floral traits associated with mating system divergence in *Mimulus*. *Evolution* 56:2138–55

Fishman L, Wyatt R. 1999. Pollinator-mediated competition, reproductive character displacement, and the evolution of selfing in *Arenaria uniflora* (Caryophyllaceae). *Evolution* 53:1723–33

Flaxman SM. 2000. The evolutionary stability of mixed strategies. *Trends Ecol. Evol.* 15:482–84

Fornoni J, Nunez-Farfan J, Valverde PL, Rausher MD. 2004. Evolution of mixed strategies of plant defense allocation against natural enemies. *Evolution* 58:1685–95

Glemin S. 2003. How are deleterious mutations purged? Drift versus nonrandom mating. *Evolution* 57:2678–87

Goodwillie C. 1999. Multiple origins of self-compatibility in *Linanthus* section *Leptosiphon* (Polemoniaceae): phylogenetic evidence from internal-transcribed-spacer sequence data. *Evolution* 53:1387–95

Goodwillie C. 2000. Inbreeding depression and mating systems in two species of *Linanthus* (Polemoniaceae). *Heredity* 84:283–93

Goodwillie C. 2001. Pollen limitation and the evolution of self-compatibility in *Linanthus* (Polemoniaceae). *Int. J. Plant Sci.* 162:1283–92

Goodwillie C, Partis KL, West JW. 2004. Transient self-incompatibility confers delayed selfing in *Leptosiphon jepsonii* (Polemoniaceae). *Int. J. Plant Sci.* 165:387–94

Grant V. 1981. *Plant Speciation*. New York: Columbia Univ.

Griffin CAM, Eckert CG. 2003. Experimental

analysis of biparental inbreeding in a self-fertilizing plant. *Evolution* 57:1513–19

Hallauer AR, Sears JH. 1973. Changes in quantitative traits associated with inbreeding in a synthetic variety of maize. *Crop Sci.* 13:327–30

Harder LD, Barrett SCH. 1995. Mating costs of large floral displays in hermaphrodite plants. *Nature* 373:512–15

Harder LD, Wilson WG. 1998. A clarification of pollen discounting and its joint effects with inbreeding depression on mating system evolution. *Am. Nat.* 152:684–95

Hauser TP, Loeschcke V. 1996. Drought stress and inbreeding depression in *Lychnis flos-cuculi* (Caryophyllaceae). *Evolution* 50:1119–26

Herlihy CR, Eckert CG. 2002. Genetic cost of reproductive assurance in a self-fertilizing plant. *Nature* 416:320–23

Herlihy CR, Eckert CG. 2004. Experimental dissection of inbreeding and its adaptive significance in a flowering plant, *Aquilegia canadensis* (Ranunculaceae). *Evolution* 58:2693–703

Heywood JS. 1991. Spatial analysis of genetic variation in plant populations. *Annu. Rev. Ecol. Syst.* 22:335–55

Heywood JS. 1993. Biparental inbreeding depression in the self-incompatible annual plant *Gaillardia pulchella* (Asteraceae). *Am. J. Bot.* 80:545–50

Holsinger KE. 1986. Dispersal and plant mating systems: the evolution of self-fertilization in subdivided populations. *Evolution* 40:405–13

Holsinger KE. 1988. Inbreeding depression doesn't matter: the genetic basis of mating system evolution. *Evolution* 42:1235–44

Holsinger KE. 1991. Mass-action models of plant mating systems: the evolutionary stability of mixed mating systems. *Am. Nat.* 138:606–22

Holsinger KE. 1996. Pollination biology and the evolution of mating systems in flowering plants. *Evol. Biol.* 29:107–49

Holsinger KE, Thomson JD. 1994. Pollen discounting in *Erythronium grandiflorum*:

mass-action estimates from pollen transfer dynamics. *Am. Nat.* 144:799–812

Husband BC, Barrett SCH. 1993. Multiple origins of self-fertilization in tristylous *Eichhornia paniculata* (Pontederiaceae): inferences from style morph and isozyme variation. *J. Evol. Biol.* 6:591–608

Husband BC, Schemske DW. 1996. Evolution of the magnitude and timing of inbreeding depression in plants. *Evolution* 50:54–70

Igic B, Bohs L, Kohn JR. 2004. Historical inferences from the self-incompatibility locus. *New Phytol.* 161:97–105

Iwasa Y. 1990. Evolution of the selfing rate and resource allocation models. *Plant Species Biol.* 5:19–30

Jain SK. 1976. The evolution of inbreeding in plants. *Annu. Rev. Ecol. Syst.* 7:469–95

Johnston MO. 1992. Effects of cross and self-fertilization on progeny fitness in *Lobelia cardinalis* and *L. siphilitica*. *Evolution* 46:688–702

Johnston MO. 1998. Evolution of intermediate selfing rates in plants: pollination ecology versus deleterious mutations. *Genetica* 102/103:267–78

Jones KN. 1994. Nonrandom mating in *Clarkia gracilis* (Onagraceae): a case of cryptic self-incompatibility. *Am. J. Bot.* 81:195–98

Kalisz S, Vogler DW. 2003. Benefits of autonomous selfing under unpredictable pollinator environments. *Ecology* 84:2928–42

Kalisz S, Vogler DW, Hanley KM. 2004. Context-dependent autonomous self-fertilization yields reproductive assurance and mixed mating. *Nature* 430:884–87

Kalisz S, Vogler D, Fails B, Finer M, Shepard E, et al. 1999. The mechanism of delayed selfing in *Collinsia verna* Scrophulariaceae. *Am. J. Bot.* 86:1239–47

Karron JD, Mitchell RJ, Holmquist KG, Bell JM, Funk B. 2004. The influence of floral display size on selfing rates in *Mimulus ringens*. *Heredity* 92:242–48

Kasagi T, Kudo G. 2003. Variations in bumble bee preference and pollen limitation among neighboring populations: comparisons between *Phyllodoce caerulea* and *Phyllodoce aleutica* (Ericaceae) along snowmelt gradients. *Am. J. Bot.* 90:1321–27

Kelly JK, Willis JH. 2002. A manipulative experiment to estimate biparental inbreeding in monkeyflowers. *Int. J. Plant Sci.* 163:575–79

Kohn JR, Barrett SCH. 1994. Pollen discounting and the spread of a selfing variant in tristylous *Eichhornia paniculata*: evidence from experimental populations. *Evolution* 48:1576–94

Kohn JR, Biardi JE. 1995. Outcrossing rates and inferred levels of inbreeding depression in gynodioecious *Cucurbita foetidissima* (Cucurbitaceae). *Heredity* 75:77–83

Kohn JR, Graham SW, Morton B, Doyle JJ, Barrett SCH. 1996. Reconstruction of the evolution of reproductive characters in Pontederiaceae using phylogenetic evidence from chloroplast DNA restriction site variation. *Evolution* 50:1454–69

Korbecka G, Klinkhamer PGL, Vrieling K. 2002. Selective embryo abortion hypothesis revisited: a molecular approach. *Plant Biol.* 4:298–310

Lande R, Schemske DW. 1985. The evolution of self-fertilization and inbreeding depression in plants. I. Genetic models. *Evolution* 39:24–40

Lande R, Schemske DW, Schultze ST. 1994. High inbreeding depression, selective interference among loci, and the threshold selfing rate for purging recessive lethal mutations. *Evolution* 48:965–78

Larson BMH, Barrett SCH. 2000. A comparative analysis of pollen limitation in flowering plants. *Biol. J. Linn. Soc.* 69:503–20

Latta R, Ritland K. 1993. Models for the evolution of selfing under alternative modes of inheritance. *Heredity* 71:1–10

Latta R, Ritland K. 1994. Conditions favoring stable mixed mating systems with jointly evolving inbreeding depression. *J. Theor. Biol.* 170:15–23

Leclerc-Potvin C, Ritland K. 1994. Modes of self-fertilization in *Mimulus guttatus* (Scrophulariaceae): a field experiment. *Am. J. Bot.* 81:199–205

Levin DA, Kerster HW. 1974. Gene flow in seed plants. *Evol. Biol.* 7:139–220

Lipow SR, Wyatt R. 2000. Single gene control of postzygotic self-incompatibility in poke milkweed, *Asclepias exaltata* L. *Genetics* 154:893–907

Lloyd DG. 1979. Some reproductive factors affecting the selection of self-fertilization in plants. *Am. Nat.* 113:67–79

Lloyd DG. 1992. Self- and cross-fertilization in plants. II. The selection of self-fertilization. *Int. J. Plant Sci.* 153:370–80

Lloyd DG, Schoen DJ. 1992. Self- and cross-fertilization in plants. I. Functional dimensions. *Int. J. Plant Sci.* 153:358–69

Loveless MD, Hamrick JL. 1984. Ecological determinants of genetic structure in plant populations. *Annu. Rev. Ecol. Syst.* 15:65–95

Lu YQ. 2000. Effects of density on mixed mating systems and reproduction in natural populations of *Impatiens capensis*. *Int. J. Plant Sci.* 161:671–81

Marshall DL, Diggle PK. 2001. Mechanisms of differential pollen donor performance in wild radish, *Raphanus sativus* (Brassicaceae). *Am. J. Bot.* 88:242–57

Marshall DL, Ellstrand NC. 1986. Sexual selection in *Raphanus sativus*: experimental data on nonrandom fertilization, maternal choice, and consequences of multiple paternity. *Am. Nat.* 127:446–61

Masuda M, Yahara T, Maki M. 2004. Evolution of floral dimorphism in a cleistogamous annual, *Impatiens noli-tangere* L. occurring under different environmental conditions. *Ecol. Res.* 19:571–80

Maynard Smith J. 1977. The sex habit in plants and animals. In *Measuring Selection in Natural Populations*, ed. FB Christiansen, TM Fenchel, pp. 265–73. Berlin: Springer-Verlag

Maynard Smith J. 1982. *Evolution and the Theory of Games*. Cambridge, UK: Cambridge Univ.

Moeller DA. 2005. Pollinator community structure and sources of spatial variation in plant-pollinator interactions in *Clarkia xantiana* ssp. *xantiana*. *Oecologia* 142:28–37

Morgan MT, Schoen DJ, Bataillon TM. 1997. The evolution of self-fertilization in perennials. *Am. Nat.* 150:618–38

Morgan MT, Wilson WG. 2005. Self-fertilization and the escape from pollen limitation in variable pollination environments. *Evolution* 59:1143–48

Nagylaki T. 1976. A model for the evolution of self-fertilization and vegetative reproduction. *J. Theor. Biol.* 58:55–58

Ornduff R. 1969. Reproductive biology in relation to systematics. *Taxon* 18:121–33

Pannell JR, Barrett SCH. 1998. Baker's law revisited: reproductive assurance in a metapopulation. *Evolution* 52:657–68

Pannell JR, Barrett SCH. 2001. Effects of population size and metapopulation dynamics on a mating system polymorphism. *Theor. Popul. Biol.* 59:145–55

Plaistow SJ, Johnstone RA, Colegrave N, Spencer M. 2004. Evolution of alternative mating tactics: conditional versus mixed strategies. *Behav. Ecol.* 15:534–42

Porcher E, Lande R. 2005a. Loss of gametophytic self-incompatibility with evolution of inbreeding depression. *Evolution* 59:46–60

Porcher E, Lande R. 2005b. Reproductive compensation in the evolution of plant mating systems. *New Phytol.* 166:673–84

Porcher E, Lande R. 2005c. The evolution of self-fertilization and inbreeding depression under pollen discounting and pollen limitation. *J. Evol. Biol.* 18:497–508

Ramsey M, Vaughton G. 1996. Inbreeding depression and pollinator availability in a partially self-fertile perennial herb *Blandfordia grandiflora* (Liliaceae). *Oikos* 76:465–74

Rathcke B, Real L. 1993. Autogamy and inbreeding depression in mountain laurel, *Kalmia latifolia* (Ericaceae). *Am. J. Bot.* 80:143–46

Rausher MD, Augustine D, VanderKooi A. 1993. Absence of pollen discounting in a genotype of *Ipomoea purpurea* exhibiting increased selfing. *Evolution* 47:1688–95

Rausher MD, Chang SM. 1999. Stabilization of mixed-mating systems by differences in the magnitude of inbreeding depression for male and female fitness components. *Am. Nat.* 154:242–48

Rigney LP. 1995. Postfertilization causes of differential success of pollen donors in *Erythronium grandiflorum* (Liliaceae): nonrandom ovule abortion. *Am. J. Bot.* 82:578–84

Ritland K. 1984. The effective proportion of self-fertilization with consanguineous mating in inbred populations. *Genetics* 106:139–52

Ritland K. 1990. Inferences about inbreeding depression based on changes of the inbreeding coefficient. *Evolution* 44:1230–41

Ritland K, Jain S. 1981. A model for the estimation of outcrossing rate and gene frequencies using *n* independent loci. *Heredity* 47:35–52

Robertson AW, Diaz A, Macnair MR. 1994. The quantitative genetics of floral characters in *Mimulus guttatus*. *Heredity* 72:300–11

Roff DA. 1997. *Evolutionary Quantitative Genetics*. New York: Chapman & Hall

Ronfort J, Couvet D. 1995. A stochastic-model of selection on selfing rates in structured populations. *Genet. Res.* 65:209–22

Routley MB, Mavraganis K, Eckert CG. 1999. Effect of population size on the mating system in a self-compatible, autogamous plant, *Aquilegia canadensis* (Ranunculaceae). *Heredity* 82:518–28

Sakai S. 1995. Evolutionarily stable selfing rates of hermaphroditic plants in competing and delayed selfing modes with allocation to attractive structures. *Evolution* 49:557–64

Sakai S, Ishii HS. 1999. Why be completely outcrossing? Evolutionarily stable outcrossing strategies in an environment where outcross pollen availability is unpredictable. *Evol. Ecol.* 1:211–22

Satake A, Sasakai A, Iwasa Y. 2001. Variable timing of reproduction in unpredictable environments: adaptation of flood plain plants. *Theor. Popul. Biol.* 60:1–15

Schemske DW, Lande R. 1985. The evolution of self-fertilization and inbreeding depression in plants. II. Empirical observations. *Evolution* 39:41–52

Schemske DW, Lande R. 1986. Mode of

pollination and selection on mating system: a comment on Aide's paper. *Evolution* 40:436

Schemske DW, Lande R. 1987. On the evolution of plant mating systems: a reply to Waller. *Am. Nat.* 130:804–6

Schoen DJ. 1983. Relative fitnesses of selfed and outcrossed progeny in *Gilia achilleifolia* (Polemoniaceae). *Evolution* 37:292–301

Schoen DJ, Brown AHD. 1991. Whole and part-flower self-pollination in *Glycine clandestine* and *G. argyrea* and the evolution of autogamy. *Evolution* 45:1651–64

Schoen DJ, Lloyd DG. 1984. The selection of cleistogamy and heteromorphic diaspores. *Biol. J. Linn. Soc.* 23:303–22

Schoen DJ, Lloyd DG. 1992. Self- and cross-fertilization in plants. III. Methods for studying modes and functional aspects of self-fertilization. *Int. J. Plant Sci.* 153:381–93

Schoen DJ, Morgan MT, Bataillon T. 1996. How does self-pollination evolve? Inferences from floral ecology and molecular genetic variation. *Philos. Trans. R. Soc. London Ser. B* 351:1281–90

Seavey SR, Bawa KS. 1986. Late-acting self-incompatibility in angiosperms. *Bot. Rev.* 52:195–219

Shaw DL, Kahler AL, Allard RW. 1981. A multilocus estimator of mating system parameters in plant populations. *Proc. Natl. Acad. Sci. USA* 78:1298–302

Spencer M, Colegrave N, Schwartz SS. 2001. Hatching fraction and timing of resting stage production in seasonal environments: effects of density dependence and uncertain season length. *J. Evol. Biol.* 14:357–67

Stebbins GL. 1974. *Flowering Plants: Evolution Above the Species Level.* Cambridge, MA: Belknap

Stephenson AG. 1981. Flower and fruit abortion: proximate causes and ultimate functions. *Annu. Rev. Ecol. Syst.* 12:253–79

Stewart HM, Stewart SC, Canne-Hilliker JM. 1996. Mixed mating system in *Agalinis neoscotica* (Scrophulariaceae) with bud pollination and delayed pollen germination. *Int. J. Plant Sci.* 157:501–8

Stone JL. 2002. Molecular mechanisms underlying the breakdown of gametophytic self-incompatibility. *Q. Rev. Biol.* 77:17–32

Takebayashi N, Morrell PL. 2001. Is self-fertilization an evolutionary dead end? Revisiting an old hypothesis with genetic theories and a macroevolutionary approach. *Am. J. Bot.* 88:1143–50

Tsitrone A, Duperron S, David P. 2003. Delayed selfing as an optimal mating strategy in preferentially outcrossing species: theoretical analysis of the optimal age at first reproduction in relation to mate availability. *Am. Nat.* 162:318–31

Ushimaru A, Kikuzawa K. 1999. Variation of breeding system, floral rewards, and reproductive success in clonal *Calystegia* species (Convolvulaceae). *Am. J. Bot.* 86:436–46

Uyenoyama MK. 1986. Inbreeding and the cost of meiosis: the evolution of selfing in populations practicing biparental inbreeding. *Evolution* 40:388–404

Uyenoyama MK, Antonovics J. 1987. The evolutionary dynamics of mixed mating systems: on the adaptive value of selfing and biparental inbreeding. In *Perspectives in Ethology*, ed. PPG Bateson, PH Klopfer, pp. 125–52. New York: Plenum

Uyenoyama MK, Waller DM. 1991a. Coevolution of self-fertilization and inbreeding depression I. Mutation-selection balance at one and two loci. *Theor. Popul. Biol.* 40:14–46

Uyenoyama MK, Waller DM. 1991b. Coevolution of self-fertilization and inbreeding depression II. Symmetric overdominance in viability. *Theor. Popul. Biol.* 40:47–77

Vallejo-Marin M, Uyenoyama MK. 2004. On the evolutionary costs of self-incompatibility: incomplete reproductive compensation due to pollen limitation. *Evolution* 58:1924–35

Venable DL. 1985. The evolutionary ecology of seed heteromorphism. *Am. Nat.* 126:577–95

Vogler DW, Das C, Stephenson AG. 1998. Phenotypic plasticity in the expression of self-incompatibility in *Campanula rapunculoides*. *Heredity* 81:546–55

Vogler DW, Kalisz S. 2001. Sex among the

flowers: the distribution of plant mating systems. *Evolution* 55:202–4

Waller DM. 1986. Is there disruptive selection for self-fertilization? *Am. Nat.* 128:421–26

Waller DM. 1993. The statics and dynamics of mating system evolution. In *The Natural History of Inbreeding and Outbreeding: Theoretical and Empirical Perspectives*, ed. NW Thornhill, pp. 97–117. Chicago: Univ. Chicago

Waser NM, Price MV. 1994. Crossing distance effects in *Delphinium nelsonii*: outbreeding and inbreeding depression in progeny fitness. *Evolution* 48:842–52

Weller SG, Ornduff R. 1977. Cryptic self-incompatibility in *Amsinckia grandiflora*. *Evolution* 31:47–51

Weller SG, Wagner WL, Sakai AK. 1995. A phylogenetic analysis of *Scheidea* and *Alsinidendron* (Caryophyllaceae: Alsinoid-eae): implications for the evolution of breeding systems. *Syst. Bot.* 20:315–37

Whitlock MC, Ingvarsson PK, Hatfield T. 2000. Local drift load and the heterosis of interconnected populations. *Heredity* 84:452–57

Wolfe LM. 1993. Inbreeding depression in *Hydrophyllum appendiculatum*: role of maternal effects, crowding, and parental mating history. *Evolution* 47:374–86

Wolfe LM, Barrett SCH. 1988. Temporal changes in the pollinator fauna of tristylous *Pontederia cordata*, an aquatic plant. *Can. J. Zool.* 66:1421–24

Wyatt R. 1988. Phylogenetics aspects of evolution of self-pollination. In *Plant Evolutionary Biology*, ed. DL Gottleib, SK Jain, pp. 109–31. New York: Chapman & Hall

Yahara T. 1992. Graphical analysis of mating system evolution in plants. *Evolution* 46:557–61

Annu. Rev. Ecol. Evol. Syst. 2005. 36:81–105
doi: 10.1146/annurev.ecolsys.36.091704.175523
Copyright © 2005 by Annual Reviews. All rights reserved
First published online as a Review in Advance on July 29, 2005

INDIRECT INTERACTION WEBS:
Herbivore-Induced Effects Through Trait Change in Plants

Takayuki Ohgushi

*Center for Ecological Research, Kyoto University, Otsu 520-2113, Japan;
email: ohgushi@ecology.kyoto-u.ac.jp*

Key Words   biodiversity, community structure, nontrophic link, plant–herbivore
interactions, trait-mediated indirect effects

■ **Abstract**   Although predation has a lethal effect on prey, mature terrestrial plants
are rarely killed by herbivores, but herbivory can change plant allelochemistry, cell
structure and growth, physiology, morphology, and phenology. This review explores
the herbivore-induced indirect effects mediated by such plant responses following herbivory in terrestrial systems. Herbivore-induced indirect effects are ubiquitous in many
plant–herbivore systems, and indirect interactions occur among temporally separated,
spatially separated, and taxonomically distinct herbivore species. Unlike interspecific
competition, herbivores can benefit each other through plant-mediated indirect effects. Herbivore-induced changes in plants occur at low levels of herbivory, which increases the likelihood of plant-mediated indirect interactions between herbivores. The
herbivore-induced indirect effects result in interaction linkages, which alter species
richness and abundance in arthropod communities. Such interaction linkages should
be depicted using indirect interaction webs, which incorporate nontrophic, indirect
links. The idea of interaction linkages by herbivore-induced indirect effects that shape
community organization and biodiversity is an important revision of the traditional
view of plant-based terrestrial food webs.

INTRODUCTION

A central issue in ecology is understanding how trophic interactions make up
food webs in various ecosystems (Berlow et al. 2004, Hunter & Price 1992, Paine
1980, Polis & Winemiller 1996). Recent studies of interactions between plants
and herbivores reveal that plants respond to herbivore damage by changes in allelochemistry, cell structure and growth, physiology, morphology, and phenology
(Karban & Baldwin 1997). Because herbivory is common and usually nonlethal
on terrestrial plants, this ensures in many ecosystems that most plants have traits
altered by herbivory. In the past, studies of trophic interactions at the ecosystem level have concentrated on how the relative abundance of biomass or energy

1543-592X/05/1215-0081$20.00

produced by one trophic level is transferred to another (Leibold et al. 1997, Oksanen et al. 1981, Polis 1999). In contrast, the consequences of ubiquitous nonlethal indirect links in plant–herbivore interactions in terrestrial systems have long been overlooked.

In this review, I argue for the prevalence of herbivore-induced plant responses in generating interaction linkages, which in turn affect herbivore community structure across trophic levels. I suggest that herbivore-induced interaction linkages have the potential to contribute greatly to the maintenance of species richness and interaction diversity in terrestrial systems. My arguments are largely restricted to herbivorous insects because they are among the richest contributors to biodiversity on the Earth, and they exhibit diverse feeding relationships with plants that produce many well-understood induced plant responses.

INTERACTION LINKAGE ON TERRESTRIAL PLANTS

In terrestrial systems, individual plant species with their associated herbivores form plant-based food chains that are interconnected with each other, producing a network of interacting species. Many studies focus on single interactions, although indirect effects can link multiple interactions in a community (Jones et al. 1998, Strauss 1997), and such interaction linkages are common in multitrophic systems (Dicke & Vet 1999, Gange & Brown 1997, Price et al. 1980). Both above- and belowground interactions are frequently influenced by indirect effects (Masters & Brown 1997, Van der Putten et al. 2001). Nevertheless, few studies have integrated the impacts of multiple indirect interactions in structuring ecological communities.

As an example of the important indirect interaction linkages, I illustrate how multiple plant–insect interactions are connected with each other on the willow *Salix miyabeana* (Figure 1). The spittlebug *Aphrophora pectoralis* is a specialist insect herbivore on the willow. In autumn, females lay eggs in the distal part of current shoots, which die within one week because of mechanical damage. This damage induces a compensatory shoot growth in the next year, producing longer shoots with a greater number of leaves (Nozawa & Ohgushi 2002). This enhanced shoot growth resulted in the increased density of 23 species of leafrolling caterpillars in early spring. After leafrolling caterpillars eclosed and left their leaf shelters, most leaf shelters were colonized by other insects, in particular, the aphid *Chaitophorus saliniger*, which is highly specialized for utilizing leafrolls (Nakamura & Ohgushi 2003). These aphids were tended by three species of ant that harvested aphid honeydew. The increased number of ants, in turn, reduced the larval survival of the leaf beetle *Plagiodera versicolora*. Direct interspecific competition is unlikely to have significant impacts on these herbivorous insects. The willow had a low level of leaf herbivory (less than 20% leaf consumption), suggesting that interspecific competition between leaf chewers is rare. Indeed, we detected a positive correlation between sap-sucking spittlebugs and aphids. Moreover, spittlebug nymphs and leaf beetle larvae are mobile so that they can avoid damaged plant tissues, even

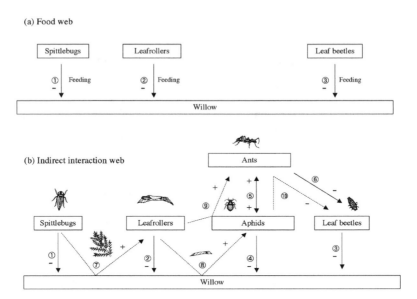

Figure 1 Comparison between a food web (*top*) and its indirect interaction web (*bottom*) of herbivorous insects on the willow *Salix miyabeana*. *Solid* and *broken lines* show direct and indirect effects, respectively. *Plus* and *minus signs* indicate positive and negative effects from an initiator to a receiver species, respectively.

if interspecific competition occurs. In contrast, spittlebug oviposition led to major indirect effects on other species. This unexpected linkage in the chain of indirect interactions indicates that such indirect effects provide an underlying mechanism responsible for a network of interactions in ecological communities.

INDIRECT EFFECTS OF HERBIVORE-INDUCED CHANGES IN PLANTS

In this section, I explore herbivore-induced indirect effects in terrestrial systems, focusing on how herbivores sharing the same host-plant have indirect interactions mediated by changes in plant characteristics. I compiled a representative set of examples of plant-mediated indirect interactions between herbivores covering a broad spectrum of plant–herbivore systems. The database was compiled by key-word searches using "indirect effects/interactions" or related concepts in articles published between 1985 and 2004 in major ecological journals including *Ecology, Oikos, Oecologia, Journal of Animal Ecology, Journal of Ecology, Functional Ecology, Ecological Entomology*, and *American Naturalist*. I also surveyed the reference sections of the papers obtained and of reviews on the topic in the *Annual Review of Ecology, Evolution, and Systematics*; the *Annual Review of Entomology*, and *Trends in Ecology and Evolution.* I included studies on indirect interactions between herbivores, including insects, mammals, fungi, and pathogens, through herbivore-induced changes in plants. I refer to the "initiator" as an herbivore that causes induced plant responses, the "receiver" as an impacted herbivore, and the "transmitter" as a plant fed upon by the initiator. Herbivore-induced indirect interactions produced by changes in plants are a type of trait-mediated indirect interaction (Abrams et al. 1996); they occur when an initiator species causes changes in traits of a transmitter species that, in turn, affects a receiver species. The cited studies meet the following criteria: (*a*) induced plant responses to initiators were directly or indirectly demonstrated, and (*b*) effects of trait changes in plants on receivers were measured by performance and/or population parameters. The final database consists of 83 pairwise interactions, which are summarized in Table 1. Detailed information on each reference is shown in Supplemental Appendix 1 (follow the Supplemental Material link from the Annual Reviews home page at http://www.annualreviews.org/).

Interactions among Temporally Separated Herbivore Species

Herbivore damage on plants often changes their nutrient status, production of defensive chemicals and volatile substances, physical defense structures of thorns, spines, and trichomes, plant architecture by compensatory regrowth, and phenology of plants including bud burst, leaf flush, and flowering onset (Karban & Baldwin 1997). These changes in plant traits following herbivory are important in determining food and habitat suitability for herbivores that subsequently utilize the same plant. These alterations in the plant are the mechanistic basis for indirect interactions between temporally separated herbivore species (Table 1).

Larvae of the ranchman's tiger moth *Platyprepia virginalis* and the western tussock moth *Orgyia vetusta* both feed on leaves of the bush lupine *Lupinus arboreus.* The former appears from February–April and the latter from May–July. Harrison & Karban (1986) demonstrated that feeding by the tiger moth larvae in early spring negatively affected the suitability of the host plant to the tussock moth larvae late in the season. Spring feeding by the tiger moth significantly reduced larval growth, pupal weight, and thus fecundity of the tussock moth. Also, it was suggested that early herbivory by the tiger moth decreased nitrogen levels in subsequently emerged leaves, which may have reduced performance of the tussock moth. Denno et al. (2000) found indirect interactions between the salt marsh-inhabiting planthoppers. Previous feeding by one planthopper species had detrimental effects on the subsequent performance and survival of the other. Prior feeding by *Prokelisia dolus* resulted in prolonged development and reduced body size in *P. marginata*, whereas development was protracted in *P. dolus* when plants were previously exposed to *P. marginata*. The mechanism of the delayed competitive effects between the two planthoppers is most likely diminished plant nutrition, because feeding by *P. dolus* significantly reduced the concentration of essential amino acids.

TABLE 1 Herbivore-induced indirect interactions between herbivorous species through changes in plant traits (see Supplemental Appendix 1, for details; follow the Supplemental Material link from the Annual Reviews home page at http://www.annualreviews.org/)

Type of interaction	Effect of initiator on receiver	Initiator→Receiver	Plant traits changed[a]	Effect on receiver[a]	Reference
Nonseparated	Negative	Caterpillar → spittlebug; aphid → aphid; planthopper → planthopper	Quality (−), leaf production (−), early senescence	Survival (−), growth (−), reproductive success (−)	Inbar et al. (1995), Karban (1986), Matsumura & Suzuki (2003)
Temporally separated	Negative	Aphid → aphid; caterpillar → caterpillar, aphid, sawfly, weevil, planthopper, leaf beetle; planthopper → planthopper; leaf beetle → leaf beetle; thrips → bumblebee; sap beetle → bumblebee	Quality (−), growth (−), defense chemicals (+), trichome (+), nectar and pollen production (−)	Survival (−), growth (−), oviposition preference (−), visitation rate (−), density (−), species richness (−)	Denno et al. (2000), Harrison & Karban (1986), Petersen & Sandström (2001), studies 5, 7–8, 10, 16–17, 20–22, 24–25, and 27–29 in Supplemental Material, Appendix 1
	Positive	Aphid → caterpillar; caterpillar → caterpillar, gall midge; flea beetle → cerambycid beetle; caterpillar, gall midge → aphid, spittlebug	Quality (+), regrowth (+)	Growth (+), density (+), species richness (+)	Danman (1989), Strauss (1991a), Williams & Myers (1984), studies 9, 12, 14–15, and 18–19 in Supplemental Material, Appendix 1

(Continued)

TABLE 1 (*Continued*)

Type of interaction	Effect of initiator on receiver	Initiator → Receiver	Plant traits changed[a]	Effect on receiver[a]	Reference
Spatially separated	Negative	Aphid → aphid; agromyzid fly → bracken chafer; click beetle → caterpillar	Quality (−), defense chemicals (+), biomass (−), seed set (−)	Growth (−), leaf consumption (−), density (−)	Bezemer et al. (2003), Masters & Brown (1992), Moran & Whitham (1990), Salt et al. (1996)
	Positive	Bracken chafer → aphid, agromyzid fly, tephritid fly; weevil, cranefly → tephritid fly	Quality (+), early flowering	Growth (+), fecundity (+), adult longevity (+)	Gange & Brown (1989), Masters & Brown (1992), Masters et al. (2001)
Temporally and spatially separated	Negative	Caterpillar → aphid, bee, syrphid fly; scarab, leaf beetle, grasshopper → caterpillar	Pollen production (−), flower number (−), flower size (−), floral tube (−), delayed flowering, phloem hydraulic pressure (−)	Visitation rate (−), time spent per flower (−), survival (−)	Johnson et al. (2002), Lehtilä & Strauss (1997), Strauss et al. (1996), study 39 in Supplemental Appendix 1
	Positive	Gall midge → leaf beetle, aphid; wireworm → honeybee, hover fly, bumblebee	Quality (+), regrowth (+), nectar production (+)	Visitation rate (+), density (+)	Nakamura et al. (2003), Poveda et al. (2003)
Temporally and taxonomically separated (pathogen–insect)	Negative	Spider mite → wilt fungus; leaf beetle → rust fungus; wilt fungus → spider mite; rust fungus → leaf beetle	Leaf quantity (−), defense chemicals (+)	Survival (−), growth (−), fecundity (−), density (−), infection (−)	Hatcher et al. (1994), Karban et al. (1987), Simon & Hilker (2003)
	Positive	Leaf beetle → rust fungus	—	Infection (+)	Simon & Hilker (2003)

Temporally and taxonomically separated (mammal–insect)	Negative	Sheep, ibex → leaf beetle; elk → sawfly	Flower number (−), fruit abundance (−), leaf quality for oviposition (−)	Density (−)	Bailey & Whitham (2003), Gómez & Gonzáles-Megías (2002)
	Positive	Moose → aphid, psyllid, leaf-miner, leaf-galler; eastern cottontail, snowshoe hare, moose, reindeer, elk → galling sawfly; beaver → leaf beetle	Vigor shoot (+), regrowth (+), defense chemicals (+), quality (+)	Defense ability (+), growth (+), density (+), species richness (+)	Danell & Huss-Danell (1985), Martinsen et al. (1998), Roininen et al. (1997), studies 50 and 54–55 in Supplemental Appendix 1
Spatially and taxonomically separated (mycorrhiza–insect)	Negative	Stem- and cone-boring caterpillar, scale → ectomycorrhizal fungus; arbuscular mycorrhizal fungus → caterpillar, tephritid fly	C/N ratio (+), defense chemicals (+)	Survival (−), growth (−), leaf consumption (−), colonization rate (−), gall size (−)	Gange & West (1994), Gehring et al. (1997), Vicari et al. (2002), studies 60 and 64 in Supplemental Appendix 1
	Positive	Arbuscular mycorrhizal fungus → caterpillar, bean beetle, aphid	Quality (+), defense chemicals (−), biomass (+)	Survival (+), growth (+), fecundity (+), pupation rate (+)	Borowicz (1997), Gange et al. (1999), Goverde et al. (2000)

(Continued)

TABLE 1 (*Continued*)

Type of interaction	Effect of initiator on receiver	Initiator → Receiver	Plant traits changed[a]	Effect on receiver[a]	Reference
Spatially and taxonomically separated (endophyte–insect)	Negative	Endophyte → aphid, caterpillar	Quality (−), defense chemicals (+), phytosterol metabolism (−)	Survival (−), growth (−), density (−), population growth rate (−)	Bultman et al. (2004), Omacini et al. (2001), Raps & Vidal (1998)
	Positive	Endophyte → aphid, grasshopper	Quality (+)	Growth (+), fecundity (+), density (+)	Gange (1996), Saikkonen et al. (1999)
Spatially and taxonomically separated (pathogen–insect)	Negative	Fungal pathogen → leaf beetle	Stem diameter (−), leaf production (−)	Survival (−), growth (−), oviposition preference (−)	Kruess (2002)
	Positive	Fungal pathogen → aphid	Quality (+), defense chemicals (+)	Growth (+), embryo development (+), density (+), population growth rate (+)	Johnson et al. (2003)
Ecosystem engineer mediated	Positive	Caterpillar → caterpillar, aphid, springtail; galling sawfly → aphid; eriophyid mite → spider mite	Leaf shelter (+)	Density (+), species richness (+)	Martinsen et al. (2000), Lill & Marquis (2003), Nakamura & Ohgushi (2003), studies 75–76 and 80–83 in Supplemental Appendix 1

[a](+), increase; (−), decrease.

In contrast, positive effects of early-attacking insects on later-emerging insects have been documented (Table 1). For instance, the fall webworm *Hyphantria cunea* feeding on leaves of the red alder from August–September, had larger pupal size and higher pupation rate on trees that were previously damaged by the western tent caterpillar *Malacosoma californicum pluviale* (Williams & Myers 1984). The heavier pupae on damaged trees resulted in a 12.5% increase in fecundity over pupae on undamaged trees.

Herbivore-induced architectural responses also generate indirect interactions between temporally separated species (Table 1). Strauss (1991a) showed that early-season bud damage by the leaf beetle *Blepharida rhois* caused subsequent production of basal vegetative shoots in the smooth sumac *Rhus glabra*. These basal shoots are the preferred oviposition site for the cerambycid stem borer *Oberea ocellata*, and stem-borer attack increased significantly after leaf beetle attack. Gall initiation by the stem gall midge *Rabdophaga rigidae* stimulates the development of lateral shoots of the willow *Salix eriocarpa* followed by a secondary leaf flush. Nakamura et al. (2003) found that lateral shoots and upper leaves on galled shoots were less tough and had a higher water and nitrogen content. As a result, density of the aphid *Aphis farinosa* was significantly higher on galled shoots than on ungalled shoots, because the aphid frequently colonized lateral shoots. Also, adults of the leaf beetles *Plagiodera versicolora* and *Smaragdina semiaurantiaca* were more abundant on galled shoots than on ungalled shoots, because they preferentially fed on young leaves produced by the secondary leaf flush.

Interactions among Spatially Separated Herbivore Species

Interactions between spatially separated insects, which share one plant but utilize different parts of it, have revealed that insects often interact significantly with each other, although the species never encounter one another directly (Table 1). These spatially separated indirect interactions could occur because tissues of an individual plant depend on a common resource budget.

Moran & Whitham (1990) described a plant-mediated interaction between two aphid species that feed on different parts of the lamb's-quarters *Chenopodium album*. One aphid *Hayhurstia atriplicis* makes leaf galls, whereas the other aphid *Pemphigus betae* feeds underground on roots. The root feeder had no significant effects on its host, but the leaf feeder severely reduced root biomass. As a result, the number of the root feeder significantly decreased, with *Pemphigus* often being eliminated entirely. Conversely, the garden chafer *Phyllopertha horticola* feeding on roots of the shepherd's purse *Capsella bursa-pastoris* improved performance of the sap-sucking aphid *Aphis fabae* (Gange & Brown 1989). The root feeder induced water stress to the host plant by a large reduction in vegetative tissue, resulting in an increase in soluble nitrogen. The enhanced host-plant quality increased growth rate and longevity, and thus fecundity of the aphid. On the other hand, the aphid affected neither the host plant nor the garden chafer. However, the garden chafer interacted with a dipteran leaf miner *Chromatomyia syngenesiae* in quite a different way on the common sow thistle *Sonchus oleraceus* (Masters & Brown 1992). Root herbivory increased pupal mass of the leaf miner and thus its fecundity, probably because of changes in host quality initiated by root feeding. In contrast, leaf herbivory reduced the growth rate of the root feeder, because leaf miner herbivory reduced root biomass considerably.

Interactions among Herbivores and Pollinators

Leaf herbivory by insects often changes considerably the quantity and/or quality of floral traits, which are of crucial importance in pollinator service (Bronstein et al. 2006, Strauss 1997). Foliar leaf damage early in the season decreases flower number, flower size, pollen production, pollen performance, and nectar production, which can affect plant relationships with pollinators (Table 1).

Strauss et al. (1996) and Lehtilä & Strauss (1997) experimentally studied how leaf damage affects plant attractiveness to pollinators in the wild radish *Raphanus raphanistrum*. Leaf damage by larvae of the white butterfly *Pieris rapae* significantly decreased the number and size of flowers. Pollinators discriminated against damaged plants by visiting such plants less frequently and by spending less time on them. Damaged plants received fewer visits by native bees than undamaged plants, probably because flower number was the main cue attracting native bees to plants. Also, syrphid flies, which were abundant pollinators, spent less time per flower on the damaged than on the undamaged plants. These studies emphasize how pollination service is largely influenced by previous herbivory, which has long been ignored in pollination ecology.

Interactions among Distantly Related Herbivore Species

Hochberg & Lawton (1990) argued that organisms in different phyla or even kingdoms may compete for the same resources, and that such interactions may be one of the most pervasive forms of interspecific competition in nature, yet still be one of the most poorly understood. As induced plant responses can influence a variety of different herbivores, the initiator and the receiver species may be related taxonomically only distantly.

INTERACTIONS AMONG HERBIVOROUS MAMMALS AND INSECTS   Mammalian browsing often affects indirectly herbivorous insects in negative or positive ways (Table 1). Danell & Huss-Danell (1985) found that herbivorous insects including aphids, psyllids, leaf miners, and leaf gallers were more abundant on birch trees of *Betula pendula* and *B. pubescence* previously browsed upon by the moose *Alces alces* than on unbrowsed trees. Browsed trees subsequently produced larger leaves with more nitrogen and chlorophyll, and this improved leaf quality resulted in higher densities of the herbivorous insects. Natural browsing by hares and moose had strongly positive effects on densities of galling insects on two host plants, *Populus balsamifera* and *Salix novaeangliae* (Roininen et al. 1997). When the mammalian browsers attacked these plants, numbers of newly developed vigorous shoots on ramets increased significantly. As a result, leaf-edge galling sawfly density increased significantly because of the improved plant quality.

The beaver *Castor canadensis* often cuts down cottonwoods *Populus* sp., removing nearly all aboveground biomass. Resprout growth from the stumps and roots of beaver-cut trees contains more phenolic glycosides and total nitrogen than normal juvenile growth. The specialist leaf beetle *Chrysomela confluens* is attracted to the resprouted growth. Martinsen et al. (1998) experimentally demonstrated the positive effects of the beaver on the leaf beetle. Beetle larvae that had fed on resprout growth were better defended against ants than those that fed on nonresprout growth, because the increased defensive chemicals in the resprout growth were sequestered and used by the beetles for their own defenses. Beetle larval development on resprout growth was also significantly faster and larval weight higher at maturity because of the increased total leaf nitrogen. Regrowth of plants following herbivory often changes plant architecture by increasing the biomass of vegetative and reproductive parts or by inducing rapid branching. When browsed heavily by mule deer and/or elk in spring, the number of inflorescences on the scarlet gilia *Ipomopsis arizonica* increased. The increased number of inflorescences, in turn, increased the density of a fruit-feeding noctuid caterpillar (Mopper et al. 1991).

INTERACTIONS AMONG MICROORGANISMS AND HERBIVOROUS INSECTS   Indirect interactions between highly unrelated organisms can also include microorganisms—such as pathogens, endophytes, and mycorrhizae—sharing a host plant (Table 1). The fungal pathogen *Verticillium dahliae* was less likely to cause symptoms of verticillium wilt on cotton seedlings previously attacked by *Tetranychus* spider mites (Karban et al. 1987). Conversely, spider mite densities decreased on seedlings infected with fungal disease, probably because of the reduced leaf tissue by fungal infection. In contrast, Johnson et al. (2003) found positive indirect effects of the fungal pathogen *Marssonina betulae* of silver birch on preference, performance, and population growth of the aphid *Euceraphis betulae*. Aphids reared on infested leaves were heavier, possessed longer hind tibiae, and displayed enhanced embryo development, compared to aphids on intact leaves. Population growth rate of aphids was also positively correlated with fungal infection. Fungal-infected leaves contained higher concentrations of free amino acids, resulting in the positive interaction between the fungus and aphids.

Systemic endophytes are well known for increasing host-plant defenses against insect herbivores and pathogenic microorganisms by producing mycotoxins (Clay 1997). Thus, insect herbivores exhibit reduced performance and/or population density on endophyte-infected plants (Table 1). However, infection by fungal endophytes may have positive effects on herbivorous insects. The sycamore aphids *Drepanosiphum platanoidis* and *Periphyllus acericola* had significantly higher densities, heavier weight, and more fecundity on infected than uninfected trees of the sycamore maple *Acer pseudoplatanus* (Gange 1996). The infected leaves had higher soluble nitrogen, which may have increased aphid performance and density.

Effects of arbuscular mycorrhizal fungi on herbivorous insects vary from negative to positive (Table 1) depending on diet breadth of insects or soil nutrients (Gange 2006). Arbuscular mycorrhizal fungi decreased herbivory by chewing and leaf-mining insects on *Plantago lanceolata* by increasing the level of the carbon-based feeding deterrents aucubin and catalpol (Gange & West 1994). In contrast, they improved performance of the aphids *Myzus ascalonicus* and *M. persicae*; adults gained greater weight and fecundity (Gange et al. 1999). On the other hand, ectomycorrhizal fungi are negatively affected by herbivorous insects (Gehring & Whitham 1994), because herbivore-induced reductions in aboveground biomass reduce the carbon-source capacity of plants to such a degree that there is insufficient carbon to meet the demands of mycorrhizal fungi.

Interactions Mediated by Insect Ecosystem Engineers

Ecosystem engineers are organisms that directly or indirectly modulate the availability of resources to other species by causing physical state changes in biotic or abiotic materials (Jones et al. 1994). Insect ecosystem engineers manipulate plants to create structural alterations that influence interactions among species. This contrasts with the trait mediation by insects discussed in previous sections where herbivory altered plant responses without causing structural changes.

Ecosystem engineering is ubiquitous on terrestrial plants. Obvious candidates include gall makers, leafrollers, case bearers, and stem borers, all of which are common insect herbivore guilds, and which provide new habitats to other herbivores and/or their natural enemies. In particular, shelter building is a very common lifestyle among the microlepidoptera and in some weevils, sawflies, and even grasshoppers. Insects that are secondary occupants of shelters can gain several benefits, including avoidance of natural enemies (Damman 1987), protection from adverse microclimates (Hunter & Willmer 1989, Larsson et al. 1997), and access to more easily eaten food (Sagers 1992) and highly nutritious food (Fukui et al. 2002).

Lawton & Jones (1995) argued that ecologists fail to recognize the role of ecosystem engineers as keystone species that exert a great influence on community organization. As we can see in the function of leafrolling caterpillars as ecosystem engineers on the willow (Figure 1), recent studies have shown that insect ecosystem engineers have the potential to greatly affect other arthropods (Marquis & Lill 2006, Table 1).

Features of Herbivore-Induced Indirect Interactions

The literature survey clearly demonstrates that herbivore-induced indirect effects through trait change in plants are widespread in many plant–herbivore systems (Supplemental Appendix 1; follow the Supplemental Material link from the Annual Reviews home page at http://www.annualreviews.org/). Although the potential importance of trait-mediated indirect effects in ecological communities has been widely accepted (Abrams et al. 1996, Strauss 1991b, Wootton 1994), they have been studied much less frequently than density-mediated indirect effects such as keystone predation, trophic cascades, and apparent competition (Holt & Lawton 1994, Menge 1995, Pace et al. 1999, Polis et al. 2000). More recently, some authors have stressed the community consequences of trait-mediated indirect interactions in herbivore–predator systems (Werner & Peacor 2003, Bolker et al. 2003). Werner & Peacor (2003) argued that ecological communities are replete with trait-mediated indirect effects that arise from phenotypic plasticity, and that these effects are quantitatively important to community dynamics. Nevertheless, trait-mediated indirect interactions have received little attention in plant–herbivore systems (but see Callaway et al. 2003).

Table 2 summarizes features of indirect effects in plant–herbivore interactions by comparing them with those found in herbivore–predator interactions. This review reveals that substantial indirect interactions caused by herbivore-induced changes in terrestrial plants frequently occur among temporally separated, spatially separated, and distantly related herbivore species. These interactions have been poorly explored for two reasons. First, the traditional view on within-trophic-level interactions has emphasized that interactions should be most prevalent among closely related species within guilds or among species that utilize the same part of a resource at the same time. Second, unlike interspecific competition, these plant-mediated indirect interactions commonly occur at low levels of herbivory resulting in underestimation of the ubiquitous indirect interactions among herbivores.

Note that herbivores sharing the same host-plant can benefit each other (47% of 83 pairwise interactions in Supplemental Appendix 1; follow the Supplemental Material link from the Annual Reviews home page at http://www.annualreviews. org/), because herbivory often enhances resource availability through improved nutritional quality and/or increased biomass of plants because of compensatory regrowth. In addition, ecosystem engineers benefit secondary users that colonize newly constructed domiciles later in the season. Despite the fact that positive interactions are ubiquitous in many ecological communities (Bruno et al. 2003, Hay et al. 2004), the beneficial interactions within the same trophic level have been largely ignored. This is because the traditional view of community ecology has largely emphasized interspecific competition as the interaction of primary importance between organisms at the same trophic level.

In terrestrial systems, trait-mediated indirect effects should predominate in plant–herbivore interactions, whereas density-mediated indirect effects should be most common in herbivore–predator interactions. This is because predators kill individuals of the lower trophic level, whereas herbivores only alter their traits. Thus, indirect effects through changes in density because of mortality by consumers occur infrequently in plant–herbivore systems. For example, Müller & Godfray (1999) suggested that indirect effects by trait mediation are less frequent than indirect effects by density mediation in aphid–parasitoid systems. There is increasing appreciation of trait-mediated indirect effects resulting from changes in prey behavior to avoid predation risk, i.e., the nonlethal effects of predators (Losey & Denno 1998, Schmitz 1998, Schmitz et al. 2004). However, there is a large difference in trait-mediated indirect effects between herbivore–predator and

TABLE 2 Comparison of features of indirect effects in plant–herbivore and herbivore–predator interactions in terrestrial systems

Type of interaction	Effect of feeding	Response of herbivore/ plant after feeding	Indirect effects involved predominantly	Within-trophic interactions at a consumer level		
				Indirect interactions among separated species	Ecosystem engineering	Positive interactions
Plant–Herbivore	Nonlethal	Trait change	Trait-mediation	Common	Common	Common
Herbivore– Predator	Lethal	Mortality	Density- mediation	Less frequent	Less frequent	Less frequent

plant–herbivore systems in terms of when herbivores or plants respond to attack by their enemies. The indirect effects mediated by changes in behavior of an herbivore prey result from the presence of a predator before feeding, whereas the plant-mediated indirect effects occur after feeding by herbivores. In other words, the trait-mediated indirect effects in plant–herbivore interactions emerge in the postfeeding process, whereas those in herbivore–predator interactions appear in the prefeeding process. In consequence, indirect interactions between herbivores via changes in plants should occur more frequently than those between predators via changes in behavior of a shared herbivore prey.

WHY ARE HERBIVORE-INDUCED INDIRECT INTERACTIONS SO COMMON IN TERRESTRIAL SYSTEMS?

The importance of interspecific interactions between herbivorous insects has long been discounted, because empirical studies often show lack of competition in nature (Lawton & Strong 1981, Seifert 1984, Strong 1984). In addition, the concentration on direct interactions has caused us to overlook the importance of widespread herbivore-induced indirect interactions on terrestrial plants. Direct interspecific competition for limited resources requires high levels of herbivory, whereas plant-mediated indirect interactions can occur at low levels of herbivory. One reason for this is that plant defenses that mediate herbivore indirect interactions are often rapidly induced at low levels of herbivory before it causes plant mortality. Conversely, heavy defoliation can actually decrease indirect interactions. Plants that are heavily exploited during outbreaks of forest defoliators, for example, cannot compensate for lost tissue. Also, habitats previously created by ecosystem engineers are hardly maintained under heavy herbivory. A lack of visible depletion of green plants, therefore, does not mean that interspecific interactions between herbivores rarely occur. Instead, limited herbivory greatly increases the likelihood of indirect interactions between herbivores mediated by changes in plant characteristics. Thus, it is inferred that plant-mediated indirect interactions between herbivores predominate at low levels of herbivory, whereas the relative importance of direct interspecific competition is apparent at high levels of herbivory.

In this context, plant-mediated indirect effects should be more common in terrestrial than in aquatic systems. In terrestrial systems, the average consumption rate by herbivores varies from 4% to 18% of aboveground plant biomass (Polis 1999), whereas in aquatic systems herbivore consumption often exceeds 50% of primary production. Indeed, primary production in aquatic systems is mainly by phytoplankton, which are killed by predation leaving an absence of organisms that can retain induced responses. The low level of herbivory in terrestrial plants therefore produces a predominance of plant-mediated indirect effects in terrestrial systems, whereas the high level of herbivory in aquatic systems produces more direct effects of grazing. Recent reviews strongly support this view that the majority of interactions between terrestrial herbivorous insects are likely to be indirect, mediated by changes in plants following herbivory (Damman 1993, Denno & Kaplan 2006, Denno et al. 1995, Masters & Brown 1997). For example, Denno et al. (1995) stressed that over half of the 145 documented cases of interspecific competition among insect herbivores involved delayed, plant-mediated competition in which previous feeding by one species induced either nutritional or allelochemical changes in the plant that adversely affected the performance of another species feeding on it later in the season.

INTEGRATING MULTIPLE INTERACTIONS INTO INDIRECT INTERACTION WEBS

In this section, I emphasize the important role of herbivore-induced indirect effects through plant traits in forming indirect interaction webs. I also illustrate that the indirect interaction web provides a conceptual tool to efficiently explore the structure and biodiversity of ecological communities by comparing the traditional food web approach.

How Does Interaction Linkage Affect Biodiversity?

Most of the Earth's biodiversity is in its interaction diversity: the tremendous variety of ways in which species are linked together into constantly interacting networks. Thus, ecologists have recognized diversity of species interactions as one of the most important components of biodiversity (Price 2002, Thompson 1996). For example, many of the adaptations and counter-adaptations of plants and their insect herbivores indicate that much of the biodiversity of the Earth results from the arms race between herbivores and their host plants (Strauss & Zangerl 2002). Temporal and spatial resource heterogeneity can increase species richness and interaction diversity in terrestrial systems (Hunter et al. 1992). Specifically, temporal and spatial heterogeneity in the food and habitat provided by terrestrial plants is greatly promoted by the feedbacks resulting from changes in plant quality and architecture in response to herbivory and the creation of physical structures by ecosystem engineers.

An understanding of interaction linkages propagated by herbivore-induced indirect effects can provide valuable insight into how a network structure of species interactions affects biodiversity in ecological communities. We are starting to examine indirect effects on biodiversity components in plant–herbivore systems (Bailey & Whitham 2002, Martinsen et al. 2000, Ohgushi 2006, Ohgushi et al. 2006, Omacini et al. 2001, Van Zandt & Agrawal 2004, Waltz & Whitham 1997). Herbivore-induced changes in terrestrial plants can generate changes that cascade upward to higher trophic levels and, thus, influence biodiversity. These bottom-up cascading effects can have repercussions through entire herbivorous insect communities and alter species richness and abundance of each species. For example, larvae of leafrollers on cottonwoods construct leaf shelters, which are later colonized by other arthropods. Martinsen et al. (2000) found four times greater species richness and seven times greater abundance of arthropods on shoots with a rolled leaf compared to adjacent shoots without leafrolls. Likewise, Lill & Marquis (2003) found that the presence of a leaf-tying caterpillar *Pseudotelphusa* sp. had a great impact on species composition of herbivorous insects on white oak *Quercus alba*. In their removal experiments, a decrease in shelter availability significantly decreased by 14–38% species richness of leaf-chewing insects. This is because positive effects of the ecosystem engineers on other arthropods that secondarily use leaf shelters later caused the increased arthropod biodiversity. Another example is the leaf-galling aphid *Pemphigus betae* and the leaf beetle *Chrysomela confluens*, which had positive and negative effects on other arthropod species on cottonwoods, respectively (Waltz & Whitham 1997). Aphid removal decreased species richness by 32% and relative abundance by 55%, respectively, because the aphids attracted various predators and parasitoids and herbivorous insects because of changes in plant quality. In contrast, the leaf beetle decreased species richness of other herbivorous insects because the leaf beetles negatively affected plants by reducing terminal shoot growth. These studies indicate that indirect interaction linkages have a significant impact on arthropod biodiversity on terrestrial plants.

Indirect Interaction Webs

Indirect effects have the potential to strongly influence biodiversity components in ecological communities by shaping a network structure of interacting species. The most complete ecological network descriptions available are food webs, a basic tool to analyze community structure (Polis & Winemiller 1996). Because food webs focus on direct trophic interactions, nontrophic interactions are not included. As I stressed, the nontrophic, herbivore-induced indirect interactions can connect herbivore species indirectly mediated by trait change in plants. Thus, plant-based terrestrial food webs that ignore nontrophic indirect links are an inadequate tool for understanding the structural organization of arthropod communities. Furthermore, the principles of trophic interactions in food webs are not of much value in understanding ecosystem engineering. To understand how multiple interactions are connected, we can use "indirect interaction webs" that include nontrophic, indirect links. The indirect interaction webs can efficiently illustrate the linkage of multiple interactions, thereby providing a tool to explore the interaction diversity in a community. Food webs alone can clarify only feeding interactions, which are a part of indirect interaction webs. In this context, Berlow et al. (2004) pointed out that the future challenge to develop the theory of food web dynamics is to incorporate nontrophic links into food web structure. Menge & Sutherland (1987) originally termed "interaction web" as a trophic structure of strong interactions. It is always a subset of the species in a food web, deriving from Paine's functional web (Paine 1980). Although recent arguments include nontrophic or indirect interactions only if they are detected as strong interactions (Menge & Branch 2001), the interaction webs are principally based on the traditional food web concept. The indirect interaction webs are an alternative that explicitly incorporate nontrophic and indirect interactions into components of traditional food webs. Food webs consist of direct trophic interactions with energy transfer, whereas indirect interaction webs include nontrophic effects without energy transfer as mediators to connect multiple interactions.

Again, let us look at an indirect interaction web illustrating the interaction linkage of herbivorous insects on the willow. A food web approach detected three independent trophic interactions (Figure 1a) consisting of spittlebugs, leafrollers, and leaf beetles that feed on plants (interactions 1, 2, and 3), each of which is temporally or spatially separated from the others. The aphids were not included in the food web because they did not directly colonize the willow in the absence of leaf shelters constructed by leafrollers. In the indirect interaction web, the following indirect interactions were added (Figure 1b): the interaction between spittlebugs and leafrollers through compensatory shoot growth (interaction 7), the interaction between leafrollers and aphids through leaf shelters (interaction 8), the interaction between leafrollers and three ant species through aphid colonies (interaction 9), and the interaction between aphids and leaf beetles through increased tending by ants (interaction 10). Because the aphids were included in this web when leaf shelters were available, three direct interactions were newly established: the interaction between aphids and willow (interaction 4), the interaction between aphids and three species of ants (interaction 5), and the interaction between ants and leaf beetles (interaction 6). Thus, the indirect interaction web revealed six direct and four indirect interactions including four positive interactions, whereas the food web approach encompassed only three negative, direct interactions. Will we find in general that direct plus indirect interaction webs increase the detection of species' influences on each other by over three times, as in the case above? The indirect interaction webs will differ greatly and depict the interaction network and diversity in ecological communities more realistically than do the traditional food webs.

FUTURE DIRECTIONS

The study of indirect effects is an increasingly rich subfield of community ecology (Wootton 2002). Indeed, there is rapidly expanding evidence to suggest the importance of herbivore-induced indirect effects as mediators of interaction linkages shaping indirect interaction webs. This subject is of great importance in understanding not only community organization but also in identifying the underlying mechanisms of maintenance of biodiversity. Thus, the study of herbivore-induced indirect effects is at a very challenging stage (Ohgushi et al. 2006). Here, I emphasize several promising directions for future research.

1. We need further evidence to determine how common and widespread herbivore-induced indirect effects are, not only in terrestrial but also in aquatic systems. As herbivore-induced indirect interactions occur at low levels of herbivory, I predict that they will be much more frequent in terrestrial systems than in pelagic systems. Note that in marine systems, seaweeds may provide

plant-mediated indirect effects because they can induce chemical defenses following herbivory (Cronin & Hay 1996, Pavia & Toth 2000).

2. We should seek out plant characteristics that provide favorable conditions for herbivore-induced indirect effects by comparing plant responses following herbivory among taxa, life histories, and life forms.

3. We need to explore herbivore-initiated interaction linkages as the important community consequences of trait-mediated indirect effects. Also, a comparison of trait-mediated indirect effects in plant–herbivore and herbivore–predator systems will contrast the two different forms of indirect effects through trait mediation.

4. Long-term studies are crucial to clarify temporal variation in herbivore-induced indirect effects. In particular, we need multigenerational studies of the population dynamics of key species that initiate indirect effects to understand how the temporal changes in indirect effects alter the structure of indirect interaction webs in ecological communities.

5. We need to know how the interaction linkages caused by nontrophic indirect effects determine community organization and biodiversity. Specifically, we should pay much attention to the positive effects of ecosystem engineering and plant compensatory growth on species richness and interaction diversity.

6. Ecologists should recognize that indirect interaction webs are a valuable tool for understanding the importance of nontrophic indirect links and interaction diversity in nature. This is because traditional food webs can rarely predict underlying mechanisms of community organization that are frequently shaped by nontrophic indirect effects.

7. We need to compare multitrophic interactions in terrestrial and aquatic systems in the context of the presence or absence of nonlethal effects that produce trait-mediated indirect effects. Because the nonlethal effects of herbivores on terrestrial plants provide a mechanistic basis for feedbacks cascading upward through trophic levels via plant-mediated indirect effects, I predict that they would be of secondary importance in pelagic systems because zooplankton has primarily lethal effects on phytoplankton.

8. Plant-mediated interactions between leaf-feeding and root-feeding insects and those between leaf- or sap-feeding insects and mycorrhizae can link above- and belowground communities (Van der Putten et al. 2001). There is increasing evidence that aboveground herbivory can change root carbon allocation, root exudation, root biomass, and morphology (Bardgett et al. 1998). Thus, the quantity and quality of organic matter input from plants damaged by herbivores have the potential to greatly influence abundance, species composition, and activity of the soil organisms in the rhizosphere by altering interactions in soil food webs.

9. Because plant responses to herbivores provide a mechanistic basis for indirect interaction linkages shaping nontrophic indirect interaction webs, an

essential question is, How do plant responses to herbivores evolve in communities that consist of diverse assemblages that interact directly and indirectly? In this context, selection pressures caused by one species can change in the presence of other species and, thus, variation in the community composition can alter the coevolutionary outcomes of interactions (Siepielski & Benkman 2004, Thompson 1994). Furthermore, we should explore the evolutionary consequences of plant-mediated indirect effects (Agrawal & Van Zandt 2003, Craig 2006). Specifically, I predict that trait-mediated indirect effects limit the potential for pairwise coevolution, and that this limitation can be seen in the diffused evolutionary arms race between plant resistance and its herbivores because multiple herbivores attack the same host-plant. On the other hand, plant-mediated indirect effects will provide valuable insights to understanding how evolutionary alterations of plant traits, in turn, affect community organization of higher trophic levels by reforming interaction linkages.

If simple systems, such as willow and its herbivorous insects discussed here, can reveal three times more interactions when both direct and indirect interactions are examined, we have a lot more ecology to study. There is also much to evaluate in terms of relative strengths of direct and indirect effects, and density- and trait-mediated effects. In addition, we have more mechanisms to understand in relation to the maintenance and increase of biodiversity. Emphasis on nontrophic and indirect effects offers great promise for enriching ecological investigations and the understanding of nature.

ACKNOWLEDGMENTS

I thank Peter Price, Tim Craig, Mark Hunter, John Thompson, Sharon Strauss, and Bob Denno for invaluable comments on this manuscript. My work was supported by the Ministry of Education, Culture, Sports, Science and Technology Grant-in-Aid for Creative Basic Research (09NP1501), Scientific Research (A-15207003), and Biodiversity Research of the 21st Century COE Program (A14).

**The *Annual Review of Ecology, Evolution, and Systematics* is online at
http://ecolsys.annualreviews.org**

LITERATURE CITED

Abrams PA, Menge BA, Mittelbach GG, Spiller DA, Yodzis P. 1996. The role of indirect effects in food webs. See Polis & Winemiller 1996, pp. 371–95

Agrawal AA, Van Zandt PA. 2003. Ecological play in the coevolutionary theatre: genetic and environmental determinants of attack by a specialist weevil on milkweed. *J. Ecol.* 91:1049–59

Bailey JK, Whitham TG. 2002. Interactions among fire, aspen, and elk affect insect diversity: reversal of a community response. *Ecology* 83:1701–12

Bailey JK, Whitham TG. 2003. Interactions

among elk, aspen, galling sawflies and insectivorous birds. *Oikos* 101:127–34

Bardgett RD, Wardle DA, Yeates GW. 1998. Linking above-ground and below-ground interactions: how plant responses to foliar herbivory influence soil organisms. *Soil Biol. Biochem.* 30:1867–78

Berlow EL, Neutel AM, Cohen JE, de Ruiter PC, Ebenman B, et al. 2004. Interaction strengths in food webs: issues and opportunities. *J. Anim. Ecol.* 73:585–98

Bezemer TM, Wagenaar R, Van Dam NM, Wäckers FL. 2003. Interactions between above- and belowground insect herbivores as mediated by the plant defense system. *Oikos* 101:555–62

Bolker B, Holyoak M, Krivan V, Rowe L, Schmitz O. 2003. Connecting theoretical and empirical studies of trait-mediated interactions. *Écology* 84:1101–14

Borowicz VA. 1997. A fungal root symbiont modifies plant resistance to an insect herbivore. *Oecologia* 112:534–42

Bronstein JL, Huxman TE, Davidowitz G. 2006. Plant-mediated effects linking herbivory and pollination. See Ohgushi et al. 2006. In press

Bruno JF, Stachowicz JJ, Bertness MD. 2003. Inclusion of facilitation into ecological theory. *Trends Ecol. Evol.* 18:119–25

Bultman TL, Bell G, Martin WD. 2004. A fungal endophyte mediates reversal of wound-induced resistance and constrains tolerance in a grass. *Ecology* 85:679–85

Callaway RM, Pennings SC, Richards CL. 2003. Phenotypic plasticity and interactions among plants. *Ecology* 84:1115–28

Clay K. 1997. Fungal endophytes, herbivores and the structure of grassland communities. See Gange & Brown 1997, pp. 151–69

Craig TP. 2006. Evolution of plant-mediated interactions among natural enemies. See Ohgushi et al. 2006. In press

Cronin CV, Hay ME. 1996. Induction of seaweed chemical defenses by amphipod grazing. *Ecology* 77:2287–301

Damman H. 1987. Leaf quality and enemy

avoidance by the larvae of a pyralid moth. *Ecology* 68:88–97

Damman H. 1989. Facilitative interactions between two lepidopteran herbivores of *Asimina. Oecologia* 78:214–19

Damman H. 1993. Patterns of interaction among herbivore species. In *Caterpillars: Ecological and Evolutionary Constraints on Foraging*, ed. NE Stamp, TM Casey, pp. 132–69. New York: Chapman & Hall

Danell K, Huss-Danell K. 1985. Feeding by insects and hares on birches earlier affected by moose browsing. *Oikos* 44:75–81

Denno RF, Kaplan I. 2006. Plant-mediated interactions in herbivorous insects: mechanisms, symmetry, and challenging the paradigms of competition past. See Ohgushi et al. 2006. In press

Denno RF, McClure MS, Ott JR. 1995. Interspecific interactions in phytophagous insects: competition reexamined and resurrected. *Annu. Rev. Entomol.* 40:297–331

Denno RF, Peterson MA, Gratton C, Cheng J, Langellotto GA, et al. 2000. Feeding-induced changes in plant quality mediate interspecific competition between sap-feeding herbivores. *Ecology* 81:1814–27

Dicke M, Vet LEM. 1999. Plant-carnivore interactions: consequences for plant, herbivore and carnivore. In *Herbivores: Between Plants and Predators*, ed. H Olff, VK Brown, RH Drent, pp. 483–520. Oxford: Blackwell Sci.

Fukui A, Murakami M, Konno K, Nakamua M, Ohgushi T. 2002. A leaf-rolling caterpillar improves leaf quality. *Entomol. Sci.* 5:263–66

Gange AC. 1996. Positive effects of endophyte infection on sycamore aphids. *Oikos* 75:500–10

Gange AC. 2006. Insect-mycorrhizal interactions: patterns, processes and consequences. See Ohgushi et al. 2006. In press

Gange AC, Bower E, Brown VK. 1999. Positive effects of an arbuscular mycorrhizal fungus on aphid life history traits. *Oecologia* 120:123–31

Gange AC, Brown VK. 1989. Effects of root

herbivory by an insect on a foliar-feeding species, mediated through changes in the host plant. *Oecologia* 81:38–42

Gange AC, Brown VK, eds. 1997. *Multitrophic Interactions in Terrestrial Systems*. London: Blackwell Sci. 448 pp.

Gange AC, West HM. 1994. Interactions between arbuscular mycorrhizal fungi and foliar-feeding insects in *Plantago lanceolata* L. *New Phytol.* 128:79–87

Gehring CA, Cobb NS, Whitham TG. 1997. Three-way interactions among ectomycorrhizal mutualists, scale insects, and resistant and susceptible pinyon pines. *Am. Nat.* 149:824–41

Gehring CA, Whitham TG. 1994. Interactions between aboveground herbivores and the mycorrhizal mutualists of plants. *Trends Ecol. Evol.* 9:251–55

Gómez JM, Gonzáles-Megías A. 2002. Asymmetrical interactions between ungulates and phytophagous insects: being different matters. *Ecology* 83:203–11

Goverde M, van der Heijden MGA, Wiemken A, Sanders IR, Erhardt A. 2000. Arbuscular mycorrhizal fungi influence life history traits of a lepidopteran herbivore. *Oecologia* 123:362–69

Harrison S, Karban R. 1986. Effects of an early-season folivrous moth on the success of a later-season species, mediated by a change in the quality of the shared host, *Lupinus arboreus* Sims. *Oecologia* 69:354–59

Hay ME, Parker JD, Burkepile DE, Caudill CC, Wilson AE, et al. 2004. Mutualisms and aquatic community structure: the enemy of my enemy is my friend. *Annu. Rev. Ecol. Evol. Syst.* 35:175–97

Hatcher PE, Paul ND, Ayres PG, Whittaker JB. 1994. Interactions between *Rumex* spp., herbivores and a rust fungus: *Gastrophysa viridula* grazing reduces subsequent infection by *Uromyces rumicis*. *Funct. Ecol.* 8:265–72

Hochberg ME, Lawton JH. 1990. Competition between kingdoms. *Trends Ecol. Evol.* 5:367–71

Holt RD, Lawton JH. 1994. The ecological consequences of shared natural enemies. *Annu. Rev. Ecol. Syst.* 25:495–520

Hunter MD, Ohgushi T, Price PW, eds. 1992. *Effects of Resource Distribution on Animal-Plant Interactions*. San Diego: Academic. 505 pp.

Hunter MD, Price PW. 1992. Playing chutes and ladders: heterogeneity and the relative roles of bottom-up and top-down forces in natural communities. *Ecology* 73:724–32

Hunter MD, Willmer PG. 1989. The potential for interspecific competition between two abundant defoliators on oak: leaf damage and habitat quality. *Ecol. Entomol.* 14:267–77

Inbar M, Eshel A, Wool D. 1995. Interspecific competition among phloem-feeding insects mediated by induced host-plant sinks. *Ecology* 76:1506–15

Johnson SN, Douglas AE, Woodward S, Hartley SE. 2003. Microbial impacts on plant-herbivore interactions: the indirect effects of a birch pathogen on a birch aphid. *Oecologia* 134:388–96

Johnson SN, Mayhew PJ, Douglas AE, Hartley SE. 2002. Insects as leaf engineers: can leaf-miners alter leaf structure for birch aphids? *Funct. Ecol.* 16:575–84

Jones CG, Lawton JH, Shachak M. 1994. Organisms as ecosystem engineers. *Oikos* 69:373–86

Jones CG, Ostfeld RS, Richard MP, Schauber EM, Wolff JO. 1998. Chain reactions linking acorns to gypsy moth outbreaks and lyme disease risk. *Science* 279:1023–26

Karban R. 1986. Interspecific competition between folivorous insects on *Erigeron glaucus*. *Ecology* 67:1063–72

Karban R, Adamchak R, Schnathorst WC. 1987. Induced resistance and interspecific competition between spider mites and a vascular wilt fungus. *Science* 235:678–80

Karban R, Baldwin IT. 1997. *Induced Responses to Herbivory*. Chicago: Univ. Chicago Press. 319 pp.

Kruess A. 2002. Indirect interaction between a fungul plant pathogen and a herbivorous beetle of the weed *Cirsium arvense*. *Oecologia* 130:563–69

Larsson S, Häggström H, Denno RF. 1997. Preference for protected feeding site by larvae of the willow-feeding leaf beetle *Galerucella lineola*. *Ecol. Entomol.* 22:445–52

Lawton JH, Jones CG. 1995. Linking species and ecosystems: organisms as ecosystem engineers. In *Linking Species and Ecosystems*, ed. CG Jones, JH Lawton, pp. 141–50. New York: Chapman & Hall

Lawton JH, Strong DRJ. 1981. Community patterns and competition in folivorous insects. *Am. Nat.* 118:317–38

Lehtilä K, Strauss SY. 1997. Leaf damage by herbivores affects attractiveness to pollinators in wild radish, *Raphanus raphanistrum*. *Oecologia* 111:396–403

Leibold MA, Chase JM, Shurin JB, Downing AL. 1997. Species turnover and the regulation of trophic structure. *Annu. Rev. Ecol. Syst.* 28:467–94

Lill JT, Marquis RJ. 2003. Ecosystem engineering by caterpillars increases insect herbivore diversity on white oak. *Ecology* 84:682–90

Losey JE, Denno RF. 1998. Interspecific variation in the escape responses of aphids: effect on risk of predation from foliar-foraging and ground-foraging predators. *Oecologia* 115:245–52

Marquis RJ, Lill JT. 2006. Effects of arthropods as physical ecosystem engineers on plant-based trophic interaction webs. See Ohgushi et al. 2006. In press

Martinsen GD, Driebe EM, Whitham TG. 1998. Indirect interactions mediated by changing plant chemistry: beaver browsing benefits beetles. *Ecology* 79:192–200

Martinsen GD, Floate KD, Waltz AM, Wimp GM, Whitham TG. 2000. Positive interactions between leafrollers and other arthropods enhance biodiversity on hybrid cottonwoods. *Oecologia* 123:82–89

Masters GJ, Brown VK. 1992. Plant-mediated interactions between two spatially separated insects. *Funct. Ecol.* 6:175–79

Masters GJ, Brown VK. 1997. Host-plant mediated interactions between spatially separated herbivores: effects on community structure. See Gange & Brown 1997, pp. 217–37

Masters GJ, Jones TH, Rogers M. 2001. Host-plant mediated effects of root herbivory on insect seed predators and their parasitoids. *Oecologia* 127:246–50

Matsumura M, Suzuki Y. 2003. Direct and feeding-induced interactions between two rice planthoppers, *Sogatella furcifera* and *Nilaparvata lugens*: effects of dispersal capability and performance. *Ecol. Entomol.* 28:174–82

Menge BA. 1995. Indirect effects in marine rocky intertidal interaction webs: patterns and importance. *Ecol. Monogr.* 65:21–74

Menge BA, Sutherland JP. 1987. Community regulation: variation in disturbance, competition, and predation in relation to environmental stress and recruitment. *Am. Nat.* 130:730–57

Menge BA, Branch GM. 2001. Rocky intertidal communities. In *Marine Community Ecology*, ed. MD Bertness, SD Gaines, ME Hay, pp. 221–51. Sunderland, MA: Sinauer

Mopper S, Maschinski J, Cobb N, Whitham TG. 1991. A new look at habitat structure: consequences of herbivore-modified plant architecture. In *Habitat Structure*, ed. SS Bell, ED McCoy, HR Mushinsky, pp. 260–80. London: Chapman & Hall

Moran NA, Whitham TG. 1990. Interspecific competition between root-feeding and leaf-galling aphids mediated by host-plant resistance. *Ecology* 71:1050–58

Müller CB, Godfray HCJ. 1999. Indirect interactions in aphid-parasitoid communities. *Res. Popul. Ecol.* 41:93–106

Nakamura M, Miyamoto Y, Ohgushi T. 2003. Gall initiation enhances the availability of food resources for herbivorous insects. *Funct. Ecol.* 17:851–57

Nakamura M, Ohgushi T. 2003. Positive and negative effects of leaf shelters on herbivorous insects: linking multiple herbivore species on a willow. *Oecologia* 136:445–49

Nozawa A, Ohgushi T. 2002. How does spittlebug oviposition affect shoot growth and bud production in two willow species? *Ecol. Res.* 17:535–43

Ohgushi T. 2006. Nontrophic, indirect interaction webs of herbivorous insects. See Ohgushi et al. 2006. In press

Ohgushi T, Craig TP, Price PW, eds. 2006. *Indirect Interaction Webs: Nontrophic Linkages through Induced Plant Traits.* Cambridge: Cambridge Univ. Press. In press

Oksanen L, Fretwell SD, Arruda J, Niemelä P. 1981. Exploitation ecosystems in gradients of primary productivity. *Am. Nat.* 118:240–61

Omacini M, Chaneton EJ, Ghersa CM, Müller CB. 2001. Symbiotic fungal endophytes control insect host-parasite interaction webs. *Nature* 409:78–81

Pace ML, Cole JJ, Carpenter SR, Kitchell JF. 1999. Trophic cascades revealed in divers ecosystems. *Trends Ecol. Evol.* 14:483–88

Paine RT. 1980. Food webs: linkage, interaction strength and community infrastructure. *J. Anim. Ecol.* 49:667–85

Pavia H, Toth GB. 2000. Inducible chemical resistance to herbivory in the brown seaweed *Ascophyllum nodosum. Ecology* 81:3212–25

Petersen MK, Sandström JP. 2001. Outcome of indirect competition between two aphid species mediated by responses in their common host plant. *Funct. Ecol.* 15:525–34

Polis GA. 1999. Why are parts of the world green? Multiple factors control productivity and the distribution of biomass. *Oikos* 86:3–15

Polis GA, Sears ALW, Huxel GR, Strong DR, Maron J. 2000. When is a trophic cascade a trophic cascade? *Trends Ecol. Evol.* 15:473–75

Polis GA, Winemiller KO, eds. 1996. *Food Webs: Integration of Patterns and Dynamics.* New York: Chapman & Hall. 472 pp.

Poveda K, Steffen-Dewenter I, Scheun S, Tscharntke T. 2003. Effects of below- and above-ground herbivores on plant growth, flower visitation and seed set. *Oecologia* 135:601–5

Price PW. 2002. Species interactions and the evolution of biodiversity. In *Plant-Animal Interactions: An Evolutionary Approach*, ed. CM Herrera, O Pellmyr, pp. 3–25. Oxford: Blackwell Sci.

Price PW, Bouton CE, Gross P, McPheron BA, Thompson JN, Weis AE. 1980. Interactions among three trophic levels: influence of plants on interactions between insect herbivores and natural enemies. *Annu. Rev. Ecol. Syst.* 11:41–65

Raps A, Vidal S. 1998. Indirect effects of an unspecialized endophytic fungus on specialized plant—herbivorous insect interactions. *Oecologia* 114:541–47

Roininen H, Price PW, Bryant JP. 1997. Response of galling insects to natural browsing by mammals in Alaska. *Oikos* 80:481–86

Sagers CL. 1992. Manipulation of host plant quality: herbivores keep leaves in the dark. *Funct. Ecol.* 6:741–43

Saikkonen K, Helander M, Faeth SH, Schulthess F, Wilson D. 1999. Endophyte-grass-herbivore interactions: the case of *Neotyphodium* endophytes in Arizona fescue populations. *Oecologia* 121:411–20

Salt DT, Fenwick P, Whittaker JB. 1996. Interspecific herbivore interactions in a high CO_2 environment: root and shoot aphids feeding on *Cardamine. Oikos* 77:326–30

Schmitz OJ. 1998. Direct and indirect effects of predation and predation risk in old-field interaction webs. *Am. Nat.* 151:327–42

Schmitz OJ, Krivan V, Ovadia O. 2004. Trophic cascades: the primacy of trait-mediated indirect interactions. *Ecol. Lett.* 7:153–63

Seifert RP. 1984. Does competition structure communities? Field studies on neotropical *Heliconia* insect communities. In *Ecological Communities: Conceptual Issues and the Evidence*, ed. DR Strong Jr, D Simberloff, LG Abele, AB Thistle, pp. 54–63. Princeton: Princeton Univ. Press

Siepielski AM, Benkman CW. 2004. Interactions among moths, crossbills, squirrels, and lodgepole pine in a geographic selection mosaic. *Evolution* 58:95–101

Simon M, Hilker M. 2003. Herbivores and pathogens on willow: do they affect each other? *Agric. For. Entomol.* 5:275–84

Strauss SY. 1991a. Direct, indirect, and cumulative effects of three native herbivores on a shared host plant. *Ecology* 72:543–58

Strauss SY. 1991b. Indirect effects in community ecology: their definition, study and importance. *Trends Ecol. Evol.* 6:206–10

Strauss SY. 1997. Floral characters link herbivores, pollinators, and plant fitness. *Ecology* 78:1640–45

Strauss SY, Conner JK, Rush SL. 1996. Foliar, herbivory affects floral characters and plant attractiveness to pollinators: implications for male and female plant fitness. *Am. Nat.* 147:1098–107

Strauss SY, Zangerl AR. 2002. Plant-insect interactions in terrestrial ecosystems. In *Plant-Animal Interactions: An Evolutionary Approach*, ed. CM Herrera, O Pellmyr, pp. 77–106. Oxford: Blackwell Sci.

Strong DR Jr. 1984. Exorcising the ghost of competition past: phytophagous insects. In *Ecological Communities: Conceptual Issues and the Evidence*, ed. DR Strong Jr, D Simberloff, LG Abele, AB Thistle, pp. 28–41. Princeton: Princeton Univ. Press

Thompson JN. 1994. *The Coevolutionary Process*. Chicago: Univ. Chicago Press. 376 pp.

Thompson JN. 1996. Evolutionary ecology and the conservation of biodiversity. *Trends Ecol. Evol.* 11:300–3

Van der Putten WH, Vet LEM, Harvey JA, Wackers FL. 2001. Linking above- and belowground multitrophic interactions of plants, herbivores, pathogens, and their antagonists. *Trends Ecol. Evol.* 16:547–54

Van Zandt PA, Agrawal AA. 2004. Community-wide impacts of herbivore-induced plant responses in milkweed (*Asclepias syriaca*). *Ecology* 85:2616–29

Vicari M, Hatcher PE, Ayres PG. 2002. Combined effect of foliar and mycorrhizal endophytes on an insect herbivore. *Ecology* 83:2452–64

Waltz AM, Whitham TG. 1997. Plant development affects arthropod communities: opposing impacts of species removal. *Ecology* 78:2133–44

Werner EE, Peacor SD. 2003. A review of trait-mediated indirect interactions in ecological communities. *Ecology* 84:1083–100

Williams KS, Myers JH. 1984. Previous herbivore attack of red alder may improve food quality for fall webworm larvae. *Oecologia* 63:166–70

Wootton JT. 1994. The nature and consequences of indirect effects in ecological communities. *Annu. Rev. Ecol. Syst.* 25:443–66

Wootton JT. 2002. Indirect effects in complex ecosystems: recent progress and future challenges. *J. Sea Res.* 48:157–72

Annu. Rev. Ecol. Evol. Syst. 2005. 36:107–24
doi: 10.1146/annurev.ecolsys.36.102403.135635
Copyright © 2005 by Annual Reviews. All rights reserved
First published online as a Review in Advance on July 29, 2005

EVOLUTIONARY HISTORY OF POALES

H. Peter Linder[1] and Paula J. Rudall[2]

[1]Institute for Systematic Botany, University of Zurich, CH-8008 Zurich, Switzerland;
email: plinder@systbot.unizh.ch
[2]Jodrell Laboratory, Royal Botanic Gardens, Kew, TW9 3DS, United Kingdom;
email: p.rudall@rbgkew.org.uk

Key Words C_4 photosynthesis, epiphytes, grasses, nutrient-poor habitats, wind
pollination

■ **Abstract**   The predominantly wind-pollinated order Poales includes about one
third of all monocot (Angiosperm) species, with c. 20,000 species dominating modern
savanna and steppe vegetation. Recent improvements in understanding relationships
within the order allow phylogenetic optimizations of habitat preferences and adaptive
character states, enabling exploration of the factors that have influenced evolution in this
successful order. Poales probably originated in the late Cretaceous in wet nutrient–poor
sunny habitats. By the Paleogene the lineage had diversified into swamps, the forest
understory, epiphytic habitats, and nutrient-poor heathlands. The Neogene saw major
diversifications of the grasses and possibly the sedges into fire-adapted vegetation in
seasonal climates and low atmospheric CO_2. Diversification into these habitats was
facilitated by morphological features such as the sympodial habit and physiological
factors that allowed frequent evolution of CO_2-concentrating mechanisms.

INTRODUCTION

The monocot order Poales, with c. 20,000 species, represents more than one third
of all monocotyledons and includes many economically significant crops. Intensive research over the past two decades has broadened the circumscription of the
order to include not only Poaceae (grasses) and Cyperaceae (sedges) and their
respective allies, but also several smaller families of hitherto more enigmatic taxonomic affinity, such as Bromeliaceae (Pineapple family), Rapateaceae (Rapatea
family), Typhaceae (Cattail family), Xyridaceae (Yellow-eyed Grasses), and Eriocaulaceae (Pipewort family). Many of these groups are ecologically dominant in
their respective habitats; for example, Poaceae are globally dominant in savannas
and steppe, Cyperaceae in wetter grassy areas, Bromeliaceae in parts of tropical
South America, and even the relatively small family Restionaceae is locally dominant in the African Cape flora. Extant families of Poales occur in virtually all
nonmarine habitats, from equator to pole, from floating waterplants to the most
severe deserts, and on most (if not all) soil types. Furthermore, there are representatives of most major life-forms—lianes, epiphytes, hydrophytes, emergent-rooted

aquatics, tall and short plants, annuals and long-lived perennials; only the tree-form is absent. There is considerable variation in species richness among different clades (Figure 1, see color insert). The most species-rich lineages are associated with sunny, seasonally arid, or generally dry (savanna and steppe grasslands and heathlands) or epiphytic habitats, and they often contain annual species, or species with CO_2-concentrating mechanisms such as C_4 photosynthesis and crassulacean acid metabolism (CAM).

Improved resolution of the relationships among the c. 18 constituent families of Poales sensu *lato*, together with clarification of subfamilial classification in some key groups, have permitted tentative dating of the origins of some of the larger clades using both molecular clocks and reference fossils. The establishment of a developmental-genetics program in grasses, prompted by the completion of the whole-sequence genome, for model organisms such as *Oryza* (rice), has greatly improved the potential for recognition and isolation of specific targeted genes from other plant species.

The improved phylogenetic background sets the framework that allows us to address questions about the selective forces and morphological innovations that promoted the considerable species richness of the larger families such as Bromeli-aceae, Cyperaceae, Restionaceae, and especially Poaceae. Factors that are commonly invoked include adaptation to wind pollination, seasonal drought, and the evolution of grazing animals, the last primarily affecting Poaceae. Here we examine the suites of morphological innovations and selective forces that both provided the foundation for the primary Poales radiation and possibly preadapted them for subsequent secondary diversification into the more derived clades. Three of the more species-rich clades are wind pollinated, and this could be a guiding force in the evolution of similar reproductive structures in different clades of Poales. Seasonal drought may have shaped aspects of the vegetative morphology and anatomy and, hence, photosynthetic physiology. We explore the hypotheses that the ancestral forms in Poales occurred in perennially wet areas and that four derived clades represent radiations into the new seasonally arid habitats that opened in the Neogene in response to global climatic change, the establishment of regular fires, and lowered levels of atmospheric CO_2.

PHYLOGENY

Early phylogenetic analyses of Poales (Dahlgren & Rasmussen 1983, Duvall et al. 1993, Kellogg & Linder 1995) employed relatively narrow sampling of taxa and/or data. More recently, several combined multigene (or multigene plus morphology) analyses of representatives of all families of Poales and related orders or all monocots have improved our understanding of the circumscription of Poales (Bremer 2002; Chase et al. 2000, 2005; Davis et al. 2004; Michelangeli et al. 2003). A broad circumscription of Poales is now widely adopted (APG 2003). The most recent and most comprehensive combined molecular analysis of monocot taxa (Chase et al. 2005) adds some further resolution within Poales that allows us to evaluate character evolution within the order and forms the basis for the tree in Figure 1. Chase et al. (2005) identified some "problem taxa"—mostly mycoheterotrophs or aquatics—that tended to destabilize tree topology and reduce support values; hence, they left these taxa out of their primary analysis. Within Poales, *Trithuria* (Hydatellaceae) represents such a problem taxon; when included, it falls near Eriocaulaceae, Mayacaceae, and Xyridaceae (i.e., in the xyrid clade). Because this placement for *Trithuria* (and alternative placement for *Mayaca*) was also retrieved by the combined analyses of Michelangeli et al. (2003) and Davis et al. (2004), we have inserted a xyrid clade in the topology shown in Figure 1. Figure 1 also includes some other taxa that were not included in the analysis of Chase et al. (2005) such as *Lyginia* and *Hopkinsia* (based on their placement in Briggs et al. 2000 and Bremer 2002) and the grass subfamilies (based on GPWG 2001).

The resulting phylogenetic hypothesis (Figure 1) has the following significant properties:

1) Three families—Bromeliaceae, Rapateaceae, and Typhaceae (including *Sparganium*)—form a basal grade (here termed the basal Poales).

2) There are four major clades of "core Poales," which we here term the graminid, restiid, cyperid, and xyrid clades, respectively. Relationships inferred between these four clades are inconsistent among analyses; for example, the xyrid clade is sometimes sister to the graminids (Bremer 2002) rather than the cyperids (Chase et al. 2005).

3) The xyrid clade (Eriocaulaceae, Hydatellaceae, Mayacaceae, Xyridaceae) is perhaps the most problematic clade of Poales in terms of both circumscription and phylogenetic position. It contains elements of several earlier proposed taxonomic groupings, including Commelinales. Xyridaceae may not be monophyletic, because *Abolboda* and *Xyris* are placed separately from the other genera in some analyses. However, the clade is largely characterized by mature ovules with very little or no nucellar tissue (in *Mayaca*, Eriocaulaceae, and some Xyridaceae), which also occurs in the graminid clade (Rudall 1997).

4) A restiid clade (Anarthriaceae, Centrolepidaceae, and Restionaceae, including *Hopkinsia* and *Lyginia*) has long been recognized, but *Ecdeiocolea* and *Flagellaria* were previously also included.

5) The graminid clade (Ecdeiocoleaceae, Flagellariaceae, Joinvilleaceae, and Poaceae) is a relatively recent circumscription; previous phylogenies placed the three small families in the restiid clade, with Poaceae as sister to the entire restiid clade. A close relationship between *Joinvillea* and Poaceae was proposed by Campbell & Kellogg (1987) and the inclusion of Ecdeiocoleaceae by Briggs et al. (2000; Bremer 2002). Placement of *Flagellaria* remains uncertain; for example, Michelangeli et al. (2003) placed it as sister to graminids plus restiids.

6) The cyperid clade (Cyperaceae, Juncaceae, Thurniaceae, and including *Prionium*) is a robust and consistent clade that has been retrieved in most analyses. It represents the former Cyperales sensu Dahlgren et al. (1985). Consistent grouping of *Thurnia* and *Prionium* led to their treatment as a single family (Bremer 2002, APG 2003).

DATING CLADES

The ancestral monocots probably evolved during the early Cretaceous and diversified into the extant major lineages during the mid- to late Cretaceous contemporaneously with other basal angiosperms (Gandolfo et al. 2000, Herendeen & Crane 1995). The earliest verifiable fossils of Poales date from the late Cretaceous (Maastrichtian) at 115 Mya (Herendeen & Crane 1995).

Dating the major nodes on the Poales phylogeny is problematic because of the limited number and phylogenetic range of fossils. Dates of nodes with no fossil evidence can be inferred by placing known fossils on a well-resolved phylogeny (Sanderson 1998). Four attempts to use this approach for Poales (Bremer 2002, Givnish et al. 2000, Janssen & Bremer 2004, Wikström et al. 2001) have produced highly discordant estimates of ages of key nodes. Perhaps the best available estimate is that of Janssen & Bremer (2004), who calibrated a detailed analysis of over 800 *rbc*L sequences using dates calculated by Bremer (2002). The ages obtained are greater than those of Bremer (2002), possibly because of more extensive sampling, which could affect rate correction calculations in nonparametric rate-smoothing methods (Linder et al. 2005). The analysis of Givnish et al. (2000), also based on *rbc*L alone but sampling fewer taxa, produced even more recent dates, whereas the analysis of Wikström et al. (2001), which has the smallest taxon sampling, returned the most recent dates. We used the dates indicated by the Janssen & Bremer analysis in Figure 1. Nodes not included in their phylogeny were calibrated, with suitable corrections, from Bremer (2002) and nodes not included in either analysis were spaced evenly between the dated nodes.

Despite recent developments in molecular dating methods (Magallón 2004, Sanderson et al. 2004), uncertainties surrounding rate correction methods (Linder et al. 2005) and difficulties in accurately constraining the age ranges of calibration nodes (Graur & Martin 2004) make confident estimation of absolute ages of nodes impossible. Node age estimation in Poales would be improved by a larger sample of genes and taxa, more critical evaluation of the possible distortions generated by the different rate correction methods, and especially by more fossil calibration points.

HABITAT EVOLUTION

Habitats

ANCESTRAL POALES   Phylogenetic optimizations of habitat ecology and associated adaptations using extant species (Figure 2a–d, see color insert) permit inferences

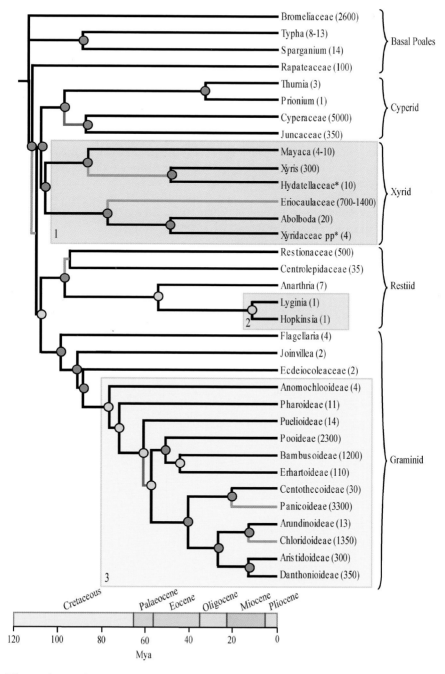

Figure 1 See legend on next page

Figure 1   Phylogram of Poales, primarily based on the topology in Chase et al. (2005), but also including a xyrid clade (1) from Davis et al. (2004), a restiid clade (2) from Bremer (2002), and infrafamilial relationships of Poaceae (3) from GPWG (2001). Clade names are followed by the number of species. Node ages marked in *blue* are from Janssen & Bremer (2004), in *yellow* from Bremer, and in *green* are spaced evenly between dated nodes. *Red* internodes indicate an increase in diversification significant at p < 0.01, and in *orange* at p < 0.05. Data sources for the families are: Anarthriaceae (Linder et al. 1998a), Bromeliaceae (Smith & Till 1998), Centrolepidaceae (Cooke 1998), Cyperaceae (Goetghebeur 1998), Ecdeiocoleaceae (Linder et al. 1998b), Eriocaulaceae (Stützel 1998), Flagellariaceae (Appel & Bayer 1998), Hydatellaceae (Hamann 1998), Joinvilleaceae (Bayer & Appel 1998), Juncaceae (Balslev 1998), Mayacaceae (Stevenson 1998), Poaceae (Clayton & Renvoize 1986, Judziewicz & Soderstrom 1989), Rapateaceae (Stevenson et al. 1998), Restionaceae (Linder et al. 1998c), Thurniaceae (Kubitzki 1998b), Typhaceae (Kubitzki 1998c), and Xyridaceae (Kral 1998).

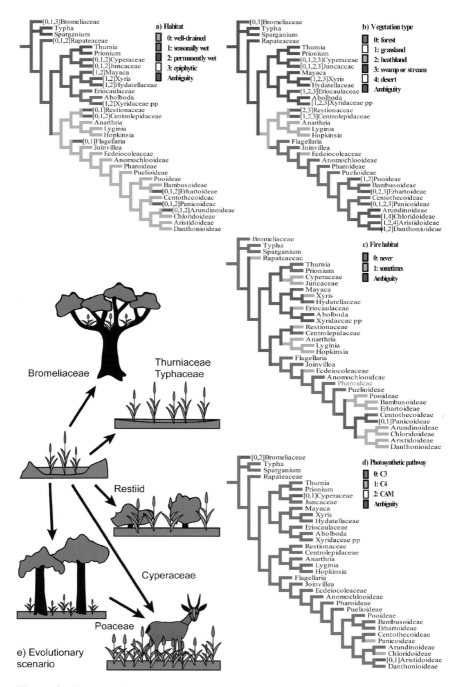

Figure 2   See legend on next page

Figure 2 Vegetative ecology. Unambiguous optimizations of different aspects of habitat preference and associated adaptations. Polymorphic taxa have their states indicated by numbers in brackets. Panels show (*a*) habitat, (*b*) vegetation type, (*c*) fire, (*d*) photosynthetic pathway, and (*e*) evolutionary scenario, indicating the evolution from wet habitats into epiphytic, seasonally wet, heathland and grassland habitats, the latter via a forest understorey habitat. Data sources for the families are: Anarthriaceae (Linder et al. 1998a), Bromeliaceae (Smith & Till 1998), Centrolepidaceae (Cooke 1998), Cyperaceae (Goetghebeur 1998), Ecdeiocoleaceae (Linder et al. 1998b), Eriocaulaceae (Stützel 1998), Flagellariaceae (Appel & Bayer 1998), Hydatellaceae (Hamann 1998), Joinvilleaceae (Bayer & Appel 1998), Juncaceae (Balslev 1998), Mayacaceae (Stevenson 1998), Poaceae (Clayton & Renvoize 1986, Judziewicz & Soderstrom 1989), Rapateaceae (Stevenson et al. 1998), Restionaceae (Linder et al. 1998c), Thurniaceae (Kubitzki 1998b), Typhaceae (Kubitzki 1998c, and Xyridaceae (Kral 1998). The optimizations were performed using Winclada (Nixon 2002).

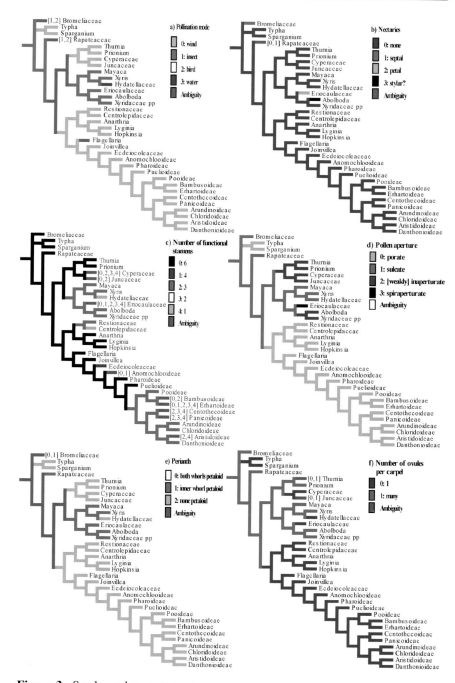

Figure 3   See legend on next page

Figure 3 Reproductive ecology. Unambiguous optimizations of different aspects of pollination biology and associated adaptations: (*a*) pollination mode, (*b*) presence and position of nectaries, (*c*) number of functional stamens, (*d*) type of pollen aperture, (*e*) perianth type, and (*f*) the number of ovules per carpel. Data sources for the families are: Anarthriaceae (Linder et al. 1998a), Bromeliaceae (Smith & Till 1998), Centrolepidaceae (Cooke 1998), Cyperaceae (Goetghebeur 1998), Ecdeiocoleaceae (Linder et al. 1998b), Eriocaulaceae (Stützel 1998), Flagellariaceae (Appel & Bayer 1998), Hydatellaceae (Hamann 1998), Joinvilleaceae (Bayer & Appel 1998), Juncaceae (Balslev 1998), Mayacaceae (Stevenson 1998), Poaceae (Clayton & Renvoize 1986, Judziewicz & Soderstrom 1989), Rapateaceae (Stevenson et al. 1998), Restionaceae (Linder et al. 1998c), Thurniaceae (Kubitzki 1998b), Typhaceae (Kubitzki 1998c), and Xyridaceae (Kral 1998). The optimizations were performed using Winclada (Nixon 2002).

regarding ancestral Poales. The Upper Cretaceous was a time of high levels of CO_2 ($\sim$1000 ppm) (Cerling et al. 1998), warm climates, and generally high rainfall. Land was largely forested; there is no evidence for the existence of grasslands or extensive open vegetation. Thus, the habitat available for sun-demanding, shallow-rooted plants lacking secondary thickening that could not compete directly with forest trees was relatively limited. It is likely that ancestral Poales grew in marshy or wet habitats, and often on nutrient-poor soils—these habitats are disadvantageous to dicot shrubs and trees. Indeed, Givnish et al. (2000) suggested that the entire commelinid (Poales and related orders) diversification began in this habitat type. Extant groups of Poales still found in these habitats are relatively species poor. They include *Abolboda* and many other Xyridaceae (Kral 1998), basal Rapateaceae (Givnish et al. 2000, 2004), basal Bromeliaceae such as *Brocchinia* (Givnish et al. 1997, 2004), Typhaceae, Thurniaceae, Hydatellaceae, and *Mayaca*. All species grow rooted in the substrate, and most bear their flowers in the air and not underwater, except some species of *Mayaca* and Hydatellaceae. Species of *Typha* and *Prionium* can still dominate wetland areas. Most of the remaining cyperid and xyrid families include at least a few species that prefer wetland habitats, though it is not clear whether these are reversals.

Ancestral members of the graminid clade invaded forest understory (Figure 2*b*); some basal graminids are still found either in forest margins as climbers (*Flagellaria*) or in deeply shaded forest understorey (Anomochlooideae, Pharoideae, Puelioideae). A curious exception is the heathland graminid family Ecdeiocoleaceae, which occurs in the semiarid Kwongan heathlands of south-western Australia (Meney & Pate 1999). Basal taxa such as *Flagellaria*, *Joinvillea*, and Ecdeiocoleaceae might offer tantalizing glimpses of ancestral Poales growing in the swampy forests and heathlands of the late Cretaceous, yet we currently know frustratingly little about their biology.

THE EPIPHYTIC DIVERSIFICATION   With their sympodial habit (growth terminated by flowering, so that vegetative growth has to continue from side shoots), monocots are apparently well adapted as epiphytes, but surprisingly few ($\sim$17) monocot families have successfully invaded this habitat and most of them are species-poor. The largest epiphytic monocot diversifications, Araceae, Bromeliaceae, and Orchidaceae, show some morphological similarities (Benzing 1990). About half of the Bromeliaceae and a few Rapateaceae (*Epiphyton*) escaped from the competition of trees and seasonal wetlands by growing on trees (epiphytically) or rocks (epilithically) and developed a complex suite of adaptations, including succulence, absorptive foliar trichomes, and CAM to survive short periods of water stress (Benzing 2000). CAM plants can absorb CO_2 during the night, when water loss is reduced. The ability to readily evolve CAM presumably preadapted the lineage for the stresses of even lower CO_2 levels during the Pliocene and Pleistocene.

THE NUTRIENT-RICH, FIRE-ADAPTED DIVERSIFICATION   The Tertiary saw major climatic changes (Zachos et al. 2001). During the Paleocene and early Eocene, temperatures gradually increased; the planet was ice-free, and the climate was probably less seasonal. A global trend toward lower temperatures set in from 50 Mya. The Oligocene was characterized by sharply lower temperatures with extensive glaciation of Antarctica (33–26 Mya), seasonal climates (Denton 1999), and possibly grass-free savanna vegetation (Wolfe 1985). The early Miocene (24–14 Mya) heralded the onset of warmer, moister climates, causing extensive deglaciation of Antarctica. Finally, from 14 Mya saw the establishment of cold winters with frost at high latitudes, and at mid-latitudes seasonal drought, possibly with regular fires (Herring 1985). Three million years ago, closure of the Panama Isthmus precipitated a rapid temperature drop and the Northern Hemisphere glaciations.

Although Poaceae have a fossil record dating back to the Paleocene, grass became a major component of the vegetation relatively recently. The earliest record for this transition, inferred from plant fossils and animal dentition and structure, is from the middle Miocene (~15 Mya) of Africa and North America, and the late Miocene (~10 Mya) of Europe and South America, while in Australia it is even later. This transition from woody to grassy vegetation could be the result of increased fires, which would tend to benefit grasses over woody plants (Keeley & Rundel 2003). In all areas, grass-dominated vegetation was initially predominantly C_3, but by 7–5 Mya C_4 grasses prevailed everywhere except Australia (Jacobs et al. 1999). Cerling (1992) suggested that C_4 dominance in East Africa dates only from the Middle Pleistocene, although the $\delta13$ isotopes of fossil bones indicate an earlier dominance of C_4 grasses.

The cyperids and, to a lesser extent, the xyrids also expanded into this new open habitat. Generally they occupy wetter habitats than the grasses within the fire-adapted vegetation. However, the picture is less clear than in the grasses, and the key may well lie in understanding the evolutionary history of *Carex*, which includes almost one third of the cyperid species.

NUTRIENT-POOR, FIRE-ADAPTED DIVERSIFICATION   Restionaceae represent an experiment in dealing with a combination of nutrient-poor soils and seasonal drought. Whereas the former selects for evergreen structures, the latter requires a reduction in the transpiration surface during the hot, dry summers. The result is loss of leaves and transfer of photosynthesis to the stems. Adaptation to regular fires (Figure 2c) selects for stems flowering in the first year; they subsequently persist and photosynthesize for several years (Linder 1991, Stock et al. 1987). Restionaceae have a long fossil record (Linder et al. 2003); plots of lineages through time indicate that diversification of the modern species groups of Restionaceae dates from the Oligocene in Africa (Linder & Hardy 2004) and the Eocene in Australia (Linder et al. 2003). Restionaceae were probably an important component of southern Mediterranean-type heathlands on nutrient-poor soils throughout much of the Tertiary. Optimizations indicate that the ancestral Restionaceae may well have been found in seasonally wet habitats, with the lineage adapting later to more arid environments (Linder 2000b), consistent with the Cenozoic global climatic changes.

Ecological Adaptations

ROOTING SYSTEMS   Several different nutrient-absorbing strategies have evolved in the Poales. Many Bromeliaceae rely on foliage for water absorption, via phytotelms (water-filled cavities) or specialized absorptive foliar trichomes that can partially replace roots. These represent a critical adaptation to habitats in which root absorption is not effective, such as in trees and on rocks, although the reduced root systems usually retain an anchorage role (Benzing 2000). Similarly, the clusters of very fine, specialized nutrient-absorbing roots of Restionaceae aid survival on the exceedingly nutrient-poor soils of the south-western tip of Africa and the Kwongan vegetation of Australia by increasing the absorbing surface (Lamont 1983). However, root structural modifications have been scarcely studied in this clade.

EVERGREEN, PERSISTENT PHOTOSYNTHETIC ORGANS   Retention of photosynthetic organs for several years can be interpreted as an adaptation to conditions of low soil fertility (Chapin 1980). This feature is well developed in Restionaceae, Thurniaceae, Bromeliaceae, Mayacaceae, and Rapateaceae, and is found in at least some species of most other families. This is most likely an ancestral condition in Poales, consistent with the interpretation of Givnish et al. (1999) that the group ancestrally occurred on nutrient-poor soils. These long-lived photosynthetic organs are defended from herbivory by spines (Bromeliaceae) and/or tannins (Restionaceae, Rapateaceae). In Restionaceae, leaves are barely developed; the main photosynthetic organs are the stems, which remain active for up to three years, resulting in a remarkable level of specialization in stem anatomy (Cutler 1969). Similar patterns are seen in some grasses and sedges that grow on nutrient-poor soils in seasonal climates (Ellis 1987). The critical nutrients, nitrogen and phosphorus, remain bound in enzyme systems in these long-lived organs rather than being released into the environment with leaf litter; they are subsequently resorbed, thus facilitating survival in nutrient-poor habitats. Most Poaceae and Cyperaceae are found on richer soils and have shorter-lived leaves. In many tropical grasses the leaves persist for only a single growing season (~6 months).

ANNUAL LIFE HISTORY   A remarkable adaptation to seasonal climates is the evolution of the annual life history, ecologically explored by Verboom et al. (2003). Annuals are relatively rare in monocots, but the strategy has evolved separately within seven families of Poales (Centrolepidaceae, Cyperaceae, Eriocaulaceae, Hydatellaceae, Juncaceae, Poaceae, Xyridaceae). However, it has not become dominant within these clades, characterizing only a few species within otherwise perennial genera. In Poaceae the shift between the annual and perennial life histories appears to be controlled by only two genes (Hu et al. 2003), suggesting that it may represent a relatively simple evolutionary transition.

SYMPODIAL GROWTH   Sympodial growth is the basal condition in the monocots and has a strong influence over the range of biologies possible in the class, permitting formation of rhizomes, bulbs, corms, and tubers (Holttum 1955). Partially because of the absence of secondary thickening, mature monocots cannot compete effectively with mature dicots in optimal environments. However, a long-lived underground organ facilitates survival of seasonal loss of leaves, whether from regular dry-season fires or frost, because at least some regeneration buds are protected. The above-ground stems that bear the inflorescence and leaves are short-lived and in many cases annual, thus minimizing the impact of defoliation. Furthermore, the smaller investment in long-lived woody organs accelerates the juvenile establishment phase, allowing re-establishment of populations after fire. For example, in Restionaceae there are two strategies for surviving regular fires: Some species store starch in a protected long-lived rhizome, others regenerate by prolific seed production and germination (Bell & Pate 1993, Pate et al. 1991).

PHOTOSYNTHETIC PATHWAYS   During the Cretaceous and again in the Eocene the atmospheric CO_2 level was several times higher than modern levels, which may be critically low for effective C_3 photosynthesis. Photorespiration can exceed photosynthesis at CO_2 levels below 500 μl^{-1} and temperatures above 25°C (Ehleringer et al. 1991, Ehleringer & Monson 1993). Both the C_4 photosynthetic pathway and CAM contain processes that concentrate CO_2 in the Rubisco-containing cells, thereby counteracting the effects of low atmospheric CO_2. Change in the photosynthetic pathway does not require new enzymes, but does necessitate both a difference in expression of existing enzymes and reorganization of the anatomy of the photosynthetic and storage tissues (Kellogg 2000). Thus, parallel evolution of C_4 in grasses and sedges, and CAM in Bromeliaceae (Figure 2d), is probably an adaptation to changes in atmospheric CO_2 concentration (Ehleringer et al. 1991) rather than to seasonal drought. However, this hypothesis is difficult to test, because the time of origin of these processes is unknown. The only evidence of past occurrence of C_4 is from changes in carbon isotope proportions in soil and herbivore bones (Cerling et al. 1998); this measures the relative abundance of these processes rather than merely their presence. Optimization on a molecularly dated tree suggests that CAM evolved only c. 17 Mya in Bromeliaceae (Givnish et al. 2004), which is consistent with both aridification and reduction on atmospheric CO_2.

C$_4$ photosynthesis evolved several times in Poaceae, but the exact number cannot be determined confidently because a few gains and many losses are almost as parsimonious as many gains (Giussani et al. 2001). Similarly, there appear to have been numerous origins of the C_4 mode in Cyperaceae (Bruhl et al. 1987, Ehleringer & Monson 1993, Stock et al. 2004) and CAM in Bromeliaceae (Benzing 2000). Thus, it is difficult to understand why C_4 photosynthesis did not evolve in several other, smaller C_3 groups of Poales such as Restionaceae and Eriocaulaceae. One possible explanation might be that in these families there has been extensive transfer of photosynthesis from the leaves to the stems, perhaps impeding the anatomical reorganization required for C_4 photosynthesis. The numerous origins in Poaceae and Cyperaceae could be accounted for by the very large number of individuals in these families, thus increasing the potential for the necessary evolutionary experimentation (Monson 2003).

EVOLUTION OF WIND POLLINATION

Pollination Modes

Insect pollination (entomophily) is undoubtably the ancestral pollination mode of Poales and predominates in basal Poales such as Bromeliaceae and Rapateaceae; bird pollination has also evolved in both these families. Within the xyrid clade, the small family Hydatellaceae is water pollinated, but all other xyrids are animal pollinated. However, four clades of Poales are wind pollinated; three of them (cyperids, graminids, and restiids) representing highly derived, species-rich groups, and a fourth (Typhaceae) representing a small, basal family of Poales. Thus, shifts from animal to wind or water pollination either occurred several (three or four) times in Poales (Figure 3*a*, see color insert), or once followed by several reversals.

Joinvillea and *Flagellaria* may represent reversals to animal pollination within wind-pollinated clades; their pollination modes are uncertain. Most texts record them as wind pollinated (Bayer & Appel 1998), but there are records of insect associations, and available evidence from both field observations and floral morphology are uncertain. Newell (1969) reported that flowers of *Joinvillea* are visited by bees and those of *Flagellaria* by ants, and hypothesized that both types of insect play a role in pollination. In *Flagellaria* the flowers are somewhat scented and the inner perianth (the perianth consists of the petals and sepals) is white and showy (Backer 1954) in contrast to those of most wind-pollinated Poales. The anthers are exserted from the flowers and mobile on the filaments, as might be expected in wind-pollinated species, but the style is short, although it is feathery.

Why did wind pollination evolve more than once within Poales, given that this pollination mode is otherwise rare in monocots (Linder 1998)? A survey of the pollination biology of *Joinvillea*, *Flagellaria*, and Hydatellaceae may help to answer this question. For example, in Hydatellaceae, the typical features of water pollination are similar to those of wind pollination: loss of morphological attributes attractive to pollinators (e.g., nectaries and attractive flowers), pollen specialization (both in morphology and scale of production), development of attributes that can trap passing pollen (such as specialized stigma morphology and inflorescence structure), unisexuality, and monoecy.

Reproductive Adaptations

INFLORESCENCES   The structure of the inflorescence (the aggregation of flowers) is arguably an important mechanistic factor in facilitation of pollen trapping in wind-pollinated plants (Friedman & Harder 2004), allowing stigmas to filter a large volume of air or causing eddies that result in deposition of pollen on strategically-placed stigmas. Little is known about concomitant modifications of stigma morphology, but inflorescence morphology, whether compact or in cone-like spikelets, can be interpreted in terms of air dynamics (Niklas 1985, 1987). However, we found no obvious correlation between inflorescence organization and wind pollination in Poales. In terms of condensation of the inflorescence, the spikelets or spikelet-like structures found in the wind-pollinated families Poaceae, Restionaceae, and Cyperaceae are comparable to similar aggregations of flowers found in the animal-pollinated Bromeliaceae and Eriocaulaceae. It is likely that such condensation of the inflorescence to a pseudanthial (flower-like) presentation in many Poales was an evolutionary precursor to wind pollination.

SEXUALITY   Unisexual flowers, whether on unisexual plants (dioecious) or bisexual plants (monoecious), are often associated with wind pollination (Linder 1998). The shift from bisexual to unisexual flowers occurred numerous times during the evolution of Poales. The only large clade with unisexual plants is the restiid lineage, but this also evolved in several other families: Typhaceae, Hydatellaceae, Cyperaceae, some Eriocaulaceae, and numerous times in Poaceae (Connor 1981). Alternatively, dioecy is frequently "leaky" and therefore readily reversible; rare instances of bisexual flowers have been recorded in Restionaceae, and there are many examples of unisexual flowers in the large families Cyperaceae and Poaceae, although the pattern in Poaceae is obscured by the frequent combination of unisexual and bisexual flowers into a single spikelet. In all cases early floral developmental stages are apparently bisexual; unisexuality is the result of the abortion of either female or male organs during ontogeny (Decraene et al. 2001, 2002; Zaitchik et al. 2000).

FLORAL NECTARIES   The presence of floral nectar is normally a good indicator of animal pollination. In monocots, nectar is predominantly produced from septal nectaries (Smets et al. 2000, Rudall 2002). By contrast, septal nectaries are rare in Poales (Figure 3b), but are present in at least one of the basal families, Bromeliaceae, which is also the only family of Poales possessing inferior ovaries. The massive labyrinthine nectaries found below the ovary chambers of the Bromeliaceae with superior ovaries are apparently more specialized than the nectaries placed between the ovary chambers of those species with ovaries below the flower, perhaps resulting in an increase in nectar production as demanded by pollinators such as birds and bats (Böhme 1988, Sajo et al. 2004). There are also a few records of septal nectaries in Rapateaceae. Venturelli & Bouman (1988) reported "open septal nectaries" in *Spathanthus* (Rapateaceae), and Givnish et al. (1999, 2000) recorded nectaries present in some species of Rapateaceae.

Because Rapateaceae and Bromeliaceae belong to the basal grade of Poales, the presence of septal nectaries may be the ancestral condition, with parallel losses occurring within Rapateaceae, the lineage leading to Typhaceae and the lineage leading to the rest of the order. Conversely, septal nectaries may have evolved de novo in Bromeliaceae and some Rapateaceae. Floral nectar is also present in some members of the xyrid clade (Eriocaulaceae and *Abolboda*), but it is produced by the perianth in Eriocaulaceae (Stützel 1986) and by stylar appendages in *Abolboda* (Stützel 1990), indicating that nectar production may have evolved independently in these lineages from ancestors that lacked septal nectaries.

PERIANTH Most Poales lack the colorful perianth (petals plus sepals) that normally indicate animal pollination. In some members of the basal grade (Bromeliaceae and Rapateaceae) and in the xyrid clade (excluding Hydatellaceae), at least the inner perianth whorl is colorful, but the outer whorl is often shorter than the inner and of a firmer texture. If we assume that a colorful perianth is the ancestral condition in the Poales, then this was lost three to five times, each time coincident with the shift to wind or water pollination (Figure 3*e*).

Perianth reductions or even losses occurred in several Poales. The perianth was reduced to bristles in Cyperaceae and Typhaceae, and to tiny lobes (lodicules) in grasses. The homologies of these lobes in grasses are still disputed, but morphological and developmental evidence indicates that they probably represent the petals (Clifford 1987, Kellogg 2000, Lawton-Rauh et al. 2000). In addition, several families (Cyperaceae, Typhaceae, Restionaceae, and Poaceae) have undergone reductions in number of perianth parts. The adaxial lodicule is absent from most derived grasses such as *Hordeum* (Pooideae), thus rendering the "typical" grass flower structurally bilaterally symmetrical (Rudall & Bateman 2004).

Shifts in perianth function have also occurred independently. In some grasses, the lodicules function to open the flower at flowering by swelling and forcing apart the bracts subtending the flowers (lemma and palea) (Clayton & Renvoize 1986). In Restionaceae the perianth appears to protect the flower; in some genera it persists around the nut, forming a wing that aids wind dispersal (Linder 1991). In both Cyperaceae and Typhaceae the bristles also aid dispersal of the nutlet.

MALE ORGANS Increased pollen to ovule ratio is normally regarded as characteristic of wind-pollinated species (Faegri & Van der Pijl 1979). This often correlates with an increase in stamen number, but in Poales there have been numerous reductions in stamen number from the ancestral condition of six stamens in two whorls of three (Figure 3*c*). Such reduction is sometimes caused by cyclic loss of a complete stamen whorl (e.g., the inner whorl in Xyridaceae and Eriocaulaceae, or the outer whorl in Restionaceae and many Poaceae) and sometimes by loss of the adaxial (back) or abaxial (front) stamens, the latter occurring independently in the xyrid, cyperid, and graminid clades—that is, in both wind- and insect-pollinated species.

Stamen number is remarkably labile in grasses, including species with one, two, three, four, or six; stamen number sometimes varies even within a single genus (e.g., *Ehrharta*). With the exception of *Anomochloa* (and *Ecdeiocolea*, the putative sister to Poaceae), the male organs of basal Poaceae are typically radially symmetrical as, for example, in *Steptochaeta*, *Pharus*, Puelioideae, and Bambusoideae, which have six stamens in two whorls. The bamboo *Ochlandra* has up to 120 stamens, presumably a derived condition. The most common stamen numbers in grasses are three (e.g., *Hordeum*) and two (e.g., *Diarrhena*), but stamen homologies are equivocal. Species with three stamens may be interpreted as possessing the entire outer stamen whorl (Hackel 1906), or three stamens in the front of the flower (Cocucci & Anton 1988). Rudall & Bateman (2004) regarded the latter interpretation as more plausible and comparable to the condition found in many other monocot groups such as Orchidaceae (Rudall & Bateman 2002). Reduction in stamen number appears an unlikely adaptive modification in wind-pollinated groups in which increase in stamen number is expected, maximizing pollen output. However, the structure of the male organs should be viewed relative to that of the entire inflorescence, which in some cases is condensed to form a pseudoflower with many protruding stamens.

POLLEN   In most monocots pollen grains are dispersed singly (as monads), but the cyperid clade is unusual in that the meiotic products remain together to form permanent tetrahedral tetrads. Dispersal of pollen in tetrads evolved three times in Poales as, for example, in some species of *Typha*, Hydatellaceae, and the cyperid clade. All are wind-pollinated groups that possess several to numerous ovules in each ovary. In some cyperids, the tetrads have become secondarily reduced to functional monads (pseudomonads) by the abortion of three of the four microspores. This happened only in clades where the ovule complement is reduced to one. Simpson et al. (2003) demonstrated that two genera of Cyperaceae have "normal" sticky pollen in monads and thus argued that animal pollination occurs in these genera. The relationship between pollen morphology, number of ovules, and pollination mode in the cyperid clade requires further evaluation.

Monocot pollen grains typically possess a single groove-like aperture (sulcus), and this is also the predominant condition in Poales, although in two wind-pollinated groups—Typhaceae and the graminid-restiid clade—the groove has been modified into a pore (Figure 3d). Porate pollen is generally associated with wind pollination (Linder 2000a). Pollen lacking apertures, or with complex spiralling apertures, is recorded in some animal-pollinated Poales, including some Bromeliaceae and several xyrids (some Xyridaceae, *Abolboda*, and Eriocaulaceae), and also a few (probably wind-pollinated) Cyperaceae (Furness & Rudall 1999).

Pollen aperture structure is diverse in the graminid clade; most species possess a thickened ring (annulus) around the aperture, which has a stopper-like lid (operculum), but Restionaceae and Centrolepidaceae show several types of aperture-rings that are not thickened (Linder & Ferguson 1985). Pores that penetrate both exine and intine layers of the pollen walls (scrobiculi) are recorded in most graminid and restiid families except Poaceae and Joinvilleaceae (Linder & Ferguson 1985). Pollen with stopper-like lids occurs in some species of *Tillandsia* (Bromeliaceae) and *Xyris* (Rudall & Sajo 1999), as well as several graminids and restiids (Anarthriaceae, Ecdeiocoleaceae, and Poaceae: Linder & Ferguson (1985). Furness & Rudall (2003) demonstrated that pollen with single groove-like apertures and with stoppers has evolved many times in monocots, including several Poales, probably due to similar environmental selection pressures, because it normally occurs in species from dry habitats, either year-round or seasonally.

FEMALE ORGANS   Reductions in both carpel and ovule number are correlated with wind pollination (Faegri & Van der Pijl 1979, Linder 1998). Pseudomonomery—the reduction in the number of fertile carpels from three to one—is found in many Poales (e.g., Poaceae, Eriocaulaceae, *Sparganium*, Restionaceae). In the wind-pollinated Poaceae and Cyperaceae, the female organ normally consists of a single carpel containing a single ovule that is located at the back of the flower, but it is not clear whether the other two carpels have been entirely suppressed or are merely sterile and lack septae between them (the pseudomonocarpellary condition, Philipson 1985). This condition seems to have evolved once in each of these families and remained fixed. Pseudomonomery evolved numerous times in Restionaceae, sometimes involving different carpels (Decraene et al. 2002; Linder 1992a,b).

Reduction in ovule number to one per carpel evolved at least five times in the Poales (Figure 3*f*). Reduced ovule number is often associated with aggregated inflorescences and pseudanthial floral presentation in which the entire inflorescence resembles a single flower. This feature may be correlated with shifts to wind pollination, in which case exceptions such as Eriocaulaceae, which has condensed pseudanthial inflorescences but is animal pollinated, would indicate reversals.

SYNTHESIS

Poales occupy a remarkably wide ecological range, but more than 80% of species belong to three ecological radiations (Figure 2*e*). Bromeliaceae, with 13% of the species, retained animal pollination. They are both biologically and morphologically divergent within the order, although they share similarities with some Rapateaceae (Givnish et al. 2004). An ecological peculiarity of this lineage is the occupation of the epiphytic habitat. A smaller diversification is that of the restiid clade, including c. 3% of the total species of Poales, into the nutrient-poor heathlands of the Southern Hemisphere. In perhaps the most recent diversification (possibly less than 15 My old), lineages from three clades (Poaceae, Cyperaceae, and some xyrids, representing a further 60% of species) evolved into seasonally inhospitable grasslands, with either severe frost or fire in the winter months. Important attributes that facilitated these apparent radiations are the sympodial growth habit and anatomical-physiological conditions that allowed frequent evolution of CO_2-concentrating mechanisms. The role of wind pollination is not clear, because this strategy is not usually associated with rapid speciation (Dodd et al. 1999). However, it could be a more effective pollination agent in dense and extensive populations in the first years after fire, when animal vectors might be in short supply.

Many questions remain about the evolutionary history of Poales. In particular, we need better fossil-based vegetation reconstructions to augment inferences based on phylogenetic optimizations. We need much more information about the biology of the some of the smaller clades, especially relatively poorly known but phylogenetically critical taxa such as *Flagellaria*, *Joinvillea*, and *Mayaca*.

ACKNOWLEDGMENTS

We are grateful to Richard Bateman, Chloé Galley, Toby Kellogg, and Timo van der Niet for critically reading the manuscript.

The *Annual Review of Ecology, Evolution, and Systematics* is online at
http://ecolsys.annualreviews.org

LITERATURE CITED

Angiosperm Phylogeny Group (APG). 2003. An update of the Angiosperm Phylogeny Group classification for the orders and families of flowering plants: APG II. *Bot. J. Linn. Soc.* 141:399–436

Appel O, Bayer C. 1998. Flagellariaceae. See Kubitzki 1998a, pp. 208–11

Backer CA. 1954. Flagellariaceae. In *Flora Malesiana*, ed. CGGJ van Steenis, pp. 245–50. Djakarta: Noordhoff-Kolff

Balslev H. 1998. Juncaceae. See Kubitzki 1998a, pp. 252–60

Bayer C, Appel O. 1998. Joinvilleaceae. See Kubitzki 1998a, pp. 249–51

Bell TL, Pate JS. 1993. Morphotypic differentiation in the SW Australian restiad *Lyginia barbata* R. Br. (Restionaceae). *Aust. J. Bot.* 41:91–104

Benzing DH. 1990. *Vascular Epiphytes. General Biology and Related Biota.* Cambridge, UK: Cambridge Univ. Press

Benzing DH. 2000. *Bromeliaceae: Profile of An Adaptive Radiation.* Cambridge, UK: Cambridge Univ. Press

Böhme S. 1988. Bromelienstudien III. Vergleichende Untersuchungen zu Bau, Lage und systematischer Verwertbarkeit der Septalnektarien von Bromeliaceen. *Trop. Subtrop. Pflanzenwelt* 62:86–89

Bremer K. 2002. Gondwanan evolution of the grass alliance of families (Poales). *Evolution* 56:1374–87

Briggs BG, Marchant AD, Gilmore S, Porter CL. 2000. A molecular phylogeny of Restionaceae and allies. See Wilson & Morrison 2000, pp. 661–71

Bruhl JJ, Stone NE, Hattersley PW. 1987. C_4 acid decarboxylation enzymes and anatomy in sedges (Cyperaceae): first record of NAD-malic enzyme species. *Aust. J. Plant Physiol.* 14:719–28

Campbell CS, Kellogg EA. 1987. Sister Group Relationships of the Poaceae. See Soderstrom et al. 1987, pp. 217–24

Cerling TE. 1992. Development of grasslands and savannas in East Africa during the Neogene. *Palaeogeogr. Palaeoclimatol. Palaeoecol.* 97:241–47

Cerling TE, Ehleringer JR, Harris JM. 1998. Carbon dioxide starvation, the development of C_4 ecosystems, and mammalian evolution. *Philos. Trans. R. Soc. Ser. B* 353:159–71

Chapin FS. 1980. The mineral-nutrition of wild plants. *Annu. Rev. Ecol. Syst.* 11:233–60

Chase MW, Fay MF, Devey DS, Maurin O, Ronsted N, et al. 2005. Multigene analyses of monocot relationships: a summary. *Aliso* In press

Chase MW, Soltis DE, Soltis PS, Rudall PJ, Fay MF, et al. 2000. Higher-level systematics of the monocotyledons: an assessment of current knowledge and a new classification. See Wilson & Morrison 2000, pp. 3–16

Clayton WD, Renvoize SA. 1986. *Genera Graminum. Grasses of the World.* London: HMSO

Clifford HT. 1987. Spikelet and floral morphology. See Soderstrom et al. 1987, pp. 21–30

Cocucci AE, Anton AM. 1988. The grass flower: Suggestions on its origin and evolution. *Flora* 181:353–62

Connor HE. 1981. Evolution of reproductive systems in the Gramineae. *Ann. Mo. Bot. Gard.* 68:48–74

Cooke DA. 1998. Centrolepidaceae. See Kubitzki 1998a, pp. 106–9

Cutler DF. 1969. Juncales. In *Anatomy of the Monocotyledons*, ed. CR Metcalfe, pp. 1–357. Oxford: Clarendon

Dahlgren RMT, Clifford HT, Yeo PF. 1985. *The Families of the Monocotyledons: Structure, Function and Evolution*. Berlin: Springer-Verlag

Dahlgren RMT, Rasmussen FN. 1983. Monocotyledon evolution: characters and phylogenetic estimation. *Evol. Biol.* 16:255–395

Davis JI, Stevenson DW, Petersen G, Seberg O, Campbell LM, et al. 2004. A phylogeny of the monocots, as inferred from rbcL and atpA sequence variation, and a comparison of methods for calculating jackknife and bootstrap values. *Syst. Bot.* 29:467–510

Decraene LPR, Linder HP, Smets EF. 2001. Floral ontogenetic evidence in support of the *Willdenowia* clade of South African Restionaceae. *J. Plant Res.* 114:329–42

Decraene LPR, Linder HP, Smets EF. 2002. Ontogeny and evolution of the flowers of South African Restionaceae with special emphasis on the gynoecium. *Plant Syst. Evol.* 231:225–58

Denton GH. 1999. Cenozoic climate change. In *African Biogeography, Climate Change, and Human Evolution*, ed. TG Bromage, F Schrenk, pp. 94–114. Oxford: Oxford Univ. Press

Dodd ME, Silvertown J, Chase MW. 1999. Phylogenetic analysis of trait evolution and species diversity variation among angiosperm families. *Evolution* 53:732–44

Duvall MR, Clegg MT, Chase MW, Clark WD, Kress WJ, et al. 1993. Phylogenetic hypotheses for the monocotyledons constructed from *rbc*L sequence data. *Ann. Mo. Bot. Gard.* 80:607–19

Ehleringer JR, Monson RK. 1993. Evolutionary and ecological aspects of photosynthetic pathway variation. *Annu. Rev. Ecol. Syst.* 24:411–39

Ehleringer JR, Sage RF, Flanagan LB, Pearcy RW. 1991. Climate change and the evolution of C_4 photosynthesis. *Trends Ecol. Evol.* 6:95–99

Ellis RP. 1987. Leaf anatomy of *Ehrharta* (Poaceae) in southern Africa: the setacea group. *Bothalia* 17:75–89

Faegri K, Van der Pijl L. 1979. *The Principles of Pollination Biology*. Oxford: Pergamon

Friedman J, Harder LD. 2004. Inflorescence architecture and wind pollination in six grass species. *Funct. Ecol.* 18:851–60

Furness CA, Rudall PJ. 1999. Microsporogenesis in Monocotyledons. *Ann. Bot.* 84:475–99

Furness CA, Rudall PJ. 2003. Apertures with lids: distribution and significance of operculate pollen in monocots. *Int. J. Plant Sci.* 164:835–54

Gandolfo MA, Nixon KC, Crepet WL. 2000. Monocotyledons: a review of their early Cretaceous record. See Wilson & Morrison 2000, pp. 44–51

Giussani LM, Cota-Sanchez JH, Zuloaga FO, Kellogg EA. 2001. A molecular phylogeny of the grass subfamily Panicoideae (Poaceae) shows multiple origins of C_4 photosynthesis. *Am. J. Bot.* 88:1993–2012

Givnish TJ, Evans TM, Pires JC, Sytsma KJ. 1999. Polyphyly and convergent morphological evolution in Commelinales and Commelinidae: evidence from rbcL sequence data. *Mol. Phylogenet. Evol.* 12:360–85

Givnish TJ, Evans TM, Zjhra ML, Patterson TB, Berry PE, Sytsma KJ. 2000. Molecular evolution, adaptive radiation, and geographic diversification in the amphiatlantic family Rapateaceae: Evidence from ndhF sequences and morphology. *Evolution* 54:1915–37

Givnish TJ, Millam KC, Evans TM, Hall JC, Pires JC, et al. 2004. Ancient vicariance or recent long-distance dispersal? Inferences about phylogeny and South American—African disjunctions in Rapateaceae and Bromeliaceae based on *ndh*F sequence data. *Int. J. Plant Sci.* 165: S35–54

Givnish TJ, Sytsma KJ, Smith JF, Hahn WJ, Benzing DH, Burkhardt EM. 1997. Molecular evolution and adaptive radiation in *Brocchinia* (Bromeliaceae: Pitcairnioideae) atop tepuis of the Guayana shield. In *Molecular Evolution and Adaptive Radiation*, ed. TJ Givnish, KJ Sytsma, pp. 259–311. Cambridge, UK: Cambridge Univ. Press

Goetghebeur P. 1998. Cyperaceae. See Kubitzki 1998a, pp. 141–90

Grass Phylogeny Work. Group (GPWG). 2001. Phylogeny and subfamilial classification of the grasses (Poaceae). *Ann. Mo. Bot. Gard.* 88:373–457

Graur D, Martin W. 2004. Reading the entrails of chickens: molecular timescales of evolution and the illusion of precision. *Trends Genet.* 20:80–86

Hackel E. 1906. Über Kleistogamie bei den Gräsern. *Österr. Bot. Z.* 56:81–88, 143–54, 180–86

Hamann U. 1998. Hydatellaceae. See Kubitzki 1998a, pp. 231–34

Herendeen PS, Crane PR. 1995. The fossil history of the monocotyledons. See Rudall et al. 1995, pp. 1–21

Herring JR. 1985. Charcoal fluxes in sediments of the North Pacific Ocean: the Cenozoic record of burning. See Sundquist & Broecker 1985, pp. 419–42

Holttum RE. 1955. Growth-habits of monocotyledons—variations on a theme. *Phytomorphology* 5:399–413

Hu FY, Tao DY, Sacks E, Fu BY, Xu P, et al. 2003. Convergent evolution of perenniality in rice and sorghum. *Proc. Natl. Acad. Sci. USA* 100:4050–54

Jacobs BF, Kingston JD, Jacobs LL. 1999. The origin of grass-dominated ecosystems. *Ann. Mo. Bot. Gard.* 86:590–643

Janssen T, Bremer K. 2004. The age of major monocot groups inferred from 800+ *rbc*L sequences. *Bot. J. Linn. Soc.* 146:385–98

Judziewicz EJ, Soderstrom TR. 1989. Morphological, anatomical, and taxonomic studies in *Anomochloa* and *Streptochaeta* (Poaceae: Bambusoideae). *Smithson. Contrib. Bot.* 68: 1–52

Keeley JE, Rundel PW. 2003. Evolution of CAM and C_4 carbon-concentrating mechanisms. *Int. J. Plant Sci.* 164:S55–77

Kellogg EA. 2000. The grasses: A case study in macroevolution. *Annu. Rev. Ecol. Syst.* 31: 217–38

Kellogg EA, Linder HP. 1995. Phylogeny of Poales. See Rudall et al. 1995, pp. 511–42

Kral R. 1998. Xyridaceae. See Kubitzki 1998a, pp. 461–69

Kubitzki K, ed. 1998a. *The Families and Genera of Vascular Plants. IV. Flowering Plants. Monocotyledons.* Berlin: Springer-Verlag

Kubitzki K. 1998b. Thurniaceae. See Kubitzki 1998a, pp. 455–57

Kubitzki K. 1998c. Typhaceae. See Kubitzki 1998a, pp. 457–61

Lamont BB. 1983. Strategies for maximizing nutrient uptake in two mediterranean ecosystems of low nutrient status. In *Mediterranean-Type Ecosystems. The Role of Nutrients,* ed. FJ Kruger, DT Mitchell, JUM Jarvis, pp. 246–73. Berlin: Springer-Verlag

Lawton-Rauh AL, Alvarez-Buylla ER, Purugganan MD. 2000. Molecular evolution of flower development. *Trends Ecol. Evol.* 15: 144–49

Linder HP. 1991. A review of the southern African Restionaceae. *Contrib. Bolus Herb.* 13:209–64

Linder HP. 1992a. The gynoecia of Australian Restionaceae: morphology, anatomy and systematic implications. *Aust. Syst. Bot.* 5:227–45

Linder HP. 1992b. The structure and evolution of the female flower of the African Restionaceae. *Bot. J. Linn. Soc.* 109:401–25

Linder HP. 1998. Morphology and the evolution of wind pollination. In *Reproductive Biology,* ed. ST Owens, PJ Rudall, pp. 123–35. Kew: R. Bot. Gard.

Linder HP. 2000a. Pollen morphology and wind pollination in Angiosperms. In *Pollen and Spores: Morphology and Biology,* ed. MM Harley, CM Morton, S Blackmore, pp. 73–88. Kew: R. Bot. Gard.

Linder HP. 2000b. Vicariance, climate change, anatomy and phylogeny of the Restionaceae. *Bot. J. Linn. Soc.* 134:159–77

Linder HP, Briggs BG, Johnson LAS. 1998a. Anarthriaceae. See Kubitzki 1998a, pp. 19–20

Linder HP, Briggs BG, Johnson LAS. 1998b. Ecdeiocoleaceae. See Kubitzki 1998a, pp. 195–97

Linder HP, Briggs BG, Johnson LAS. 1998c.

Restionaceae. See Kubitzki 1998a, pp. 425–45

Linder HP, Eldenäs P, Briggs BG. 2003. Contrasting patterns of radiation in African and Australian Restionaceae. *Evolution* 57: 2688–702

Linder HP, Ferguson IK. 1985. On the pollen morphology and phylogeny of the Restionales and Poales. *Grana* 24:65–76

Linder HP, Hardy CR. 2004. Evolution of the species-rich Cape flora. *Philos. Trans. R. Soc. London Ser. B* 359:1623–32

Linder HP, Hardy CR, Rutschmann F. 2005. Taxon sampling effects in molecular clock dating: an example from the African Restionaceae. *Mol. Phylogenet. Evol.* 35:569–82

Magallón S. 2004. Dating lineages: molecular and paleontological approaches to the temporal framework of clades. *Int. J. Plant Sci.* 165:S7–21

Meney KA, Pate JS, eds. 1999. *Australian Rushes: Biology, Identification and Conservation of Restionaceae and Allied Families.* Nedlands: Univ. West. Aust.

Michelangeli FA, Davis JJ, Stevenson DW. 2003. Phylogenetic relationships among Poaceae and related families as inferred from morphology, inversions in the plastid genome, and sequence data from the mitochondrial and plastid genomes. *Am. J. Bot.* 90:93–106

Monson RK. 2003. Gene duplication, neofunctionalization, and the evolution of C_4 photosynthesis. *Int. J. Plant Sci.* 164:S43–54

Newell TK. 1969. A study of the genus *Joinvillea* (Flagellariaceae). *J. Arnold Arbor.* 50: 527–55

Niklas KJ. 1985. The aerodynamics of wind pollination. *Bot. Rev.* 51:328–86

Niklas KJ. 1987. Pollen capture and wind-induced movement of compact and diffuse grass panicles: implications for pollination efficiency. *Am. J. Bot.* 74:74–89

Nixon KC. 2002. *WinClada* vers. 1.00.08. Published by the author, Ithaca, NY

Pate JS, Meney KA, Dixon KW. 1991. Contrasting growth and morphological characteristics of fire-sensitive (obligate seeder) and fire-resistant (resprouter) species of Restionaceae (S. Hemisphere restiads) from south-western Australia. *Aust. J. Bot.* 39: 505–25

Philipson WR. 1985. Is the grass gynoecium monocarpellary? *Am J. Bot.* 72:1954–61

Rudall PJ. 1997. The nucellus and chalaza in monocotyledons: structure and systematics. *Bot. Rev.* 63:140–81

Rudall PJ. 2002. Homologies of inferior ovaries and septal nectaries in monocotyledons. *Int. J. Plant Sci.* 163:261–76

Rudall PJ, Bateman RM. 2002. Roles of synorganisation, zygomorphy and heterotopy in floral evolution: the gynostemium and labellum of orchids and other lilioid monocots. *Biol. Rev.* 77:403–41

Rudall PJ, Bateman RM. 2004. Evolution of zygomorphy in monocot flowers: iterative patterns and developmental constraints. *New Phytol.* 162:25–44

Rudall PJ, Cribb PJ, Cutler DF, Humphries CJ, eds. 1995. *Monocotyledons: Systematics and Evolution.* Kew: R. Bot. Gard.

Rudall PJ, Sajo MG. 1999. Systematic position of *Xyris*: flower and seed anatomy. *Int. J. Plant Sci.* 160:795–808

Sajo MG, Rudall PJ, Prychid CJ. 2004. Floral anatomy of Bromeliaceae, with particular reference to the evolution of epigyny and septal nectaries in commelinid monocots. *Plant. Syst. Evol.* 247:215–31

Sanderson MJ. 1998. Estimating rate and time in molecular phylogenies: beyond the molecular clock? In *Molecular Systematics of Plants II. DNA Sequencing*, ed. DE Soltis, PS Soltis, JJ Doyle, pp. 242–64. Boston: Kluwer Acad.

Sanderson MJ, Thorne JL, Wikström N, Bremer K. 2004. Molecular evidence on plant divergence times. *Am. J. Bot.* 91:1656–65

Simpson DA, Furness CA, Hodkinson TR, Muasya AM, Chase MW. 2003. Phylogenetic relationships in Cyperaceae subfamily Mapanioideae inferred from pollen and plastid DNA sequence data. *Am. J. Bot.* 90:1071–86

Smets EF, Ronse Decraene LP, Caris P, Rudall

PJ. 2000. Floral nectaries in monocotyledons: distribution and evolution. See Wilson & Morrison 2000, pp. 230–40

Smith LB, Till W. 1998. Bromeliaceae. See Kubitzki 1998a, pp. 74–99

Soderstrom TR, Hilu KW, Campbell CS, Barkworth ME, eds. 1987. *Grass Systematics and Evolution*. Washington, DC: Smithson. Inst.

Stevenson DW. 1998. Mayacaceae. See Kubitzki 1998a, pp. 294–96

Stevenson DW, Colella M, Boom B. 1998. Rapateaceae. See Kubitzki 1998a, pp. 415–24

Stock WD, Chuba DK, Verboom GA. 2004. Distribution of South African C_3 and C_4 species of Cyperaceae in relation to climate and phylogeny. *Aust. Ecol.* 29:313–39

Stock WD, Sommerville JEM, Lewis OAM. 1987. Seasonal allocation of dry mass and nitrogen in a fynbos endemic Restionaceae species *Thamnochortus punctatis* Pill. *Oecologia* 72:315–20

Stützel T. 1986. Die epipetalen Drusen der Gattung *Eriocaulon* (Eriocaulaceae). *Beitr. Biol. Pflanz.* 60:271–76

Stützel T. 1990. "Appendices" am Gynoeceum der Xyridaceen Morphogenie, Funktion und systematische Bedeutung. *Beitr. Biol. Pflanz.* 65:275–99

Stützel T. 1998. Eriocaulaceae. See Kubitzki 1998a, pp. 197–207

Sundquist ET, Broecker WS, eds. 1985. *The Carbon Cycle and Atmospheric CO_2: Natural Variations Archean to Present*. Washington, DC: Am. Geophys. Union

Venturelli M, Bouman F. 1988. Development of ovule and seed in Rapateaceae. *Bot. J. Linn. Soc.* 97:267–94

Verboom GA, Linder HP, Stock WD. 2003. Phylogenetics of the grass genus *Ehrharta* Thunb.: evidence for radiation in the summer-arid zone of the South African Cape. *Evolution* 57:1008–21

Wikström N, Savolainen V, Chase MW. 2001. Evolution of the angiosperms: calibrating the family tree. *Proc. R. Soc. London Ser. B* 268:2211–20

Wilson KL, Morrison DA, eds. 2000. *Monocots: Systematics and Evolution*. Melbourne: CSIRO

Wolfe JA. 1985. Distribution of major vegetation types during the Tertiary. See Sundquist & Broecker 1985, pp. 357–75

Zachos J, Pagani M, Sloan L, Thomas E, Billups K. 2001. Trends, rhythms, and aberrations in global climate 65 Ma to present. *Science* 292:686–93

Zaitchik BF, LeRoux LG, Kellogg EA. 2000. Development of male flowers in *Zizania aquatica* (North American wild-rice; Gramineae). *Int. J. Plant Sci.* 161:345–51

Annu. Rev. Ecol. Evol. Syst. 2005. 36:125–46
doi: 10.1146/annurev.ecolsys.36.102403.112501
First published online as a Review in Advance on August 5, 2005

THE EVOLUTION OF POLYANDRY:
Sperm Competition, Sperm Selection, and Offspring Viability

Leigh W. Simmons

Evolutionary Biology Research Group, School of Animal Biology, The University of Western Australia, Nedlands, Western Australia 6009, Australia;
email: lsimmons@cyllene.uwa.edu.au

Key Words genetic incompatibility, intrinsic sire effects, maternal effects, multiple mating

■ **Abstract** In contrast to Bateman's principle, there is now increasing evidence that female fitness can depend on the number of mates obtained. A number of genetic benefits have been proposed for the evolution of polyandry. A meta-analysis of available data suggests that polyandry, rather than multiple mating, can have a weak but significant general effect on embryo viability, as indicated by egg hatching success. Although this effect is generally regarded as evidence in favor of the genetic incompatibility hypothesis, appropriate data that test for intrinsic sire effects on embryo viability are generally unavailable. Moreover, maternal effects that could generate the result have not been adequately controlled, and there is little unequivocal evidence to suggest that fertilization is biased toward sperm bearing genotypes that would enhance offspring viability. Greater effort is required in these areas to elucidate the mechanisms underlying observed fitness effects of polyandry.

INTRODUCTION

Bateman's principle is a well-established paradigm in evolutionary biology. Bateman argued that while a male's reproductive success should be limited only by the number of females he can inseminate, a female's reproductive success should be largely independent of the number of males with which she copulates (Bateman 1948). Given that mating incurs costs associated with exposure to disease (Thrall et al. 2000), predation (Rowe 1994), or physical harm from males (Chapman et al. 1995), females should mate only once or a few times to guarantee the fertilization of their ova. Bateman's principle provides a general explanation for why males compete for access to multiple partners and are therefore subject to sexual selection (Arnold & Duvall 1994). Nevertheless, female acceptance of matings from more than one male (polyandry) is a taxonomically widespread phenomenon (Jennions & Petrie 2000, Zeh & Zeh 2001). Polyandry has far-reaching evolutionary consequences and the adaptive significance of the behavior is currently subject to considerable debate.

One consequence of polyandry is that the sperm from two or more males can co-occur at the site of fertilization, generating competition between sperm for the fertilization of ova. Parker (1970) recognized that, for internally fertilizing species, sperm competition would favor opposing adaptations in males that both facilitate the pre-emption of sperm stored by females from previous matings and prevent the mating male's own sperm from being pre-empted by future males. In the 30 years since Parker's contribution, sperm competition has become widely recognized as a potent force of sexual selection, generating behavioral, physiological, and morphological adaptations that serve to prevent potential rivals from gaining access to females, and thus ensuring the copulating male's sperm are used for fertilization (Birkhead & Møller 1998, Simmons 2001b). Parker also recognized that there was every reason to expect selection to operate on ejaculates because these are at the front line in sperm competition. To this end, the past decade has seen the development of an extensive theoretical base with which to predict the effects of sperm competition on the characteristics of ejaculates of both internally and externally fertilizing species, from the numbers of sperm ejaculated to the morphology and behavior of individual sperm (Parker 1998, Wedell et al. 2002).

Clearly sperm competition has traditionally been viewed as an extension of male contest competition. However, researchers are increasingly aware of the importance of the female's role in fertilization events (Eberhard 1996, Thornhill 1983). Patterns of nonrandom paternity may not reflect adaptations to sperm competition at all; rather, females may have the potential to selectively fertilize their ova using sperm from males who offer greater fitness returns for their offspring. Indeed, polyandry may have its evolutionary origins in increasing female reproductive success (Sivinski 1984) despite the established dogma of Bateman's principle, and sperm selection is just one mechanism by which females could exercise "cryptic female choice" (Eberhard 1996, Simmons 2001b).

Arnqvist & Nilsson's (2000) meta-analysis of 122 experimental studies of insects found strong evidence for positive effects of polyandry on female fitness. Polyandry increased the numbers of eggs produced by females and the percentage of those eggs that hatched. This was true both for species with and without nuptial feeding. Increased egg production is a well-known response to mating in insects, arising from the transfer of hormonal stimulants in the male's ejaculate (Simmons 2001b). The significantly greater effect of polyandry on egg production in nuptial-feeding species (Arnqvist & Nilsson 2000) is perhaps not surprising, given the increased nutrient resources obtained by females through multiple mating. The effect of polyandry on hatching success did not differ between species with or without nuptial feeding and is a result that may in part be due simply to the avoidance of sperm depletion. However, a number of genetic mechanisms have been proposed that could also account for the increased hatching success of polyandrous females (Yasui 1997; Zeh & Zeh 1996, 1997). Unfortunately, distinguishing between the immediate benefits of multiple matings on a female's ability to fertilize all of her eggs and the potential genetic benefits associated with polyandry requires experiments that manipulate the degree of polyandry while controlling mating frequency, experiments that were unavailable at the time of Arnqvist & Nilsson's (2000) review. Here I provide an overview of two genetic models that predict an increased viability of offspring as a result of the overlap of sperm from different males at the site of fertilization. I provide a prospective meta-analysis of the data now available to test these models and suggest alternative explanations for the emerging patterns.

GENETIC BENEFITS OF POLYANDRY

A number of genetic benefit hypotheses have been proposed to explain the evolution of polyandry. For example, the "trading-up" hypothesis proposes that females seek extrapair copulations with males of higher genetic quality than their social mate (Kempenaers et al. 1992); the "genetic diversity" hypothesis proposes that, by increasing genetic diversity within progeny, females guard against future environmental uncertainty (Yasui 1998, 2001); and the "sexy-sperm" hypothesis proposes that, by inciting sperm competition, females produce sons that are successful in sperm competition by virtue of heritable variation in the characteristics that determine the fertilization success of their fathers (Keller & Reeve 1995). In this review I focus on the genetic incompatibility and the intrinsic male quality (or "good-sperm") models for the evolution of polyandry. Both models require sperm from multiple males to overlap at the site of fertilization so that sperm from potential sires must compete for fertilizations and/or be subject to selective use by cryptically choosing females.

Genetic Incompatibility

The genetic incompatibility hypothesis proposes that females mate with more than one male in order to hedge against incompatibilities that can arise because of selfish genetic elements (Zeh & Zeh 1996, 1997). For example, some species of maternally inherited bacteria in the genus *Wolbachia* can generate cytoplasmic incompatibility that results in complete embryo failure. When an uninfected female mates with an infected male the resulting embryos die (Stouthamer et al. 1999). If *Wolbachia*-infected males were less competitive in sperm competition, or if females could select sperm from uninfected males, multiply mating females could reduce the incidence of embryo failure. Indeed, sex ratio–distorting elements in *Drosophila* (Atlan et al. 2004, Wu 1983) and stalk-eyed flies (Wilkinson & Fry 2001) decrease sperm competition ability so that males carrying the drive element do have a lower competitive fertilization success. The superior sperm competitive abilities of drive-free males can thereby provide the necessary genetic benefits for the evolution of polyandry (Wilkinson et al. 2003).

More generally, genetic incompatibility can arise because of combinations of maternal and paternal haplotypes that produce offspring of inferior viability. For example, nonadditive interactions between alleles at one or more loci can have negative effects on an organism's fitness (Crnokrak & Roff 1995, Merilä & Sheldon 1999), leading to incompatibility. Moreover, heterozygosity is now widely recognized as being beneficial to individual fitness, and evidence for female choice of males whose haplotypes offer increased heterozygosity for offspring is growing (Brown 1997, Tregenza & Wedell 2000). An often cited example comes from work on the vertebrate Major Histocompatibility Complex (MHC). The MHC is involved in immune function, resistance to parasitic infection, and the production of odors that are important in individual recognition and mate choice (Penn 2002). Females of a number of vertebrate species ranging from mice (Yamazaki et al. 1976) to humans (Wedekind et al. 1995) have been shown to prefer potential mating partners whose MHC haplotype would result in the production of MHC heterozygous offspring that are resistant to parasites (Penn et al. 2002). For example, in the three-spined stickleback, *Gasterosteus aculeatus*, the intensity and prevalence of parasitic infections is lowest for individuals with an intermediate number of MHC class-IIB alleles (Wegner et al. 2003). Female sticklebacks appear capable of ensuring an optimal number of alleles for their offspring by choosing males with a number of MHC class-IIB alleles that complements their own (Aeschlimann et al. 2003, Reusch et al. 2001).

The deleterious effects of inbreeding arise because of a decrease in heterozygosity, or more specifically because of an increased expression of deleterious recessive mutations that result from increased homozygosity (Pusey & Wolf 1996). Thus, inbreeding avoidance can be viewed as a form of mate choice for genetic compatibility (Tregenza & Wedell 2000). Because the MHC influences the composition of odors that are used in both kin-recognition and mate choice, it has been argued that MHC-based mating preferences represent a mechanism of inbreeding avoidance (Penn 2002). Nevertheless, there may not always be mechanisms available by which females can identify males with haplotypes that would complement their own, or their precopulatory mate choices can be compromised by sexual conflict over mating decisions (Parker 1979). By mating multiple times, females have the potential to utilize postcopulatory mechanisms to ensure their eggs are fertilized by sperm with haplotypes that best complement their own through selective fertilization of eggs with sperm from compatible individuals (Zeh & Zeh 1997).

Intrinsic Male Quality

Sivinski (1984) was first to suggest that viability selection might favor the evolution of multiple mating by females. He argued that if sperm competitive ability were correlated with offspring competitive ability, polyandrous females would, on average, produce offspring of higher viability than monandrous females. Cryptic female choice could enhance such a filtering process, for example, where females provided chemically challenging environments within which sperm were forced to compete (Sivinski 1984). Sivinski recognized that sperm fitness, mediated by the haploid genotype, is unlikely to correlate strongly with fitness of the diploid organism in which the phenotype is the product of the combined sperm and egg haplotypes (Parker 1992). Although it is not yet clear whether a gamete's phenotype is controlled by its haploid genotype or by its parent's diploid genotype, there is little evidence for haploid expression in *Drosophila* (Erickson 1990). Nevertheless, fertilization success may be mediated by condition-dependent traits (Rowe & Houle 1996, Tomkins et al. 2004). For example, a male's fertilization success may depend on his ability to commit resources to the production of large numbers of sperm and/or seminal fluid products, rather than the haplotypes of individual sperm. If so, these traits should capture additive genetic variation across a large number of loci that contribute to the general viability of the organism. Yasui (1997) showed theoretically how polyandry could be maintained when a male's fertilization success reflected general viability in his so called "good-sperm" model.

The intrinsic male quality hypothesis differs from the genetic incompatibility hypothesis in that it requires heritable variation in offspring viability so that the benefits of polyandry stem solely from the genetic quality of sires rather than from an interaction between sire and dam genotypes. There is now good evidence that females can gain viability benefits for their offspring through precopulatory mate choice (Møller & Alatalo 1999). Moreover, recent studies using in vitro fertilization techniques are providing convincing evidence for paternal genetic effects on offspring viability (Sheldon et al. 2003, Wedekind et al. 2001). Therefore, if a male's fertilization success were correlated with heritable variation in viability, multiply mating females could ensure their offspring were sired by males able to bestow high viability on their offspring.

SPERM COMPETITION AND SPERM SELECTION

There is accumulating evidence suggesting that genetic incompatibility can be an important factor in determining whether a given sperm will fertilize an egg. An extreme form of incompatibility arises when closely related species occur in sympatry, and conspecific sperm and pollen precedence has been found to play an important role in reproductive isolation (Howard 1999). In the ground crickets *Allenomobius fasciatus* and *A. socius*, heterospecific crosses result in the production of hybrid offspring, indicating that heterospecific sperm are capable of fertilization. However, when females mate with both a conspecific and a heterospecific male, the conspecific male gains the majority of fertilizations (Gregory & Howard 1994). The mechanism for this postinsemination barrier to hybridization is unknown, although experimentally increasing the numbers of heterospecific sperm can overcome the homospecific advantage (Howard & Gregory 1993).

Maternal filtering of gametes is well known in flowering plants, where the risk of self-pollination can be high. Pollen competition studies have shown that self pollen is less likely to fertilize ovules than outcross pollen because the maternal tissue inhibits self pollen-tube growth (Matton et al. 1994). Self-fertilization is a general problem for hermaphroditic animals that release their gametes into the environment. In the compound ascidian *Diplosoma listerianum*, water-borne sperm enter the female gonopore where they can be stored for several weeks prior to internal fertilization. Self sperm are generally phagocytosed within the lumen of the oviduct, as are nonself sperm that share self-recognition markers with maternal tissue, resulting in a fertilization advantage for genetically dissimilar sperm (Bishop et al. 1996).

Gamete recognition systems are an important mechanism of sperm selection in broadcast spawning animals in which both males and females release their gametes into the environment (Vacquier 1998). In sea urchins, the sperm protein bindin is the principal component of the sperm acrosomal vesicle and is involved in the attachment of sperm to the egg surface and in sperm–egg fusion. There is evidence of positive selection for amino acid divergence between species, and sperm-egg recognition facilitates reproductive isolation (Levitan 2002, Metz & Palumbi 1996). In the Australian urchin *Heliocidaris erythrogramma*, the proportion of eggs fertilized in artificial spawning trials depends on an interaction between male and female genotypes (Evans & Marshall 2005), suggesting that fertilization depends on the presence of compatible sperm. Moreover, the average proportion of eggs fertilized is higher when sperm from two males are present than when only a single male's sperm are present, providing evidence for a direct fertilization benefit of polyandrous spawning. Studies of the tropical urchin *Echinometra mathaei* show that in competitive spawning a male's fertilization success depends on his bindin genotype and on the bindin genotype of females; eggs selectively fuse with sperm that share the largest number of bindin alleles (Palumbi 1999).

Gamete recognition systems also have potential to be an important mechanism for incompatibility avoidance in internally fertilizing species. Interestingly, there is some evidence to suggest that MHC haplotype may be involved in nonrandom fertilization in mice (Wedekind et al. 1996), sand lizards (Olsson et al. 2004), and tunicates (Scofield et al. 1982), in these cases favoring heterozygosity in resultant embryos. In sand lizards (Olsson et al. 1996) and sedge warblers (Marshall et al. 2003), fertilization success is greater for sperm with genetically dissimilar haplotypes to the female, and male by female interaction effects on competitive fertilization success have been reported in *Drosophila* (Clark et al. 1999, 2000), bean weevils (Wilson et al. 1997), flour beetles (Lewis & Austad 1990), and domestic fowl (Birkhead et al. 2004). However, these studies typically score paternity based on newly hatched or adult offspring genotypes so that the interaction effects observed could be caused by sperm selection or random fertilization followed by differential embryo mortality. That is, nonrandom paternity could either reflect the avoidance of genetic incompatibility or be a reflection of its consequences.

A good example of the problems associated with assigning paternity late in offspring development is seen in Calsbeek & Sinervo's (2004) study of side-blotched lizards *Uta stansburiana*. They found that paternity of hatched offspring depends on sire body size. When sires were large, females produced more sons and when sires were small, females produced more daughters. Calsbeek & Sinervo (2004) interpreted this finding as evidence for sperm selection, arguing that females selectively fertilize their eggs with Y-bearing sperm from large sires to produce sons of high fitness. Indeed, they showed that survival of sons from hatching to maturity increased with sire size, whereas survival of daughters decreased. If this pattern of sire-size–dependent mortality occurs throughout the lifespan of offspring, including the period from zygote formation through to hatching, it can explain the patterns of paternity determined from hatched offspring without invoking a process of sperm selection by females. Likewise in Evans et al.'s (2003) experiments with guppies, attractive males (those having more orange body coloration) sired more newborn young after artificial insemination with equal numbers of sperm from competing males. Although these patterns are consistent with postcopulatory female preferences for orange, the patterns of paternity could be explained by higher embryo mortality of offspring sired by less ornamented males rather than by differential fertilization success of the ornamented male's sperm.

Under the intrinsic male quality hypothesis, some males should be consistently more successful in gaining fertilizations than others across all females mated. Specifically, the hypothesis requires that fertilization success should have an additive genetic basis. There are studies examining the genetics of traits thought to be important in sperm competition, such as testis size, ejaculate volume, sperm morphology, and sperm viability (Keller & Reeve 1995, Moore et al. 2004, Pizzari & Birkhead 2002, Simmons & Kotiaho 2002). In the beetle *Onthophagus taurus*, testis size and sperm length appear to be condition-dependent, consistent with the intrinsic male quality hypothesis (Simmons & Kotiaho 2002). Nevertheless, very few studies directly examine the levels of additive genetic variance in fertilization success. Edwards (1955) demonstrated a heritable component to the probability of fertilization by mouse sperm, and heritabilities of around 30% have been reported for fertilization success in domestic fowl (Froman et al. 2002) and bulb mites (Radwan 1998). Genetic variation in fertilization success of male dung flies is also implicated by the finding that this trait evolved over 10 generations of polyandrous mating (Hosken et al. 2001). Finally, although a broad-sense heritability of 50% was found for sperm competition success in the field cricket *Teleogryllus oceanicus*, the patterns of genetic variance were consistent with maternal inheritance (Simmons 2003). When fertilization success is transmitted via female-biased mechanisms, females are less likely to acquire genetic benefits for their sons by mating with males having superior sperm competitiveness (Pizzari & Birkhead 2002).

POLYANDRY AND OFFSPRING VIABILITY

In their study of adders, *Vipera berus*, Madsen et al. (1992) found that the proportion of stillborn young produced by a female negatively correlated with the number of different males with which she had mated. Madsen et al. interpreted their findings as support for the intrinsic male quality hypothesis, suggesting that males who were successful in sperm competition bestowed high viability on their offspring. Likewise, Zeh (1997) found that multiply mated female pseudoscorpions, *Cordylochernes scorpioides*, had higher reproductive success than singly mated females; they were less likely to suffer complete brood failure and thereby produced a greater number of hatched nymphs. Zeh (1997) interpreted her findings as evidence for the genetic incompatibility hypothesis because in a second experiment, in which she allocated a single male to two different females, there was no correlation between the reproductive success of a male's mates as might be expected under the intrinsic male quality hypothesis.

Several other studies of insects and arachnids (Konior et al. 2001, Pai & Yan 2002, Watson 1991), reptiles (Olsson et al. 1994), birds (Kempenaers et al. 1999), and mammals (Hooglund 1998) reported positive correlations between the degree of female polyandry and the hatching success of eggs, or the viability and performance of hatched offspring. However, correlational studies cannot provide evidence for genetic benefits of polyandry because they confound number of mates with number of matings. Material benefits from mating could generate positive associations between mating frequency and female reproductive success. Maternal effects are a second source of potentially confounding variation. For example, some studies of social insects report higher productivity or resistance to disease for colonies headed by multiply mated queens (Hughes & Boomsma 2004, Wiernasz et al. 2004), and this has been taken as evidence for the genetic diversity hypothesis for the evolution of polyandry. However, Fjerdingstad & Keller (2004) showed that foundress queens of the ant *Lasius niger* with greater initial body weight were more likely to mate with several males during their nuptial flights, and that queen body weight, rather than mating frequency, was the best predictor of colony productivity. Thus, maternal effects can generate an apparent association between mating frequency and reproductive success in correlational studies.

Tregenza & Wedell (1998) suggested an experimental protocol with which to test the hypothesis that polyandry, rather than multiple mating, contributed to offspring viability. Using the field cricket *Gryllus bimaculatus*, they produced three groups of treatment females, all of which mated four times. In the first treatment group females mated four times with a single male, in the second group females mated twice with each of two different males, and in the third group females mated once with each of four different males. After mating, females were allowed to oviposit and the resulting offspring reared to adulthood. They found that polyandry had a significant effect on the average hatching success of eggs, which increased with the number of males mated. Many of the eggs failed to hatch, not because they were unfertilized, but because of embryo mortality. The effects of female polyandry on posthatching offspring performance were nonsignificant. These data suggest that polyandry, as opposed to multiple mating, yields significant benefits for females. Tregenza & Wedell argued that this finding was most likely due to the avoidance of genetic incompatibility on the grounds that males conferring high hatching success when the sole mate of a female did not elevate hatching success when they were one of several mates of a polyandrous female.

Tregenza & Wedell's approach has been emulated many times since, and there are now enough data to determine whether there is a general effect of polyandry on offspring performance (Table 1). These experimental tests use mostly insects, perhaps because of the ease with which experimental treatments can be conducted and the large sample sizes that can be obtained. The available data suggest that there is a small but significant general effect of polyandry on hatching success (Figure 1). The mean effect size for the viability of offspring once they had hatched is close to zero and not significant ($Z_r = 0.04 \pm 0.07$, $t_7 = 0.62$, $P = 0.55$), and there is no correlation between effect sizes on embryo viability [assumed to be reflected by hatching success, for which there is some evidence (Simmons 2001a, Tregenza & Wedell 1998)] and hatched offspring viability ($r_6 = 0.223$, P > 0.50).

Only two studies examined the benefits of polyandry while controlling for mating frequency in vertebrates. In their study of the myobatrachid frog *Crinia georgiana*, Byrne & Roberts (2000) found no effect of polyandry on egg viability ($Z_r = -0.057$) or on tadpole viability to metamorphosis ($Z_r = 0.068$; effect sizes are the means derived from measures taken in different environmental treatments). In the marsupial *Antechinus agilis*, the number of neonates that attached to the teats at birth was not greater for polyandrous females ($Z_r = -0.112$) although the offspring of polyandrous females did have a higher growth rate ($Z_r = 0.487$) (Fisher et al. 2005).

GENETIC INCOMPATIBILITY OR INTRINSIC SIRE EFFECTS?

The data in Table 1 suggest that polyandry, rather than mating frequency, can have a small but significant general effect on prehatching embryo viability. This effect of polyandry on hatching success has been taken as evidence in support of the genetic incompatibility hypothesis for the evolution of polyandry (references in Table 1). However, genetic incompatibility is generally the default interpretation based on smaller posthoc experiments and/or analyses of subsets of the data that fail to reveal consistent sire effects on hatching success across replicate females. Given the small general effect size on hatching success, the tests used to reject intrinsic sire effects in the studies detailed in Table 1 clearly lack the statistical power to adequately test for them (Ivy & Sakaluk 2005), and there is little unequivocal evidence to distinguish between the two competing hypotheses.

Tregenza & Wedell (2002) provided support for a role of genetic incompatibility in determining the benefits of polyandry in their study of inbreeding avoidance in *G. bimaculatus*. They produced three mating groups in which females mated twice: once each with two full sibling males, with two unrelated males, or with one sibling and one unrelated male. Females mated to two siblings had a significantly lower hatching success than females mated with two unrelated males, indicating a

TABLE 1 Data on the fitness consequences of polyandry derived from experimental studies in which mating frequency was controlled

Species	Number of matings: males		Fitness component	Mean ± SE (N)		Effect size (Z_r)	Notes
	Monandrous	Polyandrous		Monandrous	Polyandrous		
Dermaptera							
Euborellia plebeja	3:1	1:3	% hatch	60.0 ± 5.8 (26)	66.8 ± 5.4 (30)	0.113	1
Diptera							
Scatophaga stercoraria	2:1	1:2	% hatch	84.0 ± 2.0 (36)	78.0 ± 2.0 (36)	−0.017	2
			offspring viability	86.0 ± 3.0 (36)	88.0 ± 3.0 (36)	0.074	
Cyrtodiopsis dalmanni	3:1	1:3	% hatch	61.9 ± 5.0 (40)	73.4 ± 4.0 (32)	0.208	3
Orthoptera							
Gryllodes sigillatus	3:1	1:3	% hatch	97.1 ± 1.0 (12)	94.1 ± 2.4 (12)	−0.228	4
	5:1	1:5		97.7 ± 0.6 (12)	97.0 ± 1.0 (12)	−0.115	
			offspring viability	−5.0 ± 4.1 (12)	7.6 ± 5.7 (12)	0.360	
				−3.6 ± 3.9 (12)	7.8 ± 2.7 (12)	0.474	
Gryllus bimaculatus	4:1	2:2	% hatch	41.0 ± 3.0 (48)	47.0 ± 3.0 (24)	0.166	5
		1:4			53.0 ± 3.0 (16)	0.346	
			offspring viability	65.0 ± 2.0 (47)	69.0 ± 3.0 (23)	−0.084	
					60.0 ± 6.0 (16)	−0.127	
Teleogryllus oceanicus	2:1	1:2	% hatch	48.0 ± 2.0 (34)	55.0 ± 2.0 (64)	0.247	6
			offspring viability	28.0 ± 2.0 (34)	27.0 ± 3.0 (59)	−0.038	
Teleogryllus commodus	4:1	2:2	% hatch	30.0 ± 1.5 (20)	31.9 ± 1.5 (20)	0.138	7
		1:4			30.5 ± 2.2 (20)	0.028	
			offspring viability	46.6 ± 1.8 (20)	44.7 ± 2.4 (20)	−0.103	
					48.7 ± 4.4 (19)	0.092	
	3:1	1:3	% hatch	47.8 ± 2.9 (60)	51.2 ± 2.3 (76)	0.080	

Allonemobius socius	4:1	% hatch	19.0 ± 4.0 (24)	55.0 ± 6.0 (24)	0.669	8
		offspring viability	18.0 ± 4.0 (24)	30.0 ± 6.0 (24)	0.203	
Coleoptera						
Tenebrio molitor	4:1	% hatch	—(40)	—(43)	0.057	9
		# offspring	65.2 ± 5.5 (43)	85.4 ± 5.5 (43)	0.276	
		offspring viability	—(43)	—(43)	0.022	
Callosobruchus maculatus	2:1	egg-adult viability	91.0 ± 0.02 (30)	89.0 ± 0.0 (35)	−0.086	10
	3:1		86.0 ± 0.02 (27)	77.0 ± 0.0 (34)	−0.356	
Mecoptera						
Panorpa vulgaris	2:1	% hatch	34.0 ± 3.0 (23)	50.0 ± 7.0 (23)	0.464	11
Pseudoscorpionida						
Cordylochernes scorpioides	2:1	% aborted broods	36.0 ± 4.0 (69)	24.0 ± 3.0 (68)	0.206	12
		# offspring	181 ± 15 (72)	137 ± 13 (77)	0.235	

Effect size correlations calculated from the mean and standard deviations contrasting monandrous with polyandrous treatments and transformed to Fisher's Z_r (Rosenthal 1991). 1. Kamimura 2003; 2. Tregenza et al. 2003; 3. Baker et al. 2001; 4. Residual mean offspring viabilities controlling for female age, Ivy & Sakaluk 2005; 5. Tregenza & Wedell 1998; 6. Simmons 2001a; 7. M. Jennions, unpublished observations; 8. Fedorka & Mousseau 2002; 9. Mean values for hatching success and offspring viability were unavailable, and Z_r was calculated from the probability values for the statistical tests, Worden & Parker 2002; 10. Egg-adult viability encompasses both hatching success and offspring viability, Eady et al. 2000; 11. L. Enqvist, unpublished observation; 12. Newcomer et al. 1999.

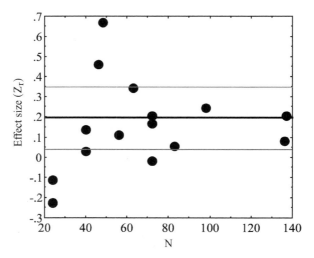

Figure 1   Plot of the size of the effect (Fisher's Z_r) of polyandry on hatching success against sample size (number of females across all treatments) using data from Table 1. The mean effect size and the 95% CIs on the mean are calculated after averaging multiple values derived from a single species and are indicated by the *bold* and *gray parallel lines*, respectively. The mean Z_r differs significantly from zero (0.192 ± 0.069, $t_{10} = 2.79$, $P = 0.019$). Data for *Callosobruchus maculatus* are not included here because hatching success and offspring viability were both incorporated into the estimated effect. Assuming the entire effect in this species was due to hatching success will reduce the mean Z_r but it remains significantly greater than zero.

cost of inbreeding. However, females mated to one full sibling and one unrelated male had the same hatching success as those that had mated with two unrelated males. Tregenza & Wedell (2002) argued that sperm from the unrelated male were used preferentially because random sperm usage would have generated a hatching success intermediate between that experienced by females mating with unrelated and sibling males. They later showed that paternity of hatched offspring from females mating with both a sibling and an unrelated male was biased toward the unrelated male, although only when the unrelated male mated first (Bretman et al. 2004). Interestingly, female *G. bimaculatus* show precopulatory preferences based on male relatedness that are consistent with Tregenza & Wedell's findings for postcopulatory preferences; females prefer to mate with unrelated males over siblings and retain the spermatophores of unrelated males longer (Simmons 1989, 1991).

Despite the compelling evidence for inbreeding avoidance benefits of polyandry in *G. bimaculatus*, similar studies from two other taxa failed to provide support for this form of genetic incompatibility generating the observed embryo viability benefits of polyandry. Using identical experimental designs, studies of the cricket *Teleogryllus commodus* (Jennions et al. 2004) and the guppy, *Poecila reticulata*

(T.E. Pitcher, unpublished observations), found that though inbreeding did reduce hatching success or number of live-born young, respectively, females mated with one sibling and one unrelated male did not have higher embryo viability than expected by random use of sperm from each male.

Only one study has attempted to test directly for intrinsic sire effects on embryo viability. In their study of the cricket *T. oceanicus*, García-González & Simmons (2005) used a paternal half-sib breeding design to estimate the amount of additive genetic variation in male effects on hatching success. They found significant sire effects, estimating the coefficient of additive genetic variation in sire-induced hatching success to be in the region of 20% and the heritability of sire-induced hatching success to be 46%. Moreover, the sire effect on hatching success was not influenced by experimentally manipulated variation in female genotypes, and there was no sire-by-dam interaction, supporting their argument that intrinsic sire effects are more important in this system than genetic incompatibilities.

MATERNAL EFFECTS?

Although the experimental design suggested by Tregenza & Wedell (1998) controls for possible benefits of multiple copulations, such as differences in sperm supply that might influence hatching success, it does not control for potential maternal effects. Maternal effects arise when females vary the amount of resources they provide for their offspring and can have profound influences on offspring performance (Mousseau & Fox 1998, Qvarnström & Price 2001). Importantly, females can vary their investment in eggs depending on the phenotype of their mating partners and thereby selectively influence the quality of offspring produced (Cunningham & Russell 2000, Sheldon 2000). In the cricket studied by Tregenza & Wedell (1998), females invest more heavily in offspring production if allowed to choose a mate rather than being allocated one (Simmons 1987). The possibility of being able to choose a mate cryptically could influence a female's investment in egg production, thereby explaining the differences in hatching success of eggs produced by polyandrous and monandrous females. Differences in maternal allocation to offspring can seriously confound studies of genetic benefits associated with female mating patterns (Sheldon 2000). For example, in their study of dung beetles, *Onthophagus taurus*, Kotiaho et al. (2003) found that genetic sire effects on male horn length were considerably inflated by differential maternal effects. In this species body size and horn length are strongly dependent on maternal provisioning, and females provide more resources for their offspring after mating with large, long-horned males.

In many insects, accessory gland products, or Acps, are incorporated into the ejaculate and transferred to the female with sperm at copulation. Acps influence female reproduction in a dose-dependent manner, often via an increase in resource allocation to egg production (Simmons 2001b). For example, the accessory glands of crickets secrete over 30 protein fractions (Kaulenas et al. 1975). Many of these proteins are transferred to the female in the ejaculate where they are converted, by prostaglandin synthetase that is secreted by the testes (Destephano & Brady 1977), into prostaglandins that stimulate an increased investment in egg production by the mated female (Stanley-Samuelson & Loher 1986, Stanley-Samuelson et al. 1986). If males show variation in the composition or potency of their Acps, this could generate variation in maternal investment in eggs and in embryo viability. Indeed, in their study of the bean weevil *Callosobruchus maculatus*, Eady et al. (2000) found that polyandrous females laid more eggs than monandrous females even though both groups received the same numbers of matings. Likewise, the number of offspring produced by polyandrous female flour beetles, *Tenebrio molitor*, was greater than for monandrous females (Worden & Parker 2002). Again, given the lack of a significant effect of mating regime on hatching success in this later study, the effect must have arisen predominantly from a greater egg-laying response from the variety of males mated in the polyandrous group. Importantly, in their study of *T. oceanicus*, García-González & Simmons (2005) found a strong and positive genetic correlation between sire effects on hatching success and sire accessory gland weight. Thus, sire effects on offspring performance may be indirect via their ability to influence maternal investment in reproduction. Polyandry could reduce the variance in quantity and/or quality of Acps acquired by females and explain the increased embryo viability of polyandrous females without invoking genetic mechanisms.

In their study of the cricket *Gryllus firmus*, Weigensberg et al. (1998) used a recessive mutation for pale eye color as a marker to examine paternal genetic effects and maternal effects induced by males, via their Acps, on the prehatching embryonic development of offspring produced by females that had mated to both a pale-eyed and a wild-type male. Within clutches there was no differences in the viability of embryos to the eye spot stage of development or in the proportion of each male's embryos that hatched. However, Weigensberg et al. (1998) did find significant genetic effects of sire identity on embryo growth at 10 days and on hatchling size; embryos sired by wild-type males were generally larger than those sired by pale-eyed males. Moreover, they also found significant nongenetic effects on embryo size; egg size at laying and embryo size at 10 days, irrespective of sire identity, were both negatively dependent on the size of the pale-eyed male. Thus, genetic sire effects and male-induced maternal effects both appear to influence embryonic growth in this cricket.

Maternal half-sib designs offer a powerful way with which to control for potentially confounding maternal effects. Using alpine whitefish, *Coregonus* sp., Wedekind et al. (2001) crossed 10 males and 10 females in all possible pairwise combinations by in vitro fertilization. They recorded early embryo mortality, which was associated with developmental problems, and late embryo mortality that was associated with a bacterial infection. They found a significant male-by-female interaction effect on early embryo mortality consistent with partial genetic incompatibilities. However, for late embryo mortality they found significant sire and dam effects but no interaction. Survival of bacterial infection was correlated with sire ornamentation, consistent with intrinsic sire effects on offspring viability that are reliably indicated by male secondary sexual characteristics. Interestingly, late embryo survival depends strongly on MHC class II genotype, but there is no indication that eggs are able to selectively fuse with sperm that would enhance embryo viability (Wedekind et al. 2004). In a similar experiment Rudolfsen et al. (2005) found strong and significant maternal effects and a strong sire-by-dam interaction effect on embryo mortality in Atlantic cod, *Gadus morhua*, indicating that in this species genetic incompatibilities may be more important than intrinsic sire effects.

The use of ionizing radiation to cull nonfocal embryos may also prove useful for controlling maternal effects. In bulb mites, *Rhizoglyphus robini*, Konior et al. (2001) found that polyandrous females produce daughters of greater fecundity. However, they did not control fully for mating frequency or for potential maternal effects. Subsequently, Kozielska et al. (2004) irradiated all but one of the mates of polyandrous females so that the offspring they produced were sired by the same male that had been mated to a monandrous female counterpart. Thus, females in both treatments were genetically monandrous. In this experiment there was no effect of "pseudopolyandry" on daughter fecundity, indicating that maternal effects associated with polyandry could not explain their previous result. Moreover, they found a positive association between sire fertilization success and the fecundity of daughters produced by both polyandrous and monandrous females, suggesting that genetically based intrinsic sire effects influence daughter performance. In contrast, the sons of "pseudopolyandrous" females had a lower sperm competitive success, suggesting that maternal effects influence son performance.

CONCLUSIONS

This review suggests that increased embryo viability may prove to be a general benefit derived from polyandry, at least in insects, and perhaps more broadly (see Stockley 2003), although the mechanisms behind the effect are far from clear. Increased effort is required in several areas before progress can be made.

Alternative explanations for the observed benefits of polyandry are not easily distinguishable. Although available data on the effects of polyandry are taken as support for genetic benefits, the alternative hypothesis that maternal effects underlie offspring reproductive performance can, in general, not yet be rejected. Research programs that control for potential maternal effects are required. A promising approach is the use of maternal half-sib designs, which are most easily accomplished with externally fertilizing species such as broadcast spawning invertebrates and externally fertilizing fish and amphibians. Recent studies aimed at elucidating the genetic benefits of precopulatory female choice using maternal half-sib designs are providing valuable information on genetic sire and sire-by-dam interaction effects (Barber et al. 2001, Sheldon et al. 2003, Wedekind et al. 2001, Welch et al. 1998).

Maternal half-sib designs are clearly more challenging with internally fertilizing species. Nevertheless, innovative use of molecular (Sheldon et al. 1997) or morphological markers (Weigensberg et al. 1998) offers an avenue for such studies. Although the use of irradiation may also prove useful for examining maternal and genetic effects on postembryonic offspring performance (Kozielska et al. 2004), the fact that it induces early embryonic mortality makes it less useful for partitioning maternal and genetic effects on hatching success.

Assuming that the observed effects of polyandry are indeed genetically based, the alternative explanations of intrinsic sire effects and genetic incompatibility are not easily distinguished with the available data. More effort must be made in examining intrinsic sire effects on hatching success, and large-scale experimental designs will be required given the rather small general effect size. Again, maternal half-sib designs offer a powerful means with which to partition the effects of sire genotypes and compatible genotypes on offspring viability. It seems likely that both genetic incompatibilities and intrinsic sire effects will prove to influence offspring performance, but it is essential to establish the relative importance of each process in a particular system as this will have implications for the strength of sexual selection (Mays & Hill 2004).

Finally, if females are to obtain genetic benefits from polyandry, fertilization success must be biased, either through superior sperm competitiveness or through sperm selection by females, toward males whose genotypes provide superior offspring fitness (Yasui 1998, Zeh & Zeh 1997). Although there is some good evidence for sperm–egg interaction effects on patterns of fertilization in broadcast spawning invertebrates, generally there is very little unequivocal evidence that paternity is biased toward males with compatible and/or intrinsically superior genotypes. The majority of studies assign paternity at a late stage of offspring development so that the patterns observed are completely confounded with the expected consequences of genetic incompatibilities and/or intrinsic sire effects on embryo viability. In order to demonstrate unequivocally that patterns of paternity arise because of biased fertilization rather than biased offspring viability, researchers need to assign paternity at a much earlier stage in offspring development, before the onset of embryo mortality. Another approach would be to assess offspring viability from particular parental genotypes independently of sperm competition trials so that patterns of paternity could be adjusted to reflect known differences in mortality prior to the point of paternity assignment (Bretman et al. 2004, Mack et al. 2002). This approach should prove particularly accurate in externally fertilizing species where eggs and sperm from the individuals to be used in sperm competition trials could be combined in a noncompetitive assay to establish the combined effects of parental genotypes and their interactions on embryo viability.

The study of sexual selection and sperm competition is undergoing a paradigm shift toward an integration of female perspectives. There is no doubt that Bateman's principle has served us well in developing a broad understanding of mating system evolution from a male perspective (Shuster & Wade 2003). However, the evidence is growing that, contrary to Bateman's principle, female fitness, like male fitness, can increase with number of mates. Time will tell just how profound an impact this will have on our understanding of mating systems, and considerable research effort is required to understand the mechanisms underlying the fitness advantage associated with female multiple mating.

ACKNOWLEDGMENTS

I thank Paco García-González and Jonathan Evans for comments on earlier drafts of the manuscript, all those authors who provided original published and/or unpublished data, and the Australian Research Council for financial support.

The *Annual Review of Ecology, Evolution, and Systematics* is online at http://ecolsys.annualreviews.org

LITERATURE CITED

Aeschlimann PB, Häberli MA, Reusch TBH, Boehm T, Milinski M. 2003. Female sticklebacks *Gasterosteus aculeatus* use self-reference to optimize MHC allele number during mate selection. *Behav. Ecol. Sociobiol.* 54:119–26

Arnold SJ, Duvall D. 1994. Animal mating systems: a synthesis based on selection theory. *Am. Nat.* 143:317–48

Arnqvist G, Nilsson T. 2000. The evolution of polyandry: multiple mating and female fitness in insects. *Anim. Behav.* 60:145–64

Atlan A, Joly D, Capillon C, Montchamp-Moreau C. 2004. *Sex-ratio* distorter of *Drosophila simulans* reduces male productivity and sperm competitive ability. *J. Evol. Biol.* 17:744–51

Baker RH, Ashwell RIS, Richards TA, Fowler K, Chapman T, Pomiankowski A. 2001. Effects of multiple mating and male eye span on female reproductive output in the stalk-eyed fly, *Cyrtodiopsis dalmanni*. *Behav. Ecol.* 12:732–39

Barber I, Arnott SA, Braithwaite VA, Andrew J, Huntingford FA. 2001. Indirect fitness consequences of mate choice in sticklebacks: offspring of brighter males grow slowly but resist parasitic infections. *Proc. R. Soc. London Ser. B* 268:71–76

Bateman AJ. 1948. Intrasexual selection in *Drosophila*. *Heredity* 2:349–68

Birkhead TR, Chaline N, Biggins JD, Burke T, Pizzari T. 2004. Nontransitivity of paternity in birds. *Evolution* 58:416–20

Birkhead TR, Møller AP, eds. 1998. *Sperm Competition and Sexual Selection*. London: Academic

Bishop JDD, Jones CS, Noble LR. 1996. Female control of paternity in the internally fertilizing compound ascidian Diplosoma listerianum. II. Investigation of male mating success using RAPD markers. *Proc. R. Soc. London Ser. B* 263:401–7

Bretman A, Wedell N, Tregenza T. 2004. Molecular evidence of post-copulatory inbreeding avoidance in the field cricket *Gryllus bimaculatus*. *Proc. R. Soc. London Ser. B* 271:159–64

Brown JL. 1997. A theory of mate choice based on heterozygosity. *Behav. Ecol.* 8:60–65

Byrne PG, Roberts JD. 2000. Does multiple paternity improve fitness of the frog *Crinia georgiana*? *Evolution* 54:968–73

Calsbeek R, Sinervo B. 2004. Within-clutch variation in offspring sex determined by differences in sire body size: cryptic mate choice in the wild. *J. Evol. Biol.* 17:464–70

Chapman T, Liddle LF, Kalb JM, Wolfner MF, Partridge L. 1995. Cost of mating in *Drosophila melanogaster* females is mediated by male accessory gland products. *Nature* 373:241–44

Clark AG, Begun DJ, Prout T. 1999. Female × male interactions in *Drosophila* sperm competition. *Science* 283:217–20

Clark AG, Dermitzakis ET, Civetta A. 2000. Nontransitivity of sperm precedence in *Drosophila*. *Evolution* 54:1030–35

Crnokrak P, Roff DA. 1995. Dominance variance: association with selectin and fitness. *Heredity* 75:530–40

Cunningham EJA, Russell AF. 2000. Egg investment is influenced by male attractiveness in the mallard. *Nature* 404:74–76

Destephano DB, Brady UE. 1977. Prostaglandin and prostaglandin synthetase in the cricket, *Acheta domesticus. J. Insect Physiol.* 23:905–11

Eady PE, Wilson N, Jackson M. 2000. Copulating with multiple mates enhances female fecundity but not egg-to-adult survival in the bruchid beetle *Callosobruchus maculatus. Evolution* 54:2161–65

Eberhard WG. 1996. *Female Control: Sexual Selection by Cryptic Female Choice.* Princeton: Princeton Univ. Press

Edwards RG. 1955. Selective fertilization following the use of sperm mixtures in the mouse. *Nature* 175:215–23

Erickson RP. 1990. Post-meiotic gene expression. *Trends Genet.* 6:264–69

Evans JP, Marshall DJ. 2005. Male-by-female interactions influence fertilization success and mediates the benefits of polyandry in the sea urchin *Heliocidaris erythrogramma. Evolution* 59:106–12

Evans JP, Zane L, Francescato S, Pilastro A. 2003. Directional postcopulatory sexual selection revealed by artificial insemination. *Nature* 421:360–63

Fedorka KM, Mousseau TA. 2002. Material and genetic benefits of female multiple mating and polyandry. *Anim. Behav.* 64:361–67

Fisher DO, Double MC, Moore BD. 2005. Number of mates and timing of mating affect offspring growth in the small marsupial, *Antichinus agilis. Anim. Behav.* In press

Fjerdingstad EJ, Keller L. 2004. Relationships between phenotype, mating behavior, and fitness of queens in the ant *Lasius niger. Evolution* 58:1056–63

Froman DP, Pizzari T, Feltmann AJ, Castillo-Juarez H, Birkhead TR. 2002. Sperm mobility: mechanisms of fertilizing efficiency, genetic variation and phenotypic relationship with male status in the domestic fowl, *Gallus gallus domesticus. Proc. R. Soc. London Ser. B* 269:607–12

García-González F, Simmons LW. 2005. The evolution of polyandry: intrinsic sire effects contribute to embryo viability. *J. Evol. Biol.* 18:1097–1103

Gregory PG, Howard DJ. 1994. A postinsemination barrier to fertilization isolates two closely related ground crickets. *Evolution* 48:705–10

Hooglund JL. 1998. Why do female Gunnison's prairie dogs copulate with more than one male? *Anim. Behav.* 55:351–59

Hosken DJ, Garner TWJ, Ward PI. 2001. Sexual conflict selects for male and female reproductive characters. *Curr. Biol.* 11:1–20

Howard DJ. 1999. Conspecific sperm and pollen precedence and speciation. *Annu. Rev. Ecol. Syst.* 30:109–32

Howard DJ, Gregory PG. 1993. Post-insemination signalling systems and reinforcement. *Philos. Trans. R. Soc. London Ser. B* 340:231–36

Hughes WOH, Boomsma JJ. 2004. Genetic diversity and disease resistance in leaf-cutting ant societies. *Evolution* 58:1251–60

Ivy TM, Sakaluk SK. 2005. Polyandry promotes enhanced offspring survival in decorated crickets. *Evolution* 59:152–59

Jennions MD, Hunt J, Graham R, Brooks R. 2004. No evidence for inbreeding avoidance through postcopulatory mechanisms in the black field cricket, *Teleogryllus commodus. Evolution* 58:2472–77

Jennions MD, Petrie M. 2000. Why do females mate multiply? A review of the genetic benefits. *Biol. Rev.* 75:21–64

Kamimura Y. 2003. Effects of repeated mating and polyandry on the fecundity, fertility and maternal behaviour of female earwigs, *Euborellia plebeja. Anim. Behav.* 65:205–14

Kaulenas MS, Yenofsky RL, Potswald HE, Burns AL. 1975. Protein synthesis by the Accessory gland of male house cricket, *Acheta domesticus*. *J. Exp. Zool.* 193:21–36

Keller L, Reeve HK. 1995. Why do females mate with multiple males? The sexually selected sperm hypothesis. *Adv. Stud. Behav.* 24:291–315

Kempenaers B, Congdon B, Boag P, Robertson RJ. 1999. Extrapair paternity and egg hatchability in tree swallows: evidence for the genetic compatibility hypothesis? *Behav. Ecol.* 10:304–11

Kempenaers B, Verheyen GR, Van den Broeck M, Burke T, Van Broekhoven C, Dhondt AA. 1992. Extra-pair paternity results from female preference for high-quality males in the blue tit. *Nature* 357:494–96

Konior M, Radwan J, Kolodziejczyk M. 2001. Polyandry increases offspring fecundity in the bulb mite. *Evolution* 55:1893–96

Kotiaho JS, Simmons LW, Hunt J, Tomkins JL. 2003. Males influence maternal effects that promote sexual selection: A quantitative genetic experiment with dung beetles *Onthophagus taurus*. *Am. Nat.* 161:852–59

Kozielska M, Krzeminska A, Radwan J. 2004. Good genes and the maternal effects of polyandry on offspring reproductive success in the bulb mite. *Proc. R. Soc. London Ser. B* 271:165–70

Levitan DR. 2002. The relationship between conspecific fertilization success and reproductive isolation among three congeneric sea urchins. *Evolution* 56:1599–609

Lewis SM, Austad SN. 1990. Sources of intraspecific variation in sperm precedence in red flour beetles. *Am. Nat.* 135:351–59

Mack PD, Hammock BA, Promislow DEL. 2002. Sperm competition ability and genetic relatedness in *Drosophila melanogaster*: similarity breeds contempt. *Evolution* 56: 1789–95

Madsen T, Shine R, Loman J, Hakansson T. 1992. Why do female adders copulate so frequently? *Nature* 355:440–41

Marshall RC, Buchanan KL, Catchpole CK. 2003. Sexual selection and individual genetic diversity in a songbird. *Proc. R. Soc. London Ser. B* 270(Suppl. 2):S248–50

Matton DP, Nass N, Clarke AE, Newbigin E. 1994. Self-incompatibility—how plants avoid illegitimate offspring. *Proc. Natl. Acad. Sci. USA* 91:1992–97

Mays HL, Hill GE. 2004. Choosing mates: good genes versus genes that are a good fit. *Trends Ecol. Evol.* 19:554–59

Merilä J, Sheldon BC. 1999. Genetic architecture of fitness and nonfitness traits: empirical patterns and development of ideas. *Heredity* 83:103–9

Metz EC, Palumbi SR. 1996. Positive selection and sequence rearrangements generate extensive polymorphism in the gamete recognition protein bindin. *Mol. Biol. Evol.* 13:397–406

Møller AP, Alatalo RV. 1999. Good-genes effects in sexual selection. *Proc. R. Soc. London Ser. B* 266:85–91

Moore PJ, Harris WE, Montrose VT, Levin D, Moore AJ. 2004. Constraints on evolution and postcopulatory sexual selection: Trade-offs among ejaculate characteristics. *Evolution* 58:1773–80

Mousseau TA, Fox CW. 1998. The adaptive significance of maternal effects. *Trends Ecol. Evol.* 13:403–7

Newcomer SD, Zeh JA, Zeh DW. 1999. Genetic benefits enhance the reproductive success of polyandrous females. *Proc. Natl. Acad. Sci. USA* 96:10236–41

Olsson M, Gullberg A, Tegelström H, Madsen T, Shine R. 1994. Can female adders multiple? *Nature* 369:528

Olsson M, Madsen T, Ujvari B, Wapstra E. 2004. Fecundity and MHC affects ejaculation tactics and paternity bias in sand lizards. *Evolution* 58:906–9

Olsson M, Shine R, Gullberg A, Madsen T, Tegelström H. 1996. Sperm selection by females. *Nature* 383:585

Pai A, Yan G. 2002. Polyandry produces sexy sons at the cost of daughters in red flour beetles. *Proc. R. Soc. London Ser. B* 269:361–68

Palumbi SR. 1999. All males are not created equal: fertility differences depend on gamete recognition polymorphisms in sea urchins. *Proc. Natl. Acad. Sci. USA* 96:12632–37

Parker GA. 1970. Sperm competition and its evolutionary consequences in the insects. *Biol. Rev.* 45:525–67

Parker GA. 1979. Sexual selection and sexual conflict. In *Sexual Selection and Reproductive Competition in Insects*, ed. MS Blum, NA Blum, pp. 123–66. London: Academic

Parker GA. 1992. Snakes and female sexuality. *Nature* 355:395–96

Parker GA. 1998. Sperm competition and the evolution of ejaculates: Towards a theory base. See Birkhead & Møller 1998, pp. 3–54

Penn DJ. 2002. The scent of genetic compatibility: sexual selection and the major histocompatibility complex. *Ethology* 108:1–21

Penn DJ, Damjanovich K, Potts WK. 2002. MHC heterozygosity confers a selective advantage against multiple-strain infections. *Proc. Natl. Acad. Sci. USA* 99:11260–64

Pizzari T, Birkhead TR. 2002. The sexually-selected sperm hypothesis: sex-biased inheritance and sexual antagonism. *Biol. Rev.* 77:183–209

Pusey A, Wolf M. 1996. Inbreeding avoidance in animals. *Trends Ecol. Evol.* 11:201–6

Qvarnström A, Price TD. 2001. Maternal effects, paternal effects and sexual selection. *Trends Ecol. Evol.* 16:95–100

Radwan J. 1998. Heritability of sperm competition success in the bulb mite, *Rhizoglyphus robini. J. Evol. Biol.* 11:321–28

Reusch TBH, Häberli MA, Aeschlimann PB, Milinski M. 2001. Female sticklebacks count alleles in a strategy of sexual selection explaining MHC polymorphism. *Nature* 414:300–2

Rosenthal R. 1991. *Meta-analytical Procedures for Social Research.* London: Sage

Rowe L. 1994. The costs of mating and mate choice in water striders. *Anim. Behav.* 48:1049–56

Rowe L, Houle D. 1996. The lek paradox and the capture of genetic variance by condition dependent traits. *Proc. R. Soc. London Ser. B* 263:1415–21

Rudolfsen G, Figenschou L, Folstad I, Nordeide JT, Søreng E. 2005. Potential fitness benefits from mate selection in the Atlantic cod (*Gadus morhua*). *J. Evol. Biol.* 18:172–79

Scofield VL, Schlumpberger JM, West LA, Weissman IL. 1982. Protochordate allorecognition is controlled by a MHC-like gene system. *Nature* 295:499–502

Sheldon BC. 2000. Differential allocation: tests, mechanisms and implications. *Trends Ecol. Evol.* 15:397–402

Sheldon BC, Arponen H, Laurila A, Crochet P-A, Merilä J. 2003. Sire coloration influences offspring survival under predation risk in the moorfrog. *J. Evol. Biol.* 16:1288–95

Sheldon BC, Merilä J, Qvarnström A, Gustafsson L, Ellegren H. 1997. Paternal genetic contribution to offspring condition predicted by size of male secondary sexual character. *Proc. R. Soc. London Ser. B* 264:297–302

Shuster SM, Wade MJ. 2003. *Mating Systems and Strategies.* Princeton: Princeton Univ. Press

Simmons LW. 1987. Female choice contributes to offspring fitness in the field cricket, *Gryllus bimaculatus* (De Geer). *Behav. Ecol. Sociobiol.* 21:313–21

Simmons LW. 1989. Kin recognition and its influence on mating preferences of the field cricket, *Gryllus bimaculatus* (De Geer). *Anim. Behav.* 38:68–77

Simmons LW. 1991. Female choice and the relatedness of mates in the field cricket, *Gryllus bimaculatus. Anim. Behav.* 41:493–501

Simmons LW. 2001a. The evolution of polyandry: an examination of the genetic incompatibility and good-sperm hypotheses. *J. Evol. Biol.* 14:585–94

Simmons LW. 2001b. *Sperm Competition and its Evolutionary Consequences in the Insects.* Princeton: Princeton Univ. Press

Simmons LW. 2003. The evolution of polyandry: patterns of genotypic variation in female mating frequency, male fertilization success and a test of the sexy-sperm hypothesis. *J. Evol. Biol.* 16:624–34

Simmons LW, Kotiaho JS. 2002. Evolution of ejaculates: patterns of phenotypic and genotypic variation and condition dependence in sperm competition traits. *Evolution* 56:1622–31

Sivinski J. 1984. Sperm in competition. In *Sperm Competition and the Evolution of Animal Mating Systems*, ed. RL Smith, pp. 86–115. London: Academic

Stanley-Samuelson DW, Loher W. 1986. Prostaglandins in insect reproduction. *Ann. Entomol. Soc. Am.* 79:841–53

Stanley-Samuelson DW, Peloquin JJ, Loher W. 1986. Egg-laying in response to prostaglandin injections in the Australian field cricket, *Teleogryllus commodus. Physiol. Entomol.* 11:213–19

Stockley P. 2003. Female multiple mating behavior, early reproductive failure and litter size variation in mammals. *Proc. R. Soc. London Ser. B* 270:271–78

Stouthamer R, Breeuwer JAJ, Hurst GDD. 1999. *Wolbachia pipientis*: Microbial manipulator or arthropod reproduction. *Annu. Rev. Microbiol.* 53:71–102

Thornhill R. 1983. Cryptic female choice and its implications in the scorpionfly *Harpobittacus nigriceps. Am. Nat.* 122:765–88

Thrall PH, Antonovics J, Dobson AP. 2000. Sexually transmitted diseases in polygynous mating systems: prevalence and impact on reproductive success. *Proc. R. Soc. London Ser. B* 267:1555–63

Tomkins JL, Radwan J, Kotiaho JS, Tregenza T. 2004. Genic capture and resolving the lek paradox. *Trends Ecol. Evol.* 19:323–28

Tregenza T, Wedell N. 1998. Benefits of multiple mates in the cricket *Gryllus bimaculatus. Evolution* 52:1726–30

Tregenza T, Wedell N. 2000. Genetic compatibility, mate choice and patterns of parentage. *Mol. Ecol.* 9:1013–27

Tregenza T, Wedell N. 2002. Polyandrous females avoid costs of inbreeding. *Nature* 415:71–73

Tregenza T, Wedell N, Hosken DJ, Ward PI. 2003. Maternal effects on offspring depend on female mating pattern and offspring environment in yellow dung flies. *Evolution* 57:297–304

Vacquier VD. 1998. Evolution of gamete recognition proteins. *Science* 281:1995–98

Watson PJ. 1991. Multiple paternity as genetic bet-hedging in female sierra dome spiders, *Linyphila litigiosa* (Linyphiidae). *Anim. Behav.* 41:343–60

Wedekind C, Chapuisat M, Macas E, Rülicke T. 1996. Non-random fertilization in mice correlates with the MHC and something else. *Heredity* 77:400–9

Wedekind C, Müller R, Spicher H. 2001. Potential genetic benefits of mate selection in whitefish. *J. Evol. Biol.* 14:980–86

Wedekind C, Seebeck T, Bettens F, Paepke AJ. 1995. MHC-dependent mate preferences in humans. *Proc. R. Soc. London Ser. B* 260: 245–49

Wedekind C, Walker M, Portmann J, Cenni B, Müller R, Binz T. 2004. MHC-linked susceptibility to a bacterial infection, but no MHC-linked cryptic female choice in whitefish. *J. Evol. Biol.* 17:11–18

Wedell N, Gage MJG, Parker GA. 2002. Sperm competition, male prudence and sperm-limited females. *Trends Ecol. Evol.* 17:313–20

Wegner KM, Kalbe M, Kurtz J, Reusch TBH, Milinski M. 2003. Parasite selection for immunogenetic optimality. *Science* 301: 1343

Weigensberg I, Carriére Y, Roff DA. 1998. Effects of male genetic contribution and paternal investment to egg and hatchling size in the cricket, *Gryllus firmus. J. Evol. Biol.* 11:135–46

Welch AM, Semlitsch RD, Gerhardt HC. 1998. Call duration as an indicator of genetic quality in male gray tree frogs. *Science* 280:1928–30

Wiernasz DC, Perroni CL, Cole BJ. 2004. Polyandry and fitness in the western harvester ant, *Pogonomyrmex occidentalis. Mol. Ecol.* 13:1601–6

Wilkinson GS, Fry CL. 2001. Meiotic drive alters sperm competitive ability in stalk-eyed flies. *Proc. R. Soc. London Ser. B* 268:2559–64

Wilkinson GS, Swallow SJ, Christenson SJ,

Madden K. 2003. Phylogeography of sex ratio and multiple mating in stalk-eyed flies from southeast Asia. *Genetica* 117:37–46

Wilson N, Tubman SC, Eady PE, Robertson G. 1997. Female genotype affects male success in sperm competition. *Proc. R. Soc. London Ser. B* 264:1491–95

Worden BD, Parker PG. 2002. Polyandry in grain beetles, *Tenebrio molitor*, leads to greater reproductive success: material or genetic benefits? *Behav. Ecol.* 12:761–67

Wu C-I. 1983. Virility deficiency and the sex-ratio trait in *Drosophila pseudoobscura*. I. Sperm displacement and sexual selection. *Genetics* 105:651–62

Yamazaki K, Boyse EA, Mike V, Thaler HT, Mathieson BJ, et al. 1976. Control of mating preferences in mice by genes in the major histocompatibility complex. *J. Exp. Med.* 144:1324–35

Yasui Y. 1997. A "good-sperm" model can explain the evolution of costly multiple mating by females. *Am. Nat.* 149:573–84

Yasui Y. 1998. The "genetic benefits" of female multiple mating reconsidered. *Trends Ecol. Evol.* 13:246–50

Yasui Y. 2001. Female multiple mating as a genetic bet-hedging strategy when mate choice criteria are unreliable. *Ecol. Res.* 16:605–16

Zeh JA. 1997. Polyandry and enhanced reproductive success in the harlequin-beetle-riding pseudoscorpion. *Behav. Ecol. Sociobiol.* 40:111–18

Zeh JA, Zeh DW. 1996. The evolution of polyandry I: intragenomic conflict and genetic incompatibility. *Proc. R. Soc. London Ser. B* 263:1711–17

Zeh JA, Zeh DW. 1997. The evolution of polyandry II: post-copulatory defences against genetic incompatibility. *Proc. R. Soc. London Ser. B* 264:69–75

Zeh JA, Zeh DW. 2001. Reproductive mode and the genetic benefits of polyandry. *Anim. Behav.* 61:1051–63

Annu. Rev. Ecol. Evol. Syst. 2005. 36:147–68
doi: 10.1146/annurev.ecolsys.36.102003.152644
First published online as a Review in Advance on August 11, 2005

INDIVIDUAL-BASED MODELING OF ECOLOGICAL AND EVOLUTIONARY PROCESSES[1]

Donald L. DeAngelis[1] and Wolf M. Mooij[2]

[1]USGS, Florida Integrated Science Centers, University of Miami, Coral Gables, Florida,
33124; email: ddeangelis@bio.miami.edu
[2]Netherlands Institute of Ecology, Center for Limnology,
The Netherlands; email: w.mooij@nioo.knaw.nl

Key Words phenotypic variation, spatially explicit, size dependent, learning, emergent properties

■ **Abstract** Individual-based models (IBMs) allow the explicit inclusion of individual variation in greater detail than do classical differential-equation and difference-equation models. Inclusion of such variation is important for continued progress in ecological and evolutionary theory. We provide a conceptual basis for IBMs by describing five major types of individual variation in IBMs: spatial, ontogenetic, phenotypic, cognitive, and genetic. IBMs are now used in almost all subfields of ecology and evolutionary biology. We map those subfields and look more closely at selected key papers on fish recruitment, forest dynamics, sympatric speciation, metapopulation dynamics, maintenance of diversity, and species conservation. Theorists are currently divided on whether IBMs represent only a practical tool for extending classical theory to more complex situations, or whether individual-based theory represents a radically new research program. We feel that the tension between these two poles of thinking can be a source of creativity in ecology and evolutionary theory.

INTRODUCTION

The past decade and a half has been a period of enormous growth of individual-based modeling (IBM) in ecology and evolutionary biology. A survey of articles with "individual-based" (or "individual-oriented") and "model" in the title, abstract, or keywords showed a linear increase from 1 in 1990 to approximately 150 in 2004. Not all papers about IBM use these keywords, and papers on the IBM approach go back at least to the1960s, so the actual number of studies that employ IBM is considerably higher. An objective of our review is to assess whether this numerical increase corresponds to real scientific progress.

---

[1]The U.S. Government has the right to retain a nonexclusive, royalty-free license in and to any copyright covering this paper.

IBM has expanded across a wide swath of ecology and evolutionary biology; it parallels the growth of related "agent-based" models of economics, social science, and artificial intelligence and of particle-based models in physics (Hockney & Eastwood 1989). An early review of IBM in ecology (Huston et al. 1988) pointed to the promise of IBM to provide a unified approach to applied and theoretical questions in ecology. However, a minireview by Grimm (1999) of IBMs in ecology (N = 50) showed that "most of the IBMs of the period 1990–1999 did not directly address general questions, but were narrower in their scope" or "pragmatic."

The field has also been reviewed in Hogeweg & Hesper (1990), DeAngelis & Gross (1992), Uchmanski & Grimm (1996), and Łomnicki (1999). In a recent book, Grimm & Railsback (2005) provide a set of guidelines for building, testing, and analyzing individual-based models. Our objective is more investigative. We ask ourselves, on the basis of the database of approximately 900 references found in our literature survey, among other sources, how has the field of IBM developed over the past decades? We classify IBMs according to their purposes and the way they take into account variation at the individual level. From this perspective, we explore and assess the current state of IBM.

What is an individual-based model? We agree with Grimm & Railsback (2005) that no absolute definition exists. In principle, IBMs simulate populations or systems of populations as being composed of discrete agents that represent individual organisms or groups of similar individual organisms, with sets of traits that vary among the agents. Each agent has a unique history of interactions with its environment and other agents. IBMs attempt to capture the variation among individuals that is relevant to the questions being addressed. Our interpretation of what constitutes an IBM is deliberately broad and inclusive. However, this interpretation is compared with a stricter definition in the Conclusions and Perspectives section below.

We provide a solid conceptual basis for our review by describing five major types of individual variation in IBMs: spatial, ontogenetic, phenotypic, cognitive, and genetic. To survey the breadth of applications of IBM, we next formulate seven groups of biological processes for which IBMs have been developed, and we highlight within each group a number of research foci. To measure the depth of the contributions of IBMs to important questions, we illustrate several areas of high concentrations of IBMs by describing papers that have been particularly influential in those areas. We offer suggestions for design and analysis of IBMs. We conclude that the explicit accounting for individual variation plays a crucial role in the development of ecological and evolutionary theory.

THE NATURE OF INDIVIDUAL VARIATION

Although a balance between parsimony and detail is essential in modeling, the growth in IBM reflects a perception that inclusion of variation among individuals is often indispensable for understanding and prediction at the population, community, and ecosystem levels. The trend toward inclusion of variation at finer resolution is already apparent in extensions of differential-equation and difference-equation models (which, for convenience, we refer to as classical models) to structure within populations, such as age or size classes, and to subpopulations within metapopulations. However, better understanding of some population and community phenomena requires zooming down to even lower levels. The level of resolution of IBMs is the individual organism, so IBMs allow accounting for differences among individuals. We suggest that it is convenient to categorize this resolution as an increasing level of mechanistic detail along each of five directions or axes. The structural characteristics along these axes are (*a*) spatial variability, local interactions, and movement; (*b*) life cycles and ontogenetic development; (*c*) phenotypic variability, plasticity, and behavior; (*d*) differences in experience and learning; and (*e*) genetic variability and evolution. Demographic stochasticity could be considered a sixth category, but we see it as an intrinsic property of any IBM and thus part of each of the above categories.

Variability in Space

Classical predator-prey models, in their functional responses, contain at least an implicit concept of the space in which interactions occur; some metapopulation models incorporate distances between patches; and reaction-diffusion models can simulate the spread of population density explicitly in space. However, classical models cannot take into account discrete individuals, which create local population nonuniformity that can affect population dynamics and ecosystem function (Durrett & Levin 1994). Myers (1976) was one of the first to show, in a spatially explicit IBM of herbivorous insects feeding on plants, that local inhomogeneity, in the form of herbivore outbreaks and extinctions that arise stochastically, could affect total-population dynamics.

Life Cycle Detail

The inclusion of age-structure or stage-structure detail has been dominated by difference-equation (matrix) models, such as Leslie models (Caswell 2000, Leslie 1945), which have proved effective at incorporating important key dynamic features that result from life cycles of organisms. However, a number of practical and theoretical questions require examination of the life cycle in finer detail, such as demographic variability, spatial structure, and short-term temporal variability. Such situations can, in principle, be handled by matrix models but then require partitioning of the population into such a large number of subclasses that IBMs become advantageous.

Phenotypic Variation and Behavior

In addition to the variation within populations caused by ontogenetic changes through the life cycle, variation also results from the unique experiences of each individual. This observation is true for plants, each of which exists on a site with its own characteristics of soil, light, water, and nutrients, as well as for animals, each of which has its own trajectory through the environment and history of feeding, disease, and exposure to other stresses. Łomnicki (1978) recognized the importance of individual phenotypic variation to populations. The earliest genre of IBMs to gain prominence, the forest gap-phase replacement models (Botkin et al. 1972, Shugart 1984), accounted for the phenotypic variations among individual trees caused by effects of shading on growth. Phenotypic variability, even within a cohort of genetically identical organisms born at roughly the same time, poses a fundamental theoretical question of how it affects population dynamics. Classical size-structure models, such as the McKendrick-von Foerster model, have been used to study variable size in populations (e.g., De Roos et al. 1992, Sinko & Streifer 1967) but have the same difficulty as matrix models in dealing with more than one or two phenotypic features, with stochasticity, or temporal autocorrelations, in changes in condition of individuals (i.e., cases in which individuals with a higher than average feeding rate one day have a higher probability of a higher than average feeding rate the next day). An additional difficult task for classical models is in dealing with complex behavioral sequences, which can occur in foraging, mating, and other activities.

Experience and Learning

Implicit learning at the population level is built into some classical models of interacting populations; for example, a predator population could increase its preference for certain prey relative to other prey types when the first type increases its proportional representation in the prey community (Steele 1974). However, learning, like phenotypic change, is a product of individual experiences. All learning involves memory, so memory of past experiences must be considered an internal state of organisms. Two general types of learning can be distinguished: learning from the environment and learning from other organisms. The latter type of learning, which results from encounters with other individuals that affect fitness, requires a game-theory approach, which can be difficult to include in classical models.

Genetics and Evolution

Genetic and evolutionary biologists have always tended to focus on the individual organism. Many of the concepts of evolutionary genetics, such as mutations, genetic drift, and founder effects, inherently involve small numbers of individuals, so that stochasticity is important. Although analytic models of dynamics are well understood for many single-locus populations, computational models are usually needed for more complex situations. IBMs are more flexible than are classical models of genetic change and can mimic real populations. IBMs have long been applied to human populations, for example, in studying the maintenance of balanced polymorphisms such as the sickle cell trait (Schull & Levin 1964).

Mechanistic detail is assumed to increase along all of these five axes. Classical structural models are usually close to the origin. IBMs, however, tend to depart substantially from the origin along more than one axis at the same time.

ECOLOGICAL AND EVOLUTIONARY PROCESSES ADDRESSED WITH IBMS

A categorization of more than 900 IBM papers with respect to the main type of biological process they deal with showed seven major groups of studies.

Movement Through Space

Models of animal local movement encompass the detailed active movement behaviors of animals and include their interactions with complex landscapes and other animals (Johnson et al. 1992) as well as the development of home ranges (South 1999). These models have been used to design and evaluate empirical studies of movement. Animal migration models take into account detection of sensory cues and the following of older, experienced members of a school, herd, or flock. Fish-migration models have revealed that the proportion of experienced members of schools is important for accurate migration to the previous year's site (Huse et al. 2002). Models of migrating birds have raised questions regarding traditional concepts of carrying capacity when applied to overwinter survival (Goss-Custard et al. 2002, Pettifor et al. 2000). Models of active animal dispersal from natal sites take into account effects of conspecifics and avoidance as well as attraction behaviors in patchy environments (McCarthy 1997). Such models can be used to determine the nature of dispersal patterns (Schwarzkopf & Alford 2002). A considerable number of studies focus on movement by water or air. Marine biologists use ocean-current models to study passive movement of ichthyo- and other plankton (Miller et al. 1998), and analogous models are used to simulate wind dispersal of seeds, especially important for grassland vegetation (Jongejans & Schippers 1999).

Formation of Patterns Among Individuals

Models of swarm/aggregation formation are used to study environmental and social forces and individual decision rules that lead to formation of swarms, flocks, schools, herds, and other groups (Flierl et al. 1999, Gueron et al. 1996, Hemelrijk & Kunz 2005, Huth & Wissel 1992, Young et al. 2001). A limited set of rules can give rise to highly organized group aggregation and movements. Settlement, territory-formation, and home-range models predict propagule settlement rates as well as the spatial patterns of territories that arise from the settling organisms (Murai et al. 1979, Tanemura & Hasegawa 1980). Learning behavior can be a factor in territory formation (Stamps & Krishnan 1999). Models of the dynamics of social networks simulate the ontogeny of caste structure in insect societies

(Hogeweg & Hesper 1983) as well as of foraging paths. Some models describe effects of sociality in vertebrates, such as development of cooperative behavior (Schank & Alberts 2000) and its effect on fitness (Vucetich et al. 1997). How animals should act in encounters with conspecifics (hawkish or dovish, selfishly or altruistically) depends on the expectation of the behavior of the other animal in the encounter. Study of this behavior is the field of game-theoretic interactions. IBM simulations that incorporate game theory are used to simulate the outcome of competing strategies (Broom & Ruxton 1998).

From Foraging and Bioenergetics to Population Dynamics

Foraging models examine the effects of variation of prey preferences among predators on frequency-dependent selection (Sherratt & Macdougall 1995) and variance in growth (Breck 1993). Complex models use artificial-life approaches to show that foraging in large flocks versus solitary foraging for birds, for example, can arise out of resource distribution (Toquenaga et al. 1995) and that spatial memory is important to optimal foraging in a complex environment (Folse et al. 1989, Nolet & Mooij 2002). A related set of models deals with spatial heterogeneity and habitat choice. The mechanisms for formation of an ideal free distribution (IFD) have been simulated by use of exploratory movements combined with memory (Regelmann 1984). Deviations from the IFD occur in spatial systems in which individuals have less than perfect knowledge (Hugie & Grand 2003, Humphries et al. 2001) or when temporal variation occurs in mortality and food availability (Railsback & Harvey 2002).

Bioenergetics models of individuals, including physiological abilities and constraints, combined with foraging models, take account of time and energy budgets and make finer-scale predictions about foraging decisions of individuals (Hölker & Breckling 2002, Kooijman 2000). Some practical and theoretical questions require attaching bioenergetics models to life-history models, as well as demographic variability, spatial structure, and environmental change (Dunham & Overall 1994, Grimm & Uchmanski, Poethke & Hovestadt 2002, Reuter & Breckling 1999, Tyre et al. 2001). An important subset of life-history models are those on fish early-life history, which focus on determinination of growth and survival of young-of-year fish. These models have been useful for understanding stock-recruitment dynamics (DeAngelis et al. 1993b, Madenjian & Carpenter 1991, Rose et al. 1999, Scheffer et al. 1995). Classical life-cycle models often leave out important phenotypic variation and have special difficulty in capturing the large variances in size and other traits that can develop through positive-feedback effects in the life cycle when growth rates or other phenotypic traits are autocorrelated from one time step to the next (DeAngelis et al. 1993a, Pfister & Stevens 2002). Phenotypic variation generated by positive feedback can lead to population-level consequences.

Kaiser (1979) was among the earliest investigators to recognize the importance of properties of individuals in population dynamics, particularly their interactions with space, in which population inhomogeneities that can have profound effects arise. Logistic growth simulated on a spatial arena produces dynamics far different from those of the classic logistic model (Berec et al. 2001, Bolker & Pacala 1997, Kaitala et al. 2001, Law et al. 2003 , Molofsky 1994, Rodd & Reznick 1997), and mate-search strategies in space affect the occurrence of the Allee effect (Berec et al. 2001). Inclusion of spatially explicit landscapes and discrete individuals has greatly extended the power and realism of metapopulation models. IBMs allow accounting for differences in individuals (e.g., age, energy levels, and dispersal ability), variation in site quality (e.g., overgrazing), population levels, and demographic stochasticity (Gustafson & Gardner 1996, Keeling 2002, Kostova et al. 2004, Ovaskainen & Hanski 2004, Rushton et al. 2000). Deviations from the classical Levins metapopulation model are observed in multispecing IBMs that result from correlations between patch occupancy and population levels (Keeling et al. 2002).

Exploitative Species Interactions

Intense interest focuses on the effects of diffusion and mixing of populations on stability and spatial pattern formation in predator-prey and host-parasitoid interactions (Cuddington & Yodzis 2002; De Roos et al. 1991; Donalson & Nisbet 1999; McCauley et al. 1993; Myers 1976; Pascual & Levin 1999; Petersen & DeAngelis 2000; Wilson et al. 1993, 1999). Smith (1991) showed that detailed movement patterns of predators and prey, such as pursuit or avoidance, can drastically affect the stability of the interaction. Microbial systems often allow model comparisons with data (Kooi & Kooijman 1994). Models of host-pathogen interactions demonstrate how the spread of parasites and disease is related to variations in individual characteristics (Jeltsch et al. 1997, Rand et al. 1995). The possible amplification of small initial size differences within a fish cohort has been shown (DeAngelis et al. 1979) in models of size-dependent predator-prey interactions. Because of the complex interactions between numbers, size distribution, and resource pool, the IBM approach has become a popular modeling tool for projecting the dynamics of fish populations in particular (Claessen et al. 2000, De Roos & Persson 2002).

Local Competition and Community Dynamics

Models of local competition of sessile organisms keep track of individual organisms either within grid cells or on a continuum. This class includes benthic or intertidal organisms (Robles & Desharnais 2003) and terrestrial plants (Law & Dieckmann 2000, Pacala 1986, Pacala & Silander 1990, Weiner et al. 2001). Models of this type show a variety of emergent phenomena, such as the development of bimodal distributions of plants from even-aged stands (Ford & Diggle 1981). A subset of these models focuses on bacterial colony growth (Kreft et al. 1998). Forest models are a particular class of local-competition models, and the earliest genre of IBMs to gain prominence were the forest gap-phase replacement models

(Botkin et al. 1972, Shugart 1984). Interactions were averaged over tree gaps (approximately 0.1 ha). Another class of forest models (e.g., Pacala et al. 1996) includes spatial locations of each tree. In recent models, physiological detail and sophisticated visualization have been increased (Parrott & Lange 2004).

Many models of community dynamics are aimed at explicating the factors that control species richness and diversity (Chave et al. 2002). In particular, the mechanisms of competition-colonization trade-offs, conspecific density-dependence effects, and niche differences all interact in a spatial context. Models also study the factors that affect community patterns and transitions between types of community (Higgins et al. 2000, Van Nes & Scheffer 2003). Food webs can be influenced by individual behaviors, such as prey behavioral responses to predator presence, the complexities of changing size distributions of populations in food webs, and the size of available habitat (Schmitz & Booth 1997).

Evolutionary Processes

The modeling of fitness and trait evolution does not always attempt to replicate the mechanisms of evolution, but instead uses computing approaches that range from simple comparisons of trait values to genetic algorithms to find optimal traits for foraging and predator avoidance, dispersal, and reproduction (Rees et al. 1999, Travis & Dytham 1998, Vos et al. 2002, Warren & Topping 2001). Other models, such as those describing coevolution of parasite or pathogen virulence and host resistance, attempt to simulate evolutionary processes (Fellowes & Travis 2000). Competition models, coupled with the capacity to evolve, have shown that the details of local interaction determine whether character displacement of competitive exclusion result (Kawata 1996). Various mechanisms of sympatric and parapatric speciation have been investigated (Dieckmann & Doebeli 1999, Gavrilets et al. 1998, Rice 1984, Savill & Hogeweg 1998, Van Doorn et al. 2004). Hybrid spatial zones can develop where speciation occurs (Sayama et al. 2003). An important evolutionary process addressed by IBMs is the maintenance of genetic diversity. Models have been used to show when balanced polymorphisms could be maintained in populations (Heuch 1978, Schull & Levin 1964). Factors affecting the threat of hybridization to diversity have been investigated (Wolf et al. 2001). The advantages of different mating systems and sexual reproduction are compared under various environmental conditions. This comparison includes dioecy versus hermaphroditism (Wilson & Harder 2003) and polygyny versus monogamy (Gronstol et al. 2003).

Management-Related Processes

IBMs have been applied in population-viability analysis for a large variety of taxa, including birds (Letcher et al. 1998), mammals (Li et al. 2003), and insects (Griebeler & Seitz 2002). Conservation models describe conservation of communities, food webs, or other groups of species at the landscape scale (Macdonald & Rushton 2003, Turner et al. 1994, Wiegand et al. 2004, Wiegand et al. 1999). Fate, effects, and risks of contaminants have been modeled for PCBs (polychlorinated biphenyls), organic toxicants, oil spills, and radiation, among other pollutants (Baveco & De Roos 1996). Invasive species such as weeds have been modeled, including the effects of plant life-history characteristics and properties of the environment (Higgins et al. 1996). Models are used to study the efficiency and sustainability of harvesting and effects of regulations, as well as to predict recruitment to stock and to analyze harvest data (Lammens et al. 2002, Whitman et al. 2004). Effects of increased landscape fragmentation, global warming, and other aspects of disturbance and environmental change on species populations and ecological systems have been modeled (Keane et al. 2001, McDonald et al. 1996, Shugart et al. 1992).

CONTRIBUTIONS OF INDIVIDUAL-BASED MODELS

The above overview illustrates the breadth of subject areas that IBMs now cover in ecology. Has the impact been deep as well?

The significance of IBM to ecology has been obvious in certain applied areas. Forest gap-phase models have been used for more than three decades and are a routine tool in forest ecology (e.g., Shugart 1984). These models are effective at projecting the influence of environmental factors on forest productivity and species composition. The commercial importance of trees and the availability of species-level data are factors in their success. The use of IBMs in fish early-life history beginning in the early 1990s and continuing today resembles the growth of forest models. Understanding the factors that affect recruitment is crucial for the fisheries industry, and IBMs are indispensable for simulating complex interactions between size, growth rate, and survival (e.g., Rose et al. 1999). A third applied area is population-viability analysis, applied to the conservation of endangered species (e.g., Letcher et al. 1998) or to the reintroduction of extinct species (Figure 1, see color insert). IBMs are efficient at modeling populations in which the factors of spatial heterogeneity, behavioral complexity, and small population-size effects (e.g., stochasticity, the Allee effect) are important.

Usage of IBM in several areas of theory has also gained prominence. One of these areas involves spatial heterogeneity and scale aspects of space to populations (e.g., Levin 1994, Levin & Durrett 1996). Spatially explicit IBMs are used to investigate how local interactions and movements of organisms manifest themselves in population-level consequences, such as stability, persistence, and coexistence. Models of local competition of sessile organisms allow investigation of the emergence of size distributions and spatial patterns (Robles & Desharnais 2003). Modeling of fish schooling and other animal-swarming behaviors is a subfield that has grown dramatically (Huse et al. 2002). Researchers can use IBMs to examine how intricate patterns of swarm-scale movement emerge from the movements of individual fish, which know the positions and orientations of only a few neighbors. Understanding the connections between rules of individuals and large-scale patterns is an underlying theme of ecological theory. The study of how speciation occurs is another subfield in which IBMs have been used with success (e.g., Gavrilets et al. 1998). Discreteness and local interactions (mating) are fundamental aspects of real populations that are captured naturally by IBMs. Speciation is an emergent property of such models.

IBMs have also played a role in interpretation of empirical data from field studies. These studies include the analysis of data on movements of animals, where radio-tagging has produced enormous amounts of information on locations or organisms through time. Models can be used to infer the rules governing the movement, although the development of statistical connections is a challenge (Mooij & DeAngelis 2003). A related subfield is the simulation of passive movements of organisms in oceanic currents. Data on settlement densities of benthic-organism propagules are often known. When linked with hydrologic circulation models, IBMs can be used to trace the paths of dispersal of propagules from sources (Miller et al. 1998).

A SAMPLE OF INDIVIDUAL-BASED MODELS

We have selected for in-depth consideration six papers as representative of areas of high concentration of IBMs. They include the most-cited papers on fish and forest IBMs by Rice et al. (1993) and Ribbens et al. (1994), respectively. More recent papers are a unifying community paper by Chave et al. (2002), a highly cited paper by Dieckmann & Doebeli (1999) on speciation, a paper by Wiegand et al. (1999) showing how landscape structure can effectively be linked to population dynamics, and a multimodel conservation study by Stephens et al. (2002).

The fish-recruitment model by Rice et al. (1993) demonstrated the importance of phenotypic variation, specifically the effects of growth-rate variation and size-dependent mortality on the size structure of a population through time. The authors modeled cohorts of initially even-sized individuals with a series of different growth-rate coefficients that stayed constant for individuals but varied among individuals. The mean growth rate of the cohort increased over a growing season because size-dependent mortality acted mostly on the smaller individuals. Both increased mean growth rate and increased growth-rate variance led to an increase in survivorship. This model was extended to allow random variation in growth rate between days, with some positive correlation between growth rates on successive days (i.e., successful foraging one day improved chances for success on the next day). With autocorrelated growth, the size-frequency distribution was substantially broader than when the growth rates were held constant over time, which demonstrates the importance of autocorrelations in addition to variation in growth rates.

The forest model by Ribbens et al. (1994) introduced a technique for calibrating a model of tree-seedling dispersal and demonstrated the effect of spatial variation in seedling recruitment on forest dynamics. The model, SORTIE, described growth, probability of survival, recruit production, and resource density on a spatial domain. Earlier IBMs of forest dynamics made simple assumptions about seedling dependence on parent distribution. The authors used data on the spatial distribution of parent trees as well as transect data on seedlings through mixed stands. These data were used to calibrate and validate a seedling-dispersal submodel. Because this analysis showed that seedling establishment can be a limiting factor in forest dynamics, the authors next included their seedling submodel in SORTIE and demonstrated that seedling-recruitment limitation could significantly affect the dynamics of a forest.

Life-cycle variation in traits provides possible mechanisms for maintaining biodiversity of sessile organisms in a diversity model by Chave et al. (2002). Such mechanisms include (*a*) life-history niche differentiation mediated by competitive trade-offs, (*b*) frequency dependence caused by species-specific pests, (*c*) recruitment limitation caused by localized dispersal, and (*d*) a speciation-extinction dynamic equilibrium mediated by stochastic drift. The relative importance of these mechanisms was assessed using a spatially explicit IBM with individual-level processes of birth, death, speciation, dispersal, and immigration from outside on a 4096×4096 lattice. Community-level patterns, such as species-area curves, relative-abundance distributions, and spatial distributions (Figure 2, see color insert) were extracted from the simulation output and compared with those in real communities. All contributed to species richness, but only life-history niche differentiation mediated by competitive trade-offs and frequency dependence caused by species-specific pests produced robust coexistence. The scale of dispersal was the most important factor affecting species area curves (steep at small scales, then more shallow, and finally steep across large scales). The model is successful because of its unifying nature in explaining many spatial patterns.

Incorporation of genetic variation in a speciation model by Dieckmann & Doebeli (1999) showed that sympatric speciation can emerge much more easily than previously expected from competition for resources. Analytic models had already been used to describe the evolution of traits in asexual species with trait value x, associated with a carrying capacity $K(x)$, which is unimodal with a maximum at the trait value x_0. Analysis predicts that in the absence of competition a species with trait x, associated with a carrying capacity $K(x)$ unimodal around x_0, will evolve toward x_0. This outcome could be evolutionarily stable, but given a sufficient level of competition at x_0, a branching occurs and two phenotypes split from the original. The authors extended this result to a sexual species. Many additive diploid loci represented the feeding preferences of individuals. The simulated population first evolved toward x_0. Under the assumption of assortative mating, on the basis of either the similarity of an ecologically important character or an ecologically neutral "marker" trait, branching occurred (Figure 3). A great advantage of this model is its simplicity and matching to the analytic result for the asexual species.

Wiegand et al. (1999) demonstrated the importance of a spatially explicit representation in examining the effects of landscape heterogeneity on populations.

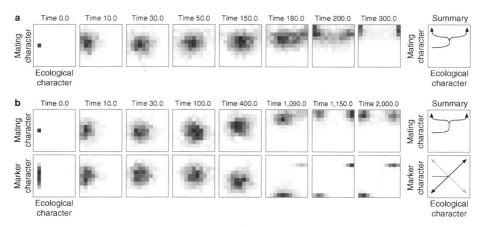

Figure 3   Evolutionary branching in sexual populations was simulated by Dieckmann & Doebeli (1999). In panel *a* the mating probabilities of individuals depend on an ecological character. In panel *b* the mating probabilities depend on a "marker" trait that is neutral with respect to ecological factors such as competition for resources. This result shows that sympatric speciation can occur, even if assortative mating is associated with an ecologically neutral marker trait. (Reprinted with permission of *Nature*.)

In a general-modeling framework, they used an algorithm to produce realistic patterns of good ($\lambda > 1$) and poor ($\lambda < 1$) habitat and matrix (avoided) cells on a 50-by-50 cell-model grid. This grid allowed them to produce various landscape types and examine effects of different levels of fragmentation. Using the European brown bear as their focus, the authors formulated rules on dispersal, habitat-selection strategies, home-range selection, mortality during dispersal, reproduction, and mortality within home ranges. One application of the model was to test earlier results on population viability based on nonspatially explicit models. An implication of earlier models was that increases in poor, or sink, habitat, while the amount of good habitat is held fixed, would result in population declines, because the models assumed that individuals would have more trouble finding the good habitat (needle in haystack effect). The results of IBM simulation by Wiegand et al. (1999) showed, however, that increases in poor habitat help the population by increasing connectivity of good habitat and that increases in poor habitat did not hinder the modeled individuals in finding good habitat.

The conservation model by Stephens et al. (2002) showed the importance of individual optimization by comparing four different approaches to a metapopulation of alpine marmots on the basis of 13 years of field data: model 1, a population-level matrix model; model 2, a matrix model based on the social group level; model 3, a spatially explicit IBM model with empirically based behavioral rules; model 4, a spatially explicit IBM with behavioral rules based on fitness optimization. In model

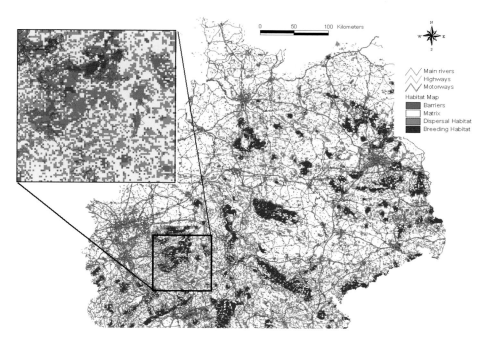

Figure 1 Detailed map of the northern half of Germany that classifies the landscape as barriers, matrix, dispersal, and breeding habitat of the Eurasian lynx. This map is used as basis for a study of the reintroduction of this species (Kramer-Schadt et al. 2004). The small inset figure shows the output of an individual-based simulation running on the landscape map: Pink areas represent home ranges occupied by adult female lynx, pink lines show dispersing subadult females, and blue lines show dispersing subadult males.

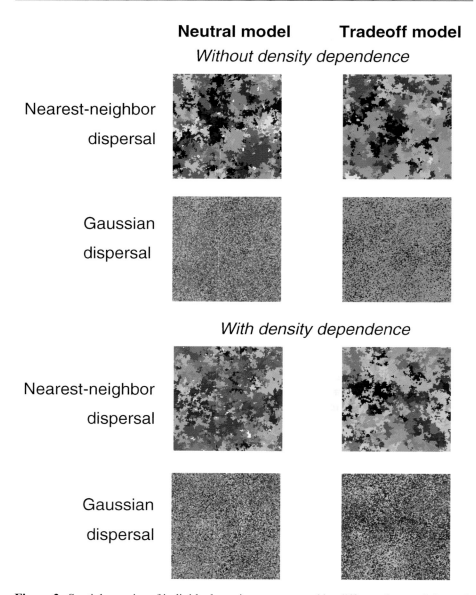

Figure 2 Spatial mosaics of individual species were created by different forms of the multispecies model (Chave et al. 2002). The grain of the patterns depends on (*a*) whether tradeoffs are made between species traits or the traits are neutral, (*b*) whether conspecific density dependence (negative effects on individual if conspecific density is high) is included, and (*c*) whether the dispersal of propagules in space is only local or is larger in scale and described by a Gaussian distribution. (Reprinted with permission of the University of Chicago Press.)

4, at each time step, the estimated optimal action by a given individual was calculated, and the individual was allowed to behave according to fitness optimization. Models 1, 2, and 4 were used to predict equilibrium population sizes well. However, only with model 4 were researchers able to predict an important contribution to an Allee effect. This Allee effect emerged from stochastic skewing of the sex ratio at low-population sizes, less efficient thermoregulation during winter hibernation, and difficulty finding mates during dispersal, even at relatively high population levels. The model is successful in elucidating important population-level effects that arise from social interactions.

DEVELOPING INDIVIDUAL-BASED MODELS

The way in which IBMs are constructed generally requires more careful consideration than is typical for classical models. The state variables of an IBM can be visualized as a table in which the rows represent the individuals and the columns represent their traits. A typical IBM keeps track of between 100 and 10,000 organisms. The table of individuals can easily be created using standard spreadsheet or database software. To add dynamics to the model, we must iterate over the individuals (see example in Figure 4). This iteration is easily done using mathematical programs. When the model reaches application stage, performance often becomes an issue. At this stage, dedicated programs written in a standard computer language such as C++ come into play. They can be optimized for the particular computational bottlenecks of a given program and are also very flexible in their input and output. A clear drawback of dedicated programs is the effort required to develop, maintain, and document them.

To get both good performance and standardization, researchers have made a number of efforts to develop flexible but high-performance frameworks for IBMs (e.g., Baveco & Lingeman 1992, Mooij & Boersma 1996). Despite the promise of these frameworks, their use tends to be limited to the developer and coworkers. We know of no framework that has been so well developed, documented, and maintained as to attain widespread use. Typically, they are built around databases that contain tables with some of the basic concepts of ecological theory, such as the individuals, the spatial units they live in, and the environmental conditions and available resources in those spatial units. To maintain a considerable flexibility in the way the actual dynamics are implemented, they often have a built-in "event handler," comparable with the ones that drive a multitasking operating system.

The focus of a developer of an individual-based model is usually on the short-term task of getting it to work, rather than on the long-term task of thorough analysis of the model. This approach runs the risk of generating a model for which a systematic analysis of its dynamics is a tedious or virtually impossible task. In this sense, the culture of IBM development is quite different from that of the culture of classical model development in theoretical ecology. In the latter, almost all the focus is either on getting analytical solutions for various aspects of the model

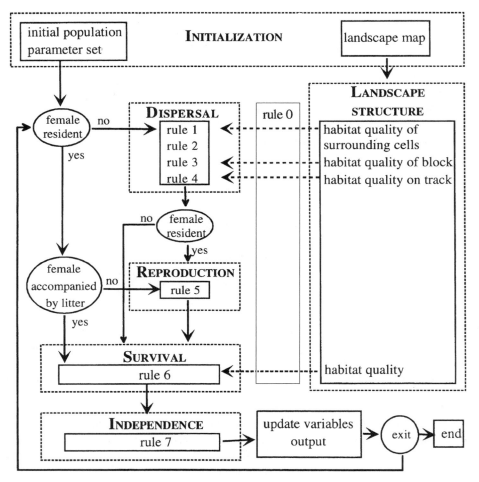

Figure 4 Flow chart for the individual-based simulation model of a bear population (Wiegand et al. 1999). The flow chart shows each of the events that can happen to a female bear during the yearly time step of the model: dispersal, reproduction, survival, and independence (from the mother). Each of these events is governed by one or more rules not shown here. This hierarchical presentation of the model results in a good overview of the general model structure and the general flow of logic. (Reprinted with permission of the University of Chicago Press.)

behavior or on a formal numerical analysis of the model, including sensitivity and bifurcation analysis. Given that an individual-based model is kept simple enough that it runs within a couple of minutes, many of the techniques applied in the analysis of classical models can also be applied to individual-based models. This aspect of the field of IBMs in ecology could use more attention.

CONCLUSIONS AND PERSPECTIVES

We can safely predict that the proportion of papers in the literature that feature IBMs will continue to increase and that, with time, new uses and aspects of IBMs in ecology and evolutionary biology will develop. We are less certain about what philosophical role IBMs will play in the future of these fields. Two general views have emerged.

One perspective is that IBMs are an extension of classical approaches (Bolker et al. 1997). From this viewpoint, use of IBMs in modeling spatial systems, life cycles, and evolution is an elaboration of the types of structures already found in some classical models. However, IBMs are more appropriate for exploring dynamic behaviors that are currently too complex for analysis, because of difficulties caused by local interactions in space, demographic stochasticity, and other types of complexity. The types of IBMs used in such explorations are as simple as possible; individuals have a minimum number of attributes and are often just points in space. Theoretical ecologists examine IBMs for indications of general relationships that can act as a starting point for the development of analytic models. They emphasize the continued building of classic mathematical theory to cover more and more complex types of ecological systems. This approach is well suited to the expansion of the understanding given by analytic models into borderlands," where analytic models cannot be formulated but where the processes are still simple enough that results of simulations can be related back to analytic results. These models at the more analytic end of the spectrum of IBMs will probably continue to have great appeal, as parsimony is a cardinal trait in science.

Nonetheless, other ecologists see in IBMs a new research program that differs radically from the classic mathematical approach to ecological theory and constitutes a new philosophical paradigm (Uchmanski & Grimm 1996). These researchers explicitly restrict the term IBM to describe models that include some threshold level of individual detail. In this view, individuals by nature have highly complex responses to their environment. Such responses include phenotypical change and learning and are thus better described by rule-based simulations than by mathematical models. Behaviors at the population and community levels should emerge from individuals behaving in an adaptive way. Future development of such models can allow individuals to have more sophisticated internal states and as much individual autonomy (an expression of variability) within the model as possible to behave and make decisions and, through mating, such models can provide the basis for evolution. Although this simulation approach generally precludes mathematical analysis, approaches such as pattern-oriented modeling (Grimm et al. 1996) provide ways of designing models and comparing model behavior with natural systems. For proponents of this strong role of IBM (e.g., Grimm & Railsback 2005), mathematical notation is not essential for rigor, as long as a model is formally documented. Grimm (1999) expressed the view that IBMs should play a much more direct role in "paradigmatic" science, that is, in making new scientific breakthroughs.

The difference in the two approaches outlined is a special case of the contrasting top-down and bottom-up views of system models. The top-down objective is to extend classic equation models into more complex domains with the assistance of IBMs. The bottom-up objective is to find the appropriate individual-level bases for the emergence of patterns at larger scales. These two viewpoints, the classical and the IBM approach, form opposite poles of thinking. We can expect that in due time a synthesis of both views will develop (e.g., Fahse et al. 1998, Wilson 1998). As we move in that direction, the stress lines between these poles can be the source of a creative tension for the fields of ecology and evolution.

ACKNOWLEDGMENTS

We thank Volker Grimm, Bart Nolet, Matthijs Vos, and Will Wilson for many useful comments. We thank Jérôme Chave, Ulf Dieckmann, Stephanie Kramer-Schadt, and Thorsten Wiegand for providing us with originals of their figures. D.L.D. received support from the Florida Integrated Science Centers. This is publication 3524 of the Netherlands Institute of Ecology (NIOO-KNAW), Center for Limnology. Both authors contributed equally to this work.

The *Annual Review of Ecology, Evolution, and Systematics* is online at
http://ecolsys.annualreviews.org

LITERATURE CITED

Baveco JM, De Roos AM. 1996. Assessing the impact of pesticides on lumbricid populations: an individual-based modelling approach. *J. Appl. Ecol.* 33:1451–68

Baveco JM, Lingeman R. 1992. An object-oriented tool for individual-oriented simulation: host parasitoid system application. *Ecol. Model.* 61:267–86

Berec L, Boukal DS, Berec M. 2001. Linking the Allee effect, sexual reproduction, and temperature-dependent sex determination via spatial dynamics. *Am. Nat.* 157:217–30

Bolker B, Pacala SW. 1997. Using moment equations to understand stochastically driven spatial pattern formation in ecological systems. *Theor. Popul. Biol.* 52:179–97

Bolker BM, Deutschman DH, Hartvigsen G, Smith DL. 1997. Individual-based modelling: What is the difference? *Trends Ecol. Evol.* 12:111

Botkin DB, Janak JF, Wallis JR. 1972. Some ecological consequences of a computer model of forest growth. *J. Ecol.* 60:849–73

Breck JE. 1993. Foraging theory and piscivorous fish: Are forage fish just big zooplankton? *Trans. Am. Fish. Soc.* 122:902–11

Broom M, Ruxton GD. 1998. Evolutionarily stable stealing: game theory applied to kleptoparasitism. *Behav. Ecol.* 9:397–403

Caswell H. 2000. *Matrix Population Models: Construction, Analysis, and Interpretation.* Sunderland, MA: Sinauer Assoc. 722 pp.

Chave J, Muller-Landau HC, Levin SA. 2002. Comparing classical community models: theoretical consequences for patterns of diversity. *Am. Nat.* 159:1–23

Claessen D, De Roos AM, Persson L. 2000. Dwarfs and giants: cannibalism and competition in size-structured populations. *Am. Nat.* 155:219–37

Cuddington K, Yodzis P. 2002. Predator-prey dynamics and movement in fractal environments. *Am. Nat.* 160:119–34

De Roos AM, Diekmann O, Metz JAJ. 1992. Studying the dynamics of structured population-models: a versatile technique and its application to Daphnia. *Am. Nat.* 139:123–47

De Roos AM, McCauley E, Wilson WG. 1991. Mobility versus density-limited predator-prey dynamics on different spatial scales. *Proc. R. Soc. London Ser. B* 246:117–22

De Roos AM, Persson L. 2002. Size-dependent life-history traits promote catastrophic collapses of top predators. *Proc. Natl. Acad. Sci. USA* 99:12907–12

DeAngelis DL, Cox DK, Coutant CC. 1980. Cannibalism and size dispersal in young-of-the-year largemouth bass: experiment and model. *Ecol. Model.* 8:133–48

DeAngelis DL, Gross LJ, eds. 1992. *Individual-Based Models and Approaches in Ecology: Populations, Communities and Ecosystems.* New York: Chapman & Hall. 525 pp.

DeAngelis DL, Rose KA, Crowder LB, Marschall EA, Lika D. 1993a. Fish cohort dynamics: application of complementary modeling approaches. *Am. Nat.* 142:604–22

DeAngelis DL, Shuter BJ, Ridgway MS, Scheffer M. 1993b. Modeling growth and survival in an age-0 fish cohort. *Trans. Am. Fish. Soc.* 122:927–41

Dieckmann U, Doebeli M. 1999. On the origin of species by sympatric speciation. *Nature* 400:354–57

Donalson DD, Nisbet RM. 1999. Population dynamics and spatial scale: effects of system size on population persistence. *Ecology* 80:2492–507

Dunham AE, Overall KL. 1994. Population responses to environmental-change-life-history variation, individual-based models, and the population-dynamics of short-lived organisms. *Am. Zool.* 34:382–96

Durrett R, Levin S. 1994. The importance of being discrete (and spatial). *Theor. Popul. Biol.* 46:363–94

Fahse L, Wissel C, Grimm V. 1998. Reconciling classical and individual-based approaches in theoretical population ecology: a protocol for extracting population parameters from individual-based models. *Am. Nat.* 152:838–52

Fellowes MDE, Travis JMJ. 2000. Linking the coevolutionary and population dynamics of host-parasitoid interactions. *Popul. Ecol.* 42:195–203

Flierl G, Grunbaum D, Levin S, Olson D. 1999. From individuals to aggregations: the interplay between behavior and physics. *J. Theor. Biol.* 196:397–454

Folse LJ, Packard JM, Grant WE. 1989. AI modelling of animal movements in a heterogeneous habitat. *Ecol. Model.* 46:57–72

Ford ED, Diggle PJ. 1981. Competition for light in a plant monoculture modelled as a spatial stochastic process. *Ann. Bot.* 48:481–500

Gavrilets S, Li H, Vose MD. 1998. Rapid parapatric speciation on holey adaptive landscapes. *Proc. R. Soc. London Ser. B* 265:1483–89

Goss-Custard JD, Stillman RA, West AD, Caldow RWG, McGorty S. 2002. Carrying capacity in overwintering migratory birds. *Biol. Conserv.* 105:27–41

Griebeler EM, Seitz A. 2002. An individual-based model for the conservation of the endangered large blue butterfly, *Maculinea arion* (Lepidoptera : Lycaenidae). *Ecol. Model.* 156:43–60

Grimm V. 1999. Ten years of individual-based modelling in ecology: What have we learned and what could we learn in the future? *Ecol. Model.* 115:129–48

Grimm V, Frank K, Jeltsch F, Brandl R, Uchmanski J, et al. 1996. Pattern-oriented modelling in population ecology. *Sci. Tot. Environ.* 183:151–66

Grimm V, Railsback SF. 2005. *Individual-Based Modeling and Ecology.* Princeton: Princeton Univ. Press. 480 pp.

Grimm V, Uchmanski J. 2002. Individual variability and population regulation: a model of the significance of within-generation density dependence. *Oecologia* 131:196–202

Gronstol GB, Byrkjedal I, Fiksen O. 2003. Predicting polygynous settlement while incorporating varying female competitive strength. *Behav. Ecol.* 14:257–67

Gueron S, Levin SA, Rubenstein DI. 1996. The dynamics of herds: from individuals to aggregations. *J. Theor. Biol.* 182:85–98

Gustafson EJ, Gardner RH. 1996. The effect of landscape heterogeneity on the probability of patch colonization. *Ecology* 77:94–107

Hemelrijk CK, Kunz H. 2005. Density distribution and size sorting in fish schools: an individual-based model. *Behav. Ecol.* 16:178–87

Heuch I. 1978. Maintenance of butterfly populations with all-female broods under recurrent extinction and recolonization. *J. Theor. Biol.* 75:115–22

Higgins SI, Bond WJ, Trollope WSW. 2000. Fire, resprouting and variability: a recipe for grass-tree coexistence in savanna. *J. Ecol.* 88:213–29

Higgins SI, Richardson DM, Cowling RM. 1996. Modeling invasive plant spread: the role of plant-environment interactions and model structure. *Ecology* 77:2043–54

Hockney RW, Eastwood JW. 1989. *Computer Simulation Using Particles.* New York: IOP Publ. 540 pp.

Hogeweg P, Hesper B. 1983. The ontogeny of the interaction structure in bumble bee colonies: a mirror model. *Behav. Ecol. Sociobiol.* 12:271–83

Hogeweg P, Hesper B. 1990. Individual-oriented modeling in ecology. *Math. Comput. Model.* 13:83–90

Hölker F, Breckling B. 2002. Influence of activity in a heterogeneous environment on the dynamics of fish growth: an individual-based model of roach. *J. Fish Biol.* 60:1170–89

Hugie DM, Grand TC. 2003. Movement between habitats by unequal competitors: effects of finite population size on ideal free distributions. *Evol. Ecol. Res.* 5:131–53

Humphries S, Ruxton GD, Van der Meer J. 2001. Unequal competitor ideal free distributions: predictions for differential effects of interference between habitats. *J. Anim. Ecol.* 70:1062–69

Huse G, Railsback S, Ferno A. 2002. Modelling changes in migration pattern of herring: collective behaviour and numerical domination. *J. Fish Biol.* 60:571–82

Huston M, DeAngelis D, Post W. 1988. New computer models unify ecological theory. *BioScience* 38:682–91

Huth A, Wissel C. 1992. The simulation of the movement of fish schools. *J. Theor. Biol.* 156:365–85

Jeltsch F, Muller MS, Grimm V, Wissel C, Brandl R. 1997. Pattern formation triggered by rare events: lessons from the spread of rabies. *Proc. R. Soc. London Ser. B* 264:495–503

Johnson AR, Wiens JA, Milne BT, Crist TO. 1992. Animal movements and population-dynamics in heterogeneous landscapes. *Landsc. Ecol.* 7:63–75

Jongejans E, Schippers P. 1999. Modeling seed dispersal by wind in herbaceous species. *Oikos* 87:362–72

Kaiser H. 1979. The dynamics of populations as the result of the properties of individual animals. *Fortschr. Zool.* 25:109–36

Kaitala V, Alaja S, Ranta E. 2001. Temporal self-similarity created by spatial individual-based population dynamics. *Oikos* 94:273–78

Kawata M. 1996. The effects of ecological and genetic neighbourhood size on the evolution of two competing species. *Evol. Ecol.* 10:609–30

Keane RE, Austin M, Field C, Huth A, Lexer MJ, et al. 2001. Tree mortality in gap models: application to climate change. *Clim. Change* 51:509–40

Keeling MJ. 2002. Using individual-based simulations to test the Levins metapopulation paradigm. *J. Anim. Ecol.* 71:270–79

Keeling MJ, Wilson HB, Pacala SW. 2002. Deterministic limits to stochastic spatial models of natural enemies. *Am. Nat.* 159:57–80

Kooi BW, Kooijman S. 1994. Existence and stability of microbial prey-predator systems. *J. Theor. Biol.* 170:75–85

Kooijman SALM. 2000. *Dynamic Energy and Mass Budgets in Biological Systems.* Cambridge, UK: Cambridge Univ. Press. 424 pp.

Kostova T, Carlsen T, Kercher J. 2004.

Individual-based spatially-explicit model of an herbivore and its resource: the effect of habitat reduction and fragmentation. *C. R. Biol.* 327:261–76

Kramer-Schadt S, Revilla E, Wiegand T, Breitenmoser U. 2004. Fragmented landscapes, road mortality and patch connectivity: modelling influences on the dispersal of Eurasian lynx. *J. Appl. Ecol.* 41:711–23

Kreft JU, Booth G, Wimpenny JWT. 1998. BacSim, a simulator for individual-based modelling of bacterial colony growth. *Microbiology* 144:3275–87

Lammens E, Van Nes EH, Mooij WM. 2002. Differences in the exploitation of bream in three shallow lake systems and their relation to water quality. *Freshw. Biol.* 47:2435–42

Law R, Dieckmann U. 2000. A dynamical system for neighborhoods in plant communities. *Ecology* 81:2137–48

Law R, Murrell DJ, Dieckmann U. 2003. Population growth in space and time: spatial logistic equations. *Ecology* 84:252–62

Leslie PH. 1945. On the use of matrices in certain population mathematics. *Biometrika* 33:213–45

Letcher BH, Priddy JA, Walters JR, Crowder LB. 1998. An individual-based, spatially-explicit simulation model of the population dynamics of the endangered red-cockaded woodpecker, *Picoides borealis*. *Biol. Conserv.* 86:1–14

Levin SA. 1994. Patchiness in marine and terrestrial systems: from individuals to populations. *Philos. Trans. R. Soc. London Ser. B* 343:99–103

Levin SA, Durrett R. 1996. From individuals to epidemics. *Philos. Trans. R. Soc. London Ser. B* 351:1615–21

Li YM, Guo ZW, Yang QS, Wang YS, Niemela J. 2003. The implications of poaching for giant panda conservation. *Biol. Conserv.* 111:125–36

Lomnicki A. 1978. Individual differences between animals and the natural regulation of their numbers. *J. Anim. Ecol.* 47:461–75

Lomnicki A. 1999. Individual-based models and the individual-based approach to population ecology. *Ecol. Model.* 115:191–98

Macdonald DW, Rushton S. 2003. Modelling space use and dispersal of mammals in real landscapes: a tool for conservation. *J. Biogeogr.* 30:607–20

Madenjian CP, Carpenter SR. 1991. Individual-based model for growth of young-of-the-year walleye: a piece of the recruitment puzzle. *Ecol. Appl.* 1:268–79

McCarthy MA. 1997. Competition and dispersal from multiple nests. *Ecology* 78:873–83

McCauley E, Wilson WG, De Roos AM. 1993. Dynamics of age-structured and spatially structured predator-prey interactions: individual-based models and population-level formulations. *Am. Nat.* 142:412–42

McDonald ME, Hershey AE, Miller MC. 1996. Global warming impacts on lake trout in arctic lakes. *Limnol. Oceanogr.* 41:1102–08

Miller CB, Lynch DR, Carlotti F, Gentleman W, Lewis CVW. 1998. Coupling of an individual-based population dynamic model of *Calanus finmarchicus* to a circulation model for the Georges Bank region. *Fish Oceanogr.* 7:219–34

Molofsky J. 1994. Population-dynamics and pattern-formation in theoretical populations. *Ecology* 75:30–39

Mooij WM, Boersma M. 1996. An object-oriented simulation framework for individual-based simulations (OSIRIS): *Daphnia* population dynamics as an example. *Ecol. Model.* 93:139–53

Mooij WM, DeAngelis DL. 2003. Uncertainty in spatially explicit animal dispersal models. *Ecol. Appl.* 13:794–805

Murai M, Thompson WA, Wellington WG. 1979. A simple computer model of animal spacing. *Res. Popul. Ecol.* 20:165–78

Myers JH. 1976. Distribution and dispersal in populations capable of resource depletion. *Oecologia* 23:255–69

Nolet BA, Mooij WM. 2002. Search paths of swans feeding on spatially autocorrelated tubers. *J. Anim. Ecol.* 71:451–62

Ovaskainen O, Hanski I. 2004. From individual behavior to metapopulation dynamics:

unifying the patchy population and classic metapopulation models. *Am. Nat.* 164:364–77

Pacala SW. 1986. Neighborhood models of plant-population dynamics. 4. Single-species and multispecies models of annuals with dormant seeds. *Am. Nat.* 128:859–78

Pacala SW, Canham CD, Saponara J, Silander JA, Kobe RK, et al. 1996. Forest models defined by field measurements: estimation, error analysis and dynamics. *Ecol. Monogr.* 66:1–43

Pacala SW, Silander JA. 1990. Field-tests of neighborhood population-dynamic models of two annual weed species. *Ecol. Monogr.* 60:113–34

Parrott L, Lange H. 2004. Use of interactive forest growth simulation to characterise spatial stand structure. *For. Ecol. Manage.* 194:29–47

Pascual M, Levin SA. 1999. From individuals to population densities: searching for the intermediate scale of nontrivial determinism. *Ecology* 80:2225–36

Petersen JH, DeAngelis DL. 2000. Dynamics of prey moving through a predator field: a model of migrating juvenile salmon. *Math. Biosci.* 165:97–114

Pettifor RA, Caldow RWG, Rowcliffe JM, Goss-Custard JD, Black JM, et al. 2000. Spatially explicit, individual-based, behavioural models of the annual cycle of two migratory goose populations. *J. Appl. Ecol.* 37:103–35

Pfister CA, Stevens FR. 2002. The genesis of size variability in plants and animals. *Ecology* 83:59–72

Poethke HJ, Hovestadt T. 2002. Evolution of density-and patch-size-dependent dispersal rates. *Proc. R. Soc. London Ser. B* 269:637–45

Railsback SF, Harvey BC. 2002. Analysis of habitat-selection rules using an individual-based model. *Ecology* 83:1817–30

Rand DA, Keeling M, Wilson HB. 1995. Invasion, stability and evolution to criticality in spatially extended, artificial host-pathogen ecologies. *Proc. R. Soc. London Ser. B* 259:55–63

Rees M, Sheppard A, Briese D, Mangel M. 1999. Evolution of size-dependent flowering in *Onopordum illyricum*: a quantitative assessment of the role of stochastic selection pressures. *Am. Nat.* 154:628–51

Regelmann K. 1984. Competitive resource sharing: a simulation model. *Anim. Behav.* 32:226–32

Reuter H, Breckling B. 1999. Emerging properties on the individual level: modelling the reproduction phase of the European robin *Erithacus rubecula*. *Ecol. Model.* 121:199–219

Ribbens E, Silander JA, Pacala SW. 1994. Seedling recruitment in forests: calibrating models to predict patterns of tree seedling dispersal. *Ecology* 75:1794–806

Rice JA, Miller TJ, Rose KA, Crowder LB, Marschall EA, et al. 1993. Growth-rate variation and larval survival: inferences from an individual-based size-dependent predation model. *Can. J. Fish. Aquat. Sci.* 50:133–42

Rice WR. 1984. Disruptive selection on habitat preference and the evolution of reproductive isolation: a simulation study. *Evolution* 38:1251–60

Robles C, Desharnais R. 2003. History and current development of a paradigm of predation in rocky intertidal communities. *Ecology* 83:1521–36

Rodd FH, Reznick DN. 1997. Variation in the demography of guppy populations: the importance of predation and life histories. *Ecology* 78:405–18

Rose KA, Rutherford ES, McDermot DS, Forney JL, Mills EL. 1999. Individual-based model of yellow perch and walleye populations in Oneida Lake. *Ecol. Monogr.* 69:127–54

Rushton SP, Barreto GW, Cormack RM, Macdonald DW, Fuller R. 2000. Modelling the effects of mink and habitat fragmentation on the water vole. *J. Appl. Ecol.* 37:475–90

Savill NJ, Hogeweg P. 1998. Spatially induced speciation prevents extinction: the evolution of dispersal distance in oscillatory predator-prey models. *Proc. R. Soc. London Ser. B* 265:25–32

Sayama H, Kaufman L, Bar-Yam Y. 2003. Spontaneous pattern formation and genetic diversity in habitats with irregular geographical features. *Conserv. Biol.* 17:893–900

Schank JC, Alberts JR. 2000. The developmental emergence of coupled activity as cooperative aggregation in rat pups. *Proc. R. Soc. London Ser. B* 267:2307–15

Scheffer M, Baveco JM, DeAngelis DL, Lammens E, Shuter B. 1995. Stunted growth and stepwise die-off in animal cohorts. *Am. Nat.* 145:376–88

Schmitz OJ, Booth G. 1997. Modelling food web complexity: the consequences of individual-based, spatially explicit behavioural ecology on trophic interactions. *Evol. Ecol.* 11:379–98

Schull WJ, Levin BS. 1964. Monte Carlo simulation: some uses in the genetic study of primitive man. In *Stochastic Models in Medicine and Biology*, ed. J Gurland, pp. 179–96. Madison, WI: Univ. Wisc. Press

Schwarzkopf L, Alford RA. 2002. Nomadic movement in tropical toads. *Oikos* 96:492–506

Sherratt TN, Macdougall AD. 1995. Some population consequences of variation in preference among individual predators. *Biol. J. Linnean Soc.* 55:93–107

Shugart HH. 1984. *A Theory of Forest Dynamics*. New York: Springer-Verlag. 278 pp.

Shugart HH, Smith TM, Post WM. 1992. The potential for application of individual-based simulation-models for assessing the effects of global change. *Annu. Rev. Ecol. Syst.* 23:15–38

Sinko JW, Streifer W. 1967. A new model for age-size structure of a population. *Ecology* 48:910–18

Smith M. 1991. Using massively-parallel supercomputers to model stochastic spatial predator-prey systems. *Ecol. Model.* 58:347–67

South A. 1999. Extrapolating from individual movement behaviour to population spacing patterns in a ranging mammal. *Ecol. Model.* 117:343–60

Stamps JA, Krishnan VV. 1999. A learning-based model of territory establishment. *Q. Rev. Biol.* 74:291–318

Steele JH. 1974. *The Structure of Marine Ecosystems*. Cambridge, MA: Harvard Univ. Press. 128 pp.

Stephens PA, Frey-Roos F, Arnold W, Sutherland WJ. 2002. Model complexity and population predictions: the alpine marmot as a case study. *J. Anim. Ecol.* 71:343–61

Tanemura M, Hasegawa M. 1980. Geometical models of territory: I. Models for synchronous and asynchronous settlement of territories. *J. Theor. Biol.* 82:477–96

Toquenaga Y, Kajitani I, Hishino T. 1995. Egrets of a feather flock together. In *Artificial Life IV*, ed. RA Brooks, P Maes, pp. 140–51. Cambridge, MA: MIT Press

Travis JMJ, Dytham C. 1998. The evolution of dispersal in a metapopulation: a spatially explicit, individual-based model. *Proc. R. Soc. London Ser. B* 265:17–23

Turner MG, Wu YA, Wallace LL, Romme WH, Brenkert A. 1994. Simulating winter interactions among ungulates, vegetation, and fire in northern Yellowstone Park. *Ecol. Appl.* 4:472–86

Tyre AJ, Possingham HP, Lindenmayer DB. 2001. Inferring process from pattern: Can territory occupancy provide information about life history parameters? *Ecol. Appl.* 11:1722–37

Uchmanski J, Grimm V. 1996. Individual-based modelling in ecology: What makes the difference? *Trends Ecol. Evol.* 11:437–41

Van Doorn GS, Dieckmann U, Weissing FJ. 2004. Sympatric speciation by sexual selection: a critical reevaluation. *Am. Nat.* 163:709–25

Van Nes EH, Scheffer M. 2003. Alternative attractors may boost uncertainty and sensitivity in ecological models. *Ecol. Model.* 159:117–24

Vos M, Flik BJG, Vijverberg J, Ringelberg J, Mooij WM. 2002. From inducible defences to population dynamics: modelling refuge use and life history changes in Daphnia. *Oikos* 99:386–96

Vucetich JA, Peterson RO, Waite TA. 1997. Effects of social structure and prey dynamics on extinction risk in gray wolves. *Conserv. Biol.* 11:957–65

Warren J, Topping C. 2001. Trait evolution in an individual-based model of herbaceous vegetation. *Evol. Ecol.* 15:15–35

Weiner J, Stoll P, Muller-Landau H, Jasentuliyana A. 2001. The effects of density, spatial pattern, and competitive symmetry on size variation in simulated plant populations. *Am. Nat.* 158:438–50

Whitman K, Starfield AM, Quadling HS, Packer C. 2004. Sustainable trophy hunting of African lions. *Nature* 428:175–78

Wiegand T, Knauer F, Kaczensky P, Naves J. 2004. Expansion of brown bears (*Ursus arctos*) into the eastern Alps: a spatially explicit population model. *Biodivers. Conserv.* 13:79–114

Wiegand T, Moloney KA, Naves J, Knauer F. 1999. Finding the missing link between landscape structure and population dynamics: a spatially explicit perspective. *Am. Nat.* 154:605–27

Wilson WG. 1998. Resolving discrepancies between deterministic population models and individual-based simulations. *Am. Nat.* 151:116–34

Wilson WG, De Roos AM, McCauley E. 1993. Spatial instabilities within the diffusive lotka-volterra system: individual-based simulation results. *Theor. Popul. Biol.* 43:91–127

Wilson WG, Harder LD. 2003. Reproductive uncertainty and the relative competitiveness of simultaneous hermaphroditism versus dioecy. *Am. Nat.* 162:220–41

Wilson WG, Harrison SP, Hastings A, McCann K. 1999. Exploring stable pattern formation in models of tussock moth populations. *J. Anim. Ecol.* 68:94–107

Wolf DE, Takebayashi N, Rieseberg LH. 2001. Predicting the risk of extinction through hybridization. *Conserv. Biol.* 15:1039–53

Young WR, Roberts AJ, Stuhne G. 2001. Reproductive pair correlations and the clustering of organisms. *Nature* 412:328–31

Annu. Rev. Ecol. Evol. Syst. 2005. 36:169–89
doi: 10.1146/annurev.ecolsys.36.102003.152617
First published online as a Review in Advance on August 11, 2005

THE INFLUENCE OF PLANT SECONDARY METABOLITES ON THE NUTRITIONAL ECOLOGY OF HERBIVOROUS TERRESTRIAL VERTEBRATES

M. Denise Dearing,[1] William J. Foley,[2] and Stuart McLean[3]

[1]Department of Biology, University of Utah, Salt Lake City, Utah 84112;
email: dearing@biology.utah.edu
[2]School of Botany and Zoology, Australian National University, Canberra ACT 0200,
Australia; email: william.foley@anu.edu.au
[3]School of Pharmacy, University of Tasmania, Hobart, Tasmania 7005, Australia;
email: Stuart.McLean@utas.edu.au

Key Words biotransformation, conditioned food aversions, conjugation, functionalization, permeability glycoprotein, regulated absorption

■ **Abstract** Plant secondary metabolites (PSMs) significantly impact the nutritional ecology of terrestrial vertebrate herbivores. Herbivores have a wide range of mechanisms (herbivore offenses) to mitigate the negative effects of PSMs. We discuss several behavioral and physiological offenses used by terrestrial vertebrates. Several newly recognized herbivore offenses such as regulated absorption and regulation of toxin intake are presented. We give a detailed description of the biotransformation system with respect to PSMs. We also summarize recent findings of plant–animal interactions for lizards, birds, and mammals. Finally, we discuss some new tools that can be applied to long-standing questions of plant–vertebrate interactions.

INTRODUCTION

Herbivores are faced with a food resource of poor quality not only because plants are low in nutrients but also because they produce plant secondary metabolites (PSMs) that have wide-ranging physiological effects from direct toxicity to digestion impairment. We review the significant impact of PSMs on the nutritional ecology of terrestrial vertebrate herbivores by beginning with a very brief introduction to PSMs. We follow with an extensive review of herbivore offenses (Karban & Agrawal 2002) including numerous behavioral and physiological mechanisms herbivores use to deal with PSMs. We finish with a short summary of the current state of research for three classes of vertebrate herbivores and novel tools for future studies. A goal of this review is to point out new avenues for research in the area of plant–animal interactions.

Plant Secondary Metabolites

Understanding the effects of PSMs on herbivore foraging may seem overwhelming given the bewildering variety of compounds. This diversity has led to different classification schemes based on chemical characteristics (e.g., terpenoid, phenolic, alkaloid), modes of action (toxins versus digestibility reducers), or plant apparency (abundance). Although these classifications have been useful in formulating some broad theories of plant defense, the more we learn, the less useful such broad classifications become (Foley et al. 1999, Foley & McArthur 1994). Most progress in the past decade in understanding how PSMs influence vertebrate herbivores has come from systems where the plant chemistry is well characterized through partnerships between chemists and ecologists (e.g., Clausen et al. 1986, Jakubas & Gullion 1990, Lawler et al. 2000, Moore et al. 2004). We believe that understanding the precise nature of the chemistry mediating plant–vertebrate interactions is extremely valuable. However, detailed chemical characterization may remain beyond many ecological studies, which will continue to consider PSMs in broad chemical groups. Readers desiring more detail on PSM chemistry are referred to Rosenthal & Berenbaum (1991).

MECHANISMS

In this section, we present three general mechanisms (avoidance, regulation, biotransformation) used by vertebrate herbivores to mitigate the effects of PSMs. Of the three sections, ecologists may be least familiar with biotransformation. We provide details of the molecular biology of biotransformation as a primer to aid ecologists attempting to access the pharmacological literature.

Avoidance

DECOMPOSITION   Herbivores that cache food before consumption may behaviorally circumvent or reduce the effects of PSMs if the compounds degrade during storage. Behavioral reduction of PSMs prior to ingestion has a number of advantages. The detoxification system is energetically demanding (Sorensen et al. 2005a). If animals can reduce the dose of toxins consumed through food storage, they may save significant quantities of energy that would be lost during detoxification (e.g., endogenous conjugates). Second, ingestion of lower doses may reduce the likelihood of the formation of toxic intermediate metabolites or free radicals. Lastly, behavioral manipulation of PSMs may enhance diet breadth by allowing herbivores to consume plants containing toxins or concentrations of toxins that they would otherwise be unable to process.

A few examples of behavioral manipulation are documented. Meadow voles (*Microtus pennsylvanicus*) snip branches from conifer trees and delay consumption for a few days, during which time the PSMs decrease (Roy & Bergeron 1990). Beavers (*Castor canadensis*) soak twigs in water for days prior to consumption.

The phenolics in the twigs are reduced through the leaching process. Beavers prefer leached twigs to fresh twigs (Muller-Schwarze et al. 2001). Pikas (*Ochotona princeps*) also manipulate the PSMs in their diet prior to consumption (Dearing 1997). In the summer, pikas preferentially collect leaves high in phenolics for consumption during the winter. During storage, the phenolic concentrations of leaves decay to acceptable levels. An added benefit of collecting leaves high in phenolics for storage is that these leaves preserve better, that is, there is less decomposition of the cache than leaves lacking phenolics.

TANNIN-BINDING SALIVARY PROTEINS (TBSP)   Many herbivores produce salivary proteins that may assuage the effects of tannins. Some tannins (condensed or hydrolyzable) react with proline-rich proteins secreted in saliva, binding to form a complex that is usually insoluble and unabsorbed (Lu & Bennick 1998). Proline has a secondary amine group that gives it a rigid conformation and disrupts the alpha-helical structure of proteins. This makes carbonyl groups in the protein available for hydrogen bonding with the phenolic group of tannins (Santos-Buelga & Scalbert 2000). The interactions are stronger with high-molecular weight condensed tannins that possess many phenolic groups.

The consequences of binding tannins with salivary proteins are twofold. Tannins are inactivated as toxins, enabling the animal to eat tannin-rich browse. The evidence that herbivores have developed specific TBSPs to bind the particular tannins found in their diet is conflicting (Hagerman & Robbins 1993, McArthur et al. 1995). However, the increased secretion of TBSPs results in protein loss in feces, which can affect nitrogen digestibility (Skopec et al. 2004) and body weight (Santos-Buelga & Scalbert 2000). Furthermore in some species with TBSPs, the rate of secretion of protein is so low as to question their role in defense against dietary tannins (McArthur et al. 1995).

PERMEABILITY GLYCOPROTEIN AND CYTOCHROME P4503A   Regulated absorption of PSMs by gut cells is a unique and only recently acknowledged herbivore offense (Sorensen & Dearing 2003, Sorensen et al. 2004). Absorption of PSMs across intestinal cells can be regulated by two molecular mechanisms, a glycoprotein transporter (permeability glycoprotein, or P-gp) and a biotransformation enzyme (cytochrome P450 3A, or CYP3A) that function either in tandem or independently to reduce the amount of PSMs absorbed. P-gp is one of a group of transporters that remove foreign substances from cells (Hoffmann & Kroemer 2004, Lin & Yamazaki 2003). It was originally discovered in tumor cells, in which it is responsible for resistance to several anticancer drugs (e.g., *Vinca* alkaloids, taxanes). P-gp is one of a large superfamily of ATP-binding cassette transporters that are highly conserved and found in all species from bacteria to mammals. P-gp is highly expressed on the apical surface of cells in the lower intestine with increasing levels from the duodenum to colon. P-gp pumps PSMs (e.g., digoxin, morphine) back into the lumen of the gut (Hoffmann & Kroemer 2004, Lin & Yamazaki 2003). P-gp acts on hydrophobic substances, usually also containing a hydrophilic region, with a tertiary nitrogen and aromatic ring. P-gp is also present in liver and renal tubular cells, where it pumps substances into the bile and urine, respectively, and is in the luminal side of the capillary endothelium that forms the blood-brain barrier (Lee & Bendayan 2004). Overall, the role of P-gp seems to be a defense against xenobiotics such as PSMs, by opposing their absorption, hastening their excretion, and protecting the brain. For example, mdr1a knockout mice that lack P-gp are more sensitive to neurotoxicity from oral ivermectin than are wild-type mice (Lin & Yamazaki 2003).

CYP3A is a biotransformation enzyme that metabolizes about 50% of all drugs and many PSMs. It is expressed in large amounts in liver cells but also in intestinal cells where it usually acts in concert with P-gp to reduce the absorption and bioavailability of PSMs (Cummins et al. 2002, Johnson et al. 2001). This joint defense is most effective against toxins that are taken in low doses and which are slowly absorbed, such as the fungal metabolite, cyclosporine (Lin & Yamazaki 2003). The K_m values of P-gp and CYP3A are <100 μM, so PSMs present in high concentrations that rapidly enter intestinal cells will saturate both transporter and enzyme, resulting in dose-dependent absorption. For very lipophilic PSMs, diffusion into intestinal cells is likely to be faster than P-gp can return them to the lumen; however, the effect of the pump will be to increase the time that the PSM will spend in the cell, where it is exposed to metabolism by CYP3A. The result is more extensive metabolism of the PSM rather than it being excreted unchanged in the feces.

Only recently have P-gps been examined within the context of plant–herbivore interactions. Several studies comparing performance of woodrats that specialize on juniper (*Neotoma stephensi*) to that of a generalist woodrat (*N. albigula*) suggest P-gps may play a role in dietary specialization. First, the specialist had lower blood levels after oral dosing with α-pinene (a juniper PSM) than the generalist, although the rates of biotransformation were not different (Sorensen & Dearing 2003). The specialist also excreted more α-pinene unchanged in the feces compared to the generalist (Sorensen et al. 2004). Lastly, in *in vitro* tests, the intestinal P-gp capacity of the specialist was significantly greater than the generalist (Green et al. 2004). Taken together, these studies suggest that the ability of *Neotoma stephensi* to specialize on juniper may be in part due to P-gp and/or CYP3A that reduce absorption of juniper toxins.

Many PSMs can act as activators, inducers, or inhibitors of P-gp and CYP3A4 (Zhou et al. 2004). Exposure to enzyme inhibitors can also trigger induction. Both P-gp and CYP3A can be induced by PSMs that bind the nuclear pregnane X receptor (PXR) of liver and intestinal cells (Dresser et al. 2003, Kullak-Ublick & Becker 2003, Moore et al. 2000). St. John's wort (a common nutraceutical used for depression in humans) decreases the bioavailability of cyclosporine and digoxin, probably through induction of P-gp and CYP3A by the flavonoids, hyperforin and hypericin (Mueller et al. 2004, Zhou et al. 2004). The potential outcome of these P-gp and CYP3A interactions may be complex in herbivores that consume plants with myriad PSMs.

MICROBIAL DETOXIFICATION   Many propose that PSMs are detoxified by microbes in foregut fermenting herbivores (e.g., ruminants, kangaroos, hoatzin). The diverse microbial populations in the foregut can perform many reactions, which can both reduce and increase the toxicity of ingested PSMs (Duncan et al. 2000, Foley et al. 1999). However, there remain too few examples to judge whether foregut microbial detoxification is an important driver of diet diversification in wild herbivores. The best example comes from agricultural systems where Jones & Megarrity (1986) showed that acquisition of a specific microorganism (*Synergistes jonesii*; Allison et al. 1992) enabled goats to consume greater quantities of *Leucaena leucocephala* containing mimosine, a toxic nonprotein amino acid. Nonetheless, agriculturalists are focusing on identifying and manipulating the microbial population of the foregut of sheep and cattle to degrade PSMs because they realize that breeding programs to eliminate PSMs from forage plants can be counter-productive (McSweeney et al. 2002).

Several recent studies (Jones et al. 2000, 2001) conclude that microbial populations from wild browsers had no greater ability to degrade tanniniferous foods than did those from domesticated cattle. This coupled with the reluctance of authorities to allow the release of genetically modified microorganisms to degrade toxic compounds (Gregg et al. 1998) suggests that enhancement of microbial detoxification is more likely to come from manipulating rumen populations rather than discovering new, highly specific organisms. In contrast, recent studies (Krause et al. 2004) have started to use community genome approaches to identify specific changes in the rumen microbiology of sheep eating tannin-rich forages, and ecologists should see rapid progress in understanding the extent of microbial response to PSM-rich diets (at least in agricultural systems) over the next few years.

Regulation of PSM Intake

Browsing mammals encounter a diverse range of PSMs in the majority of the foods they consume. Therefore, complete avoidance of PSMs in the diet is not likely to be a realistic strategy, as animals would need to exclude most available plants from their diet. The vast majority of PSMs that animals ingest are not so acutely toxic that a single bite would be lethal or severely detrimental, but large amounts of even low-potency compounds can be harmful. Consequently, animals must have mechanisms that allow them to detect and regulate their intake of PSMs to ensure that low-potency toxins do not cause damage. Understanding the sorts of mechanisms that animals use to regulate toxin intake is vital if we are to integrate the effects of PSMs with broader theories of feeding. For example, is the regulation of PSM intake part of the same mechanisms that animals use to recognize nutrient deficiencies or are there separate regulatory mechanisms imposed on normal feeding patterns? By understanding the interactions and tradeoffs between the intake of nutrients and PSMs, we can identify general patterns so that each new compound studied does not have to be treated as a special case. Below we review evidence that animals regulate PSM intake and describe potential mechanisms permitting regulation.

EVIDENCE OF REGULATED INTAKE   The evidence that the intake of PSMs is regulated comes largely from dose response studies in which purified PSMs are fed across a range of concentrations (5 to 10 times range) to captive animals. For example, both common ringtail (*Pseudocheirus peregrinus*) and brushtail (*Trichosurus vulpecula*) possums adjusted food intake on diets containing different concentrations of jensenone such that there was no difference in the daily intake of jensenone. Ruffed grouse (*Bonasa umbellus*) also regulate their intake of coniferyl benzoate (CB) over a fivefold range of dietary concentrations (Jakubas et al. 1993). However, it is worth noting that regulation does not appear to be as precise when animals feed on natural plant diets containing the same compounds. We present three mechanisms that we think animals may use to regulate their intake of PSMs.

NAUSEOUS FEEDBACK CAUSING CONDITIONED FOOD AVERSIONS   Conditioned food aversions (CFAs) have long been known to be powerful influences on the diets of herbivores (Provenza et al. 1998). CFAs arise when an animal makes an association between the taste or smell of a plant and some negative consequence—usually illness. Provenza and coworkers (e.g., Provenza et al. 2003) have been instrumental in demonstrating that CFAs influence diet choices of large domestic animals such as sheep, goats, and cattle and that CFAs are a useful practical method of protecting free-ranging livestock from being poisoned. In Provenza's view, CFAs are balanced by the possibility of animals forming conditioned preferences—they associate the flavor of food with positive consequences such as enhanced energy or protein status (Provenza et al. 1998).

In spite of the large volume of laboratory data that support a central role for CFAs in herbivore foraging, there are several caveats to consider. First, there have been few direct demonstrations that the attenuation of nauseous sensations allows herbivores to increase consumption. Lawler et al. (1998) showed that administration of the drug ondansetron [a specific serotonin ($5HT_3$) receptor antagonist that reduces nausea and vomiting in humans] led to greater intakes of the PSM jensenone in two herbivorous marsupials and attributed the effect to reduction in nauseous sensations. Similarly sheep consumed more endophyte-infected fescue when dosed with a dopamine receptor antagonist that reduces nausea and vomiting in humans (Aldrich et al. 1993).

Second, there have been few demonstrations of CFAs where both the compounds responsible for the distinctive taste of the plant and the nausea-inducing factors are naturally present in the same plant; nearly all studies rely on using the artificial agent LiCl to induce nausea. In contrast, Lawler et al. (1998) demonstrated an ecologically relevant example in *Eucalyptus* where the concentrations of the dominant *Eucalyptus* terpene, 1,8-cineole, covaried with the concentrations of a nausea-inducing agent jensenone. More examples like this would build confidence in the ecological relevance of CFAs.

Finally, the ability of animals to form CFAs to PSMs must be tested under the more complex foraging scenarios commonly experienced by free-ranging animals. For example, Duncan & Young (2002) showed that the ability of animals to make associations between illness and a particular food were reduced when all test foods were offered during the learning period, as would be the case in free-ranging animals. Caution is prescribed in extrapolating results of simple conditioning experiments carried out with captive animals to free-ranging herbivores (Duncan & Young 2002).

PRE-INGESTIVE EFFECTS: TASTE AND TRIGEMINAL STIMULATION Many studies have demonstrated that foods can be repellent to herbivores without inducing CFAs. Humans experience this when eating foods containing hot peppers. The vanilloid compound in the peppers, capsaicin, irritates trigeminal nerves in the mouth, generating a burning sensation and, if it is too intense, food consumption is reduced. However, the experience rarely leads to aversion to such dishes in the future. Similarly, it is clear that animals can use trigeminal feedback to regulate their intakes of these irritant PSMs. For example, Jakubas et al. (1993) showed that CB was a trigeminal irritant in birds and that ruffed grouse could regulate their intake of CB in aspen.

Compounds that are bitter, but not necessarily irritant, may also be effective deterrents at high concentrations and lead to thresholds in the ingestion of the PSM. For example, the intensely bitter phenolic glycoside, salicin, is an effective antifeedant in willows (*Salix* sp.) for common brushtail possums in the field. However, the effect was due to "taste" rather than postingestive effects because direct infusion of salicin into the stomach had no effect on subsequent salicin intake (Pass & Foley 2000).

Recent studies confirm that molecular differences in the vanilloid receptor between birds and mammals may explain why mammals but not birds are repelled by a hot, peppery taste (Jordt & Julius 2002). In contrast, bitter substances are significantly more repellent to birds. Recent advances in understanding the molecular basis of bitter tastes (e.g., Margolskee 2002) offer exciting opportunities to understand why some animals are more repelled by bitter compounds than others and to learn how the taste of PSMs integrates with other food components. The differences between mammals and birds in their susceptibility to different repellent PSMs is of particular interest in understanding the role of PSMs in fruit dispersal, particularly the possibility of directed dispersal of fruits by either birds or mammals (Cipollini & Levey 1997a, Tewksbury & Nabhan 2001).

FEEDBACK FROM DETOXICATION LIMITATIONS Recent experiments show that limits to the rate of detoxication of ingested PSMs can constrain feeding rates of mammalian herbivores. However, it remains unclear how animals translate physiological effects into changes in feeding behavior. For example, brushtail possums fed diets rich in benzoic acid ate more when provided with supplemental glycine because the rate of formation of the detoxified excretory product benzoyl glycine (hippuric acid) was enhanced (Marsh et al. 2005). Animals recognized whether diets contained supplemental glycine and modified their feeding accordingly. However, feeding rates did not change when brushtails were dosed with ondansetron

(Marsh et al. 2005), suggesting that nauseous feedback was not responsible for the changes in food intake induced by dietary benzoate.

Biotransformation

PSMs can be eliminated from the body by excretion or chemical change (biotransformation), or a combination of these processes. Because of the way the terrestrial kidney works to conserve water, lipophilic substances are poorly excreted. In the mammalian kidney, the plasma filtered at the glomerulus is nearly completely reabsorbed from the renal tubules, and lipophilic substances are extensively reabsorbed across the renal tubular epithelium leaving polar molecules that cannot permeate the epithelial cells to be excreted in the urine. The most lipophilic PSMs (e.g., monoterpenes) are not excreted without transformation (Boyle et al. 1999, 2000a), whereas more polar PSMs (e.g., quercetin, gallic acid) are excreted unchanged to varying extents (Schwedhelm et al. 2003, Wiggins et al. 2003).

Terrestrial animals have evolved a powerful suite of biotransformation enzymes to convert lipophilic PSMs into more polar metabolites readily excreted in urine or bile. These biotransformation enzymes are broadly categorized into two groups: functionalization (also called Phase 1), in which functional groups are introduced into metabolites, and conjugation (Phase 2), in which adducts are formed with endogenous compounds to further increase polarity. Functionalization and conjugation enzymes can work alone or in tandem, depending on the substrate. The most important are the mixed-function oxidases of the cytochrome P450 family (P450s) (Gonzalez & Nebert 1990, Guengerich 2004). These enzymes are considered to have evolved in part as a response to dietary plant toxins (Gonzalez & Gelboin 1994, Gonzalez & Nebert 1990). Although most drugs lose their activity after biotransformation, there are examples of PSMs whose toxicity is mediated by metabolites.

FUNCTIONALIZATION   The most common functionalization reactions are oxidations by P450s, although other reactions (reduction, hydrolysis) can be important, depending on the chemical structure of the PSM. P450s are a large family of enzymes with different but overlapping substrate specificities and a common mechanism of action (Guengerich 2004). They are located in the smooth endoplasmic reticulum where the membrane phospholipid provides an environment suitable for their activity. The liver is the major organ for P450 activity but extrahepatic sites, especially the gastrointestinal tract, are also important (Ding & Kaminsky 2003). P450s have a general requirement for molecular oxygen, NADPH, and cytochrome P450 reductase, and the reaction results in the introduction of an oxygen atom in the form R-H to ROH. The oxidized metabolites may be similar to the parent molecule, e.g., alcohols and acids formed from hydrocarbons such as monoterpenes (Boyle 2000b, Boyle et al. 1999) or may become unrecognizable through cleavage of the carbon skeleton (e.g., phenols and hydroxy acids formed from complex phenolics such as tannins and flavonoids) (Schwedhelm et al. 2003,

Zhou et al. 2003). Analysis of the literature on human P450s with substrates and inhibitors reveals that there are five primary P450s for drug metabolism (CYP: 2D6, 3A4, 2C, 1A2, 2B6) and four for other substances such as PSMs (CYP: 1A1, 1B1, 2E1, 2A6; Rendic 2002). Comparison of P450s from humans and the puffer fish (*Takifugu rubripes*) indicates that the overall pattern of P450 genes has been well conserved for 420 million years (Nelson 2003).

Despite the general conservation of P450 genes, the activities of P450 enzymes can vary widely between and within species due to differences in allelic forms and expression (Gonzalez & Kimura 2003, Guengerich 2002, Wilkinson 2004). For example, specialist marsupial folivores, whose diets are high in monoterpenes, are able to more extensively oxidize monoterpenes than generalist folivores (Boyle 2000a; Boyle et al. 1999, 2000b, 2001). In vitro, the P450s from marsupial folivores have greater activity in oxidizing terpenes compared to rat or human P450s (Pass et al. 2001, 2002).

Metabolism does not always detoxify PSMs as many adverse reactions are due to the formation of reactive and toxic metabolites (Gonzalez & Gelboin 1994). For example, the pyrrolizidine alkaloids (Fu et al. 2004) and aflatoxin B_1 form toxic intermediates after biotransformation.

CONJUGATION   Conjugation reactions involve the addition of an endogenous molecule directly to a PSM or a metabolite formed from a functionalization reaction. The most common conjugates are glucuronides and sulfates but glutathione may also be an important conjugate. These conjugates bind to PSMs forming highly polar products that are readily excreted in urine or bile (Shipkova et al. 2003), although they require active transport to leave the cell (Konig et al. 1999). Conjugation is more energetically costly compared with functionalization in that an endogenous compound, the conjugate, is typically excreted from the body as part of the process.

Conjugation with glucuronides occurs through any of ~50 enzymes in the mammalian superfamily of UDP-glucuronosyltransferases (UGTs). These enzymes catalyze reactions between activated UDP-glucuronic acid and the -COOH, -OH and $-NH_2$ groups of a vast array of endogenous and exogenous substrates such as steroid hormones and the phenolic metabolites of flavonoids and tannins (Miners et al. 2004, Radominska-Pandya et al. 1999). UGTs are located in the endoplasmic reticulum of liver, gut, and kidney, and substrates must be lipophilic to reach the membrane-bound enzyme. This is demonstrated by the differential glucuronidation of terpene metabolites in marsupial folivores, in which the less polar hydroxyterpenes are extensively conjugated, whereas the more polar hydroxyacid metabolites are excreted in the free form (Boyle et al. 1999, 2000).

The different forms of UGTs are proposed to have evolved in part in response to the challenge of dietary PSMs (Bock 2003). There are marked species differences in the pattern and rate of glucuronidation (Walton et al. 2001). Interestingly, the cat and related carnivores are deficient in UGT activity and are highly sensitive to toxicity of certain nondietary UGT substrates such as acetaminophen and morphine

(Court & Greenblatt 2000). In the koala (*Phascolarctos cinereus*), which specializes on *Eucalyptus*, there is negligible conjugation of terpene metabolites, which instead are extensively oxidized to enable their excretion (Boyle et al. 2000a, 2001). However, the koala does glucuronidate phenolic metabolites (McLean et al. 2003), possibly because they are not readily oxidized by P450 enzymes.

Conjugation reactions with sulfate are similarly catalyzed by a large family of cytosolic sulfotransferase enzymes (SULTs) that act on a vast array of endogenous and exogenous substrates, including phenols, benzylic alcohols, and hydroxylamines (Blanchard et al. 2004, Glatt & Meinl 2004, Kauffman 2004). The sulfonate group (SO_3^-) is transferred from 3'-phosphoadenosine 5'-phosphosulfate (PAPS) and forms a sulfate ester (SO_4) only if the acceptor is an OH group. Phenols (e.g., gallic acid) can form both sulfate and glucuronide conjugates (Yasuda et al. 2000). The proportions vary depending on the relative activities of the transferases and their cofactors (Glatt & Meinl 2004) as well as on species and dose. For example, the rabbit excretes paracetamol mainly as glucuronides, whereas the rat mainly excretes sulfates at low doses and glucuronides at high doses (Walton et al. 2001).

Glutathione (GSH) forms conjugates with electrophilic toxins, which otherwise could form adducts with nucleophilic sites on proteins and DNA leading to toxicity (Armstrong 1997, Eaton & Bammler 1999). The resulting glutathione conjugates are either excreted in bile or undergo further metabolism to mercapturic acids, which are excreted in urine. Some PSMs (e.g., formylphloroglucinols; McLean et al. 2004) bind GSH rapidly, but others (e.g., isothiocyanates from glucosinolates; Mithen et al. 2000) need catalysis by glutathione *S*-transferases (GSTs). There are several GST forms that occur both in the cytoplasm for soluble electrophiles and membranes for more lipophilic substrates (Armstrong 1997, Eaton & Bammler 1999).

Because of the serious consequences of genotoxicity, where a single molecular hit can potentially result in mutation or cancer, it is perhaps significant that there is a large overcapacity of GSH conjugating activity, both in GSH and GST concentrations (Rinaldi et al. 2002). This system protects cells from endogenous electrophiles (e.g., the fatty acid oxidation product, 4-hydroxy-2-nonenal) and PSMs (e.g., aflatoxin). PSMs can alter GST activity by inhibition (e.g., curcumin) and induction (e.g., isothiocyanates, goitrin; Eaton & Bammler 1999). The wide species differences in susceptibility to aflatoxin B_1, a secondary metabolite produced by some strains of *Aspergillus*, is due to variations in the relative activities of metabolic activation to the electrophilic aflatoxin B_1-8,9-epoxide and its detoxification by GST (Klein et al. 2000).

Although conjugation is usually considered to be a detoxifying reaction, with some substrates the glucuronide or sulfate can generate a reactive electrophile that forms adducts with proteins and DNA resulting in toxicity to the cell or genome. Examples of reactive conjugates of PSMs are the acyl glucuronide of salicylic acid (Spahn-Langguth & Benet 1992) and the sulfate ester of safrole (Kauffman 2004).

INDUCTION AND INHIBITION OF METABOLISM   The activity of enzymes and transporters can be modified by PSMs that inhibit their function or stimulate the synthesis of more active protein (a process called "induction"), or both effects in succession (Hollenberg 2002). Inhibition can be due to competition between substrates for limited enzyme or an irreversible "mechanism-based" reaction where the inhibitor is oxidized by the P450 to a reactive metabolite that binds covalently to the active site of the enzyme. One example of a mechanism-based inhibitor is 6', 7'-dihydroxybergamottin, a PSM in grapefruit juice that destroys CYP3A in enterocytes, requiring three days for recovery of enzyme activity (Greenblatt et al. 2003).

Induction of P450 enzymes occurs via several nuclear receptors. The nuclear receptors activate target genes in a similar manner to the steroid hormone receptors (Wang & LeCluyse 2003). Polycyclic aromatic hydrocarbons and dietary constituents were the first substances found to induce metabolism, and now many drugs and natural products are recognized as inducing agents (Harris et al. 2003, Hollenberg 2002).

Conjugation enzymes are inducible as well as subject to inhibition. PSMs such as hyperforin, β-napthoflavone and indole-3-carbinol induce UGTs as well as CYPs via the nuclear receptors AhR, PXR, and CAR (Bock & Kohle 2004). Induction of sulfonation is not as marked (Coughtrie & Johnston 2001) but glutathione transferases are upregulated by their electrophilic substrates (Rinaldi et al. 2002). Sulfonation is potently inhibited by the polyphenols quercetin, epigallocatechin and epicatechin (Antonio et al. 2003). Glucuronidation and glutathione conjugation do not seem to be as subject to inhibition by PSMs (Eaton & Bammler 1999, Radominska-Pandya et al. 1999), although glutathione conjugates with hydrophobic substituents (e.g., isothiocyanates) bind and inhibit the active site of GSTs (Armstrong 1997).

Because of the multiplicity of enzyme forms, PSMs do not uniformly induce or inhibit all functionalization and conjugation enzymes (Manson et al. 1997). PSMs can have complex effects on P450 regulation, depending on the substances and amounts consumed, target organ, and species (Zhou et al. 2003). There is also large individual variability in the inhibition and induction of P450 enzymes (Lin & Lu 2001) and polymorphism in transporters (Ieiri et al. 2004). The consequences for herbivores eating large amounts of PSMs that can act as inhibitors of metabolism (during the feeding session) and inducers (on the following days) will be complex interactions.

HERBIVORES AND PSMs

Below we give an overview of recent research on how PSMs influence diet selection in the major groups of terrestrial herbivores, e.g., the lizards, birds, and mammals. Some groups such as the lizards have received relatively little attention in this area compared to others such as mammals. The disparity in research among groups precludes reasonable speculation regarding the relative importance of PSMs to one group versus another. Clearly more research is necessary with respect to the comparative physiology of biotransformation mechanisms used by each group prior to the establishment of generalizations.

Lizards

Herbivory is relatively rare in lizards; only 3% of all species are herbivorous (120 species; Cooper & Vitt 2002). The diets of several herbivorous lizards are documented and several are known to eat or avoid plants with notable PSMs. For example, the chuckwalla (*Sauromalus obesus*) preferentially feeds on the flowers of creosote bush (*Larrera tridentata*) while avoiding the resinous leaves that contain high levels of toxins (Nagy 1973). Despite extensive documentation on diet selection by lizards, there are very few studies that examine the role that PSMs play in food choice.

Of the herbivorous lizards, those that have received the most attention with respect to the role that PSMs play in diet selection are the herbivorous whiptail lizards of the southern Caribbean, *Cnemidophorus murinus* and *C. arubensis*. Studies on both species revealed that whiptails avoided some PSMs such as tannins and alkaloids, but were less deterred by other PSMs such as cyanide (Dearing & Schall 1992, Schall & Ressel 1991). Furthermore cyanide was ingested at extremely high doses with no apparent effect (Schall & Ressel 1991). Caribbean whiptails are capable of discriminating plant odors from those of nonfood items (Cooper et al. 2002). The ability to detect plant chemical cues in whiptails as well as other herbivorous lizards appears to be a trait that evolved with herbivory (Cooper 2003). Herbivorous whiptails can detect differences in concentrations of PSMs, e.g., quinine. The acceptable concentration of quinine in experimental food varied seasonally and may be a function of the availability of other less toxic foods (Schall 1990). These studies indicate that PSMs influence diet selection and that herbivorous whiptails have physiological capabilities for detecting PSMs. Given that these two species of *Cnemidophorus* are not considered to be as specialized for herbivory as some of the iguanans, it is likely that the more specialized lizards also possess such traits and may have other novel traits for detecting and dealing with PSMs.

Little is known about the detoxification mechanisms that herbivorous lizards employ in processing toxins. Obviously, much more research is needed on how PSMs influence diet selection in herbivorous lizards.

Birds

Herbivory is also rare among birds, presumably due to mass tradeoffs associated with flight. However, there are a number of herbivorous birds such as geese, grouse, some parrots, and the hoatzin (*Opisthocomus hoazin*) that feed on terrestrial plants. PSMs play a significant role in diet selection of herbivorous bird species. The hoatzin, for example, is a folivore that inhabits riparian areas of South America.

Although selective in its diet, many of the plants consumed by the hoatzin contain PSMs (Dominguez-Bello et al. 1994). The hoatzin's ability to detoxify certain PSMs may be a consequence of foregut fermentation rather than biotransformation by its own enzymes. The hoatzin has an atypically enlarged crop to house symbiotic bacteria that ferment fiber. It has been proposed that these bacteria in the crop biotransform PSMs before they reach the absorptive tissue in the small intestine (Grajal et al. 1989); however, this has not yet been directly tested. Furthermore, the spectrum of compounds that these fermentative bacteria can process may be limited to alkaloids rather than tannins (Jones & Palmer 2000). Diet selection of the hoatzin is consistent with this hypothesis in that it tends to avoid tannin-rich forages and selects young leaves, which typically are defended by PSMs like alkaloids.

The ruffed grouse has likely received the most attention with respect to detoxification of PSMs. Ruffed grouse consume aspen flowers (*Populus tremuloides*) as well as leaves from a variety of evergreen shrubs and ferns. CB, a phenylpropanoid ester present in aspen, is a primary determinant of feeding (Jakubas et al. 1989). Grouse can detect CB concentration differences and feed selectively on flowers low in CB. Detoxification of CB by grouse occurs through some functionalization reactions but also extensively via conjugation with both glucuronic acid and ornithine (Jakubas et al. 1993). The energetic cost of detoxification of aspen flowers with low levels of CB, typical of that consumed by grouse in the field, was enormous. Grouse lost 10% of metabolizable energy (e.g., energy not lost in urine or feces) in simply the detoxification conjugates (glucuronic acid and ornithine) excreted. This cost is an underestimate of the total energy used in detoxification as it does not account for energy used in enzymatic reactions that attach the conjugate to the PSM or functionalization reactions. Furthermore, losses of the amino acid, ornithine, as a conjugate increased the minimum nitrogen requirement. Another study on grouse consuming leaves from several different species of shrubs and at different doses revealed that usage of conjugation pathways varies among plant species but also within species at different doses (Hewitt & Kirkpatrick 1997). Surprisingly, the shift in pathway usage within a particular food type does not appear to be related to a limitation of conjugates or saturation of the detoxification pathway (Hewitt & Kirkpatrick 1997).

PSMs are not thought to play a large role in diet selection by geese (but see Buchsbaum et al. 1984), in contrast to grouse and the hoatzin (Sedinger 1997). A possible cause of this disparity is that geese feed primarily on monocots, which are typically low in PSMs, compared to the dicots high in PSMs fed on by grouse and the hoatzin. Furthermore, monocot feeders are often believed to have lower capacities for detoxification. However, comparisons of detoxification capacities have not been conducted in herbivorous species of birds.

PSMs are thought to be instrumental in fruit selection by frugivorous birds. Because frugivores are not the subject of this review, we direct readers to a few recent works as an introduction into this active and controversial area of research (e.g., Cipollini & Levey 1997a, 1997b; Tewksbury & Nabhan 2001).

Mammals

Herbivory is widespread among mammals, which may in part account for the greater number of studies on mammals compared to lizards and birds. In the past 30 years, numerous studies have demonstrated that PSMs play a significant role in diet selection of mammalian herbivores (e.g., Berger et al. 1977, Foley et al. 1999, Reichardt et al. 1990). Furthermore, detoxification is expensive and comparable to the cost of reproduction in many mammals (Sorensen et al. 2005a). Because there have been far more studies on mammals than lizards or birds, below we synthesize what has been learned in general rather than profiling individual species of mammalian herbivores.

Studies of the role of PSMs on mammalian foraging have been largely influenced by Freeland & Janzen's seminal paper (Freeland & Janzen 1974). In this work, they wedded data from pharmacological studies on laboratory rodents to ecological studies on wild herbivores to produce theory predicting the foraging behavior of herbivores with respect to PSMs. One of their main points explained the paucity of dietary specialization among mammalian herbivores. They suggested that the default state of the mammalian biotransformation system promoted a generalist foraging strategy because this system can process limited quantities of myriad xenobiotics through numerous detoxification pathways. Such a system would force mammals to be generalist feeders to prevent exceeding the capacity of any one of the pathways. A few experimental studies support this hypothesis; herbivores eat more when offered plants with differing PSMs and process the PSMs through different pathways. Unfortunately, all the studies to date in this area have been on a single mammalian herbivore, the brushtail possum (reviewed in Marsh et al. 2005). Studies on other species are needed to evaluate whether limitations of the detoxification system explain the preponderance of dietary generalization in mammalian herbivores.

A recent hypothesis that has emerged from studies of various mammalian herbivores is that specialists use detoxification pathways that are energetically less expensive than generalist feeders. Specifically, specialists appear to rely more on functionalization pathways than on energetically costly conjugation pathways (Boyle et al. 2000a, 2001; Lamb et al. 2004). As functionalization pathways may have greater substrate specificity than conjugation pathways, a potential trade-off of specialization may be the inability to process novel toxins. Sorensen et al. (2005b) demonstrated such a tradeoff in the juniper specialist, *Neotoma stephensi*, on a novel toxin. Clearly more work on other species and PSMs is necessary to determine the generality of this tradeoff.

TOOLS FOR FUTURE RESEARCH

There is still much to understand about how PSMs mediate diet selection in terrestrial herbivores. In general, we understand little of the mechanisms used by wild herbivores and how these mechanisms provide feedback to influence diet selection. Rigorously addressing large-scale ecological or evolutionary questions requires more detail on the underlying physiological mechanisms. Below we describe two approaches that will be fruitful for future studies in plant–mammal interactions.

Genomics

The burgeoning field of genomics offers a number of tools to address the role of PSMs in nutritional ecology. The detoxification limitations hypothesis of Freeland & Janzen (Freeland & Janzen 1974) may soon be testable at the level of entire detoxification systems through the use of microarray technology. A single microarray measures gene expression of thousands of genes simultaneously. Commercial arrays profiling the detoxification system have been constructed for toxicology studies in laboratory rats, and microarrays produced for one species have been successfully used for other species (Moody et al. 2002). Thus, it may be possible to use the rat toxicology microarrays on nonmodel herbivore systems such as voles or woodrats. As the technology develops, so will the possibility and affordability of making arrays for herbivores of interest. In addition to microarrays, the rapidly growing numbers of sequencing projects facilitate studies on the evolution of detoxification genes. Questions such as how do functionalization genes change at the molecular level in specialists compared to generalists can be addressed on a large scale using a combination of database mining (e.g., GenBank) and lab work.

Standardized Functional Assays

Studies of the capacity of herbivores to eliminate dietary PSMs would be greatly assisted if we could readily measure the activity of elimination processes such as enzymes, transporters, and renal excretion. The elimination of probe drugs is used in pharmacokinetic studies, but these tests typically require procedures such as intravenous administration, serial blood sampling, and the measurement of the concentrations of drugs and their metabolites (Pelkonen 2002). These methodological complications can be avoided by using a pharmacological response such as time sleeping under anesthesia as a surrogate measure of biotransformation capability. Usually this has been done in comparative studies, such as the effect of capsaicin on hexobarbitone sleeping time (Hamid et al. 1985) and the effect of 7,8-benzoflavone on zoxazolamine paralysis.

ACKNOWLEDGMENTS

We thank Michele Skopec, Shannon Haley, Ann-Marie Torregrossa, and Ben Moore for comments on this manuscript. Our work was supported by NSF (IBN 0236402 to MMD) and the Australian Research Council (to WJF and SM). We thank Kathy Smith for assistance with manuscript preparation.

The *Annual Review of Ecology, Evolution, and Systematics* is online at
http://ecolsys.annualreviews.org

LITERATURE CITED

Aldrich CG, Rhodes MT, Miner JL, Kerley MS, Paterson JA. 1993. The effects of endophyte-infected tall fescue consumption and use of a dopamine antagonist on intake, digestibility, body temperature, and blood constituents in sheep. *J. Anim. Sci.* 71:158–63

Allison MJ, Mayberry WR, McSweeney CS, Stahl DA. 1992. *Synergistes jonesii*, gen nov, sp nov—a rumen bacterium that degrades toxic pyridinediols. *Syst. Appl. Microbiol.* 15:522–29

Antonio L, Xu J, Little JM, Burchell B, Magdalou J, Radominska-Pandya A. 2003. Glucuronidation of catechols by human hepatic, gastric, and intestinal microsomal UDP-glucuronosyltransferases (UGT) and recombinant UGT1A6, UGT1A9, and UGT2B7. *Arch. Biochem. Biophys.* 411:251–61

Armstrong RN. 1997. Structure, catalytic mechanism, and evolution of the glutathione transferases. *Chem. Res. Toxicol.* 10:2–18

Berger PJ, Sanders EH, Gardner PD, Negus NC. 1977. Phenolic plant compounds functioning as reproductive inhibitors in *Microtus montanus. Science* 195:575–77

Blanchard RL, Freimuth RR, Buck J, Weinshilboum RM, Coughtrie MWH. 2004. A proposed nomenclature system for the cytosolic sulfotransferase (SULT) superfamily. *Pharmacogenetics* 14:199–211

Bock KW. 2003. Vertebrate UDP-glucuronosyltransferases: functional and evolutionary aspects. *Biochem. Pharmacol.* 66:691–96

Bock KW, Kohle C. 2004. Coordinate regulation of drug metabolism by xenobiotic nuclear receptors: UGTs acting together with CYPs and glucuronide transporters. *Drug Metab. Rev.* 36:595–615

Boyle R. 2000. *Metabolic fate of dietary terpenes in folivorous marsupials.* PhD thesis. Univ. Tasmania, Hobart. 217 pp.

Boyle R, McLean S, Davies NW. 2000. Biotransformation of 1,8-cineole in the brushtail possum (*Trichosurus vulpecula*). *Xenobiotica* 30:915–32

Boyle R, McLean S, Foley W, Davies NW, Peacock EJ, Moore B. 2001. Metabolites of dietary 1,8-cineole in the male koala (*Phascolarctos cinereus*). *Comp. Biochem. Physiol. C* 129:385–95

Boyle R, McLean S, Foley WJ, Davies NW. 1999. Comparative metabolism of dietary terpene, p-cymene, in generalist and specialist folivorous marsupials. *J. Chem. Ecol.* 25:2109–26

Boyle R, McLean S, Foley WJ, Moore BD, Davies NW, Brandon S. 2000. Fate of the dietary terpene, p-cymene, in the male koala. *J. Chem. Ecol.* 26:1095–111

Buchsbaum R, Valiela I, Swain T. 1984. The role of phenolic compounds and other plant constituents in feeding by Canada geese in a coastal marsh. *Oecologia* 63:343–49

Cipollini ML, Levey DJ. 1997a. Secondary metabolites of fleshy vertebrate-dispersed fruits: Adaptive hypotheses and implications for seed dispersal. *Am. Nat.* 150:346–72

Cipollini ML, Levey DJ. 1997b. Why are some fruits toxic? Glycoalkaloids in *Solanum* and fruit choice by vertebrates. *Ecology* 78:782–98

Clausen TP, Reichardt PB, Bryant JP. 1986. Pinosylvin and pinosylvin methyl-ether as feeding deterrents in green alder. *J. Chem. Ecol.* 12:2117–31

Cooper WE. 2003. Correlated evolution of herbivory and food chemical discrimination in iguanian and ambush foraging lizards. *Behav. Ecol.* 14:409–16

Cooper WE, Perez-Mellado V, Vitt LJ, Budzinsky B. 2002. Behavioral responses to plant toxins by two omnivorous lizard species. *Physiol. Behav.* 76:297–303

Cooper WE, Vitt LJ. 2002. Distribution, extent, and evolution of plant consumption by lizards. *J. Zool.* 257:487–517

Coughtrie MWH, Johnston LE. 2001. Interactions between dietary chemicals and human sulfotransferases—Molecular mechanisms and clinical significance. *Drug Metab. Dispos.* 29:522–28

Court MH, Greenblatt DJ. 2000. Molecular genetic basis for deficient acetaminophen glucuronidation by cats: UGT1A6 is a pseudogene, and evidence for reduced diversity of expressed hepatic UGT1A isoforms. *Pharmacogenetics* 10:355–69

Cummins CL, Jacobsen W, Benet LZ. 2002. Unmasking the dynamic interplay between intestinal P-glycoprotein and CYP3A4. *J. Pharmacol. Exp. Ther.* 300:1036–45

Dearing MD. 1997. Effects of *Acomastylis rossii* tannins on a mammalian herbivore, the North American pika, *Ochotona princeps. Oecologia* 109:122–31

Dearing MD, Schall JJ. 1992. Testing models of optimal diet assembly by the generalist herbivorous lizard *Cnemidophorus murinus. Ecology* 7:845–58

Ding XX, Kaminsky LS. 2003. Human extrahepatic cytochromes P450: Function in xenobiotic metabolism and tissue-selective chemical toxicity in the respiratory and gastrointestinal tracts. *Annu. Rev. Pharmacol. Toxicol.* 43:149–73

Dominguez-Bello MG, Michelangeli F, Ruiz MC, Garcia A, Rodriguez E. 1994. Ecology of the folivorous hoatzin (*Opisthocomus hoazin*) on the Venezuelan plains. *Auk* 11:643–51

Dresser GK, Schwarz UI, Wilkinson GR, Kim RB. 2003. Coordinate induction of both cytochrome P4503A and MDRI by St John's wort in healthy subjects. *Clin. Pharmacol. Ther.* 73:41–50

Duncan AJ, Frutos P, Young SA. 2000. The effect of rumen adaptation to oxalic acid on selection of oxalic-acid-rich plants by goats. *Br. J. Nutr.* 83:59–65

Duncan AJ, Young SA. 2002. Can goats learn about foods through conditioned food aversions and preferences when multiple food options are simultaneously available? *J. Anim. Sci.* 80:2091–98

Eaton DL, Bammler TK. 1999. Concise review of the glutathione S-transferases and their significance to toxicology. *Toxicol. Sci.* 49:156–64

Foley WJ, Iason GR, McArthur C. 1999. Role of plant secondary metabolites in the nutritional ecology of mammalian herbivores—how far have we come in 25 years? In *Nutritional Ecology of Herbivores. 5th Int. Symp. Nutr. Herbivores,* ed. HJG Jung, GC Fahey Jr, pp. 130–209. Savoy, IL: Am. Soc. Anim. Sci.

Foley WJ, McArthur C. 1994. The effects and costs of ingested allelochemicals in mammals: an ecological perspective. In *The Digestive System in Mammals: Food, Form and Function,* ed. DJ Chivers, P Langer, pp. 370–91. Cambridge, UK: Cambridge Univ. Press

Freeland WJ, Janzen DH. 1974. Strategies in herbivory by mammals: The role of plant secondary compounds. *Am. Nat.* 108:269–89

Fu PP, Xia QS, Lin G, Chou MW. 2004. Pyrrolizidine alkaloids—Genotoxicity, metabolism enzymes, metabolic activation, and mechanisms. *Drug Metab. Rev.* 36:1–55

Glatt H, Meinl W. 2004. Pharmacogenetics of soluble sulfotransferases (SULTs). *Naunyn-Schmiedebergs Arch. Pharmacol.* 369:55–68

Gonzalez FJ, Gelboin HV. 1994. Role of human cytochromes P450 in the metabolic activation of chemical carcinogens and toxins. *Drug Metab. Rev.* 26:165–83

Gonzalez FJ, Kimura S. 2003. Study of P450 function using gene knockout and transgenic mice. *Arch. Biochem. Biophys.* 409:153–58

Gonzalez FJ, Nebert DW. 1990. Evolution of the P450-gene superfamily—Animal plant warfare, molecular drive and human genetic differences in drug oxidation. *Trends Genet.* 6:182–86

Grajal A, Strahl SD, Parra R, Dominguez MG, Neher A. 1989. Foregut fermentation in the Hoatzin, a neotropical leaf-eating bird. *Science* 245:1236–38

Green AK, Haley SL, Dearing MD, Barnes

DM, Karasov WH. 2004. Intestinal capacity of P-glycoprotein is higher in the juniper specialist, *Neotoma stephensi*, than the sympatric generalist, *Neotoma albigula*. *Comp. Biochem. Physiol. A* 139:325–33

Greenblatt DJ, von Moltke LL, Harmatz JS, Chen GS, Weemhoff JL, et al. 2003. Time course of recovery of cytochrome P450 3A function after single doses of grapefruit juice. *Clin. Pharmacol. Ther.* 74:121–29

Gregg K, Hamdorf B, Henderson K, Kopecny J, Wong C. 1998. Genetically modified ruminal bacteria protect sheep from fluoroacetate poisoning. *Appl. Environ. Microbiol.* 64:3496–98

Guengerich FP. 2002. Update information on human P450s. *Drug Metab. Rev.* 34:7–15

Guengerich FP. 2004. Cytochrome p450: What have we learned and what are the future issues? *Drug Metab. Rev.* 36:159–97

Hagerman AE, Robbins CT. 1993. Specificity of tannin-binding salivary proteins relative to diet selection by mammals. *Can. J. Zool.* 71:628–33

Hamid MR, Bachmann E, Metwally SA. 1985. Interaction of capsaicin with mixed function oxidases: *ex-vivo* and *in-vivo* studies. *J. Drug Res.* 16:29–36

Harris RZ, Jang GR, Tsunoda S. 2003. Dietary effects on drug metabolism and transport. *Clin. Pharmacokinet.* 42:1071–88

Hewitt DG, Kirkpatrick RL. 1997. Ruffed grouse consumption and detoxification of evergreen leaves. *J. Wildlife Manage.* 61:129–39

Hoffmann U, Kroemer HK. 2004. The ABC transporters MDR1 and MRP2: Multiple functions in disposition of xenobiotics and drug resistance. *Drug Metab. Rev.* 36:669–701

Hollenberg PF. 2002. Characteristics and common properties of inhibitors, inducers, and activators of CYP enzymes. *Drug Metab. Rev.* 34:17–35

Ieiri I, Takane H, Otsubo K. 2004. The MDR1 (ABCB1) gene polymorphism and its clinical implications. *Clin. Pharmacokinet.* 43:553–76

Jakubas WJ, Gullion GW. 1990. Coniferyl benzoate in quaking aspen—a ruffed grouse feeding deterrent. *J. Chem. Ecol.* 16:1077–87

Jakubas WJ, Guillion GW, Clausen TP. 1989. Ruffed grouse feeding behavior and its relationship to secondary metabolites of quaking aspen flower buds. *J. Chem. Ecol.* 15:1899–917

Jakubas WJ, Karasov WH, Guglielmo CG. 1993. Ruffed grouse tolerance and biotransformation of the plant secondary metabolite coniferyl benzoate. *Condor* 95:625–40

Johnson BM, Charman WN, Porter CJH. 2001. The impact of P-glycoprotein efflux on enterocyte residence time and enterocyte-based metabolism of verapamil. *J. Pharm. Pharmacol.* 53:1611–19

Jones RJ, Amado MAG, Dominguez-Bello MG. 2000. Comparison of the digestive ability of crop fluid from the folivorous Hoatzin (*Opisthocomus hoazin*) and cow rumen fluid with seven tropical forages. *Anim. Feed Sci. Tech.* 87:287–96

Jones RJ, Megarrity RG. 1986. Successful transfer of DHP-degrading bacteria from Hawaiian goats to Australian ruminants to overcome the toxicity of *Leucaena*. *Aust. Vet. J.* 63:259–62

Jones RJ, Meyer JHF, Bechaz FM, Stoltz MA, Palmer B, Van der Merwe G. 2001. Comparison of rumen fluid from South African game species and from sheep to digest tanniniferous browse. *Aust. J. Agric. Res.* 52:453–60

Jones RJ, Palmer B. 2000. *In vitro* digestion studies using ^{14}C-labelled polyethylene glycol (PEG) 4000: comparison of six tanniniferous shrub legumes and the grass *Panicum maximum*. *Anim. Feed Sci. Tech.* 85:215–21

Jordt SE, Julius D. 2002. Molecular basis for species-specific sensitivity to "hot" chili peppers. *Cell* 108:421–30

Karban R, Agrawal AA. 2002. Herbivore offense. *Annu. Rev. Ecol. Evol. Syst.* 33:641–64

Kauffman FC. 2004. Sulfonation in pharmacology and toxicology. *Drug Metab. Rev.* 36:823–43

Klein PJ, Buckner R, Kelly J, Coulombe RA. 2000. Biochemical basis for the extreme sensitivity of turkeys to aflatoxin B-1. *Toxicol. Appl. Pharmacol.* 165:45–52

Konig J, Nies AT, Cui YH, Leier I, Keppler D. 1999. Conjugate export pumps of the multidrug resistance protein (MRP) family: localization, substrate specificity, and MRP2-mediated drug resistance. *BBA-Biomembranes* 1461:377–94

Krause DO, Smith WJM, McSweeney CS. 2004. Use of community genome arrays (CGAs) to assess the effects of *Acacia angustissima* on rumen ecology. *Microbiology* 150:2899–909

Kullak-Ublick GA, Becker MB. 2003. Regulation of drug and bile salt transporters in liver and intestine. *Drug Metab. Rev* 35:305–17

Lamb JG, Marick P, Sorensen J, Haley S, Dearing MD. 2004. Liver biotransforming enzymes in woodrats *Neotoma stephensi* (Muridae). *Comp. Biochem. Physiol. C* 138:195–201

Lawler IR, Foley WJ, Eschler BM. 2000. Foliar concentration of a single toxin creates habitat patchiness for a marsupial folivore. *Ecology* 81:1327–38

Lawler IR, Foley WJ, Pass GJ, Eschler BM. 1998. Administration of a 5HT$_3$ receptor antagonist increases the intake of diets containing *Eucalyptus* secondary metabolites by marsupials. *J. Comp. Physiol. B* 168:611–18

Lee G, Bendayan R. 2004. Functional expression and localization of P-glycoprotein in the central nervous system: Relevance to the pathogenesis and treatment of neurological disorders. *Pharm. Res.* 21:1313–30

Lin JH, Lu AYH. 2001. Interindividual variability in inhibition and induction of cytochrome P450 enzymes. *Annu. Rev. Pharmacol. Toxicol.* 41:535–67

Lin JH, Yamazaki M. 2003. Role of P-glycoprotein in pharmacokinetics—Clinical implications. *Clin. Pharmacokinet.* 42:59–98

Lu Y, Bennick A. 1998. Interaction of tannin with human salivary proline-rich proteins. *Arch. Oral Biol.* 43:717–28

Manson MM, Ball HWL, Barrett MC, Clark HL, Judah DJ, et al. 1997. Mechanism of action of dietary chemoprotective agents in rat liver: induction of phase I and II drug metabolizing enzymes and aflatoxin B-1 metabolism. *Carcinogenesis* 18:1729–38

Margolskee RF. 2002. Molecular mechanisms of bitter and sweet taste transduction. *J. Biol. Chem.* 277:1–4

Marsh KJ, Wallis IR, Foley WJ. 2005. Detoxification rates constrain feeding in common brushtail possums (*Trichosurus vulpecula*). *Ecology.* In press

McArthur C, Sanson GD, Beal AM. 1995. Salivary proline-rich proteins in mammals— roles in oral homeostasis and counteracting diet. *J. Chem. Ecol.* 21:663–91

McLean S, Brandon S, Davies NW, Boyle R, Foley WJ, et al. 2003. Glucuronuria in the koala. *J. Chem. Ecol.* 29:1465–77

McLean S, Brandon S, Davies NW, Foley WJ, Muller HK. 2004. Jensenone: Biological reactivity of a marsupial antifeedant from *Eucalyptus. J. Chem. Ecol.* 30:19–36

McSweeney CS, Odenyo A, Krause DO. 2002. Rumen microbial responses to antinutritive factors in fodder trees and shrub legumes. *J. Appl. Anim. Res.* 21:181–205

Miners JO, Smith PA, Sorich MJ, McKinnon RA, Mackenzie PI. 2004. Predicting human drug glucuronidation parameters: Application of *in vitro* and *in silico* modeling approaches. *Annu. Rev. Pharmacol. Toxicol.* 44:1–25

Mithen RF, Dekker M, Verkerk R, Rabot S, Johnson IT. 2000. The nutritional significance, biosynthesis and bioavailability of glucosinolates in human foods. *J. Sci. Food Agric.* 80:967–84

Moody DE, Zou Z, McIntyre L. 2002. Cross-species hybridisation of pig RNA to human nylon microarrays. *BMC Genomics* 3:27

Moore BD, Wallis IR, Pala-Paul J, Brophy JJ, Willis RH, Foley WJ. 2004. Antiherbivore chemistry of *Eucalyptus*—Cues and deterrents for marsupial folivores. *J. Chem. Ecol.* 30:1743–69

Moore LB, Goodwin B, Jones SA, Wisely GB,

Serabjit-Singh CJ, et al. 2000. St. John's wort induces hepatic drug metabolism through activation of the pregnane X receptor. *Proc. Natl. Acad. Sci. USA* 97:7500–2

Mueller SC, Uehleke B, Woehling H, Petzsch M, Majcher-Peszynska J, et al. 2004. Effect of St John's wort dose and preparations on the pharmacokinetics of digoxin. *Clin. Pharmacol. Ther.* 75:546–57

Muller-Schwarze D, Brashear H, Kinnel R, Hintz KA, Lioubomirov A, Skibo C. 2001. Food processing by animals: Do beavers leach tree bark to improve palatability? *J. Chem. Ecol.* 27:1011–28

Nagy KA. 1973. Behavior, diet and reproduction in a desert lizard, *Sauromalus obesus. Copeia* 1:93–102

Nelson DR. 2003. Comparison of P450s from human and fugu: 420 million years of vertebrate P450 evolution. *Arch. Biochem. Biophys.* 409:18–24

Pass GJ, Foley WJ. 2000. Plant secondary metabolites as mammalian feeding deterrents: separating the effects of the taste of salicin from its post-ingestive consequences in the common brushtail possum (*Trichosurus vulpecula*). *J. Comp. Physiol. B* 170:185–92

Pass GJ, McLean S, Stupans I, Davies N. 2001. Microsomal metabolism of the terpene 1,8-cineole in the common brushtail possum (*Trichosurus vulpecula*), koala (*Phascolarctos cinereus*), rat and human. *Xenobiotica* 31:205–21

Pass GJ, McLean S, Stupans I, Davies NW. 2002. Microsomal metabolism and enzyme kinetics of the terpene p-cymene in the common brushtail possum (*Trichosurus vulpecula*), koala (*Phascolarctos cinereus*) and rat. *Xenobiotica* 32:383–97

Pelkonen O. 2002. Human CYPs: *In vivo* and clinical aspects. *Drug Metab. Rev.* 34:37–46

Provenza FD, Villalba JJ, Cheney CD, Werner SJ. 1998. Self-organization of foraging behaviour: From simplicity to complexity without goals. *Nutr. Res. Rev.* 11:199–222

Provenza FD, Villalba JJ, Dziba LE, Atwood SB, Banner RE. 2003. Linking herbivore experience, varied diets, and plant biochemical diversity. *Small Rum. Res.* 49:257–74

Radominska-Pandya A, Czernik PJ, Little JM, Battaglia E, Mackenzie PI. 1999. Structural and functional studies of UDP-glucuronosyltransferases. *Drug Metab. Rev.* 31:817–99

Reichardt PB, Bryant JP, Anderson BJ, Phillips D, Clausen TP, et al. 1990. Germacrone defends labrador tea from browsing by snowshoe hares. *J. Chem. Ecol.* 16:1961–70

Rendic S. 2002. Summary of information on human CYP enzymes: Human P450 metabolism data. *Drug Metab. Rev.* 34:83–448

Rinaldi R, Eliasson E, Swedmark S, Morgenstern R. 2002. Reactive intermediates and the dynamics of glutathione transferases. *Drug Metab. Dispos.* 30:1053–58

Rosenthal GA, Berenbaum MR, eds. 1991. *Herbivores: Their Interactions with Secondary Plant Metabolites.* Vol. I. *The Chemical Participants.* San Diego: Academic

Roy JR, Bergeron J-M. 1990. Branch-cutting behavior by the vole. *J. Chem. Ecol.* 16:735–41

Santos-Buelga C, Scalbert A. 2000. Proanthocyanidins and tannin-like compounds—nature, occurrence, dietary intake and effects on nutrition and health. *J. Sci. Food Agric.* 80:1094–117

Schall JJ. 1990. Aversion of whiptail lizards (*Cnemidophorus*) to a model alkaloid. *Herpetologica* 46:34–38

Schall JJ, Ressel S. 1991. Toxic plant compounds and the diet of the predominantly herbivorous whiptail lizard, *Cnemidophorus arubensis. Copeia* 1:111–19

Schwedhelm E, Maas R, Troost R, Boger RH. 2003. Clinical pharmacokinetics of antioxidants and their impact on systemic oxidative stress. *Clin. Pharmacokinet.* 42:437–59

Sedinger JS. 1997. Adaptations to and consequences of an herbivorous diet in grouse and waterfowl. *Condor* 99:314–26

Shipkova M, Armstrong VW, Oellerich M, Wieland E. 2003. Acyl glucuronide drug metabolites: Toxicological and analytical implications. *Ther. Drug Monit.* 25:1–16

Skopec MM, Hagerman AE, Karasov WH. 2004. Do salivary proline-rich proteins counteract dietary hydrolyzable tannin in laboratory rats? *J. Chem. Ecol.* 30:1679–92

Sorensen JS, Dearing MD. 2003. Elimination of plant toxins by herbivorous woodrats: revisiting an explanation for dietary specialization in mammalian herbivores. *Oecologia* 134:88–94

Sorensen JS, McLister JD, Dearing MD. 2005a. Plant secondary compounds compromise energy budgets of a specialist and generalist mammalian herbivore. *Ecology* 86:125–39

Sorensen JS, McLister JD, Dearing MD. 2005b. Novel plant secondary metabolites impact energy budget of a specialist herbivore. *Ecology* 86:140–54

Sorensen JS, Turnbull CA, Dearing MD. 2004. A specialist herbivore (*Neotoma stephensi*) absorbs fewer plant toxins than does a generalist (*Neotoma albigula*). *Physiol. Biochem. Zool.* 77:139–48

Spahn-Langguth H, Benet LZ. 1992. Acyl glucuronides revisited: is the glucuronidation process a toxification as well as a detoxification mechanism? *Drug Metab. Rev.* 24:5–47

Tewksbury JJ, Nabhan GP. 2001. Seed dispersal—Directed deterrence by capsaicin in chillies. *Nature* 412:403–4

Walton K, Dorne JL, Renwick AG. 2001. Uncertainty factors for chemical risk assessment: interspecies differences in glucuronidation. *Food Chem. Toxicol.* 39:1175–90

Wang HB, LeCluyse EL. 2003. Role of orphan nuclear receptors in the regulation of drug-metabolising enzymes. *Clin. Pharmacokinet.* 42:1331–57

Wiggins NL, McArthur C, McLean S, Boyle R. 2003. Effects of two plant secondary metabolites, cineole and gallic acid, on nightly feeding patterns of the common brushtail possum. *J. Chem. Ecol.* 29:1447–64

Wilkinson GR. 2004. Genetic variability in cytochrome P450 3A5 and in vivo cytochrome P450 3A activity: Some answers but still questions. *Clin. Pharmacol. Ther.* 76:99–103

Yasuda T, Inaba A, Ohmori M, Endo T, Kubo S, Ohsawa K. 2000. Urinary metabolites of gallic acid in rats and their radical-scavenging effects on 1,1-diphenyl-2-picrylhydrazyl radical. *J. Nat. Prod.* 63:1444–46

Zhou SF, Gao YH, Jiang WQ, Huang M, Xu AL, Paxton JW. 2003. Interactions of herbs with cytochrome P450. *Drug Metab. Rev.* 35:35–98

Zhou SF, Lim LY, Chowbay B. 2004. Herbal modulation of P-glycoprotein. *Drug Metab. Rev.* 36:57–104

Annu. Rev. Ecol. Evol. Syst. 2005. 36:191–218
doi: 10.1146/annurev.ecolsys.36.112904.151932
First published online as a Review in Advance on August 12, 2005

Biodiversity and Litter Decomposition in Terrestrial Ecosystems

Stephan Hättenschwiler,[1] Alexei V. Tiunov,[2] and Stefan Scheu[3]

[1]Center of Functional Ecology and Evolution, CEFE-CNRS, 34293 Montpellier, France;
email: stephan.hattenschwiler@cefe.cnrs.fr

[2]Institute of Ecology and Evolution, Laboratory of Soil Zoology, 119071 Moscow, Russia;
email: a·tiunov@mail.ru

[3]Institute of Zoology, University of Technology Darmstadt, 64287 Darmstadt, Germany;
email: scheu@bio.tu-darmstadt.de

Key Words   ecosystem functioning, microorganisms, nutrient cycling, soil food
web, soil-animal diversity

■ **Abstract**   We explore empirical and theoretical evidence for the functional significance of plant-litter diversity and the extraordinary high diversity of decomposer
organisms in the process of litter decomposition and the consequences for biogeochemical cycles. Potential mechanisms for the frequently observed litter-diversity effects on
mass loss and nitrogen dynamics include fungi-driven nutrient transfer among litter
species, inhibition or stimulation of microorganisms by specific litter compounds, and
positive feedback of soil fauna due to greater habitat and food diversity. Theory predicts
positive effects of microbial diversity that result from functional niche complementarity, but the few existing experiments provide conflicting results. Microbial succession
with shifting enzymatic capabilities enhances decomposition, whereas antagonistic interactions among fungi that compete for similar resources slow litter decay. Soil-fauna
diversity manipulations indicate that the number of trophic levels, species identity, and
the presence of keystone species have a strong impact on decomposition, whereas the
importance of diversity within functional groups is not clear at present. In conclusion,
litter species and decomposer diversity can significantly influence carbon and nutrient turnover rates; however, no general or predictable pattern has emerged. Proposed
mechanisms for diversity effects need confirmation and a link to functional traits for
a comprehensive understanding of how biodiversity interacts with decomposition processes and the consequences of ongoing biodiversity loss for ecosystem functioning.

INTRODUCTION

The current fast rate of biodiversity loss warrants concern for several reasons
(Wilson 1992, Chapin et al. 2000). One major consequence of decreasing diversity is associated changes in ecosystem functioning because ecosystem processes

likely depend on the presence of a specific number of functional groups, species, and genotypes of organisms (Ehrlich & Ehrlich 1981). The biodiversity crisis documented a quite unanticipated ignorance among biologists and ecologists on some basic questions: How important is biodiversity for ecosystem processes? How much biodiversity is needed to maintain ecosystem functioning? Past research typically measured aboveground plant biomass production as one variable of ecosystem functioning and its dependence on plant-species richness. Experiments have shown that, in grassland ecosystems, primary productivity is positively related to plant-species diversity (see reviews by Schläpfer & Schmid 1999, Loreau et al. 2001, Roy 2001). Much less is known about how biodiversity affects other key ecosystem processes, such as decomposition and nutrient cycling.

In terrestrial ecosystems, the above- and belowground plant-litter input constitutes the main resource of energy and matter for an extraordinarily diverse community of soil organisms connected by highly complex interactions. In terms of biomass and species numbers, the largest number of soil organisms are involved in organic matter turnover, particularly the large groups of bacteria and fungi. Recycling of carbon and nutrients during decomposition is a fundamentally important ecosystem process (Swift et al. 1979, Cadish & Giller 1997) that has major control over the carbon cycle, nutrient availability, and, consequently, plant growth and community structure (Wardle 2002, Bardgett 2005). Plant-species composition, in turn, significantly affects ecosystem nutrient cycling through plant-nutrient uptake and use, rhizosphere interactions, production of litter of specific quality, and microenvironmental changes (Hobbie 1992, Eviner & Chapin 2003). Distinguishing these different controls is essential for a mechanistic understanding of biodiversity effects on ecosystem functioning.

The role of litter diversity for the composition and activity of soil communities and processes during decomposition has rarely been studied. This circumstance is surprising because litter quality as the overriding determinant for decomposition within a given climate (Coûteaux et al. 1995, Cadish & Giller 1997) varies tremendously among species (Perez-Harguindeguy et al. 2000, Hättenschwiler 2005). Similarly, the ecosystem consequences of the diversity of soil organisms are little understood, except for some keystone species or ecosystem engineers such as earthworms, termites, and ants (Jones et al. 1994, Anderson 1995). Despite the reasonable expectation that the diversity and composition of functional groups or feeding groups are important for ecosystem processes (Setälä 2002, Heemsbergen et al. 2004), the existence and the significance of the great species diversity within functional groups is puzzling (Scheu & Setälä 2002).

A strong need exists for increased efforts to investigate interactions among litter diversity, the diversity of soil organisms, and the processes that occur during mineralization and soil organic-matter formation. In this review, we summarize current knowledge on the functional significance of litter diversity and the decomposer system that depends on the litter for decomposition processes and feedbacks to plants. We also specify what we think are the most promising areas for future research.

DOES DECOMPOSITION CHANGE WITH ALTERED LITTER DIVERSITY?

The physicochemical environment, litter quality, and the composition of the decomposer community are the three main factors controlling litter decomposition (Berg et al. 1993, Coûteaux et al. 1995, Cadish & Giller 1997). Under given environmental conditions, the remaining two factors—litter quality and decomposers—are directly related to biological diversity. Litter-decay rates differ widely among species that decompose under identical environmental conditions (Cornelissen 1996, Wardle et al. 1997). These differences in decomposition are attributed to variation in litter traits, such as leaf toughness, nitrogen, lignin, and polyphenol concentrations, and the carbon/nitrogen and lignin/nitrogen ratios and their consequences for microbial activity and substrate utilization (Berg et al. 1993, Cadish & Giller 1997, Perez-Harguindeguy et al. 2000). On the basis of the close correlation between litter quality and decomposition, litter traits can be used as predictors for decay rates across species (Aber et al. 1990) and also serve as key variables in biogeochemical models (Parton et al. 1994, Nicolardot et al. 2001). These correlations, however, are commonly determined from decomposition of single-species litters in mesh bags, from which larger soil animals are excluded. In reality, a specific litter type rarely occurs in isolation, and the important question arises as to whether data on single-species decomposition can be combined to accurately scale up to community-level decomposition and to predict ecosystem processes. Theoretically, this combination is possible only for purely additive effects (i.e., the decomposition rate of a litter mixture is calculated as the sum of the proportions of individual litter species), but it does not work if synergistic or antagonistic effects occur among litter species.

The Influence of Litter-Species Richness on Litter-Mass Loss

In sharp contrast to the large body of literature on single litter-species decomposition, data on litter mixtures and the monocultures of each species included in the test are still rare. No more than approximately 30 studies exist (Gartner & Cardon 2004), even though species-mixture effects were first investigated more than 60 years ago (Gustafson 1943). These studies show a wide range of litter-mixing effects that can be grouped according to three distinct patterns. Roughly half of all litter mixes studied showed accelerated litter-decay rates compared with what would have been predicted from monocultures of the respective species included in the mix (Gartner & Cardon 2004). These synergistic responses ranged between 1% and 65% (mean of 17%) of increased total litter-mass loss in mixes compared with the arithmetic mean of component species. In ~30% of all cases, no significant differences occurred in observed and predicted mass losses in litter mixes (i.e., additive effects), and in the remaining 20% of mixes, antagonistic effects with a slower than predicted litter decomposition were observed. Antagonistic responses ranged between 1.5% and 22%; the mean decrease was 9% (Gartner & Cardon 2004).

Apparently, synergistic interactions among litter species are twice as frequent as antagonistic interactions, and nonadditive litter-mixing effects are overall predominant, whereas purely additive responses are more the exception than the rule.

However, such a broad comparison among studies should be interpreted with great caution, and at present, one should refrain from generalizations for at least three reasons. First, although the reports on litter-species interactions cover a wide range of different ecosystems, from the high arctic to the tropical rainforest, a strong bias exists toward temperate forest studies (roughly 60% of all studies available today). Important and highly species-diverse ecosystems, such as tropical forests, are critically underrepresented (only one study by Montagnini et al. 1993), and grassland studies are similarly rare (two studies: Bardgett & Shine 1999 and Hector et al. 2000). Second, distillation of a mean mixing effect from the different studies is difficult because it reflects a momentary state in a dynamic process that has been interrupted at different stages. The high temporal resolution of CO_2 efflux measurements by McTiernan et al. (1997) showed, for example, that an initial lower CO_2 release from a *Quercus petraea/Betula pendula* leaf-litter mix was followed by a phase of higher CO_2 release, and no net difference occurred compared with the single-species treatments over the entire incubation period. Given the large differences in the duration of experiments (between 56 and 1780 days) and in experimental protocols (leaf litter exposed under artificial laboratory conditions, litterbags, or field microcosms), an average relative-mixing effect appears ecologically rather meaningless. Third, most experiments done to date have included only two or three species and compared monocultures with just one mixing treatment. This narrow range strongly limits a thorough assessment of diversity effects and a more general description of litter-mass loss as a function of litter diversity.

The available data indicate that litter-species interactions are quite common and lead to distinct decomposition trajectories that differ from those expected from monocultures of litter. However, idiosyncratic responses to increasing species richness seem to predominate (Wardle et al. 1997), which leads to the following question: Why do some mixtures decompose faster than others? The question is addressed below in the discussion of potential mechanisms involved in litter-mixing effects.

Responses of Individual Species within Litter Mixtures

In most of the past experiments, mass loss was measured in litter mixtures as a whole and compared with the predicted or expected value on the basis of single-species decomposition. This approach may mask species-specific responses to mixing litter that might well be important for decomposition processes. Individual species might behave distinctly, as was observed in most of the few studies that separated decomposition among species within mixtures (Briones & Ineson 1996, Salamanca et al. 1998, Conn & Dighton 2000, Prescott et al. 2000, Wardle et al. 2003, Hättenschwiler & Gasser 2005). Depending on the size of the effect and the variation, such species-specific responses may not be detected at the level of the whole litter mix. Observations of five different two-species mixtures by Prescott et al. (2000) indicate that contrasting mixture effects on mass loss of component species are particularly important during the initial phase of decomposition (up to 1 year) but may disappear in later stages (after 2 years). In the litter mixtures of temperate forest trees studied by Hättenschwiler & Gasser (2005), decomposition of the three most-recalcitrant litter species, *Fagus sylvatica*, *Quercus petraea*, and *Acer campestre*, increased significantly along the diversity gradient from one to six species mixtures. In contrast, no overall diversity effect occurred on the decomposition of the more rapidly decomposing species, *Carpinus betulus*, *Prunus avium*, and *Tilia platyphyllos*. The entire litter mixture collectively, or, alternatively, the presence of one or a few specific species, influences the decay rate of certain litter types and, thus, the temporal dynamics of the litter layer composition and possibly nutrient dynamics. Changes in the litter-layer composition caused by distinct diversity effects among species may alter microhabitat structure and food availability for litter-feeding animals, which, in turn, have direct or indirect consequences for the further course of decomposition. Species-specific responses to litter mixtures may actually be more common than anticipated but remained largely undetected because species were not separately analyzed.

Litter-Mixing Effects on Nitrogen Dynamics

In addition to, or independent of, mass loss, litter diversity may also influence nitrogen (N) mineralization or immobilization in decomposing litter. Gartner & Cardon (2004) concluded that in the majority of all mixtures tested (76%), non-additive nutrient dynamics have been observed that range from 100% decreased to 25% increased net N mineralization in mixtures compared with the predicted values from monocultures. In most experiments that report a change in N dynamics caused by mixing litter species, the change did not correlate with the responses in mass loss. Briones & Ineson (1996), who differentiated between species, observed an increased N release and mass loss from *Eucalyptus globulus* when mixed with *Betula pendula*. However, *B. pendula* mass loss did not change, and N release actually decreased, which largely neutralized the enhanced N loss from *E. globulus* leaf litter. Two other studies that reported a significantly higher N release from mixtures did not detect a concomitant change in CO_2 release (McTiernan et al. 1997, for the *Picea abies/Alnus glutinosa* mix) or in mass loss (Blair et al. 1990). A higher nitrogen flux from a more diverse litter than from single-species litter most likely results in higher plant N availability (Finzi & Canham 1998) that possibly increases plant growth or alters the competitive balance among species. In contrast, decreased N loss from mixtures may indicate a diminished plant N availability caused by increased N immobilization or decreased N mineralization. This condition, however, does not necessarily imply negative consequences for ecosystem properties. For example, negative litter-mixture effects on N release can help to prevent N losses from the system after disturbances. Also, mixtures may not actually decrease N availability over longer time periods but may change the timing of N release not assessed in most experiments, which typically have a relative short duration. A different pattern of N availability over time could better match plant requirements or could favor some plant species over others.

Similar to the diversity effects on litter-mass loss, the mixing of different litter species has nonadditive, largely idiosyncratic effects on N release. The currently available data suggest little or no correlation between diversity effects on mass loss and on N release. Variable effects on nutrients other than N have been reported as well (Staaf 1980, Briones & Ineson 1996), but the studies are too few for unequivocal conclusions.

MECHANISMS AND CONSEQUENCES OF LITTER-MIXTURE EFFECTS

The investigation of litter-diversity effects on decomposition is still mostly in the exploratory stage of experimental tests designed to determine if and how litter-mass loss and nutrient mineralization is changed in mixtures of different litter types. Only a few attempts have been made to identify the underlying mechanisms and to explain observed diversity effects. The data currently at hand, however, provide some insights into potential processes that are likely involved and may help to direct future research. With some overlaps, these processes may be grouped into four complexes of mechanisms: (*a*) synergistic effects caused by nutrient transfer among litter types, (*b*) stimulating or inhibiting influences of specific litter compounds, (*c*) synergistic effects that result from improved microclimatic conditions or habitat diversity in a structurally more diverse litter layer, and (*d*) synergistic or antagonistic effects that result from interactions across trophic levels.

Nutrient Transfer

Differences in chemical composition and physical properties among different litter types and their interactions is the most obvious and promising starting point from which to build and test hypotheses. Theoretical considerations and experimental evidence suggest that a nutrient-rich litter type with a low carbon/nitrogen (C/N) ratio, and, thus, a relatively fast decomposition rate, enhances the decomposition of other, poor-quality litters (Seastedt 1984, Chapman et al. 1988, Wardle et al. 1997). The rationale behind such synergistic interaction is a preferential exploitation of the high-quality litter by decomposer organisms that eventually leads to a high nutrient availability and allows nutrient transfer to the low-quality litter. Transferred nutrients, in turn, lead to a more rapid decomposition of the low-quality litter and, consequently, of the entire litter mixture. Much of the past research was motivated by this hypothesis and, because of practical considerations in forestry, aimed at answering the question of whether the addition of an easily decomposable broadleaf tree litter accelerates decomposition of a poor-quality conifer needle litter.

N transfer from the N-rich to the N-poor litter, as well as increased microbial activity accompanied by increased mass loss in the more slowly decomposing litter type, was reported by Salamanca et al. (1998) and Briones & Ineson (1996) in one of the species pairs tested. The results of this latter study additionally suggest a net transfer of other nutrients such as potassium (K), calcium (Ca), and magnesium (Mg) between litter types. Nutrient transfer by fungal hyphae or leaching alleviates nutrient limitation to poor-quality litter decomposition and, therefore, is an intuitively compelling mechanism for synergistic effects among litter species. This process is, however, rarely convincingly demonstrated. Even if apparent nutrient transfer from one to the other litter type occurs, decomposition does not necessarily change (Staaf 1980). Nutrient-rich and easily decomposing leaf litter from *Cornus florida* did not accelerate decomposition of *Pinus taeda* needle litter after 1 year of exposure in the field (Thomas 1968). In line with this finding, no synergistic effects occurred in mixtures of *Pinus ponderosa* and *Quercus gambilii*, which differed significantly in litter quality (Klemmedson 1992). In a recent test of the nutrient-transfer hypothesis that included numerous two-species mixtures, Hoorens et al. (2003) found considerable nonadditive litter-mixing effects on decomposition that, however, were not related to differences in litter chemistry. They analyzed the differences in C, N, phosphorus (P), and phenol concentrations between litter species and their relationship to differences between observed and predicted decomposition rates and found no significant correlations. These data suggest that interactions between litter species occur equally likely in chemically similar and dissimilar species (Hoorens et al. 2003) and, thus, provide evidence against the nutrient-transfer hypothesis. However, nutrient dynamics have not been measured, and whether a net transfer from one litter species to the other actually occurred and how it might be related to differences in mass loss rates is not known.

Nutrient transfer among litter species appear to be involved occasionally as a driving mechanism for litter-mixing effects, but currently available data is contradictory and suggest rather a limited importance in the determination of litter-species interactions.

Effects of Specific Compounds

Besides the variation in nutrient concentrations, litter species differ in their composition of other compounds that might inhibit or stimulate decomposition. Inhibition of microbial growth or activity by a species-specific compound can diminish, compensate, or reverse other, simultaneously operating, stimulating effects such as nutrient transfer among litter types. Polyphenols are commonly viewed as a group of secondary plant metabolites that typically inhibit decomposition. The perception of polyphenols as inhibitors, however, is far too simple, and the variety of phenolic compounds can have many different functions within the litter layer and the underlying soil (Hättenschwiler & Vitousek 2000). Even intraspecific variation in litter polyphenol concentrations can strongly influence soil processes and ecosystem functioning (Schweitzer et al. 2004). Polyphenols as regulatory compounds are critical for a better understanding of decomposition processes in general and of litter-diversity effects in particular.

The studies by Schimel and colleagues in the Alaskan taiga (e.g., Schimel et al. 1998) provide some of the most comprehensive examinations of the diversity of polyphenol effects on soil processes. Secondary succession in these forests starts with *Salix/Alnus* communities and continues to an *Alnus/Populus*, a *Populus*, and finally a *Picea alba*–dominated community. *Populus balsamifera* was found to play a key role during succession by the production of polyphenols that interfere with soil processes (Figure 1). *P. balsamifera* leaf litter releases phenolic acids that are a microbial growth substrate; this substrate leads to increased microbial N immobilization. *P. balsamifera*–specific condensed tannins, on the other hand, inhibit microbial activity that results in reduced decomposition and N mineralization rates. Even more importantly, these condensed tannins in *P. balsamifera* leaf litter suppress *Frankia* symbionts and, consequently, reduce N_2 fixation by the early successional *Alnus*. *Alnus*-specific condensed tannins have no negative influence on N_2 fixation. Taken together, the diverse effects of *Populus*-specific phenolic compounds may ultimately enhance successional dynamics and change the nitrogen availability in these ecosystems (Figure 1).

In the boreal forest ecosystem of northern Europe, observational and experimental studies have shown that the release of the species-specific phenolic

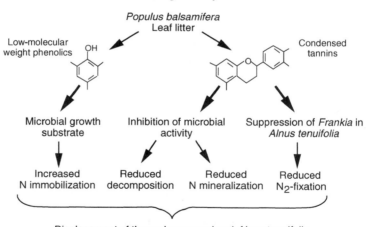

Figure 1 Schematic overview of the effects of different phenolic compounds from *Populus balsamifera* leaf litter on various soil processes and its consequences for the nitrogen cycle and successional dynamics in Alaskan taiga ecosystems. Based on research by Schimel and colleagues (e.g., Schimel et al. 1998).

compound Batatasin-III from leaf litter of the dwarf shrub *Empetrum hermaphroditum* negatively affects tree-seedling growth (Nilsson 1994). In field and laboratory experiments, the influences of allelopathy and belowground resource competition on *Pinus sylvestris* seedlings were separated (Nilsson 1994). Both allelopathy and competition independently decreased seed emergence and growth of seedlings compared with controls, and the combination of both factors led to a stronger inhibition. The negative effect of Batatasin-III on pine-seedling growth was explained by inhibition of the infection by the ectomycorrhizal fungus *Paxillus involutus* and an impaired N uptake by pine possibly caused by decreased mineralization rates. This finding documents that *E. hermaphroditum* alters soil processes by the synthesis of a specific phenolic compound in a way that its dominance is maintained in late successional stages of boreal forest ecosystems.

Even though the examples of polyphenol effects outlined above have not been studied explicitly in the context of litter-diversity effects, they suggest a strong impact of species-specific phenolic compounds on decomposition processes. A thorough analysis of the functional significance of polyphenols during decomposition of litter mixtures is lacking so far, but presumably, such an analysis will contribute to a mechanistic understanding of litter-diversity effects.

Improved Microenvironmental Conditions

A higher diverse litter layer can reasonably be assumed to be structurally richer than a monospecific litter layer. Different leaf sizes, leaf shapes, leaf-surface structures, and leaf colors all contribute to a distinct geometric organization, water-holding capacity, and radiative-energy balance in a species-rich litter layer. Such differences influence microclimatic conditions and microhabitat structure for soil animals and, therefore, have indirect consequences for decomposition.

Wardle et al. (2003) used litterbags of two adjacent compartments to study the influence of 10 different boreal forest litter species on each other's decomposition. One of the most interesting findings was the promotion of litter-mass loss and N loss of associated litter species by the presence of feather mosses (*Pleurozium schreberi* and *Hylocomium splendens*). Although feather mosses themselves are slowly decomposing, their high water-holding capacity apparently stimulated decomposition of adjacent litter species, which shows clear evidence for improved microclimatic conditions for decomposition by the presence of certain species or a particular functional group of litter species.

In a litterbag decomposition experiment that involved litter of three different broadleaf deciduous tree species, Hansen & Coleman (1998) found significantly greater microhabitat diversity and associated species richness of oribatid mites in mixed litter than in the three monocultures but found no difference in mite abundance. A similar result was obtained by Kaneko & Salamanca (1999), who observed a higher species richness of oribatid mites and a higher abundance of microarthropods in litter mixtures compared with single-species litterbags. However, the two studies are in contrast with respect to litter-mass loss. Whereas the greater faunal abundance and diversity correlated with increased mass loss in the experiment by Kaneko & Salamanca (1999), the study by Hansen & Coleman (1998) found no effect on litter-decay rate. These results might be seen as evidence for a greater importance of faunal abundance over faunal diversity for process rates.

A significantly greater initial N loss followed by a lower N immobilization was observed in three-species mixtures compared with monocultures in another litterbag study, findings that were explained by reduced fungal biomass in litter mixtures (Blair et al. 1990) (Figure 2). Moreover, fewer fungal hyphae correlated with more nematodes, apparently including fungal feeders, that might have benefited from a more diverse or microclimatically more suitable habitat in litter mixtures. Although these studies cannot fully distinguish nontrophic microenvironmental factors from trophic factors, they present convincing evidence that microenvironment-driven positive litter-species interactions contribute to a mechanistic understanding of synergistic litter-diversity effects on decomposition and nutrient dynamics.

Interactions Across Trophic Levels

Generally, the influence of soil fauna on decomposition is more difficult to quantify than that of microorganisms because it is largely an indirect effect. By regulating bacterial and fungal populations, protozoa and nematodes that make up the microfauna can alter litter decay and nutrient turnover. The mesofauna, of which springtails (Collembola) and mites (Acari) as the two main groups, have a similar function, but the saprophages among them additionally consume and process a

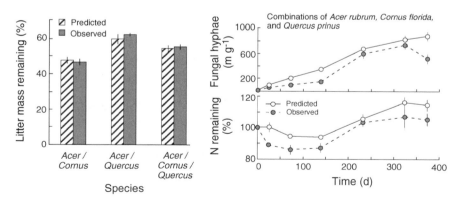

Figure 2 Comparisons of remaining litter mass after 378 days of exposure (*left*), remaining nitrogen (*bottom right*), and length of fungal hyphae (*top right*) of litter mixtures with predicted values calculated from measurements of component species. Litter mass and nitrogen are shown as percent of intial amounts, and length of hyphae is indicated in meters per gram of soil. Data points represent mean values of $n = 3$ litter bags. Reproduced with permission from Blair et al. (1990).

considerable amount of litter. Litter displacement and fragmentation and the conversion to large quantities of feces by macrofauna such as millipedes, isopods, and earthworms stimulate microbial activity and facilitate decomposition. The composition and richness of the litter layer can affect soil fauna in essentially two ways: (*a*) by shaping the microenvironment and, thus, habitat richness and patchiness, as discussed in the previous paragraph and (*b*) by providing a range of different food resources, which is the topic of the paragraph to follow.

Litter-feeding macrofauna have a tremendous impact on decomposition because they process large amounts of litter (Càrcamo et al. 2000, David & Gillon 2002) and because of their feedback on performance, activity, and community composition of microbial decomposers and smaller litter and soil fauna (Seastedt 1984, Scheu 1987, Anderson 1988, Brown 1995, Maraun et al. 1999). The saprophagous macrofauna preferentially feed on certain litter types (Zimmer & Topp 2000, Càrcamo et al. 2000, Hättenschwiler & Bretscher 2001) and are quite sensitive to changes in quality, even within a single-litter species (Hassall et al. 1987, Hättenschwiler et al. 1999). For example, isopods changed their feeding rates in particular litter species, depending on whether they did or did not have a choice among three different species (Figure 3). Consumption rates differed much less among litter species provided in monocultures than when the same species were provided in mixtures. Compared with monocultures, consumption rates of mixtures declined by factors of 1.8 for *Acer pseudoplatanus*, 3.6 for *Fagus sylvatica*, and of 4.4 for *Quercus robur* (Figure 3). On the basis of the assumption of a stimulating effect of litter

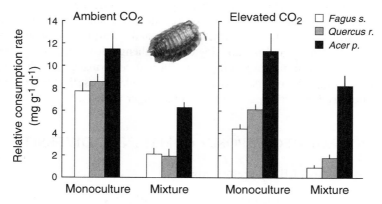

Figure 3 Relative consumption rates (in milligrams of leaf litter per grams of animal body mass per day) of the isopod *Oniscus asellus* feeding on any one of three litter species (*Fagus sylvatica*, *Quercus robur*, or *Acer pseudoplatanus*) or a mixture of all three litter species ($n = 5$ microcosms). Litter has been produced at either current ambient CO_2 (*left*) or elevated CO_2 (*right*) concentrations under otherwise identical growth conditions in the field. Total litter consumption in mixtures is the sum of the three individual columns. Data modified from Hättenschwiler & Bretscher (2001).

processing by isopods on further decay, this result suggests a faster decomposition of some preferred litter species (in this case *Acer*) compared with others when they occur together in mixtures but not when they occur in monocultures. Interestingly, when litter of the same species was produced in a CO_2-enriched atmosphere, shifts in the relative consumption of different litter species became more pronounced in both the monocultures and the mixtures (Figure 3). Litter-quality changes induced by rising atmospheric CO_2 concentration or other environmental changes, thus, can affect food selection and overall litter consumption by macrofauna. This behavior likely has consequences for decomposition and nutrient mineralization.

The influence of litter-feeding macrofauna on the decomposition of particular litter species can depend on litter-species diversity, as indicated by highly significant interactions between litter-species number and macrofauna presence in a recent field study (Hättenschwiler & Gasser 2005). For example, recalcitrant *Quercus petraea* leaf litter decomposed substantially faster with increasing litter diversity in the presence of the millipedes *Glomeris marginata/G. conspersa* (Figure 4). However, when *Glomeris* was absent, the number of associated litter species no longer influenced *Q. petraea* decomposition. Although another important litter-feeding animal, the anecic earthworm *Aporrectodea longa*, had no effect on *Q. petraea* decomposition, regardless of litter diversity, earthworm presence slowed the mass loss of the rapidly decomposing species *Prunus avium* with increasing litter-species number (Figure 4). In contrast, a significant positive relationship was seen between *P. avium* decomposition and litter-species diversity in absence of earthworms. These results clearly show that macrofauna presence can be an important driver of litter-species diversity effects.

Feedback effects between the composition and richness of litter and soil fauna appear to be important mechanisms for the understanding of how decomposition is influenced by litter diversity. Detection and quantification of such mechanisms is intellectually and methodologically challenging. In particular, litter-diversity effects on macrofauna feeding behavior and performance and the consequences for decomposition remained largely unexplored, because the litterbag method most often used in field experiments excludes larger animals.

THE ROLE OF DECOMPOSER COMMUNITY DIVERSITY

Diversity of Soil Microorganisms

Soil carbon and energy flow is mainly driven by microbial activity. The diversity of soil microorganisms is assumed to be extraordinarily high but is largely unidentified (Prosser 2002). The number of bacterial species is on the order of hundreds to thousands in 1 g of soil; total species number is estimated at 2 to 3 million (Torsvik et al. 1994, Dejonghe et al. 2001). Species diversity of soil fungi is probably only slightly less than that of bacteria (Bridge & Spooner 2001, Hawksworth 2001). One likely reason for the enormous diversity of soil microorganisms is their high fecundity combined with very short generation times and rapid growth. These

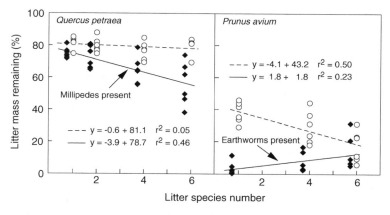

Figure 4 Remaining litter mass (percentage of initial) of a slowly decomposing species (*Quercus petraea*) (*left*) and a rapidly decomposing species (*Prunus avium*) (*right*) as a function of total litter-species number after 204 days of litter decay in the field. Data for *Q. petraea* are separated in microcosms in the presence of millipedes (*black diamonds, solid line*) and in absence of millipedes (*open circles, dashed line*). Data for *P. avium* are separated in microcosms in the presence of earthworms (*black diamonds, solid line*) and in absence of earthworms (*open circles, dashed line*). The respective regression equation along with r^2 values are indicated in the graphs. Slopes of the two regressions within litter species were significantly different. Data modified from Hättenschwiler & Gasser (2005).

factors promote a fast speciation in response to relatively small environmental changes.

Soil microbial diversity has been hypothesized to correlate positively with process rates within soils. In one of the few models that linked microbial diversity and decomposition processes, Loreau (2001) suggested that microbial diversity has a positive effect on nutrient-cycling efficiency and ecosystem processes through either greater intensity of microbial exploitation of organic matter or functional niche complementarity. Ekschmitt et al. (2001) drew similar conclusions, but few studies have been conducted to specifically address the effects of microbial diversity on process rates.

Experimental reduction in microbial diversity often did not affect gross soil processes or even increased the rate of decomposition of plant residues. After manipulation of the diversity of decomposer biota by use of chloroform fumigation, Griffiths et al. (2000) reported no consistent relationship between microbial diversity and process rates. Although nitrification, denitrification, and methane oxidation decreased along with decreasing biodiversity, plant-residue decomposition tended to be faster in pauperized soil. Similarly, decomposition of straw in fumigated and reinoculated soil consistently exceeded that in nonfumigated soil, despite reduced functional diversity of soil microorganisms (Degens 1998). In contrast, other studies found slower decomposition in response to decreased bacterial functional diversity caused by depleted uranium application (Meyer et al. 1998) or chloroform fumigation (Horwath et al. 1996). The studies that used a "synthetic" approach to create artificial microbial communities in initially sterilized substrates also gave controversial results. Salonius (1981) diluted soil suspensions to produce a gradient of microbial diversity in sterile soil microcosms. Metabolic capabilities of microbial communities were significantly reduced in less-diverse systems. In a similar experiment, no consistent effects of microbial diversity on different soil processes were found (Griffiths et al. 2001). Experiments so far provide conflicting results on the relationship between microbial diversity and rates of soil processes. However, real diversity effects could not always be separated from those introduced by disturbances. For example, chloroform fumigation selected for certain species, and diversity changed along with composition of the soil community (Griffiths et al. 2000).

Soil processes carried out by few microbial species, such as those that involve specific nutrient transformations, have been suggested to be more likely affected by shifts in diversity (Wardle 2002). Indeed, a positive correlation between over-all functional or taxonomic diversity of soil bacteria and denitrification rates was found in both laboratory and field studies (Martin et al. 1999, Griffiths et al. 2000). However, even within specific functional groups of soil bacteria (e.g., denitrifying or nitrogen-fixing bacteria) a high genetic diversity (and, thus, functional redundancy) may exist in soil and litter (Priemé et al. 2002, Widmer et al. 1999).

Does Diversity of Saprotrophic Fungi Matter?

Litter decomposition in temperate and boreal forests is mainly driven by fungal activity. Local genetic diversity of soil fungi is large, but a large portion of this diversity is present as resting stages such as conidia, spores, and inactive mycelium. The functional significance of fungal diversity can be important at small spatial scales of specific microsites within the litter layer, where only a few actively foraging hyphal tips interact with each other. The experiment by Setälä & McLean (2004) showed a clear positive effect of fungal diversity on decomposition at relatively low diversity but no influence beyond an actual diversity of 5 to 10 fungal taxa. Similarly, the decomposition of soil organic matter, and especially of pure cellulose, increased strongly with increasing number of soil fungi from monocultures to five-species mixtures (A.V. Tiunov & S. Scheu 2005a). In another experimental test, Robinson et al. (1993) reported significantly greater CO_2 release from plant litter in pairwise combinations of four fungal species compared with single-fungal-species treatments. Also, in support of positive diversity effects, Dobranic & Zak (1999) found that litter decomposition was faster at sites with high fungal diversity determined by the BIOLOG approach.

In contrast, Cox et al. (2001) documented faster pine-litter decomposition in the presence of a single fungal species compared with the same litter colonized by a diverse fungal community. In other experiments, litter decomposition by two-species or three-species mixtures of fungi did not exceed corresponding values in the best-performing monoculture (Janzen et al. 1995, Hedlund & Sjögren Öhrn 2000). Evidently, competitive interactions among fungi in a diverse community can result in reduced decomposition rates, whereas in other cases, fungal species appear to interact synergistically.

Abiotic conditions within the litter layer and litter chemical composition vary greatly. In relation to this variability, litter-decomposing fungi differ in temperature and moisture optima, and they have distinct enzymatic capabilities (Domsch et al. 1980). Niche differentiation among fungi, therefore, seems to provide a likely explanation for positive effects of fungal-species richness on litter decomposition. Common saprotrophic fungi, such as *Trichoderma*, *Mucor*, and *Rhizoctonia*, exploit spatially heterogeneous substrates by distinct strategies (Ritz 1995). The impact of fungal-species richness on litter decomposition may, therefore, be more important in heterogeneous than in homogeneous substrates, which points to the possibility of litter-diversity multiplied by fungal-diversity interactions, similar to those expected between litter diversity and litter-feeding animals discussed above. However, the only experimental test of this hypothesis to date showed larger effects of fungal diversity on decomposition of cellulose than of heterogeneous forest soil (A.V. Tiunov & S. Scheu 2005a), which suggests that facilitation rather than niche differentiation are important for interactions in species-rich fungal communities.

Interactions Between Saprotrophic and Mycorrhizal Fungi

The interaction between two main functional groups of fungi, the saprotrophic litter decomposers and the biotrophic mycorrhizal fungi, may be of greater importance for carbon and nutrient turnover than interactions within saprotrophs. Although some tree species possess arbuscular mycorrhiza, ectomycorrhizal (EM) fungi generally predominate in temperate forest ecosystems. Saprotrophic fungi (ST) gain energy from decomposing litter, whereas EM fungi receive carbon, in the form of sugars and other low-molecular-weight compounds, from their host plants. EM fungi show a considerable variation in mycelium morphology, growth pattern, enzymatic capability, and foraging strategy (Olsson et al. 2002, Lilleskov et al. 2002), but they share a wide and active set of enzymes that enables them to forage complex organic materials. Strong evidence suggests that at least some species of EM fungi gain a substantial proportion of their carbon directly from soil organic matter (Chapela et al. 2001). Competition between ST and EM fungi for nutrient and energy in the litter layer, thus, seems inevitable and may affect litter decomposition.

EM and ST basidiomycetes share many functional and structural features, including the ability for bidirectional translocation of nutrients along vegetative mycelium or particular mycelial cords and rhizomorphs (Leake et al. 2002). EM and ST fungi compete for nutrients (including organic nitrogen compounds) in forest soil, and antagonistic interactions are presumably common between these organisms (Baar & Stanton 2000, Lindahl et al. 2002). ST fungi, especially basidiomycetes, are generally more effective in breaking down dead organic matter and are almost exclusively responsible for decomposition of lignocellulose (Tanesaka et al. 1993, Colpaert & vanLaere 1996). However, because of the wide C/N ratio in most litter types, the activity of litter-decomposing fungi in temperate forests is often restricted by N availability (Ekblad & Nordgren 2002). To compensate for this deficiency, fungi translocate N from mineral soil (relatively rich in available N) to decomposing litter (Schimel & Firestone 1989, Frey et al. 2000). The activity of saprotrophs may, therefore, be strongly limited by nutrient (mainly N) sequestration by EM fungi foraging in litter or in mineral soil. Indeed, litter decomposition may be reduced in the presence of EM tree roots (Gadgil & Gadgil 1975; A.V. Tiunov, unpublished data). However, other experiments did not confirm this finding (Staaf 1988), and the interactions between ST and EM fungi remain poorly understood [cf. reviews by Leake et al. (2002) and Cairney & Meharg (2002)]. The mechanisms for either synergistic or antagonistic fungal-species interactions have been rarely addressed in a comprehensive way, which makes generalizations about the significance of fungal diversity for process rates very difficult.

The Importance of Soil-Animal Diversity

As with the microbial community, little is known about the role of animal-decomposer diversity in decomposition processes. The direct contribution of decomposer invertebrates to energy flow and carbon mineralization is low (about 10%) (Reichle et al. 1975, Schaefer 1991), whereas the direct effect on nutrient mineralization is somewhat higher (~30%) (Verhoef & Brussaard 1990, De Ruiter et al. 1993). However, the indirect effect of soil invertebrates on litter decomposition through litter fragmentation and modifications of the structure and activity of the microbial community considerably exceeds the direct effect via their own metabolism (Coleman et al. 1983, Anderson 1987, Wolters 1991). A data compilation that included 24 studies indicated that in virtually all cases, soil animals of the entire decomposer spectrum, from protists to macroarthropods, stimulated decomposition and nutrient mineralization through their effects on microorganisms (Mikola et al. 2002).

In natural ecosystems, and less so in agricultural ecosystems, the soil represents the habitat for a tremendous diversity of organisms. Moreover, soil itself is largely built through the action of animals, particularly primary and secondary decomposers (Anderson 1995, Lavelle et al. 1997, Waid 1999). Effects of soil organisms on soil processes are intimately linked to their size. Small organisms such as bacteria, fungi, and protozoa are the key drivers of energy and nutrient transformations, whereas larger decomposer organisms such as earthworms, millipedes, and isopods are the dominant habitat transformers (Lavelle et al. 1997, Anderson 2000, Scheu & Setälä 2002). These relationships suggest that at least among keystone soil-animal species, modification of the activity and structure of the microbial community, niche complementarity, and, therefore, diversity has a significant impact on decomposition and nutrient cycling. For example, evidence suggests that the diversity of earthworm species is important for microbial community composition and activity. However, in most ecosystems, earthworm diversity is comparatively low. In central Europe, for example, typically 3 to 10 species coexist. Ample evidence indicates that the different ecological groups of earthworms differentially affect the activity of soil microorganisms and decomposition processes (Shaw & Pawluk 1986, Brown et al. 2000). Experimental manipulations suggest that the loss of both functional groups and species diversity within functional groups of earthworms alters the ability of soil microorganisms to process organic substrates (Scheu et al. 2002).The functional significance of species diversity within other functional groups of soil organisms is poorly studied, and the relationship between soil-animal species diversity and soil processes remains controversial (Andrén et al. 1995, Mikola et al. 2002, Wardle 2002).

A serious difficulty for understanding the diversity–ecosystem functioning relationship in decomposer invertebrates is that knowledge on the driving forces for the evolution of soil-animal diversity is poor. The packing of animal species in soil is exceptionally dense. In forest soil, hundreds of species and thousands of individuals are concentrated in the litter layer and the uppermost mineral-soil layer the size of a footprint. Both the diversity within and the diversity between trophic groups are high. Food relationships between soil-animal species are not well studied, but some evidence shows that most taxa are food generalists rather than specialists (Anderson 1977, Petersen 2002, Maraun et al. 2003). The dominance of food generalists suggests high redundancy among soil animals, which supports evidence of a weak relationship between soil-animal diversity and ecosystem processes observed in various experiments (Bardgett & Shine 1999, Laakso & Setälä 1999, Ekschmitt et al. 2001, Cragg & Bardgett 2001). In line with these findings, the trophic structure of the decomposer community assessed with stable isotopes also indicates a high redundancy in soil-animal communities (Ponsard & Arditi 2000, Scheu & Falca 2000).

Some evidence supports the functional significance of animal diversity for soil processes (Scheu et al. 2002, Heemsbergen et al. 2004). However, studies that have explicitly addressed this question are still too few and are limited to narrow diversity gradients and low species numbers (Mikola et al. 2002). To date, the number of trophic levels and feeding guilds of soil animals (Mikola & Setälä 1998, Laakso & Setälä 1999, Setälä 2002) and the presence of certain keystone taxa (Huhta et al. 1998, Setälä 2002, Wardle 2002) appear to be more important for decomposition processes than is species diversity per se. Accordingly, Heemsbergen et al. (2004) have recently shown that functional dissimilarity rather than the number of macrofauna species drives community compositional effects on leaf litter-mass loss and soil respiration.

SOIL-DIVERSITY FEEDBACKS TO PLANTS

Changes in diversity and community structure of soil organisms likely feed back to plants and the aboveground world. On the other hand, belowground processes themselves depend on plant-community composition and diversity (Wardle 2002).

In the long-term Gisburn Forest experiment, four tree species have been planted as monocultures and all possible two-species mixtures (Chapman et al. 1988). The investigators reported significant increases in soil-fauna abundance, litter respiration, and nutrient mineralization in mixtures of *Picea abies* and *Pinus sylvestris* compared with their monocultures; a positive feedback was seen on *Picea abies* growth (Figure 5). In contrast, mixing *Picea abies* with *Alnus glutinosa* did not significantly change fauna abundance, but decreased litter respiration and mineralization compared with the predicted values of monocultures; marginally significant negative effects on *Alnus* growth and positive effects on *Picea* growth were observed. In another experiment, Nilsson et al. (1999) documented that competitive interactions among plant species change when plants are grown in humus formed from monotypic versus mixed litters, although these effects were small and tended to be idiosyncratic. In a meta-analysis of 35 studies that investigated the effects of plant litter on vegetation characteristics, Xiong & Nilsson (1999) found that plant-species richness is strongly affected by litter, which supports the notion that litter has important afterlife effects (Facelli & Pickett 1991). Despite the unequivocal

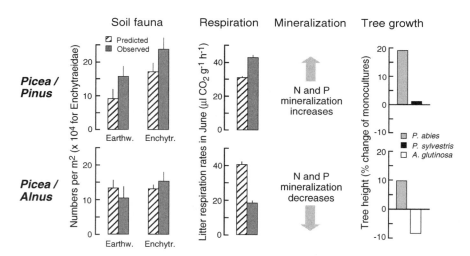

Figure 5   The consequences of mixed tree-species stands on soil-fauna abundance (numbers of individuals of earthworms and Enchytraeidea), litter-respiration rates, nitrogen (N) and phosphorus (P) mineralization rates, and their feedbacks on tree growth compared with the monocultures of the respective tree species (*n* = 3). Data for the two different tree-species mixtures *Picea abies/Pinus sylvestris* (*top*) and *Picea abies/Alnus glutinosa* (*bottom*) are shown. Soil fauna and respiration data are given as absolute numbers of actually measured values in comparison with predicted values based on measurements of monocultures. N and P mineralization are indicated as qualitative changes compared with monocultures, and tree-height growth is shown as relative changes compared with trees grown in monocultures. Data modified after Chapman et al. (1988) from the long-term Gisburn Forest experiment.

feedback of litter to plant communities, the relationship is little understood. As shown by Xiong & Nilsson (1999), litter affects plant resource competition and controls plant-community composition via the suppression of seedling establishment, particularly in early successional stages.

Soil organisms process litter that enters the detrital system not only from above ground but also from below ground. Translocation of carbon resources to below the ground in the form of roots and root exudates fuels the belowground food web and has significant implications for decomposition processes. Microorganisms are the primary recipients of this resource translocation, but they form an integral part of the soil-food web and influence the activity of litter-transforming macrofauna. Interactions between soil invertebrates and plants, mediated by soil microorganisms, are particularly numerous and important (Scheu 2001) and include, for example, grazing on mycorrhizal fungi (Klironomos & Kendrick 1995, Setälä 1995) and on plant pathogens (Curl et al. 1988, Pussard et al. 1994). Rhizosphere interactions intimately link the below- and aboveground communities. Despite their great importance, information on how these interactions affect plant growth, vegetation structure, and the aboveground food web is surprisingly limited (Van der Putten et al. 2001, Scheu & Setälä 2002, Wardle & Van der Putten 2002).

Among soil invertebrates, fungivores (Collembola, Oribatida, Nematodes) are highly abundant and usually dominate soil communities in terms of species numbers. Fungivores feed on both mycorrhizal and saprotrophic fungi, which has consequences for how these two groups of fungi interact (Tiunov & Scheu 2005b) and for nutrient transfers between plant litter, mineral soil, and plant roots (see above). Selective grazing affects fungal biomass and activity, interrupts bidirectional nutrient transfer between decomposing litter and plant roots, regulates fungal succession in decaying litter (Parkinson et al. 1979, Lussenhop 1992), and can strongly reduce mycorrhizal mycelium (Setälä 1995). However, the consequences of these grazing activities for litter decomposition are poorly studied (Sulkava & Huhta 1998). Laboratory experiments suggest that some fungivores (Collembola, Nematoda) prefer ectomycorrhizal over saprotrophic fungi (Shaw 1985, Ruess et al. 2000), but this pattern varies according to specific features of animal and fungal species. In particular, collembolans avoid toxic species of mycorrhizal basidiomycetes (Shaw 1992, Hiol et al. 1994). Furthermore, many fungivorous invertebrates function as ecosystem engineers by modifying the physical status of plant litter, and they may also feed on plant roots. Each of these trophic interactions influences plant performance, but the significance of diversity of fungivorous species is not known.

So far, we have stressed that changes in the structure of the belowground system feed back to plant community structure and the aboveground food web via modifications in plant growth, which may be viewed as bottom-up control of the plant-herbivore system. In addition, soil organisms may affect the plant-herbivore system by modifying top-down forces. Many herbivore species live within the soil at certain life stages and are integrated into the belowground food web and, thus, are subjected to belowground predation. Furthermore, some predators in their juvenile phase, such as spiders, carabid beetles, and staphylinid beetles, feed on decomposer animals. However, as adults, these predators leave the soil and forage on herbivores in the plant canopy and, thereby, foster top-down control of herbivores in their own habitat. Both processes significantly contribute to top-down control on plant herbivores and, for example, may prevent pest outbreaks. The generalist feeding habit of soil predators is an important prerequisite for this interconnection of the belowground and aboveground food web. Generalist feeding, including polyphagy, omnivory, and intraguild predation, appears to be a characteristic feature of soil predators (Scheu & Setälä 2002). This feeding habit allows switching between prey from the decomposer food web and from the aboveground system (Settle et al. 1996, Symondson et al. 2000, Halaj & Wise 2002).

In forest and agricultural ecosystems, plant-species diversity and composition may determine the susceptibility to insect outbreaks (Andow 1991, Watt 1992). Predators may reach considerably higher population densities in more diverse plant communities compared with monocultures, which increases the internal control mechanisms of prey populations. Plant-litter diversity is an important component of the positive effect of high plant-species diversity on community composition of soil predators. In agricultural systems, a clear positive correlation between the amount and composition of plant residues and the density and diversity of decomposer and predator organisms has been observed (Riechert & Bishop 1990, Mäder et al. 2002). These relationships are of key importance for successful pest management, and, therefore, a thorough understanding of trophic interactions and controls in food webs is necessary. Such understanding will only be possible if the belowground system and its key driving factors, such as the amount and diversity of litter materials, are considered in detail.

CONCLUSIONS

The literature reviewed here provides clear evidence that the diversity of litter species and decomposer organisms can significantly influence litter decomposition and nutrient mineralization and have important feedback effects to plant growth, community composition, and ecosystem functioning. However, no general relationship between species diversity and process rate has emerged so far. Responses to increasing species richness are predominantly idiosyncratic, and results are contrasting at a given level of diversity. Past research convincingly demonstrated that species identity and community composition play a critical role, but the importance of diversity per se is difficult to judge, because studies with representative gradients of species numbers, particularly for soil organisms, are rare. The characterization of functional traits that explain specific effects of a given species within the community is considered a high research priority. This effort will provide the mechanistic basis for a generalization of species-diversity effects and could promote understanding and interpretation of the conflicting results summarized here. Rather than studies to accumulate experimental evidence for diversity effects, a comprehensive approach to test hypothesized mechanisms such as nutrient transfer among litter species, or niche complementarity among soil animals, is needed to assess the functional significance of biodiversity for decomposition. Missing data have created a major gap in knowledge on interactions between substrate diversity and decomposer diversity. Although a few studies exist on how litter diversity influences microbial-community composition or soil-fauna abundance and diversity, the level of litter diversity and that of consumer diversity have not been experimentally manipulated in the same experiment. We believe that such feedback effects will elucidate the mechanisms involved in the observed processes.

ACKNOWLEDGMENTS

Financial support of A.V.T by the Alexander von Humboldt Foundation is gratefully acknowledged.

**The *Annual Review of Ecology, Evolution, and Systematics* is online at
http://ecolsys.annualreviews.org**

LITERATURE CITED

Aber JD, Melillo JM, McClaugherty CA. 1990. Predicting long-term patterns of mass loss, nitrogen dynamics, and soil organic matter formation from initial fine litter chemistry in temperate forest ecosystems. *Can. J. Bot.* 68:2201–8

Anderson JM. 1977. The organization of soil animal communities. *Ecol. Bull.* 25:15–23

Anderson JM. 1987. Interactions between invertebrates and microorganisms: noise or necessity for soil processes? In *Ecology of Microbial Communities*, ed. M Fletcher, TRG Grag, JG Jones, pp. 125–45. Cambridge, UK: Cambridge Univ. Press

Anderson JM. 1988. Spatiotemporal effects of invertebrates on soil processes. *Biol. Fertil. Soils* 6:216–27

Anderson JM. 1995. Soil organisms as engineers: microsite modulation of macroscale processes. In *Linking Species to Ecosystems*, ed. CG Jones, JH Lawton, pp. 94–106. New York: Chapman & Hall

Anderson JM. 2000. Food web functioning and ecosystem processes: problems and perceptions of scaling. In *Invertebrates as Webmasters in Ecosystems*, ed. DC Coleman, PF Hendrix, pp. 3–24. Wallingford: CAB Int.

Andow DA. 1991. Vegetational diversity and arthropod population response. *Annu. Rev. Entomol.* 36:561–86

Andrén O, Bengtsson J, Clarholm M. 1995. Biodiversity and species redundancy among litter decomposers. In *The Significance and Regulation of Soil Biodiversity*, ed. HP Collins, GP Robertson, MJ Klug, pp. 141–51. Dordrecht: Kluwer Acad.

Baar J, Stanton NL. 2000. Ectomycorrhizal fungi challenged by saprotrophic basidiomycetes and soil microfungi under different ammonium regimes in vitro. *Mycol. Res.* 104:691–97

Bardgett RD. 2005. *The Biology of Soil: A Community and Ecosystem Approach*. Oxford: Oxford Univ. Press. 253 pp.

Bardgett RD, Shine A. 1999. Linkages between plant litter diversity, soil microbial biomass and ecosystem function in temperate grasslands. *Soil Biol. Biochem.* 31:317–21

Berg B, Berg MP, Bottner P, Box E, Breymeyer A, et al. 1993. Litter mass loss rates in pine forests of Europe and eastern United States: some relationship with climate and litter quality. *Biogeochemistry* 20:127–59

Blair JM, Parmelee RW, Beare MH. 1990. Decay rates, nitrogen fluxes, and decomposer communities of single- and mixed-species foliar litter. *Ecology* 71:1976–85

Bridge P, Spooner B. 2001. Soil fungi: diversity and detection. *Plant Soil* 232:147–54

Briones MJI, Ineson P. 1996. Decomposition of eucalyptus leaves in litter mixtures. *Soil Biol. Biochem.* 28:1381–88

Brown GG. 1995. How do earthworms affect microfloral and faunal community diversity? *Plant Soil* 170:209–31

Brown GG, Barois I, Lavelle P. 2000. Regulation of soil organic matter dynamics and microbial activity in the drilosphere and the role of interactions with other edaphic functional domains. *Eur. J. Soil Biol.* 36:177–98

Cadish G, Giller KE. 1997. *Driven by Nature: Plant Litter Quality and Decomposition.* Wallingford: CAB Int. 432 pp.

Cairney JWG, Meharg AA. 2002. Interactions between ectomycorrhizal fungi and soil saprotrophs: implications for decomposition of organic matter in soils and degradation of organic pollutants in the rhizosphere. *Can. J. Bot.* 80:803–9

Càrcamo HA, Abe TA, Prescott CE, Holl FB, Chanway CP. 2000. Influence of millipedes on litter decomposition, N mineralization, and microbial communities in a coastal forest in British Columbia, Canada. *Can. J. For. Res.* 30:817–26

Chapin FS III, Zavaleta E, Eviner V, Naylor R, Vitousek PM, et al. 2000. Consequences of changing biodiversity. *Nature* 405:234–42

Chapela IH, Osher LJ, Horton TR, Henn MR. 2001. Ectomycorrhizal fungi introduced with exotic pine plantations induce soil carbon depletion. *Soil Biol. Biochem.* 33:1733–40

Chapman K, Whittaker JB, Heal OW. 1988. Metabolic and faunal activity in litters of tree mixtures compared with pure stands. *Agric. Ecosyst. Environ.* 24:33–40

Coleman DC, Reid CPP, Cole CV. 1983. Biological strategies of nutrient cycling in soil systems. In *Advances in Ecological Research*, ed. A Macfadyen, ED Ford, pp. 1–55. New York: Academic

Colpaert JV, van Laere A. 1996. A comparison of the extracellular enzyme activities of two

ectomycorrhizal and a leaf-saprotrophic basidiomycete colonizing beech leaf litter. *New Phytol.* 134:133–41

Conn C, Dighton J. 2000. Litter quality influences on decomposition, ectomycorrhizal community structure and mycorrhizal root surface acid phosphatase activity. *Soil Biol. Biochem.* 32:489–96

Cornelissen JHC. 1996. An experimental comparison of leaf decomposition rates in a wide range of temperate plant species and types. *J. Ecol.* 84:573–82

Coûteaux M-M, Bottner P, Berg B. 1995. Litter decomposition, climate and litter quality. *Trends Ecol. Evol.* 10:63–66

Cox P, Wilkinson SP, Anderson JM. 2001. Effects of fungal inocula on the decomposition of lignin and structural polysaccharides in *Pinus sylvestris* litter. *Biol. Fertil. Soils* 33:246–51

Cragg RG, Bardgett RD. 2001. How changes in soil faunal diversity and composition within a trophic group influence decomposition processes. *Soil Biol. Biochem.* 33:2073–81

Curl EA, Lartey R, Peterson CM. 1988. Interactions between root pathogens and soil microarthropods. *Agric. Ecosyst. Environ.* 24:249–61

David J-F, Gillon D. 2002. Annual feeding rate of the millipede *Glomeris marginata* on holm oak (*Quercus ilex*) leaf litter under Mediterranean conditions. *Pedobiolology* 46:42–52

Degens BP. 1998. Decreases in microbial functional diversity do not result in corresponding changes in decomposition under different moisture conditions. *Soil Biol. Biochem.* 30:1989–2000

Dejonghe W, Boon N, Seghers D, Top EM, Verstraete W. 2001. Bioaugmentation of soils by increasing microbial richness: missing links. *Environ. Microbiol.* 3:649–57

De Ruiter PC, Moore JC, Zwart KB, Bouwman LA, Hassink J, et al. 1993. Simulation of nitrogen mineralization in the belowground food webs of two winter wheat fields. *J. Appl. Ecol.* 30:95–106

Dobranic JK, Zak JC. 1999. A microtiter plate

procedure for evaluating fungal functional diversity. *Mycologia* 91:756–65

Domsch KH, Gams W, Anderson T-H. 1980. *Compendium of Soil Fungi*. Eching, Ger.: IHW-Verlag, 859 pp.

Ehrlich PR, Ehrlich AH. 1981. *Extinction: The Causes and Consequences of the Disappearance of Species*. New York: Random House. 305 pp.

Ekblad A, Nordgren A. 2002. Is growth of soil microorganisms in boreal forests limited by carbon or nitrogen availability? *Plant Soil* 242:115–22

Ekschmitt K, Klein A, Pieper B, Wolters V. 2001. Biodiversity and functioning of ecological communities—why is diversity important in some cases and unimportant in others? *J. Plant Nutr. Soil Sci.* 164:239–46

Eviner VT, Chapin FS III. 2003. Functional matrix: a conceptual framework for predicting multiple plant effects on ecosystem processes. *Annu. Rev. Ecol. Evol. Syst.* 34:487–515

Facelli JM, Pickett STA. 1991. Plant litter: Its dynamics and effects on plant community structure. *Bot. Rev.* 52:1–32

Finzi AC, Canham CD. 1998. Non-additive effects of litter mixtures on net N mineralization in a southern New England forest. *For. Ecol. Manag.* 105:129–36

Frey SD, Elliott ET, Paustian K, Peterson GA. 2000. Fungal translocation as a mechanism for soil nitrogen inputs to surface residue decomposition in a no-tillage agroecosystem. *Soil Biol. Biochem.* 32:689–98

Gadgil RL, Gadgil PD. 1975. Supression of litter decomposition by mycorrhizal roots of *Pinus radiata*. *N. Z. J. For. Sci.* 5:35–41

Gartner TB, Cardon ZG. 2004. Decomposition dynamics in mixed-species leaf litter. *Oikos* 104:230–46

Griffiths BS, Ritz K, Bardgett RD, Cook R, Christensen S, et al. 2000. Ecosystem response of pasture soil communities to fumigation-induced microbial diversity reductions: an examination of the biodiversity-ecosystem function relationship. *Oikos* 90:279–94

Griffiths BS, Ritz K, Wheatley R, Kuan HL, Boag B, et al. 2001. An examination of the biodiversity-ecosystem function relationship in arable soil microbial communities. *Soil Biol. Biochem.* 33:1713–22

Gustafson FG. 1943. Decomposition of the leaves of some forest trees under field conditions. *Plant Phys.* 18:704–07

Halaj J, Wise DH. 2002. Impact of a detrital subsidy on trophic cascades in a terrestrial grazing food web. *Ecology* 83:3141–51

Hansen RA, Coleman DC. 1998. Litter complexity and composition are determinants of the diversity and species composition of oribatid mites (Acari: Oribatidae) in litterbags. *Appl. Soil Ecol.* 9:17–23

Hassall M, Turner JG, Rands MRW. 1987. Effects of terrestrial isopods on the decomposition of different woodland leaf litter. *Oecologia* 72:597–604

Hättenschwiler S. 2005. Effects of tree species diversity on litter quality and decomposition. In *Forest Diversity and Function: Temperate and Boreal Systems*, ed. M Scherer-Lorenzen, C Körner, E-D Schulze, 176:149–64. Heidelberg: Springer-Verlag

Hättenschwiler S, Bretscher D. 2001. Isopod effects on decomposition of litter produced under elevated CO_2, N deposition and different soil types. *Global Change Biol.* 7:565–79

Hättenschwiler S, Gasser P. 2005. Soil animals alter plant litter diversity effects on decomposition. *Proc. Natl. Acad. Sci. USA* 102:1519–24

Hättenschwiler S, Vitousek PM. 2000. The role of polyphenols in terrestrial ecosystem nutrient cycling. *Trends Ecol. Evol.* 15:238–43

Hättenschwiler S, Bühler S, Körner C. 1999. Effects of elevated CO_2 on quality, decomposition and isopod consumption of tree litter. *Oikos* 85:271–81

Hawksworth DL. 2001. The magnitude of fungal diversity: the 1.5 million species estimate revisited. *Mycol. Res.* 105:1422–32

Hector A, Beale AJ, Minns A, Otway SJ, Lawton JH. 2000. Consequences of the reduction of plant diversity for litter decomposition:

effects through litter quality and microenvironment. *Oikos* 90:357–71

Hedlund K, Sjögren Öhrn M. 2000. Tritrophic interactions in a soil community enhance decomposition rates. *Oikos* 88:585–91

Heemsbergen DA, Berg MP, Loreau M, van Haj JR, Faber JH, Verhoef HA. 2004. Biodiversity effects on soil processes explained by interspecific functional dissimilarity. *Science* 306:1019–20

Hiol FH, Dixon RK, Curl EA. 1994. The feeding preference of mycophagous Collembola varies with the fungal symbiont. 5:99–103

Hobbie SE. 1992. Effects of plant species on nutrient cycling. *Trends Ecol. Evol.* 7:336–39

Hoorens B, Aerts R, Stroetenga M. 2003. Does initial litter chemistry explain litter mixture effects on decomposition? *Oecologia* 442:578–86

Horwath WR, Paul EA, Harris D, Norton J, Jagger L, Horton KA. 1996. Defining a realistic control for the chloroform fumigation-incubation method using microscopic counting and C-14-substrates. *Can. J. Soil Sci.* 76:459–67

Huhta V, Persson T, Setälä H. 1998. Functional implications of soil fauna diversity in boreal forests. *Appl. Soil Ecol.* 10:277–88

Janzen RA, Dormaar JF, McGill WB. 1995. A community-level concept of controls on decomposition processes—decomposition of barley straw by phanerochaete-chrysosporium or phlebia-radiata in pure or mixed culture. *Soil Biol. Biochem.* 27:173–79

Jones CG, Lawton JH, Shachak M. 1994. Organisms as ecosystem engineers. *Oikos* 69:373–86

Kaneko N, Salamanca EF. 1999. Mixed leaf litter effects on decomposition rates and soil microarthropod communities in an oak-pine stand in Japan. *Ecol. Res.* 14:131–38

Klemmedson JO. 1992. Decomposition and nutrient release from mixtures of Gambel oak and ponderosa pine leaf litter. *For. Ecol. Manag.* 47:349–61

Klironomos JN, Kendrick WB. 1995. Stimulative effects of arthropods on endomycorrhizas of sugar maple in the presence of decaying litter. *Funct. Ecol.* 9:528–36

Laakso J, Salminen J, Setälä H. 1995. Effects of abiotic conditions and microarthropod predation on the structure and function of soil animal communities. *Acta Zool. Fenn.* 196:162–67

Laakso J, Setälä H. 1999. Sensitivity of primary production to changes in the architecture of belowground food webs. *Oikos* 87:57–64

Lavelle P, Bignell D, Lepage M, Wolters V, Roger P, et al. 1997. Soil function in a changing world: the role of invertebrate ecosystem engineers. *Eur. J. Soil Biol.* 33:159–93

Leake JR, Donnelly DP, Boddy L. 2002. Interactions between ecto-mycorrhizal and saprotrophic fungi. In *Mycorrhizal Ecology*, ed. MGA Van der Heijden, I Sanders, pp. 345–72. Berlin: Springer-Verlag

Lilleskov EA, Hobbie EA, Fahey TJ. 2002. Ectomycorrhizal fungal taxa differing in response to nitrogen deposition also differ in pure culture organic nitrogen use and natural abundance of nitrogen isotopes. *New Phytol.* 154:219–31

Lindahl BO, Taylor AFS, Finlay RD. 2002. Defining nutritional constraints on carbon cycling in boreal forests—towards a less 'phytocentric' perspective. *Plant Soil* 242:123–35

Loreau M. 2001. Micobial diversity, producer-decomposer interactions and ecosystem processes: a theoretical model. *Proc. R. Soc. London Ser. B* 268:303–9

Loreau M, Naeem S, Inchausti P, Bengtsson J, Grime JP, et al. 2001. Biodiversity and ecosystem functioning: current knowledge and future challenges. *Science* 294:804–8

Lussenhop L. 1992. Mechanisms of microarthropod-microbial interactions in soil. *Adv. Ecol. Res.* 23:1–33

Mäder P, Fließbach A, Dubois D, Gunst L, Fried P, Niggli U. 2002. Soil fertility and biodiversity in organic farming. *Science* 296:1694–97

Maraun M, Alphei J, Bonkowski M, Buryn R, Migge S, et al. 1999. Middens of the earthworm *Lumbricus terrestris* (Lumbricidae):

microhabitats for micro- and mesofauna in forest soil. *Pedobiologia* 43:276–87

Maraun M, Migge S, Theenhaus A, Scheu S. 2003. Adding to 'the enigma of soil animal species diversity': fungal feeders and saprophagous soil invertebrates prefer similar food substrates. *Eur. J. Soil Biol.* 39:85–95

Martin TL, Trevors JT, Kaushik NK. 1999. Soil microbial diversity, community structure and denitrification in a temperate riparian zone. *Biodiv. Cons.* 8:1057–78

McTiernan KB, Ineson P, Coward PA. 1997. Respiration and nutrient release from tree leaf litter mixtures. *Oikos* 78:527–38

Meyer MC, Paschke MW, McLendon T, Price D. 1998. Decreases in soil microbial function and functional diversity in response to depleted uranium. *J. Environ. Qual.* 27:1306–11

Mikola J, Setälä H. 1998. Relating species diversity to ecosystem functioning—mechanistic backgrounds and experimental approach with a decomposer food web. *Oikos* 83:180–94

Mikola J, Bardgett RD, Hedlund K. 2002. Biodiversity, ecosystem functioning and soil decomposer food webs. In *Biodiversity and Ecosystem Functioning—Synthesis and Perspectives*, ed. M Loreau, S Naeem, P Inchausti, pp. 169–80. Oxford: Oxford Univ. Press

Montagnini F, Ramstad K, Sancho F. 1993. Litterfall, litter decomposition and the use of mulch of four indigenous tree species in the Atlantic lowlands of Costa Rica. *Agrofor. Syst.* 23:39–61

Nicolardot B, Recous S, Mary B. 2001. Simulation of C and N mineralisation during crop residue decomposition: a simple dynamic model based on the C:N ratio of the residues. *Plant Soil* 228:83–103

Nilsson M-C. 1994. Separation of allelopathy and resource competition by the boreal dwarf shrub *Empetrum hermaphroditum* Hagerup. *Oecologia* 98:1–7

Nilsson M-C, Wardle DA, Dahlberg A. 1999. Effects of plant litter species composition and diversity on the boreal forest plant-soil system. *Oikos* 86:16–26

Olsson PA, Jakobsen I, Wallander H. 2002. Foraging and resource allocation strategies of mycorrhizal fungi in a patchy environment. In *Mycorrhizal Ecology*, ed. MGA Van der Heijden, I Sanders, pp. 93–115. Berlin: Springer-Verlag

Parkinson D, Visser S, Whittaker JB. 1979. Effects of collembolan grazing on fungal colonization of leaf litter. *Soil Biol. Biochem.* 11:529–35

Parton WJ, Schimel DS, Ojima DS, Cole DV. 1994. A general model for soil organic matter dynamics. In *Sensitivity to Litter Chemistry, Texture and Management—Quantitative Modeling of Soil Forming Processes*, ed. RB Bryant, RW Arnold, pp. 137–67. Madison, WI: Soil Sci. Soc. Am.

Perez-Harguindeguy N, Diaz S, Cornelissen JHC, Venramini F, Cabido M, Castellanos A. 2000. Chemistry and toughness predict leaf litter decomposition rates over a wide spectrum of functional types and taxa in central Argentina. *Plant Soil* 218:21–30

Petersen H. 2002. General aspects of collembolan ecology at the turn of the millenium. *Pedobiologia* 46:246–60

Ponsard S, Arditi R. 2000. What can stable isotopes (delta N-15 and delta C-13) tell about the food web of soil macro-invertebrates? *Ecology* 81:852–64

Prescott CE, Zabek LM, Staley CL, Kabzerns R. 2000. Decomposition of broadleaf and needle litter in forests of British Columbia: influences of litter type, forest type, and litter mixtures. *Can. J. For. Res.* 30:1742–50

Priemé A, Braker G, Tiedje JM. 2002. Diversity of nitrate reductase (nirK and nirS) gene fragments in forest upland and wetland soils. *Appl. Environ. Microbiol.* 68:1893–900

Prosser JI. 2002. Molecular and functional diversity in soil micro-organisms. *Plant Soil* 244:9–17

Pussard M, Alabouvette C, Levrat P. 1994. Protozoa interactions with soil microflora and possibilities for biocontrol of plant pathogens. In *Soil Protozoa*, ed. JF Darbyshire, pp. 123–46. Wallingford: CAB Int.

Reichle DE, McBrager JF, Ausums S. 1975. Ecological energetics of decomposer invertebrates in a deciduous forest and total respiration budget. In *Progress in Soil Zoology, Proc. 5th Int. Soil Zool. Coll.*, ed. J Vanek, 283–91. Prague: Academia Prague

Riechert SE, Bishop S. 1990. Prey control by an assemblage of generalist predators: spiders in garden test systems. *Ecology* 71:1441–50

Ritz K. 1995. Growth responses of some soil fungi to spatially heterogeneous nutrients. *FEMS Microbiol. Ecol.* 16:269–79

Robinson CH, Dighton J, Frankland JC, Coward PA. 1993. Nutrient and carbon dioxide release by interacting species of straw-decomposing fungi. *Plant Soil* 151:139–42

Roy J. 2001. How does biodiversity control primary productivity? In *Global Terrestrial Productivity*, ed. J Roy, B Saugier, HA Mooney, pp. 169–86. San Diego: Academic

Ruess L, Garcia Zapata EJ, Dighton J. 2000. Food preferences of a fungal-feeding Aphelenchoides species. *Nematology* 2:223–30

Salamanca EF, Kaneko N, Katagiri S. 1998. Effects of leaf litter mixtures on the decomposition of *Quercus serrata* and *Pinus densiflora* using field and laboratory microcosm methods. *Ecol. Engineer.* 10:53–73

Salonius PO. 1981. Metabolic capabilities of forest soil microbial populations with reduced species diversity. *Soil Biol. Biochem.* 13:1–10

Schaefer M. 1991. Ecosystem processes: Secondary production and decomposition. In *Temperate Deciduous Forests. Ecosystems of the World*, ed. E Röhrig, B Ulrich, 7:175–218. Amsterdam: Elsevier

Scheu S. 1987. Microbial activity and nutrient dynamics in earthworm casts. Lumbricidae. *Biol. Fert. Soils* 5:230–34

Scheu S. 2001. Plants and generalist predators as links between the belowground and aboveground system. *Basic Appl. Ecol.* 2:3–13

Scheu S, Falca M. 2000. The soil food web of two beech forests (*Fagus sylvatica*) of contrasting humus type: stable isotope analysis of a macro- and a mesofauna-dominated community. *Oecologia* 123:285–96

Scheu S, Setälä H. 2002. Multitrophic interactions in decomposer communities. In *Multitrophic Level Interactions*, ed. T Tscharntke, BA Hawkins, pp. 223–64. Cambridge: Cambridge Univ. Press

Scheu S, Schlitt N, Tiunov AV, Newington JE, Jones TH. 2002. Effects of the presence and community composition of earthworms on microbial community functioning. *Oecologia* 133:254–60

Schimel JP, Firestone MK. 1989. Nitrogen incorporation and flow through a coniferous forest profile. *Soil Sci. Soc. Am. J.* 53:779–84

Schimel JP, Cates RG, Ruess R. 1998. The role of balsam poplar secondary chemicals in controlling soil nutrient dynamics through succession in the Alaskan taiga. *Biogeochemistry* 42:221–34

Schläpfer F, Schmid B. 1999. Ecosystem effects of biodiversity—a classification of hypotheses and cross-system exploration of empirical results. *Ecol. Appl.* 9:893–912

Schweitzer JA, Bailey JK, Rehill BJ, Hart SC, Lindroth RL, et al. 2004. Genetically based trait in dominant tree affects ecosystem processes. *Ecol. Lett.* 7:127–34

Seastedt TR. 1984. The role of arthropods in decomposition and mineralization processes. *Annu. Rev. Entomol.* 29:25–46

Setälä H. 1995. Growth of birch and pine seedlings in relation to grazing by soil fauna on ectomycorrhizal fungi. *Ecology* 76:1844–51

Setälä H. 2002. Sensitivity of ecosystem functioning to changes in trophic structure, functional group composition and species diversity in belowground food webs. *Ecol. Res.* 17:207–15

Setälä H, McLean MA. 2004. Decomposition rate of organic substrates in relation to the species diversity of soil saprophytic fungi. *Oecologia* 139:98–107

Settle WH, Ariawan H, Astuti ET, Cahyana W, Hakim AL, et al. 1996. Managing tropical rice pests through conservation of generalist natural enemies and alternative prey. *Ecology* 77:1975–88

Shaw C, Pawluk S. 1986. Faecel microbiology of *Octolasion tyrtaeum, Aporrectodea turgida* and *Lumbricus terrestris* and its relation to the carbon budgets of three artificial soils. *Pedobiologia* 29:377–89

Shaw PJA. 1985. Grazing preferences of *Onychiurus armatus* (Insecta: Collembola) for mycorrhizal and saprophytic fungi of pine plantations. In *Ecological Interactions in Soil*, ed. AH Fitter, D Atkinson, DJ Read, MB Usher, pp. 333–37. Oxford: Blackwell

Shaw PJA. 1992. Fungi, fungivores, and fungal food webs. In *The Fungal Community*, ed. GC Carroll, DT Wicklow, pp. 295–310. New York: Marcel Dekker

Staaf H. 1980. Influence of chemical composition, addition of raspberry leaves, and nitrogen supply on decomposition rate and dynamics of nitrogen and phosphorus in beech leaf litter. *Oikos* 35:55–62

Staaf H. 1988. Litter decomposition in beech forests: Effect of excluding tree roots. *Biol. Fertil. Soils* 6:302–5

Swift MJ, Heal OW, Anderson JM. 1979. *Decomposition in Terrestrial Ecosystems.* Berkeley: Univ. Calif. Press. 509 pp.

Sulkava P, Huhta V. 1998. Habitat patchiness affects decomposition and faunal diversity: a microcosm experiment on forest floor. *Oecologia* 116:390–96

Symondson WOC, Glen DM, Erickson ML, Liddell JE, Langdon CJ. 2000. Do earthworms help to sustain the slug predator *Pterostichus melanarius* (Coleoptera: Carabidae) within crops? Investigations using monoclonal antibodies. *Mol. Ecol.* 9:1279–92

Tanesaka E, Masuda H, Kinugawa K. 1993. Wood degrading ability of basidiomycetes that are wood decomposers, litter decomposers, or mycorrhizal symbionts. *Mycologia* 85:347–54

Thomas WA. 1968. Decomposition of loblolly pine needles with and without addition of dogwood leaves. *Ecology* 49:568–71

Tinnov AV, Scheu S. 2005a. Facilitative interactions rather than resource partitioning drive diversity-functioning relationships in laboratory fungal communities. *Ecol. Lett.* 8:618–25

Tiunov AV, Scheu S. 2005b. Arbuscular mycorrhiza and Collembola interact in affecting community composition of saprotrophic microfungi. *Oecologia* 142:636–42

Torsvik V, Goksoyr J, Daae FL, Sorheim R, Michalsen J, Solte K. 1994. Use of DNA analysis to determine the diversity of microbial communities. In *Beyond the Biomass*, ed. K Ritz, J Dighton, KE Giller, pp. 39–48. New York: John Wiley & Sons

Van der Putten WH, Vet LEM, Harvey JA, Wäckers FL. 2001. Linking above- and belowground multitrophic interactions of plants, herbivores, pathogens, and their antagonists. *Trends Ecol. Evol.* 16:547–54

Verhoef HA, Brussaard L. 1990. Decomposition and nitrogen mineralization in natural and agroecosystems: the contribution of soil animals. *Biogeochemistry* 11:175–212

Waid JS. 1999. Does soil biodiversity depend upon metabiotic activity and influences? *Appl. Soil Ecol.* 13:151–58

Wardle DA. 2002. *Communities and Ecosystems: Linking the Aboveground and Belowground Components*. Princeton, NJ: Princeton Univ. Press. 392 pp.

Wardle DA, van der Putten WH. 2002. Biodiversity ecosystem functioning and aboveground–below-ground linkages. In *Biodiversity and Ecosystem Functioning—Synthesis and Perspectives*, ed. M Loreau, S Naeem, P Inchausti, pp. 155–68. Oxford: Oxford Univ. Press

Wardle DA, Bonner KI, Nicholson KS. 1997. Biodiversity and plant litter: experimental evidence which does not support the view that enhanced species richness improves ecosystem function. *Oikos* 79:247–58

Wardle DA, Nilsson M-C, Zackrisson O, Gallet C. 2003. Determinants of litter mixing effects in a Swedish boreal forest. *Soil Biol. Biochem.* 35:827–35

Watt AD. 1992. Insect pest population

dynamics: Effects of tree species diversity. In *The Ecology of Mixed-Species Stands of Trees*, ed. MGR Cannell, DC Malcolm, PA Robertson, pp. 267–75. Oxford: Blackwell

Widmer F, Shaffer BT, Porteous LA, Seidler RJ. 1999. Analysis of nifH gene pool complexity in soil and litter at a Douglas fir forest site in Oregon cascade mountain range. *Appl. Environ. Microbiol.* 65:374–80

Wilson EO. 1992. *The Diversity of Life*. New York: WW Norton. 424 pp.

Wolters V. 1991. Soil invertebrates: Effects on nutrient turnover and soil structure—a review. *Z. Pflanzenern. Bodenk.* 154:389–402

Xiong SJ, Nilsson C. 1999. The effects of plant litter on vegetation: a meta-analysis. *J. Ecol.* 87:984–94

Zimmer M, Topp W. 2000. Species-specific utilization of food sources by sympatric woodlice (Isopoda: Oniscidea). *J. Anim. Ecol.* 69:1071–82

Annu. Rev. Ecol. Evol. Syst. 2005. 36:219–42
doi: 10.1146/annurev.ecolsys.36.102003.152620
First published online as a Review in Advance on August 12, 2005

THE FUNCTIONAL SIGNIFICANCE OF RIBOSOMAL (r)DNA VARIATION: Impacts on the Evolutionary Ecology of Organisms

Lawrence J. Weider,[1] James J. Elser,[2] Teresa J. Crease,[3] Mariana Mateos,[4] James B. Cotner,[5] and Therese A. Markow[4]

[1] Department of Zoology and Biological Station, University of Oklahoma, Norman, Oklahoma 73019; email: ljweider@ou.edu

[2] Department of Life Sciences, Division of Ecology and Organismal Biology, Arizona State University, Tempe, Arizona 85069-7100; email: j.elser@asu.edu

[3] Department of Integrative Biology, University of Guelph, Canada N1G 2W1; email: tcrease@uoguelph.ca

[4] Department of Ecology and Evolutionary Biology, University of Arizona, Tucson, Arizona 85721; email: mmateos@utep.edu; tmarkow@public.arl.arizona.edu

[5] Department of Ecology, Evolution and Behavior, University of Minnesota St. Paul, Minnesota 55108; email: cotne002@umn.edu

Key Words fitness, growth rate, intergenic spacer, microevolution, stoichiometry, ribosomal DNA

■ **Abstract** The multi-gene family that encodes ribosomal RNA (the rDNA) has been the subject of numerous review articles examining its structure and function, as well as its use as a molecular systematic marker. The purpose of this review is to integrate information about structural and functional aspects of rDNA that impact the ecology and evolution of organisms. We examine current understanding of the impact of length heterogeneity and copy number in the rDNA on fitness and the evolutionary ecology of organisms. We also examine the role that elemental ratios (biological stoichiometry) play in mediating the impact of rDNA variation in natural populations and ecosystems. The body of work examined suggests that there are strong reciprocal feedbacks between rDNA and the ecology of all organisms, from microbes to metazoans, mediated through increased phosphorus demand in organisms with high rRNA content.

INTRODUCTION

The increasingly integrative nature of science in the twentyfirst century has opened up fresh vistas in the fields of ecology, evolutionary biology, and molecular genetics, whereby a synthesis of techniques and perspectives has provided new advancements in these fields (e.g., Feder & Mitchell-Olds 2003). Herein we examine a central cellular component, the ribosome, and describe how the multi-gene family encoding ribosomal(r)RNA (the rDNA) plays a major functional role in the evolutionary ecology of organisms. Past reviews have examined the basic structure and cellular function of rDNA/rRNA (for example, Flavell 1986, Fromont-Racine et al. 2003, Moore & Steitz 2002, Sollner-Webb & Tower 1986), the evolution of rDNA among organisms (Gerbi 1985), and the use of rDNA as a molecular marker for systematics and phylogenetic reconstruction (Hillis & Dixon 1991, Mindell & Honeycutt 1990). We address the interaction between rDNA and important life-history features, especially growth rate, and how selection on rDNA variants plays an important role in the microevolutionary processes in many natural populations. We conclude by applying the emerging perspective of biological stoichiometry (Elser et al. 2000b, Sterner & Elser 2002) to examine how evolutionary forces operating on variation in rDNA can have a major effect on ecological interactions, such as competition, trophic production, and biogeochemical cycling through the central role played by ribosome biogenesis in determining growth and resource requirements.

BACKGROUND

Why study rDNA variation? Our basic thesis is that ribosome biogenesis is one of the most central processes in cellular biology from a functional perspective because of its close connections to the pace of growth and development [e.g., growth regulation mechanisms such as stringent control versus growth-related control in prokaryotes (Gourse et al. 1996), as rDNA transcription is the first step of ribosome biogenesis (Sollner-Webb & Tower 1986)]. Furthermore, rRNA synthesis represents a large energetic and nutrient sink for all growing organisms, representing $\sim$80% of cellular RNA content and 20% of total cell dry weight (Neidhardt et al. 1990). Much work in this area has focused on well-studied prokaryotes (i.e., relationship between rDNA structure and regulation and organismal function in *Escherichia coli*), but the full range of connections between rDNA structural variation and variation in rDNA expression is not completely appreciated by many researchers who study eukaryotes. In addition, the ramifications of rRNA metabolism are likely to range far beyond cell biology and will impact all organismal functions related to growth rate (e.g., life-history evolution), as well as represent a fundamental component of ecological production (Sterner & Elser 2002).

We first review the basic structure of rDNA in prokaryotes and eukaryotes; describe what is known about how transcriptional efficiencies of the rRNA-encoding genes can be influenced by variations in copy number (CN), spacer length variation, regulatory regions, and epigenetic effects; and discuss how these variations may influence growth rate. We then examine evidence for either artificial or natural selection on rDNA variants across a range of organisms. Furthermore, we will show how rDNA variations can be directly linked to the ecology of an organism via rDNA growth connections and impacts on production. Finally, because rRNA represents a critical pool of phosphorus (P) in many organisms, we link rDNA variations to the nutrient requirements of organisms, focusing on carbon:phosphorus (C:P) stoichiometry, and to the impacts of biogeochemical cycling of P as a key limiting nutrient influencing ecosystem functions.

Basic Structure of rDNA

Ribosomes are the sites of protein synthesis in all organisms. These organelles basically consist of two subunits (the large and small subunits) composed of ribosomal RNA (rRNA) and proteins and whose size (molecular weight) varies depending on systematic affinity. In prokaryotes (i.e., *E. coli*), the ribosome consists of a *30S* (small) and *50S* (large) subunit that make up the functional (*70S*) ribosome. The small subunit contains a *16S* rRNA molecule, whereas the large subunit contains a *5S* and a *23S* rRNA molecule. The genes encoding these molecules are usually closely linked and transcribed as one precursor. In addition, there are often several copies dispersed throughout the genome, but the CN is generally less than 10 in prokaryotes (Figure 1; Liao 2000).

The functional eukaryotic ribosome is also composed of a small and a large subunit. The small subunit contains a *16-18S* rRNA molecule; the large subunit contains a *5S*, *5.8S*, and a *25-28S* rRNA molecule (Figure 2). The genes encoding *5.8S*, *18S*, and *28S* rRNA are generally found together in one or more tandem arrays of a repeat unit (rDNA) that consists of the three coding regions separated by intergenic spacers (IGS). These genes are transcribed as a single large precursor that is spliced to create the mature rRNA molecules that make up the core of the ribosomal subunits (Sollner-Webb & Tower 1986).

The CN of rDNA repeats can vary from as few as one copy per haploid genome in *Tetrahymena* to hundreds or thousands of copies, and it is positively correlated with genome size (Prokopowich et al. 2000). It is thought that the multiplicity of rRNA genes is required to offset the cell's inability to amplify the original transcript via multiple rounds of translation, as occurs with mRNA. Even so, most organisms

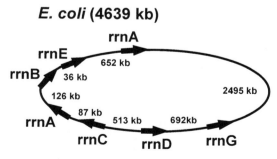

Figure 1   Diagram showing the orientation, distribution, and size, in nucleotides (expressed as kilobases—kb), of the seven rRNA operons in *E. coli* (modified from Liao 2000).

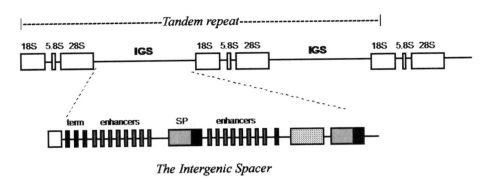

The Intergenic Spacer

Figure 2   Diagram showing the basic structure of the genes encoding for eukaryotic rRNA (the rDNA). A single tandem repeat (plus a portion of an adjoining repeat) is shown. The intergenic spacer (IGS) region is highlighted to show the location of variable numbers of subrepeats (which often contain enhancers or promoters) that greatly influence IGS length heterogeneity (modified from Sterner & Elser 2002).

possess many more copies of rDNA than are needed to meet their requirement for rRNA production because some copies remain transcriptionally inactive even at maximal growth rate (Reeder 1999).

A high degree of sequence homogeneity is observed among rDNA copies within individuals and throughout entire species despite the divergence of these sequences between species. This pattern of sequence homogeneity, which was first described in the rDNA of *Xenopus* (Brown et al. 1972), is known as concerted evolution and is thought to be the result of molecular mechanisms such as unequal crossing over, gene conversion, and gene amplification, which have collectively been termed molecular drive (Dover 1982). Even so, variation among rDNA copies within species and within individuals has been observed, with respect to both nucleotide substitutions and length heterogeneity, the latter of which is often the result of variable numbers of subrepeats in the IGS. The rDNA repeat unit is composed of regions that are under different levels of selective constraint, and thus they vary substantially in their rates of divergence among species and in their levels of intraspecific polymorphism. For example, highly conserved core regions of the rRNA genes are interspersed with variable domains or expansion segments that show high levels of divergence between species, as well as variation within populations and individuals. Similarly, the IGS tends to show much higher levels of variation than do the core regions of the rRNA genes (e.g., Pikaard 2002).

Eukaryotic rDNA is one of the best-studied multi-gene families, and the results of this work have provided important insights into the mechanisms of concerted evolution (reviewed in Elder & Turner 1995, Liao 1999). For example, studies on *Drosophila* (Polanco et al. 1998, Schlötterer & Tautz 1994), *Daphnia* (Crease & Lynch 1991), humans (Seperak et al. 1988), and plants (Copenhaver & Pikaard 1996) have provided evidence that intrachromosomal recombination is much more frequent than interchromosomal exchanges so that rDNA arrays within chromosomes tend to become homogenized into "superalleles," which then segregate within populations via sexual reproduction. In the case of some plant species with rDNA arrays on more than one chromosome, rates of exchange between the paralogous rDNA loci are so low that each array becomes homogenized for different variants (Rogers & Bendich 1987). Even so, the rDNA loci on these nonhomologous chromosomes are still more similar to one another than they are to orthologous rDNA loci in closely related species. This suggests that exchange does occur, albeit infrequently, between rDNA copies located on different chromosomes (Rogers & Bendich 1987).

Studies of rDNA variation in parthenogenetic organisms have shown that both intra- and interchromosomal exchanges occur in rDNA in the absence of meiosis and provide important opportunities for evolution within asexual lineages (Crease & Lynch 1991, Gorokhova et al. 2002, Hillis et al. 1991, Shufran et al. 2003). In addition, whereas many studies have suggested that unequal crossing over plays an important role in the generation of length variation among rDNA copies, as well as changes in overall CN of rDNA units, it is thought that this process alone cannot account for the rate of homogenization within rDNA arrays (Elder & Turner 1995; Liao 1999, 2000) and that its tendency to reduce the size of multi-gene families over time must be offset by other gene amplification mechanisms that are not yet well understood (Liao 1999, 2000; Elder & Turner 1995).

It is becoming increasingly clear that gene conversion plays a major role in the homogenization of multi-gene families and that natural selection may even favor the occurrence of initiation sites for gene conversion within the coding sequences of multi-gene families to ensure their homogenization (Liao 1999). Indeed, rates of gene conversion are estimated to be orders of magnitude higher than rates of unequal crossing over (Elder & Turner 1995, Liao 1999 and references within). In addition, Hillis et al. (1991) have shown that gene conversion in hybrid lizards can be biased and thus rapidly lead to the replacement of one variant or group of variants by another. It has also been suggested that most recombination events between members of multi-gene families on nonhomologous chromosomes occur via gene conversion (Liao 1999 and references within), an advantage of which is that such exchanges do not result in potentially deleterious gene rearrangements.

A number of other phenomena, which are beyond the scope of this review, but warrant mentioning, show how variation in rDNA structure and function can have far-reaching ecological and evolutionary consequences. For example, the presence of retrotransposable elements in rDNA (Eickbush 2002), which can lead to phenotypic abnormalities such as *bobbed* (Taylor 1923) or *abnormal abdomen* (Templeton & Rankin 1978) in *Drosophila*, can have major impacts on the fitness and survivorship of organisms (Hollocher & Templeton 1994, Templeton et al. 1993) and can be directly related to known environmental variables (e.g., temperature, rainfall) (Johnston & Templeton 1982). Likewise, work on the epigenetic phenomenon of nucleolar dominance in hybrids (Reeder 1985), whereby the rRNA genes of one parental species become transcriptionally dominant over the rRNA genes of the other parental species, has been observed in a wide variety of organisms including amphibians (i.e., *Xenopus*), insects (i.e., *Drosophila*), numerous plant species, and even among hybrid mammalian somatic cell lineages (Reeder 1985). The functional ecological bases of phenomena such as nucleolar dominance remain to be studied in detail (Pikaard 2000).

FUNCTIONAL ECOLOGY OF rDNA VARIATION

Growth Rates and Rates of Transcription

PROKARYOTES   Much of what is known about the role of rDNA in regulating rates of transcription and growth rates stems from studies done on prokaryotes, in particular, *E. coli* (Condon et al. 1995, Gourse et al. 1996, Sarmientos & Cashel 1983, Stevenson & Schmidt 1998), with much work about rate of ribosome synthesis and cellular growth rate being conducted in the 1960s (Maaløe 1969, Maaløe & Kjeldgaard 1966, Stent & Brenner 1961). The early work by Maaløe and others indicated that ribosome efficiency was relatively constant, leading many to conclude that growth rates were determined by the number of ribosomes in the cell. However, none of these experiments was performed under slow-growth conditions (Nomura 1999). Further work by Koch and colleagues (Koch 1971, Koch & Deppe 1971) demonstrated that ribosomes are in excess at low-growth rates. They argued that having unengaged ribosomes was advantageous for "shifting up" when and if substrates subsequently became available.

The rate-limiting step in ribosome synthesis at high-growth rates is the synthesis of rRNA. Gourse et al. (1996 and references within) argued that in order to make the very high number of ribosomes at high-growth rate, the promoters for the seven *E. coli* rRNA operons (Figure 1) transcribe greater than 50% of the cell's total RNA. This is remarkable given that there are approximately 2000 other operons in the cell. The central role of ribosome biogenesis in cellular function has motivated the development of a number of models to describe and explain the mechanisms responsible for regulating rRNA synthesis (Gourse et al. 1996). These models arbitrarily divide rRNA synthesis into different operating stages: (*a*) during steady-state growth at different growth rates (growth-rate-dependent control) and (*b*) during nutritional upshifts/downshifts and starvation conditions for amino acids (stringent control). Under growth-rate control of ribosome biosynthesis, free, nontranslating ribosomes inhibit the transcription of rRNA and tRNA genes (Maaløe 1969, Nomura 1999). However, under stringent control, uncharged tRNAs bound to RelA (relaxed control protein that is bound to ribosomes) cause it to produce guanosine-tetraphosphate (ppGpp), which inhibits rRNA and tRNA transcription (Nomura 1999, Wagner 1994).

EUKARYOTES   The study of rRNA transcription has been ongoing for several decades, and, in fact, rRNA genes were one of the first sets of genes to be characterized and are now among the best-studied of all eukaryotic genes (Sollner-Webb &

Tower 1986). The typical cell in a eukaryote produces approximately 10,000 different RNA species, most of which are tRNAs and mRNAs. However, nearly 50% of a cell's transcriptional capacity (closer to 80% in yeast) (Moss & Stefanovsky 2002) is directed toward the synthesis of the ~35-47S rRNA precursor of the mature ~18S, ~28S, and 5.8S RNAs of the ribosome. Even so, it is estimated that at maximum rRNA output, only about 50% of the rRNA genes are transcribed (Moss & Stefanovsky 2002). Indeed, it has been shown that yeast cells tend to alter rates of transcription of active genes rather than activate additional genes in response to increased requirements for rRNA (Reeder 1999 and references within). Even though rRNA transcription is known to be directly related to an organism's growth, development, and survivorship, it appears that the full transcriptional capacity of these genes is rarely (if ever) simultaneously engaged. The transcription of rDNA is catalyzed by RNA polymerase I (Sollner-Webb & Tower 1986), which is an efficient and highly controlled process. Each cell produces greater than a million new ribosomes each generation, which are required to support protein synthesis in the cells (e.g., for certain mammalian cells $\sim 2 \times 10^6$ ribosomes per $\sim$15 h generation time) (Sollner-Webb & Tower 1986 and references within). The life span of an average ribosome is on the order of days to weeks (Moss & Stefanovsky 2002).

Variation in rDNA CN and IGS Length Heterogeneity

PROKARYOTES   There can be a tremendous degree of variability in rDNA CN in nature and in laboratory cultures, and a number of studies have addressed the issue of whether variation in the numbers of rDNA genes (i.e., *rrn* operons) has adaptive significance in microbes. In laboratory experiments that varied temperature and shifted nutrient environments from simple to complex (i.e., varying carbon sources), Condon et al. (1995) found that increased rDNA copies increase the growth rate of *E. coli*, but only five of the seven *rrn* operons in *E. coli* (Figure 1) may actually be necessary for near-optimal, steady-state growth. However, redundant rDNAs were particularly beneficial when there was a rapid, favorable change in growth conditions (i.e., increased nutrients and/or temperature). The authors speculated that having multiple operons facilitates the surge in rRNA production induced by these new environmental conditions. Thus the significance of having seven *rrn* operons was not necessarily in supporting rapid growth rates but rather in allowing a rapid shift-up from one growth environment to a new environment.

Another important issue is whether there are any trade-offs associated with having too many copies of the rDNA operons. Similar to the findings of Condon et al. (1995), Stevenson & Schmidt (1998, 2004) found that an enhanced ability for a rapid response to favorable growth conditions occurred only when growth rates were moderate to fast. Under slow-growth-rate conditions, multiple operons appear to be nonadaptive. They noted that the presence of extra rDNA operons in the *E. coli* strain studied resulted in the overproduction of rRNA and decreased growth rates, especially under low-nutrient supply, suggesting that the regulation of rRNA synthesis was overwhelmed at slow-growth rates. This apparent disadvantage when nutrient availability is low, i.e., slow-growth-rate conditions, may be compensated by the capacity to rapidly increase growth rate in a response to an influx of nutrients. The advantage of a rapid shift in growth rate may be rarely realized in stable, low-nutrient environments. However, it may occur under more unpredictable or fluctuating conditions, which may often be the case in natural bacterial assemblages (Koch 1971).

Multiple rDNA copies are common in other prokaryotes as well. In a survey of 76 bacterial genomes, Acinas et al. (2004) found that the maximum number of rDNA operons was 15 and the minimum, and most common, was 1. Among *E. coli* strains, CN can vary widely from 1 to 7 operons as cited above. Likewise, numbers can vary widely among *Vibrio* spp. and strains (Aiyar et al. 2002, Bag et al. 1999, Heidelberg et al. 2000). Aiyar et al. (2002) showed that *Vibrio natriegens,* with doubling times of less than 10 min, manages such rapid growth through a combination of mechanisms and that one of those mechanisms is to have high rDNA CN (up to 13). This study and several others have concluded that increased rDNA CN increases the speed at which bacteria can adapt to changing environmental conditions, but a high CN does not necessarily imply a high maximal growth rate at steady state (Klappenbach et al. 2000, Stevenson & Schmidt 2004). The results of these studies suggest that there are ecological trade-offs associated with rDNA CN, with more variable and nutrient-rich environments selecting for high-copy number and less variable, nutrient-poor systems selecting for lower CNs.

Another factor related to the ecological significance of rDNA is the efficiency with which the ribosomes function and, not surprisingly, different strains seem to function at different efficiencies. In a study of several prokaryotes, Cox (2004) demonstrated that increased growth of *E. coli* and *Streptomyces* was achieved via multiple copies of *rrn* operons, but that *Streptomyces* required higher ribosome content for similar growth rates, suggesting lower per-ribosome efficiencies. Such variations have some important ecological implications. For instance, nucleic acids are phosphorus (P)-rich (9–10% by mass) compared with the remainder of major cellular biochemicals (0–3% by weight) (Sterner & Elser 2002). Thus because most of the P in growing cells is associated with nucleic acids and, particularly, ribosomes, variability in ribosome efficiency should have important ecological feedbacks by altering the stoichiometric nutrient requirements of different microbial taxa (Elser et al. 2003, Makino & Cotner 2004, Makino et al. 2003) (see below). Of particular relevance was the observation by Elser et al. (2003) that a consortium of bacteria isolated from a Minnesota lake could grow at temperature-corrected rates similar to that of *E. coli*, but they achieved those growth rates using 20–50% less RNA. Therefore, one adaptation of growth in low-nutrient environments such as lake or sea water might be to increase ribosomal efficiency.

Although counterintuitive, there may be tighter regulation of rDNA transcription when there are multiple copies in the genome (Gu et al. 2003, 2004). For instance, some rDNA operons may function more constitutively, whereas others may be induced. In *Bacillus cereus,* which has 6–10 copies of rDNA, low-temperature-tolerant strains had unique rDNA sequence signatures relative to mesophilic strains, suggesting different roles for some of the rDNA operons at low temperatures (Pruss et al. 1999). This idea was supported by a study demonstrating differential effects of particular rDNA deletions on *E. coli* growth rate and ribosome efficiency (Asai et al. 1999). Earlier work by Sarmientos & Cashel (1983) has shown that dual (tandem) promoters (*P1*, upstream; *P2*, downstream) on the *rrnA* operon of *E. coli* (Figure 1) are differentially regulated under fast- and slow-growth-rate conditions, suggesting the potential for adaptive responses in *E. coli*, and possibly in other prokaryotes.

EUKARYOTES   Among eukaryotes, variation in CN has been studied in a variety of organisms (see below). In addition, many studies have focused on the role that length heterogeneity owing to variation in the number of subrepeats in the IGS, which often contain transcription promoter and enhancer sequences, plays in influencing rates of transcription and growth rate in a range of organisms (Flavell 1986, Reeder 1984).

Substantial pioneering work on rRNA transcription has involved the frog species, *Xenopus laevis* (Reeder 1984, 1985; Reeder & Roan 1984; Reeder et al. 1983). Reeder et al. (1983) injected *X. laevis* oocytes with ribosomal gene plasmids containing variable numbers of a particular (60/81 bp) IGS subrepeat element and found that (*a*) if a long IGS and a short IGS are allowed to compete in equal molar ratios, the gene promoter attached to the long IGS always has a higher rate of transcription (i.e., competition effect), and (*b*) as the total number of 60/81 bp subrepeats in the spacer of a given plasmid increases in a reaction, there is a concomitant decrease in the total amount of transcription (i.e., sink effect). Furthermore, the authors found that regardless of orientation, the 60/81 bp repeats still conferred competitive dominance on the longer IGS types. They proposed a model to suggest that the 60/81 bp repeats serve as "attraction sites" for some (as yet) unknown factors needed to stimulate/activate the gene promoter. This suggests that multiple mechanisms related not only to absolute IGS length but also to absolute number of particular repetitive elements (i.e., 60/81 bp elements) can influence the level/amount of transcription. Thus selection might favor some optimal or intermediate IGS length and rDNA CN, or a combination of the two.

ARTIFICIAL AND NATURAL SELECTION ON rDNA VARIATION

The preceding section has highlighted some aspects related to the functional significance of rDNA variations. We now examine changes in two main features of the rDNA in response to either artificial or natural selection: (*a*) rDNA CN and (*b*) rDNA IGS length that results from variable numbers of subrepeats. For the former, we simply refer to CN, whereas for the latter, we use the term length variant (LV). Table 1 summarizes a number of studies, including some examples that we discuss below.

TABLE 1 Compilation of studies that have observed changes in rDNA structure in response to either artificial selection or putative natural selection

Organism	Type of study	Observation	References
Mixed soil bacteria	Manipulated nutritional complexity of soil medium	Bacteria with higher average rDNA operon CNs grew more rapidly	Klappenbach et al. 2000
Fungi *Neotyphodium lolii* × *Epichlöe typhina* hybrid	Monitored strains in the lab for two generations	After only two generations of single-spore purification, there were significant shifts in IGS LVs suggesting that selection may favor longer LVs	Ganley & Scott 1998
Maize *Zea mays* L.	Artificial directional selection on increased yield	Shift in predominant IGS LVs to longer variants in certain strains	Rocheford et al. 1990 Kaufman et al. 1996
Barley—wild *Hordeum spontaneum*	Monitored strain yields in field populations for 54 generations	Significant shift in frequency of IGS LVs observed; may also be linked to other loci	Saghai-Maroof et al. 1984
Barley—wild *Hordeum spontaneum*	Surveyed distribution of genetic variants across habitats	Significant associations/correlations between IGS LVs and 8 of 9 physical factors in the environment	Saghai-Maroof et al. 1990
Barley wild—*Hordeum spontaneum* cultivated—*H. vulgare*	Surveyed distribution of genetic variants across habitats	Two rDNA locus (*rrn-1* and *rrn-2*) genotypes that differed in IGS LVs were correlated with 9 physical factors in the environment	Allard et al. 1990 Zhang et al. 1990
Barley wild—*Hordeum spontaneum* cultivated—*H. vulgare*	Surveyed natural variation in rDNA LVs.	Significant associations between IGS LVs and microsite features (humidity, rainfall, edaphic factors) suggest that selection is operating along a humidity/aridity gradient	Gupta et al. 2002 Sharma et al. 2004

Barley—wild *Hordeum spontaneum* and F₁ hybrids	Artificial selection on yield and associated other reproductive traits	Yield and other traits were strongly associated with alleles at the *NOR-H3* locus, indicating that a small section of genome associated with nucleolar organizing regions has a strong effect on reproductive traits	Powell et al. 1992
Oats—wild *Avena barbata*	Natural survey of genetic variation at rDNA loci and multi-locus allozymes	Seven different habitat-genotypes/ecotypes were detected including rDNA/allozyme/morph correlations along a moisture/rainfall gradient. Ecological diversification occurred over a relatively short (150–200 generations) time span	Cluster & Allard 1995
Oats—wild *A. barbata* and other crop species	Survey of literature focusing on *A. barbata* rDNA IGS variation	Non-random distribution of IGS LVs among different species suggests that an optimal (intermediate) IGS LV may exist for many species	Jorgensen & Cluster 1988
Oats—wild *A. barbata*	Glasshouse competition experiments and quantitative genetic analyses	Follow up on earlier studies of allozyme/rDNA associations found that mesic genotype outcompeted xeric genotype: crossing experiments showed significant heritabilities for life history traits along a moisture gradient	Latta et al. 2004
Oats—cultivated *Avena sativa*	Breeding studies of cultivars	Cultivars/strains bred in controlled studies (for higher yield) had significantly longer IGS LVs than wild strains	Polanco & Pérez de la Vega 1997
Emmer wheat—wild *Triticum dicoccoides*	Natural survey of rDNA IGS LVs	Suggestive evidence that selection is operating in the IGS with weaker support for selection on total rDNA copy number	Flavell et al. 1986

(Continued)

TABLE 1 (Continued)

Organism	Type of study	Observation	References
Rice—wild and cultivated *Oryza* spp.	Monitored genetic variation in cultivated and wild populations	Significant differences in IGS LVs between cultivated and wild rice species. Cultivated African rice (*Oryza glaberrima*) showed no variation in IGS LVs compared with two wild species; suggestive of purging/purifying in IGS LVs among cultivated species?	Cordesse et al. 1990
Pitch pine *Pinus rigida* Mill.	Natural genetic variation in rDNA CN was assayed	A strong inverse correlation between rDNA CN and level of environmental stress (fire-related) suggests that high stress may be correlated with low CN	Govindaraju & Cullis 1992
Red spruce—*Picea rubens* Black spruce—*P. mariana*	Surveyed natural population genetic variation in rDNA repeat unit types and rDNA CN	Distinct rDNA alleles (repeat types) occurred across a latitudinal gradient with threefold and sixfold variation in copy number for *P. rubens* and *P. mariana*, respectively. Suggests possible role for selection (i.e., temperature, growing season length)	Bobola et al. 1992
Douglas fir *Pseudotsuga menziesii*	Natural variation in rDNA copy number across latitudinal, longitudinal and elevational gradients	Positive (weak) association between rDNA copy number and latitude (higher copy number in more northern populations) and longitudinal trend with coastal (western) populations having higher CNs	Strauss & Tsai 1988
Fruit fly *Drosophila melanogaster*	Disruptive artificial selection for increased or decreased preadult development rate	Long IGS LVs predominated in fast-developing lines, while the opposite occurred in slow-developing lines	Cluster et al. 1987

Organism	Study	Results	Reference
Fruit fly *D. melanogaster*	Monitored genetic structure of natural populations along a latitudinal range	Significant clinal shifts in distribution of IGS LVs among females across the latitudinal gradient suggests that selection exerts pressure on rDNA structure (perhaps temperature-related)	Polanco et al. 1998
Aphid/greenbug *Schizaphis graminum*	Artificial selection for insecticide (disulfoton) resistance for 200 generations (~4 years)	Control lines showed no shifts; selected lines showed shifts in IGS LVs with the loss of three LVs from the original population. Indicates the potential for large-scale genetic changes in a parthenogen	Shufran et al. 2003
Grasshopper *Dichroplus elongates*	Monitored natural genetic variation across an altitudinal gradient	The average number of RFLP rDNA variants (primarily in the IGS) per individual was significantly associated with altitude with individuals from higher altitudes expressing greater numbers of rDNA variants	Clemente et al. 2002
Water flea *Daphnia pulex*	Artificial disruptive selection on growth-related life-history features	Significant changes in predominant IGS LVs with higher proportions of longer LVs in lines exhibiting higher juvenile growth rate. Concomitant changes in percentage of RNA and phosphorus were also observed	Gorokhova et al. 2002
Frog *Xenopus laevis*	Laboratory in vitro experiments	Longer IGS LVs were more competitive than shorter LVs in a series of in vitro transcription assays	Reeder et al. 1983
Chicken several specialized strains *Gallus gallus*	Artificial selection experiments on yield and growth	Shifts in IGS regions and nucleolar (NOR) loci were related to selection for either somatic or reproductive growth	Delany & Krupkin 1999, Su & Delany 1998

Artificial Selection

The relationship between rDNA structure and important fitness-related traits (e.g., growth rate, productivity/yield) in both wild and domesticated populations has been studied extensively in crop science, where artificial selection on specific life-history traits has been conducted for a variety of species including maize (Kaufman et al. 1996, Rocheford et al. 1990), barley (Powell et al. 1992, Saghai-Maroof et al. 1984), oats (Latta et al. 2004, Polanco & Pérez de la Vega 1997), and rice (Cordesse et al. 1990). In addition, artificial selection experiments have been conducted on a few model animal species, including invertebrates such as the fruit fly *Drosophila melanogaster* (Cluster et al. 1987), the aphid/greenbug *Schizaphis graminum* (Shufran et al. 2003), and the water flea *Daphnia pulex* (Gorokhova et al. 2002), as well as vertebrate models such as the frog *X. laevis* (Reeder et al. 1983) and the chicken (Delany & Krupkin 1999, Su & Delany 1998). In general, these studies have documented significant changes in rDNA structure in response to selection on important life-history/fitness-related traits (e.g., Kaufman et al. 1996, Powell et al. 1992, Rocheford et al. 1990, Saghai-Maroof et al. 1984) (see Table 1). Furthermore, quantitative genetic studies among certain crop species (e.g., barley, wild oats) have attempted to pinpoint key ecological traits among rDNA genotypes in wild populations to elucidate the quantitative genetic underpinnings of this variation (Latta et al. 2004, Powell et al. 1992).

Among the few animal studies that have utilized artificial selection on specific traits (e.g., growth rate, life-history traits) and examined concomitant changes in rDNA, the studies by Cluster et al. (1987) and Grimaldi & Di Nocera (1988) on *D. melanogaster* are noteworthy because they show a clear response of the rDNA to directional selection. Cluster et al. (1987) selected for fast and slow development times among lines of *D. melanogaster* and noted a significant shift in the frequency of IGS LVs in the two selection regimes. A greater proportion of the faster developing lines maintained longer LVs, whereas the slower developing lines maintained shorter LVs. Subsequent work by Grimaldi & Di Nocera (1988) showed that the rate of transcriptional production of pre-rRNA was directly proportional to the number of enhancers located in the IGS. These two studies provide support for the notion that genotypes composed of longer IGS LVs may benefit from higher rDNA transcriptional rates via more enhancer and promoter sites in the subrepeat region of the IGS and thus exhibit faster development (higher growth rates).

In another study using the cyclically parthenogenetic water flea *Daphnia pulex*, Gorokhova et al. (2002) observed shifts in the frequency of IGS LVs during direct selection on a life-history character (production rate) in progeny descended from a single (clonal) stem mother. The frequency of a longer LV increased as production/fecundity rate decreased, along with concomitant shifts in other important characters such as the percentage of RNA, and body phosphorus and juvenile growth rates (see below). These data clearly show that even within a single clone of a parthenogenetic species considerable plasticity exists in the ability to shift key life-history features along with LV frequencies in response to artificial selection.

This implies that asexual organisms possess considerable flexibility to adapt to changing environmental conditions via structural mutations in the rDNA.

Finally, work with domesticated strains of chicken (Delany & Krupkin 1999, Su & Delany 1998) has shown that in some artificial selection experiments, the combination of shifts in both IGS length variation and overall rDNA CN need to be considered jointly before any predictions can be made about the outcome of the selection regime.

Natural Selection

A number of studies (Table 1) have surveyed and/or examined natural variation in rDNA CN and/or LVs in a number of organisms, including soil bacteria (Klappenbach et al. 2000), wild barley (Allard et al. 1990, Gupta et al. 2002, Saghai-Maroof et al. 1990, Sharma et al. 2004, Zhang et al. 1990), wild oats (Cluster & Allard 1995, Jorgensen & Cluster 1988), wild emmer wheat (Flavell et al. 1986), a variety of conifer species (Bobola et al. 1992, Govindaraju & Cullis 1992, Strauss & Tsai 1988), and arthropods (Clemente et al. 2002, Polanco et al. 1998), and have attempted to relate this underlying genetic variation to potentially important environmental variables.

Among prokaryotes, Klappenbach et al. (2000) found that in a phylogenetically diverse soil bacterial assemblage, species that possessed higher average CNs of rDNA genes formed colonies more quickly when exposed to nutritionally complex medium, as opposed to species that possessed lower average CNs, which responded more slowly. They concluded that rDNA CN strongly influences ecological strategies and competitive abilities among soil bacteria and potentially can be under direct natural selection.

Among eukaryotes, work on cultivated and wild barley, *Hordeum vulgare* and *Hordeum spontaneum*, respectively, has shown strong correlations between LV composition and important environmental/selective factors such as temperature and moisture availability/humidity (Allard et al. 1990, Gupta et al. 2002, Saghai-Maroof et al. 1990, Zhang et al. 1990). Allard et al. (1990) suggested that selection is acting directly on the sequence variability in the transcription units (i.e., subrepeats in the IGS), but no eco-physiological traits were assessed. In a companion paper, Zhang et al. (1990) determined that the high adaptedness associated with a few specific alleles may result from adaptively favorable nucleotide sequences in either the transcription units or the IGS and that adaptedness in barley depends more on the quality (i.e., sequence and length variation) rather than the quantity (i.e., CN) of rDNA present. Similar work on another important crop species, wild emmer wheat (*Triticum dicoccoides*), reported by Flavell et al. (1986), provides further support for the notion that natural variations at rDNA loci are significantly correlated with important environmental parameters. Govindaraju & Cullis (1992) examined rDNA CN and LVs in eight populations of pitch pine (*Pinus rigida* Mill.) associated with the Pine Barrens region of southern New Jersey. Interestingly, the authors noted an inverse relationship between rDNA CN and the level of environmental stress and proposed that strong diversifying selection is operating to influence rDNA CN in this species. Although this study is purely correlational, it suggests that rDNA CN may indeed be influenced by environmental parameters, akin to the soil microbe study of Klappenbach et al. (2000) cited above.

From the above examples and others listed in Table 1, it indeed appears that variations in rDNA CN and LVs in a broad range of organisms have ecological significance and can respond to either artificial or natural selection.

LINKING rDNA VARIATION TO ECOLOGICAL PROCESSES

A Stoichiometric Perspective

The previous sections have highlighted how structural and regulatory features of rDNA operate to affect key aspects of organism function, such as the maximal and realized rates of growth and development of diverse biota ranging from bacteria to vertebrates. In particular, we have seen that variations in rDNA structure and expression are commonly linked to the challenge of maintaining a high rate of rRNA production associated with rapid cellular proliferation. In this section we show how the functional consequences of rDNA variation extend well beyond those normally considered by cellular and evolutionary biologists. We employ the perspective of biological stoichiometry (the study of the balance of energy and multiple chemical elements in living systems) (Sterner & Elser 2002) to consider how ecological forces, such as the supply of the key limiting nutrient phosphorus (P), impinge on evolutionary change involving rDNA, as well as potential feedbacks generated by the coupling of rDNA to growth and ribosome production. In doing so we highlight some of the ecological forces that may operate to impose trade-offs on the evolution of high-growth-rate phenotypes associated with changes in the rDNA genome. Connections between growth, cellular P requirements, and RNA allocation under conditions of environmental P limitation have been established for some time in unicellular algae, dating to classic studies of Rhee and colleagues (e.g., Rhee & Gotham 1981 and references within). These connections have more recently become integrated in various emerging stoichiometric models of growth-rate regulation in photoautotrophic organisms (e.g., Ågren 2004, Klausmeier et al. 2004).

Because the associations of growth, RNA, and P in autotrophs have been covered extensively elsewhere (Frost et al. 2005, Geider & La Roche 2002, Sterner & Elser 2002), our emphasis in this section is on connections between these variables in metazoans, where the elemental composition of animal biomass, its physiological regulation, and its connections to biochemical allocations have come under close scrutiny in only the past 10 years. It is now known that the elemental composition of animal biomass, in contrast to the physiological plasticity of autotrophs, is homeostatically regulated by various physiological mechanisms around taxon- and stage-specific levels (Sterner & Elser 2002). For example, the P content of the ubiquitous crustacean zooplankter, *Daphnia*, varies between only 1.2 and 2%, with variation largely due to the stage of development (juveniles have somewhat higher

P content than adults). In contrast, another crustacean zooplankter, *Bosmina*, generally seems to have lower P content, with values between 0.6 and 0.9% (Hessen & Lyche 1991). Such studies have been extended recently to include insects and show that crustacean zooplankton and insect taxa exhibit similar ranges of variation in C:N, C:P, and N:P ratios (Elser et al. 2000b). In the following we focus on microorganisms (heterotrophic bacteria) and invertebrate animals (primarily insects and crustaceans). We choose this emphasis because in vertebrate animals the dominant form of P is in the apatite mineral that forms bone (Elser et al. 1996) and thus potential connections to rDNA variation are obscured.

Because variation in C:N:P ratios has been shown to have considerable consequences for ecological processes such as secondary production (Sterner & Schulz 1998) and consumer-driven nutrient recycling (Elser & Urabe 1999), a desire to understand the biological basis of this variation led to formulation of the growth-rate hypothesis (GRH) (Elser et al. 1996), which states that variation in organismal C:N:P ratios reflects differences in growth rate because of differential allocation to P-rich rRNA (RNA is ~9.6% P by mass) that is needed to meet the protein synthesis demands of growth. Elser et al. (2000b) extended the GRH to include its genetic basis, postulating that variations in rDNA IGS length and CN underpin variation in growth and therefore RNA allocation, and thus P content and C:N:P ratios.

So, is growth-related variation in RNA allocation sufficient to explain organism level variation in P content? A variety of recent studies indicates that this is the case, at least in microorganisms (Makino et al. 2003) and invertebrates including zooplankton (Acharya et al. 2004; Carrillo et al. 2001; Elser et al. 1996; Main et al. 1997; Vrede et al. 1998, 2002;) and insects (Schade et al. 2003). In an integrated analysis, Elser et al. (2003) showed not only that growth, RNA, and P were generally tightly coupled across all study organisms (ranging from microbes to invertebrates), but also that RNA allocations were sufficiently large such that RNA contributed on average ~50% of biomass P across the taxa reported. Thus RNA production generates a physiologically dominant pool of P in many organisms and therefore rDNA transcription itself appears to represent an ecologically and biogeochemically significant process.

This work supports the hypothesized connection among organismal growth rate, RNA allocation, and C:N:P stoichiometry proposed nearly 10 years ago (Elser et al. 1996). Thus uncovering the contributions of rDNA variations that drive variation in growth and RNA allocation will be fruitful in many areas. This empirical base has now stimulated several lines of theoretical investigation of the GRH (Ågren 2004, Klausmeier et al. 2004, Vrede et al. 2004). These papers offer different perspectives on this coupling, but their collective message is that relatively simple eco-evolutionary formulations are beginning to capture the ecological significance of RNA-related impacts on C:N:P stoichiometry and growth rate. For example, under resource-rich conditions, increased allocation to assembly machinery (i.e., RNA) results in organisms with low optimal N:P ratios (that is, organisms that are easily P limited and are poor P competitors) whereas low-resource environments favor organisms with high optimal N:P ratios (Klausmeier et al. 2004).

We now turn our attention to the question of whether connections, indeed, exist among rDNA variations and C:N:P stoichiometry. In particular, Elser et al. (2000b) proposed that there should be a generally positive association between the length of the rDNA IGS and organismal growth rate, RNA content, and P content in eukaryotes. The data collected for the explicit purpose of evaluating these connections remain quite limited; however, a number of relevant studies have appeared (discussed, in part, above). For example, Weider et al. (2004) identified IGS length variation among clones in three species of *Daphnia* and examined growth rates, RNA levels, and P contents under standardized conditions. As predicted by the GRH, there were positive correlations between RNA:DNA ratio and either growth rate or IGS length, and a significant positive correlation between IGS length and growth rate when clonal means for all three species were examined. However, no clear-cut relationships between RNA:DNA and P content were observed for any of the three species, likely because of the limited sample size and narrow range of values observed for P content and RNA:DNA ratios. However, as noted above (Gorokhova et al. 2002), IGS length in single daphnid clones can respond to artificial selection with concomitant changes in growth rate, RNA, and P content.

The aforementioned studies provide some of the first evidence that rDNA variations can be connected not only to growth itself but also to the elemental composition of living biomass. This suggests the existence of a potentially important mechanism for a trade-off in the evolution of rapid growth rate: high-growth rate appears to impose a disproportionate elevation in organismal P demands in order to maintain production of P-rich rRNA. Preliminary evidence from evolutionary/ecological studies provides some support for this claim. For example, there is a general trend for increasing growth and development rates for organisms as one moves to high latitudes (Conover & Schultz 1995), suggesting that arctic biota should be more P-rich than those in temperate and tropical regions, a pattern documented in recent studies of vascular plants (McGroddy et al. 2004, Reich & Oleksyn 2004) and in *Daphnia* (Elser et al. 2000a). Indeed, the results of feeding experiments in the latter study showed that arctic *Daphnia* were not only more P-rich than their temperate counterparts, they also were more sensitive to low P in their diets and had very low recycling rates of P. Weider et al. (2005) have recently evaluated whether stoichiometric food quality impinges on the relative success of rDNA variants. In an experiment where algal food differing in C:P ratio was externally supplied to a mixture of two allozymically different *D. pulex* clones differing in the length of their IGS, the clone with the longer IGS (and likely to have higher P requirements according to the GRH) increased dramatically relative to the clone with the shorter IGS when the external food supply was P-rich. Conversely, when food was low in P, the clone with the shorter IGS, which is likely to have lower P requirements according to the GRH, won out. In a follow-up experiment, algal food was not supplied externally, but instead algae and *Daphnia* were present together and thus nutrient recycling by the *Daphnia* could play a role. In this case, clonal coexistence of these two variants was observed in treatments of different light intensity that were intended to produce divergence in algal C:P ratios. Thus feedback mechanisms at the ecosystem level may contribute to stabilizing the co-existence of rDNA variants in nature. The above studies highlight that connections between rDNA variation, C:N:P ratios, and important ecological parameters (e.g., growth rate, competitive ability) may indeed be important and certainly warrant further study.

CONCLUDING REMARKS

In this review, we have provided evidence from a variety of organisms that selection on variation in the CN or IGS length of rDNA can have a number of substantial ecological consequences. This is because rDNA polymorphisms can affect two aspects of profound ecological importance: (*a*) growth rate, which is directly linked to ecological production, and (*b*) organismal P requirements, which contribute to stoichiometric food quality effects and the sequestering and recycling of phosphorus. Therefore, patterns of energy flow and the cycling of a key limiting nutrient (P) are simultaneously impacted by evolutionary events mediated by the functional consequences of rDNA variation. Taking such a multilevel approach as indicated above may allow us to begin linking subcellular and genetic processes (as exemplified by rDNA variation, and its impacts on ribosome biogenesis) with the evolution of major life-history traits and, ultimately, lead to a better understanding of the nature and outcome of ecological interactions in natural ecosystems.

ACKNOWLEDGMENTS

We thank K.L. Glenn, P. Jeyasingh, and M. Kyle for assistance with the literature search for this review and for comments on earlier drafts of the manuscript. We thank R.W. Sterner for first suggesting that we put together such a review, and we thank our NSF-IRCEB colleagues, in particular R.W. Sterner, J.F. Harrison, and W.F. Fagan, who helped us to formulate many of the underpinnings represented in this review. We acknowledge the support of NSF-IRCEB (grant #9977047) during the early stages of manuscript preparation, as well as support for some of the work cited herein. We thank the editors of the *Annual Review of Ecology, Evolution, and Systematics*, particularly D. Futuyma, for providing input and editorial assistance.

**The *Annual Review of Ecology, Evolution, and Systematics* is online at
http://ecolsys.annualreviews.org**

LITERATURE CITED

Acharya K, Kyle M, Elser JJ. 2004. Biological stoichiometry of *Daphnia* growth: an eco-physiological test of the growth rate hypothesis. *Limnol. Oceanogr* 49:656–65

Acinas SG, Marcelino LA, Klepac-Ceraj V,

Polz MF. 2004. Divergence and redundancy of *16S* rRNA sequences in genomes with multiple *rrn* operons. *J. Bacteriol.* 186:2629–35

Ågren GI. 2004. The C:N:P stoichiometry of

autotrophs—theory and observations. *Ecol. Lett.* 7:185–91

Aiyar SE, Gaal T, Gourse RL. 2002. rRNA promoter activity in the fast-growing bacterium *Vibrio natriegens. J. Bacteriol.* 184:1349–58

Allard RW, Saghai-Maroof MA, Zhang Q, Jorgensen RA. 1990. Genetic and molecular organization of ribosomal DNA (rDNA) variants in wild and cultivated barley. *Genetics* 126:743–51

Asai T, Condon C, Voulgaris J, Zaporojets D, Shen BH, et al. 1999. Construction and initial characterization of *Escherichia coli* strains with few or no intact chromosomal rRNA operons. *J. Bacteriol.* 181:3803–9

Bag PK, Nandi S, Bhadra RK, Ramamurthy T, Bhattacharya SK, et al. 1999. Clonal diversity among recently emerged strains of *Vibrio parahaemolyticus* O3:K6 associated with pandemic spread. *J. Clin. Microbiol.* 37:2354–57

Bobola MS, Eckert RT, Klein AS. 1992. Restriction fragment variation in the nuclear ribosomal DNA repeat unit within and between *Picea rubens* and *Picea mariana. Can. J. For. Res.* 22:255–63

Brown DD, Wensink PC, Jordan E. 1972. A comparison of the ribosomal DNAs of *Xenopus laevis* and *Xenopus mulleri*: the evolution of tandem genes. *J. Mol. Evol.* 63:57–73

Carrillo P, Villar-Argaiz M, Medina-Sanchez JM. 2001. Relationship between N:P ratio and growth rate during the life cycle of copepods: an in situ measurement. *J. Plankton Res.* 23:537–47

Clemente M, Remis MI, Vilardi JC. 2002. Ribosomal DNA variation in the grasshopper, *Dichroplus elongatus. Genome* 45:1125–33

Cluster PD, Allard RW. 1995. Evolution of ribosomal DNA (rDNA) genetic structure in colonial populations of *Avena barbata. Genetics* 139:941–54

Cluster PD, Marinkovic D, Allard RW, Ayala FJ. 1987. Correlations between development rates, enzyme activities, ribosomal DNA spacer-length phenotypes, and adaptation in *Drosophila melanogaster. Proc. Natl. Acad. Sci. USA* 84:610–14

Condon C, Liveris D, Squires C, Schwartz I, Squires CL. 1995. rRNA operon multiplicity in *Escherichia coli* and the physiological implications of *rrn* inactivation. *J. Bacteriol.* 177:4152–56

Conover DO, Schultz ET. 1995. Phenotypic similarity and the evolutionary significance of countergradient variation. *Trends Ecol. Evol.* 10:248–52

Copenhaver GP, Pikaard CS. 1996. Two-dimensional RFLP analyses reveal megabase-sized clusters of rRNA gene variants in *Arabidopsis thaliana*, suggesting local spreading of variants as the mode for gene homogenization during concerted evolution. *Plant J.* 9:273–82

Cordesse F, Second G, Delseny M. 1990. Ribosomal gene spacer length variability in cultivated and wild rice species. *Theor. Appl. Genet.* 79:81–88

Cox RA. 2004. Quantitative relationships for specific growth rates and macromolecular compositions of *Mycobacterium tuberculosis, Streptomyces coelicolor* A3(2) and *Escherichia coli* B/r: an integrative theoretical approach. *Microbiology* 150:1413–26

Crease TJ, Lynch M. 1991. Ribosomal DNA variation in *Daphnia pulex. Mol. Biol Evol.* 8:620–40

Delany ME, Krupkin AB. 1999. Molecular characterization of ribosomal gene variation within and among *NORs* segregating in specialized populations of chicken. *Genome* 42:60–71

Dover GA. 1982. Molecular drive, a cohesive model of species evolution. *Nature* 299:111–17

Eickbush TH. 2002. Repair by retrotransposition. *Nat. Genet.* 31:126–27

Elder JF, Turner BJ. 1995. Concerted evolution of repetitive DNA sequences in eukaryotes. *Q. Rev. Biol.* 70:297–320

Elser JJ, Acharya K, Kyle M, Cotner J, Makino W, et al. 2003. Growth rate—stoichiometry couplings in diverse biota. *Ecol. Lett.* 6:936–43

Elser JJ, Dobberfuhl D, MacKay NA, Schampel JH. 1996. Organism size, life history, and N:P stoichiometry: towards a unified view of cellular and ecosystem processes. *BioScience* 46:674–84

Elser JJ, Dowling T, Dobberfuhl DA, O'Brien J. 2000a. The evolution of ecosystem processes: ecological stoichiometry of a key herbivore in temperate and arctic habitats. *J. Evol. Biol.* 13:845–53

Elser JJ, Sterner RW, Gorokhova E, Fagan WF, Markow TA, et al. 2000b. Biological stoichiometry from genes to ecosystems. *Ecol. Lett.* 3:540–50

Elser JJ, Urabe J. 1999. The stoichiometry of consumer-driven nutrient cycling: theory, observations and consequences. *Ecology* 80:735–51

Feder ME, Mitchell-Olds T. 2003. Evolutionary and ecological functional genomics. *Nat. Genet.* 4:649–55

Flavell RB. 1986. The structure and control of expression of ribosomal RNA genes. *Oxford Surv. Plant Mol. Cell Biol.* 3:251–74

Flavell RB, O'Dell M, Sharp P, Nevo E, Beiles A. 1986. Variation in the intergenic spacer of ribosomal DNA of wild wheat, *Triticum dicoccoides*, in Israel. *Mol. Biol. Evol.* 3:547–58

Fromont-Racine M, Senger B, Saveanu C, Fasiolo F. 2003. Ribosome assembly in eukaryotes. *Gene* 313:17–42

Frost P, Evans-White M, Finkel Z, Jensen T, Matzek V. 2005. Are you what you eat? Physiological constraints on organismal stoichiometry in an elementally imbalanced world. *Oikos* 109:18–28

Ganley ARD, Scott B. 1998. Extraordinary ribosomal spacer length heterogeneity in a neotyphodium endophyte hybrid: implications for concerted evolution. *Genetics* 150:1625–37

Geider RJ, La Roche J. 2002. Redfield revisited: variability in C:N:P in marine microalgae and its biochemical basis. *Eur. J. Phycol.* 37:1–17

Gerbi SA. 1985. Evolution of ribosomal DNA. In *Molecular Evolutionary Genetics*, ed. RJ MacIntyre, pp. 419–517. New York: Plenum

Gorokhova E, Dowling TE, Weider LJ, Crease

TJ, Elser JJ. 2002. Functional and ecological significance of rDNA intergenic spacer variation in a clonal organism under divergent selection for production rate. *Proc. R. Soc. London Ser. B* 269:2373–79

Gourse RL, Gaal T, Bartlett MS, Appleman JA, Ross W. 1996. rRNA transcription and growth rate-dependent regulation of ribosome synthesis in *Escherichia coli. Annu. Rev. Microbiol.* 50:645–77

Govindaraju DR, Cullis CA. 1992. Ribosomal DNA variation among populations of a *Pinus rigida* Mill. (pitch pine) ecosystem: I. Distribution of copy numbers. *Heredity* 69:133–40

Grimaldi G, Di Nocera PO. 1988. Multiple repeated units in *Drosophila melanogaster* ribosomal DNA spacer stimulate rRNA precursor transcription. *Proc. Natl. Acad. Sci. USA* 85:5502–6

Gu ZL, Rifkin SA, White KP, Li WH. 2004. Duplicate genes increase gene expression diversity within and between species. *Nat. Genet.* 36:577–79

Gu ZL, Steinmetz LM, Gu X, Scharfe C, Davis RW, Li WH. 2003. Role of duplicate genes in genetic robustness against null mutations. *Nature* 421:63–66

Gupta PK, Sharma PK, Balyan HS, Roy JK, Sharma S, et al. 2002. Polymorphism at rDNA loci in barley and its relation with climatic variables. *Theor. Appl. Genet.* 104:473–81

Heidelberg JF, Eisen JA, Nelson WC, Clayton RA, Gwinn ML, et al. 2000. DNA sequence of both chromosomes of the cholera pathogen *Vibrio cholerae. Nature* 406:477–83

Hessen DO, Lyche A. 1991. Inter- and intraspecific variations in zooplankton element composition. *Arch. Hydrobiol.* 121:343–53

Hillis DM, Dixon MT. 1991. Ribosomal DNA: molecular evolution and phylogenetic inference. *Q. Rev. Biol.* 66:411–53

Hillis DM, Moritz C, Porter CA, Baker RJ. 1991. Evidence for biased gene conversion in concerted evolution of ribosomal DNA. *Science* 251:308–10

Hollocher H, Templeton AR. 1994. The

molecular through ecological genetics of *abnormal abdomen* in *Drosophila mercatorum*. *Genetics* 136:1373–84

Johnston JS, Templeton AR. 1982. Dispersal and clines in *Opuntia* breeding *Drosophila mercatorum* and *Drosophila hydei* at Kamuela, Hawaii. In *Ecological Genetics and Evolution*, ed. JSF Barker, WT Starmer, pp. 241–56. Sydney: Academic

Jorgensen RA, Cluster PD. 1988. Modes and tempos in the evolution of nuclear ribosomal DNA: new characters for evolutionary studies and new markers for genetic and population studies. *Ann. Missouri Bot. Gard.* 75: 1238–47

Kaufman B, Rocheford TR, Lambert RJ, Hallauer AR. 1996. Change in ribosomal DNA spacer-length composition in maize recurrent selection populations. 2. Analysis of BS10, BS11, RBS10, and RSSSC. *Theor. Appl. Genet.* 92:680–87

Klappenbach JA, Dunbar JM, Schmidt TM. 2000. rRNA operon copy number reflects ecological strategies of bacteria. *Appl. Environ. Microbiol.* 66:1328–33

Klausmeier CA, Litchman E, Daufresne T, Levin SA. 2004. Optimal nitrogen-to-phosphorus stoichiometry of phytoplankton. *Nature* 429:171–74

Koch AL. 1971. The adaptive responses of *Escherichia coli* to a feast and famine existence. *Adv. Microbiol. Physiol.* 6:147–217

Koch AL, Deppe CS. 1971. In vivo assay of protein synthesizing capacity of *Escherichia coli* from slowly growing chemostat cultures. *J. Mol. Biol.* 55:549–62

Latta RG, MacKenzie JL, Vats A, Schoen DJ. 2004. Divergence and variation of quantitative traits between allozyme genotypes of *Avena barbata* from contrasting habitats. *J. Ecol.* 92:57–71

Liao D. 1999. Concerted evolution: molecular mechanisms and biological implications. *Am. J. Hum. Genet.* 64:24–30

Liao D. 2000. Gene conversion drives within genic sequences: concerted evolution of ribosomal RNA genes in bacteria and archaea. *J. Mol. Evol.* 51:305–17

Maaløe O. 1969. An analysis of bacterial growth. *Dev. Biol. Suppl.* 3:33–58

Maaløe O, Kjeldgaard NO. 1966. *Control of Macromolecular Synthesis: A Study of DNA, RNA, and Protein Synthesis in Bacteria.* New York: Benjamin. 284 pp.

Main T, Dobberfuhl DR, Elser JJ. 1997. N:P stoichiometry and ontogeny in crustacean zooplankton: a test of the growth rate hypothesis. *Limnol. Oceanogr.* 42:1474–78

Makino W, Cotner JB. 2004. Elemental stoichiometry of a heterotrophic bacterial community in a freshwater lake: implications for growth- and resource-dependent variations. *Aquat. Microbiol. Ecol.* 34:33–41

Makino W, Cotner JB, Sterner RW, Elser, JJ. 2003. Are bacteria more like plants or animals? Growth rate and resource dependence of bacterial C:N:P stoichiometry. *Funct. Ecol.* 17:121–30

McGroddy ME, Daufresne T, Hedin LO. 2004. Scaling of C:N:P stoichiometry in forest ecosystems worldwide: implications of terrestrial Redfield-type ratios. *Ecology* 85:2390–401

Mindell DP, Honeycutt RL. 1990. Ribosomal RNA in vertebrates: evolution and phylogenetic applications. *Annu. Rev. Ecol. Syst.* 21:541–66

Moore PB, Steitz TA. 2002. The involvement of RNA in ribosome function. *Nature* 418:229–35

Moss T, Stefanovsky VY. 2002. At the center of eukaryotic life. *Cell* 109:545–48

Neidhardt FC, Ingraham JL, Schaechter, M. 1990. *Physiology of the Bacterial Cell: A Molecular Approach.* Sunderland, MA: Sinauer & Assoc.

Nomura M. 1999. Regulation of ribosome biosynthesis in *Escherichia coli* and *Saccharomyces cerevisiae*: diversity and common principles. *J. Bacteriol.* 181:6857–64

Pikaard CS. 2000. Nucleolar dominance: uniparental gene silencing on a multi-megabase scale in genetic hybrids. *Plant Mol. Biol.* 43:163–77

Pikaard CS. 2002. Transcription and tyranny in the nucleolus: the organization, activation, dominance and repression of ribosomal RNA genes. In *The Arabidopsis Book*, ed. CR Sommerville, EM Meyerowitz, pp. 1–23. Rockville, MD: *Am. Soc. Plant Biol.*

Polanco C, González AI, de la Fuente A, Dover GA. 1998. Multigene family of ribosomal DNA in *Drosophila melanogaster* reveals contrasting patterns of homogenization for IGS and ITS spacer regions: a possible mechanism to resolve this paradox. *Genetics* 149:243–56

Polanco C, Pérez de la Vega M. 1997. Intergenic ribosomal spacer variability in hexaploid oat cultivars and landraces. *Heredity* 78:115–23

Powell W, Thomas WTB, Thompson DM, Swanston JS, Waugh R. 1992. Association between rDNA alleles and quantitative traits in doubled haploid populations of barley. *Genetics* 130:187–94

Prokopowich CD, Gregory TR, Crease TJ. 2000. The correlation between rDNA copy number and genome size in eukaryotes. *Genome* 46:48–50

Pruss BM, Francis KP, von Stetten F, Scherer S. 1999. Correlation of *16S* ribosomal DNA signature sequences with temperature-dependent growth rate of mesophilic and psychrotolerant strains of the *Bacillus cereus* group. *J. Bacteriol.* 181:2624–30

Reeder RH. 1984. Enhancers and ribosomal gene spacers. *Cell* 38:349–51

Reeder RH. 1985. Mechanisms of nucleolar dominance in animals and plants. *J. Cell Biol.* 101:2013–16

Reeder RH. 1999. Regulation of RNA polymerase I transcription in yeast and vertebrates. *Prog. Nucleic Acid Res. Mol. Biol.* 62:293–327

Reeder RH, Roan JG. 1984. The mechanism of nucleolar dominance in *Xenopus* hybrids. *Cell* 38:39–44

Reeder RH, Roan JG, Dunaway M. 1983. Spacer regulation of *Xenopus* ribosomal gene transcription: competition in oocytes. *Cell* 35:449–56

Reich PB, Oleksyn J. 2004. Global patterns of plant leaf N and P in relation to temperature and latitude. *Proc. Natl. Acad. Sci. USA* 101:11001–6

Rhee GY, Gotham IJ. 1981. The effect of environmental factors on phytoplankton growth: light and the interactions of light with nitrate limitation. *Limnol. Oceanogr.* 26:649–59

Rocheford TR, Osterman JC, Gardner CO. 1990. Variation in the ribosomal DNA intergenic spacer of a maize population mass-selected for high grain yield. *Theor. Appl. Genet.* 79:793–800

Rogers SO, Bendich AJ. 1987. Ribosomal RNA genes in plants: variability in copy number and in the intergenic spacer. *Plant Mol. Biol.* 9:509–20

Saghai-Maroof MA, Allard RW, Zhang Q. 1990. Genetic diversity and ecogeographical differentiation among ribosomal DNA alleles in wild and cultivated barley. *Proc. Natl. Acad. Sci. USA* 87:8486–90

Saghai-Maroof MA, Soliman KM, Jorgensen RA, Allard RW. 1984. Ribosomal DNA spacer-length polymorphisms in barley: Mendelian inheritance, chromosomal location, and population dynamics. *Proc. Natl. Acad. Sci. USA* 81:8014–18

Sarmientos P, Cashel M. 1983. Carbon starvation and growth rate-dependent regulation of the *Escherichia coli* RNA promoters: Differential control of dual promoters. *Proc. Natl. Acad. Sci. USA* 80:7010–13

Schade J, Kyle M, Hobbie S, Fagan W, Elser JJ. 2003. Stoichiometric tracking of soil nutrients by a desert insect herbivore. *Ecol. Lett.* 6:96–101

Schlötterer C, Tautz D. 1994. Chromosomal homogeneity of *Drosophila* ribosomal DNA arrays suggests intrachromosomal exchanges drive concerted evolution. *Curr. Biol.* 4:777–83

Seperak P, Slatkin M, Arnheim N. 1988. Linkage disequilibrium in human ribosomal genes: implications for multigene family evolution. *Genetics* 119:943–49

Sharma S, Beharav A, Balyan HS, Nevo E, Gupta PK. 2004. Ribosomal DNA polymorphism and its association with geographical and climatic variables in 27 wild barley populations from Jordan. *Plant Sci.* 166:467–77

Shufran KA, Mayo ZB, Crease TJ. 2003. Genetic changes within an aphid clone: homogenization of rDNA intergenic spacers after insecticide selection. *Biol. J. Linn. Soc.* 79:101–5

Sollner-Webb B, Tower J. 1986. Transcription of cloned eukaryotic ribosomal RNA genes. *Annu. Rev. Genet.* 55:801–30

Stent GS, Brenner S. 1961. A genetic locus for the regulation of ribonucleic acid synthesis. *Proc. Natl. Acad. Sci. USA* 47:2005–14

Sterner RW, Elser JJ. 2002. *Ecological Stoichiometry: The Biology of Elements from Molecules to the Biosphere*. Princeton: Princeton Univ. Press. 439 pp.

Sterner RW, Schultz KL. 1998. Zooplankton nutrition: recent progress and a reality check. *Verh. Int. Verein. Limnol.* 27:3009–14

Stevenson BS, Schmidt TM. 1998. Growth rate-dependent accumulation of RNA from plasmid-borne rRNA operons in *Escherichia coli. J. Bacteriol.* 180:1970–72

Stevenson BS, Schmidt TM. 2004. Life history implications of rRNA gene copy number in *Escherichia coli. Appl. Environ. Microbiol.* 70:6670–77

Strauss SH, Tsai C-H. 1988. Ribosomal gene number variability in Douglas-fir. *J. Heredity* 79:453–58

Su MH, Delany ME. 1998. Ribosomal RNA gene copy number and nucleolar-size polymorphisms within and among chicken lines selected for enhanced growth. *Poultry Sci.* 77:1748–54

Taylor WF. 1923. *The inheritance of 'bobbed'* in Drosophila hydei. M.S. thesis, Univ. Calif., Berkeley, CA. 17 pp.

Templeton AR, Hollocher H, Johnston JS. 1993. The molecular through ecological genetics of abnormal abdomen in *Drosophila mercatorum*. V. Female phenotypic expression on natural genetic backgrounds and in natural environments. *Genetics* 134:475–85

Templeton AR, Rankin MA. 1978. Genetic revolutions and control of insect populations. In *The Screwworm Problem*, ed. RH Richardson, pp. 83–112. Austin: Univ. Texas Press

Vrede T, Andersen T, Hessen DO. 1998. Phosphorus distribution in three crustacean zooplankton species. *Limnol. Oceanogr.* 44:225–29

Vrede T, Dobberfuhl DR, Elser JJ, Kooijman SALM. 2004. The stoichiometry of production—fundamental connections among organism C:N:P stoichiometry, macromolecular composition and growth rate. *Ecology* 85:1217–29

Vrede T, Persson J, Aronsen G. 2002. The influence of food quality (P:C ratio) on RNA:DNA ratio and somatic growth rate of *Daphnia. Limnol. Oceanogr.* 47:487–94

Wagner R. 1994. The regulation of ribosomal RNA synthesis and bacterial cell growth. *Arch. Microbiol.* 161:100–9

Weider LJ, Glenn KL, Kyle M, Elser JJ. 2004. Associations among ribosomal (r)DNA intergenic spacer length variation, growth rate, and C:N:P stoichiometry in the genus *Daphnia. Limnol. Oceanogr.* 49:1417–23

Weider LJ, Makino W, Acharya K, Glenn KL, Kyle M, et al. 2005. Genotype x environment interactions, stoichiometric food quality effects, and clonal coexistence in *Daphnia pulex. Oecologia.* 143:537–47

Zhang Q, Saghai Maroof MA, Allard RW. 1990. Effects of adaptedness of variations in ribosomal DNA copy number in populations of wild barley (*Hordeum vulgare* ssp. *spontaneum*). *Proc. Natl. Acad. Sci. USA* 87:8741–45

Annu. Rev. Ecol. Evol. Syst. 2005. 36:243–66
doi: 10.1146/annurev.ecolsys.35.021103.105730
First published online as a Review in Advance on August 12, 2005

Evolutionary Ecology of Plant Adaptation to Serpentine Soils

Kristy U. Brady, Arthur R. Kruckeberg, and H.D. Bradshaw Jr.

Department of Biology, University of Washington, Seattle, Washington 98195;
email: kbrady@u.washington.edu, ark@u.washington.edu, toby@u.washington.edu

Key Words Ca:Mg, ecological genetics, edaphic, endemism, mineral nutrition, physiological ecology, serpentine tolerance, ultramafic

■ **Abstract** Plant adaptation to serpentine soil has been a topic of study for many decades, yet investigation of the genetic component of this adaptation has only recently begun. We review the defining properties of serpentine soil and the pioneering work leading to three established physiological and evolutionary mechanisms hypothesized to be responsible for serpentine tolerance: tolerance of a low calcium-to-magnesium ratio, avoidance of Mg toxicity, or a high Mg requirement. In addition, we review recent work in serpentine ecology documenting the high proportion of endemic species present, the adaptive morphologies of serpentine-tolerant plants, and the distinctive structure of serpentine communities. Studies of the physiological mechanisms proposed to confer serpentine tolerance have shown that uptake of particular ions and heavy metals varies between serpentine-tolerant and -intolerant species. Recent studies examining the genetic basis of serpentine adaptation have shown serpentine-adaptive quantitative trait loci (QTL) to have large phenotypic effects, drought tolerance to be as important as metal tolerance, and serpentine adaptation to have evolved independently multiple times within species. Investigations of plant races and species adapted to contrasting soil types have shown disparate flowering times, divergent floral morphologies, and pollen incompatibility to contribute to reproductive isolation. Finally, we propose that future studies involving serpentine systems should merge the fields of ecology, evolution, physiology, and genetics.

1. INTRODUCTION

"Nothing can be more abrupt than the change often due to diversity of soil, a sharp line dividing a pine- or heather-clad moor from calcareous hills."

—Alfred Russel Wallace (1858)

Wallace (1858) and Darwin (1859) put forth the idea that adaptation is the signature of evolution by the hand of natural selection, and that adaptation to novel environments results in the origin of new species. Though the importance of

1543-592X/05/1215-0243$20.00

adaptive evolution in shaping the Earth's diverse biota is undisputed (Schluter 2001), our understanding of the process for organisms living in their natural habitats is remarkably superficial.

Wallace (1858) recognized that plant adaptation to different soil types is evidence of the strong natural selection imposed by ecological discontinuities. Among such examples of edaphic specialization, plant adaptation to serpentine soils is a system ideal for studies in evolutionary ecology and satisfies key requirements for addressing mechanistic questions of adaptive evolution in nature. First, there is extensive literature documenting serpentine flora and systematics, natural history, ecology, and physiology. Second, differential adaptation of closely related plants to serpentine and "normal" soil is phylogenetically and geographically widespread, having evolved independently many times. Third, divergence in adaptive phenotypes is readily demonstrated via reciprocal transplants (see Figure 1). Fourth, agents of natural selection are often apparent—e.g., soil chemical and physical properties—and amenable to manipulative experiments in the field, greenhouse, and laboratory. Finally, physiological mechanisms of serpentine tolerance have been described at the whole-plant level, providing valuable clues for the discovery of adaptations at lower levels of organization (e.g., tissue, cellular, and genetic).

Figure 1   Responses of eight strains of *Achillea borealis* to serpentine soil (*upper panel*) and nonserpentine soil (*lower panel*). The serpentine strains are S-142, S-164, S-135, and S-184. Photo by A.R. Kruckeberg.

In this review, we begin with a brief description of the "serpentine problem," followed by a historical overview of research on serpentine systems, a discussion of the present state of the field, and suggestions for the directions in which future research might proceed. We propose that future studies involving serpentine systems merge the fields of ecology, evolution, physiology, and genetics, with the goal of identifying all of the genes (and mutations within them) that produce the novel phenotypes required for serpentine adaptation.

2. THE SERPENTINE SYNDROME

"It is the obdurate physical adversity of things such as peridotite bedrock which often drives life to its most surprising transformations."

—David Raines Wallace (1983)
In *The Klamath Knot*

Serpentine soils are formed by the weathering of ultramafic rocks, those igneous or metamorphic rocks comprised of at least 70% ferromagnesian, or mafic, minerals (Kruckeberg 2002). Although serpentine is present, it is by no means the exclusive mafic mineral in these soils, hence it has been acknowledged that "serpentine soil" is, in fact, a misnomer (Proctor 1999). However, this term is cemented in the field (Brooks 1987).

Serpentine soils are ubiquitous, but patchily distributed. Although some variation occurs between sites, Whittaker (1954) identified three collective traits: (*a*) poor plant productivity, (*b*) high rates of endemism, and (*c*) vegetation types distinct from those of neighboring areas. Based on these features, Whittaker (1954) divided the serpentine problem into three parts: the edaphic, the plant species-level response (autecology), and the plant community-level effect (synecology).

The edaphic factor of the serpentine problem is multifaceted, involving chemical, physical, and biotic components. Arguably, the most influential factor on plant life is the chemical one (Kruckeberg 1985). Serpentine soils are characterized by low calcium-to-magnesium ratios with Ca at significantly lower concentrations relative to surrounding areas. They also frequently contain elevated levels of heavy metals, such as iron, nickel, chromium, and cobalt, which are toxic to most plants. Serpentine soils are often deficient in essential plant nutrients such as nitrogen, potassium, and phosphorus (Brooks 1987, Gordon & Lipman 1926, Proctor & Woodell 1975, Vlamis & Jenny 1948, Walker 1954).

The physical conditions of serpentine soils also prove inhospitable for many plants. Serpentine outcrops are often steep and comparatively rocky, making them particularly vulnerable to erosion, which results in shallow soils. Silt and clay contents in serpentine soils are generally minimal. Combined, these factors yield an environment with little moisture and depressed nutrient levels (Kruckeberg 2002, Proctor & Woodell 1975, Walker 1954). Therefore, plants inhabiting these sites must often tolerate drought as well as serpentine chemical attributes.

Biologically, serpentine sites frequently host a depauperate flora compared to surrounding regions. Sparse plant cover also encourages erosion and promotes elevated soil temperatures (Kruckeberg 2002). Each of these factors poses an additional stress to plant life. Together, the chemical, physical, and biotic components of the edaphic factor produce what Jenny (1980) coined the "serpentine syndrome," a term which illustrates the fact that it is the cumulative effect of these components to which a plant must adapt.

Plant species adapted to serpentine soils often possess morphologies somewhat distinct from closely related species not adapted to serpentine sites. There are several morphological features characteristic of serpentine-tolerant species and races. First, they typically possess xeromorphic foliage, including reduced leaf size and sclerophylls. Second, the stature of serpentine-tolerant plants is significantly reduced relative to counterparts on nonserpentine soil. Finally, root systems of species growing on and off serpentine sites are often more developed on serpentine soils than on neighboring soils (Krause 1958, Pichi-Sermolli 1948, Ritter-Studnička 1968, Rune 1953).

Serpentine systems are, in part, defined by the presence of a large number of endemic species (Jenny 1980). Cuba and New Caledonia exhibit especially high levels of endemism on serpentine soils. Although serpentine regions comprise only 1% of land area in California (Kruckeberg 1985, 2002), it has been estimated that serpentine-endemic plant species represent up to 10% of the state's total endemic flora (www.biodiversityhotspots.org). Experiments have shown that serpentine-tolerant species and races are limited to serpentine soils because of their inability to compete in nonserpentine environments (Kruckeberg 1950, 1954). This suggests that along the evolutionary trajectory toward serpentine tolerance, genetic trade-offs occur, rendering the serpentine-adapted plant species or race unable to recolonize its historical habitat. Thus, serpentine-tolerant species are often endemic to serpentine regions.

The final aspect of the serpentine problem identified by Whittaker (1954) involves the structure and distribution of serpentine communities. Serpentine landscapes are perhaps most striking from afar because the sharp contrast of vegetation on serpentine and neighboring nonserpentine soils brilliantly delineates the presence of an edaphic discontinuity (see Figure 2; see color insert). In extreme cases, serpentine sites are barren, leaving no doubt of a steep ecological gradient (see Figure 3).

Each facet of the serpentine problem is important to study in order to understand the ecology and evolution of plant adaptation to serpentine soils. The details of this problem are further discussed in the comprehensive reviews by Krause (1958), Proctor & Woodell (1975), Kinzel (1982), and Brooks (1987).

Figure 3   Extremely barren serpentine slopes above De Roux Forest Camp, upper north fork Teanaway River, Wenatchee Mountains, Washington.

3. SERPENTINE TOLERANCE MECHANISMS: AN OVERVIEW

"Compared to nonserpentine species, serpentine plants typically have one or more of the following traits: greater tolerance of high Mg and low Ca levels, higher Mg requirement for maximum growth, lower Mg absorption, higher Ca absorption, and Mg exclusion from leaves."

—Tyndall & Hull (1999)
In *Savannas, Barrens, and Rock Outcrop Plant Communities of North America.*

The mechanism by which a serpentine-tolerant plant copes with elevated levels of Mg and relatively insufficient quantities of Ca in the soil is perhaps its most defining character. There are several physiological and evolutionary mechanisms hypothesized to be responsible for serpentine tolerance.

3.1. The Calcium-to-Magnesium Ratio

Loew & May (1901) first attributed the poor productivity of serpentine sites to the low Ca:Mg ratio present in serpentine substrates. From their experiments, they concluded that the Ca:Mg ratio must be at least unity for optimal growth. Vlamis & Jenny (1948) later proposed that the low concentration of Ca present in serpentine soils is the principal stress of the serpentine syndrome, and high concentrations of Mg further compound the problem by depressing the availability of Ca. This hypothesis is supported by a number of studies (Kruckeberg 1954, Vlamis 1949, Walker 1948, Walker et al. 1955) in which growth of nonserpentine plants in serpentine soil increases significantly with the addition of Ca to the soil. For example, Kruckeberg (1954) found that nonserpentine races of *Phacelia californica* (Hydrophyllaceae) could persist on serpentine soil if the soil was supplemented with Ca; however, plants survived no better on serpentine soil fertilized with N, P, and K than they did on unfertilized serpentine soil.

Walker (1948) tested this hypothesis on tomato (*Lycopersicon esculentum* Mill., var. Marglobe) and a serpentine endemic, *Streptanthus glandulosus* ssp. *pulchellus* (Brassicaceae). In this experiment, Walker grew plants in serpentine soil leached with chloride solutions containing varying concentrations of Ca and Mg and fertilized with N, P, and K. Resulting soils varied in exchangeable Ca from 5% to 80% and Mg from 94% to 19%, respectively. The soil with the highest Ca:Mg ratio resembled productive agricultural soil, whereas the lowest Ca:Mg ratio tested was lower than that found in the original serpentine soil. Growth of tomato was directly correlated with Ca levels, whereas *Streptanthus* growth remained relatively unchanged across different Ca concentrations, suggesting that this species is more tolerant of depleted Ca levels. In soils with low Ca:Mg ratios, tomato took up significantly more Mg than *Streptanthus*. These data led Walker (1948) to conclude that the poor productivity of serpentine soils is a result of the low Ca concentration and concurrent high concentration of Mg, and only those species tolerant of low Ca:Mg levels could survive on serpentine soils.

Walker et al. (1955) reached similar conclusions from their study on three crop plants and three native California serpentine endemics. Plants were grown in soils prepared using the leaching and fertilizing technique previously described. Growth of crop plants declined considerably as soil Ca levels dropped, whereas the native serpentine species were not significantly affected by changing Ca concentrations. Analysis of plant tissue indicated that, when grown in soils with low Ca:Mg ratios, native species absorbed more Ca and typically less Mg than the crop plants. Thus, Walker et al. (1955) surmised that serpentine-tolerant species survive on soils with depleted levels of Ca because they are still able to absorb sufficient quantities of Ca without taking up excessive quantities of Mg.

Vlamis (1949) grew barley and Romaine lettuce on serpentine soil collected from California and discovered that each crop displayed symptoms of disease and diminished growth. Yields improved somewhat with the addition of N (NH_4NO_3), P, and K to the soil, and significantly increased with the addition of Ca ($CaSO_4$). Disease symptoms also lessened upon addition of Ca. Plants growing in soil with supplemental Ca, N, P, and K produced the greatest yields and appeared perfectly healthy. In contrast, addition of Mg ($MgSO_4$) or K alone drastically reduced growth and exacerbated disease symptoms. Tissue analysis revealed that Ca uptake by plants was considerably reduced when Mg or K was added to the soil. Such results corroborate Vlamis & Jenny's (1948) original suggestion that plant growth on serpentine soil is most affected by the low Ca content, and the excess Mg competitively inhibits Ca uptake by the plant.

3.2. Magnesium Toxicity

Serpentine soils contain potentially toxic concentrations of Mg, leading some researchers to conclude that Mg poisoning is the primary cause of the serpentine syndrome (Brooks 1987; Brooks & Yang 1984; Marrs & Proctor 1976; Proctor 1970, 1971). Proctor (1970) analyzed the effects of elevated levels of Mg on a serpentine-tolerant race of *Agrostis canina* (Poaceae) and a nonserpentine race of *A. stolonifera*, finding the latter to be notably more susceptible to Mg toxicity than the former. Furthermore, Mg toxicity appeared contingent upon sufficiently low Ca concentrations as it subsided upon addition of Ca in both soil and water cultures. Similar conclusions were reached in a second study (Proctor 1971). When grown on serpentine soil, oat plants showed severe symptoms of toxicity, which decreased dramatically upon addition of Ca to soil. Proctor (1971) thus suggested that the serpentine syndrome results not necessarily from limited Ca in the soil, but from the excessive level of Mg, which acts antagonistically to plant uptake of Ca and results in Mg poisoning.

In analyzing tissue of a number of serpentine endemics from Zimbabwe, Brooks & Yang (1984) found the concentration of Mg in plant tissue to be inversely proportional to the concentrations of seven other nutrients: aluminum, Fe, Co, boron, manganese, P, and sodium. Interestingly, a correlation between the concentration of Ca and Mg was not found. However, these data clearly suggest that the uptake of Mg comes at a cost to the plant as the uptake of other elemental nutrients is forfeited. Therefore, Brooks & Yang (1984) proposed that the heightened level of Mg in serpentine soils and its antagonistic behavior toward other elements could be the most important factor in the serpentine syndrome.

Madhok (1965), Madhok & Walker (1969), and Grover (1960) each suggested that competitive inhibition between Ca and Mg in soils with low Ca:Mg ratios is responsible for the differential absorption of the two elements in the serpentine-endemic sunflower *Helianthus bolanderi* ssp. *exilis* (Asteraceae), and the cultivated sunflower *H. annuus*. Grover (1960) showed that the quantity of Ca taken up by *H. bolanderi* ssp. *exilis* roots increased with increasing external Ca concentrations. However, raising the external concentration of Mg inhibited the uptake of Ca. Madhok (1965) and Madhok & Walker (1969) investigated the effects of Ca and Mg levels on these two species. *H. bolanderi* ssp. *exilis* tolerated much higher concentrations of Mg than *H. annuus*, which took up more Mg and less Ca than the serpentine endemic. Likewise, Marrs & Proctor (1976) found serpentine-tolerant strains of *Agrostis stolonifera* displayed greater tolerance to elevated levels of Mg than a nontolerant strain. Furthermore, the *A. stolonifera* strain from the more extreme serpentine site tolerated higher levels of Mg than the strain from the moderate serpentine site.

Main (1970) demonstrated that tolerance to Mg is a heritable trait in *Agropyron spicatum* (Poaceae). The serpentine-tolerant race of *A. spicatum* is significantly more tolerant of high levels of Mg than the nonserpentine race. Crossing these two strains produces an F_1 hybrid that exhibits an intermediate tolerance to Mg. Brooks (1987) suggested that if a specific tolerance to Mg is necessary for plants to survive on serpentine soil and Mg tolerance is heritable, then Mg must certainly be toxic to most plants at the levels present in serpentine regions.

3.3. Magnesium Requirement

Another theory on the physiological aspect of the serpentine syndrome posits a Mg requirement. There is evidence that certain serpentine races require higher concentrations of Mg externally and, in some cases, internally than counterpart nonserpentine populations (Grover 1960; Madhok 1965; Madhok & Walker 1969; Main 1974, 1981; Marrs & Proctor 1976). Main (1974) showed that, when grown on substrates with high levels of Mg, serpentine strains of *Agropyron spicatum* maintained lower concentrations of Mg in their tissue than nonserpentine strains. Thus, Main suggested that higher concentrations of Mg in the soil were necessary in order for the serpentine strain to acquire adequate levels of the element.

In another study, Main (1981) determined that *Poa curtifolia* (Poaceae), a serpentine endemic, requires an abnormally high level of Mg both externally and internally for optimal growth. Growth of this plant is positively correlated with the concentration of Mg in shoots and negatively correlated with Ca levels. Similarly, Marrs & Proctor (1976) showed that serpentine-tolerant strains of *Agrostis stolonifera* require more Mg and contain higher internal levels of Mg than nonserpentine strains. Grover (1960) found that growth of the serpentine-endemic *Helianthus bolanderi* ssp. *exilis* correlated positively with increasing soil Mg concentrations, whereas growth of *H. annuus*, a nonserpentine species, declined precipitously under the same conditions. Madhok (1965) and Madhok & Walker (1969) revealed that *H. bolanderi* ssp. *exilis* also requires a higher internal concentration of Mg relative to *H. annuus*. However, because the serpentine species more readily takes up Ca than Mg, a high external concentration of Mg is necessary for *H. bolanderi* ssp. *exilis* to attain sufficient quantities for optimal growth. Proctor &

Woodell (1975) suggest that studies showing unusually high Mg requirements by some serpentine-tolerant species and races, especially those concentrations typically lethal to plants, could lend insight into serpentine endemism by defining the conditions that restrict serpentine-tolerant plants to serpentine soil.

3.4. Hyperaccumulation

Hyperaccumulation is another mechanism that is hypothesized to allow plants to survive on serpentine soils. Hyperaccumulators are defined as those plants which contain in their tissue more than 1000 μg/g dry weight of Ni, Co, copper, Cr, or lead, or more than 10,000 μg/g dry weight of zinc or Mn (Baker & Brooks 1989). Aside from metal tolerance, hyperaccumulation is thought to benefit the plant by means of allelopathy, defense against herbivores, or general pathogen resistance (Boyd & Jaffré 2001, Boyd & Martens 1998a, Davis et al. 2001). There are at least 400 known Ni hyperaccumulators; however, the majority of serpentine-tolerant species do not fall into this category (Proctor 1999). Therefore, this review does not further explore the phenomenon of hyperaccumulation.

3.5. Preadaptation and Cross-Tolerance

Kruckeberg (1954) described a scenario in which distinct races or species evolve through adaptation to serpentine soil because certain individuals on nonserpentine sites are somewhat preadapted to one or more characteristics of serpentine sites. As these preadapted seeds germinate and reproduce on serpentine soil, their most fit progeny survive to reproduce, and thus a discrete lineage is born through the accumulation of serpentine tolerance alleles.

Macnair (1987) suggested that colonization of areas such as mines is challenging for plants because they must evolve tolerance to toxic levels of heavy metals—e.g., Cu—as well as adapt to a suite of other edaphic restrictions, such as low nutrients or drought conditions. For this reason, it has been suggested that plants that are somewhat preadapted for any of the harsh edaphic conditions of mines successfully colonize these regions more readily (Antonovics et al. 1971, Macnair 1987). Similar inferences might be drawn regarding serpentine sites.

Analogous to the notion of preadaptation is cross-tolerance. Some unfavorable conditions are common between different harsh environments. Thus, plants adapted, for example, to the high levels of Mg present in coastal saline habitats may also be tolerant of the high levels of Mg found in serpentine soil (Proctor 1971). Kruckeberg (1954) noted that a maritime-adapted race of *Achillea borealis californica* (Asteraceae) appeared equally adapted to serpentine soil. Likewise, Proctor (1971) found maritime-adapted *Armeria maritima* (Plumbaginaceae) grew relatively well on serpentine soil, and growth of a maritime-adapted race of *Silene maritima* (Caryophyllaceae) was comparable to the serpentine-tolerant race when grown on serpentine soil.

Boyd & Martens (1998b) offer three theories to explain the "preadaptive" nature of nonserpentine populations to serpentine conditions: (*a*) high rates of gene flow from serpentine to nonserpentine populations bring serpentine tolerance alleles into the latter population, (*b*) a constitutive serpentine adaptive trait presents little or no cost to a plant, or (*c*) a serpentine adaptive trait is adaptive for more than one function.

The first hypothesis aligns with the previously described scenario proposed by Kruckeberg (1954); however, there is little data available to support this idea. Evidence for cross-tolerance lends support to the third suggestion. The second explanation seems the least likely as it would be expected that, if serpentine tolerance presented little or no cost to a plant, then serpentine-adapted plants would grow just as well as nonserpentine plants on "normal" soil. But, it has been shown that serpentine-tolerant plants do not grow as well as nonserpentine plants when both are grown together on nonserpentine soil (Kruckeberg 1950, 1954). Even in the absence of competition with nonserpentine plants, serpentine-tolerant plants seem to possess a slower intrinsic growth potential than nonserpentine plants (K.U. Brady, unpublished data). Nevertheless, Reeves & Baker (1984) suggest that metal tolerance in *Thlaspi goesingense* (Brassicaceae) is a constitutive serpentine adaptive trait and is evident only when plants are grown on soils with high metal concentrations. In their study, seeds from serpentine and calcareous populations of *T. goesingense* were germinated in a greenhouse and grown in various serpentine and artificial limestone soils, some amended with Zn. Plants from each population grown on soils with high concentrations of Ni, Co, Zn, and Mn accumulated high quantities of all metals in shoots. Plants growing on serpentine soils took up much more Mg and considerably less Ca than plants growing on nonserpentine soils, regardless of origin. However, plants from both populations performed better on nonserpentine soil with no heavy metals. The similar metal accumulation and performance of plants from each population on differing soil types suggest that metal tolerance is a constitutive trait within this species.

Boyd & Martens (1998b) reached analogous conclusions in their study on Ni tolerance in *Thlaspi montanum* var. *montanum* (Brassicaceae). Seeds collected from two serpentine and two nonserpentine populations were grown in a greenhouse in soils ranging from "normal" to high-Ni content. Plants from all four populations exhibited hyperaccumulation of Ni in a similar fashion. Thus, Boyd & Martens (1998b) concluded that Ni hyperaccumulation is a constitutive trait of *T. montanum* var. *montanum*.

Similarly, serpentine tolerance appears to be a constitutive trait of *Silene dioica* (Caryophyllaceae), which survives in serpentine soil and tolerates high Ni concentrations regardless of whether plants are from a serpentine or nonserpentine population (Westerbergh 1995, Westerbergh & Saura 1992). The authors suggest that this "preadapted" nature of *S. dioica* has allowed the species to colonize serpentine soils multiple times independently. Taylor & Levy (2002) tested three varieties of *Phacelia dubia* (Hydrophyllaceae) for preadaptation to serpentine soils and found that *P. dubia* var. *georgiana*, an endemic of granite outcrops, displayed a tolerance to lower Ca:Mg ratios than present in its native soil. The authors propose that this type of variation could represent the sort of preadaptive trait that allows a species to inhabit new environments. Likewise, inherent serpentine tolerance in *S. dioica* or metal tolerance in *Thlaspi montanum* var. *montanum* and *T. goesingense* could render each species preadapted to serpentine soils, facilitating colonization of such regions.

4. THE ECOLOGY OF SERPENTINE SYSTEMS

The ecology of serpentine systems is particularly interesting considering the high proportion of endemic plant species present, the adaptive morphologies of serpentine plants, and the distinctive structure of serpentine communities. Iturralde's (2001) analysis of plants on serpentine soils in Cuba perfectly illustrates the unique ecology of these systems. Plants exhibit various morphological adaptations associated with serpentine tolerance such as sclerophlly, microphylly, and thorny stems, as well as compact statures. The structure of these serpentine communities is characteristically open and short-statured relative to surrounding regions. A high number of endemic species and Ni hyperaccumulators were also recorded for serpentine sites in Cuba (Iturralde 2001). Disproportionate occurrence of endemic species on serpentine soils relative to total species in the surrounding region has also been documented recently in the Piacenza area of North Apennines, Italy (Vercesi 2003). Specht et al. (2001) noted corresponding features in the serpentine region of northern New South Wales, Australia. They found the vegetation on serpentine soils to be distinct from adjoining substrates and the community structure reduced in both height and basal area relative to neighboring areas.

The effects of patch size on local and regional plant diversity on serpentine sites in California have been studied at length (Harrison 1997, 1999; Harrison et al. 2001). In woody species of California serpentine chaparral, it was demonstrated that larger areas of serpentine soil (6 km^2 to 55 km^2) support higher alpha diversity (local species number) than smaller patches (0.5 ha to 3 ha), but that beta diversity (differentiation in species composition among sites) is significantly higher among small patches (Harrison 1997). Similar results were obtained in a study on the diversity of endemic serpentine herbs in the North Coast Ranges of California (Harrison 1999). However, the latter study also showed alpha diversity for all herbaceous species to be higher in small serpentine patches because of a larger number of alien species present. Harrison (1999) found that the number of alien species present in a small serpentine patch directly correlates with the soil Ca level. In contrast, the number of serpentine endemics on large patches is inversely proportional to the concentration of soil Ca. Nevertheless, even when soil conditions do not vary, small serpentine patches still host higher numbers of alien species than large patches (Harrison et al. 2001). Additionally, Gram et al. (2004) found that water availability is strongly correlated with species composition in California serpentine grasslands. Specifically, in a year with high rainfall, the number of exotic species present on rocky serpentine outcrops nearly doubled from 10 exotic species to 19. These analyses suggest that both soil properties and patch size are important factors in determining species composition in serpentine areas of California.

Batianoff & Singh (2001) investigated the differences in plant ecology and frequency of endemism between the upland and lowland landforms in the largest serpentine region in eastern Australia. They discovered that overall species richness correlates negatively with the concentration of Ni in the soil. However, they also found a greater number of endemic species in the uplands, which has higher soil Ni concentrations than the lowlands. Consequently, Batianoff & Singh (2001) concluded that edaphic conditions strongly influence species diversity and levels of endemism in this system.

A study by Cooke (1994) examined dynamics influencing species composition on serpentine and associated nonserpentine areas in the Wenatchee Mountains in the state of Washington. Results indicated substrate to have the greatest effect followed by aspect and disturbance regime. Cooke also identified several traits in serpentine-tolerant races of *Achillea* and *Senecio* (Asteraceae) as pivotal to surviving on serpentine soils. Serpentine species are much more tolerant of depleted Ca levels and elevated concentrations of Mg and Ni in the soil. To minimize water requirements and excessive water loss, serpentine plants are able to reduce water potentials to levels lower than found on nonserpentine soils, as well as keep stomata closed or nearly closed. In addition, serpentine-tolerant plants have a slower growth rate than nonserpentine species and possess morphologies adapted to drought conditions (Cooke 1994).

Several other studies have also revealed the importance of drought tolerance to survival on serpentine soil. Chiarucci (2004) identified drought stress as a more significant challenge to plants on serpentine soil in Tuscany, Italy, than the presence of heavy metals. Freitas & Mooney (1995) showed that populations of *Bromus hordeaceus* (Poaceae) growing on serpentine soil are better adapted to drought stress than *B. hordeaceus* populations growing on sandstone, as represented by both plant growth and root branching patterns. Armstrong & Huenneke (1992) examined the effect of drought on species composition in a California serpentine grassland. Although annual grasses were negatively affected by four consecutive years of drought, native bunchgrasses were little affected, likely because they possess well-developed root systems and are able to access soil moisture at greater depths. Furthermore, exotic annual grasses were more severely affected than native annual grasses. McCarten (1992) found that, in addition to mineral composition, soil depth, slope angle, and aspect are important factors influencing the structure of a California serpentine grassland by virtue of their effects on soil water content. Native perennial bunchgrasses are more commonly found on deeper soils where roots can access deep soil moisture. Native annual grasses with less-developed root systems are more often found on shallower soils. Thus, the combination of soil depth and soil chemical properties creates a patchwork of microhabitats in the serpentine grassland that results in variation in species composition and structure within a small area.

Figure 2 Yellow goldfields (*Lasthenia californica*) on serpentine (*left*) and grassland on sandstone (*right*) at Jasper Ridge, Stanford University, California. Photo by Bruce Bohm.

5. THE PHYSIOLOGICAL BASIS OF SERPENTINE TOLERANCE

As previously discussed, serpentine-tolerant plants must endure a variety of adverse chemical conditions. The adaptive mechanism(s) that confer to plants tolerance to soils with depleted quantities of Ca and high concentrations of Mg and heavy metals is still not well understood.

Lee et al. (1997) compared the foliar concentrations of nine elements (N, P, K, Ca, Mg, Ni, Cu, Co, and Cr) in 12 plant species growing on and off serpentine soil in New Zealand. In each case, the foliar concentrations of Mg and Ni were significantly higher in plants growing on serpentine soil than in conspecifics growing on nonserpentine soil. Conversely, the concentration of Ca was significantly less for plants growing on serpentine soil. In several species, the concentration of Cu, Co, and/or Cr was also significantly higher in plants growing on serpentine soil.

To gain insight into how Mg and Ca are stored in plants growing on serpentine soil, Tibbetts & Smith (1992) used *Sedum anglicum* (Crassulaceae) to analyze accumulation of Ca and Mg in the vacuole. Although *S. anglicum* is not associated with serpentine soils, the authors chose this species because it has large parenchymatous mesophyll cells, and the vacuole comprises more than 95% of cell volume. Therefore sufficient quantities of cell "sap," which essentially reflect the contents of the vacuole, can be easily obtained. In any case, *S. anglicum* can be found on soils relatively rich in Mg, and experiments revealed that the species is tolerant of a wide range of Ca:Mg ratios. However, *S. anglicum* preferentially takes up Ca over Mg. Specifically, the authors found that at a soil Ca:Mg of unity, the cell sap concentration of Ca was about twice that of Mg. Equal cellular concentrations of Ca and Mg were reached when the external Ca:Mg ratio was approximately 1:6. Interestingly, less than half of the total Mg and Ca in the leaf cell sap was in the form of free cations. Instead, chelation of Mg and Ca by soluble carboxylates such as malate, citrate, and isocitrate located in the vacuole result in the formation of metal-ligand complexes. Tibbetts & Smith (1992) thus suggest that chelation may be important for vacuolar sequestration of excessive numbers of ions that might otherwise be toxic to the cell.

Adaptive physiological differences in two races of *Lasthenia californica* (Asteraceae) found on serpentine soil have been studied extensively (Rajakaruna 2003, Rajakaruna & Bohm 1999, Rajakaruna et al. 2003c). Races A and C of *L. californica* coexist on serpentine soil at Jasper Ridge Biological Preserve, California, but inhabit soils of differing physical and chemical properties. Rajakaruna & Bohm (1999) describe race A soil as having a higher water content, percent clay, cation exchange capacity, and Na and Mg concentration than the soil race C inhabits, which has a higher Ca:Mg ratio and higher concentrations of Ca, K, and Ni. Results of these studies revealed that race A is more tolerant of ionic stresses, including high levels of Na and Mg (Rajakaruna et al. 2003c), whereas race C is more tolerant of drought stress (Rajakaruna et al. 2003b). Interestingly, uptake of both Ca and Mg was shown to be twice as high in race A plants as in race C plants. Furthermore, shoot concentrations of Ca and Mg were reported to be 127- and 28-times higher, respectively, in race A than in race C, indicating a greater tolerance of ion accumulation in the shoot in race A (Rajakaruna et al. 2003c). Conversely, ion concentrations in the roots remained approximately equal in both races, suggesting that elevated levels of ions in race A shoots are a result of both an increased rate of uptake and translocation of ions in race A. It is worth noting that these studies in conjunction with earlier research (e.g., Madhok 1965, Madhok & Walker 1969, Walker et al. 1955) suggest that the physiological basis for serpentine tolerance may involve one or more different mechanisms (Rajakaruna et al. 2003c), including ion uptake discrimination at the root level, ion translocation properties, and/or chelation.

6. THE GENETICS OF ADAPTATION TO SERPENTINE SOILS

Little is known about the process by which a serpentine-tolerant population evolves. Studies concerning genetic and adaptive differentiation in serpentine-tolerant and -intolerant races and species are necessary in order to reveal key innovations in the path to tolerance. Nyberg Berglund et al. (2001) examined the genetic differentiation of multiple serpentine populations of *Cerastium alpinum* (Caryophyllaceae) in Fennoscandia and revealed that serpentine tolerance in this species likely evolved two or more times independently. Further investigation (Nyberg Berglund et al. 2004) showed that serpentine populations of *C. alpinum* are more tolerant of elevated levels of Ni and Mg than nonserpentine populations, and that the degree of tolerance is directly correlated with the degree of saturation of Ni or Mg in the soil of the population's origin. These data suggest that serpentine tolerance in this species is locally evolved, supporting the earlier conclusion (Nyberg Berglund et al. 2001) that tolerance in *C. alpinum* has arisen more than once in this region.

Patterson & Givnish (2004) sequenced three segments of chloroplast DNA in *Calochortus* (Liliaceae) and deduced (by phylogenetic analysis using maximum parsimony) that serpentine tolerance arose seven times in this genus. Similarly, Mengoni et al. (2003) examined chloroplast genetic diversity between nine populations of the serpentine-endemic *Alyssum bertolonii* (Brassicaceae) from four serpentine regions in northern Italy. High levels of genetic differentiation were detected between populations within a region as well as between populations of different regions. Mengoni et al. suggest the high percentage of population-specific chloroplast haplotypes is evidence that serpentine tolerance within this species has arisen independently in each population.

As previously discussed, the physical state of serpentine soils produces drought conditions. Accordingly, serpentine-tolerant plants are often drought tolerant. Hughes et al. (2001) analyzed the role of drought tolerance in serpentine tolerance in the *Mimulus guttatus* (Phrymaceae) complex. Physiological responses to drought were analyzed in two serpentine-endemic species (*M. nudatus* and *M. pardalis*), two nonserpentine species (*M. marmaratus* and *M. nasutus*), and *M. guttatus*, which occurs on and off serpentine soil. Although all five species proved to be susceptible to drought, pressure-volume curves imply that the two serpentine-endemic species are more drought tolerant than the other three species, because the former possess higher hydrated osmotic pressures.

Hughes et al. (2001) also developed a population segregating for serpentine tolerance from a cross between *Mimulus nudatus*, a serpentine-endemic species, and *M. marmoratus*, a nonserpentine species. They found that serpentine tolerance in the segregating population was significantly positively correlated with drought tolerance, but not with Ni tolerance. This suggests that drought tolerance may be a more important factor than Ni tolerance in influencing adaptation to serpentine conditions in the *M. guttatus* complex. Interestingly, there seemed to be an important correlation between drought tolerance and plant size, as the most tolerant plants were also the smallest. This suggests that more compact forms may be favored in drought and serpentine conditions, seemingly explaining the characteristically low stature of serpentine communities. However, a more detailed genetic analysis is needed to determine whether the observed correlation between plant size and drought or serpentine tolerance in this system is due to linkage or pleiotropy.

A QTL mapping project by Gailing et al. (2004) identified one large QTL (or two or more tightly linked QTLs) controlling contrasting adaptive traits in serpentine-tolerant *Microseris douglasii* (Asteraceae) and serpentine-intolerant *M. bigelovii*. Adaptive traits are often genetically complex, involving multiple genes with varying effects. QTL maps provide information on the genetic architecture of a quantitative trait by identifying the number of loci involved, their map location, magnitude of effect, and mode of action. The success of QTL maps in analyzing the genetic component of adaptive differences between plant species and races is well documented (Mauricio 2001). Serpentine plants ordinarily flower and set seed earlier than conspecifics or sister taxa on nonserpentine soil in order to avoid drought conditions typical of serpentine systems. Accordingly, serpentine-tolerant *Microseris douglasii* flowers and sets seed earlier than *M. bigelovii*, which grows off serpentine soil. But flowering early comes at a cost to accumulating leaf biomass. Thus, alleles at the major QTL controlling leaf and floral bud production identified by Gailing et al. (2004) have opposite effects in *M. douglasii* and *M. bigelovii*.

7. THE EVOLUTION OF REPRODUCTIVE ISOLATION

In theory, plant populations growing on contrasting soil types could experience such strong divergent selection that subsequent genetic differentiation of the populations renders them reproductively isolated and, in extreme cases, results in ecological speciation (Schluter 2001). Kruckeberg (1986) suggests diverse edaphic substrates set the stage for ecological speciation because such geologic effects are patchily distributed and thus promote geographic isolation. Adaptation to edaphic conditions may also beget reproductive isolation indirectly via linkage or pleiotropy, or directly via the advent of pre- and postzygotic isolating mechanisms such as reinforcement to prevent hybridization (Rajakaruna & Whitton 2004, Schluter 2001).

Prezygotic isolating mechanisms between plant species or races include shifts in flowering time, a switch to primarily self-fertilization from out-crossing, and alterations in flower morphology that affect pollinator attraction and/or visitation (Macnair 1989). For example, in *Collinsia sparsiflora* (Scrophulariaceae *s.l.*), peak flowering time in serpentine and nonserpentine populations differs significantly (Wright et al. 2005). The serpentine-endemic *Mimulus nudatus* flowers earlier than serpentine-adapted populations of *M. guttatus*, the presumed progenitor of *M. nudatus* (Gardner & Macnair 2000). Furthermore, the primary pollinators of *M. guttatus* and *M. nudatus* are different owing to divergent floral morphologies. In *Lasthenia californica* disparate flowering times and pollen incompatibility reproductively isolate the two edaphic races previously described (Rajakaruna & Whitton 2004). In fact, recent phylogenetic work by Rajakaruna et al. (2003a) suggests that at Jasper Ridge Biological Preserve these races are two distinct species.

Although disparate flowering times have been confirmed for plant races and species adapted to contrasting soil types in several systems, whether shifts in phenology are strictly a result of adaptation to edaphic conditions or a method of reinforcement is not always clear (McNeilly & Antonovics 1968). Nevertheless, both processes function to reproductively isolate differentially adapted plant races or species.

8. FUTURE DIRECTIONS FOR RESEARCH ON PLANT ADAPTATION TO SERPENTINE SOILS

"When this research was in its embryonic stages, a complete solution of the problem of serpentine endemism seemed manifestly an attainable goal, wholly within the scope of this single study! Like the proverbial mirage, that goal seems elusive as, one by one, the many facets and ramifications of the problem are explored. It would be well at this time then, to take stock of just what is known about serpentine floras and what is yet to be learned."

—A.R. Kruckeberg (1950)
In *An experimental inquiry into the nature of endemism on serpentine soils*.
PhD thesis.

In contrast to our detailed knowledge of neutral evolutionary processes, where precision at the DNA sequence level is the norm, there is an almost complete lack of detailed information regarding the genetic basis of adaptive evolution for organisms living in their natural habitats. We attribute this disparity to the fact that remarkably little is known about either the ecology of highly developed model organisms in genomics (e.g., *Drosophila, Caenorhabditis, Arabidopsis, Saccharomyces*), or the genetics of any of the myriad organisms that ecologists have studied in depth. It is not a coincidence that the organisms of interest to ecologists and geneticists are different—ecologists prefer larger, longer-lived species that dominate and structure ecosystems, whereas geneticists prefer smaller, shorter-lived species that prosper in the laboratory. Yet studying adaptive evolution in nature requires the analysis of both the ecology and genetics of the organism. These two disciplines are encompassed in the field of ecological genetics, which seeks to analyze natural selection and genetic variation in concert as a means to studying adaptive evolution (Via 2002).

There are two obvious approaches to advancing the study of ecological genetics. The first is to develop a good working knowledge of the ecology of one or more of the genetic model organisms. The second is to develop powerful genetic tools for organisms and experimental systems whose ecology is already well understood. Although both approaches are worthwhile and are being pursued vigorously by many investigators worldwide, we favor the latter approach. Developing genetic and genomic resources for interesting organisms has become a major scientific enterprise; whole eukaryotic genomes can be completely sequenced in a matter of a few weeks. Ecological research in natural settings cannot match that pace; hence, we feel that it is easier to bring genetics to ecology, rather than vice versa.

Plant adaptation to serpentine soils has the potential to become a general model for ecological genetics in natural populations (Gailing et al. 2004, Pepper & Norwood 2001). Plants—because of their prolific seed production, macroscopic size, sessile growth habit, and ease of estimating darwinian fitness—are nearly uniquely suited to the large reciprocal transplant experiments that will be required for a thorough examination of adaptive genetics in nature. Reciprocal transplant experiments between serpentine populations (or species) and their spatially adjacent populations (or sister species) growing on normal soils demonstrate unequivocally that adaptation to serpentine soils has a genetic basis (Nyberg Berglund et al. 2004; Kruckeberg 1950, 1954; Rajakaruna et al. 2003b,c), but detailed genetic analyses of this adaptation have only recently been initiated (see Section 6 above).

The patchy distribution of serpentine soils in many landscapes means that adaptation often occurs on small geographic scales, with serpentine-adapted populations and their presumed nonserpentine ancestors growing adjacent to each other, but on juxtaposed soil types. The results of reciprocal transplant experiments suggest that natural selection for serpentine tolerance is strong. Strong selection favors adaptation by mutations with large phenotypic effects (Orr 1998), which improves the odds of identifying these adaptive mutations by genetic analysis. Further, as detailed above, the evolution of serpentine tolerance has occurred many times independently, even within a single species, facilitating investigation of the "repeatability" of evolutionary trajectories (Nyberg Berglund et al. 2004). Finally, the physiological basis of serpentine tolerance is likely to involve such mechanisms as ion transport, osmotic control, and temperature stress, topics that are well studied in plant genomic model systems such as *Arabidopsis thaliana* or crop species. This knowledge can be used to develop and test hypotheses about the physiology of adaptations in serpentine-tolerant plants. It is a short step from physiology to biochemistry and from there to genetics and genomics. Our current understanding of the serpentine-tolerant phenotype ("syndrome") is a somewhat vague composite of morphology, physiology, and ecology. Genetic analysis will benefit greatly from a precise description of each component of the tolerance phenotype (e.g., Ca homeostasis, Ni tolerance, drought tolerance), so that each component may be dissected genetically, then reassembled, like a mosaic, into a complete portrait of adaptation.

The physiology of ion transport is an especially promising area for more detailed study, because serpentine tolerance will universally involve adaptations to deal with the reduced availability of Ca (Hirschi 2004), high levels of Mg, and high levels of one or more heavy metals. Elemental analysis of plant tissues can now be done on tiny samples (10 μm to 50 μm in diameter) using laser ablation and mass spectrometry (Narewski et al. 2000), in principle making it feasible to measure the internal concentration of all mineral ions anywhere in the plant and compare them to the ion concentration in the soil solution (or artificial nutrient solution) in which the plant is growing. Such experiments can be carried out in a high-throughput fashion (Hirschi 2003, Salt 2004). By experimental manipulation of ion concentrations and ratios in artificial nutrient solutions, it should be possible to infer mechanistic details about ion exclusion, uptake, competition, and sequestration in various plant tissues. Indeed, many of the current hypotheses about physiological mechanisms of serpentine tolerance were developed as a result of experiments of this type, albeit in decades past when analytical methods were much less sensitive and sophisticated. Whole-plant investigations can now be followed up with electrophysiological ion flux measurements at the level of single cells (Shabala et al. 1997) down to individual ion channel proteins (Tester 1997).

Much of the selectivity in ion transport is likely to occur in the roots of serpentine-tolerant plants, yet most physiological experiments have been focused on the above-ground organs. Root growth is very sensitive to the ionic composition of the surrounding solution and has been used successfully to assay tolerance to serpentine soils (Nyberg Berglund et al. 2004) and heavy metals (Macnair 1983). Because the essential ionic features of serpentine soil can be recreated in hydroponic culture (Walker et al. 1955) and the flux of ions across roots can be measured at spatial scales comparable to the dimensions of single cells (Newman 2001, Shabala et al. 1997), root physiologists are poised to make substantial progress on the mechanisms of serpentine tolerance.

A reasonably comprehensive understanding of serpentine-tolerance physiology leads immediately to a "candidate gene" approach for identifying individual genes that contribute to serpentine tolerance in natural plant populations. The essence of the candidate gene approach is to narrow the scope of the search for adaptive genes from the whole plant genome (>25,000 genes) to a manageable handful of genes (and their alleles) whose phenotypic effects can be measured directly through labor-intensive breeding or transgenic experiments. Genes that are known or suspected to participate in the relevant physiological processes or other adaptive traits in well-studied plants—crops and genomic models such as *Arabidopsis*—become candidate genes. For example, the genes encoding many ion transport proteins in *Arabidopsis* have been identified (Hirschi 2003, Salt 2004). The orthologs of these genes can be cloned from serpentine-tolerant plants and their sister taxa, then tested for both DNA sequence variation and functional differences. Clearly, this comparative method will depend upon high-resolution phylogenies capable of resolving the sister taxon relationships of serpentine-tolerant plant populations and species. Species- and population-level phylogenies have been less commonly available in plants than in animals, but that is changing as plant molecular systematists turn to more rapidly evolving nuclear DNA sequences (Sang 2002, Small et al. 2004).

Candidate genes may also be identified in mutant screens for serpentine tolerant phenotypes in model plant systems. Intensive efforts to discover *Arabidopsis* mutants that accumulate or exclude specific ions (Hirschi 2004, Lahner et al. 2003, Salt 2004) could have obvious application to wild serpentine plant species with the same phenotype. Likewise, a mutant screen of *Arabidopsis* has shown that null alleles of *CAX1*, a Ca-proton antiporter in the tonoplast (Hirschi et al. 1996), can produce many of the growth and ion uptake phenotypes associated with serpentine-tolerant plants (Bradshaw 2005). Whether mutations in *CAX1* play a role in serpentine tolerance in natural plant populations remains to be seen.

In the absence of a well-developed physiological model for serpentine tolerance, and therefore in the absence of candidate genes, it may be wise to pursue a purely genetic approach to identifying genes that contribute to serpentine tolerance. The most general method is to genetically map and positionally clone the quantitative trait loci that are responsible for the phenotypic differences between serpentine-tolerant and nonserpentine sister taxa. Such an approach has the advantage of being unbiased; i.e., it can point to biochemical pathways or physiological mechanisms that are as yet undescribed. The QTL mapping and cloning approach (reviewed in Mauricio 2001) is initiated by making crosses between individuals from populations (or closely related species), one of which is serpentine-adapted and the other of which is not. From this F_1 hybrid generation, a segregating F_2 population is produced. For each F_2 plant, the serpentine tolerance phenotypes are measured (e.g., biomass on serpentine soil, tissue Mg concentration), and variation in these phenotypes is correlated with genome-wide molecular marker genotypes (e.g., microsatellites, single nucleotide polymorphisms) that distinguish the alleles of the serpentine-tolerant parent from those of the nonserpentine parent. The power of this method is illustrated by the results of the QTL mapping experiment in *Microseris* described previously (Gailing et al. 2004). QTL mapping gives an estimate of the number of loci involved in the fully serpentine-adapted phenotype, as well their relative magnitude of effect on each component of the tolerance phenotype. Positional cloning of QTLs is notoriously difficult (e.g., Frary et al. 2000, Fridman et al. 2000), but if selection for serpentine tolerance is strong, there should be QTLs with large phenotypic effects that make cloning possible. The first such mapping experiment confirms that these large adaptive QTLs exist (Gailing et al. 2004). Once a QTL has been cloned from a single plant species, it will be straightforward to test for its significance in other plant taxa that have evolved serpentine tolerance independently.

Although it has been more than a century since Loew & May (1901) offered a hypothesis for plant tolerance of serpentine soils, for the first time we have methods adequate for determining the phenotypes and genotypes of serpentine-tolerant plants at the level of precision needed to identify individual genes responsible for this striking adaptation. Plant adaptation to serpentine soil represents a valuable and experimentally tractable system for evolutionary ecologists and geneticists. We look forward to seeing its secrets unraveled.

ACKNOWLEDGMENTS

The authors wish to thank D. Schemske and N. Rajakaruna for their insightful comments and suggestions during the preparation of this manuscript.

**The *Annual Review of Ecology, Evolution, and Systematics* is online at
http://ecolsys.annualreviews.org**

LITERATURE CITED

Antonovics J, Bradshaw AD, Turner RG. 1971. Heavy metal tolerance in plants. *Adv. Ecol. Res.* 7:1–85

Armstrong JK, Huenneke LF. 1992. Spatial and temporal variation in species composition in California grasslands: The interaction of drought and substratum. See Baker et al. 1992, pp. 213–33

Baker AJM, Brooks RR. 1989. Terrestrial higher plants which hyperaccumulate metallic elements—A review of their distribution, ecology and phytochemistry. *Biorecovery* 1:81–126

Baker AJM, Proctor J, Reeves RD, eds. 1992. *The Vegetation of Ultramafic (Serpentine) Soils.* Andover, Engl.: Intercept. 509 pp.

Batianoff GN, Singh S. 2001. Central Queensland serpentine landforms, plant ecology and endemism. *S. Afr. J. Sci.* 97:495–500

Boyd RS, Jaffré T. 2001. Phytoenrichment of soil Ni content by *Sebertia acuminata* in New Caledonia and the concept of elemental allelopathy. *S. Afr. J. Sci.* 97:535–38

Boyd RS, Martens SN. 1998a. The significance of metal hyperaccumulation for biotic interactions. *Chemoecology* 8:1–7

Boyd RS, Martens SN. 1998b. Nickel hyperaccumulation of *Thlaspi montanum* var. *montanum* (Brassicaceae): A constitutive trait. *Am. J. Bot.* 85:259–65

Bradshaw HD. 2005. Mutations in *CAX1* produce phenotypes characteristic of plants tolerant to serpentine soils. *New Phytol.* 167:81–88

Brooks RR. 1987. In *Serpentine and Its Vegetation*, ed. TR Dudley. Portland, OR: Dioscorides. 454 pp.

Brooks RR, Yang XH. 1984. Elemental levels and relationships in the endemic serpentine flora of the Great Dyke, Zimbabwe and their significance as controlling factors for this flora. *Taxon* 33:392–99

Chiarucci A. 2004. Vegetation ecology and conservation on Tuscan ultramafic soils. *Bot. Rev.* 69:252–68

Cooke SS. 1994. *The edaphic ecology of two*

western North American composite species. PhD thesis. Univ. Wash., Seattle. 288 pp.

Darwin C. 1859. *On the Origin of Species by Means of Natural Selection, or The Preservation of Favoured Races in the Struggle for Life.* New York: Mentor. 495 pp.

Davis MA, Boyd RS, Cane JH. 2001. Host-switching does not circumvent the Ni-based defense of the Ni hyperaccumulator *Streptanthus polygaloides* (Brassicaceae). *S. Afr. J. Sci.* 97:554–57

Frary A, Nesbitt TC, Frary A, Grandillo S, van der Knaap E, et al. 2000. *fw2.2*: a quantative trait locus key to the evolution of tomato fruit size. *Science* 289:85–88

Freitas H, Mooney H. 1995. Growth responses to water stress and soil texture of two genotypes of *Bromus hordeaceus* from sandstone and serpentine soils, pg. 19. *Proc. Int. Conf. Serpentine Ecol.*, *2nd, Noumea*, 72 pp.

Fridman E, Pleban T, Zamir D. 2000. A recombination hotspot delimits a wild-species quantitative trait locus for tomato sugar content to 484bp within an invertase gene. *Proc. Natl. Acad. Sci. USA* 97:4718–25

Gailing O, Macnair MR, Bachmann K. 2004. QTL mapping for a trade-off between leaf and bud production in a recombinant inbred population of *Microseris douglasii* and *M. bigelovii* (Asteraceae, Lactuceae): a potential preadaptation for the colonization of serpentine soils. *Plant Biol.* 6:440–46

Gardner M, Macnair MR. 2000. Factors affecting the co-existence of the serpentine endemic *Mimulus nudatus* Curran and its presumed progenitor, *Mimulus guttatus* Fischer ex DC. *Biol. J. Linn. Soc.* 69:443–59

Gordon A, Lipman CB. 1926. Why are serpentine and other magnesian soils infertile? *Soil Sci.* 22:291–302

Gram WK, Borer ET, Cottingham KL, Seabloom EW, Boucher VL, et al. 2004. Distribution of plants in a California serpentine grassland: are rocky hummocks spatial refuges for native species? *Plant Ecol.* 172:159–71

Grover R. 1960. *Some aspects of Ca-Mg nutrition of plants with special reference to ser-

pentine endemism.* PhD thesis. Univ. Wash., Seattle. 136 pp.

Harrison S. 1997. How natural habitat patchiness affects the distribution of diversity in Californian serpentine chaparral. *Ecology* 78:1898–1906

Harrison S. 1999. Local and regional diversity in a patchy landscape: native, alien, and endemic herbs on serpentine. *Ecology* 80:70–80

Harrison S, Rice K, Maron J. 2001. Habitat patchiness promotes invasion by alien grasses on serpentine soil. *Biol. Conserv.* 100:45–53

Hirschi KD. 2003. Strike while the ionome is hot: making the most of plant genomics advances. *Trends Biotech.* 21:520–21

Hirschi KD. 2004. The calcium conundrum. Both versatile nutrient and specific signal. *Plant Physiol.* 136:2438–42

Hirschi KD, Zhen RG, Cunningham KW, Rea PA, Fink GR. 1996. CAX1, an H^+/Ca^{2+} antiporter from *Arabidopsis*. *Proc. Natl. Acad. Sci. USA* 93:8782–86

Hughes R, Bachmann K, Smirnoff N, Macnair MR. 2001. The role of drought tolerance in serpentine tolerance in the *Mimulus guttatus* Fischer ex DC. complex. *S. Afr. J. Sci.* 97:581–86

Iturralde RB. 2001. The influence of ultramafic soils on plants in Cuba. *S. Afr. J. Sci.* 97:510–12

Jenny H. 1980. *The Soil Resource: Origin and Behavior.* *Ecol. Stud.* 37:256–59. New York: Springer-Verlag. 377 pp.

Kinzel H. 1982. *Pflanzenökologie und Mineralstoffwechsel.* Stuttgart: Ulmer

Krause W. 1958. Andere Bodenspezialisten. In *Handbuch der Pflanzenphysiologie*, ed. G Michael, 4:758–806. Berlin: Springer-Verlag

Kruckeberg AR. 1950. *An experimental inquiry into the nature of endemism on serpentine soils.* PhD thesis. Univ. Calif., Berkeley. 154 pp.

Kruckeberg AR. 1954. The ecology of serpentine soils: A symposium. III. Plant species in relation to serpentine soils. *Ecology* 35:267–74

Kruckeberg AR. 1985. *California Serpentines: Flora, Vegetation, Geology, Soils, and Management Problems.* Berkeley: Univ. Calif. Press. 180 pp.

Kruckeberg AR. 1986. An essay: the stimulus of unusual geologies for plant speciation. *Syst. Bot.* 11:455–63

Kruckeberg AR. 2002. The influences of lithology on plant life. In *Geology and Plant Life: The Effects of Landforms and Rock Type on Plants*, pp. 160–81. Seattle/London: Univ. Wash. Press. 362 pp.

Lahner B, Gong J, Mahmoudian M, Smith El, Abid KB, et al. 2003. Genomic scale profiling of nutrient and trace elements in *Arabidopsis thaliana. Nat. Biotechnol.* 21:1215–21

Lee WG, Bannister P, Wilson JB, Mark AF. 1997. Element uptake in an ultramafic flora, Red Mountain, New Zealand. *Proc. Int. Conf. Serpentine Ecol., 2nd, Noumea*, pp. 179–86

Loew O, May DW. 1901. The relation of lime and magnesia to plant growth. *U.S. Dep. Agric. Bur. Plant Ind. Bull.* 1:1–53

Macnair MR. 1983. The genetic control of copper tolerance in the yellow monkey flower *Mimulus guttatus. Heredity* 50:283–93

Macnair MR. 1987. Heavy metal tolerance in plants: A model evolutionary system. *Trends Ecol. Evol.* 2:354–59

Macnair MR. 1989. The potential for rapid speciation in plants. *Genome* 31:203–10

Madhok OP. 1965. Magnesium nutrition of *Helianthus annuus* L. and *Helianthus bolanderi* Gray subspecies *exilis* Heiser. PhD thesis. Univ. Wash., Seattle. 124 pp.

Madhok OP, Walker RB. 1969. Magnesium nutrition of two species of sunflower. *Plant Physiol.* 44:1016–22

Main JL. 1970. *A demonstration of genetic differentiation of grass species to levels of calcium and magnesium.* PhD thesis. Univ. Wash., Seattle, 103 pp.

Main JL. 1974. Differential responses to magnesium and calcium by native populations of *Agropyron spicatum. Am. J. Bot.* 61:931–37

Main JL. 1981. Magnesium and calcium nutrition of a serpentine endemic grass. *Am. Midl. Nat.* 105:196–99

Marrs RH, Proctor J. 1976. The response of serpentine and nonserpentine *Agrostis stolonifera* L. to magnesium and calcium. *J. Ecol.* 64:953–64

Mauricio R. 2001. Mapping quantitative trait loci in plants: uses and caveats for evolutionary biology. *Nat. Rev. Genet.* 2:370–81

McCarten N. 1992. Community structure and habitat relations in a serpentine grassland in California. See Baker et al. 1992, pp. 207–11

McNeilly T, Antonovics J. 1968. Evolution in closely adjacent plant populations. IV. Barriers to gene flow. *Heredity* 23:205–18

Mengoni A, Gonnelli C, Brocchini E, Galardi F, Pucci S, et al. 2003. Chloroplast genetic diversity and biogeography in the serpentine endemic Ni-hyperaccumulator *Alyssum bertolonii. New Phytol.* 157:349–56

Narewski U, Werner G, Schulz H, Vogt C. 2000. Application of laser ablation inductively coupled mass spectrometry (LA-ICP-MS) for the determination of major, minor, and trace elements in bark samples. *Fresenius J. Anal. Chem.* 366:167–70

Newman IA. 2001. Ion transport in roots: measurements of fluxes using ion-selective microelectrodes to characterize transporter function. *Plant Cell Environ.* 24:1–14

Nyberg Berglund AB, Saura A, Westerbergh A. 2001. Genetic differentiation of a polyploid plant on ultramafic soils in Fennoscandia. *S. Afr. J. Sci.* 97:533–35

Nyberg Berglund AB, Dahlgren S, Westerbergh A. 2004. Evidence for parallel evolution and site-specific selection of serpentine tolerance in *Cerastium alpinum* during the colonization of Scandinavia. *New Phytol.* 161:199–209

Orr HA. 1998. The population genetics of adaptation: the distribution of factors fixed during adaptive evolution. *Evolution* 52:935–49

Patterson TB, Givnish TJ. 2004. Geographic cohesion, chromosomal evolution, parallel adaptive radiations, and consequent floral adaptations in *Calochortus* (Calochortaceae): evidence from a cpDNA phylogeny. *New Phytol.* 161:253–64

Pepper AE, Norwood LE. 2001. Evolution of *Caulanthus amplexicaulis* var. *barbarae* (Brassicaceae), a rare serpentine endemic plant: A molecular phylogenetic perspective. *Am. J. Bot.* 88:1479–89

Pichi-Sermolli R. 1948. Flora e vegetazione delle serpentine e delle alter ofioliti dell'alta valle del Trevere (Toscana). *Webbia* 6:1–380

Proctor J. 1970. Magnesium as a toxic element. *Nature* 227:742–43

Proctor J. 1971. The plant ecology of serpentine. II. Plant responses to serpentine soils. *J. Ecol.* 59:397–410

Proctor J. 1999. Toxins, nutrient shortages and droughts: the serpentine challenge. *Trends Ecol. Evol.* 14:334–35

Proctor J, Woodell SRJ. 1975. The ecology of serpentine soils. *Adv. Ecol. Res.* 9:255–365

Rajakaruna N. 2003. Edaphic differentiation in the *Lasthenia*: A model for studies in evolutionary ecology. *Madroño* 50:34–40

Rajakaruna N, Baldwin BG, Chan R, Desrochers AM, Bohm BA, et al. 2003a. Edaphic races and phylogenetic taxa in the *Lasthenia californica* complex (Asteraceae: Helianteae): an hypothesis of parallel evolution. *Mol. Ecol.* 12:1675–79

Rajakaruna N, Bohm BA. 1999. The edaphic factor and patterns of variation in *Lasthenia californica* (Asteraceae). *Am. J. Bot.* 86:1576–96

Rajakaruna N, Bradfield GE, Bohm BA, Whitton J. 2003b. Adaptive differentiation in response to water stress by edaphic races of *Lasthenia californica* (Asteraceae). *Int. J. Plant Sci.* 164:371–76

Rajakaruna N, Siddiqi MY, Whitton J, Bohm BA, Glass ADM. 2003c. Differential responses to Na$^+$/K$^+$ and Ca^{2+}/Mg^{2+} in two edaphic races of the *Lasthenia californica* (Asteraceae) complex: A case for parallel evolution on physiological traits. *New Phytol.* 157:93–103

Rajakaruna N, Whitton J. 2004. Trends in the evolution of edaphic specialists with an example of parallel evolution in the *Lasthenia californica* complex. In *Plant Adaptation: Molecular Biology and Ecology*, ed. QCB Cronk, J Whitton, RH Ree, IEP Taylor, pp. 103–10. Ottawa, Ont.: NRC Res. 166 pp.

Reeves RD, Baker AJM. 1984. Studies on metal uptake by plants from serpentine and nonserpentine populations of *Thlaspi goesingense* Hálácsy (Cruciferae). *New Phytol.* 98:191–204

Ritter-Studnička H. 1968. Die serpentinomorphosen der flora bosniens. *Bot. Jahrb.* 88:443–65

Roberts BA, Proctor J, eds. 1992. *The Ecology of Areas with Serpentinized Rocks: A World View*. Dordrecht: Kluwer Acad. 427 pp.

Rune O. 1953. Plant life on serpentines and related rocks in the north of Sweden. *Acta Phytogeogr. Suec.* 31:1–139

Salt DE. 2004. Update on plant ionomics. *Plant Physiol.* 136:2451–56

Sang T. 2002. Utility of low-copy nuclear gene sequences in plant phylogenetics. *Crit. Rev. Plant Biochem. Mol. Biol.* 37:121–47

Schluter D. 2001. Ecology and the origin of species. *Trends Ecol. Evol.* 16:372–80

Shabala S, Newman IA, Morris J. 1997. Oscillations in H$^+$ and Ca^{2+} ion fluxes around the elongation region of corn roots and effects of external pH. *Plant Physiol.* 113:111–18

Small RL, Cronn RC, Wendel JF. 2004. Use of nuclear genes for phylogeny reconstruction in plants. *Aust. Syst. Bot.* 17:145–70

Specht A, Forth F, Steenbeeke G. 2001. The effect of serpentine on vegetation structure, composition and endemism in northern New South Wales, Australia. *S. Afr. J. Sci.* 97:521–29

Taylor SI, Levy F. 2002. Responses to soils and a test for preadaptation to serpentine in *Phacelia dubia* (Hydrophyllaceae). *New Phytol.* 155:437–47

Tester M. 1997. Techniques for studying ion channels: an introduction. *J. Exp. Bot.* 48:353–59

Tibbetts RA, Smith JAC. 1992. Vacuolar accumulation of calcium and its interaction with magnesium availability. See Baker et al. 1992, pp. 367–73

Tyndall RW, Hull JC. 1999. Vegetation, flora, and plant physiological ecology of serpentine barrens of Eastern North America. In *Savannas, Barrens, and Rock Outcrop Plant Communities of North America*, ed. RC Anderson, JS Fralish, JM Baskin, pp. 67–82. Cambridge: Cambridge Univ. Press.

Vercesi GV. 2003. Plant ecology of ultramafic outcrops [Northern Apennines (Piacenza), Region: Emilia Romagna], pg. 34. *Proc. Int. Conf. Serpentine Ecol., 4th, Havana.* 83 pp.

Via S. 2002. The ecological genetics of speciation. *Am. Nat.* 159(Suppl.):S1–7

Vlamis J. 1949. Growth of lettuce and barley as influenced by degree of calcium saturation of soil. *Soil Sci.* 67:453–66

Vlamis J, Jenny H. 1948. Calcium deficiency in serpentine soils as revealed by absorbent technique. *Science* 107:549–51

Walker RB. 1948. *A study of serpentine soil infertility with special reference to edaphic endemism.* PhD thesis. Univ. Calif., Berkeley. 101 pp.

Walker RB. 1954. The ecology of serpentine soils: A symposium. II. Factors affecting plant growth on serpentine soils. *Ecology* 35:259–66

Walker RB, Walker HM, Ashworth PR. 1955. Calcium-magnesium nutrition with special reference to serpentine soils. *Plant Physiol.* 30:214–21

Wallace AR. 1858. On the tendency of varieties to depart indefinitely from the original type. *J. Proc. Linn. Soc. Zool.* 3:53–62

Wallace DR. 1983. *The Klamath Knot: Explorations of Myth and Evolution.* San Francisco: Sierra Club Books. 149 pp.

Westerbergh A. 1995. *Silene dioica* and its adaptation and evolution on serpentine, pg. 49. *Proc. Int. Conf. Serpentine Ecol., 2nd, Noumea.* 72 pp.

Westerbergh A, Saura A. 1992. Serpentine and the populations structure of *Silene dioica* (L.) Clairv. (Caryophyllaceae). See Baker et al. 1992, pp. 461–67

Whittaker RH. 1954. The ecology of serpentine soils: A symposium. I. Introduction. *Ecology* 35:258–59

Wright JW, Stanton ML, Scherson R. 2005. Local adaptation to serpentine and non-serpentine soils in *Collinsia sparsiflora. Evol. Ecol. Res.* In press

Annu. Rev. Ecol. Evol. Syst. 2005. 36:267–94
doi: 10.1146/annurev.ecolsys.36.102003.152636
Copyright © 2005 by Annual Reviews. All rights reserved
First published online as a Review in Advance on August 12, 2005

BIODIVERSITY-ECOSYSTEM FUNCTION RESEARCH:
Is It Relevant to Conservation?

Diane S. Srivastava[1] and Mark Vellend[2,3]

[1]Department of Zoology and Biodiversity Research Centre, University of British
Columbia, Vancouver, British Columbia, Canada V6T 1Z4; email: srivast@zoology.ubc.ca
[2]National Center for Ecological Analysis and Synthesis, Santa Barbara, California
93101-3351
[3]Current address: Departments of Botany and Zoology and Biodiversity Research Centre,
University of British Columbia, Vancouver, British Columbia, Canada V6T 1Z4;
email: mvellend@interchange.ubc.ca

Key Words   diversity, extinction, species richness, stability, ecosystem services

■ **Abstract**   It has often been argued that conserving biodiversity is necessary for
maintaining ecosystem functioning. We critically evaluate the current evidence for
this argument. Although there is substantial evidence that diversity is able to affect
function, particularly for plant communities, it is unclear if these patterns will hold for
realistic scenarios of extinctions, multitrophic communities, or larger spatial scales.
Experiments are conducted at small spatial scales, the very scales at which diversity
tends to increase owing to exotics. Stressors may affect function by many pathways,
and diversity-mediated effects on function may be a minor pathway, except in the case
of multiple-stressor insurance effects. In general, the conservation case is stronger for
stability measures of function than stock and flux measures, in part because it is easier
to attribute value unambiguously to stability and in part because stock and flux measures of functions are anticipated to be more affected by multitrophic dynamics. Nor is
biodiversity-ecosystem function theory likely to help conservation managers in practical decisions, except in the particular case of restoration. We give recommendations
for increasing the relevance of this area of research for conservation.

INTRODUCTION

The past decade has seen a flurry of ecological research on the effects of biodiversity on ecosystem functions. The biodiversity-ecosystem function (hereafter
BDEF) hypothesis posits that a reduction in biological diversity (variety of species,
genotypes, etc.) will cause a reduction in ecosystem-level processes. Although the
BDEF hypothesis has deep academic and philosophical roots (reviewed by Hector
et al. 2001, Naeem 2002), it became widely discussed by ecologists in the early
1990s as the result of seminal conferences (Schulze & Mooney 1993), international collaborations (Heywood & Watson 1995, Jones & Lawton 1994), and

research initiatives (Lubchenco et al. 1991). The studies that test the BDEF hypothesis differ from previous competition or facilitation experiments in that they encompass a much wider range of diversity treatments and consider ecosystem-level responses as interesting in their own right, rather than a mechanism for competitive exclusion (e.g., nutrient depletion) or assessing fitness (e.g., plant growth).

It is no coincidence that BDEF research emerged at a time when public interest in conservation was at a peak in Western countries. The term biodiversity, for example, was popularized by the 1992 United Nations Conference on Environment and Development in Rio de Janeiro. There are clear links between the BDEF hypothesis and conservation concerns. Many authors have argued that BDEF research will allow ecologists to predict the consequences of extinctions on ecosystem properties, many of which can be assigned economic value (e.g., carbon fixation, water purification). According to this rationale, the case for conserving biodiversity will be bolstered if reductions in biodiversity are shown to reduce ecosystem functions. Government policy documents have already begun to include general statements justifying conservation in order to maintain ecosystem functions (Government of Saskatchewan 2004). This review evaluates whether BDEF research is relevant for conservation and what would be needed to make it more relevant. Relevance could occur in two ways: either in the form of practical recommendations to conservation managers or as a more general justification to the public for conserving diversity. This review primarily considers the latter question, as this is the most common type of relevance mentioned in the BDEF literature. Our review builds on many past perspectives on this topic (Hector et al. 2001, Hooper et al. 2005, Lawler et al. 2002, Lepš 2004, Schmid & Hector 2004, Schwartz et al. 2000, Srivastava 2002) and also attempts to integrate the assessment of conservation relevance more fully into a broad theoretical framework.

Our assessment is based on the premise that using BDEF research to bolster the case for conservation requires posing a progressive sequence of questions, each question contingent on an affirmative to the previous one. The structure of the review is as follows. After defining biodiversity and ecosystem function, we evaluate assumptions about change in each variable: Is biodiversity decreasing at relevant spatial scales for experiments? Can we easily define desirable levels of ecosystem function? We then evaluate the potential for biodiversity to affect ecosystem function, asking if diversity is able to affect function, whether BDEF patterns are general, and if these patterns can be scaled-up to spatial scales relevant for conservation. We then ask if BDEF results can stand up to the addition of more ecological realism, such as differences in extinction risk between species and the inclusion of species in complex food webs. Finally we compare the importance of any BDEF effects on ecosystem function with other human-induced changes in ecosystem function. We conclude with a synthesis of our assessment of the above questions and recommendations for improving the relevance of BDEF research for conservation.

DEFINITIONS: BIODIVERSITY AND ECOSYSTEM FUNCTION

Biodiversity and ecosystem function have been defined in many ways, and the choice of definition may have consequences for linking BDEF research with conservation.

Biodiversity

Biodiversity has two fundamental levels, species diversity and genetic diversity, which are monotonic increasing functions of the number of species and genotypes, respectively, and the evenness of their relative abundances (Magurran 2004). In BDEF research, biodiversity is most often manipulated or measured as the number of species or genotypes, although a few studies have manipulated evenness (e.g., Wilsey & Polley 2002). Broader discussions of biodiversity frequently imply that any change in species composition (the identity of species present) constitutes a change in biodiversity (Chapin et al. 2000). However, in the BDEF literature, a clear distinction has been made between effects of biodiversity (as defined above) and effects of composition (e.g., Downing & Leibold 2002). In particular, BDEF researchers have attempted to remove covariance between diversity loss and composition change by randomly selecting the species that form communities of differing diversity. Although separating the effects of diversity and composition may be important in pinpointing the mechanisms underlying BDEF relationships, most realistic extinction scenarios predict directional change in composition, an issue we return to below.

Ecosystem Function

The term ecosystem function has often been used synonymously with ecosystem services, the latter defined by Daily (1997) as "the conditions and processes through which natural ecosystems, and the species that make them up, sustain and fulfill human life." Daily's (1997) list of ecosystem services includes air and water purification, maintenance of soil fertility, and aesthetic beauty. Interestingly, it also includes maintenance of biodiversity, which would be circular in BDEF studies. A more focused concept of ecosystem function is provided by Pacala & Kinzig (2002), who distinguish between three classes of ecosystem functions: stocks of energy and materials (e.g., biomass), fluxes of energy or material processing (e.g., productivity, decomposition), and the stability of rates or stocks over time. Stability of ecosystem processes has been measured as resistance to or resilience from perturbations, predictability, and the inverse of temporal variability. Many ecosystem functions measured to date fall comfortably into one of these classes, although virtually any aggregate property of an ecosystem could be considered an ecosystem function, such as invasion resistance, the community-wide prevalence of disease, or efficiency of pollination or seed dispersal.

ARE THE CONSERVATION ASSUMPTIONS BEHIND BDEF RESEARCH JUSTIFIED?

Before we discuss the premise that biodiversity loss will lead to impoverished ecosystem functioning, we examine key assumptions implicit in the BDEF case for conservation: (*a*) Diversity is being reduced at a scale relevant to ecosystem functions and (*b*) reductions in commonly measured ecosystem variables are undesirable.

Is Biodiversity Really Declining? A Balance of Extinctions and Invasions at Different Scales

BDEF studies of the last decade almost universally invoke the global decline in biodiversity as the primary impetus for research (Naeem et al. 1999). Conservation management and measurement of ecosystem functions (e.g., carbon credits) also occur on relatively large scales, from watersheds to nations. Most empirical studies, on the other hand, have been conducted at relatively small scales. Indeed, there is no doubt that biodiversity is decreasing globally (McKinney & Lockwood 1999), but the question of whether biodiversity is actually declining at the same spatial scales at which experiments or observations are conducted, or at the scales at which conservation policies are implemented, is much less certain.

When a forest is cleared and turned into an agricultural field, biodiversity declines. However, BDEF experiments are often designed to simulate extinction events in natural or semi-natural habitats, and the question of how biodiversity is changing in these systems has no such self-evident answer. A comprehensive multi-taxa review of this question (Sax & Gaines 2003) revealed the surprising answer that in many cases, particularly for plants, species diversity at local to regional scales is currently increasing through additions of exotic species that have not been fully offset by extinctions of natives. With the caveat that the trend in increasing local diversity may be system- or biome-specific, or transient, these data complicate conservation arguments based on small-scale BDEF experiments.

The fact that exotic species often increase local and regional species diversity underscores the importance of addressing the question raised by Levine and D'Antonio (1999): "Are native and exotic species all that different?" For example, of the 19 species used in the Cedar Creek experiments (e.g., Tilman et al. 2002), two (*Poa pratensis* and *Achillea millefolium*) are considered invasive weeds in many North American grasslands (Stubbendieck et al. 1994). Because a positive relationship between diversity and productivity was found in this system, could the results be used to suggest that exotic species should be encouraged? Of course, conservation efforts in natural habitats almost universally oppose the establishment and spread of non-native species, regardless of their impacts on productivity or nutrient cycling (Pimentel et al. 2000), and we are not attempting to suggest that there is any internal inconsistency in most conservation arguments—the goal is to protect native biodiversity from anthropogenic threats, which may include the introduction of non-native species. Rather, we are calling attention to a potential pitfall of using BDEF research to bolster the public case for biodiversity protection. The immediate goal of conservation is often the composition of ecological communities, not just their diversity.

Are Declines in Commonly Measured Ecosystem Variables Undesirable?

Ecosystem function is a potentially problematic concept, in part because the word function carries the unfortunate baggage of an implied purpose and an underlying assumption that particular levels can be unambiguously considered good (Lawler et al. 2002). Even when the implication of purpose is removed by defining functioning as simply showing activity (Naeem et al. 1999), the problem of value judgment remains. In intensively managed ecosystems (e.g., agriculture), it is unambiguous that high productivity is desirable, but in more natural ecosystems it is often not clear that any particular level of ecosystem function is good or bad (Vandermeer et al. 2002). Furthermore, the same ecosystem function may be valued very differently in different contexts. High productivity is often not desired in lake management but would be when managing a forest for carbon credits. In a purely scientific endeavor, the issue can be sidestepped by simply asking how ecosystem-level variables depend on biodiversity. However, any conservation implication of the BDEF hypothesis assumes that we would like to maintain or enhance the functioning of ecosystems, so a particular target level of functioning is often implied (e.g., high productivity is good), although rarely stated explicitly. Nor is it generally stated in BDEF studies which ecosystem functions are more important than others, critical to assessing the overall impact of declining diversity. We would like to emphasize that these points are not a criticism of academic BDEF research, but rather a call for more explicit considerations of what is meant by ecosystem function when deriving implications for conservation.

THE THREE PHASES OF BIODIVERSITY RESEARCH

We distinguish between three sequential questions that BDEF research needs to answer in the affirmative before concluding that conserving biodiversity is valuable in terms of maintaining ecosystem functions. Phase 1 of BDEF research addresses whether diversity loss is able to reduce ecosystem functions. This question, which has dominated BDEF research for the last decade, concerns the existence and sign of the arrow between diversity and ecosystem function in Figure 1. Phase 2 research asks if loss in diversity is likely to reduce ecosystem functions and, in Figure 1, refers to the net effect of the arrows linking the stressor with ecosystem function via diversity and composition changes. Phase 3 research asks if biodiversity loss is an important pathway by which ecosystem functions are reduced. In Figure 1b and 1c, Phase 3 research refers to the relative importance (arrow width) of the indirect

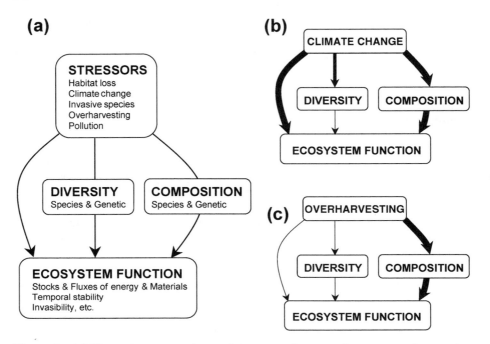

Figure 1 (*a*) There are many pathways between environmental stressors and ecosystem functions. Simply showing a positive effect of diversity on ecosystem function is insufficient evidence that reducing a stressor will lead to improvements in ecosystem functioning. The relative importance of each pathway (indicated by arrow width in *b,c*) may differ between specific stressors, in this case (*b*) climate change and (*c*) overharvesting.

effects of a stressor on ecosystem function, as mediated by diversity, versus other effects of stressors on ecosystem function (direct or via composition).

PHASE 1: IS SPECIES LOSS ABLE TO REDUCE FUNCTION?

There are two questions we can ask of Phase 1 research that are particularly important for conservation. Are patterns between biodiversity and ecosystem function truly general across systems and methods? Only a general pattern can be used to make a consistent argument for conserving biodiversity. Can results from small-scale BDEF studies be scaled-up to the larger spatial scales of conservation applications? As argued above, conservation policy deals with much larger spatial scales than the norm for ecological experiments.

How General are BDEF Relationships?

There has been heated debate about the generality of patterns between biodiversity and ecosystem function (Kaiser 2000). Earlier reviews of BDEF studies include

TABLE 1 Potential mechanisms behind positive effects of diversity on function in single-trophic level systems

Stocks and flux types of ecosystem functions

Niche complementarity: Niche differentiation between species or genotypes allows diverse communities or populations to be more efficient at exploiting resources than depauperate ones, leading to greater productivity and retention of nutrients within the ecosystem (Antonovics 1978, Loreau 2000, Tilman et al. 1997).

Functional facilitation: A positive effect of one species on the functional capability of another will lead to an increase in function in more diverse communities. Examples include promotion of net consumption rates in diverse aquatic insect assemblages through modifications of hydrology and detrital particle size (Cardinale et al. 2002, Heard 1994).

Sampling effect: Also called positive selection effect (Loreau 2000). This effect combines probability theory with species-sorting mechanisms. When there is positive covariance between the competitive ability of a species and its per capita effect on ecosystem function, the probability of including a dominant, functionally important species will increase with diversity (Holt & Loreau 2002, Huston 1997, Ives et al. 2005, Tilman et al. 1997).

Dilution effect: Lower densities of each species or genotype in high-diversity communities may reduce the per capita effects of specialized enemies such as pathogens (e.g., via reduced transmission efficiency) or predators (e.g., via reduced searching efficiency). In essence, specialized enemies create frequency-dependent selection among species or genotypes (e.g., Mitchell et al. 2002).

Stability of ecosystem functions

Insurance effects: Species that are redundant in functional roles or capacity respond differently to stressors, allowing maintenance of net community function after perturbation (Ives et al. 1999, Yachi & Loreau 1999).

Portfolio effect: Independent fluctuations of many individual species may show lower variability in aggregate than fluctuations of any one species (Doak et al. 1998, Tilman et al. 1998), much as a diversified stock portfolio represents a more conservative investment strategy than would any single stock. This effect does not require any interactions between species.

Compensatory dynamic effects: Negative temporal covariance between species abundances create lower variance in their aggregate properties, such as total biomass (Tilman et al. 1998).

a 20-study review of Schwartz et al. (2000), a 49-study review by Schmid et al. (2002), a broad overview of the field (Hooper et al. 2005), as well as reviews of specific functions (Jolliffe 1997, Levine & D'Antonio 1999) and specific systems (Covich et al. 2004, Loreau et al. 2002). We compiled a total of 100 BDEF studies (see Supplemental Table 1; follow the Supplemental Material link from the Annual Reviews home page at http://www.annualreviews.org/). Of these studies, 71% found a positive effect of diversity on at least one ecosystem function. This strong support for the BDEF hypothesis should be tempered by several caveats. First, many studies examined multiple ecosystem functions, and it is possible that the more functions are examined, the more likely at least one function will exhibit a positive response to diversity. If we randomly choose one function per study, then only 62% of studies found positive BDEF effects. Second, positive effects of diversity on ecosystem functions were more often log-linear relationships (53%) than linear (39%), with 8% showing other patterns (see also Schmid et al. 2002, Schwartz et al. 2000). Several authors (Schwartz et al. 2000, Wardle 2002) have argued that log-linear BDEF functions imply that many species can be lost from a system before there is much decrease in function. Other authors speculate that increasing spatial or temporal scale will change asymptotic patterns to linear (Lawler et al. 2002, Tilman 1999). It is useful to understand why BDEF studies differ in results. Various authors have pointed out that the effect of species diversity on function depends on the system studied and the function measured (Duffy 2003, Schmid et al. 2002, Schwartz et al. 2000, Wardle 2002). For example, within grassland experiments, 74% of studies showed a positive BDEF effect on primary productivity, but only 44% reported a similar effect on decomposition. Overall, studies were less likely to find positive diversity effects on stocks and flux measures of functions (67% found at least one positive effect) than measures of stability (73%) or invasion resistance (73%). System effects are also important. Half (51%) of the studies used grassland communities, mainly north-temperate, of which 78% found at least one positive BDEF effect. Only 63% of studies in other systems found at least one positive BDEF effect, suggesting that ecologists are disproportionately looking for an effect in systems where it is likely to occur.

Research design likely also explains some of the variance in results. The empirical studies making up Phase 1 research include both observational evidence and experiments. All experiments require communities of differing richness, but these have been created by either artificially assembling communities, removing species, or through perturbation (e.g., fertilization). Furthermore, species could be lost at random, according to species sensitivities or following a set order (nested). Several authors have shown that the likelihood of detecting a BDEF effect depends on these design choices (Allison 1999, Mikola et al. 2002, Zavaleta & Hulvey 2004).

The strength of BDEF effects may also depend on the level of diversity manipulated (genetic, species, or functional group). When species are clustered into functional groups, functional diversity effects generally tend to outweigh species diversity effects (Heemsbergen et al. 2004, Hooper et al. 2002, Petchey et al. 2004b). Relatively few Phase 1 studies have examined the effects of genetic diversity within species (Bell 1990; Boles et al. 2004; Booth & Grime 2003; Hughes & Stachowicz 2004; Madritch & Hunter 2002–2004; Weltzin et al. 2003), and it is difficult to generalize about results.

Phase 1 BDEF research has been useful in outlining potential mechanisms behind positive effects of diversity on function (Table 1), although it is often difficult to link patterns unequivocally to mechanisms (Ives et al. 2005, Wardle 2002). Note that most mechanisms proposed to date apply largely to diversity within trophic levels, although Ives et al. (2005) suggest that they may extend to predator diversity effects on prey density. Below we describe the complexity that arises when one considers multiple trophic levels.

Can We Scale-Up Results from Local BDEF Experiments?

The size of experimental units in BDEF research varies from Petri dishes (e.g., Naeem & Li 1997) to field plots of almost 100 m^2 (e.g., Tilman et al. 2002). Observational evidence includes sites up to several hectares in size (e.g., Aoki 2003). By contrast, conservation policies are often implemented at scales ranging from a few hectares to many square kilometers. Can we scale-up from small-scale experiments to large-scale conservation implications?

Several theoretical studies have outlined potential mechanisms linking BDEF relationships at different scales (Bond & Chase 2002; Cardinale et al. 2000, 2004; Loreau & Mouquet 1999; Loreau et al. 2003; Peterson et al. 1998; Srivastava 2002; Tilman 1999). Based on the scaling of species richness with area, Tilman (1999) argued that the 11 species needed to maintain function at the local (0.5 m^2) scale implied that 232 species needed to be maintained at the regional (1 km^2) scale. Unfortunately, species-area relationships tell us that species richness tends to be correlated across different scales but do not explain why. Without a mechanism, it is difficult to predict the exact effect of reducing regional diversity on local diversity. By contrast, species-saturation and metacommunity theory predict mechanisms by which reductions in regional richness may affect local richness. Saturation theory predicts that regional extinctions will reduce local richness only when the regional species pool (or dispersal from it) is already relatively depauperate (Cornell & Lawton 1992). At high levels of regional richness, communities should be saturated with immigrating species, and local diversity will be limited more by interspecific interactions. Thus if local diversity influences local ecosystem function, regional extinctions should result in reduced levels of ecosystem function when local communities are not saturated with species (Ruesink & Srivastava 2001, Srivastava 2002). However, the degree of saturation in most ecological communities remains unclear (Hillebrand & Blenckner 2002, Shurin & Srivastava 2005).

Metacommunity models consider a landscape composed of many local-scale patches, with dispersal among them (Leibold et al. 2004). If we assume that in a given patch some species are better than others at carrying out a function, then at the scale of this patch, increasing diversity may have a positive effect on function through an increased probability of including functionally important species (sampling effect: Table 1); however, there may also be a negative effect when additional species reduce, through interspecific competition, the abundance of functionally important species (Bond & Chase 2002, Loreau et al. 2003). At larger spatial scales, patches become increasingly different in environmental conditions and, consequently, species composition (Gilbert & Lechowicz 2004). This could have several outcomes. First, species that do not contribute to ecosystem function in one patch may become functionally important in other patches. Thus if species with a high level of functional importance in a given patch type are favored to occur in that patch type, high regional diversity could increase the overall magnitude (Cardinale et al. 2004, Loreau et al. 2003) or variety (Bond & Chase 2002) of ecosystem function at the regional level. For example, in a system of wet and dry habitat patches, a diverse species pool will allow for greater productivity than a species pool lacking in drought-adapted species. Second, because species may be stochastically lost from individual patches, dispersal between patches may be needed both to rescue the patch population (Cardinale et al. 2004, Loreau et al. 2003) and to maintain regional richness (Loreau et al. 2003). It is important to note that this body of theory is as much about incorporating beta diversity (regardless of spatial scale) as it is about increasing spatial scale per se.

There are as yet no empirical studies that directly test the above hypotheses relating regional extinctions and local function. However, some empirical studies suggest that regional processes are important in maintaining both diversity and ecosystem function (Symstad et al. 2003). Increasing propagule pressure (i.e. increasing dispersal from the regional pool) increases productivity and diversity in grasslands under some conditions (Foster & Dickson 2004, Foster et al. 2004) but not others (Foster & Dickson 2004, Foster et al. 2004, Wilsey & Polley 2003). Isolation of habitat fragments (i.e., reduced dispersal from the regional pool) is linked to reductions in both diversity and function in studies of pollinators and coffee production (De Marco & Coelho 2004), parasitoids and pest suppression (Kruess & Tscharntke 1994), and microarthropods and secondary production (Gonzalez & Chaneton 2002). Unfortunately, it is often difficult to isolate (see Figure 1) the effect of diversity change from other effects of fragmentation on function (but see Gonzalez & Chaneton 2002), a point we return to below.

In sum, theory suggests that when local diversity is constrained by immigration from the regional species pool, changes in regional diversity could lead to changes in ecosystem function. At present, it is unknown whether this would actually happen, even in an experimental system. Thus it is too early to derive any significant conservation implications concerning links between regional-scale diversity and ecosystem function, although this line of inquiry seems promising.

PHASE 2: IS REALISTIC SPECIES LOSS LIKELY TO REDUCE FUNCTION?

The first generation of experiments (Phase 1) explored the ecological conditions under which change in diversity is able to affect ecosystem function. This is an interesting academic question and will probably continue to be actively investigated for some time to come. However, in order to answer this question rigorously, experiments have often had to make simplifying assumptions, such as random extinction of species and single trophic-level communities. In terms of conservation, therefore, the important follow-up question (Phase 2) is whether extinctions of species or loss of genetic diversity will likely lead to change in ecosystem functioning. We first ask whether BDEF relationships still hold for more realistic patterns of diversity loss. We then ask how BDEF relationships will be affected by extinctions occurring at, or cascading through, multiple trophic levels.

Incorporating Non-Random Species Loss

After some of the first BDEF experiments (Naeem et al. 1994, Tilman & Downing 1994) were criticized for confounding species number with species composition (Aarssen 1997, Huston 1997, Huston & McBride 2002, Wardle 2002), researchers have been careful to assign composition randomly at each richness level. We refer to such experiments as random-loss experiments. By biased-loss we mean the non-random association of certain species with richness levels. The results of random-loss BDEF experiments could most easily be used to predict the effects of species extinctions when such extinctions naturally occur independently of species traits (Lepš 2004, Srivastava 2002). However, species extinctions do not appear to be random in either the geological record or modern time. Extinctions of fossil species are often correlated with traits such as long generation time, large body size, small geographic range, and low local density (McKinney 1997), although the strength of these correlations differs between mass extinction events and periods of background extinction (Jablonski 2001). The same suite of traits generally also characterizes modern species that have gone or are on the verge of going extinct (Lawton 1995, McKinney 1997).

Given that extinction is biased to certain types of species, can we still use the results of random-loss experiments to predict likely effects of real extinctions on ecosystem functions? If there is little difference in predicted BDEF relationships between random-loss and realistic biased-loss scenarios, random-loss experiments might still give a "general expectation in the absence of detailed predictions" (Hector et al. 2001: p. 625). However, the few studies to date that have contrasted ecosystem effects of realistic extinctions with those of random extinctions have found marked differences. These studies examine effects of marine invertebrate extinctions on bioturbation (Solan et al. 2004), terrestrial vertebrate extinctions on risk of Lyme Disease (Ostfeld & LoGiudice 2003), grassland species loss on invasion resistance (contrast Zavaleta & Hulvey 2004 with Dukes 2001) and primary production (Schläpfer et al. 2005), as well as simulations with theoretical food webs (Ives & Cardinale 2004).

These differences between real and random extinctions can be understood, in part, by examining trait correlations. If the traits that cause species to be extinction-prone are positively correlated with the traits that cause species to be functionally important, then the first species to be lost will initially cause a disproportionately large reduction in functioning (Hooper et al. 2002, Lavorel & Garnier 2002). For example, large-bodied species often have higher extinction risk, in part, because large-body size is correlated with other extinction-prone traits: small population sizes, long generation times, and high trophic positions (Cardillo & Bromham 2001, Gaston & Blackburn 1995, Gonzalez & Chaneton 2002, McKinney 1997). In the marine invertebrate study (Solan et al. 2004) described above, large-bodied species are also among the most effective bioturbators. In the Lyme disease study (Ostfeld & LoGiudice 2003), large-bodied species such as deer and raccoons tend not to transmit the Lyme disease agent to ticks and so reduce disease prevalence. More generally, simulations of multiple data sets show that when extinctions are biased toward animal species with large-body sizes, there is a disproportionate loss of function-effect traits (Petchey & Gaston 2002). The opposite pattern may also dominate. If the traits that cause species to be extinction-prone are negatively correlated with traits that cause species to be functionally important, then the first few species to be lost will have little initial effect on ecosystem functions. For example, species at low abundance are disproportionately likely to go extinct, as they are more sensitive to environmental, demographic, and genetic stochasticity (Lawton 1995, McKinney 1997). However, such rare species often have the smallest effects on ecosystem functioning simply because they contribute very little to total community abundance (but see Lyons & Schwartz 2001). For example, in the marine invertebrate study (Solan et al. 2004), bioturbation rates could be largely maintained despite loss of rare species, as long as one of the most abundant and largest species, a brittlestar, remained. A more common pattern is for large-bodied species to have low abundance (Brown 1995), and it will be more challenging to predict the net effects of losing such species.

The above discussion relates to initial decreases in function. Longer-term effects of biased extinctions on function also depend on the strength of interspecific interactions, and whether niche complementarity, facilitation, or sampling effects (Table 1) predominate (Gross & Cardinale 2005). The potential of the community to compensate is affected by the traits of the remaining species that are, of course, also biased in biased-loss scenarios (Gross & Cardinale 2005, Solan et al. 2004). Removal experiments in grasslands demonstrate that different segments of the plant community have different capacities for functional compensation (Lyons & Schwartz 2001, Smith & Knapp 2003). For example, the study by Smith & Knapp (2003) shows that removal of the rarest species can be compensated by increased growth of the dominant species, whereas reductions in the density of the dominant species cannot be compensated by rare species.

One of the clearest extinction patterns is the preferential loss of top trophic levels. Top predators are recorded to be among the first species to go extinct in marine (Myers & Worm 2003, Pauly et al. 1998), freshwater (Petchey et al. 2004a), and terrestrial systems (Didham et al. 1998, Kruess & Tscharntke 1994). Experimental perturbations of model systems also often result in preferential loss of species at high trophic levels (Gilbert et al. 1998, Petchey et al. 1999). The reasons behind this preferential loss of predators include life-history correlates, food web dynamics, and human behavior (Holt et al. 1999, Pauly et al. 1998, Petchey et al. 2004a). Extinction of species at top trophic levels may, in turn, have disproportionately strong impacts on ecosystem functions. For example, the effects on plant biomass of removing a top trophic level (Shurin et al. 2002) have been argued to be comparable to the effects of drastic reductions in plant diversity (Duffy 2003). It should be noted that in the studies summarized in these meta-analyses, removal of the top trophic level was generally achieved by removing one predator species, whereas the decrease in plant diversity involved a mean ($\pm$SD) loss of 10.9 $\pm$ 8.6 species (Duffy 2003), suggesting that the effect of losing a single species was greater when the species originated from a higher trophic level (Duffy 2003). To date, we know of only two experiments appropriate for testing this hypothesis. In seagrass mesocosms, producer (eelgrass, macroalgae, epiphyte) biomass tended to be 17 to 33% more strongly affected by loss of the only predator species than by the loss of a single grazer species (data from Duffy et al. 2005). By contrast, in plant-aphid-predator mesocosms, aphid suppression was equally affected ($\pm$15% change) when either a plant or predator species was removed (data from Aquilino et al. 2005). The capacity for functional compensation may be reduced for extinctions occurring at higher trophic levels simply because there are fewer species at the same trophic level that could potentially compensate (Duffy 2002, Vinebrooke et al. 2003). That is, species richness is inversely related to trophic level (Duffy 2002, Petchey et al. 2004a, Vinebrooke et al. 2003).

Although species at high trophic levels may strongly affect various ecosystem functions, the direction and mechanism of such effects are often difficult to predict without full knowledge of the food web (Petchey et al. 2004a, Thébault & Loreau 2003). In linear food webs, extinctions may initiate trophic cascades such that extinctions at adjacent trophic levels can have opposite effects on basal trophic levels (Carpenter & Kitchell 1993). In simulations of more complex food webs, the impact of extinctions on ecosystem biomass depends on food web shape, connectance and interaction strengths (Ives & Cardinale 2004, Petchey et al. 2004a, Thébault & Loreau 2003). We now turn to a more detailed consideration of multitrophic effects of extinctions on ecosystem function.

Returning Species to Food Webs

Species are at risk throughout food webs. The *Exxon Valdez* oil spill in Alaska, for example, affected organisms throughout the coastal marine food web, from seaweed to orcas (Paine et al. 1996). However, the majority of BDEF experiments have considered only extinctions within a single trophic level, although there is an increasing number of multitrophic studies (reviewed by Duffy 2003, Duffy et al. 2005, Schmid et al. 2002). Whether this emphasis on single trophic level experiments and theory prevents ecologists from applying their results to real, multitrophic food webs depends on the answers to the following three questions:

1. When multiple extinctions occur at different trophic levels, do they have opposing or synergistic effects on ecosystem function? Theoretically, we might expect that ecosystem functions will be affected differently by extinctions at different trophic levels, but the direction of this interaction is unclear. To illustrate this, let us envisage the simplest possible multitrophic system, one with just producers and consumers. To make things even simpler, let us assume reductions in producer diversity lead to reductions in primary productivity, and ask whether this reduction in function is offset or exacerbated

by reduced consumer diversity (for more dynamic scenarios see Thébault & Loreau 2003). Loss of all consumers could increase primary production when producers experience a trade-off between growth rate and palatability, such that consumers lead to the dominance of slow-growing inedible species (Fox 2004b, Norberg 2000, Thébault & Loreau 2003). Alternatively, loss of all consumers could reduce primary productivity when herbivory acceler-ates nutrient cycling between plant matter and soil (Hik & Jefferies 1990, McNaughton et al. 1997). Reductions in consumer diversity that do not lead to loss of the entire trophic level can increase primary productivity, for ex-ample, in cases where consumers previously complemented each other in consumption efficiency (Norberg 2000, Sommer et al. 2001), or where high diversity allowed the most efficient predator species to persist (Fox 2004b, Holt & Loreau 2002), a type of sampling effect (Denoth et al. 2002, Ives et al. 2005). Alternatively, reductions in consumer diversity can reduce primary productivity, for example, in cases where the efficiency of diverse predator assemblages was reduced by intraguild predation or interference (Finke & Denno 2004). Theoretically, therefore, consumer extinctions could either amplify or buffer the effects of producer extinctions. Only a few experi-ments have addressed interactive effects of extinctions between two trophic levels, and results are predictably mixed. In aquatic microcosms, effects of algal richness on primary productivity and community biomass have been shown to depend also on the richness of other trophic levels (bacteria × al-gal richness effects, Naeem et al. 2000; herbivore × algal richness effects, Gamfeldt et al. 2005). However, in another aquatic microcosm experiment, algal productivity is independent of both algal richness and herbivore rich-ness, although there are transient effects (Fox 2004a). In Swedish grasslands, a strong relationship between plant richness and biomass accrual depends on removal of all insects from the system (Mulder et al. 1999). However, BDEF relationships have been found in other grassland experiments despite not actively excluding insects (e.g., Tilman et al. 1996). A tritrophic study in seagrass mesocosms found eelgrass and epiphyte biomass to be strongly affected by a grazer diversity by predator presence interaction (Duffy et al. 2005). By contrast, a tritrophic study in crop-insect mesocosms found no interaction between plant diversity and predacious insect diversity on sup-pression of a herbivore, the pea aphid (Aquilino et al. 2005).

2. Can extinctions cascade through food webs and thus impact ecosystem func-tion at other levels? Changes in diversity at one trophic level have often been implicated in changes in diversity at other trophic levels, either higher or lower (Dyer & Letourneau 2003, Hunter & Price 1992, Siemann 1998). The proposed mechanisms behind extinction cascades vary with the trophic role of the lost species, that is, they depend on how composition change covaries with diversity loss. In the case of predator-mediated coexistence, loss of a predator species can lead to decreases in prey diversity (Paine 1966). Re-ductions in diversity at basal trophic levels may also cause loss of species at

higher trophic levels through reduction in the types of resources, net amount of resources, or number of facilitative interactions (Siemann 1998, van der Heijden et al. 1998). Indirect effects can allow such extinction cascades to propagate through multiple levels in food webs (Borrvall et al. 2000, Dyer & Letourneau 2003, Morris et al. 2004). However, there is little consensus as to the conditions favoring extinction cascades. Studies with simulated food webs have found extinction cascades to be more likely with either many (Lundberg et al. 2000) or few species per trophic level (Borrvall et al. 2000) and either high (Pimm 1982) or low connectance between species (Dunne et al. 2002). The presence of cascading extinctions is important from a conservation perspective, as it suggests that the effects of losing a single species may ultimately have consequences for multiple species and their associated ecosystem functions (van der Heijden et al. 1998).

3. Do the mechanisms behind BDEF relationships depend on multitrophic interactions, including indirect effects? Several empirical studies have found that changes in diversity at one trophic level affect functions performed by a different (unmanipulated) trophic level, clearly requiring multitrophic interactions (Cardinale et al. 2003, Duffy 2003, Duffy et al. 2005, Finke & Denno 2004, Naeem et al. 2000, van der Heijden et al. 1998). Indirect effects are a special subset of multitrophic interactions that involve linked pairwise interactions. In several multitrophic BDEF experiments, indirect effects determined effects of diversity on function (Cardinale et al. 2003, Downing & Leibold 2002, Duffy et al. 2003, Finke & Denno 2004, Mikola & Setälä 1998). For example, diverse assemblages of natural enemies were more efficient in suppressing pea aphids because of apparent facilitation involving a second aphid species (Cardinale et al. 2003). Initial theoretical examinations show that indirect effects may modify BDEF relationships, although less so than previously expected (Ives et al. 2005). Increased diversity can stabilize ecosystem function through a variety of mechanisms. Several mechanisms do not require multitrophic interactions (see Table 1). However, recent simulations of food webs suggest that multitrophic interactions can create additional mechanisms to stabilize communities, for example through omnivory (McCann & Hastings 1997) or the buffering of strong interactions by weak interactions (McCann et al. 1998, McCann 2000). Low mean interaction strengths tend to stabilize food webs (Jansen & Kokkoris 2003) and, in simulations of community assembly, characterize more diverse communities (Kokkoris et al. 1999). Although the last observation suggests that extinctions result in less stable food webs and hence ecosystem functions, it should be cautioned that this conclusion is likely highly dependent on the exact pattern of extinction (Ives & Cardinale 2004, Ives et al. 2000).

In summarizing these three questions about multitrophic systems, we found that multiple extinctions at various trophic levels can affect a given ecosystem function either synergistically or antagonistically. Such multiple extinctions may occur not only through a multitrophic stressor (such as pollution or habitat loss), but also because single extinctions can result in further cascading extinctions throughout food webs. Consequently, multitrophic interactions can have strong effects on the relationship between diversity and function, but the direction of these effects is highly context dependent (Thébault & Loreau 2003). In practice, it will be difficult to predict such BDEF effects in all but the simplest food webs. Note that many multitrophic BDEF experiments report unexpected or idiosyncratic results (Downing & Leibold 2002; Duffy et al. 2001, 2005; Mikola & Setälä 1998; Naeem et al. 2000; Norberg 2000). Thus multitrophic extinctions and interactions weaken the BDEF argument for conserving diversity in order to maintain certain levels of ecosystem stocks and fluxes. By contrast, in terms of stability of function, multitrophic effects may—at least under some extinction scenarios—exacerbate the destabilizing effects of extinctions (Ives & Cardinale 2004, McCann 2000).

PHASE 3: IS SPECIES LOSS THE MOST IMPORTANT PATHWAY TO REDUCED FUNCTION?

Simulating realistic patterns of extinction in real food webs brings us closer to connecting BDEF research and conservation concerns, but ultimately we need to consider the underlying causes of biodiversity change and their direct effects on ecosystem function (Gonzalez & Chaneton 2002, Schmid & Hector 2004, Srivastava 2002, Wardle 2002, Figure 1). The five top reasons for changes in biodiversity are habitat loss and conversion, climate change, invasive species, pollution and enrichment, and over-harvesting (Diamond 1989, Parmesan & Yohe 2003, Sala et al. 2000). Each of these may affect ecosystem functioning directly or indirectly via changes in the biota. Indirect biotic effects on ecosystem function can be further divided into effects of composition (including shifts in abundance) and effects of diversity per se (Figure 1). If the effects of diversity per se on ecosystem function are minimal relative to the other effects, it will be difficult to justify the conservation of biodiversity on the grounds that reductions will lead to a loss of ecosystem functioning. A full examination of different drivers of biodiversity loss is beyond the scope of this review. Here we contrast two scenarios for which we can make preliminary assessments of the likely importance of the three pathways to changes in ecosystem function.

Over-harvesting generally results in the selective removal of particular species, with relatively minor direct effects on ecosystem function. For example, pelagic fisheries often focus on particular target species, and apart from an immediate reduction in the biomass of fish, direct effects on ecosystem functions such as stocks or fluxes of nutrients are likely to be minor. However, ecosystem function may be greatly altered via a cascade of effects through the food web. For example, Estes et al. (1998) have suggested that declines in fish stocks off the shores of Alaska ultimately led to the collapse of kelp forests (and associated primary production). Declining fish stocks are thought to have led to declines in pinniped populations, which, in turn, caused killer whales to shift their primary food source from pinnipeds to sea otters, the main regulator of kelp-eating sea urchins. This scenario suggests a strong effect of over-harvesting on the abundances of species in the food web (i.e., a composition effect), which ultimately reduced productivity in kelp forests. Direct effects of fishing on kelp production, and indirect effects via changes in diversity, were likely quite minor.

Climate change will have profound direct effects on ecosystem function. Rates of energy and material flux depend fundamentally on temperature and moisture, both predicted to show significant changes over the next century (Intergovernmental Panel on Climate Change 2001). Thus even if species ranges were unresponsive to climate change, ecosystem function at all spatial scales would be dramatically altered. Species ranges have consistently shifted in response to climate change in the past (Delcourt & Delcourt 1991), and many species already show clear signals of shifts in response to contemporary climate change (Parmesan & Yohe 2003, Root et al. 2003). Species composition at particular sites is, therefore, certain to be altered dramatically, with consequent effects on ecosystem function. Encroachment of grasslands by shrubs, for example, should greatly alter carbon dynamics (Jackson et al. 2002). Whereas global diversity is most likely to decrease with climate change (Thomas 2004), regional and local diversity seem just as likely to increase as decrease depending on geographic location. It is difficult to imagine that changes in diversity resulting from climate change will translate into changes in ecosystem function that are measurable relative to large direct effects of climate change and indirect effects via changes in composition.

In these two scenarios of biodiversity change, one seems likely to show a relatively weak direct effect on ecosystem function (over-harvesting), the other a relatively strong effect (climate change). Both may show strong indirect effects via composition, but neither show strong indirect effects via diversity. Strong indirect effects via diversity change also seem unlikely for other drivers. For example, diversity effects accounted for only a fraction of the total change in ecosystem function following experimental habitat fragmentation (6 %; Gonzalez & Chaneton 2002) or warming (<33%; O. Petchey personal communication in Srivastava 2002). However, we realize that biased-extinction scenarios represent simultaneous changes in both composition and diversity and that our separation of composition and diversity effects may be tenuous in many real situations.

It is also important to note that compositional change depends fundamentally on the existence of diversity to begin with. If diversity is reduced by a driver, the range of possible responses of the ecosystem to subsequent drivers may be compromised; that is, biodiversity provides a kind of insurance in the face of different kinds of environmental perturbation (Ives et al. 1999, Naeem 1998, Yachi & Loreau 1999). Many studies indicate that diversity promotes resilience or resistance of ecosystem functions to stressors (McCann 2000) (see Supplemental Table 1; follow the Supplemental Material link from the Annual Reviews home page at http://www.annualreviews.org/) by allowing species composition to respond to changing conditions. For example, high diversity grassland plots at Cedar Creek were more resistant to drought because they contained key drought-tolerant species that increased in biomass as other species declined (Tilman 1996). A few studies have shown that communities experiencing high levels of one stressor have both reduced diversity and reduced resistance to a subsequent stressor (Griffiths et al. 2000, Tilman & Downing 1994), although it is difficult to unambiguously attribute resistance to diversity-mediated insurance effects. In general, such multiple-stressor insurance effects will be strongest when species tolerances to one stressor are positively correlated with tolerances to a subsequent stressor (see Vinebrooke et al. 2004 for a more comprehensive discussion). The corollary is that effects of multiple stressors will depend critically on the identity of the stressors. There are many examples of non-additive effects of multiple stressors on ecosystem states (Folt et al. 1999, Payette et al. 2000, Vinebrooke et al. 2004); it would be instructive to revisit these examples to test specifically for insurance effects (see also Vinebrooke et al. 2004).

CONCLUSIONS

Ecological and evolutionary research provide the underpinning for our understanding of the natural world and, therefore, our ability to manage and protect it. However, particular sub-fields or studies within these disciplines may or may not have direct conservation implications. For example, it is undisputable that the functional characteristics of organisms play a central role in regulating ecosystem processes (Wardle 2002). However, as pointed out by Hooper et al. (2005), this fact was established by more than 50 years of research in ecosystem ecology long before the recent spate of BDEF studies. The key question here is whether BDEF research of the last decade or so has conservation implications above and beyond those derived from earlier work. Posing the question in this way distinguishes our approach from some previous efforts to derive conservation implications from BDEF research, which focused more broadly to include long-standing research topics such as the difference between monocultures and polycultures in agriculture or the functional importance of individual species (e.g., Chapin et al. 2000, Hooper et al. 2005).

As outlined in the introduction to this review, BDEF research may have conservation implications if (*a*) it improves our ability to manage particular ecosystems or (*b*) if it bolsters the public case for general conservation of biodiversity. The last decade of fervent activity in BDEF research was motivated by the latter possibility. Here we present four conclusions based on our review of the BDEF literature (Table 2) and our assessment of its conservation relevance.

BDEF research is most useful for conservation managers in cases where humans directly control plant diversity and where high levels of ecosystem functions are unambiguously preferred (e.g., in restoration projects). Restoration ecologists are frequently concerned with reclaiming severely disturbed sites, in which case high levels of functions such as primary productivity and nutrient retention are

TABLE 2   Summary of the main conclusions of this review

ARE THE CONSERVATION ASSUMPTIONS BEHIND BDEF RESEARCH JUSTIFIED?

Biodiversity is declining at global scales, but not always at the smaller spatial scales at which policies are implemented or experiments conducted.

Target levels are often unclear for stocks and flux types of functions. However, stability of functions is often desirable.

PHASE 1: IS SPECIES LOSS ABLE TO REDUCE FUNCTION?

Although many studies show positive effects of biodiversity on function, others do not. Generality is obscured by differences between systems, functions measured, level of diversity manipulated, as well as experimental approaches.

It is not yet clear how results from small-scale BDEF experiments can be scaled-up to larger spatial scales relevant to conservation.

PHASE 2: IS SPECIES LOSS LIKELY TO REDUCE FUNCTION?

Realistic patterns of species extinctions can result in BDEF effects different from those predicted by Phase 1 random-loss experiments.

Preferential loss of species at top trophic levels may have particularly strong ecosystem effects.

Ecosystem effects of extinctions in multitrophic food webs will be difficult to predict because of numerous indirect effects and the likelihood of simultaneous or cascading extinctions through multiple trophic levels.

PHASE 3: IS SPECIES LOSS IMPORTANT IN REDUCING FUNCTION?

Anthropogenic drivers of extinction will also directly affect ecosystem function. Indirect effects of such drivers on ecosystem function, via diversity effects independent of composition, may often be minor by comparison.

Extinctions caused by one stressor may reduce the potential for compositional shifts in response to a second stressor, reducing stability of ecosystem functioning. Such insurance effects depend on the particular stressors involved, and covariance in species tolerances.

---

preferred. BDEF studies in temperate grasslands have clearly demonstrated that diverse plantings of species generally show elevated levels of productivity and nutrient uptake. In some cases, diverse plant communities are also more stable in the face of perturbations. For example, restoration of seagrass beds is most effective when diverse genotypes are planted, as genetic diversity increases resistance to goose grazing (Hughes & Stachowicz 2004). However, apart from restoration, BDEF research has little to offer in the way of practical advice for conservation managers. Many conservation managers pursue diversity as a goal unto itself. When certain ecosystem functions are desired in the course of reserve management (e.g., reduction of invasive species biomass), it can be more effective to identify functionally important types of species (e.g., biocontrol agents or native competitors) than simply to advocate increased diversity. We now shift our attention to the question of whether BDEF research bolsters the general case for conservation.

It is easier to make the general case for conservation implications of positive biodiversity effects on ecosystem stability than on stocks or fluxes of energy and materials. Many stock and flux measures of ecosystem functions are desirable in certain contexts but not others. For example, high productivity is desirable when growing forests for carbon credits but is undesirable in many lakes. By contrast, human society and economic systems generally depend on the stability of ecosystems and their functions. Although many Phase 1 studies demonstrated positive relationships between biodiversity and stocks or fluxes of energy and materials (particularly in the case of grasslands), our review indicates that once multitrophic interactions are considered explicitly it becomes challenging to predict the response of such stocks and fluxes to loss of diversity. By contrast, studies of both single trophic and multitrophic systems suggest that loss of diversity is likely to precipitate some reduction in ecosystem resistance or resilience. Furthermore, when diversity effects on stock and flux types of ecosystem functions are compared with the direct effects of stressors or species composition, diversity effects are expected to be of minor importance. However, maintaining diversity may be an important insurance strategy, ensuring the option of composition change in response to future stressors. Therefore, the case for conserving diversity in order to maintain function is generally weaker with respect to stocks and fluxes than it is for their stability.

The validity of any function-based argument for conserving diversity will depend critically on the anticipated pattern of extinctions. Compositional change is an inevitable consequence of extinctions and cannot be ignored in ecologists' predictions about the functional consequences of future extinctions. Ironically, the random-loss BDEF experiments that proved so powerful in isolating functional effects of diversity from those of composition are insufficient to yield predictions about likely biased-loss extinction scenarios. This will be particularly true when extinctions result in changes in trophic structure.

In neither the case of ecosystem stocks and fluxes nor that of stability has BDEF research yet been able to create a convincing general argument for conservation in complex systems, either because important questions have not yet been resolved or because of context-dependency of results (Table 2). This does not mean that the BDEF case for conservation cannot be strengthened in the future. Within the past few years, several studies have started to add ecological realism to BDEF research by simulating realistic extinction scenarios, using multitrophic communities and invoking larger-scale processes. We interpret this as a positive sign that the transition from Phase 1 to Phase 2 of BDEF research has begun and that the results of experiments are starting to become more relevant to conservation. However, we are not presently in a position to say that "incorporating diversity effects into policy and management is essential" (Hooper et al. 2005) with respect to maintaining ecosystem function, unless diversity is defined so broadly as to mean the biotic community (as in figure 1 of Hooper et al. 2005).

If BDEF research is to fulfill its mandate of predicting ecosystem consequences of the current and impending loss of biodiversity, we suggest that future BDEF research (*a*) use realistic scenarios of extinctions, either by trait-analysis of endangered species or by utilizing real stressors; (*b*) expand to systems underrepresented in BDEF research, particularly multitrophic systems; (*c*) extrapolate results and theory to the larger spatial scales relevant to diversity loss and conservation; (*d*) reconcile the role of invasive species as both a response (invasibility as a function) and manipulated (contributing to biodiversity) variable; (*e*) contrast the importance of biodiversity effects on function with other pathways between stressors and functions (direct effects, effects via composition); and (*f*) in the case of multiple stressors, explore the insurance effects of diversity on future composition shifts.

ACKNOWLEDGMENTS

This manuscript was improved by comments from Bradley Cardinale, Andrew Gonzalez, Robin Naidoo, and Mary O'Connor, as well as by many ecologists at the University of British Columbia, especially Andrew MacDougall and Jordan Rosenfeld. We appreciated the thoughtful editorial comments of Daniel Simberloff. Thanks to the National Center for Ecological Analysis and Synthesis (NCEAS) for providing a space for discussions. D.S.S. was supported by the Natural Sciences and Engineering Research Council (Canada); M.V. by NCEAS, which is funded by the National Science Foundation (Grant #DEB-0072909), the University of California, and its Santa Barbara campus.

The *Annual Review of Ecology, Evolution, and Systematics* is online at
http://ecolsys.annualreviews.org

LITERATURE CITED

Aarssen LW. 1997. High productivity in grassland ecosystems: effected by species diversity or productive species? *Oikos* 80:183–84

Allison GW. 1999. The implications of experimental design for biodiversity manipulations. *Am. Nat.* 153:26–45

Antonovics J. 1978. The population genetics of species mixtures. In *Plant Relations in Pastures*, ed. JR Wilson, pp. 223–52. Melbourne, Australia: CSIRO

Aoki I. 2003. Diversity-productivity-stability relationship in freshwater ecosystems: Whole-systemic view of all trophic levels. *Ecol. Res.* 18:397–404

Aquilino KM, Cardinale BJ, Ives AR. 2005. Reciprocal effects of host plant and natural enemy diversity on herbivore surpression: an empirical study of a model tritrophic system. *Oikos* 108:275–82

Bell G. 1990. The ecology and genetics of fitness in *Chlamydomonas* 2. The properties of mixtures of strains. *Proc. R. Soc. London Ser. B* 240:323–50

Boles BR, Thoendal M, Singh PK. 2004. Self-generated diversity produces "insurance effects" in biofilm communities. *Proc. Natl. Acad. Sci. USA* 101:16630–35

Bond EM, Chase JM. 2002. Biodiversity and ecosystem functioning at local and regional spatial scales. *Ecol. Lett.* 5:467–70

Booth RE, Grime JP. 2003. Effects of genetic impoverishment on plant community diversity. *J. Ecol.* 91:721–30

Borrvall C, Ebenman B, Jonsson T. 2000. Biodiversity lessens the risk of cascading extinction in model food webs. *Ecol. Lett.* 3:131–36

Brown JH. 1995. *Macroecology*. Chicago: Univ. Chicago Press. 269 pp.

Cardillo M, Bromham L. 2001. Body size and

risk of extinction in Australian mammals. *Conserv. Biol.* 15:1435–40

Cardinale BJ, Harvey CT, Gross K, Ives AR. 2003. Biodiversity and biocontrol: emergent impacts of a multi-enemy assemblage on pest suppression and crop yield in an agroecosystem. *Ecol. Lett.* 6:857–65

Cardinale BJ, Ives AR, Inchausti P. 2004. Effects of species diversity on the primary productivity of ecosystems: extending our spatial and temporal scales of inference. *Oikos* 104:437–50

Cardinale BJ, Nelson K, Palmer MA. 2000. Linking species diversity to the functioning of ecosystems: on the importance of environmental context. *Oikos* 91:175–83

Cardinale BJ, Palmer MA, Collins SL. 2002. Species diversity enhances ecosystem functioning through interspecific facilitation. *Nature* 415:426–29

Carpenter SR, Kitchell JF, eds. 1993. *The Trophic Cascade in Lakes*. Cambridge, UK: Cambridge Univ. Press. 385 pp.

Chapin FS, Zavaleta ES, Eviner VT, Naylor RL, Vitousek PM, et al. 2000. Consequences of changing biodiversity. *Nature* 405:234–42

Cornell HV, Lawton JH. 1992. Species interactions, local and regional processes, and limits to the richness of ecological communities: a theoretical perspective. *J. Anim. Ecol.* 61:1–12

Covich AP, Austen MC, Barlocher F, Chauvet E, Cardinale BJ, et al. 2004. The role of biodiversity in the functioning of freshwater and marine benthic ecosystems. *BioScience* 54:767–75

Daily GC. 1997. *Nature's Services: Societal Dependence on Natural Ecosystems*. Washington, DC: Island

De Marco P, Coelho FM. 2004. Services performed by the ecosystem: forest remnants influence agricultural cultures' pollination and production. *Biodivers. Conserv.* 13:1245–55

Delcourt HR, Delcourt PA. 1991. *Quaternary Ecology: A Paleoecological Perspective*. New York: Chapman & Hall

Denoth M, Frid L, Myers JH. 2002. Multiple agents in biological control: improving the odds? *Biol. Control* 24:20–30

Diamond JM. 1989. The present, past and future of human-caused extinctions. *Philos. Trans. R. Soc. London Ser. B* 325:469–77

Didham RK, Lawton JH, Hammond PM, Eggleton P. 1998. Trophic structure stability and extinction dynamics of beetles (Coleoptera) in tropical forest fragments. *Philos. Trans. R. Soc. London Ser. B* 353: 437–51

Doak DF, Bigger D, Harding EK, Marvier MA, O'Malley RE, Thomson D. 1998. The statistical inevitability of stability-diversity relationships in community ecology. *Am. Nat.* 151:264–76

Downing AL, Leibold MA. 2002. Ecosystem consequences of species richness and composition in pond food webs. *Nature* 416:837–41

Duffy JE. 2002. Biodiversity and ecosystem function: the consumer connection. *Oikos* 99: 201–19

Duffy JE. 2003. Biodiversity loss, trophic skew and ecosystem functioning. *Ecol. Lett.* 6: 680–87

Duffy JE, Macdonald KS, Rhode JM, Parker JD. 2001. Grazer diversity, functional redundancy, and productivity in seagrass beds: an experimental test. *Ecology* 82:2417–34

Duffy JE, Richardson JP, Canuel EA. 2003. Grazer diversity effects on ecosystem functioning in seagrass beds. *Ecol. Lett.* 6:637–45

Duffy JE, Richardson JP, France KE. 2005. Ecosystem consequences of diversity depend on food chain length in estuarine vegetation. *Ecol. Lett.* 8:301–9

Dukes JS. 2001. Biodiversity and invasibility in grassland microcosms. *Oecologia* 126:563–68

Dunne JA, Williams RJ, Martinez ND. 2002. Network structure and biodiversity loss in food webs: robustness increases with connectance. *Ecol. Lett.* 5:558–67

Dyer LA, Letourneau D. 2003. Top-down and bottom-up diversity cascades in detrital vs. living food webs. *Ecol. Lett.* 6:60–68

Estes JA, Tinker MT, Williams TM, Doak DF. 1998. Killer whale predation on sea otters linking oceanic and nearshore ecosystems. *Science* 282:473–76

Finke DL, Denno RF. 2004. Predator diversity dampens trophic cascades. *Nature* 429:407–10

Folt CL, Chen CY, Moore MV, Burnaford J. 1999. Synergism and antagonism among multiple stressors. *Limnol. Oceanogr.* 44: 864–77

Foster BL, Dickson TL. 2004. Grassland diversity and productivity: the interplay of resource availability and propagule pools. *Ecology* 85:1541–47

Foster BL, Dickson TL, Murphy CA, Karel IS, Smith VH. 2004. Propagule pools mediate community assembly and diversity-ecosystem regulation along a grassland productivity gradient. *J. Ecol.* 92:435–49

Fox JW. 2004a. Effects of algal and herbivore diversity on the partitioning of biomass within and among trophic levels. *Ecology* 85:549–59

Fox JW. 2004b. Modelling the joint effects of predator and prey diversity on total prey biomass. *J. Anim. Ecol.* 73:88–96

Gamfeldt L, Hillebrand H, Jonsson PR. 2005. Species richness changes across two trophic levels simultaneously affect prey and consumer biomass. *Ecol. Lett.* 8:696–703

Gaston KJ, Blackburn TM. 1995. Birds, body-size and the threat of extinction. *Philos. Trans. R. Soc. London Ser. B* 347:205–12

Gilbert B, Lechowicz MJ. 2004. Neutrality, niches, and dispersal in a temperate forest understory. *Proc. Natl. Acad. Sci.USA* 101: 7651–56

Gilbert F, Gonzalez A, Evans-Freke I. 1998. Corridors maintain species richness in the fragmented landscapes of a microecosystem. *Proc. R. Soc. London Ser. B* 265:577–82

Gonzalez A, Chaneton EJ. 2002. Heterotroph species extinction, abundance and biomass dynamics in an experimentally fragmented microecosystem. *J. Anim. Ecol.* 71:594–602

Government of Saskatchewan. 2004. *Caring for Natural Environments: A Biodiversity Action Plan for Saskatchewan's Future.* Saskatchewan, Can: Saskatchewan Biodiversity Interagency Steer. Comm.

Griffiths BS, Ritz K, Bardgett RD, Cook R, Christensen S, et al. 2000. Ecosystem response of pasture soil communities to fumigation-induced microbial diversity reductions: an examination of the biodiversity-ecosystem function relationship. *Oikos* 90:279–94

Gross K, Cardinale BJ. 2005. The functional consequences of random versus ordered species extinctions. *Ecol. Lett.* 8:409–18

Heard SB. 1994. Pitcher-plant midges and mosquitoes: a processing chain commensalism. *Ecology* 75:1647–60

Hector A, Joshi J, Lawler SP, Spehn EM, Wilby A. 2001. Conservation implications of the link between biodiversity and ecosystem functioning. *Oecologia* 129:624–28

Heemsbergen DA, Berg MP, Loreau M, van Haj JR, Faber JH, Verhoef HA. 2004. Biodiversity effects on soil processes explained by interspecific functional dissimilarity. *Science* 306:1019–20

Heywood VH, Watson RT, eds. 1995. *Global Biodiversity Assessment.* Cambridge, UK: Cambridge Univ. Press. 1140 pp.

Hik DS, Jefferies RL. 1990. Increases in the net aboveground primary production of a salt-marsh forage grass—a test of the predictions of the herbivore-optimization model. *J. Ecol.* 78:180–95

Hillebrand H, Blenckner T. 2002. Regional and local impact on species diversity—from pattern to process. *Oecologia* 132:479–91

Holt RD, Lawton JH, Polis GA, Martinez ND. 1999. Trophic rank and the species-area relationship. *Ecology* 80:1495–504

Holt RD, Loreau M. 2002. Biodiversity and ecosystem functioning: the role of trophic interactions and the importance of system openness. In *The Functional Consequences of Biodiversity: Empirical Progress and Theoretical Extensions*, ed. AP Kinzig, SW Pacala, D Tilman, pp. 246–63. Princeton, NJ: Princeton Univ. Press

Hooper DU, Chapin FS, Ewel JJ, Hector A, Inchausti P, et al. 2005. Effects of biodiversity on ecosystem functioning: a consensus of current knowledge. *Ecol. Monogr.* 75:3–35

Hooper DU, Solan M, Symstad A, Díaz S, Gessner MO, et al. 2002. Species diversity, functional diversity, and ecosystem functioning. See Loreau et al. 2002, pp. 195–208

Hughes AR, Stachowicz JJ. 2004. Genetic diversity enhances the resistance of a seagrass ecosystem to disturbance. *Proc. Natl. Acad. Sci. USA* 101:8998–9002

Hunter MD, Price PW. 1992. Playing chutes and ladders: heterogeneity and the relative roles of bottom-up and top-down forces in natural communities. *Ecology* 73:724–32

Huston MA. 1997. Hidden treatments in ecological experiments: re-evaluating the ecosystem function of biodiversity. *Oecologia* 110:449–60

Huston MA, McBride AC. 2002. Evaluating the relative strengths of biotic versus abiotic controls on ecosystem processes. See Loreau et al. 2002, pp. 47–60

Intergovernmental Panel on Climate Change. 2001. *Climate Change 2001: Impacts, Adaptation, and Vulnerability.* Cambridge, UK: Cambridge Univ. Press

Ives AR, Cardinale BJ. 2004. Food-web interactions govern the resistance of communities after non-random extinctions. *Nature* 429: 174–77

Ives AR, Cardinale BJ, Snyder WE. 2005. A synthesis of subdisciplines: predator-prey interactions, and biodiversity and ecosystem functioning. *Ecol. Lett.* 8:102–16

Ives AR, Gross K, Klug JL. 1999. Stability and variability in competitive communities. *Science* 286:542–44

Ives AR, Klug JL, Gross K. 2000. Stability and species richness in complex communities. *Ecol. Lett.* 3:399–411

Jablonski D. 2001. Lessons from the past: evolutionary impacts of mass extinctions. *Proc. Natl. Acad. Sci. USA* 98:5393–98

Jackson RB, Banner JL, Jobbágy EG, Pockman WT, Wall DH. 2002. Ecosystem carbon loss with woody plant invasion of grasslands. *Nature* 418:623–26

Jansen VAA, Kokkoris GD. 2003. Complexity and stability revisited. *Ecol. Lett.* 6:498–502

Jolliffe PA. 1997. Are mixed populations of plant species more productive than pure stands? *Oikos* 80:595–602

Jones CG, Lawton JH, eds. 1994. *Linking Species and Ecosystems.* New York: Chapman & Hall

Kaiser J. 2000. Rift over biodiversity divides ecologists. *Science* 289:1282–83

Kokkoris GD, Troumbis AY, Lawton JH. 1999. Patterns of species interaction strength in assembled theoretical competition communities. *Ecol. Lett.* 2:70–74

Kruess A, Tscharntke T. 1994. Habitat fragmentation, species loss, and biological control. *Science* 264:1581–84

Lavorel S, Garnier E. 2002. Predicting changes in community composition and ecosystem functioning from plant traits: revisiting the Holy Grail. *Funct. Ecol.* 16:545–56

Lawler SP, Armesto JJ, Kareiva P. 2002. How relevant to conservation are studies linking biodiversity and ecosystem functioning? In *The Functional Consequences of Biodiversity: Empirical Progress and Theoretical Extensions*, ed. AP Kinzig, SW Pacala, D Tilman, pp. 294–313. Princeton, NJ: Princeton Univ. Press

Lawton JH. 1995. Population dynamic principles. In *Extinction Rates*, ed. JH Lawton, RM May, pp. 147–63. Oxford, UK: Oxford Univ. Press

Leibold MA, Holyoak M, Mouquet N, Amarasekare P, Chase JM, et al. 2004. The metacommunity concept: a framework for multiscale community ecology. *Ecol. Lett.* 7:601–13

Lepš J. 2004. What do the biodiversity experiments tell us about consequences of plant species loss in the real world? *Basic Appl. Ecol.* 5:529–34

Levine JM, D'Antonio CM. 1999. Elton revisited: a review of evidence linking diversity and invasibility. *Oikos* 87:15–26

Loreau M. 2000. Biodiversity and ecosystem functioning: recent theoretical advances. *Oikos* 91:3–17

Loreau M, Mouquet N. 1999. Immigration and the maintenance of local species diversity. *Am. Nat.* 154:427–40

Loreau M, Mouquet N, Gonzalez A. 2003. Biodiversity as spatial insurance in heterogeneous landscapes. *Proc. Natl. Acad. Sci. USA* 100:12765–70

Loreau M, Naeem S, Inchausti P, eds. 2002. *Biodiversity and Ecosystem Functioning: Synthesis and Perspectives.* Oxford, UK: Oxford Univ. Press. 294 pp.

Lubchenco J, Olson AM, Brubaker LB, Carpenter SR, Holland MM, et al. 1991. The sustainable biosphere initiative: an ecological research agenda. *Ecology* 72:371–412

Lundberg P, Ranta E, Kaitala V. 2000. Species loss leads to community closure. *Ecol. Lett.* 3:465–68

Lyons KG, Schwartz MW. 2001. Rare species loss alters ecosystem function—invasion resistance. *Ecol. Lett.* 4:358–65

Madritch MD, Hunter MD. 2002. Phenotypic diversity influences ecosystem functioning in an oak sandhills community. *Ecology* 83:2084–90

Madritch MD, Hunter MD. 2003. Intraspecific litter diversity and nitrogen deposition affect nutrient dynamics and soil respiration. *Oecologia* 136:124–28

Madritch MD, Hunter MD. 2004. Phenotypic diversity and litter chemistry affect nutrient dynamics during litter decomposition in a two species mix. *Oikos* 105:125–31

Magurran AE. 2004. *Measuring Biological Diversity.* Oxford, UK: Blackwell

McCann K, Hastings A. 1997. Re-evaluating the omnivory-stability relationship in food webs. *Proc. R. Soc. London Ser. B.* 264:1249–54

McCann K, Hastings A, Huxel GR. 1998. Weak trophic interactions and the balance of nature. *Nature* 395:794–98

McCann KS. 2000. The diversity-stability debate. *Nature* 405:228–33

McKinney ML. 1997. Extinction vulnerability and selectivity: combining ecological and paleontological views. *Annu. Rev. Ecol. Syst.* 28:495–516

McKinney ML, Lockwood JL. 1999. Biotic homogenization: a few winners replacing many losers in the next mass extinction. *Trends Ecol. Evol.* 14:450–53

McNaughton SJ, Banyikwa FF, McNaughton MM. 1997. Promotion of the cycling of diet-enhancing nutrients by African grazers. *Science* 278:1798–800

Mikola J, Salonen V, Setala H. 2002. Studying the effects of plant species richness on ecosystem functioning: Does the choice of experimental design matter? *Oecologia* 133:594–98

Mikola J, Setälä H. 1998. Relating species diversity to ecosystem functioning: mechanistic backgrounds and experimental approach with a decomposer food web. *Oikos* 83:180–94

Mitchell CE, Tilman D, Groth JV. 2002. Effects of grassland plant species diversity, abundance, and composition on foliar fungal disease. *Ecology* 83:1713–26

Morris RJ, Lewis OT, Godfray HCJ. 2004. Experimental evidence for apparent competition in a tropical forest food web. *Nature* 428:310–13

Mulder CPH, Koricheva J, Huss-Danell K, Hogberg P, Joshi J. 1999. Insects affect relationships between plant species richness and ecosystem processes. *Ecol. Lett.* 2:237–46

Myers RA, Worm B. 2003. Rapid worldwide depletion of predatory fish communities. *Nature* 423:280–83

Naeem S. 1998. Species redundancy and ecosystem reliability. *Conserv. Biol.* 12:39–45

Naeem S. 2002. Ecosystem consequences of biodiversity loss: the evolution of a paradigm. *Ecology* 83:1537–52

Naeem S, Chapin FS, Costanza R, Ehrlich PR, Golley FB, et al. 1999. Biodiversity and ecosystem functioning: maintaining natural life support processes. *Issues Ecol.* 4:1–12

Naeem S, Hahn DR, Schuurman G. 2000. Producer-decomposer co-dependency influences biodiversity effects. *Nature* 403:762–64

Naeem S, Li SB. 1997. Biodiversity enhances ecosystem reliability. *Nature* 390:507–9

Naeem S, Thompson LJ, Lawler SP, Lawton JH, Woodfin RM. 1994. Declining biodiversity can alter the performance of ecosysems. *Nature* 368:734–37

Norberg J. 2000. Resource-niche complementarity and autotrophic compensation determines ecosystem-level responses to increased cladoceran species richness. *Oecologia* 122:264–72

Ostfeld RS, LoGiudice K. 2003. Community disassembly, biodiversity loss, and the erosion of an ecosystem service. *Ecology* 84: 1421–27

Pacala S, Kinzig AP. 2002. Introduction to theory and the common ecosystem model. In *Functional Consequences of Biodiversity: Empirical Progress and Theoretical Extensions*, ed. AP Kinzig, SW Pacala, D Tilman, pp. 169–74. Princeton, NJ: Princeton Univ. Press

Paine RT. 1966. Food web complexity and species diversity. *Am. Nat.* 100:65–75

Paine RT, Ruesink JL, Sun A, Soulanille EL, Wonham MJ, et al. 1996. Trouble on oiled waters: lessons from the Exxon Valdez oil spill. *Annu. Rev. Ecol. Syst.* 27:197–235

Parmesan C, Yohe G. 2003. A globally coherent fingerprint of climate change impacts across natural systems. *Nature* 421:37–42

Pauly D, Christensen V, Dalsgaard J, Froese R, Torres F Jr. 1998. Fishing down marine food webs. *Science* 279:860–63

Payette S, Bhiry N, Delwaide A, Simard M. 2000. Origin of the lichen woodland at its southern range limit in eastern Canada: the catastrophic impact of insect defoliators and fire on the spruce-moss forest. *Can. J. For. Res.* 30:288–305

Petchey OL, Downing AL, Mittelbach GG, Persson L, Steiner CF, et al. 2004a. Species loss and the structure and functioning of multitrophic aquatic systems. *Oikos* 104:467–78

Petchey OL, Gaston KJ. 2002. Extinction and the loss of functional diversity. *Proc. R. Soc. London Ser. B* 269:1721–27

Petchey OL, Hector A, Gaston KJ. 2004b. How do different measures of functional diversity perform? *Ecology* 85:847–57

Petchey OL, McPhearson PT, Casey TM, Morin PJ. 1999. Environmental warming alters food-web structure and ecosystem function. *Nature* 402:69–72

Peterson G, Allen CR, Holling CS. 1998. Ecological resilience, biodiversity, and scale. *Ecosystems* 1:6–18

Pimentel D, Lach L, Zuniga R, Morrison D. 2000. Environmental and economic costs of nonindigenous species in the United States. *BioScience* 50:53–65

Pimm SL. 1982. *Food Webs*. London/New York: Chapman & Hall. 219 pp.

Root TL, Price JT, Hall KR, Schneider SH, Rosenzweig C, Pounds JA. 2003. Fingerprints of global warming on wild animals and plants. *Nature* 421:57–60

Ruesink JL, Srivastava DS. 2001. Numerical and per capita responses to species loss: mechanisms maintaining ecosystem function in a community of stream insect detritivores. *Oikos* 93:221–34

Sala EO, Chapin FS III, Amnesto JJ, Berlow E, Bloomfield J, et al. 2000. Global biodiversity scenarios for the year 2100. *Science* 287:1770–74

Sax DF, Gaines SD. 2003. Species diversity: from global decreases to local increases. *Trends Ecol. Evol.* 18:561–66

Schläpfer F, Pfisterer AB, Schmid B. 2005. Non-random species extinction and ecosystem functioning. *J. Appl. Ecol.* 42:13–24

Schmid B, Hector A. 2004. The value of biodiversity experiments. *Basic Appl. Ecol.* 5:535–42

Schmid B, Joshi J, Schläpfer F. 2002. Empirical evidence for biodiversity-ecosystem functioning relationships. In *Functional Consequences of Biodiversity: Empirical Progress and Theoretical Extensions*, ed. AP Kinzig, SW Pacala, D Tilman, pp. 120–50. Princeton, NJ: Princeton Univ. Press

Schulze ED, Mooney HA. 1993. *Biodiversity and Ecosystem Function*. Berlin: Springer-Verlag

Schwartz MW, Brigham CA, Hoeksema JD,

Lyons KG, van Mantgem PJ. 2000. Linking biodiversity to ecosystem function: implications for conservation ecology. *Oecologia* 122:297–305

Shurin JB, Borer ET, Seabloom EW, Anderson K, Blanchette CA, et al. 2002. A cross-ecosystem comparison of the strength of trophic cascades. *Ecol. Lett.* 5:785–91

Shurin JB, Srivastava DS. 2005. New perspectives on local and regional diversity: beyond saturation. In *Metacommunities*, ed. M Holyoak, MA Leibold, RD Holt, pp. 399–417. Chicago: Univ. Chicago Press

Siemann E. 1998. Experimental tests of effects of plant productivity and diversity on grassland arthropod diversity. *Ecology* 79:2057–70

Smith MD, Knapp AK. 2003. Dominant species maintain ecosystem function with non-random species loss. *Ecol. Lett.* 6:509–17

Solan M, Cardinale BJ, Downing AL, Engelhardt KAM, Ruesink JL, Srivastava DS. 2004. Extinction and ecosystem function in the marine benthos. *Science* 306:1177–80

Sommer U, Sommer F, Santer B, Jamieson C, Boersma M, et al. 2001. Complementary impact of copepods and cladocerans on phytoplankton. *Ecol. Lett.* 4:545–50

Srivastava DS. 2002. The role of conservation in expanding biodiversity research. *Oikos* 98:351–60

Stubbendieck J, Friisoe GY, Bolick MR. 1994. *Weeds of Nebraska and the Great Plains*. Lincoln, NE: Nebraska Dept. Agriculture

Symstad AJ, Chapin FS, Wall DH, Gross KL, Huenneke LF, et al. 2003. Long-term and large-scale perspectives on the relationship between biodiversity and ecosystem functioning. *BioScience* 53:89–98

Thébault E, Loreau M. 2003. Food-web constraints on biodiversity-ecosystem functioning relationships. *Proc. Natl. Acad. Sci. USA* 100:14949–54

Thomas CDEA. 2004. Extinction risk from climate change. *Nature* 427:145–48

Tilman D. 1996. Biodiversity: population versus ecosystem stability. *Ecology* 77:350–63

Tilman D. 1999. Ecology—diversity and production in European grasslands. *Science* 286:1099–100

Tilman D, Downing JA. 1994. Biodiversity and stability in grasslands. *Nature* 367:363–65

Tilman D, Knops J, Wedin D, Reich P. 2002. Plant diversity and composition: effects on productivity and nutrient dynamics of experimental grasslands. See Loreau et al. 2002, pp. 21–35.

Tilman D, Lehman CL, Bristow CE. 1998. Diversity-stability relationships: statistical inevitability or ecological consequence? *Am. Nat.* 151:277–82

Tilman D, Lehman CL, Thomson KT. 1997. Plant diversity and ecosystem productivity: theoretical considerations. *Proc. Natl. Acad. Sci. USA* 94:1857–61

Tilman D, Wedin D, Knops J. 1996. Productivity and sustainability influenced by biodiversity in grassland ecosystems. *Nature* 379:718–20

van der Heijden MGA, Klironomos JN, Ursic M, Moutoglis P, Streitwolf-Engel R, et al. 1998. Mycorrhizal fungal diversity determines plant biodiversity, ecosystem variability and productivity. *Nature* 396:69–72

Vandermeer J, Lawrence D, Symstad A, Hobbie S. 2002. Effect of biodiversity on ecosystem functioning in managed ecosystems. See Loreau et al. 2002, pp. 221–33

Vinebrooke RD, Cottingham KL, Norberg J, Scheffer M, Dodson SI, et al. 2004. Impacts of multiple stressors on biodiversity and ecosystem functioning: the role of species co-tolerance. *Oikos* 104:451–57

Vinebrooke RD, Schindler DW, Findlay DL, Turner MA, Paterson M, Milis KH. 2003. Trophic dependence of ecosystem resistance and species compensation in experimentally acidified lake 302S (Canada). *Ecosystems* 6:101–13

Wardle DA. 2002. *Communities and Ecosystems: Linking the Aboveground and Belowground Components*. Princeton, NJ: Princeton Univ. Press. 392 pp.

Weltzin JF, Muth NZ, Von Holle B, Cole PG. 2003. Genetic diversity and invasibility: a test using a model system with a novel experimental design. *Oikos* 103:505–18

Wilsey BJ, Polley HW. 2002. Reductions in grassland species evenness increase dicot seedling invasion and spittle bug infestation. *Ecol. Lett.* 5:676–84

Wilsey BJ, Polley HW. 2003. Effects of seed additions and grazing history on diversity and productivity of subhumid grasslands. *Ecology* 84:920–31

Yachi S, Loreau M. 1999. Biodiversity and ecosystem productivity in a fluctuating environment: the insurance hypothesis. *Proc. Natl. Acad. Sci. USA* 96:1463–68

Zavaleta ES, Hulvey KB. 2004. Realistic species losses disproportionately reduce grassland resistance to biological invaders. *Science* 306:1175–77

Annu. Rev. Ecol. Evol. Syst. 2005. 36:295–317
doi: 10.1146/annurev.ecolsys.35.021103.105715
First published online as a Review in Advance on August 17, 2005

CONSEQUENCES OF THE CRETACEOUS/PALEOGENE MASS EXTINCTION FOR MARINE ECOSYSTEMS

Steven D'Hondt

Graduate School of Oceanography, University of Rhode Island, Narragansett Bay Campus, Narragansett, Rhode Island 02882; email: dhondt@gso.uri.edu

Key Words biogeochemistry, carbon, evolution, recovery, Tertiary

■ **Abstract** One of the greatest mass extinctions in Earth's history occurred at the end of the Cretaceous era, sixty-five million years (Myr) ago. Considerable evidence indicates that the impact of a large asteroid or comet was the ultimate cause of this extraordinary event. At the time of mass extinction, the organic flux to the deep sea collapsed, and production of calcium carbonate by marine plankton radically declined. These biogeochemical processes did not fully recover for a few million years. The drastic decline and long lag in final recovery of these processes are most simply explained as consequences of open-ocean ecosystem alteration by the mass extinction. If this explanation is correct, the extent and timing of marine biogeochemical recovery from the end-Cretaceous event was ultimately contingent on the extent and timing of open-ocean ecosystem recovery. The biogeochemical recovery may in turn have created new evolutionary opportunities for a diverse array of marine organisms.

INTRODUCTION

The end-Cretaceous mass extinction is one of the largest known extinction events in Earth's history (Sepkoski 1996). At the genus level, the total diversity of fossilized marine genera declined by nearly 50% from the end-Cretaceous Maastrichtian stage [~71 to 65 Mya] to the earliest Paleogene Danian stage (~65 to 61 Mya) (Sepkoski 1996; absolute age estimates from Berggren et al. 1995). Entire categories of organisms completely disappeared. In the marine realm, these categories included mosasaurs, sauropterygians (plesiosaurs and pliosaurs), ammonites, and heterohelicid planktic foraminifera. In the terrestrial realm, these categories included all forms of nonavian dinosaurs (Fastovsky & Sheehan 2005).

This event has historically been called the Cretaceous/Tertiary (K/T) mass extinction. However, the International Commission on Stratigraphy (ICS) does not recognize the Tertiary as a formal unit of Earth's history. In this review, I follow their lead and refer to the end-Cretaceous event as the Cretaceous/Paleogene (K/Pg) mass extinction.

1543-592X/05/1215-0295$20.00

Over the years, the end-Cretaceous extinction has been attributed to many ultimate causes including a supernova (Russell & Tucker 1971), flood-basalt eruption (Officer & Drake 1985), and the impact of a large asteroid or comet (Alvarez et al. 1980, Smit & Hertogen 1980). The vast majority of evidence favors the impact model.

The evidence for a large-body impact at the end of the Cretaceous is overwhelming (D'Hondt 1994, Smit 1999). Evidence of the impact is so widespread and clear that the ICS defines the K/Pg boundary as the base of the impact-debris layer, as exemplified by the iridium-bearing clay layer at El Kef, Tunisia (Cowie et al. 1989). The ICS has recently estimated the age of the K/Pg boundary to be 65.5 Mya. Herein I use an age estimate of 65 Mya for the K/Pg boundary because most paleontological and paleoceanographic studies of the last decade calibrated their data to a standard timescale that placed the boundary at 65 Mya.

Details of the K/Pg extinction remain a subject of occasional debate in the paleontological literature. However, the fossils observed in uppermost Cretaceous strata often demonstrate a striking pattern of occurrence; the more abundant, widely distributed, and often studied the fossil taxon, the more closely its final occurrence is associated with the impact-debris horizon. Rare but charismatic macrofossils, such as ammonite shells or articulated vertebrate bones, are rarely found close to the debris horizon (e.g., Marshall & Ward 1996). Abundant and widespread Maastrichtian fossils, such as the remains of calcareous nannoplankton (Pospichal 1994), planktic foraminifera (e.g., D'Hondt et al. 1996, Molina et al. 1998) and, in Denmark, cheilostome bryozoans (Håkansson and Thomsen 1999), generally occur very close below the debris horizon.

Local distributions of relatively rare fossils such as ammonites or dinosaur remains have occasionally been statistically analyzed to assess the possible rate and timing of end-Cretaceous extinction. The results of those analyses are consistent with the interpretation of mass extinction at the time of impact (e.g., Fastovsky & Sheehan 2005, Marshall & Ward 1996).

The coincidence of extinction and impact is particularly striking for planktic foraminifera, whose tests (skeletons) are abundant in upper Cretaceous marine limestones and carbonate oozes throughout the world. All latest Cretaceous morphospecies of planktic foraminifera occur just below the impact-debris horizon at one location or another (D'Hondt et al. 1996). Only two or three of these species consistently occur more than a few centimeters or tens of centimeters above the impact horizon in normally deposited marine sediments. The scattered tests of other Late Cretaceous species in lower Paleogene sediments exhibit carbon isotopic signatures, preservational states, and fossil distributions that indicate they were reworked from preimpact sediments (Huber 1996, Huber et al. 2002, Kaiho & Lamolda 1999, Zachos et al. 1992).

The proximate cause (or causes) of the end-Cretaceous extinction remains unknown. Advocates of the impact-extinction hypothesis (and their few but stalwart opponents) have proposed many potential causes (Kring 2000). These include global darkness (Alvarez et al. 1980), global cooling (Toon et al. 1982), acid rain (Lewis et al. 1982), rapid thermal radiation (Emiliani 1980), and heavy metal poisoning (Erickson & Dickson 1987). The relative importance of these respective causes is model dependent and presumably varied from one environment to another. In marine environments, darkness and (perhaps) cooling may have been important causes of end-Cretaceous extinction. Brief thermal radiation is a highly unlikely cause of widespread marine extinction because the thermal inertia of the upper ocean is great. Acid rain is also an unlikely cause of the marine extinction because the acid-buffering capacity of the upper ocean exceeds most estimates of acid production by the end-Cretaceous impact (D'Hondt et al. 1994a).

Whatever the proximate cause of the end-Cretaceous extinction, it greatly affected the diversity of life on Earth. In one sense, diversity never recovered; the organisms that disappeared will never reappear. In another sense, diversity recovered, but took a very long time; analyses of stage-level (multi-million-year) data compilations suggest that evolutionary rates, measured as rates of genus origination, lagged rates of peak extinction by millions of years (Kirchner & Weil 2000, Sepkoski 1998).

SELECTIVITY OF THE END-CRETACEOUS MASS EXTINCTION

Studies of marine fossil assemblages in distant locations show that the diversity of Danian communities was consistently lower than that of Maastrichtian communities. For example, fossil deposits in Upper Maastrichtian sediments of New Jersey (United States) contain diverse remains of abundant bivalves, gastropods, ammonites, sharks, bony fish, turtles, mosasaurs, crocodiles, and an occasional brachiopod (Gallagher 2003). In contrast, fossil remains in the overlying Danian sediments make up a depauperate assemblage where the most abundant fossils are sponges, brachiopods, and solitary corals (each limited to a single species) (Gallagher 2003). The Danian molluscs are much smaller, less abundant, and far less diverse than their Maastrichtian counterparts. Occasional remains of crocodiles and lamnid sharks provide the only evidence of higher predators (Gallagher 2003). Studies of macroinvertebrate faunas in Denmark (Håkansson & Thomsen 1999) and of molluscan taxa in the Gulf Coast (United States) region [Alabama (Jones et al. 1987) and Texas (Hansen et al. 1993)] have similarly identified large decreases in diversity across the K/Pg boundary.

Some groups of organisms were far more resistant to the K/Pg mass extinction than others. Taxa that preferentially survived the mass extinction included benthic foraminifera (Alegret & Thomas 2004, Culver 2003) and fossilizable dinoflagellates (Brinkhuis et al. 1998, Wendler & Willems 2002). Microorganisms that preferentially survived the mass extinction generally exhibit life-history characteristics that might have enabled them to wait out periods of extreme stress to open-ocean ecosystems. These characteristics include benthic lifestyles and the ability to form cysts (Brinkhuis et al. 1998, Kitchell et al. 1986, Wendler & Willems 2002). In contrast, the marine microorganisms most affected by the K/Pg mass extinction, planktic foraminifera and coccolithophorids, generally lack these life-history characteristics. Planktic foraminifera may be particularly susceptible to extinction when driven to very low population densities because, unlike dinoflagellates, diatoms, calcareous nannoplankton, and many radiolaria and benthic foraminifera, they can only reproduce sexually (D'Hondt et al. 1996).

Macrofossil groups that preferentially survived the mass extinction also exhibit consistent ecological characteristics. For example, several studies have inferred that detritus-feeding marine and terrestrial organisms were less susceptible than other organisms to end-Cretaceous extinction (Arthur et al. 1987, Sheehan & Hansen 1986). This trophic selectivity and the life-history selectivity described in the previous paragraph are consistent with the proximate cause of extinction being a drastic but short-lived decrease in photosynthesis brought about by a brief period of impact-induced global darkness (Arthur et al. 1987, Sheehan & Hansen 1986).

Taxonomic patterns of macrofossil extinction selectivity vary considerably from region to region (Hansen et al. 1993, McClure & Bohonak 1995). This variability is consistent with variation in selection factors from region to region and ecosystem to ecosystem (McClure & Bohonak 1995). It is also consistent with chance having played a significant role in taxonomic survival. Within some taxonomic groups, survival was so low that selective extinction is hard to distinguish from a null hypothesis of chance survival. On a local scale, these taxa include molluscs in the Brazos region of Texas (Hansen et al. 1993). On a global scale, they include planktic foraminifera.

Traits of species or higher taxa are sometimes considered to protect taxa against background extinction. Some of these properties—population size of species (Lockwood 2003), species-level geographic distributions, and species richness of clades (Jablonski 2001)—conferred no significant resistance to the K/Pg mass extinction. Geographic distribution may have played some role in genus-level survival of the K/Pg mass extinction, at least among bivalves (Jablonski 2003) and calcareous dinoflagellates (Wendler & Willems 2002). Nonetheless, it did not consistently protect organisms against the end-Cretaceous mass extinction. For example, almost all genera of Late Cretaceous planktic foraminifera were driven to extinction, despite being globally distributed. In at least some cases, mundane organismal properties visibly trumped geographic distribution in conferring resistance to the K/Pg mass extinction. A global study of echinoids showed that adult feeding strategies (omnivory and fine organodetritus deposit feeding) are strongly correlated to survival of the extinction event but geographic distribution is not (Smith & Jeffrey 1998).

BIOGEOCHEMICAL COLLAPSE AND RECOVERY AFTER THE MASS EXTINCTION

The mass disappearance of species and genera at the K/Pg boundary was an extraordinary event. Its potential consequences went well beyond biological diversity into the environmental realm. Per-area rates of deep-sea carbonate sedimentation drastically dropped at the time of mass extinction (e.g., Zachos & Arthur 1986). The flux of organic matter to the deep seafloor declined tremendously at the same time (e.g., Hsü et al. 1982, Zachos et al. 1989). These processes did not recover for millions of years (Arthur et al. 1987; D'Hondt et al. 1998a,b). Understanding of the extinction's ecological consequences requires a closer examination of these processes.

PRODUCTION OF CARBONATE MICROFOSSILS

Calcium carbonate ($CaCO_3$) microfossils produced by nannoplankton (principally coccolithophorids) and planktic foraminifera blanket approximately 70% of the deep seafloor in the modern ocean. These microfossils compose an unusually low percentage of sediments deposited during the Danian stage. Throughout the world ocean, the $CaCO_3$ fraction of deep-sea sediments crashed to nearly zero at the very end of the Cretaceous and did not completely return to pre-extinction values for three or four million years (Figure 1A).

Paleoceanographic studies have consistently shown that the Danian carbonate crash resulted from a radical and long-lasting decrease in the rate of $CaCO_3$ accumulation (D'Hondt et al. 1996, Zachos & Arthur 1986). This decrease in carbonate accumulation occurred in every ocean basin (Zachos & Arthur 1986) (Figure 1B). It began at the time of K/Pg impact and microfossil mass extinction (D'Hondt et al. 1996). Carbonate accumulation rates remained unusually low for more than a million years (Zachos & Arthur 1986). Present data suggest that carbonate accumulation may have remained unusually low for about four million years (D'Hondt et al. 1998b) (Figure 1B).

A large decrease in $CaCO_3$ accumulation can result from either (a) a large decrease in production of $CaCO_3$ microfossils or (b) a large increase in the dissolution of $CaCO_3$ microfossils by deep water. Records of microfossil preservation indicate that the unusually low carbonate accumulation rates of the Danian stage resulted from unusually low rates of microfossil production. For example, at moderate levels of carbonate dissolution, foraminiferal tests are visibly fragmented. Fragmentation of foraminiferal tests is much lower in sediments deposited during the postextinction interval of low $CaCO_3$ accumulation than in the sediments deposited before or after (Figure 1C). Fragmentation remained unusually low for about three million years (Figure 1C).

The unusually low rate of postextinction $CaCO_3$ accumulation principally resulted from the rate of calcareous nannofossil (fine carbonate) accumulation

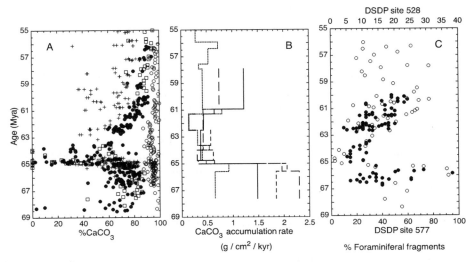

Figure 1 Patterns of K/Pg carbonate sedimentation at widely distant open-ocean sites: (*A*) Sedimentary carbonate concentrations (% CaCO₃) at Central Pacific DSDP Site 577 (*white circles*), South Atlantic DSDP Site 528 (*black circles*), South Atlantic DSDP Site 527 (*white squares*), and Caribbean ODP Site 1001A (*crosses*). (*B*) Mean CaCO₃ accumulation rates at Site 577 (*dotted line*), Site 528 (*line with short dashes*), Site 527 (*line with long dashes*), and Site 1001A (*solid line*). (*C*) Percent foraminiferal fragments [number of planktic foraminiferal fragments/(number of planktic foraminifera + fragments)*100]. Note the pronounced decreases in the properties exhibited by all three panels at 65 Mya (the K/Pg boundary) and their final recoveries at approximately 62–61 Mya. Age assignments are based on magnetic reversal data (Bleil 1985, Chave 1984, Sigurdsson et al. 1997), the K/Pg and Paleocene/Eocene (P/E) boundaries (Sigurdsson et al. 1997, D'Hondt et al. 1998a), the magnetic reversal timescale of Cande & Kent (1995), and age estimates for the K/Pg boundary and the P/E boundary from Berggren et al. (1995). Accumulation rates immediately above and below the K/Pg boundary (within paleomagnetic reversal interval 29R) are based on precessional cycle counts of D'Hondt et al. (1996). Except at the K/Pg boundary where precessional control precisely identifies the changes in mean accumulation rates, the step-like nature of the accumulation rate estimates in panel *B* results from the positions of the chronostratigraphic tie points used to calculate the rates. Actual changes in accumulation rates may be more gradual and may occur at any time between tie points.

decreasing by a factor of four or more (D'Hondt et al. 1996, Zachos & Arthur 1986). Foraminifera constitute a very small fraction of the total carbonate in upper Cretaceous sediments (Figure 2A). Furthermore, despite the extraordinary level of end-Cretaceous planktic foraminiferal extinction, the average rate of foraminiferal accumulation decreased by a factor of two or less across the mass extinction horizon

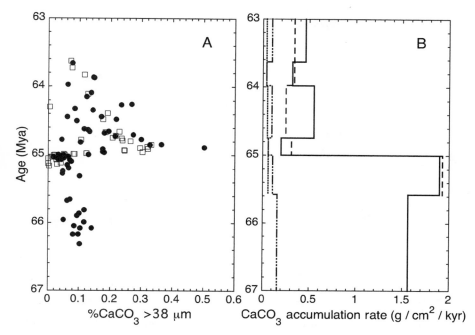

Figure 2   Carbonate constituents and their accumulation rates: (*A*) Percent CaCO₃ >38 μm (foraminiferal tests and their fragments) at South Atlantic DSDP Sites 527 (*white squares*) and 528 (*black circles*). (*B*) Accumulation of CaCO₃ in the >38 μm fraction at Sites 527 (*dotted line*) and 528 (*line of alternating dots and dashs*), and the <38 μm fraction, principally composed of calcareous nannofossils, at Sites 527 (*dashed line*) and 528 (*solid line*). Age assignments and uncertainties in timing of changes in accumulation rates are as in Figure 1. Note the pronounced change at the K/Pg boundary (65 Mya) in both panels.

(Figure 2*B*). Consequently, the low carbonate accumulation rates in Danian deep sea sediments principally resulted from a radical and long-lasting decrease in CaCO₃ production by calcareous nannoplankton.

THE FLUX OF ORGANIC MATTER TO DEEP WATER

Four lines of evidence indicate that the organic flux to the global deep ocean radically decreased at the time of the K/Pg impact and microfossil mass extinction: The first line of evidence is the improved preservation of planktic foraminifera in Danian sediments (Figure 1*C*). The second line of evidence is a tremendous decrease in the carbon isotopic (δ^{13}C) difference between planktic foraminifera and benthic foraminifera at sites throughout the world ocean (D'Hondt et al. 1998a; Stott & Kennett 1989; Zachos et al. 1989, 1992) (Figure 3*A*,*B*). The third line of evidence is convergence of the carbon isotopic signatures of benthic foraminiferal

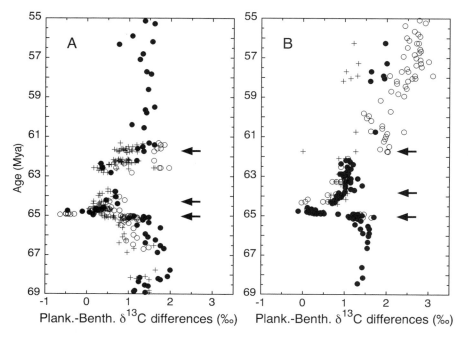

Figure 3 Carbon isotopic proxies for organic flux to the deep sea: (*A*) δ^{13}C differences between planktic foraminifera and benthic foraminifera at South Atlantic DSDP Site 528; (*B*) δ^{13}C differences between planktic foraminifera and benthic foraminifera at central Pacific DSDP Site 577. Age assignments are as in Figure 1. The black circles represent δ^{13}C differences between fine carbonate (principally calcareous nannofossils) and benthic foraminifera, the white circles represent differences between planktic foraminifera that dwelt near the sea surface and benthic foraminifera, and the crosses represent differences between deeper water (thermocline) planktic foraminifera and benthic foraminifera. Isotopic data are from D'Hondt et al. (1998b) and S. D'Hondt and J.C. Zachos (unpublished data). Note the sharp negative shift in both records at 65 Mya (the K/Pg boundary) (*bottom arrow* in each panel), the partial recovery of the δ^{13}C differences within the following Myr (*middle arrow* in each panel), and the eventual recovery of both records at approximately 62 Mya (*top arrow* in each panel).

tests from the Atlantic and Pacific oceans (Stott & Kennett 1989, Zachos et al. 1992). The fourth line of evidence is collapse of the δ^{13}C gradient between infaunal and epifaunal benthic foraminifera (Zachos et al. 1989).

A fuller understanding of these lines of evidence requires a brief discussion of the oceanic carbon system. The concentration of dissolved inorganic carbon (DIC = HCO_3^- + CO_3^{2-} + CO_2) in surface seawater is approximately in equilibrium with the atmosphere and is inversely proportional to temperature. Consequently, DIC concentrations in the surface ocean are lowest in the warm subtropical waters (1900 μmol/kg) and highest in the cold high-latitude ocean (2200 μmol/kg)

(Broecker & Peng 1982). Deep water DIC concentrations depend on (*a*) the atmospheric equilibration in the surface-ocean region(s) of deep water formation and (*b*) the oxidation of sinking organic matter from the surface ocean.

In seawater, CO_2 released by oxidation of organic matter increases the solubility of $CaCO_3$ by decreasing pH and carbonate ion (CO_3^{2-}) concentration (Pilson 1998). Consequently, the sinking and decay of organic carbon renders deep water corrosive to $CaCO_3$ microfossils. The decreased fragmentation of planktic foraminifera in lowermost Paleogene sediments indicates that the postextinction ocean was less corrosive to carbonate microfossils than the pre-extinction ocean for about three million years (Figure 1*C*). This in turn suggests that the decay of organic carbon in deep waters and, by implication, the flux of organic carbon to deep water were unusually low for the same interval of time.

Carbon 12 constitutes 98.89% of the stable carbon in the world (Faure 1986). The remaining 1.11% is ^{13}C. The ratio of stable carbon isotopes in environmental samples is typically expressed in parts per thousand (‰) relative to a standard value, using δ notation:

$$\delta^{13}C = [(^{13}C/^{12}C)_{sample} - (^{13}C/^{12}C)_{standard}]/(^{13}C/^{12}C)_{standard} * 1000.$$

The conventional standard value is the $^{13}C/^{12}C$ ratio of the PDB standard, a sample of fossil belemnites from the Cretaceous PeeDee Formation of South Carolina (United States) (Faure 1986).

The $\delta^{13}C$ of dissolved carbon in the modern deep ocean is 1‰ to 2‰ more negative than the $\delta^{13}C$ of dissolved carbon in the surface ocean (Kroopnick 1980, Kroopnick et al. 1970) (Figure 4). This difference ultimately results from the flux of organic carbon from the surface ocean to the deep sea. The $\delta^{13}C$ of marine plankton is about 20‰ more negative than the $\delta^{13}C$ of the surface ocean (Anderson & Arthur 1983). Consequently, deep water oxidation of sinking organic matter from the surface ocean renders deep-ocean $\delta^{13}C$ more negative than surface-ocean $\delta^{13}C$ by adding DIC with the $\delta^{13}C$ of marine plankton to the deep ocean. The $\delta^{13}C$ of DIC in the modern deep ocean is least negative in the North Atlantic basin, which contains relatively young deep water, and most negative in the North Pacific basin, which contains deep water that has accumulated DIC from organic oxidation for many hundreds of years (Figure 4).

Because foraminifera derive most of the carbonate in their tests from DIC in their environment, $\delta^{13}C$ differences between surface water DIC and deep water DIC are approximated by $\delta^{13}C$ differences between the tests of planktic foraminifera that live in the surface ocean and the tests of benthic foraminifera that live on the seafloor (epibenthic foraminifera). Similarly, $\delta^{13}C$ differences between seawater DIC in the thermocline (tens to hundreds of meters beneath the sea surface) and deep water DIC are approximated by $\delta^{13}C$ differences between thermocline-dwelling planktic foraminifera and the tests of epibenthic foraminifera.

Carbon isotopic differences between planktic and benthic foraminiferal tests of the latest Cretaceous (Figure 3*A,B*) are roughly equivalent to $\delta^{13}C$ differences between surface water DIC and deep water DIC in the modern ocean (Figure 4). This equivalence indicates that the distribution of oceanic DIC in the pre-extinction

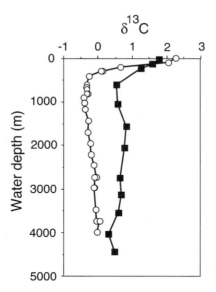

Figure 4 δ^{13}C of dissolved inorganic carbon in the modern North Pacific [Geosecs Site, SIO data (*white circles*)] and South Atlantic [Site 103 data (*black squares*)]. Data from Kroopnick et al. (1970) and Kroopnick (1980).

Cretaceous was much as it is today. For this to be so, patterns of organic-matter oxidation in the deep ocean and, by inference, the flux of organic matter to the deep ocean, must have been similar to what it is today.

This carbon isotopic system changed radically at the time of end-Cretaceous extinction, when the δ^{13}C values of planktic and benthic foraminiferal tests converged (Figure 3A,B) (D'Hondt et al. 1998a; Stott & Kennett 1989; Zachos et al. 1989, 1992). The δ^{13}C difference between planktic tests and benthic tests did not fully recover to pre-extinction values until approximately three million years after the extinction event (Figures 3A,B) (Adams et al. 2004, D'Hondt et al. 1998a). The K/Pg convergence of planktic and benthic δ^{13}C signals provides strong evidence that the impact and mass extinction coincided with a radical decrease in the flux of organic matter to the deep sea. Coeval convergence of benthic foraminiferal δ^{13}C signals from distant ocean basins similarly indicates a radical decrease in the flux of organic matter to the deep ocean (Stott & Kennett 1989, Zachos et al. 1992). The long delay in recovery of planktic to benthic δ^{13}C differences indicates that the flux of organic matter to deep water did not recover for about three million years. The Danian difference between planktic and benthic δ^{13}C signals recovered in two stages (Adams et al. 2004, D'Hondt et al. 1998a). The first recovery stage is marked by continuous relaxation toward an interval of low but relatively stable δ^{13}C differences between planktic and benthic foraminifera. The second stage is defined by a discontinuous adjustment to approximately pre-extinction δ^{13}C differences about three million years after the end-Cretaceous impact and mass extinction. Like the postextinction changes in $CaCO_3$ preservation and the end-Cretaceous collapse of planktic to benthic $\delta^{13}C$ differences, the two-stage recovery of planktic to benthic $\delta^{13}C$ differences occurred in both the Atlantic and Pacific Oceans (Figure 3).

ECOLOGICAL MODELS OF THE POSTEXTINCTION OCEAN

Two models have been proposed to explain the interval of low organic flux and low nannofossil production that followed the end-Cretaceous impact. It was first interpreted as the result of unusually low marine biological productivity throughout the postextinction ocean (e.g., Hsü et al. 1982). This interpretation is usually described as the Strangelove Ocean model, following Broecker & Peng (1982), who called an ocean with no life a "Strangelove ocean." The alternative model, proposed by D'Hondt et al. (1998b), holds that marine productivity was relatively high but the fraction of total productivity that sank to the deep sea was relatively low during the multi-million year interval of low organic flux. This interpretation has been called the Living Ocean model (e.g., Adams et al. 2004).

Interpretations of the Danian postextinction ocean as a Strangelove ocean (e.g., Hsü & McKenzie 1985) implicitly redefine a Strangelove ocean as one where life is present but biological productivity is low. On very short timescales, such an ocean would naturally have resulted from at least one hypothetical consequence of the K/Pg impact—a large reduction in sunlight at Earth's surface owing to high concentrations of impact dust (Alvarez et al. 1980) and sulfuric aerosol released by vaporization of anhydrite at the impact site (Brett 1992). Such a reduction in sunlight may have greatly reduced photosynthesis for a short period of time after the impact. However, quantitative models of large impact events and their consequences suggest that dust and aerosols would have been swept from the atmosphere in less than a year (Kring 2000).

Maintenance of low phytoplankton production on significantly longer timescales is difficult to envision (Arthur et al. 1987). Marine phytoplankton have typical doubling times of hours to days. Given such doubling times, a multi-million-year decrease in phytoplankton production would require some key environmental property, such as availability of a biologically limiting nutrient, to instantly decrease at the K/Pg boundary and remain anomalously low for a few million years. However, no physical consequences of large impacts are predicted to last as long as a million years by quantitative models. In fact, very few environmental consequences of large impacts have been inferred to last beyond a decade. Furthermore, the physical consequences of large impacts are not obviously linked to the long-term state of biologically limiting properties, such as marine nutrient availability.

D'Hondt et al. (1998b) addressed these issues by hypothesizing that marine phytoplankton production quickly returned after the impact and mass extinction. In this Living Ocean model, low organic flux to the deep ocean was principally a consequence of ecosystem structure in the postextinction ocean. The proportion of organic production that sinks from the surface ocean is controlled by several ecosystem properties, including the ratio of phytoplanktic respiration to photosynthesis; phytoplankton size (only the largest plankton sink rapidly enough to settle to deep waters); the ability of phytoplankton to aggregate into larger particles; and the size, trophic efficiency, and gut structure of animals (which may repackage biomass into large particles that sink rapidly).

Given these factors, many rearrangements of the open-ocean ecosystem could decrease the flux of organic matter to the deep sea. These possible rearrangements include a decrease in the mean size of marine phytoplankton (increased dominance of marine production by picoplankton), a decrease in the abundance of colonial or aggregating phytoplankton, a decrease in the mean photosynthesis/respiration ratio of phytoplankton, a decrease in the mean size of pelagic grazers, or a shift in dominance from grazers that create fecal pellets (fish) to grazers that do not (e.g., jellyfish). Any or all of these changes could result from mass extinction. In such altered ecosystems, an increased fraction of total production would be shunted through the microbial food web. By allowing essential nutrients to remain in easily remineralized forms (such as tiny microbially grazed plankton) in the euphotic zone, such changes may also increase rates of nutrient recycling and biomass production in the near-surface ocean.

Application of this model to the Danian postextinction ocean suggests that dissolved nutrient availability and biological productivity may have been relatively high in the near-surface ocean throughout the three-million-year interval of reduced organic flux to the deep sea. If so, final recovery of the organic flux to the deep sea at the end of this interval would have increased the rate at which nutrients were lost from the near-surface ocean in sinking organic matter. It would have also reduced the rate of nutrient recycling in the near-surface ocean. Both of these effects would have shifted large portions of the ocean to an oligotrophic state that more closely resembled the pre-extinction ocean (and the modern ocean).

FOSSIL EVIDENCE OF POSTEXTINCTION ECOSYSTEM STRUCTURES AND RECOVERY

Microfossil Evidence of Open-Ocean Ecosystem Structure and Recovery

Records of fossil occurrences have rarely been placed in the context of the biogeochemical and sedimentological data that define the long lag in postextinction recovery of organic flux and nannofossil production. Because the carbon isotopic records are principally derived from foraminiferal calcite, planktic foraminiferal records allow the most straightforward comparison of the biological and biogeochemical responses to the K/Pg event.

Planktic foraminiferal occurrence data show that faunal turnover was nearly 100% throughout the world at the time of mass extinction (e.g., D'Hondt et al.

1996, Molina et al. 1998, Smit 1977, Troelsen 1957). They also provide intriguing evidence of faunal succession during the early and final stages of biogeochemical recovery. The survivor *Guembelitria cretacea* and its immediate descendents (Olsson et al. 1999) dominated assemblages during the earliest postextinction interval (D'Hondt et al. 1996, Gerstel et al. 1987), when planktic to benthic δ^{13}C differences were lowest. As the first stage of δ^{13}C recovery proceeded, descendents of a different survivor [*Hedbergella monmouthensis* (Olsson et al. 1999)] came to dominate planktic foraminiferal assemblages. This group dominated planktic foraminiferal communities until the time of final δ^{13}C recovery at the end of the Danian, when a third group became dominant. The *Morozovella* and *Acarinina* species that dominated this last assemblage have been interpreted as descendents of the third survivor taxon, *Hedbergella holmdelensis* (Olsson et al. 1999). This group went on to dominate tropical and temperate planktic foraminiferal assemblages for millions of years. Its close coincidence with the final δ^{13}C recovery indicates that the ascendance of this group to dominance closely coincided with the final recovery of the organic flux to the deep sea.

The genera that dominated the first two postextinction assemblages were generally distributed throughout the open ocean. However, their isotopic signatures and the geographic distribution of their relative abundances suggest that each genus in these assemblages inhabited a slightly different niche, in terms of seasonality, water depth and, possibly, a weak degree of photosymbiont reliance (D'Hondt & Zachos 1993). The genera that came to dominate planktic foraminiferal assemblages at the time of final δ^{13}C recovery differed in two critical ways from the principal members of the earlier assemblages. First, their tests were generally much larger. Second, their carbon and oxygen isotopic signatures indicate that they relied very strongly on photosymbionts for nutrition (D'Hondt et al. 1994b, Norris 1996). In both regards, these postrecovery taxa resembled members of modern oligotrophic communities (D'Hondt et al. 1994b, Norris 1996) and many members of the preextinction communities (D'Hondt & Zachos 1998, Houston & Huber 1998).

Records of planktic foraminiferal test size provide additional evidence of ecosystem structure during the long interval of biogeochemical recovery (Schmidt et al. 2004). For example, at South Atlantic Site 528 and Equatorial Pacific Site 577, tiny tests made up an unusually high fraction of the mass of planktic foraminiferal calcite during the long postextinction interval of low organic flux to the deep sea (Figure 5A). In contrast, large foraminiferal tests were common before the extinction event, were rare throughout the long interval of low organic fluxes, and returned to pre-extinction abundance at the approximate time of final biogeochemical recovery (Figure 5B). The relative abundance of tiny tests in the earliest Paleogene sediments becomes even more striking if the individual masses of large and small tests are considered. For example, a 212-μm diameter test weighs nearly an order of magnitude more than a 106-μm diameter test (D'Hondt et al. 1994b) and approximately two orders of magnitude more than a 38-μm diameter test. Consequently, an earliest Paleogene sample with 70% of its mass composed of 38-μm to 106-μm tests and 7% of its mass composed of >212 μm tests contains roughly

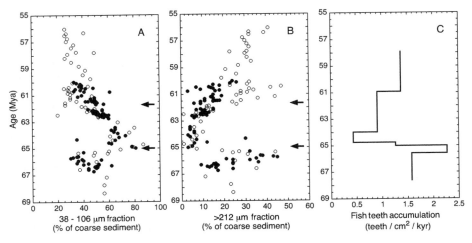

Figure 5 Records of planktic foraminiferal test diameter and fish teeth accumulation. Planktic foraminiferal tests comprise most of the coarse (>38 μm) sediment at Site 528 (*black circles*) and Site 577 (*white circles*): (*A*) Percent of coarse sediment in the 38- to 106-μm diameter fraction. (*B*) Percent of coarse sediment in the >212-μm diameter fraction. (*C*) Mean rates of fish teeth accumulation at South Atlantic DSDP Site 527. In panels *A* and *B*, the lower arrow marks the K/P event horizon and the upper arrow marks the approximate time of final recovery of the carbon isotope system (see Figure 3). Accumulation rates are based on fish teeth counts from Shackleton et al. (1984) and the chronostratigraphic data described in the Figure 1 caption. Age assignments and uncertainties in timing of changes in accumulation rates are as in Figure 1.

1000 times as many tiny tests as large tests. Given these mass differences and the relative constancy of foraminiferal calcite accumulation across the mass extinction horizon (Figure 2*B*), the flux of individual foraminiferal tests to the seafloor was very high over at least the first two million years of the postextinction interval.

Numerous studies have documented tremendous mass extinction of calcareous nannofossils at the K/Pg boundary (e.g., Pospichal 1994). However, few studies have closely examined the postextinction recovery of calcareous nannofossil assemblages. Study of Antarctic ODP Site 738 showed that successive changes in assemblage composition occurred in the first few million years after the extinction (Wei & Pospichal 1991). No studies have closely examined the relationship, if any, between calcareous nannofossil assemblages, CaCO₃ accumulation, and organic fluxes to the deep sea throughout the entire postextinction interval of biogeochemical recovery.

Fossil occurrences of silica-secreting plankton, such as diatoms and radiolaria, are generally scarcer and more rarely examined than the fossil occurrences of carbonate-secreting planktic foraminifera and calcareous nannofossils. Fortunately, diatom frustules and radiolarian tests are relatively abundant in upper

Cretaceous and lower Paleogene marine sediments of eastern New Zealand (Hollis et al. 1995). In these sediments, the broad composition of radiolarian assemblages radically changed at the K/Pg boundary (Hollis et al. 1995). At several sites in this region, concentrations of silica normalized to titanium or aluminum (and inferred to be composed of radiolarian tests and diatom frustules) increased at the K/Pg boundary and remained high for at least one or two million years (Hollis et al. 2003). These patterns of silica concentration have been interpreted as evidence of an increase in biosiliceous productivity across the K/Pg boundary, followed by high biosiliceous productivity for at least the first 1.5 million years after the mass extinction (Hollis et al. 2003). Whether the final recovery of organic flux to the deep sea coincided with significant changes in diatom and/or radiolarian assemblages remains to be determined.

Macrofossil evidence of open-ocean ecosystem structure is rare because macrofossils are rare in deep-sea sediments and are almost never recovered in deep-sea cores. However, some macroorganisms (particularly fish) leave a microfossil record that can be used to infer some characteristics of open-ocean ecosystems. At least one such record spans the entire multi-million-year interval of carbon-system collapse and recovery. This record, from South Atlantic DSDP Site 527, shows that the average fish teeth accumulation rate declined at the time of mass extinction, increased within about a million years, and increased again at the approximate time of final organic flux recovery (Figure 5C).

Macrofossil Evidence of Ecosystem Structure and Recovery in Shallow Marine Communities

Lower Paleogene records of marine macrofossils are very patchily distributed. Consequently, the precise timing and geographic extent of postextinction changes in marine macroinvertebrate and vertebrate communities are poorly constrained. The Brazos region of Texas (United States) may be the only location where marine macroinvertebrate faunal succession has been directly compared with the earliest stages of postextinction planktic foraminiferal succession (Hansen et al. 1993). At that location, the postextinction succession of macroinvertebrate assemblages closely coincides with the postextinction succession of planktic foraminiferal assemblages. A pronounced bloom of a microherbivorous gastropod occurred during the earliest Paleogene reign of the guembelitriid-dominated foraminiferal assemblage. Deposit-feeding molluscs grew to dominate the macroinvertebrate assemblages as the descendents of *H. monmouthensis* grew to dominate the planktic foraminiferal community (Hansen et al. 1993). Comparison of these foraminiferal records with records at open-ocean sites suggests that this ecologic ascendance of deposit-feeding molluscs approximately coincided with the initiation of the first stage of open-ocean carbon system recovery.

Similarly detailed records of macroorganismal assemblage succession are known from few other localities. Succession aside, available data suggest that broad patterns of ecologic dominance in postextinction communities were somewhat different at other localities. In particular, remains of suspension-feeding organisms are relatively more abundant in Danian sediments of Alabama (United States) (Bryan & Jones 1989), New Jersey (United States) (Gallagher 2003), and Denmark (Håkansson & Thomsen 1999) than in the Brazos, Texas, sediments. The categories of organisms that dominate these fossil assemblages differ considerably from region to region. To some extent, these compositional differences may result from sub-million-year differences in the ages of the assemblages being compared (Hansen et al. 1993). However, they are also contingent on the different environments and pre-extinction assemblages of the different localities.

These contingencies can be illustrated by brief examination of the K/Pg record in Denmark. The K/Pg communities of the Gulf Coast and New Jersey inhabited shallow-water siliciclastic sedimentary environments. In contrast, the Danish communities inhabited—and, in the Maastrichtian, created—a shallow water carbonate environment. In Denmark, bryozoan fossils dominate Maastrichtian assemblages, but are almost entirely absent from sediments deposited during the first several tens of kyrs (tens of thousands of years) after the end-Cretaceous extinction (Håkansson & Thomsen 1999). The lowermost Danian sediments of this region are nearly devoid of macrofossils. The fossils present in this zone are dominated by skeletal elements from two genera of crinoids; they also contain fragments from a few echinoid taxa and very rare cheilostome bryozoa. Bryozoans returned to faunal dominance within a few hundred kyrs after the mass extinction (Håkansson & Thomsen 1999), during the first stage of open-ocean carbon system recovery. This bryozoan-dominated recovery fauna was much less diverse taxonomically and ecologically than the bryozoan-dominated pre-extinction fauna (Håkansson & Thomsen 1999).

Paleogene records of marine macrofossils are not known in enough detail to determine whether marine macroinvertebrate ecosystems significantly changed at the time of final carbon-system recovery. The few available records are consistent with close linkage of benthic recovery to carbon system recovery. The bryozoan-dominated ecosystem and its shallow-water carbonate habitat disappeared from Denmark by the end of the Danian (61.7 Mya) to be replaced by clastic sediments (Håkansson & Thomsen 1999). In New Jersey, macrofossil diversity returned to pre-extinction levels by the Thanetian stage (58.7 to 55.8 Mya) (Gallagher 2003). However, the exact timing of these events (the disappearance of the Danish carbonate communities, the recovery of macrofossil diversity on the New Jersey shelf) and the exact timing of final ecosystem recovery in other shallow marine environments remain largely unconstrained. Consequently, their relationship to final recovery of the marine carbon system also remains to be determined.

Fossil Evidence and Ecological Models of the Postextinction Ocean

Many studies have interpreted the composition of Danian macrofossil assemblages to be consistent with a Strangelove (low-productivity) model of the postextinction ocean (e.g., Arthur et al. 1987, Gallagher 2003, Hansen et al. 1993).

The composition of these assemblages is also generally consistent with a Living Ocean model. For example, Gallagher (2003) interpreted the postextinction abundance of sponges and brachiopods in the Atlantic Coastal Plain (United States) to be consistent with a low-productivity (Strangelove) ocean where plankton are small and scarce. In making this argument, Gallagher (2003) states that sponges and brachiopods are "minimalist" organisms that specialize in filtering fine organic matter from large volumes of water. These properties are also consistent with a Living Ocean model of the postextinction ocean, where productivity is normal (or even high) and plankton are abundant but tiny. The nearly four-million-year interval of low fish-tooth accumulation at Site 527 (Figure 5) is similarly consistent with either (*a*) reduced food availability in a very long-lasting postextinction Strangelove ocean or (*b*) reduced transfer of biomass to relatively large grazers and higher predators in a normal-productivity ocean where a relatively higher fraction of biomass is shunted through the microbial community.

Other results are inconsistent with a low-productivity (Strangelove) model of the Danian ocean but very consistent with a Living Ocean model. Most notably, the New Zealand records of siliceous microfossil abundance have consistently been interpreted as evidence of regionally high primary production (e.g., Hollis et al. 2003). Planktic foraminiferal data are also more readily consistent with a Living Ocean model than a Strangelove Ocean model (D'Hondt et al. 1998a). The abundance of planktic foraminifera with tiny tests and low photosymbiont reliance required an abundance of suitably tiny prey throughout the long interval of low organic flux to the deep sea (Figure 5). Furthermore, the large, highly photosymbiotic individuals that characterized planktic foraminiferal assemblages after the final recovery of organic fluxes are fossil analogues to modern taxa that are highly adapted for oligotrophic conditions (with scarce nutrients and relatively scarce but large prey) (D'Hondt et al. 1994b, Norris 1996).

BIOGEOCHEMICAL AND BIOLOGICAL RECOVERY FROM MASS EXTINCTION

If the Living Ocean model of the Danian ocean is correct, the rapid and long-lasting K/Pg changes in planktic carbonate production and the organic flux to the deep sea were direct consequences of the mass extinction. Furthermore, if this model is correct, the marine biogeochemical recovery from the K/Pg event was a direct consequence of the biological recovery from the mass extinction. For example, the two stages of recovery of the organic flux to the deep sea should be closely linked to stages of ecosystem recovery.

In evaluating this possibility, three scenarios must be considered. In the first scenario, recovery of the marine carbon system was contingent on ecological recovery. For example, the evolution of one or more key components of the ecosystem (such as large phytoplankton or large grazing zooplankton) may have returned the marine carbon system to its pre-extinction state. In the second scenario, ecological recovery was contingent on biogeochemical recovery. Here, final recovery of the marine carbon system sparked biological diversification by creating a broad range of new opportunities for other organisms. It is conceivable that both the first and second scenarios apply, with positive feedback between the ecological recovery and the biogeochemical recovery. In the third scenario, the composition of marine ecosystems and the state of the marine biogeochemical system are largely independent of each other. This scenario corresponds to a Strangelove Ocean scenario where the long delays in environmental recovery are consequences of unknown multi-million-year physical or chemical consequences of the K/Pg impact. In this scenario, no direct link is necessarily expected between the timing of biological recovery and the timing of environmental recovery.

The relative timing of biological events and the biogeochemical events suggests that the two categories of events were closely linked. The mass extinction directly coincided with the collapse of the organic flux to deep water (e.g., Hsü et al. 1982) and the drastic decline in nannofossil production (e.g., Zachos et al. 1986). Furthermore, as described in the preceding subsections, the first stages of recovery in diverse marine communities roughly coincided with the first stage of carbon-flux recovery. This stage of recovery occurred during the first million years after the extinction, as marine ecosystems were reassembled by a patchwork of migration and evolution. Also as described in preceding subsections, the final stage of organic-flux recovery globally coincided with pronounced turnover of planktic foraminiferal assemblages a few million years after the mass extinction. More strikingly, this final stage of recovery of the organic flux to the deep sea was closely followed by rapid diversification of the previously monospecific morozovellid and acarinid lineages (H.K. Coxall, S. D'Hondt, J.C. Zachos, manuscript submitted). For many subsequent millions of years, these clades constituted the most species-rich components of planktic foraminiferal assemblages (Olsson et al. 1999). A recent study suggests that clades of calcareous nannofossils underwent similar rapid diversification shortly after the final recovery of the organic flux to the deep sea (Fuqua & Bralower 2004). These relationships between final carbon-system recovery and planktic diversification support the second scenario outlined above.

Present data are insufficient to determine whether other groups of marine organisms underwent significant radiation at the time of final organic-flux recovery. However, analyses of stage-level (multi-million-year) data compilations suggest that evolutionary rates, measured as rates of genus origination, lagged the K/Pg extinction by millions of years (Kirchner & Weil 2000, Sepkoski 1998). This long lag in taxonomic recovery is not unique to the early Paleogene. Similarly long lags followed the other major mass extinctions (Sepkoski 1998). A long lag in recovery of the marine carbon cycle is also not unique to the early Paleogene. Carbon isotopic records from marine sediments throughout the world demonstrate that the largest known mass extinction (at the end of the Permian, 250 million years ago) altered the marine carbon cycle for approximately one million years after the extinction event (D'Hondt et al. 2000, Holser et al. 1989).

The delays in evolutionary recovery from mass extinctions have been attributed to the internal dynamics of diversification (Erwin 2001, Kirchner & Weil 2000, Sepkoski 1998). Some of those dynamics may well have played out in the biogeochemical arena. For example, if the Living Ocean model of the Danian postextinction ocean is correct, the final recovery of the organic flux to the deep sea would have stripped nutrients from the surface ocean and driven it toward more broadly oligotrophic conditions. In doing so, this final recovery of the marine carbon system may have created new evolutionary opportunities for a diverse array of marine organisms.

ACKNOWLEDGMENTS

During my studies of the biological and environmental consequences of the K/Pg event, I have benefited greatly from discussions and collaborations with many colleagues, including J.C. Zachos, P. Schultz, D.E. Fastovsky, P. Sheehan, R.K. Olsson, J.J. Sepkoski Jr., J.W. King, T.D. Herbert, M.A. Arthur, H. Sigurdsson, M. Lindinger, M.E.Q. Pilson, E. Molina, D.A. Kring, P. Donaghay, H. Coxall, and numerous others. I thank E. Nadin, G. Hoke, and M. Neissingh for generating much of the data shown in Figures 1 and 5. I thank M. Smith for helping to draft the figures and A. Pariseault for helping in ways too numerous to describe. This manuscript was greatly improved by S. Wing's thoughtful review. Financial support was provided by the U.S. National Science Foundation Division of Earth Sciences.

The *Annual Review of Ecology, Evolution, and Systematics* is online at
http://ecolsys.annualreviews.org

LITERATURE CITED

Adams JB, Mann ME, D'Hondt S. 2004. The Cretaceous-Tertiary extinction: modeling carbon flux and ecological response. *Paleoceanography* 19, PA1002, doi:10.1029/2002PA000849

Alegret L, Thomas E. 2004. Benthic foraminifera and environmental turnover across the Cretaceous/Paleogene boundary at Blake Nose (ODP Hole 1049C, Northwestern Atlantic). *Palaeogeogr. Palaeoclimatol. Palaeoecol.* 208:59–83

Alvarez LW, Alvarez W, Asaro F, Michel HV. 1980. Extraterrestrial cause for the Cretaceous-Tertiary extinction. *Science* 208: 1095–108

Anderson TF, Arthur MA. 1983. Stable isotopes of oxygen and carbon and their application to sedimentologic and paleoenvironmental problems. In *Stable Isotopes in Sedimentary Geology*, ed. MA Arthur, TF Anderson, IR Kaplan, J Veizer, LS Land, pp. 1-1–1-151. Tulsa: SEPM (Society for Sedimentary Geology) Short Course No. 10.

Arthur MA, Zachos JC, Jones DS. 1987. Primary productivity and the Cretaceous/ Tertiary boundary event in the oceans. *Cretaceous Res.* 8:43–45

Berggren WA, Kent DV, Swisher CC III, Aubry M-P. 1995. A revised Cenozoic geochronology and chronostratigraphy. In *Geochronology, Time Scales and Global Stratigraphic Correlation*, ed. WA Berggren, DV Kent, M-P Aubrey, J Hardenbol, 54:129–212. Tulsa: SEPM (Society for Sedimentary Geology) Special Publication

Bleil U. 1985. The magnetostratigraphy of

Northwest Pacific sediments, Deep Sea Drilling Project Leg 86. *Deep Sea Drilling Proj. Initial Rep.* 6:441–58

Brett R. 1992. The Cretaceous-Tertiary extinction: a lethal mechanism involving anhydrite target rocks. *Geochim. Cosmochim. Acta* 56:3603–6

Brinkhuis H, Bujak JP, Smit J, Versteegh GJM, Visscher H. 1998. Dinoflagellate-based sea-surface temperature reconstructions across the Cretaceous-Tertiary boundary. *Palaeogeogr. Palaeoclimatol. Palaeoecol.* 141:67–83

Broecker WS, Peng T-H. 1982. *Tracers in the Sea.* Lamont-Doherty Geological Observatory. Palisades, NY: Columbia Univ. Press. 620 pp.

Bryan JR, Jones DS. 1989. Fabric of the Cretaceous-Tertiary marine macrofaunal transition at Braggs, Alabama. *Palaeogeogr. Palaeoclimatol. Palaeoecol.* 69:279–301

Cande SC, Kent DV. 1995. Revised calibration of the magnetic polarity timescales for the Late Cretaceous and Cenozoic. *JGR* 100: 6093–95

Chave AD. 1984. Lower Paleocene-Upper Cretaceous Magnetostratigraphy, sites 525, 527, 528, and 529, Deep Sea Drilling Project Leg 74. *Deep Sea Drilling Proj. Initial Rep.* 74: 525–32

Cowie JW, Zieger W, Remane J. 1989. Stratigraphic commission accelerates progress, 1984–1989. *Episodes* 112:79–83

Culver SJ. 2003. Benthic foraminifera across the Cretaceous-Tertiary (K-T) boundary: a review. *Mar. Micropaleontol.* 47:177–226

D'Hondt S. 1994. The evidence for a meteorite impact at the Cretaceous-Tertiary boundary. In *Extinction and the Fossil Record*, ed. E Molina, pp. 75–96. Zaragoza, Spain: Univ. Zaragoza

D'Hondt S, Donaghay P, Zachos JC, Luttenberg D, Lindinger M. 1998a. Organic carbon fluxes and ecological recovery from the Cretaceous-Tertiary mass extinction. *Science* 282:276–79

D'Hondt S, Herbert TD, King J, Gibson C. 1996. Planktic foraminifera, asteroids, and marine production: death and recovery at the Cretaceous-Tertiary boundary. In *New Developments Regarding the K/T Event and Other Catastrophes in Earth History*, ed. GT Ryder, DE Fastovsky, S Gartner, pp. 303–17. Boulder, CO: Geol. Soc. Am.

D'Hondt S, King J, Galbrun B, Bralower TJ. 1998b. Recovery of carbonate accumulation after the Cretaceous/Paleogene impact. *Eos, Am. Geophys. Union Trans.* 79(17):S172

D'Hondt S, Pilson MEQ, Sigurdsson H, Hanson A, Carey S. 1994a. Surface-water acidification and extinction at the Cretaceous-Tertiary boundary. *Geology* 22:983–86

D'Hondt S, Zachos JC. 1993. On stable isotopic variation and earliest Paleocene planktonic foraminifera. *Paleoceanography* 8(4):527–47

D'Hondt S, Zachos JC. 1998. Cretaceous foraminifera and the evolutionary history of planktic photosymbiosis. *Paleobiology* 24: 512–23

D'Hondt S, Zachos JC, Bowring S, Hoke G, Martin M, et al. 2000. Permo/Triassic events and the carbon isotope record of Meishan, China. *Geol. Soc. Am.* 32(7):A368 (Abstr.)

D'Hondt S, Zachos JC, Schultz G. 1994b. Stable isotopic signals and photosymbiosis in Late Paleocene planktic foraminifera. *Paleobiology* 20(3):391–406

Emiliani C. 1980. Death and renovation at the end of the Mesozoic. *Eos* 61:505–6

Erickson DJ III, Dickson SM. 1987. Global trace-element biogeochemistry at the K/T boundary, oceanic and biotic response to a hypothetical meteorite impact. *Geology* 15: 1014–17

Erwin DH. 2001. Lessons from the past: biotic recoveries from mass extinctions. *Proc. Natl. Acad. Sci. USA* 98(10):5399–403

Fastovsky DE, Sheehan PM. 2005. The extinction of the dinosaurs in North America. *GSA Today* 15(3):4–10

Faure G. 1986. *Principles of Isotope Geology.* New York: Wiley & Sons. 2nd. ed.

Fuqua LM, Bralower TJ. 2004. *Evolutionary events and phytoplankton recovery after the K/T mass extinction.* Eos Trans. Am.

Geophys. Union 85(47), Fall Meet. Suppl., (Abstr. PP11B-0570)

Gallagher WB. 2003. Oligotrophic oceans and minimalist organisms: collapse of the Maastrichtian marine ecosystem and Paleocene recovery in the Cretaceous-Tertiary sequence of New Jersey. *Netherlands J. Geosci./Geol. Mijnbouw* 82(3):225–31

Gerstel J, Thunell R, Ehrlich R. 1987. Danian faunal succession; planktonic foraminiferal response to a changing marine environment. *Geology* 15:665–68

Håkansson E, Thomsen E. 1999. Benthic extinction and recovery patterns at the K/T boundary in shallow-water carbonates, Denmark. *Palaeogeogr. Palaeoclimatol. Palaeoecol.* 154:67–85

Hansen TA, Upshaw B III, Kauffman EG, Gose W. 1993. Patterns of molluscan extinction and recovery across the Cretaceous-Tertiary boundary in east Texas; report on new outcrops. *Cretaceous Res.* 14:685–706

Hollis CJ, Rodgers KA, Parker RJ. 1995. Siliceous plankton bloom in the earliest Tertiary of Marlborough, New Zealand. *Geology* 23:835–38

Hollis CJ, Rodgers KA, Strong CP, Field BD, Rogers KM. 2003. Paleoenvironmental changes across the Cretaceous/Tertiary boundary in the northern Clarence valley, southeastern Marlborough, New Zealand. *N. Z. J. Geol. Geophys.* 46:209–34

Holser WT, Schoenlaub H-P, Attrep M, Boeckelmann J, Klein P, et al. 1989. A unique geochemical record at the Permian/Triassic boundary. *Nature* 337:39–44

Houston RM, Huber BT. 1998. Evidence of photosymbiosis in fossil taxa? Ontogenetic stable isotope trends in some Late Cretaceous planktonic foraminifera. *Mar. Micropaleontol.* 34:29–46

Hsü KJ, He Q, McKenzie JA, Weissert H, Perch-Nielsen K, et al. 1982. Mass mortality and its environmental and evolutionary consequences. *Science* 216:249–56

Hsü KJ, McKenzie J. 1985. A "Strangelove" ocean in the earliest Tertiary. The carbon cycle and atmospheric CO_2: natural variations Archean to Present, ed. WS Broecker, ET Sundquist. *Am. Geophys. Union Monogr.* 32:487–92

Huber BT. 1996. Evidence for planktonic foraminifer reworking versus survivorship across the Cretaceous-Tertiary boundary at high latitudes. In *The Cretaceous-Tertiary Event and Other Catastrophes in Earth History*, ed. G Ryder, D Fastovsky, S Gartner, pp. 319–34. Boulder, CO: Geol. Soc. Am. Spec. Pap. 307. 569 pp.

Huber BT, MacLeod KG, Norris RD. 2002. Abrupt extinction and subsequent reworking of Cretaceous planktonic foraminifera across the Cretaceous-Tertiary boundary: Evidence from the subtropical North Atlantic. In *Catastrophic Events and Mass Extinctions: Impacts and Beyond*, ed. C Koeberl, KG MacLeod, pp. 277–89. Boulder, CO: Geol. Soc. Am. Spec. Pap. 356

Jablonski D. 2001. Lessons from the past: evolutionary impacts of mass extinctions. *Proc. Natl. Acad. Sci. USA* 98(10):5393–98

Jablonski D. 2003. The interplay of physical and biotic factors in macroevolution. In *Evolution on Planet Earth: The Impact of the Physical Environment*, ed. A Lister, L Rothschild, pp. 235–52. New York: Academic

Jones DS, Mueller PA, Bryan JR, Dobson JP, Channell JET, et al. 1987. Biotic, geochemical, and paleomagnetic changes across the Cretaceous/Tertiary boundary at Braggs, Alabama. *Geology* 15:311–15

Kaiho K, Lamolda M. 1999. Catastrophic extinction of planktonic foraminifera at the Cretaceous/Tertiary boundary evidenced by stable isotopes and foraminiferal abundance at Caravaca, Spain. *Geology* 27:355–58

Kirchner JW, Weil A. 2000. Delayed biological recovery from extinctions throughout the fossil record. *Nature* 404:177–80

Kitchell JA, Clark DL, Gombos AM. 1986. Biological selectivity of extinction: a link between background and mass extinction. *Palaios* 1:504–11

Kring DA. 2000. Impact events and their effect on the origin, evolution, and distribution of life. *GSA Today* 10(8):1–7

Kroopnick P. 1980. The distribution of ^{13}C in the Atlantic Ocean. *Earth Planet. Sci. Lett.* 49:469–84

Kroopnick P, Deuser WG, Craig H. 1970. Carbon 13 measurements on dissolved inorganic carbon at the North Pacific 1969 Geosecs Station. *J. Geophys. Res.* 75:7668–71

Lewis JS, Watkins GH, Hartman H, Prinn RG. 1982. Chemical consequences of major impact events on earth. In *Geological Implications of Impacts of Large Asteroids and Comets on the Earth*, ed. LT Silver, PH Schultz, 190:215–21. Boulder, CO: Geol. Soc. Am.

Lockwood R. 2003. Abundance not linked to survival across the end-Cretaceous mass extinction: patterns in North American bivalves. *Proc. Natl. Acad. Sci. USA* 100:2478–82

Marshall CR, Ward PD. 1996. Sudden and gradual molluscan extinction in the latest Cretaceous of Western European Tethys. *Science* 274:1360–63

McClure M, Bohonak AJ. 1995. Non-selectivity in extinction of bivalves in the Late Cretaceous of the Atlantic and Gulf Coastal Plain of North America. *J. Evol. Biol.* 8:779–94

Molina E, Arenillas I, Arz JA. 1998. Mass extinction in planktic foraminifera at the Cretaceous/Tertiary boundary in subtropical and temperate latitudes. *Bull. Soc. Geol. Fr.* 169(3):351–63

Norris RD. 1996. Symbiosis as an evolutionary innovation in the radiation of Paleocene planktic foraminifera. *Paleobiology* 22(4):461–80

Officer CB, Drake CL. 1985. Terminal Cretaceous environmental events. *Science* 227:1161–67

Olsson RK, Hemleben C, Berggren WA, Huber BT, eds. 1999. *Atlas of Paleocene Planktonic Foraminifera. Smithsonian Contributions to Paleobiology 85*. Washington, DC: Smithsonian Inst. 252 pp.

Pilson MEQ. 1998. *An Introduction to the Chemistry of the Sea*. Upper Saddle River, NJ: Prentice Hall

Pospichal JJ. 1994. Calcarous nannofossils at the K-T boundary, El Kef: no evidence for stepwise, gradual or sequential extinctions. *Geology* 22:99–102

Russell D, Tucker WH. 1971. Supernovae and the extinction of the dinosaurs. *Nature* 229:553–54

Schmidt DN, Thierstein HR, Bollmann J. 2004. The evolutionary history of size variation of planktic foraminiferal assemblages in the Cenozoic. *Palaeogeogr. Palaeoclimatol. Palaeoecol.* 212:159–80

Sepkoski JJ Jr. 1993. Ten years in the library: new data confirm paleontological patterns. *Paleobiology* 19(1):43–51

Sepkoski JJ Jr. 1996. Patterns of Phanerozoic extinctions: a perspective from global databases. In *Global Events and Event Stratigraphy*, ed. OH Walliser, pp. 35–52. Berlin: Springer

Sepkoski JJ Jr. 1998. Rates of speciation in the fossil record. *Proc. R. Soc. London. Ser. B* 353:315–26

Shackleton NJ and members of the shipboard scientific party. 1984. Accumulation rates in Leg 74 sediments, *Deep Sea Drilling Proj. Initial Rep.* 74:621–44

Sheehan PM, Hansen TA. 1986. Detritus feeding as a buffer to extinction at the end of the Cretaceous. *Geology* 14:868–70

Sigurdsson H, Leckie RM, Acton GD, Abrams LJ, Bralower TJ, et al. 1997. *Ocean Drilling Program Proc. Initial Rep.* Vol. 165. College Station, TX: ODP

Smit J. 1977. Discovery of a planktonic foraminiferal association between the Abathomphalus mayaroensis zone and the globigerina eugubina zone at the Cretaceous/Tertiary boundary in the Barranco del Gredero (Caravaca, SE Spain), A preliminary report I and II. *Koninkl. Nederlandse Akad. Wetensch. Proc. Ser. B* 80:280–301

Smit J. 1999. The global stratigraphy of the Cretaceous Tertiary boundary impact ejecta. *Annu. Rev. Earth Planet. Sci.* 27:75–113

Smit J, Hertogen J. 1980. An extraterrestrial event at the Cretaceous-Tertiary boundary. *Nature* 285:198–200

Smith AB, Jeffrey CH. 1998. Selectivity of extinction among sea urchins at the end of the Cretaceous period. *Nature* 392:69–71

Stott LD, Kennett JP. 1989. New constraints on early Tertiary palaeoproductivity from carbon isotopes in foraminifera. *Nature* 342(6249):526–29

Toon OB, Pollack JB, Ackerman TP, Turco RP, McKay CP, Liu MS. 1982. Evolution of an impact-generated dust cloud and its effects on the atmosphere. In *Geological Implications of Impacts of Large Asteroids and Comets on the Earth*, ed. LT Silver, PH Schultz, 190:187–200. Boulder, CO: Geol. Soc. Am.

Troelsen JC. 1957. Some planktonic foraminifera of the type Danian and their stratigraphic importance. In *Studies in Foraminifera*, ed. AR Loeblich Jr, and collaborators, 215:125–32. Washington, D.C.: US Natl. Mus. Bull.

Wei W, Pospichal JJ. 1991. Danian calcareous nannofossil succession at Site 738 in the southern Indian Ocean. *Ocean Drill. Program Proc. Sci. Results.* 119:495–12

Wendler J, Willems H. 2002. Distribution pattern of calcareous dinoflagellate cysts across the Cretaceous-Tertiary boundary (Fish Clay, Stevns Klint, Denmark): implications for our understanding of species-selective extinction. In *Catastrophic Events and Mass Extinctions: Impacts and Beyond*, ed. C Koeberl, KG MacLeod, 356:265–75. Boulder, CO: Geol. Soc. Am.

Zachos JC, Arthur MA. 1986. Paleoceanography of the Cretaceous-Tertiary boundary event: inferences from stable isotopic and other data. *Paleoceanography* 1(1):5–26

Zachos JC, Arthur MA, Dean WE. 1989. Geochemical evidence for suppression of pelagic marine productivity at the Cretaceous/Tertiary boundary. *Nature* 337(5):61–64

Zachos JC, Arthur MA, Thunell RC, Williams DF, Tappa EJ. 1985. Stable isotope and trace-element geochemistry of carbonate sediments across the Cretaceous/Tertiary boundary at Deep Sea Drilling Project Hole 577, Leg 86. *Deep Sea Drilling Proj. Initial Rep.* 86:513–32

Zachos JC, Aubry M-P, Berggren WA, Ehrendorfer T, Heider F. 1992. Magneto-biochemostratigraphy across the Cretaceous/Paleogene boundary at ODP Site 750A, Southern Kerguelen Plateau. *Ocean Drilling Program Proc. Sci. Results* 120(2):961–77

Annu. Rev. Ecol. Evol. Syst. 2005. 36:319–44
doi: 10.1146/annurev.ecolsys.36.102003.152614
First published online as a Review in Advance on August 17, 2005

LANDSCAPE ECOLOGY: What Is the State of the Science?

Monica G. Turner

*Department of Zoology, University of Wisconsin, Madison, Wisconsin 53706;
email: turnermg@wisc.edu*

Key Words   disturbance, fragmentation, spatial heterogeneity, spatial pattern,
succession

■ **Abstract**   Landscape ecology focuses on the reciprocal interactions between spatial pattern and ecological processes, and it is well integrated with ecology. The field has grown rapidly over the past 15 years. The persistent influence of land-use history and natural disturbance on contemporary ecosystems has become apparent. Development of pattern metrics has largely stabilized, and they are widely used to relate landscape pattern to ecological responses. Analyses conducted at multiple scales have demonstrated the importance of landscape pattern for many taxa, and spatially mediated interspecific interactions are receiving increased attention. Disturbance remains prominent in landscape studies, and current research is addressing disturbance interactions. Integration of ecosystem and landscape ecology remains challenging but should enhance understanding of landscape function. Landscape ecology should continue to refine knowledge of when spatial heterogeneity is fundamentally important, rigorously test the generality of its concepts, and develop a more mechanistic understanding of the relationships between pattern and process.

INTRODUCTION

Scientists have observed and described heterogeneity (complexity or variability in a system property of interest in space and time) (Li & Reynolds 1995) in ecological systems for a very long time. However, an explicit focus on understanding spatial heterogeneity—revealing its myriad abiotic and biotic causes and its ecological consequences—emerged in the 1980s as landscape ecology developed and spatial data and analysis methods became more widely available. Since then, progress in landscape ecology has been substantial and rapid, and its concepts and methods are now widely used in many branches of ecology. Landscape ecological approaches are not limited to land, but are also applied in aquatic and marine ecosystems (e.g., Bell et al. 1999, Ward et al. 2002). Research in landscape ecology has enhanced understanding of the causes and consequences of spatial heterogeneity and how they vary with scale and has influenced management of both natural and human-dominated landscapes.

1543-592X/05/1215-0319$20.00

Most generally, a landscape is an area that is spatially heterogeneous in at least one factor of interest (Turner et al. 2001). This flexible definition is applicable across scales and adaptable to different systems. Landscape ecology, a term coined by the German biogeographer Carl Troll and elaborated in 1950 (Troll 1950), arose from the European traditions of regional geography and vegetation science and was motivated by the new perspective offered by aerial photography. Landscape ecology has since been defined in various ways (Pickett & Cadenasso 1995, Risser et al. 1984, Turner 1989, Turner et al. 2001, Urban et al. 1987), but common to all definitions is a focus on understanding the reciprocal interactions between spatial heterogeneity and ecological processes. Nonetheless, landscape ecology has developed with two distinct approaches that, although not mutually exclusive, have led to some confusion about its scope. Landscape ecology often emphasizes large areas or regions and includes humans and their activities, which reflects a strong European tradition. The focus of landscape ecology is more anthropocentric in Europe and aligned closely with land planning (e.g., Bastian 2001, Opdam et al. 2002). However, landscape ecology also encompasses the causes and consequences of spatial pattern at variable spatial scales defined by the organism or process of interest, which reflects traditions in North America and Australia. Thus, streambeds may be considered landscapes for stream invertebrates (Palmer et al. 2000), and spatial heterogeneity in soils may be characterized at very fine scales relevant to individual plants or even microbes. These diversities in approach and tradition are both contrasting and complimentary (Wu & Hobbs 2002) and an inherent part of the field.

The rapid development of landscape ecology in the past two decades suggests that a review of the field is timely, albeit daunting. The number of landscape ecology articles published each year has increased exponentially since the early 1990s (Turner 2005). Reviews have been published for particular areas of landscape ecology, such as quantitative analyses of spatial pattern (e.g., Gustafson 1998, Haines-Young & Chopping 1996, Hargis et al. 1998, Li & Reynolds 1995) and disturbance dynamics (e.g., Foster et al. 1998, Perry 2002), and several synthetic articles have catalyzed progress (e.g., Pickett & Cadenasso 1995, Wiens 1999, Wu & Hobbs 2002). An edited volume of early foundation papers in landscape ecology provides access to the intellectual foundations of the field and lists the numerous books on landscape ecology published in the past decade (Wiens et al. 2005). Here, I emphasize developments in landscape ecology since my 1989 review (Turner 1989) and use a similar organization for context and comparison. My focus is primarily on contributions of landscape ecology to basic ecological understanding rather than to land management. I identify general concepts, highlight contemporary areas of inquiry, and suggest future research directions.

Several general themes are implicit throughout this review. First, understanding scale (Levin 1992, Wiens 1989) has been and remains closely aligned with landscape ecology. As ecology moved to broader scales and embraced heterogeneity, an understanding of the profound effects of grain, extent, and level of organization on analyses was crucial. Second, landscape ecology addresses both basic and applied questions and moves easily between these realms; indeed, the demand for landscape science in resource management has been quite high (Liu & Taylor 2002). Third, the use of multiple approaches, including historical or remotely sensed data, field measurements, experimental model systems, and simulation modeling, is the norm in landscape studies; the interplay of models and data has been characteristic of the field.

CAUSES OF LANDSCAPE PATTERN

Landscape patterns result from complex relationships among multiple factors, many of which are well known. The abiotic template includes climate, which strongly controls biogeographic patterns, and landform, which produces patterns of physical relief and soil development (e.g., Parker & Bendix 1996). Biotic interactions—such as competition, herbivory, and predation—and the role of keystone species or ecosystem engineers are played out on the abiotic template and influence species assemblages. Disturbance and succession are key drivers of spatial and temporal heterogeneity; many disturbances have a strong climate forcing and may interact with landform. Finally, the ways in which humans use the land are key drivers of landscape pattern (Riitters et al. 2002). These causes have been well described for many systems, yet explaining and predicting landscape patterns remains surprisingly difficult. Current questions focus on understanding landscape legacies and multiple drivers and their interactions, and on forecasting future landscapes.

Landscape Legacies

What aspects of current landscape patterns are explained by past land use or disturbance, and for how long do such influences persist? All landscapes have a history. Paleoecologists have elucidated long-term changes in the biota, but the rise of environmental history (e.g., Cronon 1983, Russell 1997) and recognition that history might explain contemporary patterns emerged more recently (e.g., Foster 1992, but see also Wells et al. 1976). In areas of northeastern France deforested during the Roman occupation and farmed during 50 to 250 AD, species richness and plant communities still varied—2000 years later—with the intensity of former agriculture (Dupouey et al. 2002). In central Massachusetts, historical land use predicted forest overstory composition well in 1992, even though other major natural disturbances occurred after land use ceased (Motzkin et al. 1999). The persistent influence of land-use history in explaining the vegetation and biogeochemical characteristics of contemporary ecosystems has become increasingly apparent (Compton & Boone 2000, Foster 2002, Goodale & Aber 2001).

Natural disturbances can also leave legacies that persist for decades to centuries. For example, stand-replacing fire is the dominant disturbance in the coniferous forest landscape of Yellowstone National Park, Wyoming. Using a chronosequence approach, Kashian et al. (2005a,b) found detectable effects of historic fires on stand density and growth rate for nearly two centuries following those fires. In tropical forests of Puerto Rico, current vegetation patterns were influenced by both historical land use and hurricanes (Foster et al. 1999). Thus, the legacies of land use and disturbance can be remarkably persistent, and integrating this history with current understanding remains an important goal. We must consider the future legacies of today's landscape patterns: What variables will be most affected, and for how long? Enhanced understanding of long-term landscape development is important for both explaining the present and looking to the future.

Multiple Drivers and Their Interactions

Understanding the relative importance of different factors (and their roles at multiple scales) in producing landscape patterns is another important challenge. Most studies have focused on a dominant driver rather than on the multiple drivers that together generate spatial pattern; interactions among the varied drivers remain poorly understood, in part because they are difficult to study. Urban et al. (2002) addressed landscape patterns of vegetation in Sequoia-Kings Canyon National Park, California. These authors recognized explicitly that spatial autocorrelation in ecological data, coupled with strong patterns of correlation among environmental factors (such as the gradients governed by elevation), makes the varied agents that produce vegetation patterns difficult to disentangle.

A number of studies have related landscape patterns to variable sets that include both biophysical and socioeconomic factors or their surrogates. Interactions between land ownership and landscape position have emerged as strong determinants of land-cover patterns and changes (Mladenoff et al. 1993, Spies et al. 1994, Wear & Bolstad 1998). Black et al. (2003) assessed the role of several economic, demographic, cultural, climatic, topographic, and geologic factors in forest spatial-pattern changes (from the 1930s to the 1990s) across an 800,000-km^2 area in the interior northwest United States. Their results nicely illustrated how social-system factors are imposed on biophysical factors to generate pattern change in the study of landscape. Furthermore, the scales of response and explanatory variables often did not correspond; broad-scale factors related to land-ownership systems, economic market structures, and cultural-value systems appeared in all significant models, regardless of the response scale, and biophysical parameters related to growing conditions at the site moderated or exacerbated changes (Black et al. 2003).

Future Landscape Patterns

Forecasting future landscape patterns remains a challenging task in which the suite of drivers of landscape pattern and their interactions must be considered. The exploration of alternative scenarios and their ecological implications is particularly important in applied landscape ecology (e.g., White et al. 1997). Empirical models that use a set of independent variables to explain past land-use changes have been informative, although extrapolation of those models to the future is problematic.

A widely used approach is based on logistic regression in which the likelihood of a particular land-cover transition is estimated and simulated into the future (Wear et al. 1996, Wear & Bolstad 1998). Because the transition probabilities in these models can be influenced by many factors (e.g., elevation, distance to roads or market center, population density, and patch size), they potentially have better predictive power than simple Markov models when run in a spatial framework.

Spatially explicit simulation models are the primary tools for exploring plausible future landscape patterns and processes. For example, interactions among fire, windthrow, forest harvesting, and tree-species dynamics were explored for a 500,000-ha heterogeneous landscape of the upper Midwest United States, by use of the spatially explicit, stochastic model LANDIS (He & Mladenoff 1999). Costanza et al. (2002) developed a spatially explicit, process-based model of the 2352-km^2 Patuxent River watershed in Maryland. The model addressed the effects of both the magnitude and spatial patterns of human settlements and agricultural practices on hydrology, plant productivity, and nutrient cycling in the landscape. Such broad-scale, spatially explicit models highlight the complex nature of landscape responses. Balancing the trade-offs between the simplicity of general models and the complexity of more realistic spatial models remains a challenge.

QUANTIFYING LANDSCAPE PATTERNS

Landscape Metrics

The quantification of spatial heterogeneity is necessary to elucidate relationships between ecological processes and spatial patterns; thus, the measurement, analysis, and interpretation of spatial patterns receive much attention in landscape ecology. A wide array of metrics for landscape composition (what and how much is present, such as the number and amount of different habitat types) and configuration (how those classes are arranged spatially) were developed for categorical data. Excellent software packages are readily available; FRAGSTATS (McGarigal & Marks 1995) is used most widely. Some metrics have also been integrated into existing geographic information system (GIS) software (e.g., Patch Analyst in Arc/View). Importantly, spatial pattern analysis is a tool rather than a goal of its own, and the objectives or questions driving any analysis must be specified a priori; this specification must include the qualities of pattern to be represented and why.

A variety of issues associated with interpreting pattern metrics are now well understood by practitioners (Gustafson 1998, Haines-Young & Chopping 1996, Li & Wu 2004, Turner et al. 2001). For example, different results are obtained by analyzing different classifications of the same data (Gustafson 1998) or using different patch-definition rules. Many metrics are sensitive to changes in the grain size (spatial resolution) of the data or the extent (area) of the study landscape (e.g., Wickham & Riitters 1995), and numerous correlations occur among landscape-pattern indices (Cain et al. 1997, Riitters et al. 1995). Composition, particularly the proportions of cover types on the landscape, influences the values of many metrics (Gardner et al. 1987, Gustafson & Parker 1992, Tischendorf 2001). No single metric can adequately capture the pattern on a given landscape, and several suggestions have been made for a meaningful set of metrics that minimize redundancy while capturing the desired qualities (Riitters et al. 1995). Comparisons made among landscapes, with different data types or through time must now routinely account for these known complexities.

Despite numerous calls for improved linkages, the relationship between processes that create patterns and the patterns themselves still is not readily apparent. Krummel et al. (1987) suggested that simple, rectilinear shapes of forest patches indicated human influences in shaping landscape patterns. Numerous authors have shown that dispersed clear-cuts in forested landscapes produce distinctive landscape patterns with high patch and edge densities and small patch areas (e.g., Spies et al. 1994). The habitat loss and fragmentation associated with human land use in many regions is also well described in landscape ecology and conservation biology (e.g., Heilman et al. 2002, Riitters et al. 2000, Saunders et al. 1991). Nonetheless, no general framework exists that permits a particular spatial pattern to be linked to specific generating factors. Current research is developing a more rigorous statistical interpretation of spatial pattern analysis that rekindles the attempt to link processes with patterns and addresses several persistent challenges.

Building upon the tradition of neutral landscape models (Gardner et al. 1987, With & King 1997), Fortin et al. (2003) explored the spatial realization of simple stochastic processes on a landscape and interpreted the resulting patterns using landscape metrics. Landscape patterns were generated by independent variation of two parameters: one that represents composition (the amount of a given habitat) and one that represents configuration (its arrangement, represented here by the amount of spatial autocorrelation) of a single habitat type. Inspection of the pair-wise scatterplots between seven landscape metrics revealed that many relationships were not linear, and several were not even monotonic (Figure 1). Thus, the expectation of linear relationships among landscape metrics that has been implicit in most previous studies may be misleading.

The statistical properties and behavior of many pattern metrics remain poorly understood. Because the distributions of landscape metrics are not known, expected values and variances are not available for statistical comparisons to be made among multiple observations of a particular metric (Li & Wu 2004, Remmel & Csillag 2003, Turner et al. 2001). Remmel & Csillig (2003) used the approach of Fortin et al. (2003) to develop neutral landscape models based on composition and configuration. They generated confidence intervals for landscape metric values by collecting their empirical distribution over a series of landscapes that were simulated using values of the two parameters estimated from the observed landscape. If the confidence intervals between two landscapes overlap, then the landscapes do not differ for the given metric (Figure 2). This approach lends much greater rigor to studies that seek to identify differences among landscapes or to detect changes through time in a given landscape. As the authors conclude, testing whether two landscape metrics differ significantly should become a standard approach.

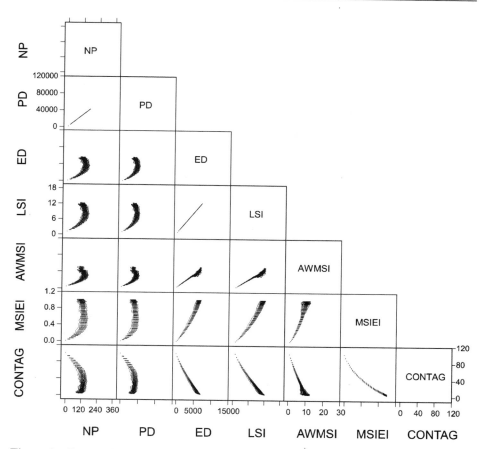

Figure 1   Scatter plots of seven landscape metrics derived from 1000 simulated binary landscapes with high autocorrelation. Abbreviations: NP, number of patches; PD, patch density; ED, edge density; LSI, landscape shape index; AWMSI, area-weighted mean shape index; MSIEI, modified Simplson's evenness index; CONTAG, contagion. The relationships are not monotonic and suggest that relationships among landscape metrics may be nonlinear. Reproduced with permission from Fortin et al. (2003).

Despite their limitations, landscape metrics remain widely used and useful. Mapped distributions of metric values (rather than the original categorical data from which they were derived) can also offer new perspectives on spatial variation across regions (Riitters et al. 2000). For example, replicate locations that share some qualities of spatial pattern are often difficult to identify; mapped distributions of metrics can be used to stratify sites for empirical study appropriately when some aspect of landscape pattern is an independent variable. Mapped patterns may also identify higher-order information not easily discernible from tabular summaries (Riitters et al. 2000).

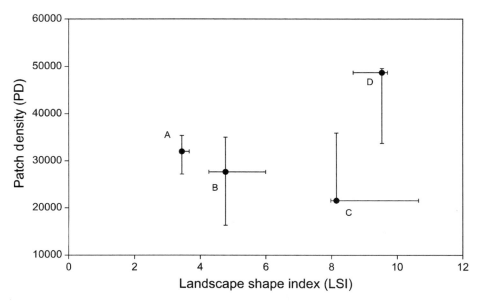

Figure 2   The 99% statistical confidence intervals for measures of patch density (PD) and landscape shape index (LSI) for four landscapes (*A–D*) near Prince George, British Columbia, Canada. *Solid circles* are actual values measured from each landscape, and confidence intervals are derived from 100 realized simulations. PD did not differ significantly among the landscapes, but LSI discriminated landscapes. Reproduced with permission from Remmel & Csillag (2003).

Spatial Statistics

Methods that treat continuous rather than discrete variation in space are receiving increased attention; the landscape metrics described above use categorical data, but spatial heterogeneity may also be continuous. Spatial statistics (Rossi et al. 1992) use the continuous distribution of a quantity of interest and do not require categorization. To illustrate the distinction, forest cover could be represented categorically (as forest or nonforest) or continuously (by tree density). Ecosystem process rates (e.g., net ecosystem production, nitrogen or carbon mineralization, and respiration) also vary continuously and, thus, may be especially amenable to analysis using spatial statistics. These methods do not depend on patch definitions or boundaries; however, additional assumptions, such as stationary of variance in the data or isotropy, may apply.

Spatial statistics are applied somewhat less in landscape ecology than are the methods based on discrete space, but they serve several important purposes. First, the spatial structure (i.e., autocorrelation) of a variable might be quantified using spatial statistics so that sampling or data analyses can avoid locations that are spatially autocorrelated or build that structure into the study. Second, variability and the scale of spatial structure can serve as the response of interest, and spatial statistics offers efficient sampling designs to assess this (e.g., Burrows et al. 2002). However, such studies are still far from routine. Pastor et al. (1998) tested three geostatistical models of the spatial distribution of available browse, annual browse consumption, conifer basal area, and soil-nitrogen availability on Isle Royale, Michigan. Their results suggested that dynamic interactions between moose foraging and plant communities produce characteristic spatial patterns of vegetation and soil properties. For the Luquillo Experimental Forest, Puerto Rico, Wang et al. (2002) examined the spatial correlations of soil properties and environmental factors to better understand the controls on biogeochemical processes within ecosystems. They hypothesized varying degrees of spatial structure in soil organic carbon, soil moisture, and soil bulk density along gradients of elevation, slope, and aspect. Cross-correlograms indicated that soil organic carbon was correlated positively with elevation at separation distances less than 3000 m and negatively at separation distances greater than 6000 m. Fraterrigo et al. (2005) also used spatial structure as a response variable; they hypothesized and detected a change in the spatial structure of soil nutrients with historic land use. Results of such analyses may elucidate mechanisms that underpin observed patterns or suggest relationships between environmental heterogeneity and process rates of interest. Nonetheless, use of spatial statistics as responses in ecological studies still presents some challenges. For example, interpretation of semivariograms calculated for replicated spatial units is not straightforward and neither is comparison of results from different models (e.g., spherical, sinusoidal, and exponential).

Spatial statistics also offer methods for interpolating spatial patterns from point data. Kriging and cokriging, which includes environmental covariates, are used to predict values in locations where measurements have not been made. However, when Bolstad et al. (1998) compared methods for predicting vegetation patterns throughout a basin, they found that multiple regression may be stronger than cokriging if the relationships between predictor and response variables are understood.

Because the data in landscape studies are almost always spatial, spatial statistics can and should be used in conjunction with classical statistics, such as regression and analysis of variance, to determine and correct for spatial autocorrelation of errors (residuals). Statistical software packages have incorporated methods to detect and correctly model the spatial covariance structure of data, and ecological studies are beginning to implement these methods (e.g., Schwartz et al. 2003). The assumption of independent errors is important in classical statistics, and failure to account for spatial autocorrelation may lead to false conclusions (Lichstein et al. 2002).

ORGANISMS IN HETEROGENEOUS LANDSCAPES

Populations exist in spatially heterogeneous environments, and the review by Wiens (1976) may mark the beginning of a landscape approach to population dynamics. How interactions within and among populations create spatial patterns in species distributions is well developed within population ecology. For example, competition and predator-prey dynamics may produce spatial patterns in the distribution of organisms even when the underlying environment is homogeneous (Durrett & Levin 1994). How organisms create spatial patterns through spatially explicit feeding relationships and physical alterations of the environment, along with how populations respond to complex patterns and actual landscapes, is addressed in landscape ecology. Considerable overlap occurs with metapopulation biology (Hanski & Gilpin 1997) in questions and approaches.

Effects of Organisms on Landscape Heterogeneity

Although the response of organisms to landscape heterogeneity dominates research on organism-landscape interactions, the role of the biota in creating heterogeneity has also been recognized. "Ecosystem engineers" (Jones et al. 1997) are noted as key sources of heterogeneity in a variety of systems. For example, bison (*Bison bison*) drive heterogeneity patterns within the North American prairie (Knapp et al. 1999). The nightly feeding forays of hippopotami (*Hippopotamus amphibius*) create a maze of trails and canals that are movement corridors for water as well as many other species (Naiman and Rogers 1997). Despite these key examples, the role of organisms as sources of spatial heterogeneity has been somewhat understudied in landscape ecology.

Effects of Landscape Heterogeneity on Organisms

Landscape ecologists emphasize how organisms use resources that are spatially heterogeneous and how they live, reproduce, disperse, and interact in landscape mosaics. The context for much of this work has been to understand how altered landscape patterns affect the distribution, persistence, and abundance of species, often in the face of land management controversies [e.g., Northern Spotted Owls (*Strix occidentalis caurina*) (Murphy & Noon 1992)]. Effects of habitat loss and fragmentation have received much attention (e.g., Andrén 1994, Fahrig 2003, Haila 2002, Saunders et al. 1991).

Much has been learned from studies that have evaluated factors that explain variation in the presence or abundance or organisms in the landscape. Patch size has a strong effect on edge and interior species but is negligible for generalist species (Bender et al. 1998). However, local habitat conditions may be inadequate to explain species presence or abundance; a significant effect of boundary shape or characteristics of the surrounding landscape—usually referred to as landscape context (Mazerolle & Villard 1999)—may be present. For example, empirical studies of butterfly taxa in landscapes with naturally isolated meadows demonstrated that the matrix that surrounds patches could influence their effective isolation (Ricketts 2001). In an experiment that controlled for patch area, Tewksbury et al. (2002) found that pollination and seed dispersal, two key plant–animal interactions, were facilitated by the presence of corridors that connect habitat patches. Murphy & Lovett-Doust (2004) argue for an integration of metapopulation and landscape-ecological approaches for understanding regional dynamics in plants, emphasizing notions of connectivity and context to describe components of variability in the landscape from a species-specific perspective.

Analyses conducted at multiple scales have demonstrated the importance of landscape context for a wide range of taxa (e.g., Lindenmayer et al. 1999, Pearson et al. 1995, Steffan-Dewenter et al. 2002, Stoner & Joern 2004), although the influence may be less if the focal habitat is abundant and well connected (e.g., Miller et al. 2004a). Many studies have also demonstrated that habitat connectivity is scale dependent; that is, whether a given pattern of habitat is connected depends on the mobility of the species and the pattern of the habitat (Goodwin & Fahrig 2002, Vos et al. 2001). Organisms may respond to multivariate habitat heterogeneity at multiple scales, and identification of the factors and scales that best explain variation in the presence or abundance of organisms remains a key goal in landscape ecology.

Disentangling the effects of landscape composition (what and how much is there) and landscape configuration (how is it spatially arranged) on populations is an important area of current research (Fahrig 1997, McGarigal & Cushman 2002). In their review of 134 published fragmentation studies, McGarigal & Cushman (2002) concluded that the ecological mechanisms and effects of habitat fragmentation on populations remain poorly understood. Evidence is mounting for a primary effect of composition and secondary effect of configuration. Field studies of forest-breeding birds conducted in 94 landscapes of 10×10 km each found a consistent positive relationship between forest cover and the distribution of forest-breeding birds but weaker and variable effects of forest fragmentation (Trzcinski et al. 1999). In a study of the incidence of herbaceous species in deciduous forests of south Sweden, Dupré & Ehrlén (2002) found that habitat quality was more important than habitat configuration. Moreover, the importance of habitat configuration varied with life history; species that were habitat specialists and clonal perennials that produced fewer seeds were more likely to be affected negatively by patch isolation. Animal-dispersed species were more negatively affected by small stand size than were species dispersed by other mechanisms (Dupré & Ehrlén 2002).

Simulation studies also suggest that changes in landscape composition are likely to have a greater effect on population persistence than are changes in landscape configuration. Fahrig & Nuttle (2005) hypothesize that landscape configuration will be important only if configuration has a large effect on among-patch movements and among-patch movements have a large effect on population survival. Results of a modeling study by Flather & Bevers (2002) found that, over a broad range of habitat amounts and arrangements, population size was largely determined by the abundance of habitat. However, habitat configuration became important in landscapes with low habitat abundance, in which dispersal mortality became important. King & With (2002) obtained similar results in which spatial pattern was important for poorly dispersed species that occurred in landscapes with low habitat abundance.

Landscape ecology has also focused attention on developing more sophisticated habitat assessments for the distribution of biota. One approach maintains the simple categorization of suitable versus unsuitable habitat but reassesses the landscape for different taxa using rules and scales appropriate for each species or functional group (e.g., Addicott et al. 1987, Pearson et al. 1999). Knight & Morris (1996) used evolutionary theories to document how density-dependent habitat selection and habitat variation could be applied to identify habitats in landscapes. Statistical methods such as resource-selection functions (Manly et al. 2002) that are based on logistic regression provide multivariate and continuous assessments of habitat selection by different taxa that can be evaluated across a range of scales. These analyses employ a used versus available design and are frequently conducted across multiple scales (e.g., Boyce et al. 2003). Studies of this sort clearly demonstrate that the same landscape may look very different to different species and underscore the importance of an organism-centered view of landscape heterogeneity (Wiens 1989).

Recent studies identify important situations in which the patch-based framework simply does not apply and suggest the need for a broader conceptual framework of spatial pattern. For example, the dendritic metapopulations that characterize fish and other species constrained to disperse within river-creek systems are not well represented by either a linear or a two-dimensional representation of spatial structure and metapopulation dynamics (Fagan 2002). Fagan (2002) combined a simple geometric model with a metapopulation model and empirical data to explore the consequences of dendritic landscapes. Depending on dispersal details, the connectivity patterns of dendritic landscapes could either enhance or reduce metapopulation persistence compared with linear systems. Furthermore, the specific location of fragmentation events becomes especially important in the dendritic systems.

A recent call for the integration of landscape ecology and population genetics (Manel et al. 2003) suggests opportunities for new insights about how geographical and environmental features structure genetic variation and for reconstruction of the spatial movements and spread of populations. In particular, landscape genetics may yield new insights regarding the spread of invasive species and native species, such as top predators, that are recovering from earlier extirpation and dispersing in heterogeneous landscapes (Lucchini et al. 2002, Reuness et al. 2003) or responding to landscape change (e.g., Keyghobadi et al. 1999).

The spatial implications of trophic cascades suggest important effects of spatial heterogeneity on species interactions. In fragmented forests of the Pacific Northwest, elevated densities of deer mice (*Peromyscus maniculatus*) in clear-cuts were associated with reduced recruitment of trillium (*Trillium ovatum*) because of increased seed predation (Tallmon et al. 2003). In a sophisticated study of predator-prey dynamics, With et al. (2002) determined how landscape structure affected the ability of two species of ladybird beetle (Coleoptera: Coccinellidae) to track aphid populations in experimental landscapes that differed in the abundance and fragmentation of red clover (*Trifolium pratense*). A compelling finding from this study was that thresholds in landscape structure can be perpetuated across trophic levels, producing similar thresholds in the distribution of pest populations and suggesting a mechanistic link between individual movements and population-level phenomena that affect predator-prey interactions in fragmented landscapes. The effects of predator-herbivore-plant relationships on spatial variability in plant communities is also intriguing. Top predators may influence their herbivore prey populations numerically, by reducing population size, or behaviorally, by influencing patterns of habitat use. When spatially variable, these top-down effects may ultimately influence the landscape vegetation patterns (Schmitz et al. 2000).

More broadly, need exists for addressing community dynamics in heterogeneous landscapes. Opdam et al. (2003) noted that a major gap in studies of population persistence in heterogeneous landscapes is the lack of methods to transfer studies on single species to generalized knowledge about the relation between landscape pattern and biodiversity. Most studies have indeed focused on single species, or perhaps functional groups, yet understanding species assemblages, especially in changing landscapes, bears further study.

LANDSCAPE HETEROGENEITY AND DISTURBANCE

Studies of disturbance and succession continue to generate new understanding about the interactions between ecological processes and landscape pattern. A disturbance is "any relatively discrete event in time that disrupts ecosystem, community, or population structure and changes resources, substrate availability, or the physical environment" (Pickett & White 1985). Disturbances often result in "open space" and, through their gradients of severity, introduce complex spatial heterogeneity. Furthermore, the occurrence or effects of disturbance may depend on the system's state before the disturbance occurred. Thus, disturbances are particularly interesting in landscape ecology because they both respond to and create spatial heterogeneity at multiple scales.

Enhanced understanding of landscape disturbance dynamics underlies the important conceptual shift that recognized dynamic equilibria and nonequilibrium systems in ecology (Perry 2002, Turner et al. 1993, Wu & Loucks 1995). Indeed, Wu & Loucks (1995) argued that the past inability to incorporate heterogeneity and multiple scales into quantitative expressions of stability led, in part, to the failure of the classical equilibrium paradigm in ecology. From a nonequilibrium perspective, stochastic events such as disturbance alter system state and trajectory and are integral to the system. Equilibrium is but one of several outcomes, and it may be apparent only at certain scales (Turner et al. 1993). Empirical studies of several landscapes found marked fluctuations in landscape composition (e.g., Baker 1989), particularly when disturbances were large and infrequent (Moritz 1997, Turner et al. 1993). The steady-state mosaic, in which sites are in different stages of succession but the landscape proportions of successional stages remain constant, was found to apply only in some cases.

Effects of Disturbance on Landscape Heterogeneity

Disturbances produce a mosaic of disturbed versus undisturbed areas and complex spatial variation in severities within disturbed areas. Foster et al. (1998) used several examples from large, infrequent disturbances to illustrate the diverse spatial patterns that can result. Single-disturbance types have been reasonably well studied, but less is known about interacting disturbances or whole disturbance regimes; this feature is an important thrust of current research. Paine et al. (1998) suggested that particular co-occurrences or sequences of different disturbances could produce ecological surprises or qualitative shifts in the system state. For example, the composition of the southern boreal forest changed substantially in a century in response to climate-driven changes in fire frequency, forest fragmentation, and logging. They suggest that understanding the ecological synergisms among disturbances is basic to future environmental management (Paine et al. 1998).

Studies designed to understand the combined contingent effects of multiple disturbances are promising. Interactions between fire and spruce-beetle (*Dendroctonus rufipennis*) outbreaks over more than a century were studied in a 2800-km^2 landscape by Bebi et al. (2003). Results showed that fire history had the greatest effect on stand susceptibility to spruce-beetle outbreak. Radeloff et al. (2000) also found that interactions between disturbances, here jack pine–budworm defoliation and salvage logging, substantially changed landscape heterogeneity in the pine barrens of northwest Wisconsin, and they hypothesized that the presettlement landscape patterns were shaped by interactions between insect defoliators and fire.

Changing disturbance regimes is another important area of current research. Because disturbances are such important agents of pattern formation in landscapes, changes in their frequency, intensity, or extent may well alter landscape structure. However, how much do disturbance regimes need to shift before landscape patterns are altered qualitatively? The answer is not known, yet it assumes increasing importance in the context of global change. Many disturbances, such as fires, floods, and hurricanes, have a strong climate forcing, and development pressure is increasing in many disturbance-prone sites (e.g., Hansen et al. 2002). How altered landscapes will themselves influence disturbance regimes is not known.

Landscape management often relies, either implicitly or explicitly, on an understanding of disturbance regimes. Management may attempt to mimic spatial and temporal patterns of disturbance or seek to maintain or return a landscape to its historic range of variability (HRV) (Landres et al. 1999). Considerable discussion has occurred about the use of the timing and spatial patterns of natural disturbances as a model for human activities (e.g., Attiwill 1994). This approach implicitly assumes that ecological processes will be better maintained in this way, and current management of the Ontario Crown Lands, Canda, offers an excellent example of implementing these concepts at a broad scale (Perera et al. 2000).

Effect of Landscape Heterogeneity on Disturbance

Assessment of the role of landscape heterogeneity on the spread of disturbance was identified at an early workshop as one of the key questions in landscape ecology (Risser et al. 1984) and was the theme of the first U.S. landscape ecology symposium. A number of studies have now documented significant influences of landscape heterogeneity on the spread or effects of disturbances. Effects of hurricanes, wind events, and fires can vary with spatial location on the landscape; researchers have frequently found a strong influence of landform on these effects. For example, the severity of hurricanes on vegetation varies with the exposure of the sites (Boose et al. 1994). In vast, relatively unlogged forests of coastal Alaska, Kramer et al. (2001) documented a spatially predictable windthrow gradient that contrasted sharply with the prior emphasis on gap-phase disturbances in these forests. In the southern Appalachian Mountains, changes in land use and land cover are often concentrated at the low to mid elevation, sheltered positions near streams that coincide with species-rich cove hardwood forests (Turner et al. 2003, Wear & Bolstad 1998). Thus, landscape position can influence susceptibility to disturbance and, hence, the spatial heterogeneity of disturbance severity.

Jules et al. (2002) investigated the role of heterogeneity in governing the spread of an invasive disease (a fatal root pathogen, *Phytophthora lateralis*) on a patchily distributed conifer, Port Orford cedar (*Chamaecyparis lawsoniana*). Their study showed that cedar populations along creeks crossed by roads were more likely to be infected than were those on creeks without road crossings; furthermore, the pathogen spread farther if it was vectored along roads. Studies have shown a strong influence of patch size and juxtaposition on incidence of Lyme disease by alteration of the community composition of vertebrate hosts and the abundance of larval ticks (*Ixodes scapularis*) (Allan et al. 2003). Indeed, the integration of landscape ecology and epidemiology may offer new approaches for understanding emerging infectious diseases and the effects of global change on vector-borne diseases (Kitron 1998).

In general, landscape position influences disturbance when the disturbance has a distinct directionality or locational specificity such that some locations are exposed more than others. However, the disturbance also must be of moderate intensity such that it can respond to gradients in the landscape—for example, fires burning under extreme conditions may show little variation in effects with landscape position (Moritz 1997). Accordingly, no predictable effect of landscape position is seen when the disturbance has no directionality, such as the smaller gap-forming down-bursts in the upper midwestern United States (Frelich & Lorimer 1991) or when intensity is extremely high (Moritz 1997).

ECOSYSTEM PROCESSES IN HETEROGENEOUS LANDSCAPES

The interface of ecosystem and landscape ecology is less developed than are the previous research areas, despite a tradition in Eastern Europe (e.g., Ryszkowski et al. 1999) and stronger connections during the early development of landscape ecology in North America. Ecosystem ecology has largely considered fluxes of matter and energy in the absence of a spatial context, and landscape ecology has had less focus on ecosystem processes. Recent studies suggest that spatial variability in some ecosystem processes may be of similar magnitude to temporal variation (e.g., Burrows et al. 2002, Turner et al. 2004), and efforts to explain and predict such variation are increasing. The importance of transfers among patches, which represent losses from donor ecosystems and subsidies to recipient ecosystems, for the long-term sustainability of ecosystems is also now acknowledged (Chapin et al. 2002, Reiners & Driese 2004). The patterns, causes, and consequences of spatial heterogeneity for ecosystem function are recognized as a current research frontier in both landscape ecology and ecosystem ecology (Lovett et al. 2005).

Progress has been made in the effort to recognize and explain spatial heterogeneity in ecosystem process rates that are either measured or simulated at many points. The role of landscape position has been elucidated (Soranno et al. 1999), and regional variation in a variety of stocks and processes (soil organic matter or carbon, denitrification, and net nitrogen mineralization rates) has been explored (e.g., Burke et al. 2002, Groffman et al. 1992). For example, the relationship between soil nitrogen mineralization and both biotic and abiotic factors was analyzed and mapped for the midwestern Great Lakes region of the United States (Fan et al. 1998). In the Greater Yellowstone Ecosystem, spatial patterns of aboveground net primary production were predicted by elevation and cover type (Hansen et al. 2000).

Broad conceptual frameworks have considered the conditions under which spatial pattern, or particular aspects of spatial pattern, should influence a lateral flux. Wiens et al. (1985) proposed a framework for lateral fluxes that included the factors that determine boundary locations, how boundaries affect the movement of materials over an area, and how imbalances in these transfers in space can affect landscape configuration. Weller et al. (1998) explored how and why different riparian buffer configurations would vary in their ability to intercept nutrient fluxes that move from a source ecosystem to an aquatic system. Loreau et al. (2003) developed a metaecosystem framework by extending metapopulation models to represent fluxes of matter or energy. Simulation models ranging from simple representations (e.g., Gardner et al. 1992) to complex, process-based spatial models (e.g., Costanza et al. 2002) have also been employed to identify the aspects of spatial configuration that could enhance or retard a lateral flux. Strayer et al. (2003b) proposed a useful conceptualization of model complexity relative to inclusion of spatial and temporal heterogeneity. However, a general understanding of lateral fluxes in landscape mosaics has remained elusive, despite promising conceptual frameworks developed for semiarid systems (e.g., Tongway & Ludwig 2001).

Many empirical studies have taken a comparative approach using integrative measurements, such as nutrient concentrations in aquatic ecosystems, as indicators of how spatial heterogeneity influences the end result of lateral fluxes (Strayer et al. 2003a). Most of these studies focus on nutrients, such as nitrogen or phosphorus, related to surface water quality (e.g., Jordan et al. 1997, Soranno et al. 1996). Variation in topography, the amount of impervious surfaces (e.g., pavement), and the extent of agricultural and urban land uses have all been related to the concentration or loading of nutrients in waters. For example, landscape heterogeneity explained from 65% to 86% of the variation in nitrogen yields to streams in the U.S. Mid-Atlantic region (Jones et al. 2001). However, the particular aspects of spatial heterogeneity that are significant or the spatial scales over which that influence is most important have varied among studies (Gergel et al. 2002). The lack of consistency among the comparative studies may arise, in part, from the absence of mechanistic understanding about how materials actually flow horizontally across heterogeneous landscapes.

CONCLUSIONS AND FUTURE DIRECTIONS

Landscape ecology is now well integrated with the ecological sciences. Consideration of spatial heterogeneity, which requires a conscious decision about whether and how to include it, has become the norm. Reasonable consensus about the approaches for quantifying landscape structure exists, but general relationships between generative processes and resulting patterns remain elusive. However, there is now an extensive library of empirical studies that explore ecological responses to landscape patterns. The multiple approaches suggested at the close of my 1989 article have in fact been widely applied in landscape studies.

Applications represent an important test of new conceptual frameworks and knowledge, and the demand for applied landscape ecology remains high (Liu & Taylor 2002). Networks of conservation areas based on the understanding of multispecies responses to landscape mosaics are providing a basis for long-term landscape planning in Europe (e.g., Bruinderink et al. 2003). In Ontario, Canada, legislation mandating that Crown forests be managed to keep landscape patterns consistent with long-term norms emerged directly from new understanding of landscape heterogeneity and disturbance (Perera et al. 2000). Forest-harvest strategies now incorporate consideration of the spatial landscape dynamics and the effects on a variety of species (Boutin & Herbert 2002). Predictions of invasive species (With 2002) and water quality (Strayer et al. 2003a) require consideration of landscape patterns.

Spatial extrapolation offers another mechanism for rigorous testing of the relationships between spatial patterns and processes (Miller et al. 2004b). Whether based on empirical observation or simulation modeling, the prediction of pattern (and the associated uncertainties) followed by testing with independent data or through cross validation offers a powerful way to evaluate current understanding. Understanding of the causes and effects of spatial heterogeneity will be enhanced by closely examining the conditions under which spatial extrapolation fails or succeeds (Miller et al. 2004b). Scaling remains challenging, despite numerous calls for progress during the past 15 years. However, scaling rules that integrate the scale dependency of patterns and processes in ways that organisms scale their responses to these patterns and processes are promising (Ludwig et al. 2000).

Where are the key directions for future research in landscape ecology? Interaction is a key theme related to several current research areas: interactions among multiple drivers that generate spatial patterns, particularly biophysical and socioeconomic factors; interactions among different kinds and scales of disturbances; and interactions among trophic levels in landscape mosaics. Landscape ecology should lead the next generation of studies taking a more comprehensive look at ecological dynamics in heterogeneous landscapes. A compelling need for expanding the temporal horizon of landscape studies also exists. Paleoecological studies provide critical context for understanding landscape dynamics, and historical dynamics shape current landscapes and may constrain future responses. Contemporary land-use patterns are creating future legacies, yet these potential legacies remain poorly understood.

Enhanced understanding of the ecological importance of spatial nonlinearities and thresholds remains an important research challenge. If important nonlinearities or thresholds are present among interacting variables, then landscape patterns may be even more difficult to predict, and unexpected changes in the state of an ecosystem or landscape may ensue (Frelich & Reich 1999, Groffman et al. 2005). Critical thresholds in spatial pattern have been suggested from theoretical and empirical studies, but a prediction of when a system is nearing a threshold is still difficult to make (Groffman et al. 2005).

Much remains to be learned about ecosystem processes in heterogeneous landscapes. The successful integration of ecosystem ecology and landscape ecology should produce a much more complete understanding of landscape function than has been developed to date. A landscape perspective still offers a prime opportunity for linking populations and ecosystem processes and services (Lundberg & Moberg 2003); organisms exist in heterogeneous space, and they use, transform, and transport matter and energy. Augustine & Frank (2001) demonstrated an influence of grazers on the distribution of soil nitrogen properties at every spatial scale from individual plants to landscapes. Seagle (2003) hypothesized that the juxtaposition of land uses with different forage-nutrient concentrations interacts nonlinearly with deer behavior to effect nutrient transport of sufficient magnitude to alter ecosystem nutrient budgets. Herbivore-mediated changes in forest composition have been shown to have important implications for patterns of nutrient cycling (Pastor et al. 1998, 1999). Studies have also identified the role of piscivores in the transportation of nutrients derived from aquatic ecosystems to terrestrial ecosystems through their foraging patterns (e.g., Naiman et al. 2002). Considering habitat use and movement patterns of species in a spatial context provides a wealth of opportunities to enhance the linkage between species and ecosystems and enhance functional understanding of landscape mosaics.

In conclusion, landscape ecology has matured. As noted by Wiens (1999), the discipline draws strength from the distinctiveness of its approach—its emphasis on spatial patterns and relationships, scaling, heterogeneity, boundaries, and flows of energy and materials in space. The themes of landscape ecology—reciprocal interactions between pattern and process, heterogeneity, scaling, critical thresholds, and boundaries and flows—have enriched the discipline of ecology. Landscape ecology should continue to refine knowledge of when spatial heterogeneity is fundamentally important in ecology (and, thus, the inverse, when it can be ignored), rigorously test the generality of its conceptual frameworks, and focus on developing a more mechanistic understanding of the reciprocal relationships between pattern and process.

ACKNOWLEDGMENTS

I sincerely thank Dean Anderson, James Forrester, Jennifer Fraterrigo, David Mladenoff, Dan Tinker, and John Wiens for constructive reviews of this manuscript. The ideas presented in this article have emerged from many collaborations and research projects, and I gratefully acknowledge my colleagues and laboratory group. This manuscript would not have been possible without research support from the National Science Foundation (Ecology, Ecosystems, Biocomplexity, and Long-term Ecological Research Programs), U.S. Department of Agriculture, U.S. Environmental Protection Agency (STAR Program), National Geographic Society, and the Andrew W. Mellon Foundation.

The *Annual Review of Ecology, Evolution, and Systematics* is online at
http://ecolsys.annualreviews.org

LITERATURE CITED

Addicott JF, Aho JM, Antolin MF, Padilla DK, Richardson JS, Soluk DA. 1987. Ecological neighborhoods: scaling environmental patterns. *Oikos* 49:340–46

Allan BF, Keesing F, Ostfeld RS. 2003. Effect of forest fragmentation on Lyme disease risk. *Conserv. Biol.* 17:267–72

Andrén H. 1994. Effects of habitat fragmentation on birds and mammals in landscapes with different proportions of suitable habitat. *Oikos* 71:355–66

Attiwill PM. 1994. The disturbance of forest ecosystems: the ecological basis for conservative management. *Forest Ecol. Manag.* 63:247–300

Augustine DJ, Frank DA. 2001. Effects of migratory grazers on spatial heterogeneity of soil nitrogen properties in a grassland ecosystem. *Ecology* 82:3149–62

Baker WL. 1989. Landscape ecology and nature reserve design in the Boundary Waters Canoe Area, Minnesota. *Ecology* 70:23–35

Bastian O. 2001. Landscape ecology: towards a unified discipline? *Landsc. Ecol.* 16:757–66

Bebi P, Kulakowski D, Veblen TT. 2003. Interactions between fire and spruce beetles in a subalpine Rocky Mountain forest landscape. *Ecology* 84:362–71

Bell SS, Robbins BD, Jensen SL. 1999. Gap dynamics in a seagrass landscape. *Ecosystems* 2:493–504

Bender DJ, Contreras TA, Fahrig L. 1998. Habitat loss and population decline: a meta-analysis of the patch size effect. *Ecology* 79:517–33

Black AE, Morgan P, Hessburg PF. 2003. Social and biophysical correlates of change in forest landscapes of the interior Columbia Basin, USA. *Ecol. Appl.* 13:51–67

Bolstad PV, Swank W, Vose J. 1998. Predicting Southern Appalachian overstory vegetation with digital terrain data. *Landsc. Ecol.* 13:271–83

Boose ER, Foster DR, Fluet M. 1994. Hurricane impacts to tropical and temperate forest landscapes. *Ecol. Monogr.* 64:369–400

Boutin S, Hebert D. 2002. Landscape ecology and forest management: developing an effective partnership. *Ecol. Appl.* 12:390–97

Boyce MS, Mao JS, Merrill EH, Fortin D, Turner MG, et al. 2003. Scale and heterogeneity in habitat selection by elk in

Yellowstone National Park. *EcoScience* 10:421–31

Bruinderink GG, Van Der SluisT, Lammertsma D, Opdam P, Pouwels R. 2003. Designing a coherent ecological network for large mammals in northwestern Europe. *Conserv. Biol.* 17:549–57

Burke IC, Lauenroth WK, Cunfer G, Barrett JE, Mosier A, et al. 2002. Nitrogen in the central grasslands region of the United States. *BioScience* 52:813–23

Burrows SN, Gower ST, Clayton MK, Mackay DS, Ahl DE, et al. 2002. Application of geostatistics to characterize leaf area index (LAI) from flux tower to landscape scales using a cyclic sampling design. *Ecosystems* 5:667–79

Cain DH, Riitters K, Orvis K. 1997. A multiscale analysis of landscape statistics. *Landscape Ecol.* 12:199–212

Chapin FS III, Matson PA, Mooney HA. 2002. *Principles of Terrestrial Ecosystem Ecology.* New York: Springer-Verlag. 436 pp.

Compton JE, Boone RD. 2000. Long-term impacts of agriculture on soil carbon and nitrogen in New England forests. *Ecology* 81:2314–30

Costanza R, Voinov A, Boumans R, Maxwell T, Villa F, et al. 2002. Integrated ecological economic modeling of the Patuxent River watershed, Maryland. *Ecol. Monogr.* 72:203–32

Cronon W. 1983. *Changes in the Land: Indians, Colonists and the History of New England.* New York: Hill & Wang. 241 pp.

Dupouey JL, Dambrine E, Laffite JD, Moares C. 2002. Irreversible impact of past land use on forest soils and biodiversity. *Ecology* 83:2978–84

Dupré C, Ehrlén J. 2002. Habitat configuration, species trains and plant distributions. *J. Ecol.* 90:796–805

Durrett R, Levin SA. 1994. Stochastic spatial models: a user's guide to ecological applications. *Philos. Trans. R. Soc. London Ser. B.* 343:329–50

Fagan WF. 2002. Connectivity, fragmentation, and extinction risk in dendritic metapopulations. *Ecology* 83:3243–49

Fahrig L. 1997. Relative effects of habitat loss and fragmentation on population extinction. *J. Wildl. Manag.* 61:603–10

Fahrig L. 2003. Effects of habitat fragmentation on biodiversity. *Annu. Rev. Ecol. Evol. Syst.* 34:487–15

Fahrig L, Nuttle WK. 2005. Population ecology in spatially heterogeneous environments. In *Ecosystem Function in Heterogeneous Landscapes*, ed. GM Lovett, CG Jones, MG Turner, KC Weathers. New York: Springer-Verlag. In press

Fan W, Randolph JC, Ehman JL. 1998. Regional estimation of nitrogen mineralization in forest ecosystems using Geographic Information Systems. *Ecol. Appl.* 8:734–47

Flather CH, Bevers M. 2002. Patchy reaction-diffusion and population abundance: the relative importance of habitat amount and arrangement. *Am. Nat.* 159:40–56

Fortin M-J, Boots B, Csillag F, Remmel TK. 2003. On the role of spatial stochastic models in understanding landscape indices. *Oikos* 102:203–12

Foster DR. 1992. Land-use history (1730–1990) and vegetation dynamics in central New England, USA. *J. Ecol.* 80:753–72

Foster DR. 2002. Insights from historical geography to ecology and conservation: lessons from the New England landscape. *J. Biogeogr.* 29:1269–75

Foster DR, Fluet M, Boose ER. 1999. Human or natural disturbance: landscape-scale dynamics of the tropical forests of Puerto Rico. *Ecol. Appl.* 9:555–72

Foster DR, Knight DH, Franklin JF. 1998. Landscape patterns and legacies resulting from large infrequent forest disturbances. *Ecosystems* 1:497–510

Fraterrigo J, Turner MG, Pearson SM, Dixon P. 2005. Effects of past land use on spatial heterogeneity of soil nutrients in Southern Appalachian forests. *Ecol. Monogr.* 75:215–30

Frelich LE, Lorimer CG. 1991. Natural disturbance regimes in hemlock-hardwood forests of the upper Great Lakes region. *Ecol. Monogr.* 61:145–64

Frelich LE, Reich PB. 1999. Neighborhood effects, disturbance severity, and community stability in forests. *Ecosystems* 2:151–66

Gardner RH, Milne BT, Turner MG, O'Neill RV. 1987. Neutral models for the analysis of broad-scale landscape patterns. *Landsc. Ecol.* 1:19–28

Gardner RH, Dale VH, O'Neill RV, Turner MG. 1992. A percolation model of ecological flows. In *Landscape Boundaries: Consequences for Biotic Diversity and Ecological Flow*, ed. AJ Hansen, F Di Castri, 259–69. New York: Springer-Verlag

Gergel SE, Turner MG, Miller JR, Melack JM, Stanley EH. 2002. Landscape indicators of human impacts to river-floodplain systems. *Aquat. Sci.* 64:118–28

Goodale CL, Aber JD. 2001. The long-term effects of land-use history on nitrogen cycling in northern hardwood forests. *Ecol. Appl.* 11:253–67

Goodwin BJ, Fahrig L. 2002. How does landscape structure influence landscape connectivity? *Oikos* 99:552–70

Groffman PM, Baron JS, Blett T, Gold AJ, Goodman I, et al. 2005. Ecological thresholds: the key to successful environmental management or an important concept with no practical application? *Ecosystems.* In press

Groffman PM, Tiedje TM, Mokma DL, Simkins S. 1992. Regional-scale analysis of denitrification in north temperate forest soils. *Landsc. Ecol.* 7:45–54

Gustafson EJ. 1998. Quantifying landscape spatial pattern: What is the state of the art? *Ecosystems* 1:143–56

Gustafson EJ, Parker GR. 1992. Relationships between landcover proportion and indices of landscape spatial pattern. *Landsc. Ecol.* 7:101–10

Haila Y. 2002. A conceptual genealogy of fragmentation research: from island biogeography to landscape ecology. *Ecol. Appl.* 12:321–34

Haines-Young R, Chopping M. 1996. Quantifying landscape structure: a review of landscape indices and their application to forested landscapes. *Prog. Phys. Geog.* 20:418–45

Hansen AJ, Rotella JJ, Kraska MPV, Brown D. 2000. Spatial patterns of primary productivity in the greater Yellowstone ecosystem. *Landsc. Ecol.* 15:505–22

Hansen AJ, Rasker R, Maxwell B, Rotella JJ, Johnson JD, et al. 2002. Ecological causes and consequences of demographic change in the new west. *BioScience* 5:151–62

Hanski IA, Gilpin ME, eds. 1997. *Metapopulation Biology.* New York, NY: Academic. 512 pp.

Hargis CD, Bissonette JA, David JL. 1998. The behavior of landscape metrics commonly used in the study of habitat fragmentation. *Landscape Ecol.* 13:167–86

He HS, Mladenoff DJ. 1999. Spatially explicit and stochastic simulation of forest-landscape fire disturbance and succession. *Ecology* 80:81–99

Heilman GE, Strittholt JR, Slosser NC, Dellasala DA. 2002. Forest fragmentation of the conterminous United States: assessing forest intactness through road density and spatial characteristics. *BioScience* 52:411–22

Jones CG, Lawton JH, Shachak M. 1997. Positive and negative effects of organisms as physical ecosystem engineers. *Ecology* 78:1946–57

Jones KB, Neale AC, Nash MS, Van Remortel RD, Wickham JD, et al. 2001. Predicting nutrient and sediment loadings to streams from landscape metrics: a multiple watershed study from the United States mid-Atlantic region. *Landscape Ecol.* 16:301–12

Jordan TE, Correll DL, Weller DE. 1997. Relating nutrient discharges from watersheds to land use and streamflow variability. *Water Resour. Res.* 33:2579–90

Jules ES, Kauffman MJ, Ritts WD, Carroll AL. 2002. Spread of an invasive pathogen over a variable landscape: a nonnative root rot on Port Orford cedar. *Ecology* 83:3167–81

Kashian DM, Turner MG, Romme WH. 2005a. Changes in leaf area and stemwood increment with stand development in Yellowstone National Park: relationships between forest stand structure and function. *Ecosystems.* 8:48–61

Kashian DM, Turner MG, Romme WH, Lorimer CJ. 2005b. Variability and convergence in stand structure with forest development on a fire-dominated landscape. *Ecology.* 86:643–54

Keyghobadi N, Roland J, Strobeck C. 1999. Influence of landscape on the population genetic structure of the alpine butterfly *Parnassius smintheus* (Papilionidae). *Mol. Ecol.* 8:1481–95

King AW, With KA. 2002. Dispersal success on spatially structured landscapes: When do spatial pattern and dispersal behavior really matter? *Ecol. Model.* 147:23–39

Kitron U. 1998. Landscape ecology and epidemiology of vector-borne diseases: tools for spatial analysis. *J. Med. Entomol.* 35:435–45

Knapp AK, Blair JM, Briggs JM, Collins SL, Hartnett DC, et al. 1999. The keystone role of bison in North American tallgrass prairie—Bison increase habitat heterogeneity and alter a broad array of plant, community, and ecosystem processes. *BioScience* 49:39–50

Knight TW, Morris DW. 1996. How many habitats do landscapes contain? *Ecology* 77:1756–64

Kramer MG, Hansen AJ, Taper ML, Kissinger EJ. 2001. Abiotic controls on long-term windthrow disturbance and temperate rain forest dynamics in southeast Alaska. *Ecology* 82:2749–68

Krummel JR, Gardner RH, Sugihara G, O'Neill RV, Coleman PR. 1987. Landscape patterns in a disturbed environment. *Oikos* 48:321–24

Landres PB, Morgan P, Swanson FJ. 1999. Overview of the use of natural variability concepts in managing ecological systems. *Ecol. Appl.* 9:1179–88

Levin SA. 1992. The problem of pattern and scale in ecology. *Ecology* 73:1943–83

Li H, Reynolds JF. 1995. On definition and quantification of heterogeneity. *Oikos* 73:280–84

Li HB, Wu JG. 2004. Use and misuse of landscape indices. *Landsc. Ecol.* 19:389–99

Lichstein JW, Simons TR, Shriner SA, Franzreb KE. 2002. Spatial autocorrelation and autoregressive models in ecology. *Ecol. Monogr.* 72:445–63

Lindenmayer DB, Cunningham RB, Pope ML, Donnelly CF. 1999. The response of arboreal marsupials to landscape context: a large-scale fragmentation study. *Ecol. Appl.* 9:594–611

Liu J, Taylor WW, eds. 2002. *Integrating Landscape Ecology into Natural Resource Management.* Cambridge, UK: Cambridge Univ. Press. 480 pp.

Loreau M, Mouquet N, Holt RD. 2003. Metaecosystems: a theoretical framework for a spatial ecosystem ecology. *Ecol. Lett.* 6:673–79

Lovett GM, Jones CG, Turner MG, Weathers KC, eds. 2005. Ecosystem function in heterogeneous landscapes. New York: Springer-Verlag. In press

Lucchini V, Fabbri E, Marucco F, Ricci S, Boitani L, et al. 2002. Noninvasive molecular tracking of colonizing wolf (*Canis lupus*) packs in the western Italian Alps. *Mol. Ecol.* 11:857–68

Ludwig JA, Wiens JA, Tongway DJ. 2000. A scaling rule for landscape patches and how it applies to conserving soil resources in savannas. *Ecosystems* 3:84–97

Lundberg J, Moberg F. 2003. Mobile link organisms and ecosystem functioning: implications for ecosystem resilience and management. *Ecosystems* 6:87–98

Manel S, Schwartz MK, Luikart G, Taberlet P. 2003. Landscape genetics: combining landscape ecology and population genetics. *Trends Ecol. Evol.* 18:189–97

Manly BFJ, McDonald LL, Thomas DL, McDonald TL, Erickson WP. 2002. *Resource Selection by Animals: Statistical Design and Analysis for Field Studies.* Dordrect, The Netherlands: Kluwer Acad. 221 pp.

Mazerolle MJ, Villard MA. 1999. Patch characteristics and landscape context as predictors of species presence and abundance: a review. *Ecoscience* 6:117–24

McGarigal K, Cushman SA. 2002. Comparative evaluation of experimental approaches to the study of habitat fragmentation effects. *Ecol. Appl.* 12:335–45

McGarigal K, Marks BJ. 1995. FRAGSTATS. Spatial analysis program for quantifying landscape structure. *USDA For. Serv. Gen. Tech. Rep.* PNW-GTR-351

Miller JR, Dixon MD, Turner MG. 2004a. Response of avian communities in large-river floodplains to environmental variation at multiple scales. *Ecol. Appl.* 14:1394–10

Miller JR, Turner MG, Smithwick EAH, Dent CL, Stanley EH. 2004b. Spatial extrapolation: the science of predicting ecological patterns and processes. *BioScience* 54:310–20

Mladenoff DJ, White MA, Pastor J, Crow TR. 1993. Comparing spatial pattern in unaltered old-growth and disturbed forest landscapes. *Ecol. Appl.* 3:294–306

Moritz MA. 1997. Analyzing extreme disturbance events: fire in Los Padres National Forest. *Ecol. Appl.* 7:1252–62

Motzkin G, Wilson GP, Foster DR, Arthur A. 1999. Vegetation patterns in heterogeneous landscapes: the importance of history and environment. *J. Veg. Sci.* 10:903–20

Murphy HT, Lovett-Doust J. 2004. Context and connectivity in plant metapopulations and landscape mosaics: Does the matrix matter? *Oikos* 105:3–14

Murphy DD, Noon BR. 1992. Integrating scientific methods with habitat conservation planning—reserve design for northern spotted owls. *Ecol. Appl.* 2:3–17

Naiman RJ, Rogers KH. 1997. Large animals and system level characteristics in river corridors. *BioScience* 47:521–29

Naiman RJ, Bilby RE, Schindler DE, Helfield JM. 2002. Pacific salmon, nutrients, and the dynamics of freshwater and riparian ecosystems. *Ecosystems* 5:399–17

Opdam P, Foppen R, Vos C. 2002. Bridging the gap between ecology and spatial planning in landscape ecology. *Landsc. Ecol.* 16:767–79

Opdam P, Verboom J, Pouwels R. 2003. Landscape cohesion: an index for the conservation potential of landscapes for biodiversity. *Landsc. Ecol.* 18:113–26

Paine RT, Tegner MJ, Johnson EA. 1998. Compounded perturbations yield ecological surprises. *Ecosystems* 1:535–45

Palmer MA, Swan CM, Nelson K, Silver P, Alvestad R. 2000. Streambed landscapes: evidence that stream invertebrates respond to the type and spatial arrangement of patches. *Landsc. Ecol.* 15:563–76

Parker KC, Bendix J. 1996. Landscape-scale geomorphic influences on vegetation patterns in four environments. *Phys. Geogr.* 17: 113–41

Pastor J, Dewey B, Moen R, Mladenoff DJ, White M, et al. 1998. Spatial patterns in the moose-forest-soil ecosystem on Isle Royale, Michigan, USA. *Ecol. Appl.* 8:411–24

Pastor J, Cohen Y, Moen R. 1999. Generation of spatial patterns in boreal forest landscapes. *Ecosystems* 2:439–50

Pearson SM, Turner MG, Drake JB. 1999. Landscape change and habitat availability in the southern Appalachian highlands and the Olympic Peninsula. *Ecol. Appl.* 9:1288–04

Pearson SM, Turner MG, Wallace LL, Romme WH. 1995. Winter habitat use by large ungulates following fires in northern Yellowstone National Park. *Ecol. Appl.* 5:744–55

Perera AH, Euler DL, Thompson ID, eds. 2000. *Ecology of a Managed Terrestrial Landscape: Patterns and Processes of Forest Landscapes in Ontario.* Vancouver, BC: UBC Press. 336 pp.

Perry GLW. 2002. Landscapes, space and equilibrium: shifting viewpoints. *Prog. Phys. Geog.* 26:339–59

Pickett STA, Cadenasso ML. 1995. Landscape ecology: spatial heterogeneity in ecological systems. *Science* 269:331–34

Pickett STA, White PS, eds. 1985. *The Ecology of Natural Disturbance and Patch Dynamics.* New York: Academic. 472 pp.

Radeloff VC, Mladenoff DJ, Boyce MS. 2000. Effects of interacting disturbances on landscape patterns: budworm defoliation and salvage logging. *Ecol. Appl.* 10:233–47

Reiners WA, Driese KL. 2004. *Transport Processes in Nature.* Cambridge, UK: Cambridge Univ. Press. 302 pp.

Reuness EK, Stenseth NC, O'Donoghue M,

Boutin S, Ellegren H, et al. 2003. Ecological and genetic sptial structuring in the Canada lynx. *Nature* 425:69–72

Remmel TK, Csillag F. 2003. When are two landscape pattern indices significantly different? *J. Geogr. Syst.* 5:331–51

Ricketts TH. 2001. The matrix matters: effective isolation in fragmented landscapes. *Am. Nat.* 158:87–99

Riitters K, Wickham J, O'Neill R, Jones B, Smith E. 2000. Global-scale patterns of forest fragmentation. *Conserv. Ecol.* 4(2):3. http://www.consecol.org/vol4/iss2/art3/

Riitters KH, Wickham JD, O'Neill RV, Jones KB, Smith ER, et al. 2002. Fragmentation of continental United States forests. *Ecosystems* 5:815–22

Riitters KH, O'Neill RV, Hunsaker CT, Wickham JD, Yankee DH, et al. 1995. A factor analysis of landscape pattern and structure metrics. *Landsc. Ecol.* 10:23–40

Risser PG, Karr JR, Forman RTT. 1984. Landscape ecology: directions and approaches. *Spec. Publ. No.2*, Ill. Nat. Hist. Surv., Champaign, Ill.

Rossi RE, Mulla DJ, Journel AG, Franz EH. 1992. Geostatistical tools for modeling and interpreting ecological spatial dependence. *Ecol. Monogr.* 62:277–314

Russell EB. 1997. *People and the Land Through Time.* New Haven: Yale Univ. Press. 306 pp.

Ryszkowski L, Bartoszewicz A, Kedziora A. 1999. Management of matter fluxes by biogeochemical barriers at the agricultural landscape level. *Landsc. Ecol.* 14:479–92

Saunders D, Hobbs RJ, Margules CR. 1991. Biological consequences of ecosystem fragmentation: a review. *Conserv. Biol.* 5:18–32

Schmitz OJ, Hambäck PA, Beckerman AP. 2000. Trophic cascades in terrestrial systems: a review of the effects of carnivore removals on plants. *Am. Nat.* 155:141–53

Schwartz MK, Mills LS, Ortega Y, Ruggiero LF, Allendorf FW. 2003. Landscape location affects genetic variation of Canada lynx (*Lynx canadensis*). *Mol. Ecol.* 12:1807–16

Seagle SW. 2003. Can deer foraging in multiple-use landscapes alter forest nitrogen budgets? *Oikos* 103:230–34

Soranno PA, Hubler SL, Carpenter SR, Lathrop RC. 1996. Phosphorus loads to surface waters: a simple model to account for spatial pattern of land use. *Ecol. Appl.* 6:865–78

Soranno PA, Webster KE, Riera JL, Kratz TK, Baron JS, et al. 1999. Spatial variation among lakes within landscapes: ecological organization along lake chains. *Ecosystems* 2:395–410

Spies TA, Ripple WJ, Bradshaw GA. 1994. Dynamics and pattern of a managed coniferous forest landscape in Oregon. *Ecol. Appl.* 4:555–68

Steffan-Dewenter I, Munzenberg U, Burger C, Thies C, Tscharntke T. 2002. Scale-dependent effects of landscape context on three pollinator guilds. *Ecology* 83:1421–32

Stoner KJL, Joern A. 2004. Landscape versus local habitat scale influences to insect communities from tallgrass prairie remnants. *Ecol. Appl.* 14:1306–20

Strayer DL, Beighley RE, Thompson LC, Brooks A, Nilsson C, et al. 2003a. Effects of land cover on stream ecosystems: roles of empirical models and scaling issues. *Ecosystems* 6:407–23

Strayer DL, Ewing HA, Bigelow S. 2003b. What kind of spatial and temporal details are required in models of heterogeneous systems? *Oikos* 102:654–62

Tallmon DA, Jules ES, Radke NJ, Mills LS. 2003. Of mice and men and trillium: cascading effects of forest fragmentation. *Ecol. Appl.* 13:1193–1203

Tewksbury JJ, Levey DJ, Haddad NM, Sargent S, Orrock JL, et al. 2002. Corridors affect plants, animals, and their interactions in fragmented landscapes. *Proc. Natl. Acad. Sci. USA* 99:12923–62

Tischendorf L. 2001. Can landscape indices predict ecological processes consistently? *Landsc. Ecol.* 16:235–54

Tongway DJ, Ludwig JA. 2001. Theories on the origins, maintenance, dynamics and functioning of banded landscapes. In *Banded Vegetation Patterning in Arid and Semiarid*

Environments: Ecological Processes and Consequences for Management. ed. D Tongway, C Valentin, J Segheri, pp. 20–31. New York: Springer-Verlag

Troll C. 1950. Die geographisched Landschaft und ihre Erforschung. *Studium Generale* 3:163–81

Trzcinski MK, Fahrig L, Merriam G. 1999. Independent effects of forest cover and fragmentation on the distribution of forest breeding birds. *Ecol. Appl.* 9:586–93

Turner MG. 1989. Landscape ecology: the effect of pattern on process. *Annu. Rev. Ecol. Syst.* 20:171–97

Turner MG. 2005. Landscape ecology in North America: past, present and future. *Ecology* 86:1967–74

Turner MG, Gardner RH, O'Neill RV. 2001. *Landscape Ecology in Theory and Practice.* New York: Springer-Verlag. 401 pp.

Turner MG, Pearson SM, Bolstad P, Wear DN. 2003. Effects of land-cover change on spatial pattern of forest communities in the southern Appalachian Mountains (USA). *Landscape Ecol.* 18:449–64

Turner MG, Romme WH, Gardner RH, O'Neill RV, Kratz TK. 1993. A revised concept of landscape equilibrium: disturbance and stability on scaled landscapes. *Landsc. Ecol.* 8:213–27

Turner MG, Tinker DB, Romme WH, Kashian DM, Litton CM. 2004. Landscape patterns of sapling density, leaf area, and aboveground net primary production in postfire lodgepole pine forests, Yellowstone National Park (USA). *Ecosystems* 7:751–75

Urban D, Goslee S, Pierce K, Lookingbill T. 2002. Extending community ecology to landscapes. *Ecoscience* 9:200–12

Urban DL, O'Neill RV, Shugart HH. 1987. Landscape ecology. *BioScience* 37:119–27

Vos CC, Verboom J, Opdam PFM, Ter Braak CJF. 2001.Toward ecologically scaled landscape indices. *Am. Nat.* 157:24–41

Wang H, Hall CAS, Cornell JD, Hall MHP. 2002. Spatial dependence and the relationship of soil organic carbon and soil moisture in Luquillo experimental forest. *Landsc. Ecol.* 17:671–84

Ward JV, Malard F, Tockner K. 2002. Landscape ecology: a framework for integrating pattern and process in river corridors. *Landsc. Ecol.* 17 (Suppl.):35–45

Wear DN, Bolstad P. 1998. Land-use changes in southern Appalachian landscapes: spatial analysis and forecast evaluation. *Ecosystems* 1:575–94

Wear DN, Turner MG, Flamm RO. 1996. Ecosystem management with multiple owners: landscape dynamics in a Southern Appalachian watershed. *Ecol. Appl.* 6:1173–88

Weller DE, Jordan TE, Correll DL. 1998. Heuristic models for material discharge from landscapes with riparian buffers. *Ecol. Appl.* 8:1156–69

Wells TC, Sheail J, Ball DF, Ward LK. 1976. Ecological studies on the Porton Ranges: relationships between vegetation, soils and land-use history. *J. Ecol.* 64:589–26

White D, Minotti PG, Barczak MJ, Sifneos JC, Freemark KE, et al. 1997. Assessing risks to biodiversity from future landscape change. *Conserv. Biol.* 11:349–60

Wickham JD, Riitters KH. 1995. Sensitivity of landscape metrics to pixel size. *Int. J. Remote Sens.* 16:3585–94

Wiens JA. 1999. The science and practice of landscape ecology. In *Landscape Ecological Analysis: Issues and Applications,* ed. JM Klopatek, RH Gardner, pp. 372–83. New York: Springer-Verlag

Wiens JA. 1976. Population responses to patchy environments. *Annu. Rev. Ecol. Syst.* 7:81–120

Wiens JA. 1989. Spatial scaling in ecology. *Funct. Ecol.* 3:385–97

Wiens JA, Crawford CS, Gosz JR. 1985. Boundary dynamics—a conceptual framework for studying landscape ecosystems. *Oikos* 45:421–27

Wiens JA, Moss MR, Turner MG, Mladenoff DJ, eds. 2005. *Foundation Papers in Landscape Ecology.* New York: Columbia Univ. Press. In press

With KA. 2002. The landscape ecology of

invasive spread. *Conserv. Biol.* 16:1192–03

With KA, King AW. 1997. The use and misuse of neutral landscape models in ecology. *Oikos* 79:219–29

With KA, Pavuk DM, Worchuck JL, Oates RK, Fisher JL. 2002. Threshold effects of landscape structure on biological control in agroecosystems. *Ecol. Appl.* 12:52–65

Wu J, Hobbs RJ. 2002. Key issues and research priorities in landscape ecology: an idiosyncratic synthesis. *Landsc. Ecol.* 17:355–65

Wu J, Loucks OL. 1995. From balance of nature to hierarchical patch dynamics: a paradigm shift in ecology. *Q. Rev. Biol.* 70:439–66

Annu. Rev. Ecol. Evol. Syst. 2005. 36:345–72
doi: 10.1146/annurev.ecolsys.36.091704.175531
Copyright © 2005 by Annual Reviews. All rights reserved

ECOLOGY AND EVOLUTION OF APHID-ANT INTERACTIONS

Bernhard Stadler[1] and Anthony F.G. Dixon[2]

[1]*Bayreuth Institute for Terrestrial Ecosystem Research, University of Bayreuth, Bayreuth D-95440, Germany; email: bernhard.stadler@bitoek.uni-bayreuth.de*
[2]*School of Biological Sciences, University of East Anglia, Norwich NR4 7TJ, United Kingdom; email: a.f.dixon@uea.ac.uk*

Key Words mutualism, aphid-ant relationship, cost-benefit analysis

■ **Abstract** Aphids and ants are two abundant and highly successful insect groups, which often live in the same habitat and therefore are likely to interact with one another. Whether the outcome of such an interaction is a predator-prey or mutualistic one is dependent on what each partner has to offer relative to the needs of the other. Consequently, understanding why some aphids enter mutualistic interactions with ants is dependent on understanding the physiological, ecological, and evolutionary traits of both partners. This includes an appreciation of the spatial, temporal, and taxonomic context in which mutualistic interactions developed. In this review, we use aphid-ant interactions to illustrate the whole range of interactions from antagonistic to mutualistic as well as to identify the processes affecting the degree of association and in particular the context within which such interactions evolved. The constraints of establishing and maintaining beneficial interactions between aphids and ants is addressed from a cost-benefit perspective. Prospects for future research are identified to further the understanding of the patterns and processes associated with aphid-ant relationships.

INTRODUCTION

In many ecology textbooks, *mutualism* is defined as an interaction between two species that is beneficial for both species (Begon et al. 1999, Boucher et al. 1982, Krohne 1998). However, assuming that it is individuals that benefit, then mutualism is when organisms reciprocally and positively affect the individual fitness or per capita growth rate of their partners (Addicott 1985, Bartlett 1961). Or more simply, mutualism is a reciprocally beneficial relationship between organisms (Herre et al. 1999). There are several reviews of the ecology of mutualism (Boucher et al. 1982, Bronstein 1994b, Bronstein & Barbosa 2002, Buckley 1987a, Connor 1995, Hoeksema & Bruna 2000). These give general accounts of the conditions under which such beneficial associations might develop, e.g., the type of environment in which they are likely to evolve and the constraints. When dealing with mutualism, there is a tendency to describe general patterns across different taxonomic groups and omit details and mechanisms that appear of little importance for the larger picture. Seemingly, there are many similarities in the biologies of the partners of ants (e.g., aphids, coccids, membracids, scale insects, and lycaenid butterflies), notably the production of honeydew/nectar, which makes it tempting to look for broad, general patterns (c.f. Delabie 2001) at the risk of ignoring mechanisms that shape different degrees of association. In this review, however, we focus exclusively on ant-aphid interactions, which were last reviewed some time ago (Buckley 1987a, Way 1963). These reviews are still cited in ecological textbooks as a paradigm for mutualistic associations between insects (e.g., Begon et al. 1999, Hölldobler & Wilson 1990). Over the past two decades, however, considerable progress has been made in gathering information on the physiological adaptations, the ecological context, and the evolutionary constraints acting on both partners when entering an association. This information challenges the widely held view that the interactions between these two groups of organisms are always positive. This information needs to be incorporated into any overview of the subject if one is to understand the costs and benefits of the associations between aphids and ants.

The many papers published on mutualistic relationships over the past two decades led to predictions about the conditions that favor the evolution of reduced antagonism and mutualism (Boucher et al. 1982, Bronstein 1994a, Bronstein 2001, Hoeksema & Bruna 2000, Holland et al. 2002, Stachowicz 2001). For example, it is suggested that facultative mutualism and those involving third partners should be more variable than obligate mutualism. In addition, mutualisms that depend on one or both partners being abundant are more likely to vary with conditions, obligate mutualism must include mechanisms limiting the abundance of partners, and highly specialized mutualisms should be rare. However, there is no general theory that predicts the outcome of species interactions ranging from mutualistic to antagonistic. This is possibly because there is no common currency for measuring the costs and benefits for both partners. However, there is a need to understand the physiological and ecological costs and benefits at the individual level to identify the structuring forces in the antagonism-mutualism continuum. Aphids and ants are characterized by distinctive population and genetic structures, which highlight the different temporal, spatial, and evolutionary aspects that are likely to affect these relationships.

The aim of this review is to use aphid-ant interactions to (a) illustrate the whole range of interactions from antagonistic to mutualistic, (b) identify the mechanisms regulating the degree of association, and, in particular, (c) identify the context in which such interactions evolved. The evolution and maintenance of beneficial interactions between aphids and ants is addressed from the perspective of both partners, and the research needed to further the understanding of the conflict of interests in aphid-ant interactions is highlighted.

FEATURES OF THE PARTNERS

In the next two sections we briefly summarize the main features of aphids and ants and the potential costs and benefits associated with myrmecophily.

Aphids

Aphids most likely evolved some 280 mya in the Carboniferous (Heie 1967), and there are now about 4000 species worldwide (Eastop 1973, Remaudière & Remaudière 1997). They characteristically have several parthenogenetic generations during summer, a single sexual generation in autumn, and overwinter as eggs. The parthenogenetic, iteroparous mode of reproduction associated with the telescoping of generations, in which aphid embryos start developing in their grandmother and develop to an advanced stage inside their mother (Dixon 1998, Kindlmann & Dixon 1989), results in rapid multiplication and facilitates the exploitation of short-lived resources. In addition, many species are highly polyphenic, with winged individuals specialized more for dispersal than reproduction and unwinged individuals more for reproduction than dispersal. That is, they show division of labor. Their prodigious rates of increase are unparalleled in other herbivorous insects. Aphids feed on phloem sap, which is typically rich in sugars but low in nitrogen (N). As a consequence, aphids need to ingest large volumes of phloem sap—most of which is excreted as honeydew (Dixon 2004, Stadler et al. 1998, Zoebelein 1954). Feeding on plant sap is a very old way of obtaining food and dates back to the Early Devonian (Labandeira 1997). There is a clear North-South gradient in species richness, with relatively few species in the tropics (Dixon et al. 1987). This geographic pattern is attributed to the small fat reserves, large investment in offspring, and high host-plant specificity of aphids. These life-history attributes greatly limit the amount of time aphids can spend searching for host plants. This in association with a high plant diversity in the tropics means that very few plants there are abundant enough to host aphids (Dixon 2004, Dixon et al. 1993a, Dixon et al. 1993b).

Ants

Ants have survived since the upper Cretaceous, some 100 mya, and now comprise some 8000 species (Hölldobler & Wilson 1990). They show division of labor and eusociality, which enables colonies to sustain high population densities for long periods of time (Hölldobler & Wilson 1990, Wilson 1987). Like an aphid colony, an ant colony is an almost exclusively female society. Central to the success of ant societies is that the members of each colony are divided into reproductive and nonreproductive worker castes. Workers are able to forage over vast areas relative to their size but operate from a fixed base (nest) to which they need to return. The nests of mound-building ants in temperate regions often last for many decades and have a strong influence on their immediate environment. For example, herbivore pressure on trees may be significantly reduced around ant nests, resulting in "green islands" (Laine & Niemela 1980) with higher survival rates of trees and a strong shift in herbivore community structure (Skinner & Whittaker 1981, Wellenstein 1980). Ants inhabit a broad geographic range with the highest species richness in the tropics, where they are the dominant insect group in terms of biomass (Fittkau & Klinge 1973, Wilson 1990). Not all ant taxa have developed trophobiotic relationships with aphids: Only workers of the subfamilies *Formicinae* and *Dolichoderinae* and a few species in the genus *Myrmica* and *Tetramorium* of the *Myrmicinae* collect honeydew (Kunkel et al. 1985, Nixon 1951). Baltic amber fossils indicate that associations between ants and aphids date back at least to the early Oligocene (Wheeler 1914). However, mutualistic interactions with ants do not seem to be constrained to particular aphid taxa as, although not analyzed phylogenetically, the frequency of ant attendance does not vary strikingly between groups. This probably suggests multiple origins of myrmecophily (Bristow 1991, Stadler et al. 2003).

In view of their abundance and territoriality, their burrowing and mound construction, the relative stability of their populations, as well as their feeding habits and aggressiveness, ants are a group of insects that almost any other insect group is likely to encounter at some stage. The outcome of these encounters can be positive or negative.

Benefits and Costs

Before considering the interactions between ants and aphids, it is important to have some idea of the benefits and costs of the mutualism for both partners. Benefits for one partner may entail costs for the other. Therefore, the tension within these systems and the relative magnitude of the resulting conflict/shared interests determines whether a relationship is positive or negative. At the risk of oversimplification, ants are very active and many forage over great distances, so energy for foraging is likely to be a very important limiting factor. Aphids have to process very large quantities of phloem sap to sustain their very high growth rates, so honeydew is often likely to be abundant and available for fueling ant foraging. However, because phloem sap contains very little amino nitrogen and aphids are very good at assimilating most of it, honeydew is unlikely to be a source of N for ants. In addition to being a fuel for foraging, honeydew may also be stored and used to tide ants over periods of adverse conditions. A good example of this is the storing of coccid honeydew by honey-pot ants (Gullan & Kosztarab 1997). There is no example of this involving ants and aphids, but it could exist.

The cost for ants is that they need to monopolize, collect, transport, and pass honeydew to their nest mates, which involves morphological and behavioral adaptations. However, the biggest cost is likely to be that associated with being dependent on aphids for fuel for foraging. This is particularly so for obligate myrmecophily, involving one species of ant and aphid. Although the distribution of the aphid *Stomaphis quercus* (L.) on oak trees is limited to those that grow within the territories of the ant *Lasius fuliginosus* (Latreille) (Goidanich 1956), it is unknown whether the distribution and abundance of the ant is dependent on that of the aphid. The expectation, however, is that ants should rarely be dependent on a single aphid species because this would put them at great risk of extinction.

Aphids are soft bodied and have little defense against natural enemies other than avoidance. Therefore, it is likely that a major benefit of ant attendance for aphids is protection. In habitats where aphids are at particular risk of attack from natural enemies, a high incidence of ant attendance is predicted. Most aphids that are ant attended are gregarious. Clearly, this is advantageous for ants because it results in the sources of energy being concentrated in a few places rather than scattered throughout their territory. However, in being gregarious, aphids become more attractive to natural enemies, which could put an upper limit on the size of ant-attended aphid colonies.

The cost for aphids appears to be mainly one of producing large quantities of high-quality honeydew to attract ants. It is well established that facultatively attended aphids increase their rate of honeydew production when attended by ants (Nixon 1951). Therefore, if unattended aphids feed at an optimum rate for the assimilation of amino nitrogen, then a faster rate is likely to adversely affect their feeding efficiency and consequently their rate of growth. If aphids are obligately ant attended, then another cost for aphids is the effect this has on their distribution. Good examples of this are the oak aphid, *S. quercus*, and the thyme aphid, *Aphis serpylli* Koc, both of which have markedly more restricted distributions than their host plants because they are dependent on ants being present in the habitat (Hopkins & Thacker 1999, Hopkins et al. 1998). This cost-benefit perspective is adopted in the remainder of the review.

PATTERNS IN APHID-ANT INTERACTIONS

Given the ecological success of both ants and aphids, that aphids excrete energy-rich honeydew, and that ants aggressively defend resources, it is difficult to understand why so few species of aphid have evolved a close relationship with ants. For example, only a quarter of the aphid species in the Rocky Mountain region are attended by ants (Bristow 1991), and in Central Europe, about one third are obligate myrmecophiles (Stadler & Dixon 1998b, Tizado et al. 1993). More than a dozen hypotheses have been proposed to account for the variability in aphid-ant relationships and the low proportion of attended species. First among them is the plant permissive hypothesis (Bristow 1991), which suggests that host-plant quality plays a critical role in determining the attractiveness of aphids for ants. Both variation in the quality of different hosts or different parts of the same host plant may affect the quality of honeydew, which is either more or less attractive to ants. However, this hypothesis ignores the fact that several species of aphids may feed on the same host plant or even plant organ but have very different degrees of associations with ants and assumes that the quality of phloem sap determines that of the honeydew. Recent studies have found that feeding on woody plant parts is positively associated with ant attendance, whereas mobility, feeding in isolation, and having winged adults are negatively associated with ant attendance (Dixon 1998, Stadler et al. 2003).

The literature provides little support for the idea that a single hypothesis can account for the diversity of aphid-ant relationships. Fitting aphid-ant relationships into a continuum from highly mutualistic to antagonistic and identifying the physiological, ecological, and evolutionary constraints provide a conceptual framework for studying the opportunities and constraints shaping such relationships (Figure 1). Figure 1 also provides a road map for this review. The role of the evolutionary constraints has not been tested with the same rigor and is unlikely to be of the same importance, but these constraints may affect the direction of change in the mutualism/antagonism balance. A mutualistic relationship is expected if aphids and ants achieve higher population growth rates and larger colony sizes (Figure 1a). Aphids must be able to modify honeydew to make it more attractive to ants, which must be able to modify their foraging behavior so that they can effectively harvest this energy-rich resource.

Other ecological effects are well documented. For example, hygienic services like the collection of sugary excreta and protection from fungal infections (Figure 1b) are thought to favor mutualism. However, these are likely to be secondary because unattended aphids effectively dispose of their honeydew and are apparently not so prone to fungal infections associated with honeydew. The fragmentation of habitats and plant-related factors, such as abundance, wide distribution, and quality, are thought to facilitate mutualism because these attributes result in an increase in aphid abundance, which increases the probability of their being encountered by ants. Put another way, obligate mutualisms are unlikely to evolve between rare species. Lastly, morphological and behavioral adaptations in ants are likely to facilitate the successful establishment of mutualistic relationships (Figure 1c). Selection for an antagonistic relationship is likely if the costs, such as the production of high-quality honeydew and the absence of ant partners in many habitats or inaccessibility of suitable hosts, associated with ant attendance, are high (Figure 1d). If plants compete with aphids for the services of mutualists [e.g., via extrafloral nectaries (EFNs)] or if the chemical protection of the host plant affects the ants (e.g., honeydew containing secondary plant metabolites), an intimate relationship is less likely to develop (Figure 1f). Similarly, if the mortality due to natural enemies is low or if specialized predators or parasitoids exploit ant-attended aphids (Kaneko 2002, Völkl 1992) (Figure 1f), costs might outweigh benefits. These factors and associated mechanisms are addressed in more detail below.

It is difficult to assess the relative importance of these factors for aphid-ant relationships because of the poor understanding of their temporal dynamics or the interactions between factors. Very few studies last longer than 30 days (i.e., one aphid generation) even though the relative impact of the factors varies with time. For example, although protection is thought to be important, even obligate myrmecophiles, such as *Symydobius oblongus* (von Heyden) and *Metopeurum fuscoviride* Stroyan, are unattended on more than 25% of the occassions when surveyed during the course of a season (Stadler 2004, Stadler & Dixon 1999). Thus, short-term studies are unlikely to identify the dynamic aspects of the costs and benefits associated with ant attendance.

Features associated with:

Mutualism ← → **Antagonism**

Physiological

a)
- High growth rates—at least temporarily (El-Ziady & Kennedy 1956, Flatt & Weisser 2000, Stadler & Dixon 1999, Stadler et al. 2002)
- Ability to make the sugar composition of honeydew more suitable for ants (Fischer & Shingleton 2001, Kiss 1981, Völkl et al. 1999, Yao & Akimoto 2001)
- Honeydew is a waste product needing little further investment
- Energy source for high activity tempo foragers (Davidson 1998, Oster & Wilson 1978)

d)
- Cost of producing high-quality/large amounts of honeydew (Fischer & Shingleton 2001, Katayama & Suzuki 2002, Yao & Akimoto 2001, Yao & Akimoto 2002, Yao et al. 2000)
- Changing nutritional requirements of ants during their life cycle e.g. need for less honeydew (Hölldobler & Wilson 1990, Katayama & Suzuki 2002, Pontin 1978, Rosengren & Sundström 1991, Sudd & Sudd 1985)

Ecological

b)
- Hygienic services (Buckley 1987b, Nixon 1951)
- Protection by ants (Addicott 1978a, Cushman & Addicott 1989, Katayama & Suzuki 2002, Morris 1992, Nixon 1951, Offenberg 2000, Sakata & Hashimoto 2000, Tilles & Wood 1982, Way 1963)
- Habitat fragmentation (Braschler & Baur 2003, Braschler et al. 2003, Gotelli & Ellison 2002b, Wilson et al. 2003)
- Abundance of aphids
- Distribution of aphids
- High quality of host (Breton & Addicott 1992b, Bristow 1991)
- Gregariousness (Stadler et al. 2003)

e)
- Predation by ants (Cushman & Addicott 1989, Cushman & Whitham 1991, Engel et al. 2001, Katayama & Suzuki 2002, Offenberg 2001, Sakata 1994, Sakata 1995, Sakata & Hashimoto 2000, Yao & Akimoto 2001)
- Competition for mutualists (Addicott 1978a, Cushman & Addicott 1989, Engel et al. 2001, Fischer et al. 2001, Offenberg 2001, Sakata & Hashimoto 2000)
- Fitness costs (Stadler 2004, Stadler & Dixon 1998a, Stadler et al. 2002)
- Low predictability of C resource
- Alternative sugar sources (e.g. EFNs) (Becerra 1989, Cushman & Addicott 1989, Dreisig 1988, Offenberg 2000, Offenberg 2001, Sakata 1995, Sakata & Hashimoto 2000)
- Chemical defense of plants affects honeydew and ants (Vrieling et al. 1991, Wink & Witte 1991)

Evolutionary

c)
- Proventriculus for storing honeydew (Eisner 1957)
- Extensible gaster further facilitates the storage of honeydew (Taylor 1978)
- Overcome initial defense or aggressiveness (Doebeli & Knowlton 1998, Sakata 1994)

f)
- Low predator/parasitoid pressure (association costs) (Stadler et al. 2001)
- Exploiters of mutualism (e.g. other aphids)
- Exploiters of mutualism (e.g. specialised parasitoids, predators) (Bartlett 1961, Bronstein 2001, Mackauer & Völkl 1993, Sloggett & Majerus 2003, Völkl 1992)

Figure 1 Factors that influence the strength and direction of the associations between aphids and ants. Temporal or spatial aspects of these factors are rarely documented.

PROCESSES SHAPING APHID-ANT INTERACTIONS

Next, we will focus on the adaptations aphids and ants developed to exploit each other and the constraints they encounter in doing so. The combination of physiological, morphological, and ecological boundary conditions ultimately determines whether the net outcome of a relationship is antagonistic or mutualistic.

Adaptations of Aphids and Ants

Papers on aphid-ant relationships often suggest that ant attendance has a positive effect on aphids, resulting in larger colonies (El-Ziady & Kennedy 1956, Flatt & Weisser 2000, Skinner & Whittaker 1981), lower mortality rates (Way 1963), and removal of sticky honeydew (Nixon 1951). However, such general statements do not account for the fact that different aphid species have developed associations with ants ranging from close (obligate) through occasional (facultative) to avoidance (unattended). Only recently have the benefits aphids are assumed to derive from ant attendance been questioned and critically examined. The results reveal a number of physiological costs associated with close associations. In particular, one needs to ask whether increases in feeding and excretion actually benefit aphids, in particular those that are not closely associated with ants (see, e.g., El-Ziady & Kennedy 1956). Aphids do not simply tap into the phloem elements of plants and passively regulate the flow of plant sap through their bodies, which is sometimes collected by ants. They actively modify the composition of the sap to avoid dehydration (Fisher et al. 1984, Rhodes et al. 1996, Rhodes et al. 1997), which may also make it more attractive to ants and could have significant metabolic costs. It is possible that osmoregulation is a preadaptation for forming an association with ants. For example, the aphid *Tuberculatus quercicola* (Matsumura) incurs significant costs when attended by *Formica yessensis* Forel because it produces smaller and less fecund adults than when unattended (Yao et al. 2000). Ant-attended aphids excrete smaller droplets of honeydew at a higher rate, and the honeydew contains significantly higher concentrations of amino acids (Yao & Akimoto 2002), sucrose, and trehalose (Yao & Akimoto 2001) than that of unattended aphids. Although the concentration of amino acids in honeydew is increased, possibly as a consequence of the effect of the increased flow of phloem sap on assimilation, its significance for ants needs to be established. It is suggested that in producing large quantities of honeydew aphids have less N for growth and reproduction (Stadler & Dixon 1998a). *Chaitophorus* spp. can even modify the sugar composition of their honeydew, with those that are more closely attended (*Chaitophorus populialbae*, *Chaitophorus populeti*) able to reduce the melezitose content when unattended (Fischer & Shingleton 2001). In contrast, *Chaitophorus tremulae*, which is less often associated with ants, does not show this response. That is, the production of large quantities of honeydew, which is attractive to ants, is likely to be costly for aphids because they have to feed faster and increase the rate of converting simple to complex sugars.

It is difficult to demonstrate the fitness cost for aphids. Contrary to the claim of El-Ziady & Kennedy (1956), *Aphis fabae cirsiiacanthoides* Scop., feeding on *Cirsium arvense* (L.) Scop., incurs significantly higher costs when attended by *Lasius niger* L. It has a significantly lower mean relative growth rate, produces fewer offspring, takes longer to reach maturity, and invests less in gonads (Stadler & Dixon 1998a). The facultative myrmecophile species feeding on tansy (*Tanacetum vulgare* L.) also have significantly lower potential growth rates when attended by *L. niger* (Stadler et al. 2002). Therefore, there is increasing evidence that aphids can adapt their physiology in response to ant attendance, and this is costly. That is, there is a need to add a new dimension to the classical textbook paradigm of ant-aphid relationships.

Essential adaptations for ants involved in aphid-ant relationships are morphological changes to the proventriculus and gaster, which enable them to carry large quantities of honeydew. In the subfamilies *Formicinae* and *Dolichoderinae*, the development of a proventriculus enables them to gather large quantities of fluid rich in carbohydrate from plants or Homoptera (Davidson 1997, Davidson et al. 2004, Eisner 1957). This digestive organ is situated posterior to the crop and regulates the flow of food. The proventricular bulb pumps liquid from the crop into the midgut and prevents the posterior flow. Only in the *Dolichoderinae* and *Formicinae* does the proventriculus passively occlude the passage of food through the gut, which allows the associated musculature to be reduced (Eisner 1957). In this way ants are able to control the movement of honeydew from the crop, the "social stomach," into the midgut, where it is digested. In addition, to store large quantities of honeydew and carry it to the nest, ants need the ability to expand and contract their gaster with changes in the volume of honeydew collected (Kunkel et al. 1985, Taylor 1978).

The Ecological Framework

If one is to understand mutualistic relationships, it is necessary to address the ecological context in which both partners interact. It is likely that ants evaluate the nutritive quality and associated mortality risks of food patches, i.e., the trade-off between colony growth and forager survival. The predictions of optimal foraging theory are generally well supported, when animals face simple decisions, such as maximizing net energy returns by choosing a particular prey (Stephens & Krebs 1986). It is now appreciated that foragers are more flexible and face more complex decisions in nature, which affect their lifetime reproductive success (Clark & Mangel 2000, Houston et al. 1988, Houston & McNamara 1988, Houston & McNamara 1999). Information on the environment and physiological condition of the organism making the decisions can be incorporated into optimal foraging models.

Ants do not simply collect honeydew but are also subject to predation risks and adjust their foraging activity accordingly (Carroll & Janzen 1973). For example, when food is offered to *Lasius pallitarsis* (Provancher) in patches where the risk of predation by *Formica subnuda* Emery varies, *L. pallitarsis* spends less time foraging in the patches with *F. subnuda*, even though these patches contain high-quality resources (Nonacs & Dill 1991). Furthermore, the use of a high-quality patch depends on the magnitude of the difference in terms of growth between feeding in risky and safe patches: The greater the benefit of feeding in a risky patch, the more likely it is exploited. Thus, ant workers are capable of evaluating risks and rewards, and they forage in a way that maximizes colony fitness (Nonacs & Calabi 1992, Nonacs & Dill 1990). The trail marking by scouts, which recruit nest mates to an aphid colony, seems to be based on the rate at which they ingest honeydew (Mailleux et al. 2000). Thus, it is likely that the number of foragers recruited to a resource is related to the total production of honeydew or aphid colony size (de Biseau & Pasteels 1994, Mailleux et al. 2003, Völkl et al. 1999).

This simplistic view becomes more complicated if the spatial context in which aphid-ant relationships evolved is taken into consideration. There are numerous reports that aphid colonies close to ant nests experience higher attendance than those further away (Boucher et al. 1982, Bradley & Hinks 1968, Pontin 1978). However, even this tendency changes during the course of a season because aphids colonize trees further and further away from the main nest (Scheurer 1967). Aphids living in fragmented grassland are more frequently visited than those in unfragmented plots (Braschler et al. 2003). It is suggested that protection from predators may positively affect ant-attended colonies. However, unattended colonies are also more numerous in fragmented habitats, which indicates that fragmentation per se is more important than the services supplied by ants. Fragmentation is not only associated with an increase in aphid numbers but also a change in the composition of ant communities. For example, the polygynous ant *Formica aquilonia* Yarrow is more abundant in old forests and large fragments of forests and the monogynous *Formica lugubris* Zett. in young forests, small fragments of old-forest, and the edges of forests (Punttila 1996). Boreal ant communities, i.e., those in spruce-dominated forests, are influenced by succession more than the species of aphid they attend. There is good evidence that ant communities can affect the composition of aphid communities. In forest and grassland, apparent competition, the sharing of natural enemies (Bishop & Bristow 2001, Müller & Godfray 1999, Rott et al. 1998), or competition for mutualists is implicated in this (Addicott 1978a, Cushman & Addicott 1989).

Information on aphid ecology, mortality risks, ant foraging, recruitment behavior, and the spatial context, referred to above, is best summarized in a conceptual model, which predicts when and under what environmental conditions mutualistic relationships between aphids and ants are likely to evolve (Figure 2). It is reasonable to assume that with increase in aphid population size the energy supply for the ants will increase (Figure 2a). The energy gain is greater if the aphid species attended by the ants produce more or better quality honeydew. The more honeydew available per unit of foraging time/area the more likely ants will focus on the most productive colonies because fewer workers are required to collect the same amount of energy. Less productive colonies further away from a nest are likely to be abandoned first. As a consequence of the decline in the foraging activity for honeydew, with fewer workers protecting distant aphid colonies, the mortality of aphids due to natural enemies might increase (Figure 2b). The risk to aphids of death caused by predators can be decreased through dispersal, leading to smaller colony sizes, which are too small to be attractive to natural enemies (Kindlmann & Dixon 1993), or through recruiting more ants by producing more attractive honeydew (Kiss 1981). This is likely to result in competition for mutualists, which may be the principal factor increasing the costs for myrmecophilic aphids (Addicott 1978a, Cushman & Addicott 1989, Offenberg 2001, Sakata & Hashimoto 2000). Competition between myrmecophilous aphids for mutualists is a "red queen game" as the need to produce attractive honeydew suggests that alternative resources should affect aphid-ant relationships, e.g., in situations where aphids feed on plants with EFNs or food bodies. However, the results are equivocal with some studies reporting no competition between EFNs and honeydew-producing Homoptera (Delclaro & Oliveira 1993), whereas others do find negative interactions (Engel et al. 2001, Koptur 1991, Sakata & Hashimoto 2000). In addition, currently the effects of EFNs on aphid-ant interactions do not discriminate between different mechanisms or cause and effect in these associations (Offenberg 2000). In any case, closer associations with ants are likely to reduce the risk of mortality due to the activity of natural enemies (Figure 2b). Obligate myrmecophiles are probably more successful in attracting ants and benefit from being better protected than facultative myrmecophiles.

As noted above, the production of large quantities of honeydew entails metabolic costs. This is likely to be true for both obligate and facultative myrmecophiles. If this is correct, then there should be a negative correlation between the intensity of tending by ants and the cost of being a myrmecophile (Figure 2c). For a particular level of attendance, the overall costs should be higher in patches with a low mortality risk. This is because there is an investment in defense via ants, but the investment does not yield any interest in low mortality risk patches. Several recent studies have demonstrated that aphids are able to perceive mortality risks, e.g., *Acyrthosiphon pisum* Harris produces winged morphs in the presence of natural enemies such as ladybirds (*Coccinella septempunctata* L.) (Dixon & Agarwala 1999, Weisser et al. 1999), hoverfly larvae [*Episyrphus balteatus* (De Geer)], lacewing larvae [*Chrysoperla canea* (Stephens)] (Kunert & Weisser 2003), or parasitoids (*Aphidius ervi* Haliday) (Sloggett & Weisser 2002). This occurs either after contact with a predator or with chemical cues left by ladybirds. We hypothesize that ant-attended aphids do not respond in this way to the presence of ladybird larvae (Dixon & Agarwala 1999).

For aphids, the relationships between the different model parameters can be reduced to a simple ratio of the mortality risk (m) relative to the cost of being a myrmecophile (c). The expectation is that if $m/c < 1$ the aphids should not develop a relationship with ants, if $m/c \approx 1$ a facultative association is possible, and if $m/c > 1$ obligate myrmecophily is likely. This ratio reflects the trade-off between the

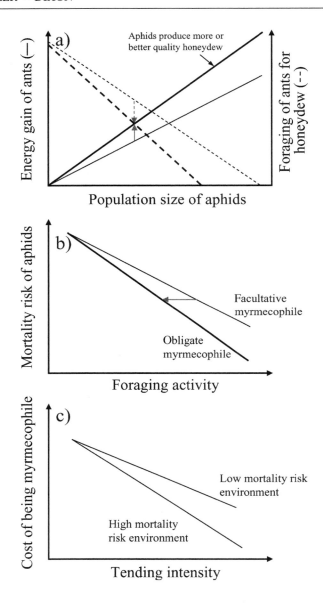

mortality suffered by a clonal population and the cost of producing more or better honeydew in response to a specific foraging activity and environment. The unit of measurement for these parameters must be related to the fitness of an aphid clone, which indicates the importance of using a common currency when measuring the effects of ant attendance. Currently, it is unknown whether the relationships shown in Figure 2*a–c* are linear.

Evolutionary Constraints

Once ants evolved ways of collecting and processing honeydew, it is likely they were able to forage larger areas and become more abundant. Oster & Wilson (1978) distinguish between "high tempo" and "low tempo" ants. They assume that a positive correlation exists between behavioral tempo, colony size, and polymorphism. Foraging over a large rather than a small area might impose a higher mortality risk, which may account for a higher turnover rate of workers. However, little attention was paid to the type of food collected, and they predicted that homopteran-tending ants would be low tempo foragers. This conflicts with the observed variation in foraging by ants associated with aphids, and diet analyses also show a more complex picture with a high variability in the food types used (Blüthgen et al. 2003). It is more likely that the remarkable range in worker activity is associated with the price of honeydew, which is dependent on the distance of the colony (gasoline station) from the ant's nest and the time required to collect the honeydew. Ants that are able to build their nest, or parts of it, close to such resources should have a selective advantage over those that are restricted in where they can build their nests. The formation of subsidiary nests by *Formica rufa* L. leads to a network of intercommunicating nests with up to 200 meters between the main and peripheral nests (Gösswald 1941, Rosengren & Pamilo 1983). That is, polydomy could be the typical nest structure of species collecting large quantities of honeydew (Davidson 1997, Davidson 1998).

Another difficulty in the evolution of a mutualistic association is the initial response of aphids to ants. It is claimed that mutualism evolved out of a predacious/parasitic relationship, e.g., through the hosts' ability to terminate exploitation by a predator/parasite (Johnstone & Bshary 2002). Others claim that individuals in populations with a mutualistic strategy survive environmental stress better than

←

Figure 2   Expected associations between ant and aphid features in mutualistic relationships. (*a*) Population size of myrmecophiles is negatively correlated with the foraging activity of ants (*dashed lines*) and positively correlated with energy gain (*solid lines*). For a particular population size, those species of aphid that produce large quantities of honeydew are a better source of energy (*solid arrow*) because the ants spend less time foraging for honeydew (*dashed arrow*). (*b*) With decline in foraging activity, the mortality risk increases for ant-attended aphids. The expectation is that the rate of decline is higher for obligate myrmecophiles. Thus, to reduce this mortality risk, a facultative myrmecophile needs to increase the quality or quantity of honeydew it produces, which may lead to a closer relationship with ants. (*c*) Associated with the degree of myrmecophily and foraging activity, there are costs, which depend on the mortality risk associated with particular habitats. For a particular tending level, those patches with a low mortality risk are costly for myrmecophiles because their investment in defense (production of high-quality honeydew) exceeds the benefits afforded by protection when natural enemy pressure is low.

those with an antagonistic strategy (Stachowicz 2001). However, a major evolutionary challenge for the development of mutualism is likely to have been the ability of aphids to survive encounters with ants (Doebeli & Knowlton 1998), especially as they are aggressive and ecologically dominant. Aphids, however, are a poor food resource (e.g., Toft 1995) because of their high sugar content, and aphids are unlikely to be a preferred prey as long as alternative prey is available and ants do not become specialist predators of aphids [but see *Lasius flavus* (Pontin 1978)].

Mutualistic relationships also tend to be exploited by specialist predators and parasitoids or by cheaters (Bronstein 2001). For example, a number of coccinellid larvae are known to have evolved protective structures like waxy wool, chemical mimicry, or inconspicuous movements. The wax covered larvae of *Scymnus interruptus* (Goeze) and *Scymnus nigrinus* Kugelann survive attacks by *L. niger* and *Formica polyctena* more often than larvae without wax (Völkl & Vohland 1996) and reach higher densities in ant-attended colonies, indicating ants probably afford the larvae protection against their natural enemies. Similarly, the wax covered larvae of *Stator sordidus* Horn, *Hyperaspis congressus* (Bartlett 1961), and *Platynaspis luteorubra* Goeze (Völkl 1995) are reported to be protected. Larvae and adults of *Coccinella magnifica* Redtenbacher use behavioral and chemical defense to avoid attacks by *F. rufa* (Sloggett & Majerus 2003, Sloggett et al. 1998). Aphid parasitoids, such as *Lysiphlebus cardui* (Marshall), *Lysiphlebus hirticornis* Mackauer, *Paralipsis eicoae* Fr. Smith, and *Paralipsis enervis* (Nees), have evolved chemical and behavioral mimicry, which enable them to exploit ant-attended aphid colonies without being attacked (Mackauer & Völkl 1993, Takada & Hashimoto 1985, Völkl 1992, Völkl et al. 1996). It is, however, not clear how often such relationships have evolved, what the relative costs are for these specialist aphid predators and parasitoids, or what kind of selection pressure they exert on aphid-ant relationships.

In addition, aphid-ant relationships may also be exploited by other species of aphids. As indicated above, the association with ants does entail costs. Therefore, to obtain these attendance benefits without paying the cost would be beneficial for an aphid. Although conceivable, there is no direct evidence that aphids are able to exploit established aphid-ant relationships, for example, by benefiting from protection services on a shared host plant. The indirect evidence is that on the leaves of birch (*Betula pendula* Roth) the co-occurrence of *Betualphis brevipilosa* Börner with *Callipterinella calliptera* (Hartig), which is ant attended, is more frequent than expected by chance, whereas co-occurrence with the unattended *Euceraphis betula* (Koch) is random (Hajek & Dahlsten 1986). Therefore, not only interspecific competition for ants might act against the development of mutualistic relationships but also specialized natural enemies and opportunistic aphids.

When considering aphid-ant relationships, one needs to appreciate how selection might act. Aphid colonies largely consist of clones; ants are socially organized with a high genetic relatedness, and as a consequence, selection in both taxa does not operate on individuals but on the colony or genet to which an individual belongs. Therefore, a useful approach is to attempt to understand how a clone, rather than an individual, should invest in defense and reproduction. An interesting case is aphid soldiers, which evolved in the closely related families Pemphigidae and Hormaphididae (Aoki 1977, Stern & Foster 1997). In a number of social aphid species, it is known that on both their primary and secondary host plants the aphids may have different means of protection. The tropical aphid *Cerataphis fransseni* (Hille Ris Lambers), for example, are attended by various species of ants on the secondary host but not on the primary host where this aphid produces galls (Stern et al. 1995). Similarly, *Pseudoregma sundanica* (van der Goot) has two defense strategies: It is obligately attended by ants, that is, it exploits their protection services and produces sterile soldiers. The level of investment in soldiers is adjusted in response to ant attendance, with soldier production and ant attendance being negatively correlated (Shingleton & Foster 2000). This response is rapid and leads to a significant change in caste structure. However, only if soldier production is directly and inversely proportional to the incidence of ant attendance are the costs the same. Investment in ants and/or soldiers might be just two alternative ways of defense, equally costly, but with different payoffs in different environments. Therefore, a major challenge is to identify costs and benefits at the level of a clone, even if only the fitness of individuals can be studied. Preliminary molecular evidence on phylogenetic patterns in aphid-ant relationships indicates that ant tending is an evolutionarily labile trait, which was evolved and lost several times. For example, ant attendance in the genus *Chaitophorus* evolved at least five times; that is, it is relatively "easy" for an aphid lineage to "gain and lose tending" (Shingleton & Stern 2003). A similar conclusion was drawn from a comparative analysis of ant attendance of 112 species of aphids and 103 species of lycaenids (butterflies) from Europe, using morphological and ecological traits, such as size, feeding site, colony structure, or host characteristics. Overall, in both insect groups, relationships with ants were only slightly (10%) associated with environmental and ecological traits. Most of the variation in the degree of ant attendance in aphids is explained at the subfamily level and least at the species level, whereas for the lycaenids most is explained at a higher taxonomic level, such as a tribe (Stadler et al. 2003). This suggests that aphids are more flexible in their associations with ants at the species level, entering and leaving associations with ants whenever it is advantageous. The most likely reason for the difference is that honeydew is a waste product, whereas lycaenids have to invest in morphological structures such as nectar glands, the structure and function of which are less easy to modify.

Little is known about the indirect costs of aphid-ant relationships. Indirect costs occur when aphids develop specific adaptations fostering a closer association with ants (e.g., aggregation behavior or reduction in defensive structures, such a siphunculi, or loss of defensive behavior), which become maladaptive when ants are not available. These are the "association costs." For example, aphids that feed on the woody parts of a tree need long stylets to reach the deeply located phloem elements. Because it takes longer to withdraw long than short stylets, the aphids are at greater risk of attack from natural enemies. This is likely to favor a closer relationship with ants to exploit their protective function (Dixon 1998, Shingleton et al. 2005). As a consequence, host plants growing where an appropriate ant partner is not available cannot be exploited. A restriction in host availability is likely to lead to lower abundance and increased risk of local extinction. As the area where conditions are suitable for both partners is likely to be less than that potentially suitable for the aphid selection, pressures should operate against strong dependence on ants to minimize the association costs. That is, nonattendance and facultative myrmecophily rather than obligate associations should prevail, which is indeed the case for most aphids, coccids, and lycaenids (Bristow 1991, Pierce et al. 2002).

MUTUALISM IN A DYNAMIC CONTEXT

It is puzzling that some aphids are closely associated with ants, whereas closely related species and other species living on the same host plant are not. This is independent of whether the host is a tree, herbaceous plant, or grass. There are few if any exceptions to the general finding that on abundant, long-lived plants there are species of aphids that show obligate, facultative, or no association with ants. Because of the multiplicity of strategies, one must be extremely careful not to oversell particular hypotheses, such as the plant permissive hypothesis or competition for mutualists, resulting in statements that one strategy is superior to (or more successful than) another. Many of the factors that are thought to influence the relationships between aphids and ants (Figure 1) were revealed by experiments carried out in a constant environment. This could lead to erroneous conclusions. For example, the idea of competition for mutualists, although well documented, is supported by few field studies. That is, conclusions derived from laboratory experiments done under specific environmental conditions demonstrate just one of a range of possible outcomes and do not explain the coexistence of different species of aphids and their association with ants on a shared host plant. Below we describe some good examples that incorporate a dynamic approach to ant-aphid associations not apparent in the dichotomy of Figure 1.

One notable exception to the static approach is the pioneering work of John Addicott and coworkers who studied the fireweed-aphid-ant system in great detail. They conclusively demonstrated that the ecological framework, such as alternative sugar resources, affects the strength of competitive interactions with intra- and interspecific neighbors for the services of two species of ants (Addicott 1978a, Addicott 1985). In addition, they clearly demonstrated that the positive effect of ants on aphid population growth rates is density dependent. At low aphid densities, ant attendance reduced the probability of extinction, whereas it had either no effect or even increased the probability of a decline at high aphid densities (Breton & Addicott 1992a). The concept of ecological neighborhood (Addicott et al. 1987), which is the area of activity or influence of organisms within the temporal and spatial context, stresses the idea that the scales at which observations and experiments are made are important for the interpretation of the results. There is no single ecological neighborhood for any given organism, and it is important to address and report the scales to facilitate comparisons between studies and between empirical and theoretical postulates. One example is given below.

The immediate environment of an aphid is its host plant and position on the plant. Plant quality changes during the course of a growing period, for example, with the age of the plant tissue the aphids feed on (Johnson et al. 2003), with growing conspecific or interspecific herbivore load, and via induced plant defense (Karban & Baldwin 1997). Although the effects of host plants on aphid-ant relationships might not be easy to identify (Breton & Addicott 1992b), the general expectation is that aphids do best on high-quality plants. As a consequence, not only is the local distribution of ants' nests in a habitat important for myrmecophiles (top-down effect) (see above) but also the distribution and seasonal changes in their host plants (bottom-up effects). Most plants consist of a mosaic of different tissues in different developmental stages (Gill et al. 1995). Therefore, it is important for an aphid to exploit the highest quality patches where their reproduction is highest. Feeding on host tissues of suboptimal quality, e.g., when confined to certain plant organs by attending ants, carries costs. This view emphasizes reproduction and fitness costs and benefits as determinants of the outcome of aphid-ant interactions. This is different from the plant permissive hypothesis because it favors a bottom-up control, which means that plant quality determines honeydew quality and thus the level of association in ant-aphid mutualisms.

Another study of the dynamic nature of ant-aphid associations combined field and laboratory experiments on the four co-occurring aphids on tansy, in which plant quality and access of ants were manipulated, to determine the relative fitness consequences for aphids showing different levels of associations with ants. In the absence of natural enemies and ants, the obligate myrmecophile, *M. fuscoviride* Stroyan, had the lowest potential growth rate on high-quality plants (supplemented with N), whereas the facultative myrmecophiles, *Aphis fabae* Scop. and *Brachycaudus cardui* (L.), and especially the unattended species, *Macrosiphoniella tanacetaria* (Kaltenb.), did significantly better. Low plant quality (no N addition) usually adversely affected the performance of the unattended species and facultative myrmecophiles, indicating the importance of high-quality plants for high aphid growth rates. Thus, a combination of host and degree of mutualism differentially affects the fitness of these aphids in different environments (Stadler et al. 2002). In the field, the realized population growth rates are less than the potential population growth rates because of fluctuations in abiotic conditions and predator pressure. On high-quality plants with ants, the aphid *M. fuscoviride* had the highest realized fitness, whereas on low-quality plants, the colonies became extinct probably because the small colonies that developed on these plants were not attractive to ants. Under these conditions, the mobile *M. tanacetaria* did better. In addition, the relative frequency of high- and low-quality plants and the presence of ants are likely to influence the outcome (Stadler 2004). This heterogeneity in patch quality might be the reason why *M. tanacetaria* is so mobile, with most colonies surviving for only 1–2 weeks (Massonnet et al. 2002). Therefore, it is suggested that a combination of top-down and bottom-up forces affect aphids within a season and determine the advantage of mutualism. Different species experience different advantages and disadvantages from mutualistic relationships in a changing environment, which allows different degrees of association with ants and a simultaneous exploitation of a shared host species.

As indicated above, plant quality affects aphid-ant interactions in many ways. As a consequence, external influences on plant quality might also cascade through to ant-aphid mutualisms. For example, in addition to the N content of plant tissue (Mattson 1980), secondary plant compounds affect aphid performance (Montllor 1991, Pickett et al. 1992) and might also be expected to affect aphid-ant interactions. Evidence is slowly accumulating that this is indeed the case. For instance, several aphid species, such as *Macrosiphum albifrons* Essig and *Aphis genistae* Scop., are known to sequester quinolizidine alkaloids, reaching 4 mg/g fresh weight in *A. genistae* and 1.8 mg/g in *M. albifrons* (Wink & Witte 1991). It is suggested that these aphids are able to exploit the chemical defense compounds of plants for their own defense against natural enemies. *A. genistae* and *Aphis cytisorum* are facultatively associated with ants, and the honeydew collected by *L. niger* workers from colonies of *A. cytisorum* contain on average 45 μg/g fresh weight of quinolizidine alkaloids (Szentesi & Wink 1991). However, it is unknown whether these secondary compounds affect the ants. The interactions between ragwort, *Senecio jacobaea* L., the aphid, *Aphis jacobaeae* Schrank, the moth, *Tyria jacobaeae* L., and ants are associated with variation in the pyrrolizidine alkaloid content of the host plant. In this system, aphids colonize plants with low-alkaloid concentrations and avoid those with high concentrations (Vrieling et al. 1991). Plants without aphids are often defoliated by *T. jacobaeae* because of the absence of ants. If not defoliated in the absence of ants, then they are likely to have a high pyrrolizidine alkaloid content, which protects them from the moth. It is suggested that the above interactions maintain the genetic variation in the alkaloid content of *S. jacobaea*. However, it is unknown whether ants actively select aphid colonies that produce honeydew with a low-alkaloid content.

For plants, the variation in availability of resources, particularly N, may determine their investment in secondary compounds and ultimately aphid-feeding behavior (Coley et al. 1985, Herms & Mattson 1992, Kainulainen et al. 1996). Application of fertilizers to plants often increases the performance of the aphids they host as the result of either direct effects (Grüber & Dixon 1988) and/or indirect effects, e.g., a concurrent increase in tending and patrolling by ants (Strauss 1987). In addition, the application of fertilizer often results in a decrease in the concentration of secondary compounds, such as phenolics (Kainulainen et al. 1996), but the effect of this on aphid-ant relationships is less clear. Soil invertebrates such as protozoa, collembola, and earthworms can also affect the N and secondary compound content of plants. For example, the soil biota act via the host plants on cereal aphids [*Sitobion avenae* (F.) and *Myzus persicae* (Sulzer)], but the effects are different for different groups of soil organisms. Protozoa and collembola, but not earthworms, significantly increase aphid performance (Bonkowski et al. 2001, Scheu et al. 1999). This indicates that indirect effects, such as the grazing of bacteria by protozoa and fungi by collembola, might have a pronounced effect on nutrient availability, plant growth, and aphid performance. It is, however, not clear whether soil biota significantly affects mutualisms between aphids and ants, but it is conceivable. The conclusion is that factors that are patch specific (bottom-up) are likely to influence both partners and might affect the development of mutualistic relationships. It is increasingly recognized that multitrophic/multispecies effects on mutualistic interactions are important (Bronstein & Barbosa 2002), but the mechanisms shaping the strength of the interactions are difficult to identify, and the common patterns in different systems even more so. The take-home message is that those parts of a habitat in which aphids can achieve high growth rates and establish mutualistic relationships, especially close associations, with ants is probably considerably smaller than the potential habitat. Experimental and theoretical studies of aphid-ant associations are more likely to be successful if a patch-oriented perspective of coevolution (Thompson 1994) is adopted to explain the breadth of the associations between aphids and ants.

CONCLUSIONS AND FRONTIERS IN APHID-ANT INTERACTIONS

The study of aphid-ant interactions is hampered by many factors that influence the outcome of this relationship. For example, bottom-up effects like the distribution and abundance of the host plants of the aphids or community composition of soil invertebrates are likely to cascade through the system and either directly affect the growth rates of aphids or indirectly via plant nutrients or secondary plant compounds. In a similar way, top-down effects of natural enemies are likely to shape the strength of aphid-ant relationships. As a consequence, we should not expect a high degree of specialization in these interactions as most mutual associations are best characterized as temporary. This is probably true for many partners of ants. For example, even though there are spectacular instances of highly specialized associations between ants and aphids, coccids, membracids, and lycaenids, the majority are often facultative and unspecific (Pierce et al. 2002, Bristow 1991, Fiedler 1991, Delabie 2001), and the number of myrmecophilous species declines from the tropics to the temperate regions (Fiedler 1998).

Given their multitrophic nature, the outcome of these relationships needs to be followed for a complete season or several seasons to appreciate the full range of costs and benefits for both partners. Laboratory experiments done under constant conditions can only provide an indication of the constraints acting on both partners and need to be paralleled by field investigations whose duration is sufficiently long to encompass the life cycles of both partners. It is important to determine fitness costs and benefits in different environments. Although it is experimentally challenging to simultaneously determine intrinsic rates of increase, it is more rewarding to compare and understand the costs and benefits for different species in different environments.

Our current understanding of costs is still rather rudimentary. For example, at the physiological level, there is little or no information on the change in the composition of honeydew and associated adaptation costs for different generations of aphids in responding to ant attendance, when both colony size and quality of the host plants vary. Clearly, there is need to determine the extent to which foraging activity is associated with sugar:protein imbalances in the diet of ants, and their effect on aphid communities. In terms of ecology, there is a growing awareness that spatial variability affects the distribution and abundance of both partners and the effects of bottom-up and top-down forces (Blossey & Hunt-Joshi 2003, Edson 1985, Gotelli & Ellison 2002a, Morris 1985, Müller & Godfray 1999). This indicates that a more spatially explicit or metapopulation perspective may be more appropriate (Addicott 1978b, Albrecht & Gotelli 2001, Edson 1985). The relative importance of mutualistic, neutral, and antagonistic interactions between aphids and ants and their relative role in community structure and species diversity are beginning to be addressed (Wimp & Whitham 2001). Still further in the future, but no less important, is the need to address the effects of the mutualistic and antagonistic interactions of these two dominant groups of insects on nutrient cycling and ecosystem functioning (Loreau 1995, Stadler et al. 1998).

In the immediate future, there are a number of interesting questions that should be addressed, either experimentally or theoretically. For example, do the population dynamics of myrmecophiles and nonmyrmecophiles differ, and if so, in what way? How do costs and benefits vary with colony size and during the course of a season? How can costs and benefits be determined considering the different genetic structures of the clonal and socially organized partners? Are there differences in the dispersal rates/patterns of myrmecophiles and nonmyrmecophiles? How does patch size affect ant attendance, and what are the costs for ants in attending aphids? How does the spatial and temporal variability in plant phenology/quality affect the distribution and abundance of aphids showing different degrees of associations with ants? What is the relative importance, in a temporal and spatial context, of the factors depicted in Figure 1 for the outcome of mutualistic relationships? In what way do plant secondary compounds or the belowground community structure affect aphid-ant relationships?

The traditional view is that ants are in control of the interaction with aphids. However, many aphid species do not compete for the services of ants and appear to have a range of options to cope with ant partners that are an unpredictable and unreliable resource. Considering the different life cycles (e.g., parthenogenetic reproduction versus social organization) and the different selection pressures that are associated with these features, there is ample opportunity for both partners to exploit each other. For a clonal fast-reproducing organism, for example, there might be little cost involved if ants prey on a few individuals. However, tempo foragers (Oster & Wilson 1978) need to monopolize a source of energy within their foraging area to avoid conflicts with conspecifics or other ant species. Therefore, it is likely that adaptive changes in the physiology, behavior, and genetic structure of the partners results in an increase in their fitness. Aphid-ant relationships are easy to manipulate and an ideal system for defining the driving forces in the ecology and evolution of antagonistic/mutualistic relationships.

ACKNOWLEDGMENTS

We would like to thank Aaron Ellison and Shin-ichi Akimoto for providing helpful comments on the manuscript. Financial support was given from the German Ministry for Research and Technology (Fördernummer: BMBF No. PT BEO 51-0339476D).

The *Annual Review of Ecology, Evolution, and Systematics* is online at http://ecolsys.annualreviews.org

LITERATURE CITED

Addicott JF. 1978a. Competition for mutualists—aphids and ants. *Can. J. Zool.* 56:2093–96

Addicott JF. 1978b. The population dynamics of aphids on fireweed: a comparison of local populations and metapopulations. *Can. J. Zool.* 56:2554–64

Addicott JF. 1985. Competition in mutualistic systems. In *The Biology of Mutualism*, ed. DH Boucher, pp. 217–47. New York: Oxford Univ. Press

Addicott JF, Aho JM, Antolin MF, Padilla DK, Richardson JS, Soluk DA. 1987. Ecological neighborhoods: scaling environmental patterns. *Oikos* 49:340–46

Albrecht M, Gotelli NJ. 2001. Spatial and temporal niche partitioning in grassland ants. *Oecologia* 126:134–41

Aoki S. 1977. *Colophina clematis* (Homoptera, Pemphigidae), an aphid species with "soldiers." *Kontyû* 45:276–82

Bartlett BR. 1961. The influence of ants upon parasites, predators, and scale insects. *Ann. Entomol. Soc. Am.* 54:543–51

Becerra JXI. 1989. Extrafloral nectaries: a defense against ant-homoptera mutualisms? *Oikos* 55:276–80

Begon ME, Harper JL, Townsend CR. 1999. *Ecology.* Oxford: Blackwell Sci. 1068 pp.

Bishop DB, Bristow CM. 2001. Effect of Allegheny mound ant (Hymenoptera: Formicidae) presence on homopteran and predator populations in Michigan jack pine forests. *Ann. Entomol. Soc. Am.* 94:33–40

Blossey B, Hunt-Joshi TR. 2003. Belowground herbivory by insects: influence on plants and aboveground herbivores. *Annu. Rev. Entomol.* 48:521–47

Blüthgen N, Gebauer G, Fiedler K. 2003. Disentangling a rainforest food web using stable isotopes: dietary diversity in a species-rich ant community. *Oecologia* 137:426–35

Bonkowski M, Geoghegan IE, Birch ANE, Griffiths BS. 2001. Effects of soil decomposer invertebrates (protozoa and earthworms) on an above-ground phytophagous insect (cereal aphid) mediated through changes in the host plant. *Oikos* 95:441–50

Boucher DH, James S, Keeler KH. 1982. The ecology of mutualism. *Annu. Rev. Ecol. Syst.* 13:315–47

Bradley GA, Hinks JD. 1968. Ants aphids and jack pine in Manitoba. *Can. Entomol.* 100:40–50

Braschler B, Baur B. 2003. Effects of experimental small-scale grassland fragmentation on spatial distribution, density, and persistence of ant nests. *Ecol. Entomol.* 28:651–58

Braschler B, Lampel G, Baur B. 2003. Experimental small-scale grassland fragmentation alters aphid population dynamics. *Oikos* 100:581–91

Breton LM, Addicott JF. 1992a. Density-dependent mutualism in an aphid-ant interaction. *Ecology* 73:2175–80

Breton LM, Addicott JF. 1992b. Does host plant quality mediate aphid-ant mutualism. *Oikos* 63:253–59

Bristow CM. 1991. Why are so few aphids anttended? See Huxley & Cutler 1991, pp. 104–19

Bronstein JL. 1994a. Conditional outcomes in mutualistic interactions. *Trends Ecol. Evol.* 9:214–17

Bronstein JL. 1994b. Our current understanding of mutualism. *Q. Rev. Biol.* 69:31–51

Bronstein JL. 2001. The exploitation of mutualism. *Ecol. Lett.* 4:277–87

Bronstein JL, Barbosa P. 2002. Multitrophic/multispecies mutualistic interactions: the role of non-mutualists in shaping and mediating mutualisms. In *Multitrophic Level Interactions*, ed. T Tscharntke, BA Hawkins, pp. 44–66. Cambridge: Cambridge Univ. Press

Buckley R. 1987a. Ant-plant-homopteran interactions. *Adv. Ecol. Res.* 16:53–85

Buckley RC. 1987b. Interactions involving plants, Homoptera, and ants. *Annu. Rev. Ecol. Syst.* 18:111–35

Carroll CR, Janzen DH. 1973. Ecology of foraging by ants. *Annu. Rev. Ecol. Syst.* 4:231–57

Clark C, Mangel M. 2000. *Dynamic State Variable Models in Ecology.* Oxford: Oxford Univ. Press. 289 pp.

Coley PD, Bryant JP, Chapin FS. 1985. Resource availability and plant antiherbivore defense. *Science* 230:895–99

Connor RC. 1995. The benefits of mutualism: a conceptual framework. *Biol. Rev.* 70:427–57

Cushman JH, Addicott JF. 1989. Intra- and interspecific competition for mutualists: ants as a limited and limiting resource for aphids. *Oecologia* 79:315–21

Cushman JH, Whitham TG. 1991. Competition mediating the outcome of a mutualism—protective services of ants as a limiting resource for membracids. *Am. Nat.* 138:851–65

Davidson DW. 1997. The role of resource imbalances in the evolutionary ecology of tropical arboreal ants. *Biol. J. Linn. Soc.* 61:153–81

Davidson DW. 1998. Resource discovery versus resource domination in ants: a functional mechanism for breaking the trade-off. *Ecol. Entomol.* 23:484–90

Davidson DW, Cook SC, Snelling RR. 2004. Liquid-feeding performances of ants (Formicidae): ecological and evolutionary implications. *Oecologia* 139:255–66

de Biseau JC, Pasteels JM. 1994. Regulated food requirement through individual behaviour of scouts in the ant, *Myrmica sabuleti* (Hymenoptera: Formicidae). *J. Insect Behav.* 7:767–77

Delabie JHC. 2001. Trophobiosis between Formicidae and Hemiptera (Sternorrhyncha and Auchenorrhyncha): an overview. *Neotrop. Entomol.* 30:501–16

Delclaro K, Oliveira PS. 1993. Ant-homoptera interaction: Do alternative sugar sources distract tending ants? *Oikos* 68:202–6

Dixon AFG. 1998. *Aphid Ecology.* London: Chapman & Hall. 300 pp.

Dixon AFG. 2004. *Insect Herbivore-Host Dynamics: Tree Dwelling Aphids.* Cambridge: Cambridge Univ. Press

Dixon AFG, Agarwala BK. 1999. Ladybird-induced life-history changes in aphids. *Proc. R. Soc. London Ser. B* 266:1549–53

Dixon AFG, Horth S, Kindlmann P. 1993a. Migration in insects: cost and strategies. *J. Anim. Ecol.* 62:182–90

Dixon AFG, Kindlmann P, Lepš J, Holman J. 1987. Why there are so few species of aphids, especially in the tropics. *Am. Nat.* 129:580–92

Dixon AFG, Kundu R, Kindlmann P. 1993b. Reproductive effort and maternal age in iteroparous insects using aphids as a model group. *Funct. Ecol.* 7:267–72

Doebeli M, Knowlton N. 1998. The evolution of interspecific mutualism. *Proc. Natl. Acad. Sci. USA* 95:8676–80

Dreisig H. 1988. Foraging rates of ants collecting honeydew or extrafloral nectar, and some possible constraints. *Ecol. Entomol.* 13:143–54

Eastop VF. 1973. Deductions from the present day host plants of aphids and related insects. In *Insect/Plant Relationships,* ed. HF Emden, pp 157–78. London: 6th Symp. R. Entomol. Soc. London

Edson J. 1985. The influence of predation and resource subdivision on the coexistence of goldenrod aphids. *Ecology* 66:1736–43

Eisner T. 1957. A comparative morphological study of the proventriculus of ants (Hymenoptera: Formicidae). *Bull. Mus. Comp. Zool.* 116:429–90

El-Ziady S, Kennedy JS. 1956. Beneficial effects of the common garden ant, *Lasius niger* L. on the black bean aphid, *Aphis fabae* Scopoli. *Proc. R. Entomol. Soc. London Ser. A* 31:61–65

Engel V, Fischer MK, Wäckers FL, Völkl W. 2001. Interactions between extrafloral nectaries, aphids and ants: Are there competition effects between plant and homopteran sugar sources? *Oecologia* 129:577–84

Fiedler K. 1991. Systematic, evolutionary, and ecological implications of myrmecophily within the Lycaenidae (Insecta: Lepidoptera: Papilionidae). *Bonn. Zool. Monogr.* 31:1–210

Fiedler K. 1998. Geographical patterns in life-history traits of Lycaenid butterflies—ecological and evolutionary implications. *Zoology* 100:336–47

Fischer MK, Hoffmann KH, Völkl W. 2001. Competition for mutualists in an ant-homopteran interaction mediated by hierarchies of ant attendance. *Oikos* 92:531–41

Fischer MK, Shingleton AW. 2001. Host plant and ants influence the honeydew sugar composition of aphids. *Funct. Ecol.* 15:544–50

Fisher DB, Wright JP, Mittler TE. 1984. Osmoregulation by the aphid *Myzus persicae*: a physiological role for honeydew oligosaccharides. *J. Insect Physiol.* 30:387–93

Fittkau EJ, Klinge H. 1973. On biomass and trophic structure of the central Amazonian rain forest ecosystem. *Biotropica* 5:2–14

Flatt T, Weisser WW. 2000. The effects of mutualistic ants on aphid life history traits. *Ecology* 81:3522–29

Gill DE, Perkins LCS, Wolf JB. 1995. Genetic mosaicism in plants and clonal animals. *Annu. Rev. Ecol. Syst.* 26:423–44

Goidanich A. 1956. *Stomaphis quercus* and ants. *Bull. Inst. Entomol. Univ. Bologna* 23:93–131

Gösswald K. 1941. Rassenstudien an der roten Waldameise *Formica rufa* L. auf systematischer, ökologischer, physiologischer und biologischer Grundlage. *Z. Angew. Entomol.* 28:62–124

Gotelli NJ, Ellison AM. 2002a. Assembly rules for New England ant assemblages. *Oikos* 99:591–99

Gotelli NJ, Ellison AM. 2002b. Biogeography at a regional scale: determinants of ant species density in New England bogs and forests. *Ecology* 83:1604–9

Grüber K, Dixon AFG. 1988. The effect of nutrient stress on development and reproduction in an aphid. *Entomol. Exp. Appl.* 47:23–30

Gullan PJ, Kosztarab M. 1997. Adaptations in scale insects. *Annu. Rev. Entomol.* 42:23–50

Hajek AE, Dahlsten DL. 1986. Coexistence of three species of leaf-feeding aphids (Homoptera) on *Betula pendula. Oecologia* 68:380–86

Heie OE. 1967. Studies on fossil aphids (Homoptera: Aphidoidea). *Spolia Zool. Musei Hauniensis* 26:1–274

Herms DA, Mattson JW. 1992. The dilemma of plants: to grow or defend. *Q. Rev. Biol.* 67:283–335

Herre EA, Knowlton N, Mueller UG, Rehner SA. 1999. The evolution of mutualisms: exploring the paths between conflict and cooperation. *Trends Ecol. Evol.* 14:49–53

Hoeksema JD, Bruna EM. 2000. Pursuing the big questions about interspecific mutualism: a review of theoretical approaches. *Oecologia* 125:321–30

Holland N, DeAngelis DL, Bronstein JL. 2002. Population dynamics and mutualism: functional responses of costs and benefits. *Am. Nat.* 159:231–44

Hölldobler B, Wilson EO. 1990. *The Ants*. Cambridge, MA: Harvard Univ. Press. 732 pp.

Hopkins GW, Thacker JI. 1999. Ants and habitat specificity in aphids. *J. Insect Conserv.* 3: 25–31

Hopkins GW, Thacker JI, Dixon AFG. 1998. Limits to the abundance of rare species: an experimental test with a tree aphid. *Ecol. Entomol.* 23:386–90

Houston AI, Clark C, McNamara JM, Mangel M. 1988. Dynamic models in behavioural and evolutionary ecology. *Nature* 332:29–34

Houston AI, McNamara JM. 1988. A framework for the functional analysis of behaviour. *Behav. Brain Sci.* 11:117–63

Houston AI, McNamara JM. 1999. *Models of Adaptive Behaviour*. Cambridge, UK: Cambridge Univ. Press. 378 pp.

Huxley CR, Cutler DF, eds. 1991. *Ant-Plant Interactions*. Oxford: Oxford Univ. Press

Johnson SN, Elston DA, Hartley S. 2003. Influence of host plant heterogeneity on the distribution of a birch aphid. *Ecol. Entomol.* 28:533–41

Johnstone RA, Bshary R. 2002. From parasitism to mutualism: partner control in asymetric interactions. *Ecol. Lett.* 5:634–39

Kainulainen P, Holopainen J, Palomäki V, Holopainen T. 1996. Effects of nitrogen fertilization on secondary chemistry and ectomycorrhizal state of Scots pine seedlings and on growth of grey pine aphid. *J. Chem. Ecol.* 22:617–36

Kaneko S. 2002. Aphid-attending ants increase the number of emerging adults of the aphid's primary parasitoid and hyperparasitoids by repelling intraguild predators. *Entomol. Sci.* 5:131–46

Karban R, Baldwin IT. 1997. *Induced Responses to Herbivory*. Chicago, IL: Univ. Chicago Press

Katayama N, Suzuki N. 2002. Cost and benefit of ant attendance for *Aphis craccivora* (Hemiptera: Aphididae) with reference to aphid colony size. *Can. Entomol.* 134:241–49

Kindlmann P, Dixon AFG. 1989. Developmental constraints in the evolution of reproductive strategies: telescoping of generations in parthenogenetic aphids. *Funct. Ecol.* 3:531–37

Kindlmann P, Dixon AFG. 1993. Optimal foraging in ladybird beetles (Coleoptera: Coccinellidae) and its consequences for their use in biological control. *Eur. J. Entomol.* 90: 443–50

Kiss A. 1981. Melezitose, aphids and ants. *Oikos* 37:382

Koptur S. 1991. Extrafloral nectaries of herbs and trees: modeling the interaction with ants and parasitoids. See Huxley & Cutler 1991, pp. 213–30

Krohne DT. 1998. *General Ecology*. Belmont, CA: Wadsworth. 722 pp.

Kunert G, Weisser WW. 2003. The interplay between density- and trait-mediated effects in predator-prey interactions: a case study in aphid wing polymorphism. *Oecologia* 135: 304–12

Kunkel H, Kloft WJ, Fossel A. 1985. Die Honigtau-Erzeuger des Waldes. In *Waldtracht und Waldhonig in der Imkerei*, ed. WJ Kloft, H Kunkel, pp. 48–265. Munich: Ehrenwirth

Labandeira CC. 1997. Insect mouth parts: ascertaining the palaeobiology of insect feeding strategies. *Annu. Rev. Ecol. Syst.* 28:153–93

Laine KJ, Niemela P. 1980. The influence of ants on the survival of mountain birches during an *Oporinia autumnata* (Lep. Geometridae) outbreak. *Oecologia* 47:39–42

Loreau M. 1995. Consumers as maximizers of matter and energy flow in ecosystems. *Am. Nat.* 145:22–42

Mackauer M, Völkl W. 1993. Regulation of aphid populations by aphiid wasps: Does parasitoid foraging behaviour or hyperparasitism limit impact? *Oecologia* 94:339–50

Mailleux AC, Deneubourg JL, Detrain C. 2003. Regulation of ants' foraging to resource productivity. *Proc. R. Soc. London Ser. B* 270: 1609–16

Mailleux AC, Jean-Louis D, Detrain C. 2000. How do ants assess food volume? *Anim. Behav.* 59:1061–69

Massonnet B, Simon JC, Weisser WW. 2002. Metapopulation structure of the specialized herbivore *Macrosiphoniella tanacetaria* (Homoptera, Aphididae). *Mol. Ecol.* 11: 2511–21

Mattson JW. 1980. Herbivory in relation to plant nitrogen content. *Annu. Rev. Ecol. Syst.* 11:119–61

Montllor CB. 1991. The influence of plant chemistry on aphid feeding behavior. In *Insect Plant Interactions*, Vol. 3, ed. E Bernays, pp. 125–73. Boston: CRC

Morris DW. 1985. Natural selection for reproductive optima. *Oikos* 45:290–92

Morris WF. 1992. The effects of natural enemies, competition, and host plant water availability on an aphid population. *Oecologia* 90: 359–65

Müller CB, Godfray HCJ. 1999. Predators and mutualists influence the exclusion of aphid species from natural communities. *Oecologia* 119:120–25

Nixon GEJ. 1951. *The Association of Ants with Aphids and Coccids.* London: Commonwealth Inst. Entomol. 36 pp.

Nonacs P, Calabi P. 1992. Competition and predation risk: their perception alone affects ant colony growth. *Proc. R. Soc. London Ser. B* 249:95–99

Nonacs P, Dill LM. 1990. Mortality risk vs. food quality trade-offs in a common currency: ant patch preferences. *Ecology* 71: 1886–92

Nonacs P, Dill LM. 1991. Mortality risk versus food quality trade-offs in ants: patch use over time. *Ecol. Entomol.* 16:73–80

Offenberg J. 2000. Correlated evolution of the association between aphids and ants and the association between aphids and plants with extrafloral nectaries. *Oikos* 91:146–52

Offenberg J. 2001. Balancing between mutualism and exploitation: the symbiotic interaction between Lasius ants and aphids. *Behav. Ecol. Sociobiol.* 49:304–10

Oster GF, Wilson EO. 1978. *Caste and Ecology in the Social Insects.* Princeton: Princeton Univ. Press

Pickett JA, Wadhams LJ, Woodcock CM, Hardie J. 1992. The chemical ecology of aphids. *Annu. Rev. Entomol.* 37:67–90

Pierce NE, Braby MF, Heath A, Lohman DJ, Mathew J, et al. 2002. The ecology and evolution of ant association in the Lycaenidae (Lepidoptera). *Annu. Rev. Entomol.* 47:733–71

Pontin AJ. 1978. The number and distribution of subterranean aphids and their exploitation by the ant *Lasius flavus* (Fabr.). *Ecol. Entomol.* 3:203–7

Punttila P. 1996. Succession, forest fragmentation, and the distribution of wood ants. *Oikos* 75:291–98

Remaudière G, Remaudière M. 1997. *Catalogue of the World's Aphididae.* Paris: Institut National de la Recherche Agronomique. 473 pp.

Rhodes JD, Croghan PC, Dixon AFG. 1996. Uptake, excretion and respiration of sucrose and amino acids by the pea aphid *Acyrthosiphon pisum. J. Exp. Biol.* 199:1269–76

Rhodes JD, Croghan PC, Dixon AFG. 1997. Dietary sucrose and oligosaccharide synthesis in relation to osmoregulation in the pea aphid, *Acyrthosiphon pisum. Physiol. Entomol.* 22:373–79

Rosengren R, Pamilo P. 1983. The evolution of polygyny and polydomy in mound-building *Formica* ants. *Acta Entomol. Fenn.* 42:65–77

Rosengren R, Sundström L. 1991. The interaction between red wood ants, Cinara aphids, and pines. A ghost of mutualism past? See Huxley & Cutler 1991, pp. 81–91

Rott AS, Müller CB, Godfray HCJ. 1998. Indirect population interactions between two aphid species. *Ecol. Lett.* 1:99–103

Sakata H. 1994. How an ant decides to prey on or to attend aphids. *Res. Popul. Ecol.* 36:45–51

Sakata H. 1995. Density-dependent predation of the ant *Lasius niger* (Hymenoptera: Formicidae) on two attended aphids *Lachnus tropicalis* and *Myzocallis kuricola* (Homoptera: Aphididae). *Res. Popul. Ecol.* 37:159–64

Sakata H, Hashimoto Y. 2000. Should aphids attract or repel ants? Effect of rival aphids and extrafloral nectaries on ant-aphid interactions. *Res. Popul. Ecol.* 42:171–78

Scheu S, Theenhaus A, Jones TH. 1999. Links between the detritivore and the herbivore system: effects of earthworms and Collembola on plant growth and aphid development. *Oecologia* 119:541–51

Scheurer S. 1967. Populationsdynamische Beobachtungen an auf Pinus lebenden Lachniden während des Jahres 1965. *Waldhygiene* 7:7–22

Shingleton AW, Foster WA. 2000. Ant tending influences soldier production in a social aphid. *Proc. R. Soc. London Ser. B* 267: 1863–68

Shingleton AW, Stern DL. 2003. Molecular phylogenetic evidence for multiple gains or losses of ant mutualism within the aphid genus *Chaitophorus. Mol. Phylogenet. Evol.* 26:26–35

Shingleton AW, Stern DLS, Foster WA. 2005. The origin of a mutualism: a morphological trait promoting the evolution of ant-aphid mutualisms. *Evolution* 59:921–26

Skinner GJ, Whittaker JB. 1981. An experimental investigation of interrelationships between the wood-ant (*Formica rufa*) and some tree canopy herbivores. *J. Anim. Ecol.* 50:313–26

Sloggett JJ, Majerus MEN. 2003. Adaptations of *Coccinella magnifica*, a myrmecophilous coccinellid to aggression by wood ants (*Formica rufa* group). II. Larval behaviour, and ladybird oviposition location. *Eur. J. Entomol.* 100:337–44

Sloggett JJ, Weisser WW. 2002. Parasitoids induce production of the dispersal morph of the pea aphid, *Acyrthosiphon pisum. Oikos* 98:323–33

Sloggett JJ, Wood RA, Majerus M. 1998. Adaptation of *Cocinella magnifica* Redtenbacher, a myrmecophilous coccinellid, to aggression by wood ants (*Formica rufa* group). I. Adult behavioral adaptation, its ecological context and evolution. *J. Insect Behav.* 11:889–904

Stachowicz JJ. 2001. Mutualism, facilitation, and the structure of ecological communities. *BioScience* 51:235–46

Stadler B. 2004. Wedged between bottom-up and top-down processes: aphids on tansy. *Ecol. Entomol.* 29:106–16

Stadler B, Dixon AFG. 1998a. Costs of ant attendance for aphids. *J. Anim. Ecol.* 67:454–59

Stadler B, Dixon AFG. 1998b. Why are obligate mutualistic interactions between aphids and ants so rare? In *Aphids in Natural and Managed Ecosystems*, ed. JM Nieto Nafria, AFG Dixon, pp. 271–8. Leon, Spain: Univ. Leon

Stadler B, Dixon AFG. 1999. Ant attendance in aphids: why different degrees of myrmecophily? *Ecol Entomol.* 24:363–69

Stadler B, Dixon AFG, Kindlmann P. 2002. Relative fitness of aphids: effects of plant quality and ants. *Ecol. Lett.* 5:216–22

Stadler B, Fiedler K, Kawecki TJ, Weisser WW. 2001. Costs and benefits for phytophagous myrmecophiles: when ants are not always available. *Oikos* 92:467–78

Stadler B, Kindlmann P, Šmilauer P, Fiedler K. 2003. A comparative analysis of morphological and ecological characters of European aphids and lycaenids in relation to ant attendance. *Oecologia* 135:422–30

Stadler B, Michalzik B, Müller T. 1998. Linking aphid ecology with nutrient fluxes in a coniferous forest. *Ecology* 79:1514–25

Stephens DW, Krebs JR. 1986. *Foraging Theory.* Princeton: Princeton Univ. Press

Stern DL, Aoki S, Kurosu DU. 1995. The life-cycle and natural history of the tropical aphid *Cerataphis fransseni* (Homoptera, Aphididae, Hormaphidinae), with reference to the evolution of host alternation in aphids. *J. Nat. Hist.* 29:231–42

Stern DL, Foster WA. 1997. The evolution of sociality in aphids: a clone's-eye of view. In *The Evolution of Social Behavior in Insects and Arachnids*, ed. J Choe, B Crespi, pp. 150–65. Cambridge, UK: Cambridge Univ. Press.

Strauss SY. 1987. Direct and indirect effects of host-plant fertilization on an insect community. *Ecology* 68:1670–78

Sudd JH, Sudd ME. 1985. Seasonal changes

in the response of wood-ants (*Formica lugubris*) to sucrose baits. *Ecol. Entomol.* 10: 89–97

Szentesi A, Wink M. 1991. Fate of quinolizidine alkaloids through 3 trophic levels - *Laburnum anagyroides* (Leguminosae) and associated organisms. *J. Chem. Ecol.* 17:1557–73

Takada H, Hashimoto Y. 1985. Association of the root aphid parasitoids *Aclitus sappaphis* and *Paralipsis eikoae* (Hymenoptera, Aphidiidae) with the aphid-attending ants *Pheidole fervida* and *Lasius niger* (Hymenoptera, Formicidae). *Kontyu* 53:150–60

Taylor RW. 1978. *Nothomyrmecia macrops*: a living-fossil ant rediscovered. *Science* 201: 979–85

Thompson JN. 1994. *The Coevolutionary Process*. Chicago, IL: Univ. Chicago Press. 376 pp.

Tilles DA, Wood DL. 1982. The influence of carpenter ant (*Camponotus modoc*) (Hymenoptera: Formicidae) attendance on the development and survival of aphids (*Cinara* spp.) (Homoptera: Aphididae) in a giant sequoia forest. *Can. Entomol.* 114:1133–42

Tizado EJ, Tinaut A, Nieto Nafria JM. 1993. Relationships between ants and aphids in the province of Leon (Spain) (Hym: Formicidae; Hom: Aphiddiae). *Vie Milieu* 43:63–68

Toft S. 1995. Value of the aphid *Rhopalosiphum padi* as food for cereal spiders. *J. Appl. Ecol.* 32:552–60

Völkl W. 1992. Aphids or their parasitoids: Who actually benefits from ant-attendance? *J. Anim. Ecol.* 61:273–81

Völkl W. 1995. Behavioral and morphological adaptations of the Coccinellid, *Platynaspis luteorubra*, for exploiting ant-attended resources (Coleoptera: Coccinellidae). *J. Insect Behav.* 8:653–70

Völkl W, Liepert C, Birnbach R, Hübner G, Dettner K. 1996. Chemical and tactile communication between the root aphid parasitoid *Paralipsis enervis* and trophobiotic ants: consequences for parasitoid survival. *Experientia* 52:731–38

Völkl W, Vohland K. 1996. Wax covers in larvae of two *Scymnus* species: Do they enhance cocinellid larval survival? *Oecologia* 107:498–503

Völkl W, Woodring J, Fischer M, Lorenz MW, Hoffmann KH. 1999. Ant-aphid mutualisms: the impact of honeydew production and honeydew sugar composition on ant preferences. *Oecologia* 118:483–91

Vrieling K, Smit W, Vandermeijden E. 1991. Tritrophic interactions between aphids (*Aphis jacobaeae* Schrank), ant species, *Tyria jacobaeae* L., and *Senecio jacobaea* L. lead to maintenance of genetic variation in pyrrolizidine alkaloid concentration. *Oecologia* 86:177–82

Way MJ. 1963. Mutualism between ants and honeydew-producing Homoptera. *Annu. Rev. Entomol.* 8:307–44

Weisser WW, Braendle C, Minoretti N. 1999. Predator-induced morphological shift in the pea aphid. *Proc. R. Soc. London Ser. B* 266: 1175–81

Wellenstein G. 1980. Auswirkungen hügelbauender Waldameisen der *Formica rufa*-Gruppe auf forstschädliche Raupen und das Wachstum der Waldbäume. *Z. Angew. Entomol.* 89:144–57

Wheeler NA. 1914. The ants of the Baltic amber. *Schr. Phys.-Ökon. Ges. Königsberg* 55: 1–142

Wilson EO. 1987. Causes of ecological success: the case of the ants. *J. Anim. Ecol.* 56:1–9

Wilson EO. 1990. *Success and Dominance in Ecosystems: The Case of Social Insects*. Nordbünte: Ecol. Inst. 104 pp.

Wilson WG, Morris WF, Bronstein JL. 2003. Coexistence of mutualists and exploiters on spatial landscapes. *Ecol. Monogr.* 73:397–413

Wimp GM, Whitham TG. 2001. Biodiversity consequences of predation and host plant hybridization on an aphid-ant mutualism. *Ecology* 82:440–52

Wink M, Witte L. 1991. Storage of quinolizidine alkaloids in *Macrosiphum albifrons* and *Aphis genistae* (Homoptera: Aphididae). *Entomol. Gen.* 15:237–54

Yao I, Akimoto S. 2001. Ant attendance

changes the sugar composition of the honeydew of the drepanosiphid aphid *Tuberculatus quercicola. Oecologia* 128:36–43

Yao I, Akimoto S. 2002. Flexibility in the composition and concentration of amino acids in honeydew of the drepanosiphid aphid *Tuberculatus quercicola. Ecol. Entomol.* 27:745–52

Yao I, Shibao H, Akimato S. 2000. Costs and benefits of ant attendance to the drepanosiphid aphid *Tuberculatus quercicola. Oikos* 89:3–10

Zoebelein G. 1954. Versuche zur Feststellung des Honigtauertrages von Fichtenbeständen mit Hilfe von Waldameisen. *Z. Angew. Entomol.* 36:358–62

Annu. Rev. Ecol. Evol. Syst. 2005. 36:373–97
doi: 10.1146/annurev.ecolsys.36.102003.152622
First published online as a Review in Advance on August 17, 2005

EVOLUTIONARY CAUSES AND CONSEQUENCES OF IMMUNOPATHOLOGY

Andrea L. Graham, Judith E. Allen, and Andrew F. Read

*Institutes of Evolution, Immunology & Infection Research, School of Biological Sciences,
University of Edinburgh, Edinburgh, Scotland EH9 3JT;
email: andrea.graham@ed.ac.uk, j.allen@ed.ac.uk, a.read@ed.ac.uk*

Key Words   defense, ecological immunology, host-parasite evolution, resistance, virulence

■ **Abstract**   Immune responses can cause severe disease, despite the role immunity plays in defending against parasitism. Indeed, immunopathology is a remarkably common cause of disease and has strong impacts upon both host and parasite fitness. Why has immune-mediated disease not been eliminated by natural selection? What constraints might immunopathology impose upon the evolution of resistance? In this review, we explore two major mechanistic causes of immunopathology in mammals and consider how such disease may have influenced immune system design. We then propose hypotheses that could explain the failure of natural selection to eliminate immunopathology. Finally, we suggest how the evolution of strategies for parasite virulence and host resistance may be shaped by this "double-edged sword" of immunity. Future work may reveal whether immunopathology constrains the evolution of resistance in all host taxa.

1. INTRODUCTION

Inappropriate immune responses can have profound fitness effects. Most obviously, failure to control parasite proliferation may be detrimental to host fitness. But immune responses also damage hosts if responses are too strong, involve the wrong parasite-killing mechanism, or are elicited by the wrong antigens, including those on the host's own cells (leading to autoimmunity) or on innocuous substances such as food (leading to allergy). Such immune-mediated diseases are termed immunopathology.

On the face of it, immunopathology conflicts with Darwinism: Organisms should not self-harm. Evolutionary biologists spent much of the past century analyzing—and for the most part explaining—other traits that appeared maladaptive. But unlike altruism and the peacock's tail, immunopathology causes human disease. Immunopathology is thus a surprising omission from evolutionary biology to date, given the humanitarian importance as well as intellectual appeal of understanding apparently maladaptive immune responses.

In this review, we start by explaining the causes of two common classes of immunopathology: Type 1 and Type 2. We then argue that, although immunopathology probably helped to shape the immune system, the failure of natural selection to eliminate immune-mediated disease demands evolutionary explanation. Finally, we contend that several areas of evolutionary research could be reshaped by a full appreciation of immunopathology. For example, immunopathology can increase the costs of both parasite virulence and host defense, thereby altering selection on these traits and leading to different evolutionary optima from those predicted under the assumption that immunopathology does not occur (e.g., Figure 1*a* and 1*b*; the theory behind this is discussed in more detail in Sections 4.1 and 4.2).

1.1. Severe Infectious Disease Usually Involves Immunopathology

During infection, immunopathology can be difficult to distinguish from more direct effects of parasites, but the distinction is real. Consider 10 tropical diseases accorded high priority by the World Health Organization (WHO). These diseases, the most deadly of which are tuberculosis and malaria, account for billions of infections and nearly 3 million deaths per year. They also have profound sublethal effects (WHO 2004). Critically, all 10 are at least partly immunopathological (Table 1), and hosts with the most severe symptoms do not necessarily harbor the most parasites.

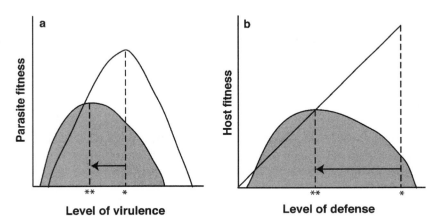

Figure 1   Optimal strategies for parasite virulence (*a*) and host defense (*b*) may be altered by immunopathology. For example, (*a*) if immunopathology increases virulence without increasing transmission—e.g., by killing the host—we predict selection for decreased parasite virulence (**) compared to that predicted under an assumption of no immunopathology (*), and (*b*) if high investment in defense is associated with immunopathology, we predict lower optimal levels of defense (**) than those predicted in the absence of immunopathology (*). In both diagrams, the gray curve represents the fitness function in the case of immunopathology.

In tuberculosis, for example, the immune response that clears bacteria also recruits fluid and cells into the air spaces of the lung (Bekker et al. 2000). Similarly, the immunological molecules that control malaria replication exacerbate disease (Akanmori et al. 2000); in mice, 10% of malarial anemia is explained by immunological exuberance rather than parasite-mediated destruction of red blood cells (Graham et al. 2005). Beyond the WHO top 10, immunopathology is manifest in common diseases of tropical and temperate residents alike. For example, influenza induces much more immunological activity than is necessary to clear the virus, and it is the excess that does most of the damage to the lung (Hussell et al. 2001, Xu et al. 2004). Immunopathology may also have delayed fitness consequences, reducing lifespan in people prone to strong immune responses (Finch & Crimmins 2004). It is said that Chagas disease (#7 in Table 1) long debilitated and eventually killed Charles Darwin (Adler 1997), so infection-induced immunopathology may have even claimed one of evolutionary biology's finest minds.

This dual effect of the immune system—fighting infection while causing immunopathology—is driven by two broad classes of mechanisms, cytotoxicity and tissue remodelling. Both are required for resolution of the diverse infections encountered over a lifetime, but each causes disease if immoderate. Cytotoxic immune responses, for example, can spiral out of control to kill host as well as parasite cells (Pfeffer 2003). Tissue remodelling is essential if the immune system is to sequester parasites or their eggs, but excessive deposition of collagen and subsequent hardening cause organs to become blocked, stiffened and, ultimately, dysfunctional (Wynn 2004). As we explain below, beyond some threshold, the parasite-controlling function of the immune system ends and host tissue damage begins.

1.2. Humans and Mice as Model Systems

Throughout this review, we focus upon humans and mice. These are the most intensively studied hosts, so they are the best-characterized models for the biological phenomenon of misdirected defense. We do expect that our general arguments will apply to other mammals, other vertebrates, invertebrates, and maybe even to plants. Such extrapolation is, however, not immediately possible: We are unaware of any attempt to determine the importance of immunopathology in infectious disease severity in "natural" mammalian-parasite interactions, let alone in nonmammalian systems. When data like those in Table 1 become available for diseases that afflict animals other than ourselves and our domesticated or laboratory mammals, it may become apparent that immunopathology is actually rare in nature.

Could the bulk of immunopathology in human populations and laboratory mice be a consequence of novel environments or novel parasites? For people, the novelty may include the plethora of parasites acquired after human populations became dense enough to sustain their transmission. Studies of wild animals could determine whether novel conditions are necessary for immunopathology, and we look forward to such work. In the meantime, we note that a selective factor can be important even if few affected individuals are observed. For instance, risk of injury has

TABLE 1   Estimated fitness effects of immunopathology in 10 tropical diseases accorded high priority by the WHO

Disease[a]	Infectious agent	People infected[b(Source)]	Deaths in 2002[c]	Age at greatest risk[(Source)]	Is there evidence that severe cases are immunopathological?[(Source)]
Tuberculosis	*Mycobacterium tuberculosis*	1,900,000,000[(Dye et al.1999)]	1,566,000	20–30 years[(Daniel et al. 2004)]	Yes, faulty responses prolong disease[(Hirsch et al. 1996)] and lung damage is independent of bacterial load[(Ehlers et al. 2001)]
Malaria	*Plasmodium* species	300,000,000[(WHO 2004)]	1,272,000	<5 years[(Snow et al. 1999)]	Yes, unregulated immune responses increase disease severity[(Akanmori et al. 2000, Dodoo et al. 2002, Li et al. 2003, Omer et al. 2003)]
Leishmaniasis	*Leishmania* species	12,000,000[(WHO 2000)]	51,000	<20 years[(Saran et al. 1989)]	Yes, immune responses increase the size of skin lesions[(Louzir et al. 1998)] and damage the liver[(Satoskar et al. 2000)]
Sleeping sickness	*Trypanosoma brucei*	300,000[(WHO 2000)]	48,000	>20 years[(Abaru 1985)]	Yes, unregulated responses[(Magez et al. 2004)] damage the central nervous system[(Hunter & Kennedy 1992, Maclean et al. 2004)]
Dengue	Dengue viruses	50,000,000[(WHO 2000)]	19,000	<15 years[(Gubler 1998)]	Yes, secondary responses cause hemorrhagic fever[(Mongkolsapaya et al. 2003)] and 11% of liver damage[(Libraty et al. 2002)]
Schistosomiasis	*Schistosoma* species	200,000,000[(WHO 2004)]	15,000	>20 years[(Booth et al. 2004)]	Yes, liver and urinary tract damage are immune-mediated[(Booth et al. 2004, Hesse et al. 2004, Hoffmann et al. 2002, Wamachi et al. 2004)]
Chagas disease	*Trypanosoma cruzi*	16,000,000[(Moncayo 1992)]	14,000	>20 years[(Jorge et al. 2003)]	Yes, damage to heart muscle is due to inflammation[(Andrade 1999, Holscher et al. 2000)] not parasite load[(Soares et al. 2001)]
Leprosy	*Mycobacterium leprae*	750,000[(Sasaki et al. 2001)]	6000	20–35 years[(Scollard 1993)]	Yes, inappropriate immune responses damage nerve cells[(Khanolkar–Young et al. 1995, Spierings et al. 2001)]
Lymphatic filariasis	*Wuchereria & Brugia* species	120,000,000[(Michael & Bundy 1997)]	0	>40 years[(Leang et al. 2004)]	Yes, elephantiasis is associated with immunological hyper-responsiveness[(Sartono et al. 1997)]
Onchocerciasis	*Onchocerca volvulus*	17,000,000[(WHO 1995)]	0	>20 years[(Murdoch et al. 2002)]	Yes, corneal opacity and skin damage are immune-mediated[(Hall & Pearlman 1999, Stewart et al. 1999)]

[a]We investigated 10 diseases prioritized by the Special Program for Tropical Disease Research, a global research partnership convened by the World Health Organization (WHO) (http://www.who.int/tdr/index.html). The diseases are high priority because they affect millions of people and yet receive only scant dollars and scientific attention. We obtained the annual mortality attributable to each infection and the age group most affected by severe disease. Then we evaluated the field and laboratory evidence that immunopathology has a role in severe cases. We were specifically interested in whether disease symptoms were in excess of those attributable to within-host parasite density. Immunopathology was implicated in severe cases of all 10 diseases.

[b]Estimated number of people infected at any given time. The lower limits of the WHO's estimated range of number of cases is shown.

[c]Mortality data are based upon the World Health Report (WHO 2004), which often underestimates cause-specific mortality rates (cf. van der Werf et al. 2003). This table is thus conservative in its estimates of mortality. Furthermore, for reasons of space, we have omitted quantitative estimates of sublethal fitness effects of these diseases, though such estimates are available (WHO 2004).

undoubtedly posed significant selective pressures on avian flight, yet birds with crash injuries are extremely rare (Cuthill & Guilford 1990). To determine the selective pressure imposed by immunopathology in nature, it may be necessary to provoke hyper-responsiveness experimentally in wild animals. It is our contention that such experiments will confirm that, as in people and mice, immunopathology has severe fitness effects. Moreover, as we discuss below, many aspects of the highly orchestrated vertebrate immune system can be understood as adaptations to reduce immunopathology. Unless these are highly atypical, immunopathology must have had a major role in the evolution of immunity in general. From what we know of human diseases (e.g., Table 1), immunopathology can bring selective pressures upon hosts that rival selection due to parasitism itself.

2. MECHANISTIC AND EVOLUTIONARY CAUSES OF IMMUNOPATHOLOGY

Immunopathology is prevalent, but how might it constrain the evolution of resistance, and why has it not been eliminated by natural selection? Two major classes of parasites, microparasites (mostly intracellular) and macroparasites (mostly extracellular), are matched by two major types of immunity (Abbas et al. 1996): Type 1 responses are mainly cytotoxic and are essential to killing microparasites, whereas Type 2 responses enable tissue remodelling to combat macroparasites. Cytokines are the molecules that enable these Type 1 versus Type 2 antiparasitic (effector) mechanisms. Here, we focus upon tumor necrosis factor alpha (TNF-α) and interleukin 13 (IL-13), which characteristically function in these two different types of response. For both cytokines, the context and quantity in which they are produced determine their protective efficacy. A constrained ability of the immune system to regulate these cytokines may be the prime evolutionary explanation of immunopathology.

2.1. Control of Microparasites Versus Type 1 Immunopathology: TNF-α

TNF-α is at once the most protective and most pathological of cytokines. It is critical to the induction of nearly all immune responses (Pfeffer 2003) and, along with other Type 1 cytokines, is important for control of microparasites. To kill these parasites as well as tumors (hence its name, tumor necrosis factor), TNF-α has a broad array of functions, including recruitment of cells to the site of infection, activation of phagocytic cells to release toxic chemicals, and even direct killing of target cells (Tracey & Cerami 1994). Such powerful antiparasitic activities, if not moderated or properly directed, result in severe host tissue damage.

The deadliest example of a TNF-α response gone wrong is that of septic shock. In this case, TNF-α is released in enormous quantities into the bloodstream following escape of bacteria from the gut into the blood (i.e., sepsis). The TNF-α

alters blood vessel walls to allow blood fluid, clotting factors, and cells to enter the tissues, as would be appropriate if TNF-α were released in a localized site of infection. Systemic release, however, causes shock—decreased blood volume and multiorgan failure. Mice deficient in TNF-α readily survive a level of sepsis that would kill a normal mouse but succumb to minor bacterial infections (Pfeffer et al. 1993). Such hosts avoid immunopathology but fail to control parasite replication.

TNF-α's dual role is also well documented in malaria (Akanmori et al. 2000, Dodoo et al. 2002, Li et al. 2003, Omer et al. 2003). TNF-α does protect against disease by killing infected red blood cells (RBCs). Yet TNF-α is associated with changes to blood vessels of the brain, which can lead to coma and death during cerebral malaria (Hunt & Grau 2003). In milder malaria, too, as parasites rupture out of the RBCs, the host responds with bursts of TNF-α that lead to fever and malaise. Hosts must therefore balance the need to kill parasites against the dangers of excess Type 1 cytokine. We have found quantitative evidence in support of this balanced optimum in mice with malaria: The least anemic were those who made Type 1 responses of intermediate magnitude (Figure 2, based upon data in Graham et al. 2005). Among the other WHO top 10, TNF-α also has dual roles in sleeping sickness (Maclean et al. 2004), Chagas (Holscher et al. 2000), and leprosy (Khanolkar-Young et al. 1995). TNF-α can even be purely pathological, as in autoimmunity (Pfeffer 2003).

Variation in TNF-α responsiveness—and, more generally, the magnitude of infection-induced inflammatory responses (Terry et al. 2000, van de Vosse et al. 2004)—has a polymorphic genetic basis. Though the effects of TNF-α promoter polymorphism are tied up with the major histocompatibility complex (MHC) in terms of chromosomal location, TNF-α polymorphisms affect disease outcome even when its effects are dissected from those of MHC (Daser et al. 1996). Genetic heterogeneity for TNF-α expression has particularly been shown to influence the likelihood of severe anemia and cerebral symptoms during malaria (Bayley et al. 2004). We thus have reason to expect that heterogeneity in the expression of TNF-α-mediated pathology is heritable and can evolve. In Section 3.3 below, we consider why natural selection might not have eliminated immunopathological expression of TNF-α.

2.2. Coping with Macroparasites Versus Type 2 Immunopathology: IL-13

Very different cytokines are involved in the Type 2 response elicited by large, extracellular macroparasites, such as the worms that infect over a third of the world's population (Chan 1997). Along with other Type 2 cytokines, IL-13 is stereotypically induced by helminths, despite the vastly different physiologies and life histories of, for example, nematodes versus trematodes. IL-13 helps to expel worms from the intestines (Finkelman et al. 2004) and destroy tissue-dwelling worms such as filarial nematodes (Maizels et al. 2004). But IL-13, like TNF-α, can cause disease when its quantities are not modulated.

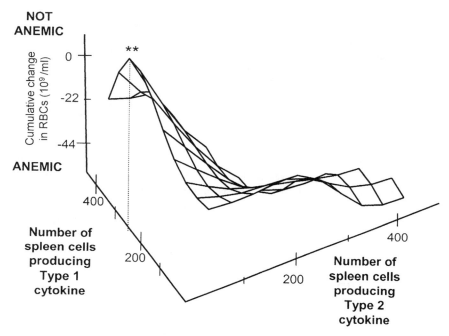

Figure 2   The optimal immune response to malaria balances parasite killing against immunopathology. During studies of malaria-filaria coinfection (Graham et al. 2005), we were able to identify quantitatively the immune response that minimized the severity of malarial symptoms. We used this system because concomitant filarial infection extends the range of immune responses that laboratory mice mount against malaria. We found that the healthiest (here, least anemic) mice were those that made Type 1 immune responses of intermediate magnitude. The observed optimum—i.e., the immune response that minimized cumulative loss of red blood cell (RBC) density—is indicated by ** (whole-model $P < 0.005$, with significant linear and quadratic functions of Type 1 cytokine).

The importance of getting the right IL-13 balance is illustrated by the tissue remodelling induced by schistosomiasis (#6 in Table 1). Schistosomes live in blood vessels and release thousands of eggs per day. A large proportion of these become lodged in host tissues, particularly the liver. Eggs release tissue-damaging toxins, and the immune system protects the liver by encapsulating the egg in an orderly arrangement of cells and molecules called a granuloma (Hoffmann et al. 2002). IL-13 is essential for the creation of this granuloma, providing the framework for extracellular matrix deposition and onward structural strengthening via collagen deposition (Wynn 2004). As tissue remodelling proceeds, more and more collagen is recruited. Here, immoderate IL-13 can cause fibrosis, filling liver tissue with so much collagen that it can no longer perform blood purification. It is this fibrotic immune response that leads to host death (Hoffmann et al. 2002). IL-13 can also go wrong in the absence of infection: It is becoming increasingly apparent that IL-13-mediated fibrosis is responsible for severe diseases such as asthma, where the immune system is trying, unsuccessfully, to contain foreign objects (Wills-Karp 2004). To avoid immunopathology, production of IL-13 must be tightly regulated in organs such as the lung or liver.

The IL-13 gene, like the TNF-α gene, is highly polymorphic in people around the world (Tarazona-Santos & Tishkoff 2005). Functionally, IL-13 polymorphisms help to control predisposition to at least 2 of the WHO top 10: severe versus mild schistosomiasis (Dessein et al. 2004) and onchocerciasis of the skin (Hoerauf et al. 2002). Overall, Type 1 versus Type 2 cytokine bias is also genetically controlled and is predictive of how well a host fights different infections (Mitchison et al. 2000). Again, the genetic raw material for evolutionary change appears to be present. Indeed, we suggest below that the risk of immunopathology may have shaped immune responsiveness.

3. CONSEQUENCES OF IMMUNOPATHOLOGY FOR EVOLUTION OF THE IMMUNE SYSTEM

The fact that some of the most potent antiparasitic molecules have profound destructive power has probably helped to shape regulatory pathways in the evolution of mammalian immunity. Several design features of the immune system are indicative of the importance of avoiding immunopathology. We present these here and then move on to assess hypotheses to explain why evolution has not succeeded in fully eliminating immunopathology.

3.1. Evolution of Type 2 Effector Responses to Minimize Immunopathology

The evolutionary reasons for a Type 1 response, as outlined above, are readily apparent: Without it, we die from overwhelming microparasitic infection. The reasons for the evolution of the Type 2 response are less obvious. Yet IL-13 and other Type 2 cytokines have multifaceted roles suggestive of strong selection pressures posed by immunopathology.

It may seem, from the foregoing discussion of IL-13, that the reason we need a Type 2 response is to fight worms. However, Type 1 responses (e.g., macrophages activated by cytokines such as TNF-α) are also capable of destroying worms (Rodriguez-Sosa et al. 2004, Thomas et al. 1997). We propose that Type 1 responses are not normally used this way because of the immunopathology that goes with large-scale TNF-α-like responses. The benefits of controlling rampantly proliferating microparasites must outweigh the costs of self-harm. But macroparasites are not immediately threatening—worms tend to induce lower case fatality rates than do microparasitic infections (WHO 2004), and even mice totally deficient in adaptive immunity survive worm infections, albeit at much higher parasite burdens (Urban et al. 1995). Thus, better-targeted control is possible, and the Type 2 response may have evolved as a safer alternative to the damaging Type 1 response. Importantly, cytokine cross-regulation ensures that strong Type 2 responses inhibit Type 1 responses (Abbas et al. 1996). The central theme in the evolution of Type 2 immunity to macroparasites may thus have been avoidance of Type 1 immunopathology.

There is a further means by which IL-13 minimizes the fitness effects of worm infection. Worms cause wounds, due to their size, motility, and necessity to eat host tissue. For example, feeding by common intestinal parasites such as hookworms induces wounds with strong fitness effects: Beyond the dangers of internal bleeding, unhealed worm bites in the intestinal wall might lead to gut leakage and, ultimately, sepsis. More generally, many intestinal worms have stages that migrate through the lung, which means that billions of people worldwide (Chan 1997) harbor lung-migrating parasites. Lung damage induced by migrating worms is appropriately repaired (McNeil et al. 2002), and it is becoming clear that IL-13 helps to direct such healing (Wynn 2004). Which activities of IL-13 arose first is open to question, but the evidence suggests that Type 2 cytokines function as much to prevent pathology as to kill macroparasites.

3.2. Moderation in All Things: A Major Role for Adaptive Immunity

Cytokines that, in excess, lead to immunopathology are produced during both innate and adaptive immune responses. Beyond the mutual inhibition between Type 1 and Type 2 cytokines (Abbas et al. 1996) that can minimize immunopathology, critical additional control is provided by regulatory T cells of the adaptive immune system. These immunomodulatory cells dampen responses that are too damaging to self-tissue (Mills 2004) by producing cytokines, particularly transforming growth factor beta (TGF-β) and interleukin 10 (IL-10), that inactivate effector cells—e.g., by switching off the production of toxic molecules by phagocytes (Mills 2004). Regulatory T cells can also increase the activation threshold at which Type 1 or Type 2 cytokines are produced (Abbas et al. 2004), thereby preventing immunopathological levels of activation from being achieved at all.

Important roles for regulatory T cells have been demonstrated for many of the WHO top 10, including malaria (Hisaeda et al. 2004), leishmaniasis (Sacks & Anderson 2004), schistosomiasis (Hesse et al. 2004), filariasis (Taylor et al. 2005), and onchocerchiasis (Satoguina et al. 2002). The critical importance of modulatory adaptive immunity is especially well illustrated in malaria, where regulatory T cells, IL-10, and TGF-β are required for host survival (Hisaeda et al. 2004, Li et al. 2003, Omer et al. 2003). Without these modulators, hosts die of malarial immunopathology, as outlined in Section 2.1 above. T regulatory cells also protect against other forms of immunopathology, including Type 1 autoimmune attack of the central nervous system in multiple sclerosis (Viglietta et al. 2004) and Type 2 allergic responses to airborne particles in asthma (Maizels et al. 2004). Indeed, in human populations, genetic polymorphisms in both IL-10 (Moore et al. 2001) and TGF-β (Gentile et al. 2003) are associated with differential predisposition to infectious, autoimmune, and allergic diseases. Immunopathology may have imposed strong selection pressure for the immune system to use these pathways to eliminate cells that are causing harm to self, even well after selection against self-reactivity in the thymus early in development.

It is tempting to speculate that the adaptive immune system may have evolved expressly to use antigen-specific receptors to focus and direct the response only where needed and to control the production of potentially destructive cytokines such as TNF-α and IL-13. The greatest defense against the fitness effects of infection may indeed be moderation (Figure 3).

3.3. Why Has Evolution Not Eliminated Immunopathology?

Despite these antipathology design features of the immune system, immunopathology still dramatically reduces the fitness of people and mice (Table 1). Why has selection not eliminated such self-harm? Below, we outline several explanatory hypotheses; these are not mutually exclusive but instead may combine to cause immunopathology to persist. The relevant explanation will almost certainly vary from system to system. Importantly, we should be able to detect which explanation holds.

First, immunopathology may be retained by a balance between costs and benefits of fighting parasites. Frank (2002) suggested that polymorphism at antigen recognition loci might be maintained by processes beyond mutation-selection balance. Such processes probably maintain polymorphism at cytokine loci as well. For example, occasionally excessive TNF-α or IL-13 responses may be unavoidable consequences of useful parasite-killing or tissue-remodelling mechanisms. Natural selection may favor high responsiveness as the default option immediately following infection, to ensure control of parasites despite the risk of immunopathology. Such decision rules should themselves select for mechanisms to moderate responses (e.g., regulatory T cells) and focus them on known threats (e.g., memory responses). Frank (2002) called for mathematical analysis to clarify the necessary conditions for stable polymorphism between high- and low-response tendencies. We second that call, adding that consideration of response efficacy (via Type 1, Type 2, and modulatory cytokines), plus empirical work to quantify costs and benefits, are also critical to that analysis.

Second, immunopathology may result from the constrained ability of the immune system to achieve optima. The design flaws may consist in poor modulatory control, signaling delays, or other mechanisms. For example, hosts are unable to simultaneously mount the two mutually antagonistic types of response. Some immunopathology may thus be a consequence of the Type 1–2 cross-regulation intrinsic to immune functioning, a proposition that has generated testable hypotheses about coinfection (Graham 2002). Still, invoking such trade-offs to explain immunopathology just shifts the problem back a level. What makes the regulatory

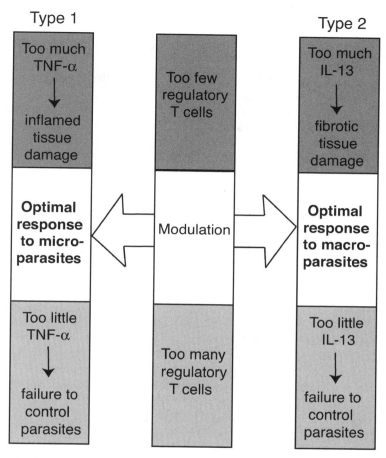

Figure 3   To control the negative fitness effects of infection, the magnitude of Type 1 responses to microparasites and Type 2 responses to macroparasites must be controlled. For example, TNF-α hypo- and hyper-responsiveness against microparasites are equally detrimental to host fitness, and the optimal region in the left-hand bar may correspond to the optimal antimalaria response identified in Figure 2 (e.g., 200–300 spleen cells producing Type 1 cytokines). T regulatory cells are critical to achieving such optima. For a mechanistic view, see figure 5 of Mills (2004).

mechanisms insufficient? Why are signals indicating self-harm so slow to arrive? The implication is that more precise immunological control is more expensive than immunopathology. Just how these engineering costs arise remains to be determined.

Third, immunopathology may be expressed mostly in cases where selection has not had time to act. It may be that novel parasites or environments experienced by modern humans, for instance, impose substantial selection on genetic determinants of immunopathology, but there has not been sufficient time to make the response to this selection detectable. Further, parasites, by virtue of shorter generation times, may have a coevolutionary grace period during which host responses are suboptimal (Behnke et al. 1992).

Fourth, lateness of onset could result in reduced selection against immunopathology. As with other diseases of senescence, selection against late-life immunopathological disease may be weak, whether infection is current (Table 1) or summed over a lifetime (Finch & Crimmins 2004). This hypothesis can be rejected, however, for early-onset immunopathologies.

Finally, selection on parasite genotypes may be an evolutionary cause of immunopathology—for example, when immune evasion or immunopathology itself is good for parasite transmission (explored in Section 4.1, below). A positive association between immunopathology and transmission might favor parasite genotypes that manipulate host immunity. Such manipulations are common among helminths (Maizels et al. 2004), protozoa (Sacks & Sher 2002), and viruses (Tortorella et al. 2000), and immunomodulatory products of parasites are genetically polymorphic in at least some species (Behnke et al. 1992, Britton et al. 1995, Yatsuda et al. 2001). It is critical, for both biomedical and evolutionary studies, to determine whether parasites or hosts benefit most from modulated immune responses and, relatedly, whether hosts are being manipulated into immunopathology.

The development of evolutionary explanations of other apparently maladaptive traits has proven highly productive, expanding evolutionary theory while revealing new properties about the traits in question. Evolutionary explanations of immunopathology may well do the same, revealing insight into the evolution of complex systems while successfully predicting the occurrence and severity of immunopathology, as well as the clinical and evolutionary consequences of medical interventions designed to alleviate it. With such studies, the interests of global health management and evolutionary biology would converge.

4. CONSEQUENCES OF IMMUNOPATHOLOGY FOR EVOLUTIONARY RESEARCH

We propose that at least three research areas in host-parasite evolutionary biology would benefit from a focused inclusion of immunopathology: the evolution of virulence, the evolution of resistance, and the evolution of immunogenetic polymorphism (e.g., in the MHC). Other research areas might likewise benefit, but we are most familiar with the aims and methods of these three. We argue that they are also representative of active research in host-parasite evolutionary biology—the combined literature on just the evolution of virulence and ecological immunology accounts for about one third of the host-parasite evolution literature (on all host taxa) searchable by Pub Med and Web of Science. We end this section by addressing empirical obstacles that all research areas must overcome to incorporate immunopathology.

4.1. Evolution of Parasite Virulence

Immunopathology can alter the costs and benefits of parasite virulence. Virulence in evolutionary terms is defined as the negative impact of infection upon host fitness. As such, virulence encompasses damage due to direct effects of parasites as well as damage due to infection-induced immunopathology. The two sources of virulence can be difficult to distinguish in practice, but they may have very different evolutionary implications.

Parasite strategies for host exploitation are predicted to evolve toward greater virulence when high replication rates and high virulence are associated with high transmission rates (Frank 1996). If immunity acts solely to reduce parasite replication, then selection is predicted to increase parasite virulence (Gandon et al. 2001). The underlying assumption that virulence and transmission are positively correlated has strong empirical support, at least in some disease systems (Mackinnon & Read 2004). But when virulence is determined independently of parasite density— for example, if immunopathology decouples transmission and virulence—then the costs and benefits of virulence are likely to be altered (Lipsitch & Moxon 1997).

Parasites might benefit from immunopathology if it is accompanied by increased transmission. In tuberculosis, for example, there is good evidence that immunopathological necrosis of the lung enhances transmission (Ehlers et al. 2001, Kaushal et al. 2002). Similarly, tissue remodelling around schistosome eggs facilitates their passage into the environment to complete the life cycle (Doenhoff 1998). Immunopathology is also associated with increased transmission of dengue (Gagnon et al. 1999, Mongkolsapaya et al. 2003) and chronicity of leishmaniasis (Sacks & Anderson 2004). Immunopathology thus seems to aid transmission of at least 4 of the WHO top 10. In these cases, the assumed positive relationship between virulence and transmission would be upheld even at the immunopathological extreme of the spectrum, and the qualitative (if not quantitative) predictions of basic theory may hold.

It is equally possible, however, that immunopathology is bad for parasite fitness, increasing virulence without increasing transmission. In this scenario, immunopathology increases the cost/benefit ratio of parasite virulence. This would qualitatively alter the predicted trajectory of the evolution of virulence (illustrated in Figure 1a), possibly selecting for decreased virulence via decreased replication or immunogenicity. For at least 3 of the WHO top 10, inducing severe immunopathology appears to bring no benefit to the parasite. Lymphatic filariasis, for example, is most virulent (i.e., causes elephantiasis) when nontransmissible (Behnke et al. 1992, Sartono et al. 1997). Severe cases of sleeping sickness are associated with inflammatory reactions to parasites in the central nervous system (Hunter & Kennedy 1992), and it may be excess TNF-α that breaches the blood-brain barrier (Maclean et al. 2004). Can this migration possibly facilitate transmission to the tsetse fly? In malaria, cytokines that, if unchecked, cause severe disease (Akanmori et al. 2000, Dodoo et al. 2002) can also block parasite transmission (Karunaweera et al. 1992). As these examples demonstrate, the effects of immunopathology upon virulence-transmission relationships may not conform to the positive correlation assumed in basic theory.

Several theoretical studies support the idea that pathological immune responses can alter the evolution of virulence. The important acknowledgment that virulence is a coevolutionary issue (beginning with van Baalen 1998) led to the insight that the evolutionary trajectory of virulence strongly depends on whether hosts develop resistance at all, and whether resistance is qualitative (i.e., each host is either susceptible or resistant) or quantitative (Gandon & Michalakis 2000, Gandon et al. 2002). Costs of immunological up-regulation further alter the evolution of virulence (Alizon & van Baalen 2005, Day & Burns 2003, Restif & Koella 2003), but only by distinguishing between direct and indirect effects of parasites were Alizon & van Baalen (2005) able to consider closely the role of immunopathology. Their provocative results lend formal support to the intuitive arguments above: The relative amount of damage that immune responses do to hosts versus to parasites determines the evolutionarily stable level of virulence (Alizon & van Baalen 2005).

A theoretical study to explore how immunopathology impacts the evolution of virulence might usefully add host heterogeneity. Baseline immunopathology (accounting for the proportion of mild disease that is due to immune hyper-reactivity) would be complemented by additional immunopathology experienced only by a proportion of hosts (accounting for the excess immunopathology that kills, as in each of the WHO top 10; Table 1). Such a model could assess whether there is a threshold proportion of damage due to immunopathology above which immune responses rather than parasites dominate dynamics. It might also predict the influence of mild versus severe immunopathology on the evolution of virulence. Better still, its parameters would be measurable and, thus, the model testable. It is our contention that in both theoretical and empirical work on the evolution of virulence, a role for immunopathology should at least be assessed before it is omitted from study.

4.2. Evolution of Resistance

Immunopathology is arguably the highest cost of immune defense. As such, it should be considered in studies of ecological (Sheldon & Verhulst 1996) or evolutionary (Lochmiller & Deerenberg 2000) immunology. To life history theorists, immunity, like foraging behavior or territorial defense, is just another trait whose costs and benefits are traded off against other fitness determinants (e.g., frequency with which nestlings are fed) (Sheldon & Verhulst 1996). Most work in ecological immunology focuses on whether costs of immunity can explain heterogeneity among hosts in their level of defense (reviewed by Schmid-Hempel 2003). Essentially, if immunity were cheap, then all hosts would be predicted to respond vigorously to infection. If, on the other hand, immunity were expensive, then hosts with differing budgets at their disposal and differing allocation priorities would differ in their investment in defense. In the costly defense scenario, immunoheterogeneity is unsurprising.

Several studies have indeed demonstrated fitness costs of immune defense (Ilmonen et al. 2000, Råberg & Stjernman 2003), but physiological costs have proven more difficult to detect (Lochmiller & Deerenberg 2000). Part of the problem may be that only energetic physiological costs have received substantial attention in this literature. In an explicit test of the energetic costs of immunity that showed lower basal metabolic rates in mice with higher adaptive immunity, Råberg et al. (2002) concluded that the evolution of immunity was probably not constrained by energy. Some studies do concede that another probable physiological cost of the immune system is immunopathology—Schmid-Hempel (2003) even proposed that "self-reactivity" is greatly underrated in ecological immunology—but very few measure how benefits of defense trade off against immunopathological costs. Studies investigating the optimal strength of response that balances the benefits of parasite killing against risks of immunopathology (e.g., Borghans et al. 1999, Råberg et al. 1998, Segel & Bar-Or 1999, Wu et al. 1996) are generally conducted outside of ecological immunology.

Most studies instead assume that greater numbers of immunological cells or molecules confer greater fitness [e.g., Nunn et al. (2000), with reservations registered by ourselves (Read & Allen 2000)]. Although there is empirical support for the notion that more is better (Biozzi et al. 1984, Luster et al. 1993), there is also substantial evidence to the contrary (Table 1; Figure 2). Exuberant immune responses can lead to complete parasite clearance and yet host death via immune-mediated organ damage, not energetic collapse.

Inclusion of immunopathological costs in ecological immunology models might predict relatively low optimal levels of defense (illustrated in Figure 1*b*). Still, arguments about condition-dependent costs of defense [e.g., a lower cost and thus higher optimal magnitude of response in high-quality individuals (Getty 1998)] as well as context-dependent costs (e.g., increasing benefit of immunity with increased exposure to infection) can apply to immunopathology. For example, the cost of defense may be energetic for nutritionally stressed individuals, whereas high-quality individuals with lots to invest in immunity may be more prone to immunopathological costs. This idea should be testable in many systems.

Theoretical studies confirm that immunopathology should receive greater attention in this field. For example, the efficacy of immune responses is a key determinant of the optimal level of investment (van Boven & Weissing 2004). More compellingly, theory that explicitly sets out to minimize the sum of parasite-induced and immunopathological damage can help to explain circumstances when the immune system should choose one response (e.g., TNF-α) over another (e.g., IL-13) (Shudo & Iwasa 2001) and lends formal support to the idea (in Section 3.1, above) that Type 2 immunity exists primarily to prevent immunopathology. Theory even predicts when immunomodulation should begin: just after parasite replication plateaus (Shudo & Iwasa 2004). These predictions should be testable, as should the further prediction that immunomodulation requires serial overshooting (Shudo & Iwasa 2004). In future, costs of resistance would ideally be broken into energetic and immunopathological parts. The extent to which covariance of the two determines evolutionary predictions would be of great interest.

Intriguingly, mathematical models that account for immunopathology predict incomplete clearance of parasites as the optimal way for a host to spend its resources (Medley 2002, Shudo & Iwasa 2004), according with empirical and verbal models (Behnke et al. 1992). Essentially, the costs of immunopathology can come to outweigh the benefits of parasite clearance. If subsequent research shows that the tolerable parasite burden is quantitatively predictable, then evolutionary biology will have contributed enormously to biomedicine. It may be possible to predict, for instance, the worm burden that is worth expelling despite the risk of immunopathology. This threshold is likely to vary across host species, sex, and condition, as well as parasite species—a rich area for experimental and field tests. Only by incorporating immunopathology can ecological immunology make such a contribution.

4.3. Evolution of Antigen Recognition Polymorphism

Perfect recognition of parasite antigens does not preclude immunopathology. Explicitly genetic theories of defense that are based upon recognition of parasites by the immune system aim to interpret immunoheterogeneity (Hedrick 2002)—to explain high diversity of MHC alleles, for example, via heterozygote and/or rare allele advantage (Apanius et al. 1997, Borghans et al. 2004, McClelland et al. 2003). Such approaches have unmasked selection pressures on genes that encode parasite-recognizing proteins (Frank 2002, Schmid-Hempel 2003). For instance, a parabolic relationship between parasite load and the number of stickleback MHC alleles has provided evidence of balancing selection and suggests that there are limits to the benefits of MHC allelic diversity (Wegner et al. 2003).

However, heterogeneity in the ability of hosts to recognize parasites does not always explain the distribution of disease (Hill 1998). The immune system must not only recognize parasite antigens but it must also choose the appropriate number and type of parasite-killing mechanisms (i.e., modulated Type 1 versus Type 2 responses; Figure 3). Defined antigens elicit differentially protective immune responses in mice with matched recognition capabilities, and similar processes operate in human hosts (Frank 2002). The recognition and effector steps of an immune response are therefore equally important: Mounting the wrong type or magnitude of response, even against the right antigen, can be very detrimental to host fitness (Graham 2002). Even in cases where MHC explains a good deal of variance in fitness, incorporating immune effector function may do still more.

Indeed, data from biomedical genetics studies support the notion that cytokines can rival or surpass the importance of MHC genotype in determining the outcome of infection (Behnke et al. 1992, Hill 1998, Mitchison et al. 2000). For the WHO top 10 (Table 1), allelic variants at MHC loci help to explain susceptibility to tuberculosis, malaria, dengue, and leprosy (Hill 1998). Inclusion of cytokine polymorphism further explains susceptibility to those diseases and adds leishmaniasis and schistosomiasis (Hill 1998) as well as onchocerciasis (Hoerauf et al. 2002) to the list. Such synergy between recognition and cytokine profiles explains the outcome of many other infectious diseases (Daser et al. 1996). For example, MHC Class II promoter polymorphism affects the type of effector mechanisms preferentially enabled by a host; this has led to the intriguing suggestion that beyond recognition capability, heterozygote advantage may consist in flexibility or fine-tuning of parasite killing (Mitchison et al. 2000).

Recognition processes thus combine with effector mechanisms to determine whether hosts clear infection entirely, control parasite replication yet permit transmission, and/or generate immunopathology. These outcomes have different evolutionary implications and merit further study (Frank 2002). Integration of the influences of the efficacy and specificity of responses is certainly an important step toward a "unified defense theory" (Jokela et al. 2000), but evolutionary studies as yet largely omit investigation of immune response efficacy. With this review of how cytokines determine protection versus pathology and thus alter evolutionary trajectories, we hope to equip all sides for theoretical and empirical progress.

4.4. Empiricism

To optimize empirical studies of the evolutionary causes and consequences of immunopathology, the main innovation needed is quantification of immune-mediated disease. This is not trivial: There is unlikely to be one measure applicable to all host-parasite interactions. Direct measurements would ideally be made—for example, the diameter of immunopathological lesions of the liver or lung (Doenhoff 1998) or the number of T cells targeting uninfected host cells (Gagnon et al. 1999). Such measurements may be system-specific and technically difficult to obtain, but just a single, well-chosen marker of immunity (such as titer of a key cytokine) would be an important advance, if the immunology were analyzed alongside parasite density as predictors of host and/or parasite fitness. When carefully chosen (Read & Allen 2000), even nonspecific measures are remarkably good at predicting resistance to infection (Biozzi et al. 1984, Luster et al. 1993). Similarly, rough estimates of success at killing parasites are better than ignoring immune efficacy.

An immediately promising experimental direction would be to take advantage of the reagents and methods available to dissect immune responses in mice and farm animals. Heterogeneity among hosts (observable even among laboratory mice) makes it possible to statistically disentangle fitness effects of high parasite density versus excess cytokine production (Figure 2) (Graham et al. 2005). It is also possible to examine the fitness consequences of artificial selection for high versus low levels of immunological activity in agricultural systems (e.g., the work of Magnusson et al. 1999). Such data could reveal the true shape of the net benefit curve for defense (thereby verifying or falsifying Figure 1b). Is the assumption of more-is-better largely correct? Studies of model animals should also help to define immunological measures that reliably predict fitness and provide guidance for choosing measures in nonmodel systems.

We hope that the experiments would be paralleled by analogous field studies in human and wild animal populations. Already, genetic and epidemiological studies of immunopathology are performed in human populations (Akanmori et al. 2000, Booth et al. 2004, Dessein et al. 2004, Dodoo et al. 2002, Maclean et al. 2004); analogous studies in wild mammals, especially rodents, are feasible. Moreover, human studies could generate much valuable data for evolutionary ecologists if measures of disease transmission were included. In wild animal populations where individual host fitness and life histories are well characterized—e.g., in mammals (Clutton-Brock & Pemberton 2004) or birds (Sheldon et al. 2003)—measurement of immunogenetic polymorphisms or cytokine levels [e.g., TNF-α in birds (Erf 2004)] would lead to substantial advances. With proper design, vaccination studies even make it possible to detect directional versus stabilizing selection on immune responsiveness in natural populations (Råberg & Stjernman 2003). The possibilities for improving current empirical practice to better understand evolutionary immunopathology appear substantial.

5. OUTLOOK

Only via concerted evolutionary and mechanistic study will we come to understand causation sufficiently well to predict the occurrence of immunopathology and its impact upon host and parasite evolution. Given the wide range of infections that induce immunopathology, understanding self-harm is essential to elucidating how natural selection acts on host and parasite genotypes in their major arena of interaction, the immune system.

Evolutionary analysis of immunopathology should open up new avenues of research. Not least is a systematic analysis of how medical interventions might alter the contribution of the host to the severity of infectious disease. For example, might vaccination increase immune efficiency by minimizing immunopathology—shortening the relatively damaging innate phase of a response and/or shortening the entire response via precisely targeted parasite killing? Or might vaccination instead boost immune responsiveness to more pathology-prone heights (Alizon & van Baalen 2005)? The evolutionary implications of these ecological effects of immunopathology could readily be explored (e.g., Gandon et al. 2001). A cost-benefit analysis of immunity might also quantitatively inform the rational treatment of fevers, which are often immunopathological. Finally, as we have reviewed elsewhere (Graham 2002), coinfections prevalent in natural populations may impose conflicting selection pressures on immune responsiveness. Do medical treatments that minimize the prevalence of coinfection predictably alter the odds of immunopathology? Short of clinical trials, we currently have no way of knowing the effects of, for example, antihelminthics on malarial disease burdens.

Our arguments about immunopathology may also apply to diseases of nonvertebrate hosts. Invertebrates share many immunological traits with vertebrates, making use of similar parasite-killing mechanisms—for example, an earthworm homologue of TNF-α can kill protozoa (Olivares Fontt et al. 2002), and there is evidence that the snail hosts of schistosomes, like the human hosts, use fibrosis to fight worms (Zhang & Loker 2004). Given these similarities, plus the progress and promise of research in ecological immunology of invertebrates (Rolff & Siva-Jothy 2003), we may soon understand how evolutionary immunopathology operates in invertebrate as well as vertebrate taxa. As with vertebrates (Read & Allen 2000), the relationship between, for example, hemocyte number and immune efficacy is not necessarily linear. Fitness could be reduced at both the low end, where individuals are undefended against parasitism, and at the high end, where individuals experience immunopathology. Indeed, fruit flies (Brandt et al. 2004) and beetles (B. Sadd & M.T. Siva-Jothy, manuscript submitted) appear prone to immune-mediated disease. If generalizations about immunopathology extend to invertebrates and beyond, deeper understanding of the costs and benefits of defense will begin to emerge.

ACKNOWLEDGMENTS

A.L.G. is an Early Career Fellow of The Leverhulme Trust and the University of Edinburgh's School of Biological Sciences. Our empirical work is supported by the U.K. Biotechnology and Biological Sciences Research Council, the Medical Research Council, The Leverhulme Trust, and The Wellcome Trust. We are very grateful to Adam Balic, Sylvain Gandon, Drew Harvell, Tracey Lamb, Tom Little, Gráinne Long, Margaret Mackinnon, and Lars Råberg for helpful discussions and/or comments on the manuscript.

LITERATURE CITED

Abaru DE. 1985. Sleeping sickness in Busoga, Uganda, 1976–1983. *Trop. Med. Parasitol.* 36:72–76

Abbas AK, Lohr J, Knoechel B, Nagabhushanam V. 2004. T cell tolerance and autoimmunity. *Autoimmun. Rev.* 3:471–75

Abbas AK, Murphy KM, Sher A. 1996. Functional diversity of helper T lymphocytes. *Nature* 383:787–93

Adler J. 1997. The dueling diagnoses of Darwin. *JAMA* 277:1275–77

Akanmori BD, Kurtzhals JAL, Goka BQ, Adabayeri V, Ofori MF, et al. 2000. Distinct patterns of cytokine regulation in discrete clinical forms of *Plasmodium falciparum* malaria. *Eur. Cytokine Netw.* 11:113–18

Alizon S, van Baalen M. 2005. Emergence of a convex trade-off between transmission and virulence. *Am. Nat.* 165:155–67

Andrade ZA. 1999. Immunopathology of Chagas disease. *Mem. Inst. Oswaldo Cruz* 94:71–80

Apanius V, Penn D, Slev PR, Ruff LR, Potts WK. 1997. The nature of selection on the major histocompatibility complex. *Crit. Rev. Immunol.* 17:179–224

Bayley JP, Ottenhoff THM, Verweij CL. 2004. Is there a future for TNF promoter polymorphisms? *Genes Immun.* 5:315–29

Behnke JM, Barnard CJ, Wakelin D. 1992. Understanding chronic nematode infections: Evolutionary considerations, current hypotheses and the way forward. *Int. J. Parasitol.* 22:861–907

Bekker LG, Moreira AL, Bergtold A, Freeman

S, Ryffel B, Kaplan G. 2000. Immunopathologic effects of tumor necrosis factor alpha in murine mycobacterial infection are dose dependent. *Infect. Immun.* 68:6954–61

Biozzi G, Mouton D, Stiffel C, Bouthillier Y. 1984. A major role of the macrophage in quantitative genetic regulation of immunoresponsiveness and antiinfectious immunity. *Adv. Immunol.* 36:189–234

Booth M, Mwatha JK, Joseph S, Jones FM, Kadzo H, et al. 2004. Periportal fibrosis in human *Schistosoma mansoni* infection is associated with low IL-10, low IFN-γ, high TNF-α, or low RANTES, depending on age and gender. *J. Immunol.* 172:1295–303

Borghans JA, Beltman JB, De Boer RJ. 2004. MHC polymorphism under host-pathogen coevolution. *Immunogenetics* 55:732–39

Borghans JA, Noest AJ, De Boer RJ. 1999. How specific should immunological memory be? *J. Immunol.* 163:569–75

Brandt SM, Dionne MS, Khush RS, Pham LN, Vigdal TJ, Schneider DS. 2004. Secreted bacterial effectors and host-produced eiger/TNF drive death in a *Salmonella*-infected fruit fly. *PLoS Biol.* 2:e418

Britton C, Moore J, Gilleard JS, Kennedy MW. 1995. Extensive diversity in repeat unit sequences of the cDNA encoding the polyprotein antigen/allergen from the bovine lungworm *Dictyocaulus viviparus*. *Mol. Biochem. Parasitol.* 72:77–88

Chan MS. 1997. The global burden of intestinal nematode infections—fifty years on. *Parasitol. Today* 13:438–43

Clutton-Brock TH, Pemberton JM, eds. 2004. *Soay Sheep: Dynamics and Selection in an Island Population.* Cambridge, UK: Cambridge Univ. Press. 396 pp.

Cuthill IC, Guilford T. 1990. Perceived risk and obstacle avoidance in flying birds. *Anim. Behav.* 40:188–90

Daniel OJ, Salako AA, Oluwole FA, Alausa OK, Oladapo OT. 2004. HIV sero-prevalence among newly diagnosed adult pulmonary tuberculosis patients in Sagamu. *Niger J. Med.* 13:393–97

Daser A, Mitchison H, Mitchison A, Muller

B. 1996. Non-classical-MHC genetics of immunological disease in man and mouse. The key role of pro-inflammatory cytokine genes. *Cytokine* 8:593–97

Day T, Burns JG. 2003. A consideration of patterns of virulence arising from host-parasite coevolution. *Evol. Int. J. Org. Evol.* 57:671–76

Dessein A, Kouriba B, Eboumbou C, Dessein H, Argiro L, et al. 2004. Interleukin-13 in the skin and interferon-γ in the liver are key players in immune protection in human schistosomiasis. *Immunol. Rev.* 201:180–90

Dodoo D, Omer FM, Todd J, Akanmori BD, Koram KA, Riley EM. 2002. Absolute levels and ratios of proinflammatory and antiinflammatory cytokine production *in vitro* predict clinical immunity to *Plasmodium falciparum* malaria. *J. Infect. Dis.* 185:971–79

Doenhoff MJ. 1998. Granulomatous inflammation and the transmission of infection: schistosomiasis—and TB too? *Immunol. Today* 19:462–67

Dye C, Scheele S, Dolin P, Pathania V, Raviglione MC. 1999. Consensus statement. Global burden of tuberculosis: estimated incidence, prevalence, and mortality by country. WHO Global Surveillance and Monitoring Project. *JAMA* 282:677–86

Ehlers S, Benini J, Held HD, Roeck C, Alber G, Uhlig S. 2001. $\alpha\beta$ T cell receptor-positive cells and interferon-γ, but not inducible nitric oxide synthase, are critical for granuloma necrosis in a mouse model of mycobacteria-induced pulmonary immunopathology. *J. Exp. Med.* 194:1847–59

Erf GF. 2004. Cell-mediated immunity in poultry. *Poult. Sci.* 83:580–90

Finch CE, Crimmins EM. 2004. Inflammatory exposure and historical changes in human life-spans. *Science* 305:1736–39

Finkelman FD, Shea-Donohue T, Morris SC, Gildea L, Strait R, et al. 2004. Interleukin-4- and interleukin-13-mediated host protection against intestinal nematode parasites. *Immunol. Rev.* 201:139–55

Frank SA. 1996. Models of parasite virulence. *Q. Rev. Biol.* 71:37–78

Frank SA. 2002. *Immunology and Evolution of Infectious Disease.* Princeton, NJ: Princeton Univ. Press. 348 pp.

Gagnon SJ, Ennis FA, Rothman AL. 1999. Bystander target cell lysis and cytokine production by dengue virus-specific human CD4$^+$ cytotoxic T-lymphocyte clones. *J. Virol.* 73:3623–29

Gandon S, Mackinnon MJ, Nee S, Read AF. 2001. Imperfect vaccines and the evolution of pathogen virulence. *Nature* 414:751–56

Gandon S, Michalakis Y. 2000. Evolution of parasite virulence against qualitative or quantitative host resistance. *Proc. R. Soc. London Ser. B* 267:985–90

Gandon S, van Baalen M, Jansen VAA. 2002. The evolution of parasite virulence, superinfection, and host resistance. *Am. Nat.* 159:658–69

Gentile DA, Doyle WJ, Zeevi A, Howe-Adams J, Kapadia S, et al. 2003. Cytokine gene polymorphisms moderate illness severity in infants with respiratory syncytial virus infection. *Hum. Immunol.* 64:338–44

Getty T. 1998. Handicap signalling: when viability and fecundity do not add up. *Anim. Behav.* 56:127–30

Graham AL. 2002. When T-helper cells don't help: Immunopathology during concomitant infection. *Q. Rev. Biol.* 77:409–33

Graham AL, Lamb TJ, Read AF, Allen JE. 2005. Malaria-filaria coinfection in mice makes malarial disease more severe unless filarial infection achieves patency. *J. Infect. Dis.* 191:410–21

Gubler DJ. 1998. Dengue and dengue hemorrhagic fever. *Clin. Microbiol. Rev.* 11:480–96

Hall LR, Pearlman E. 1999. Pathogenesis of onchocercal keratitis (river blindness). *Clin. Microbiol. Rev.* 12:445–53

Hedrick PW. 2002. Pathogen resistance and genetic variation at MHC loci. *Evol. Int. J. Org. Evol.* 56:1902–8

Hesse M, Piccirillo CA, Belkaid Y, Prufer J, Mentink-Kane M, et al. 2004. The pathogenesis of schistosomiasis is controlled by cooperating IL-10-producing innate effector and regulatory T cells. *J. Immunol.* 172:3157–66

Hill AVS. 1998. The immunogenetics of human infectious diseases. *Annu. Rev. Immunol.* 16:593–617

Hirsch CS, Hussain R, Toossi Z, Dawood G, Shahid F, Ellner JJ. 1996. Cross-modulation by transforming growth factor β in human tuberculosis: suppression of antigen-driven blastogenesis and interferon γ production. *Proc. Natl. Acad. Sci. USA* 93:3193–98

Hisaeda H, Maekawa Y, Iwakawa D, Okada H, Himeno K, et al. 2004. Escape of malaria parasites from host immunity requires CD4$^+$ CD25$^+$ regulatory T cells. *Nat. Med.* 10:29–30

Hoerauf A, Kruse S, Brattig NW, Heinzmann A, Mueller-Myhsok B, Deichmann KA. 2002. The variant Arg110Gln of human IL-13 is associated with an immunologically hyper-reactive form of onchocerciasis (sowda). *Microbes Infect.* 4:37–42

Hoffmann KF, Wynn TA, Dunne DW. 2002. Cytokine-mediated host responses during schistosome infections: walking the fine line between immunological control and immunopathology. *Adv. Parasitol.* 52:265–307

Holscher C, Mohrs M, Dai WJ, Kohler G, Ryffel B, et al. 2000. Tumor necrosis factor alpha-mediated toxic shock in *Trypanosoma cruzi*-infected interleukin 10-deficient mice. *Infect. Immun.* 68:4075–83

Hunt NH, Grau GE. 2003. Cytokines: accelerators and brakes in the pathogenesis of cerebral malaria. *Trends Immunol.* 24:491–99

Hunter CA, Kennedy PG. 1992. Immunopathology in central nervous system human African trypanosomiasis. *J. Neuroimmunol.* 36:91–95

Hussell T, Pennycook A, Openshaw PJ. 2001. Inhibition of tumor necrosis factor reduces the severity of virus-specific lung immunopathology. *Eur. J. Immunol.* 31:2566–73

Ilmonen P, Taarna T, Hasselquist D. 2000. Experimentally-activated immune defense in female pied flycatchers results in reduced breeding success. *Proc. R. Soc. London Ser.* B 267:665–70

Jokela J, Schmid-Hempel P, Rigby MC. 2000. Dr. Pangloss restrained by the Red Queen—Steps towards a unified defence theory. *Oikos* 89:267–74

Jorge MT, Macedo TA, Janones RS, Carizzi DP, Heredia RA, Acha RE. 2003. Types of arrhythmia among cases of American trypanosomiasis, compared with those in other cardiology patients. *Ann. Trop. Med. Parasitol.* 97:139–48

Karunaweera ND, Carter R, Grau GE, Kwiatkowski D, Del Giudice G, Mendis KN. 1992. Tumour necrosis factor-dependent parasite-killing effects during paroxysms in non-immune *Plasmodium vivax* malaria patients. *Clin. Exp. Immunol.* 88:499–505

Kaushal D, Schroeder BG, Tyagi S, Yoshimatsu T, Scott C, et al. 2002. Reduced immunopathology and mortality despite tissue persistence in a *Mycobacterium tuberculosis* mutant lacking alternative sigma factor, SigH. *Proc. Natl. Acad. Sci. USA* 99:8330–35

Khanolkar-Young S, Rayment N, Brickell PM, Katz DR, Vinayakumar S, et al. 1995. Tumour necrosis factor-alpha (TNF-α) synthesis is associated with the skin and peripheral nerve pathology of leprosy reversal reactions. *Clin. Exp. Immunol.* 99:196–202

Leang R, Socheat D, Bin B, Bunkea T, Odermatt P. 2004. Assessment of disease and infection of lymphatic filariasis in Northeastern Cambodia. *Trop. Med. Int. Health* 9:1115–20

Li C, Sanni LA, Omer F, Riley E, Langhorne J. 2003. Pathology of *Plasmodium chabaudi chabaudi* infection and mortality in IL-10-deficient mice are ameliorated by anti-tumor necrosis factor alpha and exacerbated by anti-transforming growth factor β antibodies. *Infect. Immun.* 71:4850–56

Libraty DH, Endy TP, Houng HS, Green S, Kalayanarooj S, et al. 2002. Differing influences of virus burden and immune activation on disease severity in secondary dengue-3 virus infections. *J. Infect. Dis.* 185:1213–21

Lipsitch M, Moxon RE. 1997. Virulence and transmissibility of pathogens: what is the relationship? *Trends Microbiol.* 5:31–37

Lochmiller RL, Deerenberg C. 2000. Trade-offs in evolutionary immunology: just what is the cost of immunity? *Oikos* 88:87–98

Louzir H, Melby PC, Ben Salah A, Marrakchi H, Aoun K, et al. 1998. Immunologic determinants of disease evolution in localized cutaneous leishmaniasis due to *Leishmania major. J. Infect. Dis.* 177:1687–95

Luster MI, Portier C, Pait DG, Rosenthal GJ, Germolec DR, et al. 1993. Risk assessment in immunotoxicology. II. Relationships between immune and host resistance tests. *Fundam. Appl. Toxicol.* 21:71–82

Mackinnon MJ, Read AF. 2004. Virulence in malaria: an evolutionary viewpoint. *Philos. Trans. R. Soc. London Ser. B* 359:965–86

Maclean L, Chisi JE, Odiit M, Gibson WC, Ferris V, et al. 2004. Severity of human African trypanosomiasis in East Africa is associated with geographic location, parasite genotype, and host inflammatory cytokine response profile. *Infect. Immun.* 72:7040–44

Magez S, Truyens C, Merimi M, Radwanska M, Stijlemans B, et al. 2004. P75 tumor necrosis factor-receptor shedding occurs as a protective host response during African trypanosomiasis. *J. Infect. Dis.* 189:527–39

Magnusson U, Wilkie B, Artursson K, Mallard B. 1999. Interferon-alpha and haptoglobin in pigs selectively bred for high and low immune response and infected with *Mycoplasma hyorhinis. Vet. Immunol. Immunopathol.* 68:131–37

Maizels RM, Balic A, Gomez-Escobar N, Nair M, Taylor MD, Allen JE. 2004. Helminth parasites—masters of regulation. *Immunol. Rev.* 201:89–116

McClelland EE, Penn DJ, Potts WK. 2003. Major histocompatibility complex heterozygote superiority during coinfection. *Infect. Immun.* 71:2079–86

McNeil KS, Knox DP, Proudfoot L. 2002. Antiinflammatory responses and oxidative stress in *Nippostrongylus brasiliensis*-induced pulmonary inflammation. *Parasite Immunol.* 24:15–22

Medley GF. 2002. The epidemiological consequences of optimisation of the individual host immune response. *Parasitology* 125:S61–70

Michael E, Bundy DA. 1997. Global mapping of lymphatic filariasis. *Parasitol. Today* 13:472–76

Mills KH. 2004. Regulatory T cells: friend or foe in immunity to infection? *Nat. Rev. Immunol.* 4:841–55

Mitchison NA, Muller B, Segal RM. 2000. Natural variation in immune responsiveness, with special reference to immunodeficiency and promoter polymorphism in class II MHC genes. *Hum. Immunol.* 61:177–81

Moncayo A. 1992. Chagas disease: epidemiology and prospects for interruption of transmission in the Americas. *World Health Stat. Q.* 45:276–79

Mongkolsapaya J, Dejnirattisai W, Xu XN, Vasanawathana S, Tangthawornchaikul N, et al. 2003. Original antigenic sin and apoptosis in the pathogenesis of dengue hemorrhagic fever. *Nat. Med.* 9:921–27

Moore KW, de Waal Malefyt R, Coffman RL, O'Garra A. 2001. Interleukin-10 and the interleukin-10 receptor. *Annu. Rev. Immunol.* 19:683–765

Murdoch ME, Asuzu MC, Hagan M, Makunde WH, Ngoumou P, et al. 2002. Onchocerciasis: the clinical and epidemiological burden of skin disease in Africa. *Ann. Trop. Med. Parasitol.* 96:283–96

Nunn CL, Gittleman JL, Antonovics J. 2000. Promiscuity and the primate immune system. *Science* 290:1168–70

Olivares Fontt E, Beschin A, Van Dijck E, Vercruysse V, Bilej M, et al. 2002. *Trypanosoma cruzi* is lysed by coelomic cytolytic factor-1, an invertebrate analogue of tumor necrosis factor, and induces phenoloxidase activity in the coelomic fluid of *Eisenia foetida foetida*. *Dev. Comp. Immunol.* 26:27–34

Omer FM, de Souza JB, Riley EM. 2003. Differential induction of TGF-β regulates proinflammatory cytokine production and determines the outcome of lethal and nonlethal *Plasmodium yoelii* infections. *J. Immunol.* 171:5430–36

Pfeffer K. 2003. Biological functions of tumor necrosis factor cytokines and their receptors. *Cytokine Growth Factor Rev.* 14:185–91

Pfeffer K, Matsuyama T, Kundig TM, Wakeham A, Kishihara K, et al. 1993. Mice deficient for the 55 kd tumor necrosis factor receptor are resistant to endotoxic shock, yet succumb to *L. monocytogenes* infection. *Cell* 73:457–67

Råberg L, Grahn M, Hasselquist D, Svensson E. 1998. On the adaptive significance of stress-induced immunosuppression. *Proc. R. Soc. London Ser. B* 265:1637–41

Råberg L, Stjernman M. 2003. Natural selection on immune responsiveness in blue tits *Parus caeruleus*. *Evol. Int. J. Org. Evol.* 57:1670–78

Råberg L, Vestberg M, Hasselquist D, Holmdahl R, Svensson E, Nilsson JA. 2002. Basal metabolic rate and the evolution of the adaptive immune system. *Proc. R. Soc. London Ser. B* 269:817–21

Read AF, Allen JE. 2000. The economics of immunity. *Science* 290:1104–5

Restif O, Koella JC. 2003. Shared control of epidemiological traits in a coevolutionary model of host-parasite interactions. *Am. Nat.* 161:827–36

Rodriguez-Sosa M, Saavedra R, Tenorio EP, Rosas LE, Satoskar AR, Terrazas LI. 2004. A STAT4-dependent Th1 response is required for resistance to the helminth parasite *Taenia crassiceps*. *Infect. Immun.* 72:4552–60

Rolff J, Siva-Jothy MT. 2003. Invertebrate ecological immunology. *Science* 301:472–75

Sacks D, Anderson C. 2004. Re-examination of the immunosuppressive mechanisms mediating non-cure of *Leishmania* infection in mice. *Immunol. Rev.* 201:225–38

Sacks D, Sher A. 2002. Evasion of innate immunity by parasitic protozoa. *Nat. Immunol.* 3:1041–47

Saran R, Sharma MC, Sen AB. 1989. Quantitative grading of *Leishmania donovani* amastigotes related to age of kala-azar patients. *J. Commun. Dis.* 21:262–64

Sartono E, Kruize YC, Kurniawan A, Maizels RM, Yazdanbakhsh M. 1997. Depression of antigen-specific IL-5 and IFN-γ responses in human lymphatic filariasis as a function of clinical status and age. *J. Infect. Dis.* 175: 1276–80

Sasaki S, Takeshita F, Okuda K, Ishii N. 2001. *Mycobacterium leprae* and leprosy: a compendium. *Microbiol. Immunol.* 45:729–36

Satoguina J, Mempel M, Larbi J, Badusche M, Loliger C, et al. 2002. Antigen-specific T regulatory-1 cells are associated with immunosuppression in a chronic helminth infection (onchocerciasis). *Microbes Infect.* 4: 1291–300

Satoskar AR, Rodig S, Telford SR 3rd, Satoskar AA, Ghosh SK, et al. 2000. IL-12 gene-deficient C57BL/6 mice are susceptible to *Leishmania donovani* but have diminished hepatic immunopathology. *Eur. J. Immunol.* 30:834–39

Schmid-Hempel P. 2003. Variation in immune defence as a question of evolutionary ecology. *Proc. R. Soc. London Ser. B* 270:357–66

Scollard DM. 1993. Time and change: new dimensions in the immunopathologic spectrum of leprosy. *Ann. Soc. Belg. Med. Trop.* 73:5–11

Segel LA, Bar-Or RL. 1999. On the role of feedback in promoting conflicting goals of the adaptive immune system. *J. Immunol.* 163:1342–49

Sheldon BC, Kruuk LEB, Merila J. 2003. Natural selection and inheritance of breeding time and clutch size in the collared flycatcher. *Evol. Int. J. Org. Evol.* 57:406–20

Sheldon BC, Verhulst S. 1996. Ecological immunology: Costly parasite defences and trade-offs in evolutionary ecology. *Trends Ecol. Evol.* 11:317–21

Shudo E, Iwasa Y. 2001. Inducible defense against pathogens and parasites: optimal choice among multiple options. *J. Theor. Biol.* 209:233–47

Shudo E, Iwasa Y. 2004. Dynamic optimization of host defense, immune memory, and post-infection pathogen levels in mammals. *J. Theor. Biol.* 228:17–29

Snow RW, Craig M, Deichmann U, Marsh K. 1999. Estimating mortality, morbidity and disability due to malaria among Africa's nonpregnant population. *Bull. WHO* 77:624–40

Soares MBP, Silva-Mota KN, Lima RS, Bellintani MC, Pontes-de-Carvalho L, Ribeiro-dos-Santos R. 2001. Modulation of chagasic cardiomyopathy by interleukin-4: dissociation between inflammation and tissue parasitism. *Am. J. Pathol.* 159:703–9

Spierings E, de Boer T, Wieles B, Adams LB, Marani E, Ottenhoff TH. 2001. *Mycobacterium leprae*-specific, HLA class II-restricted killing of human Schwann cells by CD4+ Th1 cells: a novel immunopathogenic mechanism of nerve damage in leprosy. *J. Immunol.* 166:5883–88

Stewart GR, Boussinesq M, Coulson T, Elson L, Nutman TB, Bradley JE. 1999. Onchocerciasis modulates the immune response to mycobacterial antigens. *Clin. Exp. Immunol.* 117:517–23

Tarazona-Santos E, Tishkoff SA. 2005. Divergent patterns of linkage disequilibrium and haplotype structure across global populations at the interleukin-13 (IL13) locus. *Genes Immun.* 6:53–65

Taylor MD, Le Goff L, Harris A, Malone E, Allen JE, Maizels RM. 2005. Removal of regulatory T cell activity reverses hyporesponsiveness and leads to filarial parasite clearance *in vivo. J. Immunol.* 174:4924–33

Terry CF, Loukaci V, Green FR. 2000. Cooperative influence of genetic polymorphisms on interleukin 6 transcriptional regulation. *J. Biol. Chem.* 275:18138–44

Thomas GR, McCrossan M, Selkirk ME. 1997. Cytostatic and cytotoxic effects of activated macrophages and nitric oxide donors on *Brugia malayi. Infect. Immun.* 65:2732–39

Tortorella D, Gewurz BE, Furman MH, Schust DJ, Ploegh HL. 2000. Viral subversion of the immune system. *Annu. Rev. Immunol.* 18: 861–926

Tracey KJ, Cerami A. 1994. Tumor necrosis

factor: a pleiotropic cytokine and therapeutic target. *Annu. Rev. Med.* 45:491–503

Urban JF Jr, Maliszewski CR, Madden KB, Katona IM, Finkelman FD. 1995. IL-4 treatment can cure established gastrointestinal nematode infections in immunocompetent and immunodeficient mice. *J. Immunol.* 154:4675–84

van Baalen M. 1998. Coevolution of recovery ability and virulence. *Proc. R. Soc. London Ser. B* 265:317–25

van Boven M, Weissing FJ. 2004. The evolutionary economics of immunity. *Am. Nat.* 163:277–94

van der Werf MJ, de Vlas SJ, Brooker S, Looman CW, Nagelkerke NJ, et al. 2003. Quantification of clinical morbidity associated with schistosome infection in sub-Saharan Africa. *Acta Trop.* 86:125–39

van de Vosse E, Hoeve MA, Ottenhoff TH. 2004. Human genetics of intracellular infectious diseases: molecular and cellular immunity against mycobacteria and salmonellae. *Lancet Infect. Dis.* 4:739–49

Viglietta V, Baecher-Allan C, Weiner HL, Hafler DA. 2004. Loss of functional suppression by CD4$^+$ CD25$^+$ regulatory T cells in patients with multiple sclerosis. *J. Exp. Med.* 199:971–79

Wamachi AN, Mayadev JS, Mungai PL, Magak PL, Ouma JH, et al. 2004. Increased ratio of tumor necrosis factor-α to interleukin-10 production is associated with *Schistosoma haematobium*-induced urinary-tract morbidity. *J. Infect. Dis.* 190:2020–30

Wegner KM, Kalbe M, Kurtz J, Reusch TBH, Milinski M. 2003. Parasite selection for immunogenetic optimality. *Science* 301:1343

WHO. 1995. *Onchocerciasis and its Control: Report of a WHO Expert Committee on Onchocerciasis Control. Rep. 852.* Geneva: WHO

WHO. 2000. *Report on Global Surveillance of Epidemic-Prone Infectious Diseases,* Dep. Commun. Dis. Surveill. Response. Geneva: WHO

WHO. 2004. *World Health Report.* Geneva: WHO

Wills-Karp M. 2004. Interleukin-13 in asthma pathogenesis. *Immunol. Rev.* 202:175–90

Wu J, Longmate JA, Adamus G, Hargrave PA, Wakeland EK. 1996. Interval mapping of quantitative trait loci controlling humoral immunity to exogenous antigens: evidence that non-MHC immune response genes may also influence susceptibility to autoimmunity. *J. Immunol.* 157:2498–505

Wynn TA. 2004. Fibrotic disease and the T_H1/T_H2 paradigm. *Nat. Rev. Immunol.* 4:583–94

Xu L, Yoon H, Zhao MQ, Liu J, Ramana CV, Enelow RI. 2004. Cutting edge: pulmonary immunopathology mediated by antigen-specific expression of TNF-α by antiviral CD8$^+$ T cells. *J. Immunol.* 173:721–25

Yatsuda AP, De Vries E, Vieira Bressan MC, Eysker M. 2001. A *Cooperia punctata* gene family encoding 14 kDa excretory-secretory antigens conserved for trichostrongyloid nematodes. *Parasitology* 123:631–39

Zhang SM, Loker ES. 2004. Representation of an immune responsive gene family encoding fibrinogen-related proteins in the freshwater mollusc *Biomphalaria glabrata*, an intermediate host for *Schistosoma mansoni. Genetics* 341:255–66

Annu. Rev. Ecol. Evol. Syst. 2005. 36:399–417
doi: 10.1146/annurev.ecolsys.36.102003.152629
Copyright © 2005 by Annual Reviews. All rights reserved
First published online as a Review in Advance on August 17, 2005

THE EVOLUTIONARY ECOLOGY OF GYNOGENESIS

Ingo Schlupp

*Department of Zoology, University of Oklahoma, Norman,
Oklahoma 73019; email: schlupp@ou.edu*

Key Words   maintenance of sex, *Poecilia formosa*, pseudogamy

■ **Abstract**   Most metazoans engage in recombination every generation. In theory this is associated with considerable cost, such as the production of males, so that asexual organisms, which do not pay this cost, should be able to invade populations of sexuals. Some asexuals depend on sperm of sexual males to trigger embryogenesis, a reproductive mode called gynogenesis. The genetic information of males is typically not used. Theory predicts that such mating complexes are short-lived and highly unstable. Sperm dependency is not only the defining feature of the biology of gynogenetic metazoans, it is also a major puzzle in evolutionary biology. Organisms that apparently combine disadvantages of both sexuality and asexuality are a serious challenge to theory. A number of questions about these systems are still unresolved.

CONCEPTUAL BACKGROUND: WHY SEX?

The evolution and maintenance of sexual reproduction is still one of the major unresolved problems in evolutionary biology (Redfield 1994, West et al. 1999). Once evolved, recombination must persist against invading asexuals (Doncaster et al. 2000), which do not show meiosis and recombination. During sexual reproduction any given individual pays a cost: It reduces the number of genes passed on to the next generation by half [cost of meiosis (Williams 1975)]. Additionally, for sexual reproduction two sexes are needed to restore the parental state, mostly diploidy. Hence, females have to produce males [cost of males (Maynard Smith 1978)] and mate with them. The costs also include finding a mate and risks associated with mating, like sexually transmitted diseases or increased predation. This contrasts with clonal reproduction in which the whole genome is transmitted unchanged and only females are produced. Parthenogenesis should be the more successful strategy—at least in the short term. In the long run, clonal organisms also appear to pay a cost: They accumulate deleterious mutations that cannot be purged without recombination and rare beneficial mutations cannot be combined in the same individual (Muller 1964, Williams 1975). Several theories have been proposed to explain the persistence of sexual reproduction in spite of the costs (Kondrashov 1993, West et al. 1999), but no single theory has unanimous support. Currently, one of the most widely accepted, and arguably best supported, ideas to explain the maintenance of recombination is the Red Queen hypothesis (Hamilton 1980, Ladle 1992, Van Valen 1973). It proposes that recombination results in genetically diverse offspring which, contrary to the uniform offspring of asexuals, are difficult targets for parasites and diseases (Bell 1982).

Essentially, we still lack an understanding of how sexual organisms can keep asexuals at bay given the hypothesized short-term advantages of asexuality (Agrawal 2001, West et al. 1999). One approach to this problem is to directly compare closely related sexual and asexual species. Such animals are unsuitable to study the origin of recombination, but questions concerning the maintenance of sex can be addressed.

WHAT IS GYNOGENESIS?

Parthenogenesis in metazoans is the development of an adult organism using only maternal genetic information. Parthenogenesis is not uncommon in metazoans and, in many species, parthenogenesis and sexual reproduction coexist. They may alternate throughout the life cycle as in waterfleas (*Daphnia*) (Suomalainen et al. 1987), or exist simultaneously as in hymenopterans, which produce males from unfertilized eggs and females from fertilized eggs (Suomalainen et al. 1987). In the related phenomenon of androgenesis, eggs develop after the disintegration of the maternal chromosomes, using only the remaining paternal genes. Natural androgenesis is extremely rare [for example, in *Bacillus rossius*, a stick insect (Mantovani & Scali 1992)]. Natural parthenogenetic development can be spontaneous or it may need to be induced by sperm. This is called pseudogamy, or gynogenesis (Figure 1). The "correct" ploidy, often diploidy, of the organism is achieved either by absence of meiosis (apomixis) or by one of several mechanisms that restore ploidy after meiosis. A common mechanism is fusion of the egg cell with one of the polar bodies (Suomalainen et al. 1987). In gynogenesis, the sperm serves

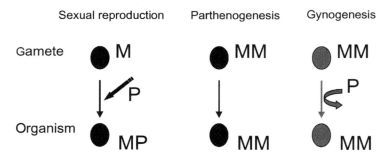

Figure 1   Schematic representation of sexual reproduction, parthenogenesis, and gynogenesis. M refers to maternal contributions; P to paternal contributions.

solely to trigger embryogenesis, but further contribution of the sperm is usually absent (for exceptions, see below).

In hybridogenesis, another form of sperm-dependent clonal reproduction, haploid egg cells are formed, which fuse with the sperm nucleus. During gametogenesis, the male chromosomes are removed from the egg and only the clonal part of the genome is passed on. This is called hemiclonal because only the maternal half of the genome is clonal. Hybridogenesis is found in several vertebrates and invertebrates (Beukeboom & Vrijenhoek 1998). Examples from vertebrates include the waterfrog complex (*Rana esculenta*) (Hellriegel & Reyer 2000, Reyer et al. 1999, Tunner 1979), and a group of fish (*Poeciliopsis* complex) (Vrijenhoek 1994, 1998).

Gynogenesis differs from sexual reproduction because the female genome is passed on unchanged. It is similar to sexual reproduction because it also requires males. For sperm-dependent unisexuals this means they need to use sperm from a heterospecific male. Due to this physiological dependency, gynogenesis has been called "sperm parasitism" (Hubbs 1964). Gynogens are interesting models because they combine disadvantageous traits from both sexuality and asexuality. Understanding how they evolve is especially revealing. Gynogenesis is being investigated at three levels: (*a*) the level of genetics because organisms without recombination have to deal with genetic decay, (*b*) the level of ecology to understand how stability in associations of sexuals and gynogens evolves and is maintained, and (*c*) the level of behavior because mating decisions of the associated males may be crucial to the stability of such mating systems.

EVOLUTIONARY ORIGINS OF GYNOGENESIS

Gynogenesis occurs in a number of taxa. Gynogens are not species in the sense of the biospecies concept (Mayr 1967), leading to difficulties in taxonomy (Cole 1985). On the one hand, they do not reproduce sexually in the sense of the definition, but on the other hand, they are usually morphologically, ecologically, and genetically distinct. Nonetheless, knowing the number of gynogentic (or parthenogenetic) species is difficult as the numbers of "forms" counted vary. As most gynogens are thought to have had multiple origins and coexist in multiple, genetically identifiable lineages, it is very difficult to decide which should be counted as a distinct entity. Excellent lists of gynogens and other parthenogenetic metazoans are provided in Beukeboom & Vrijenhoek 1998, Suomalainen et al. 1987, and Vrijenhoek et al. 1989. It appears that gynogenesis has evolved multiple times independently in several taxonomic groups. In one genus, *Poeciliopsis*, both gynogenesis and hybridogenesis have been described (Schultz 1967). Gynogens and parthenogens are likely to be overlooked by classical methods used to catalogue biodiversity. This is highlighted by the recent description of a cryptic clonal lineage of loach, *Misgurnus anguillicaudatus* (Morishima et al. 2002), which was detected using flow cytometry. Nonetheless, it is curious that so few gynogenetic taxa have been described.

Gynogenesis is absent in some major groups of vertebrates, like mammals. It is thought that genomic imprinting, an epigenetic process responsible for differences between homologous chromosomes depending on parental origin (Georgiades et al. 2001), plays a role in the absence of gynogenesis in mammals.

Hybrid Origin

Most gynogenetic vertebrates are of hybrid origin (Beukeboom & Vrijenhoek 1998, Vrijenhoek et al. 1989). One important consequence of a hybrid origin is that such species start out with extremely high heterozygosity and are therefore often called "frozen F_1." Molecular tools offer a powerful approach to identify hybrid origins (Alves et al. 1997, Dawley et al. 1987, Schartl et al. 1995b, Turner et al. 1990).

The Amazon molly (*Poecilia formosa*), a livebearing fish, seems to represent a case of a single hybrid origin (D. Möller, J. Parzefall & I. Schlupp, in preparation). Hubbs & Hubbs (1932) suggested that Amazon mollies have a hybrid origin. They proposed *P. sphenops* and *P. latipinna* to be the parental species. Later it became clear that *P. mexicana* was the maternal ancestor, but at the time, what is now known as *P. mexicana* was not recognized as a separate species. The Amazon molly was originally described by the French ichthyologist Girard in 1859 (Girard 1859), but it was not until 1932 that Laura and Carl Hubbs recognized that Amazons are unisexual (Hubbs & Hubbs 1932). Supporting molecular evidence for the hybrid origin was provided using genetic methods (Avise et al. 1991, Schartl et al. 1995b, Turner 1982). Multiple origins are known, e.g., for *Ambystoma* salamanders (Bogart 1989, Hedges et al. 1992), and triploid *Poeciliopsis* (Quattro et al. 1992, Vrijenhoek 1993).

For gynogens of hybrid origin it is usually assumed that they acquire all characteristic features of gynogenesis at once. Two scenarios for the evolution of gynogenesis with hybrid origin seem possible:

1) *Single step.* Here multiple evolutionary changes occur simultaneously. This includes disruption of meiosis, discarding the second sex, and several other features. Although this seems to be the most widely accepted scenario, it is not clear how exactly this complex change happens. However, if this hypothesis is correct, it may explain why the *de novo* formation of gynogens seems so rare: If only one of the multiple changes fails to function, the successful formation of a unisexual fails. Apparently the conditions under which the Amazon molly arose are not repeatable. Numerous attempts to create unisexual F_1's using various populations of the putative parental species invariably led to sexually reproducing F_1 (Dries 2003, Schartl et al. 1991, Schlupp et al. 1992). Such F_1 hybrids have been repeatedly used in research (Dries 2003, Schartl et al. 1991, Schlupp et al. 1992) but comparisons with true Amazon mollies should be made with caution, because the sexual lineages used to produce the F_1 have had an independent evolution over many generations. The repeated successful formation of hybridogenetic fishes (*Poeciliopsis*;

Schultz 1973; Vrijenhoek 1993, 1994) and triploid planthoppers of the genus *Muelleriana* (Drosopoulos 1978) are in agreement with this hypothesis, as no intermediate stages were reported.

2) *Multiple steps.* A sexual F_1 might be formed first, potentially with a strongly biased sex ratio according to Haldane's rule. Subsequently an additional evolutionary step may lead to loss of sexuality and selection might then favor the gynogen.

Nonhybrid Origin

Several invertebrate gynogens are probably not of hybrid origin. In these cases, an elevation of ploidy, through autopolyploidization, apparently led to mostly triploid forms. The only vertebrate that seems to be of nonhybrid origin is *Carassius auratus gibelio*. A recent report noted sexual reproduction within the gynogenetic species (Zhou et al. 2000), and it is presently difficult to estimate whether this is a recent evolutionary event or an overlooked old phenomenon.

Age of Gynogens

Gynogens are conventionally seen as evolutionarily short lived. Age estimates of up to 100,000 years have been published for some of the gynogenetic vertebrates (Avise et al. 1991, Hedges et al. 1992, Schartl et al. 1995b, Spolsky et al. 1992). These estimates are often based on the molecular clock and have large confidence intervals. This was referred to by Dries (2003), who argued that Amazon mollies, for which several age estimates ranging between 10,000 and 100,000 years have been published (Avise et al. 1991, Schartl et al. 1995b), might be as young as the lower confidence limit (8900 years). Dries (2003) correctly pointed out that in this case not enough time may have elapsed for evolutionary changes. One might also argue, however, that Amazon mollies are as old as the upper confidence limit (600,000 years), which should have been sufficient time for evolution. Nonetheless, the roughly 100,000 years that have been mentioned for individual lineages of *Ambystoma* (Spolsky et al. 1992) and *P. formosa* (Schartl et al. 1995b) are old under models of mutational meltdown based on the accumulation of deleterious mutations (Gabriel & Bürger 2000, Gabriel et al. 1993). Their ages pale, however, compared to those of the so-called "asexual scandals" (Judson & Normark 1996), especially bdelloid rotifers (Poinar & Ricci 1992) and ostracods (Judson & Normark 1996; Maynard Smith 1978, 1992), which are several millions of years old.

CONSEQUENCES OF SPERM DEPENDENCY

Sperm dependency has two main consequences: (*a*) it provides a mechanism for the introgression of paternal DNA into the clonal genome of gynogens, and (*b*) the gynogens cannot ecologically out-compete their hosts. Their physiological sperm dependency leads to behavioral and ecological interactions with males and females of a host species.

Genetic Consequences of Sperm Dependency

The need for sperm to trigger embryogenesis results in an intimate association of egg and sperm. One result of this is the occasional introgression of paternal DNA into the clonal genome of gynogens. Paternal DNA may be incorporated as a heritable component into the genome. This has been documented in salamanders (*Ambystoma laterale* complex; Bogart 1989, Bogart & Licht 1986, Bogart et al. 1989). In Amazon mollies, two genetic interactions between asexuals and sexuals have been described, formation of triploids and introgression of subgenomic amounts of DNA.

TRIPLOIDY   The inclusion of the complete sperm genome leads to triploid lineages (Schultz & Kallman 1968; Lamatsch et al. 2000a,b; Nanda et al. 1995). Fertile triploids have been reported from natural populations (Lamatsch et al. 2000a,b, but laboratory-produced triploids were sterile (Nanda et al. 1995, Schultz & Kallman 1968). Ploidy levels higher than 3N have been described for *Ambystoma,* where it can reach 5N. In *Ambystoma*, there is an inverse relationship between ploidy and fitness (Lowcock et al. 1991). In Amazon mollies, triploids are thought to be restricted to the Rio Purification drainage in Mexico (Balsano et al. 1989), and it is unclear why triploids do not have a wider geographical distribution. A recent study suggested that triploidy arose only once within the Amazon mollies (Lampert et al. 2005), whereas it arose multiple times independently in the *Poeciliopsis* complex (Mateos & Vrijenhoek 2005). The overall morphology, including size, does not allow the identification of triploid Amazons in the field, suggesting they might be more widespread than currently appreciated.

In many complexes, a present sperm donor species has been involved in the hybridization that lead to the gynogenetic hybrid. Consequently, the additional genome is provided by a species that has already contributed to the hybrid. In triploid Amazon mollies, the parental species have been *P. latipinna* and *P. mexicana*, and in the triploids typically a *P. mexicana* genome has been added (Balsano et al. 1989). Triploids with an added *P. latipinna* genome are extremely rare (Balsano et al. 1989). This allows study not only of the effects of elevated ploidy, but also of the interactions of an ancient genome with a recently recombined one.

In other species, triploids have an added genome from a species not involved in the hybridization. Especially in *Ambystoma*, the situation is very complex and paternal chromosome sets have been added from several species (Hedges et al. 1992). Behavior of triploid Amazon mollies, fitness consequences, and the coexistence remain to be studied, but good data on distribution and ecology are available from intensive field studies (Balsano et al. 1989). In triploid Amazon mollies, laboratory-born males have been reported to produce sperm, although spermatogenesis was abnormal. Breeding these males produces no offspring with sexual females, but males were capable of triggering embryogenesis in Amazon mollies (Lamatsch et al. 2000a,b).

MICROCHROMOSOMES   A second mechanism for adding paternal genetic material is the inclusion of only a small part of the sperm genome. Microchromosomes are very small, centromere bearing parts of a paternal chromosome. Their addition to a clonal genome leads to aneuploid lineages. In Amazon mollies, stable, microchromosome-bearing lineages (Schartl et al. 1995a, 1997) have been described both from the laboratory and the field. Triploid lineages can also bear microchromosomes. Given the size of the microchromosome, only limited amounts of new genetic material will be added this way. Whether this has important evolutionary consequences is still unclear, but clearly possible: In a laboratory-produced lineage of Amazon mollies bearing microchromosomes originating from "Black Mollies" (an all-black strain, which is a hybrid of several species of *Poecilia*), some individuals develop melanoma, which is unknown in Amazon mollies otherwise, because they lack the cell type (macromelanophores) for this cancer (Schartl et al. 1997). The gene coding for the tumor is located on the microchromosome.

Both mechanisms of stable introgression give us insights into the mechanism of egg activation. It appears that the sperm nucleus enters the egg cell and that the paternal chromosomes are being destroyed. If this process is halted before the entire chromosome is digested, it results in a microchromosome.

Stability of Mating Complexes

Asexual females have an intrinsic advantage over sexual females: They do not have to produce male offspring. Assuming equal reproductive output (which is true for *P. mexicana* and *P. formosa*: I. Schlupp & A. Taebel-Hellwig, in preparation), theoretically the number of gynogens in a population should grow much faster as compared to the sexuals. At the same time, the sex ratio of all females (combining gynogens and sexuals) to males becomes increasingly female biased. If no mechanisms counter this, eventually the sexuals in the population go extinct. As a direct consequence, the gynogens will also go extinct. Two mechanisms have been suggested to explain stability in complexes of gynogens and sexuals: behavioral regulation and ecological differences.

ECOLOGICAL DIFFERENCES   Stable coexistence can be expected if gynogens and sexuals use different ecological niches. Gynogens of hybrid origin differ genetically from both parental species, which predicts differences in niche usage (Vrijenhoek 1994). However, the physiological dependence on sperm forces the gynogens into close ecological proximity of the sexuals. The conditions for stable coexistence of sexuals and gynogens have been reviewed by Vrijenhoek (1994). His *Frozen Niche Variation* model (Vrijenhoek 1979) is very well supported. It argues that clones of multiple origin may use resources underutilized by the sexual parental species. Case & Taper (1986) argued that the broader niche width of the sexuals is important for stable coexistence. They also argued that the intrinsic advantage of the asexuals is not always twofold. A related argument was presented by Doncaster et al. (2000). In a model studying the hybridogenetic frog (*R. esculenta*) (Hellriegel & Reyer 2000), a combination of factors was found to allow for stable coexistence: female mating preferences, female fecundity, larval performance, and dispersal were important. Hakoyama & Iwasa (2004) argued that parasites might also significantly contribute to stability.

On a small scale, instability and extinction might go undetected, but a recent study (Heubel 2004) suggested that this dynamic might actually be present in nature. Heubel (2004) found that frequencies of gynogens in six mixed populations of *P. formosa* and *P. latipinna* increased over the reproductive season, apparently according to different population growth rates of the two types of females. However, environmental instability caused by the winter, floods, draughts, or other environmental perturbations prevent the system to reach climax (*mosaic-cycle hypothesis*) (Hutchinson 1961), which in this particular system would be the extinction of the sexuals. It would also lead to the extinction of the gynogens. Other mechanisms for stability are also likely. There could be a significant reduction of the gynogen's fitness at any stage of their life history, but to date, few studies are available. In a field study of the population biology of Amazon and Sailfin mollies (Heubel 2004), equal reproductive output (I. Schlupp & A. Taebel-Hellwig, in preparation), absence of differences in parasite loads (Tobler & Schlupp 2005), and no difference in the degree of filial cannibalism (C. Hubbs & I. Schlupp, in preparation) indicate a remarkable ecological similarity of the females.

Earlier models argued that niche separation is not required to explain stable coexistence (Moore 1975, Moore & McKay 1971, Stenseth et al. 1985) and highlighted the importance of behavioral decisions, especially male choice and male sperm allocation.

BEHAVIORAL REGULATION   Behavioral mechanisms may be critical in regulating the coexistence of gynogens and their sexual hosts. Gynogenetic females have to obtain access to sperm, but males do not gain from matings with gynogens. Consequently, good candidates for stabilizing mechanisms include male behavior, especially male choice. Theoretically it would be easiest for gynogens to obtain sperm from males that release sperm for external fertilization. However, several gynogenetic species have internal fertilization. Females require intimate contact with males to obtain sperm. The Amazon molly uses the two species that were involved in the hybridization, *P. latipinna* and *P. mexicana* (Schlupp et al. 2002). A third host species, *P. latipunctata*, has recently been described (Niemeitz et al. 2002). In this case, the third host may have made secondary contact with the Amazon molly only recently.

Internal fertilization is a severe constraint for the females—something other unisexuals with external fertilization do not face (like, e.g., the Crucian carp, *Carassius*) (Hakoyama & Iguchi 2002). From the male's perspective, these matings are wasted because they have no fitness-relevant offspring with gynogens.

This should promote strong discrimination resulting in the extinction of the gyno-gen. Obviously males do mate with gynogens, raising the question, How can this be explained? A connected question is whether gynogens show adaptations countering male discrimination (Dries 2003, Schlupp et al. 1991). An example for such adaptations and the underlying evolutionary "arms-race" might be sexual mimicry. This was described in the hybridogenetic fish *Poeciliopsis* (Lima et al. 1996), where asexual females mimic the color pattern around the genital opening found in their sexual competitors.

So far three hypotheses have been offered to explain heterospecific matings.

1) Males may make mistakes and/or use a "best of a bad job" strategy (Kawecki 1988). They might be unable to distinguish between the two types of females or have only weak preferences. However, numerous studies have reported male discrimination in the sperm donors of the Amazon mollies (Dries 2003, Schlupp et al. 1998), as well as in other species like bark beetles (*Ips*) (Loyning & Kirkendall 1996). In addition, one study showed reproductive character displacement in male *P. latipinna* (Gabor & Ryan 2001). Males from sympatric populations were better able to distinguish between gynogenetic and sexual females.

   Interestingly, a small number of studies were unable to find male preferences, either in general (Balsano et al. 1981, Balsano et al. 1985, Heubel 2004, Woodhead 1985) or under certain circumstances (Schlupp et al. 1991). The latter study provided an explanation for the lack of preference: Females that were receptive were more attractive to males, independent of species. Male preferences may be influenced by the females' sexual cycle (Parzefall 1973): Females are receptive only for a few days during each monthly cycle. Therefore, only a small proportion of the females of a population are fertile at any given time. Males are probably under selection not to forsake any potential matings (Schlupp & Ryan 1997). Alternatively, mate choice under certain environmental conditions might become so costly—especially time-consuming—that males will do better by indiscriminately mating with every female. Heubel & Schlupp (2005) identified natural turbidity as potential "visual noise" limiting male mating behavior.

   The methods used to study female preferences and male mate choice (Jennions & Petrie 1997) differ, and it is unclear whether this interacts with the results reported. In several studies, full interaction of all participating individuals was allowed (e.g., Gabor & Ryan 2001, Ryan et al. 1996, Schlupp et al. 1991). This approach provides a relatively high degree of biological realism, as it mimics the natural interactions leading to actual insemination. In these studies, male choice is usually scored using male copulation attempts as proxy for actual sperm transfer. The disadvantage is that the outcome of such interactions only partly reflects male mating preferences, partly female choice, and (depending on the design) partly female-female interactions. A recent study indicated that not all copulation attempts do result in sperm transfer (Schlupp & Plath 2005).

Another approach to measuring male preferences is to separate the choosing male from the stimulus females. Using such a setup will limit the amount of information available to the choosing male. The female stimuli may be presented live (Niemeitz et al. 2002) or as video-playback (Körner et al. 1999, Landmann et al. 1999), as well as simultaneously or sequentially (K.U. Heubel & I. Schlupp, manuscript submitted). A study conducted by K.U. Heubel & I. Schlupp (manuscript submitted) raised the possibility that seasonality might be an important component in male choice: Males were choosier during the peak of the reproductive season.

Nonetheless, in the majority of studies, males were found to prefer conspecific females, although not under all circumstances. So far, male mating behavior was mainly studied at the behavioral level, without examining the possibility of differential sperm transfer. A recent study (Aspbury & Gabor 2004) showed that males of one of the host species (*P. latipinna*) of the Amazon molly prime more sperm after being visually exposed to a conspecific female as compared to a *P. formosa* female. This demonstrated that males can adjust their investment in sperm, but did not determine if males actually transfer less or no sperm in matings with the Amazon molly. This was addressed in a recent study (Schlupp & Plath 2005), which found that behavioral preferences match the sperm transferred to females. Amazon mollies received less sexual behavior and also less sperm.

2) Male choice may be frequency- or density-dependent: The basic idea is that males should decline matings with the most common female phenotype, thereby selectively favoring the rare females. Males, however, should not selectively mate with gynogens when frequencies of sexuals are high. Several studies addressed this problem (Moore & McKay 1971, Stenseth et al. 1985).

3) Males may gain from mating with the gynogens: This hypothesis has been proposed by Schlupp et al. (1994). In several species, females do not always exercise independent mate choice, but may be influenced by observing the mating decisions of other females (reviewed by Brooks 1998, Westneat et al. 2000). Schlupp et al. (1994) showed that females of *P. latipinna* (Sailfin mollies) copied the mating preference of Amazon mollies, thereby providing an indirect benefit for males. Subsequently, K.U. Heubel & I. Schlupp (manuscript in preparation) showed that female *P. mexicana* (Atlantic mollies) also copy the mate choice of Amazon mollies and that Amazon mollies copy each other. Essentially, all females from sympatry copy each other, including the sexual females (K.U. Heubel & I. Schlupp, manuscript in preparation; Witte & Ryan 1998, 2002). It is noteworthy that Witte & Ryan (2002) demonstrated that mate copying also takes place in the field. For mate copying to work, males must visibly interact with Amazons, but it does not require actual sperm transfer. It can also be

argued that male mating behavior is a signal of male presence. It is not clear why sexual females imitate the gynogens. This is in agreement, however, with the overall ecological similarity of the two types of females.

FEMALE–FEMALE INTERACTIONS   A more indirect consequence of sperm dependency is that gynogens also have to interact with sexual females. This goes beyond typical intraspecific interactions. Amazon mollies, e.g., and sexual *P. latipinna* can distinguish each other (Schlupp & Ryan 1996). They prefer to associate with conspecifics, but trade off this preference for a larger group. This preference for larger groups facilitates the formation of mixed shoals (Schlupp & Ryan 1996), which set the stage for all interspecific interactions. Females may also compete over resources. In this context, aggression between sexuals and gynogens has been described (Foran & Ryan 1994, Schlupp et al. 1991), with the gynogens being more aggressive than the sexuals.

EVOLUTIONARY POTENTIAL OF GYNOGENS

Microevolution

So far, studies comparing the behavior of gynogens and their sexual hosts have found very few differences. The female preference for large males, e.g., in Amazon mollies (Marler & Ryan 1997), did not differ from the same preference exhibited by the sexual females. The host males and the gynogens may be locked into a coevolutionary cycle with adaptations and counter-adaptations that are difficult to detect. Interesting exceptions are two studies on visual preferences in Amazon mollies. When given a choice between a host male (*P. latipinna*) with or without a novel ornament, Amazon mollies and Sailfin mollies differed. Amazon mollies showed no preference for the novel trait, but the sexual females did. In a second case, sexual *P. mexicana* preferred a novel trait, but the Amazon mollies did not (Schlupp et al. 1999). In another study, also using the Amazon molly, the gynogens rejected images of males with signs of a parasitic infection, but sexual females (*P. latipinna* and *P. mexicana*) did not show any preference (Tobler & Schlupp 2005). It is not certain, however, that differences between gynogens and sexuals indicate evolution of the Amazons, because alternately the sexual species might have evolved. Equally, differences between gynogens and F_1 hybrids cannot establish evolution of gynogens (Dries 2003, Schlupp et al. 1992).

Macroevolution

Macroevolutionary events, like speciation or adaptive radiation, are not known from gynogens. In the *Poeciliopsis* system, the formation of a sexual species from a hemiclonal lineage has been reported (Vrijenhoek 1993).

One problem associated with assessing the evolutionary potential of gynogens is that many events are very likely to go undetected: suppose a triploid gynogen, with two genomes of parental species A, A^+ and one of parental species B, randomly looses one genome. If the lost genome is the rare one (B), the resulting genotype (AA^+) is probably not distinguishable from the true sexual and furthermore probably less fit owing to the accumulated deleterious mutations.

Cryptic Sex?

Mutations may be the main source of genetic diversity in gynogens, but they are clearly not the only source. Mutations alone would allow for a considerable evolutionary potential if mutation rates were high enough and population sizes were large. It is unclear how frequent the *de novo* addition of genetic material is and what the evolutionary consequences are. In Amazon mollies, triploids found in the Rio Purification seem to have a single origin. It has been proposed that adding genetic material to a clonal genome can influence the rate of genetic decay and prolong the gynogen's evolutionary lifespan (Schartl et al. 1995a). Although there is not enough evidence on this subject yet, it is interesting to speculate if this qualifies as a form of rare, cryptic sex and whether less sex is as good as recombination of the whole genome every generation (Green & Noakes 1995). Of course, gynogens would still be dependent on a sexual species.

Red Queen

The Red Queen Hypothesis states that an immediate advantage of sexual reproduction lies in the production of genetically diverse offspring that provide a "moving target" for parasites. Common genotypes will quickly be targeted by parasites and experience reduced fitness. This predicts coevolutionary cycles of hosts and parasites in sexual species. It also predicts higher rates of infection of asexuals, when asexuals and sexuals have comparable exposure to parasites (Vrijenhoek 1994). This has been tested using Crucian carp (*Carassius auratus*): Here, diploid sexuals had fewer parasites than triploid gynogens (Hakoyama et al. 2001), which may play an important role in the stability of this complex (Hakoyama & Iwasa 2004). In Amazon mollies, comparing several parasites, Tobler & Schlupp (2005) did not find consistent differences between the gynogens and the syntopic sexual species, *P. latipinna*.

OPEN QUESTIONS

Sperm dependency is the defining feature of gynogens and a major puzzle in evolutionary biology. Organisms that apparently combine disadvantages of both sexuality and asexuality are a challenge to theory. A number of open questions are still associated with this.

1) Two major patterns of the taxonomical distribution of asexuals lack a convincing explanation. Gynogenesis and sperm dependency occur in some

fishes and amphibians, whereas all parthenogenetic reptiles do not require sperm. Within taxa that occasionally give rise to asexuals, like teleosts, large groups show no signs of asexuals, although hybrids are common (Dowling & Secor 1997). In East African cichlids, hybridization is hypothesized to form new species (Seehausen 2004), but no unisexuals have been detected. Relevant questions about the conditions for the *de novo* generation of asexual hybrids can be addressed experimentally. Is it a certain genetic distance that allows for the formation of asexuals? It is likely that many asexuals go undetected, especially sperm-dependant species, because they are observed and collected in the field together with males. Only careful behavioral and genetic studies can identify them as gynogens.

A related set of questions has to do with the first generations after the *de novo* formation. What selection pressures shape the new species? In the *Poeciliopsis* group, gynogenesis and hybridogenesis coexist, but it is unclear how this situation has evolved.

2) A better understanding of the consequences of sperm dependency is needed: It is still not clear what the costs and benefits associated with sperm dependency are. In many species, males provide direct benefits to females (Andersson 1994). In gynogens indirect benefits cannot play a role, because the males are in a different gene pool, but direct benefits might in theory ameliorate the cost of mating for asexual females (Neiman 2004).

Gynogenesis must be maintained against both the already present sexuals and against a potentially arising sperm-independent mutant. In other words, why does the occasional truly parthenogenetic mutant not spread in the population? If the cost of gynogenesis relative to parthenogenesis is low, the advantage of a newly arisen parthenogenetic mutant may be insignificant. It is important to note that gynogens and other parthenogens have already committed a major investment in female traits like morphology, so that the added cost of engaging in matings may be relatively small. Interestingly, in some parthenogenetic species, sexual behavior is still present. The best documented cases are female–female pseudocopulations in parthenogenetic *Cnemidophorus* lizards that enhance reproductive output in the asexuals (Crews & Fitzgerald 1980, Crews et al. 1986).

Many of these questions can be resolved by long-term field studies of ecology and life-history, such as studies published by Balsano et al. (1989), Heubel (2004), and Hubbs (1964), combined with mathematical models of population dynamics.

3) Behavior is context dependent. There are at least two categories of environmental variables that interact with behavior: (*a*) social context, i.e., other individuals nearby, and (*b*) abiotic conditions. Recent studies have established the idea of communication networks (McGregor & Peake 2000). This conceptual framework considers individual behavior in the presence of audiences. In many animals, signaling and sexual behavior takes place

in close proximity of other individuals of the same species. These individuals may use information conveyed in interactions to their own advantage. This can be illustrated with mate copying (see above). In a mixed group of Amazon mollies and Sailfin mollies, the two types of females can be either the audience or the actor, leading to a large number of possible interactions. All of these can occur in rapid sequence. In nature these interactions may additionally be influenced by other behaviors, such as predator avoidance, male–male interactions (Schlupp & Ryan 1997), and female–female interactions (Foran & Ryan 1994), leading to very complex social interaction networks (Matos & Schlupp 2005). An example for an environmental variable influencing behavior is turbidity: Male choice is hampered under turbid conditions (K.U. Heubel & I. Schlupp, 2005). Especially male choice, which can be critical for the stability of gynogenetic mating systems, should be studied using a communication network approach.

Once a gynogenetic lineage has evolved it may remain gynogenetic or evolve a different mode of reproduction. The lineage may evolve into sperm-independent parthenogens (not observed), or hybridogens (not observed, although the transition from hybridogenesis to gynogenesis, or vice versa, has been reported in the hybridogenetic *Poeciliopsis*) (Vrijenhoek 1994), or finally into a sexual species. Evolution of an asexual into a sexual species is likely to go undetected, but has been reported for *Poeciliopsis* (Vrijenhoek 1989, 1993). Given the relatively young phylogenetic age of most gynogens, the genes coding for male functions are likely to be intact. In the Amazon molly, it was possible to induce male phenotypes by administering androgens (Schartl et al. 1991, Schlupp et al. 1992, Turner & Steeves 1989). Furthermore, spontaneous masculinization does occur. Lamatsch et al. (2000a,b) reported that males spontaneously occurred in a triploid lineage. These males had functional sperm and sired offspring with Amazon molly females, but not with sexual females.

4) Our knowledge of the exact mechanisms of fertilization and interactions of sperm and egg is still limited (Jun & Yigui 1991). In most species conclusive experiments showing that sperm is actually the trigger are lacking. As in parthenogenetic lizards (*Cnemidophorus*), copulatory stimuli might play an important role. Another problem in this context has to do with the acquisition of host species. Are all host species equally efficient? In most cases of gynogens of hybrid origin one of the parental species is a sperm donor. Additional hosts have been acquired in Amazon mollies (Niemeitz et al. 2002) and *Ambystoma* (Bogart 1989). It is unknown how these relationships evolve, especially in gynogens of nonhybrid origin.

ACKNOWLEDGMENTS

I am very grateful to the students and colleagues from the groups in Austin, Würzburg, Zürich, and Hamburg who accompanied my research. K. Heubel,

M. Plath, H.-U. Reyer, M. Schartl, and M. Tobler kindly provided comments on this manuscript. I wish to thank DAAD, DFG, and NATO for funding my research.

The *Annual Review of Ecology, Evolution, and Systematics* is online at
http://ecolsys.annualreviews.org

LITERATURE CITED

Agrawal AF. 2001. Sexual selection and the maintenance of sexual reproduction. *Nature* 411:692–95

Alves MJ, Coelho MM, Collares-Pereira MJ, Dowling TE. 1997. Maternal ancestry of the *Rutilus alburnoides* complex (Teleostei, Cyprinidae) as determined by analysis of cytochrome b sequences. *Evolution* 51:1584–92

Andersson M. 1994. *Sexual Selection*. Princeton: Princeton Univ. Press

Aspbury A, Gabor CR. 2004. Discriminating males alter sperm production between species. *Proc. Natl. Acad. Sci. USA* 101:15970–73

Avise JC, Trexler J, Travis J, Nelson WS. 1991. *Poecilia mexicana* is the recent female parent of the unisexual fish *Poecilia formosa. Evolution* 45(6):1530–33

Balsano JS, Kucharski K, Randle EJ, Rasch EM, Monaco PJ. 1981. Reduction of competition between bisexual and unisexual females of *Poecilia* in northeastern Mexico. *Environ. Biol. Fishes* 6:39–48

Balsano JS, Randle EJ, Rasch EM, Monaco PJ. 1985. Reproductive behavior and the maintenance of all-female *Poecilia. Environ. Biol. Fishes* 12:251–63

Balsano JS, Rasch EM, Monaco PJ. 1989. The evolutionary ecology of *Poecilia formosa* and its triploid associate. In *Ecology and Evolution of Livebearing Fishes (Poeciliidae)*, ed. GK Meffe, FF Snelson, pp. 277–98. Englewood Cliffs, NJ: Prentice Hall

Bell G. 1982. *The Masterpiece of Nature*. London: Croom Helm

Beukeboom LW, Vrijenhoek RC. 1998. Evolutionary genetics and ecology of sperm-dependent parthenogenesis. *J. Evol. Biol.* 11:755–82

Bogart JP. 1989. A mechanism for interspecific gene exchange via all—female salamander hybrids. See Dawley & Bogart 1989, pp. 170–79

Bogart JP, Elinson RP, Licht LE. 1989. Temperature and sperm incorporation in polyploid salamanders. *Science* 246:1032–34

Bogart JP, Licht LE. 1986. Reproduction and the origin of polyploids in hybrid salamanders of the genus *Ambystoma. Can. J. Genet. Cytol.* 28:605–17

Brooks R. 1998. The importance of mate copying and cultural inheritance of mating preferences. *Trends Ecol. Evol.* 13:45–46

Case MJ, Taper TJ. 1986. On the coexistence and coevolution of asexual and sexual competitors. *Evolution* 40: 366–87

Cole CJ. 1985. Taxonomy of parthenogenetic species of hybrid origin. *Syst. Zool.* 34:359–63

Crews D, Fitzgerald KT. 1980. Sexual behavior in parthenogenetic lizards *Cnemidophorus. Proc. Natl. Acad. Sci. USA* 77:499–502

Crews D, Grassman M, Lindzey J. 1986. Behavioral facilitation of reproduction in sexual and unisexual whiptail lizards. *Proc. Natl. Acad. Sci. USA* 83:9547–50

Dawley RM, Bogart JP, eds. 1989. *Evolution and Ecology of Unisexual Vertebrates*. Albany: NY State Mus.

Dawley RM, Schultz RJ, Goddard KA. 1987. Clonal reproduction and polyploidy in unisexual hybrids of *Phoxinus eos* and *Phoxinus neogaeus* (Pisces; Cyprinidae). *Copeia* 1987:275–83

Doncaster CP, Pound GE, Cox SJ. 2000. The ecological cost of sex. *Nature* 404:281–85

Dowling TE, Secor CL. 1997. The role of hybridization and introgression in the diversification of animals. *Annu. Rev. Ecol. Syst.* 28:593–619

Dries LA. 2003. Peering through the looking glass at a sexual parasite: Are Amazon mollies Red Queens? *Evolution* 57:1387–96

Drosopoulos S. 1978. Laboratory synthesis of a pseudogamous triploid "species" of the genus *Muellerianella* (Homoptera, Delphacidae). *Evolution* 32:916–20

Foran CM, Ryan MJ. 1994. Female-female competition in a unisexual/bisexual complex of mollies. *Copeia* 2:504–8

Gabor CR, Ryan MJ. 2001. Geographical variation in reproductive character displacement in mate choice by male sailfin mollies. *Proc. R. Soc. Biol. Sci. B* 268:1063–70

Gabriel W, Bürger R. 2000. Fixation of clonal lineages under Muller's ratchet. *Evolution* 54:1116–25

Gabriel W, Lynch M, Bürger R. 1993. Muller's ratchet and mutational meltdowns. *Evolution* 47:1744–57

Georgiades P, Watkins M, Burton GJ, Ferguson-Smith AC. 2001. Roles for genomic imprinting and the zygotic genome in placental development. *Proc. Natl. Acad. Sci. USA* 98:4522–27

Girard C. 1859. Ichtyological notices. *Proc. Acad. Nat. Sci. Phila.* 11:113–22

Green RF, Noakes DLG. 1995. Is a little bit of sex as good as a lot? *J. Theor. Biol.* 174:87–96

Hakoyama H, Iguchi KI. 2002. Male mate choice in the gynogenetic-sexual complex of crucian carp, *Carassius auratus. Acta Ethol.* 4:85–90

Hakoyama H, Iwasa Y. 2004. Coexistence of a sexual and an unisexual form stabilized by parasites. *J. Theor. Biol.* 226:185–94

Hakoyama H, Nishimura T, Matsubara N, Iguchi KI. 2001. Differences in parasite load and nonspecific immune reaction between sexual and gynogenetic forms of *Carassius auratus. Biol. J. Linn. Soc.* 72:401–7

Hamilton WD. 1980. Sex versus non-sex versus parasite. *Oikos* 35:282–90

Hedges SB, Bogart JP, Maxon LR. 1992. Ancestry of unisexual salamanders. *Nature* 356:708–10

Hellriegel B, Reyer H-U. 2000. Factors influencing the composition of mixed populations of a hemiclonal hybrid and its sexual host. *J. Evol. Biol.* 13:906–18

Heubel KU. 2004. *Population ecology and sexual preferences in the mating complex of the unisexual Amazon molly Poecilia formosa (GIRARD, 1859).* Diss. thesis. Univ. Hamburg, Hamburg. 142 pp.

Heubel KU, Schlupp I. 2005. Turbidity affected association behaviour in male sailfin mollies (*Poecilia latipinna*). *J. Fish Bio.* In press

Hubbs C. 1964. Interactions between bisexual fish species and its gynogenetic sexual parasite. *Bull. Tex. Mem. Mus.* 8:1–72

Hubbs CL, Hubbs LC. 1932. Apparent parthenogenesis in nature in a form of fish of hybrid origin. *Science* 76:628–30

Hutchinson JMC. 1961. The paradox of plankton. *Am. Nat.* 95:137–45

Jennions MD, Petrie M. 1997. Variation in mate choice and mating preferences: A review of causes and consequences. *Biol. Rev. Camb. Philos. Soc.* 72:283–327

Judson OP, Normark BB. 1996. Ancient asexual scandals. *Trends Ecol. Evol.* 11:41–46

Jun D, Yigui J. 1991. Studies on the phenomena of stimulating the development of sperm nucleus during oocyte maturation in two types of fishes. *Acta Hydrobiol. Sin.* 15:289–94

Kawecki TJ. 1988. Unisexual-bisexual breeding complexes in Poeciliidae: why do males copulate with unisexual females? *Evolution* 42:1018–23

Kondrashov AS. 1993. Classification of hypotheses on the advantage of amphimixis. *J. Hered.* 84:372–87

Körner KE, Lütjens O, Parzefall J, Schlupp I. 1999. The role of experience in mating preferences of the unisexual Amazon molly. *Behaviour* 136:257–68

Ladle RJ. 1992. Parasites and sex: Catching the Red Queen. *Trends Ecol. Evol.* 7:405–8

Lamatsch DK, Nanda I, Epplen JT, Schmid M, Schartl M. 2000a. Unusual triploid males in a microchromosome-carrying clone of the Amazon molly, *Poecilia formosa. Cytogenet. Cell Genet.* 91:148–56

Lamatsch DK, Steinlein C, Schmid M, Schartl M. 2000b. Noninvasive determination of genome size and ploidy level in fishes by flow cytometry: detection of triploid *Poecilia formosa. Cytometry* 39:91–95

Lampert KP, Lamatsch DK, Epplen JT, Schartl M. 2005. Evidence for monophyletic origin of triploid clones of the Amazon molly, *Poecilia formosa. Evolution* 59:881–89

Landmann K, Parzefall J, Schlupp I. 1999. A sexual preference in the Amazon molly, *Poecilia formosa. Environ. Biol. Fishes* 56:325–31

Lima NRW, Kobak CJ, Vrijenhoek RC. 1996. Evolution of sexual mimicry in sperm-dependent all-female forms of *Poeciliopsis* (Pisces: Poeciliidae). *J. Evol. Biol.* 9:185–203

Lowcock LA, Griffith H, Murphy RW. 1991. The *Ambystoma-laterale-jeffersonianum* complex in Central Ontario: ploidy structure, sex ratio, and breeding dynamics in a bisexual-unisexual community. *Copeia* 1991:87–105

Loyning MK, Kirkendall LR. 1996. Mate discrimination in a pseudogamous bark beetle (Coleoptera: Scolytidae): Male *Ips acuminatus* prefer sexual to clonal females. *Oikos* 77:336–44

Mantovani B, Scali V. 1992. Hybridogenesis and androgenesis in the stick-insect *Bacillus rossius-grandii benazii* (Insecta, Phasmatodea). *Evolution* 46:783–96

Marler CA, Ryan MJ. 1997. Origin and maintenance of a female mating preference. *Evolution* 51:1244–48

Mateos M, Vrijenhoek RC. 2005. Independent origins of allotriploidy in the fish genus *Poeciliopsis. J. Hered.* 96:32–39

Matos RJ, Schlupp I. 2005. Performing in front of an audience—Signalers and the social environment. In *Animal Communication Networks*, ed. PK McGregor, pp. 63–83. Cambridge, MA: Cambridge Univ. Press

Maynard Smith J. 1978. *The Evolution of Sex.* Cambridge, UK: Cambridge Univ. Press

Maynard Smith J. 1992. Age and the unisexual lineage. *Nature* 356:661–62

Mayr E. 1967. *Artbegriff und Evolution.* Hamburg/Berlin: Parey

McGregor PK, Peake TM. 2000. Communication networks: social environments for receiving and signaling behaviour. *Acta Ethol.* 2:71–81

Moore WS. 1975. Stability of small unisexual-bisexual populations of *Poeciliopsis. Ecology* 56:791–808

Moore WS, McKay FE. 1971. Coexistence in unisexual-bisexual breeding complexes of *Poeciliopsis* (Pisces: Poeciliidae). *Ecology* 52:791–99

Morishima K, Horie S, Yamaha E, Arai K. 2002. A cryptic clonal line of the loach *Misgurnus anguillicaudatus* (Teleostei: Cobitidae) evidenced by induced gynogenesis, interspecific hybridization, microsatellite genotyping and multilocus DNA fingerprinting. *Zool. Sci.* 19:565–75

Muller HJ. 1964. The relation of recombination to mutational advance. *Mutat. Res.* 1:2–9

Nanda I, Schartl M, Feichtinger W, Schlupp I, Parzefall J, Schmid M. 1995. Chromosomal evidence for laboratory synthesis of a triploid hybrid between the gynogenetic teleost *Poecilia formosa* and its host species. *J. Fish Biol.* 47:619–23

Neiman M. 2004. Physiological dependence on copulation in parthenogenetic females can reduce the cost of sex. *Anim. Behav.* 67:811–22

Niemeitz A, Kreutzfeldt R, Schartl M, Parzefall J, Schlupp I. 2002. Male mating behaviour of a molly, *Poecilia latipunctata*: a third host for the sperm-dependent Amazon molly, *Poecilia formosa. Acta Ethol.* 5:45–49

Parzefall J. 1973. Attraction and sexual cycle of poeciliids. In *Genetics and Mutagenesis of Fish*, ed. JH Schröder, pp. 177–83. Berlin: Springer-Verlag

Poinar GO Jr, Ricci C. 1992. Bdelloid rotifers in Domenican amber: Evidence for parthenogenetic continuity. *Experientia* 48:408–10

Quattro JM, Avise JC, Vrijenhoek RC. 1992. Mode of origin and sources of genotypic diversity in triploid gynogenetic fish clones (*Poeciliopsis*: Poeciliidae). *Genetics* 130:621–28

Redfield RJ. 1994. Male mutation rates and the cost of sex for females. *Nature* 369:145–47

Reyer H-U, Frei G, Som C. 1999. Cryptic female choice: Frogs reduce clutch size when amplexed by undesired males. *Proc. R. Soc. Biol. Sci. B* 266:2101–7

Ryan MJ, Dries LA, Batra P, Hillis DM. 1996. Male mate preferences in a gynogenetic species complex of Amazon mollies. *Anim. Behav.* 52:1225–36

Schartl A, Hornung U, Nanda I, Wacker R, Müller-Hermelink HK, et al. 1997. Susceptibility to the development of pigment cell tumors in a clone of the Amazon molly, *Poecilia formosa*, introduced through a microchromosome. *Cancer Res.* 57:2993–3000

Schartl M, Nanda I, Schlupp I, Wilde B, Epplen JT, et al. 1995a. Incorporation of subgenomic amounts of DNA as compensation for mutational load in a gynogenetic fish. *Nature* 373:68–71

Schartl M, Schlupp I, Schartl A, Meyer M, Nanda I, et al. 1991. On the stability of dispensable constituents of the eukaryotic genome stability of coding sequences versus truly hypervariable sequences in a clonal vertebrate, the amazon molly, *Poecilia formosa*. *Proc. Natl. Acad. Sci. USA* 88:8759–63

Schartl M, Wilde B, Schlupp I, Parzefall J. 1995b. Evolutionary origin of a parthenoform, the Amazon molly *Poecilia formosa*, on the basis of a molecular genealogy. *Evolution* 49:827–35

Schlupp I, Marler C, Ryan MJ. 1994. Benefit to male sailfin mollies of mating with heterospecific females. *Science* 263:373–74

Schlupp I, Nanda I, Döbler M, Lamatsch DK, Epplen JT, et al. 1998. Dispensable and indispensable genes in an ameiotic fish, the Amazon molly *Poecilia formosa*. *Cytogenet. Cell Genet.* 80:193–98

Schlupp I, Parzefall J, Epplen JT, Nanda I,

Schmid M, Schartl M. 1992. Pseudomale behaviour and spontaneous masculinization in the all-female teleost *Poecilia formosa* (Teleostei: Poeciliidae). *Behaviour* 122:88–104

Schlupp I, Parzefall J, Schartl M. 1991. Male mate choice in mixed bisexual/unisexual breeding complexes of *Poecilia* (Teleostei: Poeciliidae). *Ethology* 88:215–22

Schlupp I, Parzefall J, Schartl M. 2002. Biogeography of the unisexual Amazon molly, *Poecilia formosa*. *J. Biogeogr.* 29:1–6

Schlupp I, Plath M. 2005. Male mate choice and sperm allocation in a sexual/asexual mating complex of *Poecilia* (Poeciliidae, Teleostei). *Biol. Lett.* 1:166–68

Schlupp I, Ryan MJ. 1996. Mixed-species shoals and the maintenance of a sexual-asexual mating system in mollies. *Anim. Behav.* 52:885–90

Schlupp I, Ryan MJ. 1997. Male sailfin mollies (*Poecilia latipinna*) copy the mate choice of other males. *Behav. Ecol.* 8:104–7

Schlupp I, Waschulewski M, Ryan MJ. 1999. Female preferences for naturally-occurring novel male traits. *Behaviour* 136:519–27

Schultz RJ. 1967. Gynogenesis and triploidy in the viviparous fish *Poeciliopsis*. *Science* 157:1564–67

Schultz RJ. 1973. Unisexual fish: Laboratory synthesis of a "species." *Science* 179:180–81

Schultz RJ, Kallman KD. 1968. Triploid hybrids between the all-female teleost *Poecilia formosa* and *Poecilia sphenops*. *Nature* 219:280–82

Seehausen O. 2004. Hybridization and adaptive radiation. *Trends Ecol. Evol.* 19:198–206

Spolsky CM, Phillips CA, Uzzell T. 1992. Antiquity of clonal salamander lineages revealed by mitochondrial DNA. *Nature* 356:706–8

Stenseth NC, Kirkendall LR, Moran N. 1985. On the evolution of pseudogamy. *Evolution* 39:294–307

Suomalainen E, Saura A, Lokki J. 1987. *Cytology and Evolution in Parthenogenesis*. Boca Raton, FL: CRC

Tobler M, Schlupp I. 2005. Parasite in sexual and asexual mollies (*Poecilia*, Poeciliidae, Teleostei): a case for the Red Queen? *Biol. Lett.* In press

Tunner HG. 1979. Heterosis in the common European waterfrog. *Naturwissenschaften* 66:268–69

Turner BJ. 1982. The evolutionary genetics of a unisexual fish, *Poecilia formosa*. In *Mechanisms of Speciation*, ed. C Barigozzi, pp. 265–305. New York: Liss

Turner BJ, Elder JFJ, Laughlin TF, Davis WP. 1990. Genetic variation in clonal vertebrates detected by simple—sequence DNA fingerprinting. *Proc. Natl. Acad. Sci. USA* 87:5653–57

Turner BJ, Steeves HRI. 1989. Induction of spermatogenesis in an all-female fish species by treatment with an exogeneous androgen. See Dawley & Bogart 1989, pp. 113–22

Van Valen L. 1973. A new evolutionary law. *Evol. Theory* 1:1–30

Vrijenhoek RC. 1979. Factors affecting clonal diversity and coexistence. *Am. Zool.* 19:787–89

Vrijenhoek RC. 1989. Genetic and ecological constraints on the origins and establishment of unisexual vertebrates. See Dawley & Bogart 1989, pp. 24–31

Vrijenhoek RC. 1993. The origin and evolution of clones versus the maintenance of sex in *Poeciliopsis. J. Hered.* 84:388–95

Vrijenhoek RC. 1994. Unisexual fish: model systems for studying ecology and evolution. *Annu. Rev. Ecol. Syst.* 25:71–96

Vrijenhoek RC. 1998. Animal clones and diversity. *BioScience* 48:617–28

Vrijenhoek RC, Dawley RM, Cole CJ, Bogart JP. 1989. A list of known unisexual vertebrates. See Dawley & Bogart 1989, pp. 19–23

West SA, Lively CM, Read AF. 1999. A pluralist approach to sex and recombination. *J. Evol. Biol.* 12:1003–12

Westneat DF, Walters A, McCarthy TM, Hatch MI, Hein WK. 2000. Alternative mechanisms of nonindependent mate choice. *Anim. Behav.* 59:467–76

Williams GC. 1975. *Sex and Evolution.* Princeton: Princeton Univ. Press

Witte K, Ryan MJ. 1998. Male body length influences mate-choice copying in the sailfin molly *Poecilia latipinna. Behav. Ecol.* 9:534–39

Witte K, Ryan MJ. 2002. Mate choice copying in the sailfin molly, *Poecilia latipinna*, in the wild. *Anim. Behav.* 63:943–49

Woodhead AD. 1985. Aspects of the mating behavior of male mollies (*Poecilia* spp.). *J. Fish Biol.* 27:593–601

Zhou L, Wang Y, Gui J-F. 2000. Genetic evidence for gonochoristic reproduction in gynogenetic silver crucian carp (*Carassius auratus gibelio* Bloch) as revealed by RAPD assays. *J. Mol. Evol.* 51:498–506

Annu. Rev. Ecol. Evol. Syst. 2005. 36:419–44
doi: 10.1146/annurev.ecolsys.36.091704.175535

MEASUREMENT OF INTERACTION STRENGTH IN NATURE

J. Timothy Wootton[1] and Mark Emmerson[2]

[1]Department of Ecology and Evolution, University of Chicago, Chicago, Illinois 60637;
email: twootton@uchicago.edu
[2]Department of Zoology, Ecology, and Plant Sciences, University College Cork, Cork,
Ireland; email: M.Emmerson@ucc.ie

Key Words   allometry, energy flow, food web, modeling, scale

■ **Abstract**   Understanding and predicting the dynamics of multispecies systems generally require estimates of interaction strength among species. Measuring interaction strength is difficult because of the large number of interactions in any natural system, long-term feedback, multiple pathways of effects between species pairs, and possible nonlinearities in interaction-strength functions. Presently, the few studies that extensively estimate interaction strength suggest that distributions of interaction strength tend to be skewed toward few strong and many weak interactions. Modeling studies indicate that such skewed patterns tend to promote system stability and arise during assembly of persistent communities. Methods for estimating interaction strength efficiently from traits of organisms, such as allometric relationships, show some promise. Methods for estimating community response to environmental perturbations without an estimate of interaction strength may also be of use. Spatial and temporal scale may affect patterns of interaction strength, but these effects require further investigation and new multispecies modeling frameworks. Future progress will be aided by development of long-term multispecies time series of natural communities, by experimental tests of different methods for estimating interaction strength, and by increased understanding of nonlinear functional forms.

INTRODUCTION

All living organisms interact with other species through a variety of mechanisms, and these interactions can strongly shape the structure and function of ecological communities and ecosystems (Elton 1927, Paine 1966). To quantify the strengths of species interactions, identify the patterns that occur between species, and determine the mechanisms that cause interactions to vary across space and time in natural ecosystems are challenging but important goals for several reasons. First, because species interactions can strongly affect ecosystems, being able to predict the consequences of changes in species composition in response to the ongoing processes of biodiversity change is desirable. Ecosystems provide a range of goods and services that are important to human populations. An understanding of how the provision of these services is affected by extinctions and alien introductions is important to have because species interactions often mediate how changes in the physical and chemical environment, such as those expected from global change and human activity, play out through the ecosystem (Schindler et al. 1987, Wootton et al. 1996). Because species are directly and indirectly connected to each other through a complex web of interactions (Elton 1927, Paine 1966), impacts that affect one species in the system can ramify to other system components through multiple direct and indirect pathways that may be of different sign or strength (Holbrook & Schmitt 2004, Schoener 1993, Wootton 1992). If we make predictions about the consequences of environmental impacts without knowledge of the strengths of species interactions, our predictions become indeterminate for any ecosystem with a reasonable degree of complexity (Yodzis 1988). Second, development of a general understanding of how ecological communities are structured can benefit from an analysis of the general properties of multispecies models (May 1973, Pimm 1982). Knowledge of the pattern of interaction strengths in natural ecosystems can help to guide the development of appropriate multispecies models. Here, we review recent developments in our understanding of interaction strength, emphasizing the links to empirical studies.

INTERACTION-STRENGTH CONCEPTS

What is interaction strength? Although the concept seems intuitive, recent reviews highlight the wide diversity of definitions (Berlow et al. 2004, Laska & Wootton 1998), which are shaped by the interests and goals of individual investigators and, to some extent, by the empirical data available. Most studies of interaction strength have focused on consumer-resource interactions in food webs, although an estimate of the strengths of nontrophic interactions such as direct mutualism and interference competition is also needed.

The discovery from field experiments that local deletion of species from ecosystems can generate variable results both in the magnitude of the response and in the extent to which impacts ramify through the community (Paine 1980, Menge et al. 1994, Morin et al. 1988) has motivated recent interest in interaction strength. These results lead naturally to an interaction-strength concept based on the consequences of species deletion. Although intuitively appealing, this definition has shortcomings, because the results of deletion experiments are context dependent in several ways. Experimental results can be a function of experimental duration (Davidson et al. 1984, but see Menge 1997) because the consequences of interactions, particularly via indirect pathways of interaction (Abrams et al. 1996, Schoener 1993, Wootton 1994), take some time to establish. Second, the impacts depend on the density of the focal species, which could vary widely and further lead to variable interaction-strength estimates. Third, because indirect effects may be a component of the net effect of one species on another, estimates of the strength of interaction among species are likely to be strongly dependent on species composition, which further generates highly variable estimates for the same interaction. Finally, estimates of the overall impact of a species over a period of time can involve multiple mechanisms, and recovery of the parameters that describe the mechanisms that underlie the net long-term effects of one species on another is often difficult.

Development of a precise definition of interaction strength may facilitate the linking of multispecies theoretical frameworks with empirical information, which requires indices that are compatible with the structure of theory. Building these links would constrain model behavior and, thereby, increase our predictive ability. Several indices have been derived from models of multiple interacting species. One is the per capita interaction strength, the short-term effect of one individual on an individual of another species. Several authors have advocated focusing on this concept because it a priori should be a more consistent measure than the aggregate response of the system to a species removal, because it is defined without reference to equilibrium conditions, and because all other metrics of interaction strength can be derived from this measure (Laska & Wootton 1998, Paine 1992). A second possible index, Jacobian matrix elements, contains the partial derivatives of the system, which describe responses to a small, pulsed perturbation of a single species from equilibrium (Bender et al. 1984, May 1973, Schoener 1993). In some cases, these elements are standardized by the self-limiting effect of the perturbed species [i.e., intraspecific interactions (May 1973)]. This index has played a prominent role in theoretical analyses of multispecies systems, largely because it determines the stability properties of multispecies models. Conceptually, it represents the direct effect of a single individual of a focal species on the total population of another species at equilibrium. A third theoretical index, elements of the inverted matrix of per capita interaction strengths, describes the predicted change in the equilibrium conditions of the system to a sustained addition of individuals of a focal species (Yodzis 1988). Hence, it incorporates the net effects of an individual of one species on another through both direct and indirect pathways. Although similar to empirical experiments that delete species from a system over an extended period, it differs because complete species deletion changes the structural properties of the system, which can lead to different outcomes (Dambacher et al. 2003a,b; Laska & Wootton 1998). A fourth theoretical measure, the removal matrix, therefore, compares changes in the abundance of a species at equilibrium before and after the removal of that species from the system. These interaction-strength concepts, each useful in various contexts, behave differently; sometimes they even change in opposite ways as per capita parameters change (Laska & Wootton 1998). Therefore, investigators must explicitly identify which theoretical interaction-strength concept they are considering in their work and understand its implications.

Other metrics of interaction strength have been considered on the basis of the types of information available in various empirical studies. These metrics include magnitudes of energy or elemental flow through ecosystems (e.g., Hall et al. 2000), the magnitudes of path coefficients in structural equation models (Gough & Grace 1999, Wootton 1994), and the transitions among sessile species in Markov chain models (Wootton 2001). The latter two methods can identify strong interactions in a field context in some cases but do not generate the interaction-strength parameters in standard dynamical models; hence they will not be considered further here. Measures of energy flow through ecosystems, used extensively in ecosystem models, have been criticized as inaccurately portraying interaction strength (Paine 1980, Rafaelli & Hall 1996). Most energy-flow models are structured with flow magnitudes independent of consumer abundance or biomass (i.e., the models are donor controlled), which limits their applicability to many communities where consumers have an impact on prey. However, to discard energy-flow information in the assessment of interaction strength may be premature for two reasons. First, energy-flow magnitudes probably reflect the interaction strength of resources on consumers, whereas critics of energy-flow metrics are interested in effects of consumers on resources. Second, energy-flow patterns can be cast into per capita consumer impacts by scaling them to consumer densities (de Ruiter et al. 1995, Hall et al. 2000, Moore et al. 1993, Wootton 1997). Whether this approach works experimentally is not well tested, however.

Whether interaction strength is best represented as a constant or as a function has received periodic attention (Abrams 2001, Ruesink 1998, Sarnelle 2003). Many theoretical (May 1973) and empirical studies (Gause 1934, Paine 1992, Wootton 1997) implicitly treat interaction strength as a single number. This treatment is partly a consequence of the substantial role in ecology played by the generalized Lotka-Volterra equations, which assume constant per capita interaction strengths of one species on another. The use of a constant-interaction coefficient is further reinforced by other theoretical analyses focused on system behavior in proximity to an equilibrium point (May 1973, Yodzis 1988). Here, responses to small perturbations are likely to approximate linear behavior even in nonlinear systems. Theoretical arguments suggest that interaction strength can be a function that depends on the densities both of the interacting species and of other members of the ecological community (Abrams 1983, 2001). Functional responses of consumers on prey may be a nonlinear function of prey density because of handling time (Holling 1965, Ruesink 1998, Sarnelle 2003). Also, species may modify the per capita interaction between individuals of other species, either by changing the traits of the interacting species ("trait-mediated indirect interactions") (Abrams et al. 1996, Werner 1992) or by changing the environmental context in which an interaction occurs (Crowder & Cooper 1982, Wootton 1992).

Several approaches have been adopted in response to nonlinear interactions. One approach takes nonlinearities as given but only focuses on a single parameter within the interaction-strength function, such as maximum interaction strength (Berlow et al. 1999, Reusink 1998). Although this approach provides some information about the system, a full parameterization of the interaction-strength function would be desirable (Abrams 2001), and interaction strength typically might be far from its maximum. Hence, a second approach would estimate all parameters in all plausible nonlinear functions that describe interaction strength in an ecosystem (Abrams 2001). Although conceptually ideal, this approach is extremely difficult to implement at present for several reasons. First, we still have a rudimentary understanding of which of the infinite possible forms of nonlinear interaction-strength functions we should attempt to apply a priori, particularly for interaction modifications. Identification of the most appropriate functional forms requires more extensive empirical exploration. Second, this approach substantially increases the number of parameters to be estimated, which increases the difficulty of an already challenging task and expands the potential for error propagation in multispecies models (Ludwig & Walters 1985). An alternative approach starts with the simplest possible characterization of interaction strength and introduces additional complexity as empirical investigation warrants. Although logistically more feasible, it remains a huge challenge and carries the risk that unanticipated, abrupt changes in system behavior will arise after large perturbation to the system as a result of unidentified nonlinearities. This approach characterizes most recent empirical attempts to quantify interaction strength (Moore et al. 1993, Paine 1992, Rafaelli & Hall 1996, Wootton 1997). The debate over which approach is more effective remains open. For example, Abrams (2001) criticized many empirical studies for use of linear estimates of interaction strength, yet in some cases, linear estimates have predicted reasonably well the effects of independent experiments (Pfister 1995, Schmitz 1997, Wootton 1997, but see Ruesink 1998, Sarnelle 2003).

ISSUES AND APPROACHES FOR DIRECT MEASUREMENT OF INTERACTION STRENGTH

With the recent increased interest in estimation of interaction strength in natural communities, several authors have examined in some depth issues related to effective synthesis of field studies with established theoretical frameworks. Aside from the problem of variable definitions of interaction strength described above, several studies have explored the behavior of commonly used empirical interaction-strength metrics after they were applied to simulated data sets and have identified other key difficulties (Bender et al. 1984, Berlow et al. 1999, Laska & Wootton 1998). First, empirical indices that were not explicitly developed to interface with multispecies mathematical models rarely estimate actual per capita interaction strength (Berlow et al. 1999, Laska & Wootton 1998). Second, empirical data can only be taken over discrete time intervals, whereas most ecological theory is based on instantaneous changes; hence, a mismatch in timescale can develop. In particular, if empirical measurements are taken over too long a time interval, indirect effects and density-dependent feedbacks can develop, which influence estimates of interaction strength between directly interacting species (Bender et al. 1984, Berlow et al. 1999, Laska & Wootton 1998). Hence, interaction-strength measures must be made over short time intervals. This requirement can be problematic empirically because changes in populations over short intervals are likely to be slight and, therefore, can be strongly affected or obscured by measurement error and stochastic processes.

Third, Berlow et al. (1999) found that interaction-strength metrics derived from models with linear per capita interaction strengths (Osenberg et al. 1997, Wootton 1997) did not accurately estimate one component (maximum interaction strength) of models with nonlinear interaction-strength functions. One may view this result in two ways. Finding that methods explicitly designed for one theoretical framework fail to estimate a parameter in a different theoretical framework is not particularly surprising; hence, such an analysis may not be a fair test of the approach. A more appropriate test would first derive the proper interaction-strength metrics for the new model form and then implement the necessary experimental treatments to estimate it, as was originally done for the linear case. The alternative perspective is that we may not know a priori the correct functional form of interaction strength and, therefore, may use the metric that represents the simplest theoretical case by default. In this situation, knowing the biases of the metric when plausible alternative functions are in fact operating may be useful.

With these considerations in mind, we identify four basic strategies to estimate interaction strength: field experimentation, laboratory experimentation, observational approaches, and analysis of system dynamics.

Field Experiments

Field experiments have been advanced as the most accurate way to assess interaction strength (Bender et al. 1984, Berlow et al. 1999, Paine 1992). As noted above, such experiments must be short term to estimate interaction strength well. A key difficulty of this approach is the logistical challenge of fully implementing it, because doing so generally requires as many treatments as the number of parameters in a community model.

Paine (1992) applied field experiments to a guild of rocky intertidal grazers that feed on the kelp *Alaria marginata*. This author constructed arenas from which grazers were removed or that contained a single grazer species and measured the change in kelp population size after 8 months. By use of an interaction-strength metric that standardized the difference in kelp abundance between treatments and controls by grazer and kelp density, the author found a distribution of interaction strength skewed toward a few strong and many weak impacts. This study has stimulated many recent attempts to measure interaction strength in the field, but it illustrates several of the general pitfalls inherent in estimation of interaction strength. The interaction-strength metric used matches theoretical parameters only under the limited condition of a two-species system at equilibrium (Berlow et al. 1999, Laska & Wootton 1998), and whether equilibrium conditions were reached is uncertain. Additionally, statistically significant positive effects of putative grazers on the kelp were identified, which suggests that indirect interactions with other species over the duration of the experiment affected the per capita estimates; hence, the consistency of such measures is doubtful under different community configurations.

Raffaelli & Hall (1996) summarized results of field experiments for an estuarine food web that involved effects of large predators on invertebrate prey. Like

Paine (1992), they found a skewed distribution of consumer effects; most interaction strengths were weak and only a few effects were strong. They also found net positive effects for some putative consumers, which suggests that indirect interactions may have affected their estimates. Alternatively, sampling error coupled with small populations of consumers and prey may generate positive but statistically nonsignificant results. The authors also noted that interactions with high per capita effects did not necessarily translate into large overall impacts on the system because of predator rarity, which illustrates the difference in behavior of different interaction-strength metrics.

Laboratory Experiments

Short-term laboratory experiments may be used to provide estimates of interaction strength (Abrams 2001, Vandermeer 1969). Laboratory experiments have the advantage that species interactions can be examined in isolation from other species, and often treatments and conditions can be more easily replicated, which allows a more precise estimate of interaction strength between species pairs. These experiments are usually done over smaller scales, which increases the tractability, but whether estimates of interaction strength derived from these studies can be readily transferred to field conditions is unclear (Skelly 2002, Taylor et al. 2002). One compromise approach is to carry out small-scale mesocosm experiments in the field (e.g., Fagan & Hurd 1994, Reusink 1998, Sarnelle 2003, Wilbur & Fauth 1994).

Levitan (1987) used numerous laboratory experiments that estimated nutrient uptake and excretion, mortality, consumption rates, and resource-dependent reproduction to characterize a planktonic lake food web that encompassed two classes of phytoplankton, three zooplankton genera, and externally managed trout populations. The author then analyzed the resulting model for stability properties and the contributions of particular interactions to stability. The model, one of the most thorough to date that involves both competitive and consumer–resource interactions across a food web, suggested that the system was fairly stable. Resource differentiation by both the phytoplankton and the zooplankton were the most important interactions that affected stability. An analysis of the interaction-strength estimates illustrates (*a*) expected asymmetries between consumers and their resources, in which consumers have stronger effects on their prey, and (*b*) an overall skewed distribution in the magnitude of interactions. No field experiments are available to verify the model predictions and evaluate the approach, but the study illustrates the substantial effort required to parameterize a multispecies model with an experimental approach. The study still required an abstraction of the system to a minimum number of species suspected of being important.

Schmitz (1997) carried out an experimental study on an old-field assemblage to explore the ability of multispecies models to predict the consequences of chronic perturbations. This author estimated interaction strengths in a laboratory setting by measuring short-term feeding rates among grasshopper and plant species, nutrient uptake rates of plant species, and competitive interactions among plants. He then carried out a longer-term field experiment by manipulation of nutrients and grazers to determine how well his parameterized model predicted net effects on other species. The author found good qualitative agreement between predicted and observed patterns but some deviations from quantitative predictions. Interestingly, a generally consistent pattern to the deviations occurred, such that predicted plant responses tended to be approximately an order of magnitude greater than observed responses in nutrient manipulations and grazer-exclusion experiments, which suggests that some key parameter was either missed or not estimated accurately. These differences may also have been affected because predictions were based on the inverse community matrix, whereas the experimental manipulation of grasshoppers involved a complete species deletion, which changes the structure of the system and makes analysis by use of the removal matrix more appropriate. One limitation of this study was the inability to evaluate dynamics over more than one generation, which is likely to be a general problem in the use of laboratory experiments.

Observational Approaches

Observational information, such as rates of feeding, abundance, body size, and life-history rates, when appropriately transformed, can be used to estimate interaction strength (Hall et al. 2000, Moore et al. 1993, Rafaelli & Hall 1996, Wootton 1997, Yodzis & Innes 1992). In some cases, this approach may be logistically easier because experiments are not necessarily required.

Wootton (1997) used a direct observational approach to estimate the strengths of interactions of bird consumers on intertidal invertebrate prey. Because of the moderate size of both predators and prey species, in this case, direct observation of rates of feeding, diet composition, and abundances of predators and prey was feasible. Data were combined to obtain metrics with units appropriate to per capita interaction strength and yielded a range of estimated consumer effects on resources. As in other studies, a skewed distribution of interaction strength showed many weak and few strong per capita effects. More surprising, when other metrics of interaction strength were derived, they also showed a skewed pattern of distribution, although the identity of species involved in strong effects changed as a result of varying densities. For a subset of the estimates, experimental manipulations were available and were consistent with observationally derived estimates, which suggests that the approach is valid. The analysis centered on a focal group of avian species and only evaluated consumer impacts on resources; consequently, the results do not necessarily reflect community-wide patterns of interaction strength in the system. Inferences about the general pattern of relative interaction strength of prey on consumers probably could be inferred from the data, when it is combined with estimates of shell-free biomass of individual prey items, because consumption rates are likely to scale reasonably well with per capita effects of prey on consumers. Finally, although the direct observation approach plausibly could be extended to additional interactions that involve large prey and predators (e.g., whelks or seastars), this method would be much more difficult or impossible to apply to other interactions, such as limpets grazing on microalgae or sessile species competing for space.

A second observational approach incorporates energetics with diet-composition information. Moore et al. (1993) estimated per capita interaction strength of consumer–resource interactions in soil food webs by documentation of diets, body size, and biomass of component taxa. The authors then incorporated prior information on consumption efficiency of different taxa from laboratory studies and used size-specific metabolism relationships to estimate predicted energy demand by a population of a given biomass and body size, under the assumption that the community was at equilibrium. This approach yielded estimates of per capita (or per biomass) effects both of predators on prey and of prey on predators, and it potentially generated one of the most complete model parameterizations of a system to date. Subsequently, these data have been used to explore the distribution of interaction strength within the food web and its impacts on system stability (de Ruiter et al. 1995, Neutel et al. 2002). Key findings included differences in magnitude between reciprocal interaction strengths and nonrandom arrangement of interaction strengths within the food web that appear to stabilize the whole food web. As more comprehensive estimates become available, a comparison of these results with other systems to assess their generality will be interesting and informative (e.g., Levitan 1987).

Observational approaches tied to the structure of ecological theory can provide information on interaction strength in field situations without numerous experimental manipulations. The approach is limited by the researcher's ability to make direct observations of rates of interactions, by the way it ignores potentially important species differences when allometric relationships are used to estimate key physiological parameters, by the limited applicability of the approach to nontrophic species interactions (e.g., interference competition and mutualism), and by the assumption of equilibrium made in energetics models. Additionally, energetics approaches may be unable to estimate important nonlinearities in interaction-strength functions (but see Yodzis & Innes 1992). Finally, parameterizations that use the energetics approach have not been independently evaluated by experimental manipulations, and this prospect remains a ripe area for future research.

Analysis of System Dynamics

A fourth approach, which fits models to system dynamics over time, may provide an efficient method for estimating interaction strength (Elkinton et al. 1996; Ives et al. 1999, 2003; Laska & Wootton 1998; Pfister 1995; Seifert & Siefert 1976). Most ecological theory is based on changes in abundance or biomass over time; hence, fitting these models to system dynamics is a natural approach to estimation of model parameters.

Laska & Wootton (1998) evaluated this approach by use of simulation studies based partially on model fits to the dynamics of a rocky intertidal system.

They found good agreement between estimates of interaction strength generated from limited segments of the simulated time series and actual parameters. The accuracy of the approach increased when observations were made on time intervals sufficiently spaced that temporal correlation in the time series was low. Unlike experimental studies, which require an experiment for each parameter to be estimated, effective parameter estimation requires only zero to two pulsed experimental manipulations if a sufficiently long time series is available and if identifiable disturbance events are exploited. The analysis could be extended in several ways to better evaluate the effectiveness of the time-series approach. First, the authors only simulated linear models and, thus, did not explore the power to detect nonlinear dynamics. Second, their simulations did not investigate process and measurement error, which could reduce the power of the approach. For example, Pascual & Karieva (1996) used dynamic model fitting to analyze data from the seminal studies by Gause (1934) that estimated interaction strength and found that, with observation error, the parameter range that described competitive interactions among *Paramecium* species was very wide. These studies also highlighted that when the goal is estimation of dynamical parameters, incorporation of experimental treatments that cover a range of initial conditions is much more effective than implemention of the highly replicated experiments that involve the few treatment levels that experimental ecologists typically use.

Ives et al. (1999, 2003) used multispecies dynamic data describing planktonic communities in north temperate lakes to parameterize models of the system and evaluate the processes affecting system stability. They further developed the approach by illustrating appropriate maximum-likelihood model fitting, integrating environmental covariates into the approach, and considering the separate effects of process and measurement error. The authors estimated per capita interaction strengths and used them to identify key processes that affect stability and evaluated the role of various direct and indirect pathways involved in responses to manipulation of fish planktivory. Unlike in most studies, the magnitudes of interaction strengths estimated in this study were only weakly skewed. This result may have arisen for several reasons. First, interactions with phytoplankton were not directly modeled. Therefore, many interactions were abstract competitive effects among zooplankton, whereas most other studies examined consumer–resource interactions. Second, the form of model estimated by Ives et al. (1999, 2003) follows the Gompertz equation, where per capita rates of change are functions of natural logs of species densities, which implies multiplicative effects of species. This difference raises a key issue: What type of discrete time approximation should be used in modeling a system? Many other studies have used Ricker-type models, in which the per capita population change is related to initial abundances of species, following the assumption of additive species effects in Lotka-Volterra models. A third approach is to estimate instantaneous rates of change by fitting some arbitrary function, such as separate linear, polynomial, or exponential functions, to a small subset of points around a focal data point and calculating the average slope of the function at the focal data point (e.g., Ellner et al. 2002). More systematic exploration of the functions that provide the best fit to empirical data and assessment of the implications that these functions have for community dynamics and stability are needed.

Treatment of error in development of multispecies models from time series also requires further attention. Measurement error can bias parameter estimates in many contexts but is particularly problematic in time series because it introduces serial correlation into the data set. Serial correlation can be easily misinterpreted as an important process, such as density dependence (Ives et al. 2003). Ives et al. (2003) and others (e.g., de Valpine & Hastings 2002, Pascual & Karieva 1996) present methods to incorporate measurement error into time series analyses. Measurement errors can be estimated from the data as free parameters with these approaches but, consequently, require substantial increases in the data needed to maintain the power of the approach. An easier and effective alternative is to directly estimate measurement error during the course of empirical studies and insert it into statistical models, but ecologists rarely do this (Carpenter et al. 1994).

Process error also needs further attention in development of the time-series approach. The ability to estimate process error with time-series methods is appealing because stochastic processes have potentially important impacts on ecological systems (e.g., Chesson & Huntly 1997). Most analyses of time-series models assume that process error is an additive source of noise in growth rates, usually normally distributed. This feature need not be the structure of process error, however. For example, variation in the environment might change interaction coefficients and lead to

$$dN_i/dt \, N_i\Big[r_i + \Big(\sum_{j=1:s}[f_{ij}(N_1 \ldots N_s) + e_{ij}]N_j \Big) + e_i \Big],$$

where e_{ij} is the process error associated with a particular interaction between species i and species j, f_{ij} is the function describing per capita interactions between species i and j, and N_i is the abundance of species i. Hence, the effects of process error might interact with species densities. Incorporation of such error structure is undesirable because it increases the number of parameters to estimate, but the point where process error enters the system might have important consequences. For example, the temporal storage-effect mechanism for coexisting competitors (Chesson & Huntly 1997) involves variation in both the abundance of competitors and the intensity of the competitive interaction. The implication of inserting process error in different positions on theoretical predictions needs further exploration.

The use of system dynamics to fit multispecies models is a promising approach that needs further empirical and theoretical evaluation, including a better integration of mechanistic and statistical theory. Aside from its natural links to theory, the approach may be more powerful than direct experimentation because it requires fewer experimental manipulations of the system and can more easily probe short-term dynamics to avoid contamination of the estimates by indirect effects. Probing dynamics works best for systems that explore a wide range of possible parameter space (e.g., starting away from equilibrium), such as when experimental manipulations are imposed or when well-defined, pulsed disturbances have affected the system. In general, the approach also requires studies of longer duration to generate sufficient dynamical data. At present, the multispecies, multitrophic time series of sufficient length needed for this approach are very rare and may be hard to obtain for some systems. Development of such data sets should be an important priority for ecology to better explore these methods and related methods that exploit ecological dynamics.

ALLOMETRIC APPROACHES TO ESTIMATION OF INTERACTION STRENGTH

Given the difficulties of direct estimation of interaction strength in the field, identification of general relationships between interaction strength and more readily measured properties of species would be convenient. Because the field measurement of interaction strength is still in its infancy, the existence of useful generalizations is uncertain, but some recent results are tantalizing.

Body size is one general property of organisms that might scale with interaction strength (Elton 1927, Yodzis & Innes 1992); therefore, investigating allometric relationships with interaction strength may be profitable. Several allometric relationships relevant to various interaction-strength concepts or parameters in multispecies models in general are already well established. For example, metabolic rates and movement rates scale strongly with body size (Calder 1996, Harestad & Bunnell 1979, Peters 1983) and potentially affect organism death rates, encounter rates, and assimilation efficiencies. Foraging theory also suggests that consumption rate should be affected by the size of interacting species (e.g., Schluter 1982, Werner & Hall 1974). Several empirical data sets are available for a preliminary analysis of allometric relationships (Emmerson & Raffaelli 2004, Sala & Graham 2002, Wootton 1997). When data from all three studies are combined, they tend to fall around the same relationship (Figure 1), which suggests a generally predictable pattern of interaction strength. However, some inconsistencies are present in the pattern. Specifically, consumption rates of kelp grazers (Sala & Graham 2002) and crabs (Emmerson & Rafaelli 2004) decrease with prey:predator size ratio, whereas bird consumption rates (Wootton 1997) increase. Because bird predation involved a wider range of prey:predator size ratios, these interactions dominate the overall relationship. The basis for these discrepancies is unknown, but they could arise as a result of a unimodal function in combination with trait differences between consumer and prey classes, or from differences in how body-size gradients were established (experimentally in the lab versus naturally in the field; single-prey species versus the natural range of prey species).

The feasibility of allometrically derived interaction coefficients were assessed by Emmerson et al. (2005), who also used allometric relationships between body size and population density to estimate equilibrium population sizes in an estuary and a stream. Jacobian matrices were calculated by combination of the estimated

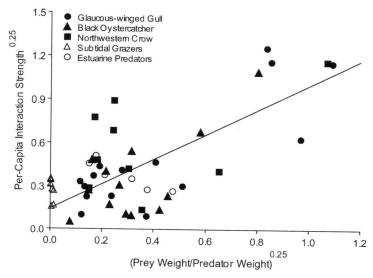

Figure 1   Relationship between per capita interaction strength and the ratio of prey to predator body size (fourth-root transformed data) from studies of avian rocky intertidal predators (Wootton 1997), subtidal grazers (Sala & Graham 2002), and estuarine predators (Emmerson & Raffaelli 2004). Points for estuarine predators illustrate the range of values observed in Emmerson & Rafaelli (2004), not specific data points in the experiment. The regression line presented is for intertidal bird predators only [$PCIS^{0.25} = 0.14 + 0.85 \times$ (prey/predator mass)$^{0.25}$, $R^2 = 0.50$].

interaction coefficients with estimated population densities and then assessed for stability. The resulting food-web parameterizations were stable, even though the webs contained 88 and 32 species, respectively.

INTERACTION STRENGTH-FREE APPROACHES

Several authors have suggested that studies focused on species-specific processes should be abandoned in favor of aggregate (macroecological) level properties of communities and ecosystems (Brown 1995, Lawton 1999), because of the difficulties associated with the estimation of interaction strength and the indeterminate predictions of models with species interactions. Such coarse description of ecological communities, however, may not be relevant in addressing many pressing questions, particularly those of importance to conservation. These questions usually concern the consequences of environmental impacts on, or extinction of, particular species, which requires consideration of the well-documented variation in responses and impacts by different species. Nevertheless, identification of ways in which species-specific responses can be determined without resorting to esti-mation of interaction strengths would be very helpful. Several approaches have been advanced in which variation in interaction strength is ignored.

Neutral theories of biodiversity (Hubbell 2001) have recently been advanced that do not consider variation in species interaction strength. These theories posit that species in competition for a limiting resource are ecologically identical, and that variation in species abundances are primarily the result of random-walk processes, speciation rates, and the effects of dispersal limitation in a spatially structured system. These theories are appealing because they can generate real-istic species-abundance and species-area distributions in ecological communities from first principles of population biology, and initial studies suggested that their predictions might apply to a wide range of communities (Hubbell 2001). A weak-ness of these analyses, however, is that the models are tuned to the data with several unmeasured parameters (McGill 2003). Stronger tests that use informa-tion on species identity have been developed, in which neutral models are tested against system dynamics in spatially distributed areas (Clark & McLachlan 2003) or parameterized with one data set and then tested against novel, independent data (Adler 2004, Wootton 2005). In these tests, the neutral models have not fared well, which indicates that, although their underlying mechanisms may play an important role in shaping community structure, accounting for variation in species interaction strength is crucial to making accurate predictions.

Analyses of linkage patterns in empirical food webs without reference to dif-ferences in the strength of the interactions that generate the links (e.g., Cohen & Briand 1984, Pimm et al. 1991) have been carried out extensively, but these studies generally have not attempted to make specific predictions about experimental or environmental impacts. Several recent approaches have revealed that topological characteristics of interaction webs can play a large role in the determination of ecosystem structure. Hence, perturbations that change these characteristics may generate predictable system responses.

Loop analysis does not account for interaction strength but predicts system re-sponse to perturbations near equilibrium on the basis of the qualitative patterns of species interactions in a community web (Levins 1975, Puccia & Levins 1985). Un-til recently, loop analysis received modest attention because its predictions quickly became indeterminate in the absence of additional information on per capita inter-action strength. Advances, however, may substantially expand its scope in making useful predictions. Dambacher et al. (2003a,b) introduced methods for quantifying the strength and complexity of different feedback pathways in relatively complex systems and used simulation studies based on systems with empirically estimated interaction strengths to evaluate the performance of their methods. They found that by quantifying feedback patterns, they could gain substantial insight into the predictability of species response to system impacts. Therefore, even if all com-ponents of a complex system do not respond predictably to a perturbation, loop analysis may still provide valuable predictive information. This method has not been tested experimentally in the field, but may offer substantial promise if its assumptions of approximately linear interaction strength and additivity of species interactions are not too badly violated.

Other analyses have also suggested that web architecture may have predictable effects (Pimm 1982). For example, Solé & Montoya (2001) investigated the effects of the loss of highly connected species from complex real food webs relative to the loss of species at random from these same webs. They found that food webs subject to random species loss tend to persist, but food webs that lose highly connected species become highly fragmented and experience many secondary extinctions. Empirical information may support this pattern. For example, in rocky intertidal systems of the northeastern Pacific, experimental deletion of the highly omnivorous starfish *Pisaster ochraceus* causes large responses in the rest of the system, whereas deletion of the specialist limpet *Acmaea mitra* has minor effects (Paine 1980). Whether the impact of a species generally correlates with its effect on overall linkage patterns remains to be determined, and to our knowledge, no one has undertaken a direct experimental test of this prediction. Montoya et al. (2005) used the inverse community matrix derived from estimates of Emmerson et al. (2005) to assess the relative importance of species connectance within food webs and found a negative relationship between linkage density of each species and the mean net indirect effect of that species on other species in the communities. The relative importance of specialist versus generalist species within food webs to other species could be tested experimentally in other systems to determine the validity of these findings.

INTERACTION-STRENGTH DISTRIBUTIONS: IMPLICATIONS AND ORIGINS

Empirical studies to date that have estimated per capita interaction strength among a number of species pairs have generally found distributions skewed toward many weak interactions and few strong interactions. This pattern is seen in the population-level interactions (Jacobian elements) documented by de Ruiter et al. (1995), which clearly show a strong skew toward weaker interactions for both the negative effects of predators on prey and the positive effects of prey on predators (Figure 2). This prevalent pattern is also seen for a range of studies that have quantified per capita effects in marine (Paine 1992, Raffaelli & Hall 1996, Sala & Graham 2002, Wootton 1997) (Figure 3A–D), freshwater (Levitan 1987) (Figure 3F), and terrestrial (Fagan & Hurd 1994) (Figure 3E) ecosystems. Interaction-strength distributions of non-trophic interactions have received less attention. Strengths of pollinator effects on plants seem to exhibit a skewed distribution on a per-interaction basis (Pettersson 1991, Schemske & Horvitz 1984), but whether these correspond to per capita interactions is unclear because pollinator abundance was unknown and indices of pollination success (pollen deposition, fruit set) may not translate directly to population dynamics. Despite the large amount of literature on interspecific competition, surprisingly few studies quantify the distribution of interaction strength

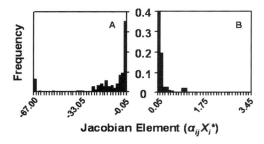

Figure 2 Frequency distribution of interaction strengths (Jacobian elements) from de Ruiter et al. (1995) soil food webs. (A) Negative effects of predators on prey. (B) Positive effects of prey on predators. Both positive and negative effects are skewed toward weak interaction strengths. Note difference in scale between consumer and resource effects.

among direct (nonexploitative) competitors in multispecies systems. Roxburgh & Wilson (2000) report the Jacobian matrix for a seven-species plant community, and the distribution of competitive interaction strengths exhibits a skewed distribution.

Such a regular pattern of skewed interaction strengths raises the question of how the shape of these distributions affects community dynamics and stability, and why such regular patterns might arise. Early theoretical studies of stability in multispecies webs assumed normal distributions of interaction strength and found that diverse webs tended to be unstable (May 1973, Pimm 1982), but recent work suggests that skewed interaction-strength distributions exhibit higher persistence under certain conditions, particularly when omnivory and multiprey satiation occur (Emmerson & Yearsley 2004, McCann et al. 1998). The relationship between the empirical and theoretical results is uncertain, however, because different interaction-strength units have generally been used: per capita interaction strengths (empirical) and Jacobian elements (theoretical). Hence, further studies exploring alternative interaction-strength metrics would be beneficial.

The effect of interaction strength on stability raises the question of how such interaction-strength distributions arise. One possibility is that community assembly processes favor interaction-strength distributions that convey stability. This proposition has been explored in several theoretical studies. In some cases, species interaction strengths were assigned at random, and simulations were carried out to determine the characteristics of interaction strength that lead to long-term persistence. Emmerson & Yearsley (2004) studied randomly parameterized Lotka-Volterra food webs with four species and found that the subset of locally stable communities had emergent skewed interaction-strength distributions, and weaker interactions were present in omnivorous loops. An alternative approach is to carry out simulations while continuously adding species with random characteristics. Kokkoris et al. (1999) studied Lotka-Volterra competition systems of

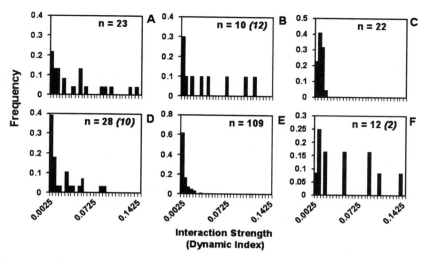

Figure 3   Frequency distribution of absolute per capita effects of consumers on their prey. Coefficients were calculated as $ln(C/E \times P)$, where C and P are the abundance of prey and consumers, respectively, in control treatments, and E is the prey abundance in predator removal treatments (Osenberg et al. 1997, Wootton 1997). Each distribution was calculated using the raw data obtained from the studies detailed in (*A*) Paine (1992), (*B*) Raffaelli & Hall (1996), (*C*) Sala & Graham (2002), (*D*) Wootton (1997), (*E*) Fagan & Hurd (1994), and (*F*) Levitan (1987). These data represent some of the most comprehensive currently available in the literature. All distributions are skewed toward many weak interactions and few strong interations, a pattern that is consistent across all of the systems studied. Numbers on the graph represent sample size (number of interactions) and numbers in parentheses indicate the number of interaction strengths found above the scale of the graphs.

40 species and found that skewed interaction-strength distributions developed over time. Drossel et al. (2004) studied assembly of food webs assuming a range of interaction-strength functional forms but placed no constraints on food-web configuration or species number. They found that resulting assembled communities exhibited skewed interaction-strength distributions. They also reported that food webs of realistic structure required mechanisms in which species focused on consuming the most profitable prey in the system. Further study of the outcome of community assembly processes would be beneficial to clarify the generality and causes of conflicting patterns of food-web structure. Such contrasting patterns occur in models with and without food-web constraints. However, the generality of the conclusion that stable communities exhibit distributions skewed toward weak interactions across studies suggests that it may be robust.

EFFECTS OF SCALE

Temporal and spatial scale may affect estimates of interaction strength in several ways, but their implications are only starting to be explored because of the difficulty of estimating interaction-strength functions at even a single scale. Indirect effects and density-dependent feedback can change interaction-strength estimates over long timescales of measurement. Disparate timescales among different species and different types of ecological processes may also affect interaction strength. For example, deaths, growth, and behavioral changes occur fairly continuously through time, whereas births are often seasonally pulsed. Such differences may affect key aspects of interaction strength, such as encounter rates among individuals of different species, and may favor biomass- rather than abundance-based approaches to species interactions. Similarly, different components of food webs have different generation times, which may affect the intensity of interactions through time. For example, in freshwater systems, algae may have generation times on the order of days, arthropod grazers have generation times on the order of weeks to months, and fish predators have generation times on the order of years. Hence, species operating on a fast timescale can quickly respond numerically to changes in the strength of species interactions, whereas species operating on longer timescales cannot. Asymmetries in timescales must be addressed by developing models that integrate fast processes over the timescale of slower processes or species, which can lead to nonlinear interactions. Additionally, integrating functions to account for disparate timescales can generate asymmetries in the functional form that describes the interaction for each of the interacting species (e.g., Nicholson & Bailey 1935).

One shortcut for dealing with disparate timescales of dynamically interacting entities is to assume that faster-scale processes quickly come to equilibrium (MacArthur 1972, Schaffer 1981), such that the interactions involving slower processes primarily experience these equilibrium conditions. MacArthur (1972) showed that resources generated locally under logistic growth at fast timescales produce logistic interaction-strength functions for consumers in which consumer dynamics are not explicitly linked to resource abundances, but are controlled by critical resource-rate parameters, which yield mechanistic interpretations of "intrinsic" population growth rate and carrying capacity. Similarly, for systems supported by nutrients derived outside the system, the effect of nutrients on plants generates a nonlinear, semi–ratio-dependent (DeAngelis et al. 1975) interaction-strength function, which is independent of nutrient concentrations but is controlled by critical nutrient-rate parameters. Because these functions collapse into equations involving a single trophic level in some cases, highly disparate timescales may allow abstraction to fewer component species, potentially simplifying the problem of dealing with all species in the system at once (Schaffer 1981). An understanding of how species operating at other timescales affect parameters within a population is essential, however, to predicting how a community will respond to environmental impacts. Additionally, when interaction strength is a nonlinear function (e.g., a saturating type II functional response), multitimescale models generate much less tractable results.

Shifting the focus from abundance to biomass may help researchers to avoid disparate timescale issues (Hall et al. 2000, Moore et al. 1993, Reusink 2000, Wootton & Power 1993, Yodzis & Innes 1992). Pulsed reproductive bouts generate delayed changes in abundance that reflect integrated patterns of prior consumption, whereas energy is stored immediately as biomass at the same temporal scale as consumption. Modeling biomass dynamics rather than abundance dynamics typically requires several key assumptions, such as larger individuals will have equivalent consumption rates as several smaller individuals of equal biomass, and larger prey individuals will be encountered at rates as high as the encounter of several small individuals. These assumptions require careful empirical investigation. Chalcraft & Resitarits (2004) provided an initial assessment by varying size structure of predatory fish and frog tadpole prey and found that, although the assumptions were met qualitatively, the relationships were affected by density-dependent processes.

Spatial scale can also affect our perception of interaction strength in several ways. Spatial scale can affect interaction strength through changes in habitat heterogeneity. If the performance of interacting species is habitat dependent, then interaction strength may change with scale. For example, competing species often perform better in different habitats, which leads to habitat shifts and exclusion of each species by the other from its preferred habitat (e.g., Connell 1961, Grace & Wetzel 1981, Pacala & Roughgarden 1985). In this case, small-scale studies in one habitat would estimate strong interspecific interaction strengths among competitors consistent with competitive exclusion, but studies at larger scales that include multiple habitats would estimate lower interspecific interaction strengths, consistent with competitive coexistence. A similar situation would arise where habitats differ in structural cover, and habitat structure alters consumer–resource interactions (e.g., Werner et al. 1983). The way in which species perceive their physical environment and interact with it and other species depends on body size. As we already noted, the home range of a species is positively correlated with body size, and so size effects may also affect the way in which interactions are modified over varying spatial scales.

Spatial scale might also affect interaction strength because of localized interactions. Ultimately, all species impacts arise via interactions among individuals, and closer individuals are more likely to interact with each other than are distant individuals. Hence, as spatial scale increases, the mass action assumptions inherent in many models of communities may be violated, which affects interaction-strength parameters by generation of patchy population structure. In these situations, spatially explicit individual-based models may be required to adequately model multispecies systems. Pascual et al. (2001, 2002) explored this issue theoretically by use of an individual-based predator-prey system, and then examined how spatially explicit dynamics affected the form of mean field-population models at larger scales. Interestingly, they found that the individual-based models generated mean field dynamics with similar form to standard consumer–resource models. The per capita interaction strengths in these models, however, depended on spatially relevant parameters such as individual dispersal distance; lower dispersal tended to weaken interspecific interaction strength. The generality of these results needs to be explored for other types of interactions and, more systematically, for changes in habitat scale, but may indicate that localized interactions can be adequately handled by standard interaction-strength functions.

Spatial-scale effects on interaction strength have been reported in several empirical studies. For example, Skelly (2002) and Taylor et al. (2002) compared per capita interaction strengths between species in small-scale mesocosms and large-scale plots imbedded in natural ecosystems and found variation between experimental venues. Wooster (1994) reported a meta-analysis of experiments that probed consumer–resource interactions and found changes in the magnitude of predator impacts with experimental plot size. Because manipulations were carried out by use of different methods, however, other differences in experimental venues than scale could have caused the documented differences. Experiments in which all aspects except scale remain constant are needed in future empirical evaluations of this question (e.g., Englund & Cooper 2003). In many cases, the scale-dependent patterns appear to arise because of variation in the openness of experimental arenas to migration: Larger arenas will tend to be less affected by populations outside the manipulated areas because of perimeter to area effects, and this may dilute or enhance the total effect of a manipulated species, depending on the relative mobility of the interacting species (Englund & Olsson 1996).

INTERACTION-STRENGTH LINKS TO ECOSYSTEM FUNCTION

Much interest has developed in understanding how biodiversity affects aggregate properties of ecosystems (Naeem et al. 1994, Petchey et al. 2004). Ultimately, the mechanisms by which species generate patterns in ecosystem properties depend in part on the characteristics of species interactions. To our knowledge no explicit experimental investigation has linked ecosystem function with quantified per capita effects of species interactions, even though niche overlap and complementary resource utilization among competitors are regularly invoked to explain increased ecosystem function with increased species richness.

Most studies that investigate biodiversity and ecosystem function only consider single trophic-level systems, whereas most natural ecosystems have multiple trophic levels. A few theoretical and empirical studies have been carried out to investigate the effects of diversity on the functioning of multitrophic ecosystems (Downing & Leibold 2002, Fox 2004, Petchey et al. 2004, Thébault & Loreau 2004). Uncertainty remains, however, in the role that interaction strength plays in the generation of the patterns in these studies. Thébault & Loreau (2004) investigated the effects of introduction of herbivorous consumers to a system of competing plants by use of linear per capita effects. Their models suggest that both the strength and arrangement of interactions causes ecosystem-level effects within the community. Incorporating studies of species interactions into systems that probe the phenomenological association between biodiversity and ecosystem function would help clarify their relationship.

EVOLUTION AND INTERACTION STRENGTH

Interaction strengths depend in part on the traits of interacting organisms and represent potentially strong selective pressures; hence, they should evolve over time. Coevolution concerns itself with the evolution of interaction strength, and its literature is sufficiently large that we cannot adequately review it here (see Abrams 2000, Thompson 1998). One particularly exciting, and challenging, area of investigation is the link between coevolution and ecological dynamics. Although evolution is often assumed to proceed slowly, several studies show that it can proceed on an ecological timescale (e.g., Grant & Grant 1989, Lee 1999). Furthermore, adaptive phenotypic plasticity may evolve under some circumstances, which creates nonlinear interaction-strength functions and trait-mediated indirect effects when viewed at a species-wide level (Tollrain & Harvell 1999, Werner 1992). Such trait changes could affect ecological stability. For example, as consumer pressure increases, prey may evolve more resistant traits, which reduces interaction strength, but when consumer pressure is low, costs of resistance may favor susceptible genotypes. Such a situation could help stabilize ecological dynamics. Concurrent consumer evolution, however, might introduce cyclical time lags, thereby destabilizing systems. Although a challenge, some progress in modeling linked ecological and evolutionary dynamics has been made (Abrams & Matsuda 1996, Marrow et al. 1996, Saloniemi 1993), and it confirms some of these contingencies. Experimental studies with bacteria-phage systems demonstrate empirically that evolutionary changes in interaction strength can affect ecological dynamics (Bohannan & Lenski 2000).

CONCLUSIONS

Estimation of interaction strength in ecosystems and linkage of these estimates to theoretical frameworks remains a daunting challenge but is needed in many cases to aid in the understanding of ecological structure and function and the prediction of responses to environmental impacts. General rules, such as characteristic patterns of interaction-strength distributions across communities and within interaction webs, and relationships with species traits such as body size may be of use in the future, but their identification will require more extensive estimates from field systems. A key need is the development of comprehensive long-term, multispecies, multitrophic time series for a group of focal ecosystems. These data would allow more comprehensive analyses of system dynamics to estimate interaction strength, as well as provide insight into other ecological questions and processes. Interaction strength–free approaches may also provide some insight as new methods become more available to analyze the effects of changing the structure of interaction webs. These methods, and several approaches, such as energetics, for estimating interaction strength also need to be experimentally tested in field situations. Empirical studies that incorporate interaction-strength estimates, body-size relationships, and ecosystem measures into highly resolved interaction webs will be valuable for clarifying the relative roles of interaction strength and web architecture on system behavior. The diverse approaches employed by ecologists illustrate the possibilities of bringing a range of techniques to bear on the challenging problems that society charges contemporary ecologists to address. Use of multiple approaches may enable us to distill some semblance of reality from natural complex systems.

The *Annual Review of Ecology, Evolution, and Systematics* is online at
http://ecolsys.annualreviews.org

LITERATURE CITED

Abrams PA. 1983. Arguments in favor of higher order interactions. *Am. Nat.* 121:887–91

Abrams PA. 2000. The evolution of predator-prey interactions: theory and evidence. *Annu. Rev. Ecol. Syst.* 31:79–105

Abrams PA. 2001. Describing and quantifying interspecific interactions: a commentary on recent approaches. *Oikos* 94:209–18

Abrams PA, Matsuda H. 1996. Fitness minimization and dynamic instability as a consequence of predator-prey coevolution. *Evol. Ecol.* 10:167–86

Abrams PA, Menge BA, Mittelbach GG, Spiller D, Yodzis P. 1996. The role of indirect effects in food webs. In *Food Webs: Dynamics and Structure*, ed. G Polis, K Winemiller, pp. 371–95. New York: Chapman & Hall

Adler PB. 2004. Neutral models fail to reproduce observed species-area and species-time relationships in Kansas grasslands. *Ecology* 85:1265–72

Bender EA, Case TJ, Gilpin ME. 1984. Perturbation experiments in community ecology: theory and practice. *Ecology* 65:1–13

Berlow EL, Navarrete SA, Briggs CJ, Power ME, Menge BA. 1999. Quantifying variation in the strengths of species interactions. *Ecology* 80:2206–24

Berlow EL, Neutel A-M, Cohen JE, de Ruiter PC, Ebenman BO, et al. 2004. Interaction strengths in food webs: issues and opportunities. *J. Anim. Ecol.* 73:585–98

Bohannan BJM, Lenski RJ. 2000. Linking genetic change to community evolution: insights from studies of bacteria and bacteriophage. *Ecol. Lett.* 3:362–77

Brown JH. 1995. *Macroecology.* Chicago: Univ. Chicago Press. 269 pp.

Calder WA III. 1996. *Size, Function, and Life History.* New York: Dover. 431 pp.

Carpenter SR, Cottingham KL, Stow CA. 1994. Fitting predator-prey models to time series with observation errors. *Ecology* 75:1254–64

Chalcraft D, Resetarits WJ Jr. 2004. Metabolic rate models and the substitutability of predator populations. *J. Anim. Ecol.* 73:323–32

Chesson P, Huntley N. 1997. The roles of harsh and fluctuating conditions in the dynamics of ecological communities. *Am. Nat.* 150:519–53

Clark JS, McLachlan JS. 2003. Stability of forest biodiversity. *Nature* 423:635–38

Cohen JE, Briand F. 1984. Trophic links of community food webs. *Proc. Natl. Acad. Sci. USA* 81:4105–9

Connell JH. 1961. The influence of interspecific competition and other factors on the distribution of the barnacle *Chthamalus stellatus*. *Ecology* 42:710–23

Crowder LG, Cooper WE. 1982. Habitat structural complexity and the interaction between bluegills and their prey. *Ecology* 63:1802–13

Dambacher JM, Li HW, Rossignol PA. 2003a. Qualitative predictions in model ecosystems. *Ecol. Model.* 161:79–93

Dambacher JM, Luh H-K, Li HW, Rossignol PA. 2003b. Qualitative stability and ambiguity in model ecosystems. *Am. Nat.* 161:876–88

Davidson DW, Inouye RS, Brown JH. 1984. Granivory in a desert ecosystem: experimental evidence for indirect facilitation of ants by rodents. *Ecology* 65:1780–86

DeAngelis DL, Goldstein RA, O'Neill RV. 1975. A model for trophic interaction. *Ecology* 56:881–92

de Ruiter PC, Neutel A, Moore JC. 1995. Energetics, patterns of interactions strengths, and stability in real ecosystems. *Science* 269:1257–60

de Valpine P, Hastings A. 2002. Fitting population models incorporating process noise and observation error. *Ecol. Monogr.* 72:57–76

Downing AL, Leibold MA. 2002. Ecosystem consequences of species richness and composition in pond food webs. *Nature* 416:837–41

Drossel B, McKane AJ, Quince C. 2004. The impact of nonlinear functional responses on the long-term evolution of food web structure. *J. Theor. Biol.* 229:539–48

Elkinton JS, Healy WM, Bounaccorsi JP, Boettner GH, Hazzard AM, et al. 1996. Interactions among gypsy moths, white-footed mice, and acorns. *Ecology* 77:2332–42

Ellner SP, Seifu Y, Smith RH. 2002. Fitting population dynamic models to time-series data by gradient matching. *Ecology* 83:2256–70

Elton C. 1927. *Animal Ecology*. Reprinted 2001, Chicago: Univ. Chicago Press. 204 pp.

Emmerson MC, Montoya JM, Woodward G. 2005. Body size, interaction strength and food web dynamics. In *Dynamic Food Webs: Multispecies Assemblages, Ecosystem Development, and Environmental Change*, ed. PC de Ruiter, V Wolters, JC Moore. New York: Academic. In press

Emmerson MC, Raffaelli D. 2004. Predator-prey body size, interaction strength and the stability of a real food web. *J. Anim. Ecol.* 73:399–409

Emmerson M, Yearsley JM. 2004. Weak interactions, omnivory and emergent food-web properties. *Proc. Natl. Acad. Sci. USA* 271:397–405

Englund G, Cooper SD. 2003. Scale effects and extrapolation in ecological experiments. *Adv. Ecol. Res.* 33:161–213

Englund G, Olsson T. 1996. Treatment effects in a stream fish enclosure experiment: influence of predation rate and prey movements. *Oikos* 77:519–28

Fagan WF, Hurd LE. 1994. Hatch density variation of a generalist arthropod predator: population consequences and community impact. *Ecology* 75:2022–32

Fox JW. 2004. Modelling the joint effects of predator and prey diversity on total prey biomass. *J. Anim. Ecol.* 73:88–96

Gause GF. 1934. *The Struggle for Existence*. Reprinted 1964, New York: Hafner. 163 pp.

Gough L, Grace JB. 1999. Effects of environmental change on plant species density: comparing predictions with experiments. *Ecology* 80:882–90

Grace JB, Wetzel RG. 1981. Habitat partitioning and competitive displacement in cattails *Typha*: experimental field studies. *Am. Nat.* 118:463–74

Grant BR, Grant PR. 1989. *Evolutionary Dynamics of a Natural Population: The Large Cactus Finch of the Galápagos*. Chicago: Univ. Chicago Press. 350 pp.

Hall RO Jr, Wallace BJ, Eggert SL. 2000. Organic matter flow in stream food webs with reduced detrital resource base. *Ecology* 81:3445–63

Harestad AS, Bunnell FL. 1979. Home range and body weight—a reevaluation. *Ecology* 60:389–402

Holbrook SJ, Schmitt RJ. 2004. Population dynamics of a damselfish: effects of a competitor that also is an indirect mutualist. *Ecology* 85:979–85

Holling CS. 1965. The functional response of predators to prey density and its role in mimicry and population regulation. *Mem. Entomol. Soc. Can.* 45:1–60

Hubbell SP. 2001. *The Unified Theory of Biodiversity and Biogeography*. Princeton, NJ: Princeton Univ. Press. 375 pp.

Ives AR, Carpenter SR, Dennis B. 1999. Community interaction webs and zooplankton responses to planktivory manipulations. *Ecology* 80:1405–21

Ives AR, Dennis B, Cottingham KL, Carpenter SR. 2003. Estimating community stability and ecological interactions from time-series data. *Ecol. Monogr.* 73:301–30

Kokkoris GD, Troumbis AY, Lawton JH. 1999. Patterns of species interaction strength in assembled theoretical competition communities. *Ecol. Lett.* 2:70–74

Laska MS, Wootton JT. 1998. Theoretical concepts and empirical approaches to measuring interaction strength. *Ecology* 79:461–76

Lawton JH. 1999. Are there general laws in ecology? *Oikos* 84:177–92

Lee CE. 1999. Rapid and repeated invasions of fresh water by the copepod *Eurytemora affinis*. *Evolution* 53:1423–34

Levins R. 1975. Evolution in communities near equilibrium. In *Ecology and Evolution of Communities*, ed. M Cody, J Diamond, pp. 16–50. Cambridge, MA: Harvard Univ. Press

Levitan C. 1987. Formal stability analysis of a planktonic freshwater community. In *Predation: Direct and Indirect Impacts on Aquatic Communities*, ed. WC Kerfoot, A Sih, pp. 71–100. Hanover, NH: Univ. Press New Engl.

Ludwig D, Walters CA. 1985. Are age-structured models appropriate for catch-

effort data? *Can. J. Fish. Aquat. Sci.* 42:1066–72

MacArthur RH. 1972. *Geographical Ecology*. New York: Harper & Row. 269 pp.

May RM. 1973. *Stability and Complexity in Model Ecosystems*. Princeton, NJ: Princeton Univ. Press. 265 pp.

McCann K, Hastings A, Huxel GR. 1998. Weak trophic interactions and the balance of nature. *Nature* 395:794–97

McGill BJ. 2003. Strong and weak tests of macroecological theory. *Oikos* 102:679–85

Menge BA. 1997. Detection of direct versus indirect effects: Were experiments long enough? *Am. Nat.* 149:801–23

Menge BA, Berlow EL, Blanchette CA, Navarrete SA, Yamada SB. 1994. The keystone species concept: variation in interaction strength in a rocky intertidal habitat. *Ecol. Monogr.* 64:249–86

Montoya JM, Emmerson MC, Solé RV, Woodward G. 2005. Perturbations and indirect effects in complex food webs. In *Dynamic Food Webs: Multispecies Assemblages, Ecosystem Development, and Environmental Change*, ed. PC de Ruiter, V Wolters, JC Moore. New York: Academic. In press

Montoya JM, Sole RV. 2003. Topological properties of food webs: from real data to community assembly models. *Oikos* 102:614–22

Moore JC, de Ruiter PC, Hunt HW. 1993. The influence of productivity on the stability of real and model ecosystems. *Science* 261:906–08

Morin PJ, Lawler SP, Johnson EA. 1988. Competition between aquatic insects and vertebrates: interaction strength and higher order interactions. *Ecology* 69:1401–09

Marrow P, Dieckmann U, Law R. 1996. Evolutionary dynamics of predator-prey systems: an ecological perspective. *J. Math. Biol.* 34:556–78

Naeem S, Thompson LJ, Lawler SP, Lawton JH, Woodfin RM. 1994. Declining biodiversity can alter the performance of ecosystems. *Nature* 368:734–37

Neutel A-M, Heesterbeek JAP, de Ruiter PC.

2002. Stability in real food webs: weak links in long loops. *Science* 296:1120–23

Nicholson AJ, Bailey VA. 1935. The balance of animal populations. Part I. *Proc. Zool. Soc. London* 3:551–98

Osenberg CW, Sarnelle O, Cooper SD. 1997. Effect size in ecological experiments: the application of biological models in meta-analysis. *Am. Nat.* 150:798–812

Pacala SW, Roughgarden J. 1985. Population experiments with the *Anolis* lizards of St-Maarten and St-Eustatius, Netherlands Antilles. *Ecology* 66:129–41

Paine RT. 1966. Food web complexity and species diversity. *Am. Nat.* 100:65–75

Paine RT. 1980. Food webs: linkage, interaction strength and community infrastructure. *J. Anim. Ecol.* 49:667–85

Paine RT. 1992. Food-web analysis through field measurement of per capita interaction strength. *Nature* 355:73–75

Pascual M, Mazzega P, Levin SA. 2001. Oscillatory dynamics and spatial scale: the role of noise and unresolved pattern. *Ecology* 82:2357–69

Pascual M, Roy M, Franc A. 2002. Simple temporal models for ecological systems with complex spatial patterns. *Ecol. Lett.* 5:412–19

Pascual MA, Kareiva P. 1996. Predicting the outcome of competition using experimental data: maximum likelihood and Bayesian approaches. *Ecology* 77:337–49

Petchey OL, Downing AL, Mittelbach GG, Persson L, Steiner CF, et al. 2004. Species loss and the structure and functioning of multitrophic aquatic systems. *Oikos* 104:467–78

Peters RH. 1983. *The Ecological Implications of Body Size.* Cambridge, UK: Cambridge Univ. Press. 329 pp.

Pettersson MW. 1991. Pollination in a guild of fluctuating moth populations: option for unspecialization in *Silene vulgaris. J. Ecol.* 79:591–604

Pfister CA. 1995. Estimating competition coefficients from census data: a test with field manipulations of tidepool fishes. *Am. Nat.* 146:271–91

Pimm SL. 1982. *Food Webs.* Reprinted 2002, Chicago: Univ. Chicago Press. 219 pp.

Pimm SL, Lawton JH, Cohen JE. 1991. Food web patterns and their consequences. *Nature* 350:669–74

Puccia CT, Levins R. 1985. *Qualitative Modeling of Complex Systems: An Introduction to Loop Analysis and Time Averaging.* Cambridge, MA: Harvard Univ. Press. 259 pp.

Rafaelli DG, Hall SJ. 1996. Assessing the relative importance of trophic links in food webs. In *Food Webs: Integration of Patterns and Dynamics*, ed. G Polis, K Winemiller, pp. 185–91. New York: Chapman & Hall

Raimondi PT, Forde SE, Delph LF, Lively CM. 2000. Processes structuring communities: evidence for trait-mediated indirect effects through induced polymorphisms. *Oikos* 91:353–61

Reusink JL. 1998. Variation in per capita interaction strength: thresholds due to nonlinear dynamics and nonequilibrium conditions. *Proc. Natl. Acad. Sci. USA* 95:6843–47

Reusink JL. 2000. Intertidal mesograzers in field microcosms: linking laboratory feeding rates to community dynamics. *J. Exp. Mar. Biol. Ecol.* 248:163–76

Roxburgh SH, Wilson JB. 2000. Stability and coexistence in a lawn community: mathematical prediction of stability using a community matrix with parameters derived from competition experiments. *Oikos* 88:395–408

Sala E, Graham MH. 2002. Community-wide distribution of predator–prey interaction strength in kelp forests. *Proc. Natl. Acad. Sci. USA* 99:3678–83

Saloniemi I. 1993. A coevolutionary predator-prey model with quantitative characters. *Am. Nat.* 141:880–96

Sarnelle O. 2003. Nonlinear effects of an aquatic consumer: causes and consequences. *Am. Nat.* 161:478–96

Schaffer WM. 1981. Ecological abstraction the consequences of reduced dimensionality in ecological models. *Ecol. Monogr.* 51:383–402

Schemske DW, Horvitz CC. 1984. Variation among floral visitors in pollination ability:

a precondition for mutualism specialization. *Science* 225:519–21

Schindler DW, Mills KH, Malley DF, Findlay DL, Shearer JA, et al. 1987. Long-term ecosystem stress: the effects of years of experimental acidification on a small lake. *Science* 228:1395–401

Schluter D. 1982. Seed and patch selection by Galapagos Ecuador ground finches *Geospiza* in relation to foraging efficiency and food supply. *Ecology* 63:1106–20

Schmitz OJ. 1997. Press perturbations and the predictability of ecological interactions in a food web. *Ecology* 78:55–69

Schoener TW. 1993. On the relative importance of direct versus indirect effects in ecological communities. In *Mutualism and Community Organization*, ed. H Kawanabe, JE Cohen, K Iwasaki, pp. 365–415. Oxford, UK: Oxford Univ. Press

Seifert RP, Seifert FH. 1976. A community matrix analysis of *Heliconia* insect communities. *Am. Nat.* 110:461–83

Skelly DK. 2002. Experimental venue and estimation of interaction strength. *Ecology* 83:2097–101

Solé RV, Montoya JM. 2001. Complexity and fragility in ecological networks. *Proc. R. Soc. Lond. B-Biol. Sci.* 268:2039–45

Taylor BW, McIntosh AR, Peckarsky BL. 2002. Reach-scale manipulations show invertebrate grazers depress algal resources in streams. *Limnol. Oceanog.* 47:893–99

Thébault E, Loreau M. 2004. Food-web constraints on biodiversity and ecosystem functioning relationships. *Proc. Natl. Acad. Sci. USA* 100:14949–54

Thompson JN. 1988. Variation in interspecific interactions. *Annu. Rev. Ecol. Syst.* 19:65–87

Tollrain R, Harvell CD. 1999. *The Ecology and Evolution of Inducible Defenses*. Princeton, NJ: Princeton Univ. Press. 383 pp.

Vandermeer JH. 1969. The competitive structure of communities: an experimental approach with protozoa. *Ecology* 50:362–71

Werner EE. 1992. Individual behavior and higher-order species interactions. *Am. Nat.* 140:5–32 (Suppl.)

Werner EE, Gilliam JF, Hall DJ, Mittelbach GG. 1983. An experimental test of the effects of predation risk on habitat use in fish ecology. *Ecology* 64:1540–48

Werner EE, Hall DJ. 1974. Optimal foraging and the size selection of prey by the bluegill sunfish *Lepomis macrochirus*. *Ecology* 55:1042–52

Wilbur HM, Fauth JE. 1994. Experimental aquatic food webs interactions between two predators and two prey. *Am. Nat.* 135:176–204

Wooster D. 1994. Predator impacts on stream benthic prey. *Oecologia* 99:7–15

Wootton JT. 1992. Indirect effects, prey susceptibility, and habitat selection: impacts of birds on limpets and algae. *Ecology* 73:981–91

Wootton JT. 1994. Predicting direct and indirect effects: an integrated approach using experiments and path analysis. *Ecology* 75:151–65

Wootton JT. 1997. Estimates and test of per capita interaction strength: diet, abundance, and impact of intertidally foraging birds. *Ecol. Monogr.* 67:45–64

Wootton JT. 2001. Prediction in complex ecological communities: analysis of empirically derived Markov models. *Ecology* 82:580–98

Wootton JT. 2005. Field parameterization and experimental test of the neutral theory of biodiversity. *Nature* 433:309–12

Wootton JT, Parker MS, Power ME. 1996. Effects of disturbance on river food webs. *Science* 273:1558–61

Wootton JT, Power ME. 1993. Productivity, consumers, and the structure of a river food chain. *Proc. Natl. Acad. Sci. USA* 90:1384–87

Yodzis P. 1988. The indeterminacy of ecological interactions as perceived through perturbation experiments. *Ecology* 69:508–15

Yodzis P, Innes S. 1992. Body size and consumer-resource dynamics. *Am. Nat.* 139:1151–75

Annu. Rev. Ecol. Evol. Syst. 2005. 36:445–66
doi: 10.1146/annurev.ecolsys.36.102003.152633
First published online as a Review in Advance on September 16, 2005

MODEL SELECTION IN PHYLOGENETICS

Jack Sullivan[1,2] and Paul Joyce[2,3]

[1]*Department of Biological Sciences, University Idaho, Moscow, Idaho 83844-3051;
email: jacks@uidaho.edu*
[2]*Initiative in Bioinformatics and Evolutionary Studies (IBEST), University of Idaho,
Moscow, Idaho 83844*
[3]*Department of Mathematics, University of Idaho, Moscow, Idaho 83844-1103;
email: joyce@uidaho.edu*

Key Words   AIC, BIC, decision theory, likelihood ratio, statistical phylogenetics

■ **Abstract**   Investigation into model selection has a long history in the statistical literature. As model-based approaches begin dominating systematic biology, increased attention has focused on how models should be selected for distance-based, likelihood, and Bayesian phylogenetics. Here, we review issues that render model-based approaches necessary, briefly review nucleotide-based models that attempt to capture relevant features of evolutionary processes, and review methods that have been applied to model selection in phylogenetics: likelihood-ratio tests, AIC, BIC, and performance-based approaches.

INTRODUCTION

In this review, we assume the well-known view first voiced by Box (1976) that all models are wrong, but some are useful. After a brief introduction, we discuss alternatives for evaluating the adequacy of the chosen model. Finally, we illustrate how each of the traditional approaches to model selection fit well within the framework of decision theory (DT) and that DT facilitates an understanding of the goals and assumptions of these approaches.

The Importance of Models

Phylogenetic analysis is entering the genomics era, and as tools for surveying genomes (e.g., expressed sequence tags, single-nucleotide polymorphisms, genome sequencing, etc.) become more widely available, phylogenetic studies at all levels, from intraspecific phylogeography to the tree of life, will increasingly use data from multiple-gene loci. Concurrent with the advent of phylogenomics is the application of phylogenies to an ever-widening array of disciplines. For example, statistical phylogenetics have been permitted as evidence in a criminal court recently in which a Louisiana physician was convicted of infecting his former girlfriend with HIV from one of his HIV-positive patients (Metzker et al. 2002), and phylogenetic testing has been used recently to refute the hypothesis that contaminated polio vaccine was the origin of the AIDS epidemic (Worobey et al. 2004).

Applying the emerging wealth of data to such an array of issues, however, presents difficulties because multiple loci are likely to be evolving under very different constraints and, therefore, may be subject to diverse substitution processes. One must, therefore, decide how best to account for the diversity of substitution processes in model-based phylogeny estimation, even for potential partitions in a single-gene data set. In our review of model choice in phylogenetics, we begin by introducing first the importance of probabilistic models in science generally, and then in the particular case of phylogenetics.

Models in Science

Statistical models allow scientists to exceed a mere description of their data and extend to proposing and testing general principles that can explain the data. Thus, statistical models add precision to the formulation of a scientific hypothesis and provide a rigorous means by which to assess the evidence for or against a hypothesis by providing a context for making predictions. Statistical models and methods are therefore ubiquitous in science.

Interestingly, the founder of modern statistics, R.A. Fisher, discovered the likelihood principle and invented maximum likelihood (ML) (Fisher 1958) primarily to answer questions related to evolutionary genetics. However, he did most of his work before the discovery of DNA, and, thus, he focused on quantitative genetics. Fisher's paradigm has been the centerpiece of data analysis throughout much of science in general, and much of biology in particular (e.g., Johnson & Omland 2004), but application of the ML principle and its explicit modeling approach has been slow in coming to phylogenetics. This delay was caused partly by the computational complexity of the problem and partly by an antithetical attitude of some systematists toward statistical approaches (e.g., Siddall & Kluge 1997). Computational difficulties have been ameliorated by a number of advances in theory and implementation (e.g., Huelsenbeck & Ronquist 2001, Swofford 1998), and philosophical objections have not proved sufficiently compelling to the broader community of systematics to halt the advance of model-based approaches to phylogenetics. Thus, the fact that Fisher's methodology is now dominating the field of phylogenetic biology, particularly in the analysis of molecular data, seems particularly appropriate to us.

Models in Phylogenetics

The necessity of models in molecular phylogenetics and evolution was recognized in the first comparative analyses of DNA sequence data (e.g., Brown et al. 1982, Jukes & Cantor 1969). Sequence divergence is roughly linear with time only shortly after a divergence event. The cause of this deviation from linearity is multiple substitutions at the same site (i.e., multiple hits), and the earliest molecular evolutionary studies attempted to accommodate multiple hits in estimating the number of substitutions that have occurred since two sequences diverged from a common ancestor by use of explicit models (Jukes & Cantor 1969).

Furthermore, the consequence of ignoring multiple substitutions was also recognized early: underestimation of the number of substitutions that have occurred since two sequences last shared a common ancestor. More importantly, however, this underestimation is not uniform. Long branches (and large genetic distances) will be underestimated disproportionately more than will short branches and genetic distances (e.g., Gillespie 1986). Some of the implications of this nonuniform underestimation are well studied [e.g., long-branch attraction (LBA) (Felsenstein 1978)], but the effect of model choice on data exploration seems to be less appreciated.

MODELS IN EXPLORING DATA VIA SATURATION PLOTS   The recognition that multiple hits can occur led to the concept of substitutional saturation (e.g., Brown et al. 1982), which is still of concern to many molecular phylogeneticists. However, because most studies lack the fossil data that Brown et al. (1982) and others have used to establish the x-axis in early saturation plots, most assessments of saturation use some measure of pairwise genetic distance on the x-axis as a proxy for time. Some other aspect of molecular evolution, say, the absolute number of transitions, is then plotted on the y-axis to make inferences about the relationship between that variable and genetic distance. Such plots are frequently used as exploratory tools with which to understand the processes that have generated a data set of interest (e.g., López-Fernández et al. 2005) and are frequently used to justify decisions about data elimination (e.g., Han & Ro 2005).

However, for the x-axis to be at all meaningful, estimates of genetic distances for use as the x-axis must be based on a model of evolution that estimates multiple substitutions adequately. If an underparameterized model is used, genetic distances will be undercorrected and will underestimate the actual number of substitutions disproportionately more for large distances than for small distances (e.g., Golding 1983). The effect that this error will have on saturation plots is simple to predict; the x-axis will be compressed nonuniformly and use of overly simple models in saturation plots (or even worse, use of uncorrected p-distances) will obfuscate understanding of the processes of molecular evolution.

This problem is common and is illustrated in Figure 1. These plots were generated from the COI data of Cicero & Johnson (2001), who used them (along with data from Cyt b, ND2, and ND3) to estimate phylogenetic relationships among *Empidonax* flycatchers. They illustrated a linear relationship apparent between third-position transitions in the original saturation plots by use of p-distances (figure 3 in Cicero & Johnson 2001), and this apparent linearity was used to justify inclusion of those sites in an equally weighted parsimony analysis, whereas other data were eliminated (not shown). However, a plot based on the HKY + I + Γ distances (Figure 1A), which the authors chose for ML analysis by application of the hierarchical likelihood-ratio test (LRT, see below), leads to very different

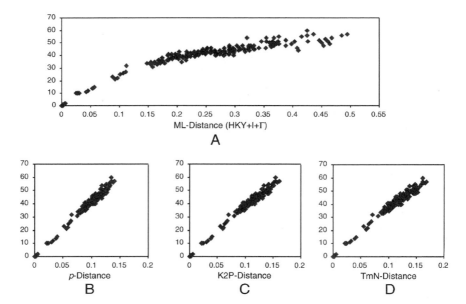

Figure 1   The effect of model choice on data exploration. Data are from Cicero & Johnson (2001); the *y*-axis is absolute number of third-position transitions, and the *x*-axis is genetic distance corrected by use of various models. Modeltest was used by the orignal authors to select HKY $+$ I $+$ Γ.

conclusions regarding the pervasiveness of multiple transitions at third-codon positions in their COI data than does their plot based on *p*-distances (Figure 1*B*). Their conclusions about the prevalence of multiple hits that involve third-position transitions in this data set are spurious and the result of use of a poorly chosen model in the saturation plots. Furthermore, many studies have used models such as the Kimura two-parameter (K2P) model (Kimura 1980) or Tamura-Nei distances (Tamura & Nei 1993) to calculate genetic distances for the *x*-axis in saturation plots. However, as is shown in Figure 1*C* and 1*D*, use of neither of these simple models as the *x*-axis in saturation plots results in detection of multiple substitutions that involve third-position transitions in the *Empidonax* COI data. Clearly, model choice has a dramatic effect on exploration of data.

THE EFFECT OF UNDERESTIMATION OF MULTIPLE SUBSTITUTIONS IN PHYLOGENY
Felsenstein (1978) was the first to point out that the underestimation of multiple hits can result in inconsistent estimation of phylogeny if the (unknown) true tree contains long branches separated by a short internal branch. This result is caused by the well-studied phenomenon of LBA and is the result of precisely the same underestimation of evolutionary change (number of substitutions) described above. Huelsenbeck & Hillis (1993) examined the performance of many methods across

a variety of tree shapes and demonstrated that accurate estimation of phylogenies is difficult, regardless of method, under the conditions that Felsenstein (1978) had described. This conclusion led them to dub that region of tree space (where two long branches are separated by a short internal branch) the Felsenstein zone. In subsequent studies, investigators have demonstrated via simulations that the underestimation of nucleotide substitutions associated with overly simplified models leads to LBA and inconsistent estimation in the Felsenstien zone, even when ML is used (e.g., Gaut & Lewis 1995, Sullivan & Swofford 2001). Furthermore, a few studies have demonstrated that use of inadequate likelihood models can lead to LBA in real data sets (e.g., Anderson & Swofford 2004, Sullivan & Swofford 1997).

The large body of simulation studies show that the shape of the underlying true tree has an enormous impact on the importance of model choice. In the ideal case (Figure 2), the underlying tree shape is such that all existing methods estimate phylogeny accurately; ML estimation is very robust to violations of model assumptions, and model choice is not critical (e.g., Sullivan & Swofford 2001). However, model choice is critical in the Felsenstein zone (Figure 2), and that observation is widely accepted.

Although perhaps not a widely appreciated, biases associated with violation of model assumption may favor the true tree. Specifically, if long terminal branches are adjacent to a short internal branch [termed the Farris zone by Siddall (1998) and the inverse Felsenstein zone by Swofford et al. (2001)] (Figure 2), the underestimation of long terminal branches will result in overestimation of the short internal branch and cause the most biased methods (such as parsimony and ML under an oversimplified model) to recover the true tree with high confidence and with very little data (Bruno & Halpern 1999, Siddall 1998, Sullivan & Swofford 2001, Swofford et al. 2001, Yang 1997). In fact, the most overly simplified method of phylogenetic estimation will be the most efficient (Sullivan & Swofford 2001). Some have suggested that this bias might be a useful attribute of methods such as parsimony and ML under simplistic models (Siddall 1998, Yang 1997). However, others have suggested that this bias is caused by misinterpretation of convergent substitutions as synapomorphies and should be avoided (Bruno & Halpern 1999, Sullivan & Swofford 2001, Swofford et al. 2001). Model choice is therefore critical here as well.

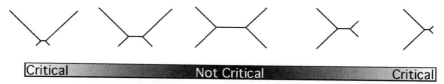

Figure 2 The effect of topology on robustness. At the center of the continuum, phylogenetics signal is strong and model choice is not critical (i.e., maximum likelihood is robust to violations of model assumptions). In the Felsenstein zone (*left*), model selection is critical, as is also the case for the inverse Felsenstein zone (*right*).

Although estimation of topology may not always be compromised by use of overly simple models, estimation of nodal support certainly is. This outcome has been demonstrated for nonparametric bootstrap values (Buckley & Cunningham 2002), parametric bootstrap tests of a priori hypotheses (Buckley 2002), and Bayesian posterior probabilities (Erixon et al. 2003, Huelsenbeck & Rannala 2004, Lemmon & Moriarty 2004). Furthermore, although most simulation studies have focused on the four-taxon cases shown in Figure 2, any large phylogeny can reasonably be expected to contain subtrees from across the continuum. That is, unless one has some assurance that no terribly short and no terribly long branches exist anywhere in the phylogeny that one is attempting to estimate, choice of an overly simple model is likely to impinge negatively on phylogeny estimation.

Overparameterization

Given the potential problems associated with overly simplistic models, an obvious reaction would be to always use the most complex model available. Indeed, use of the most complex model available has been advocated at times, at least for Bayesian estimation (e.g., Huelsenbeck & Rannala 2004). However, in general, this approach seems like a poor strategy. Although an increase in the number of parameters will always increase the fit between model and data (i.e., increase the likelihood), if that increase is simply the result of parameterizing stochastic variation, nothing is gained. With increased use of mutlilocus data for phylogeny estimation, the temptation will inevitably arise to partition data excessively. Such overparameterization can result in nonidentifiability of parameters because of a loss of degrees of freedom (Rannala 2002). Furthermore, Buckley et al. (2001) examined the performance of several models with regard to branch-length estimation from a data set containing 25 sequences of three mtDNA genes (COI, A6, and tRNA$_{Asp}$) from *Maoricicada* and two outgroups. They found that both GTR $+ I + \Gamma$ and GTR $+ \Gamma$ models (applied to all sites) provided better estimates of branch lengths than did a 10-class, site-specific rates (SSR) model (GTR $+$ SSR$_{10}$), despite the fact that the SSR model is more parameter rich and has a better likelihood. Models with the best likelihood score are not guaranteed to produce the best estimates of branch lengths from finite data and, by extension, should not necessarily be expected to perform best in phylogeny estimation.

This suggestion by Huelsenbeck & Rannala (2004) was generated by the fact that, when they simulated data under a simple Jukes-Cantor (JC) model, they were able to estimate nodal probabilities accurately by estimating with an overparameterized GTR $+ \Gamma$, even with sequences as short as 100 nt. This result is encouraging, but the recommendation based on that conclusion should be tempered somewhat for two reasons. First, the simulation conditions are very artificial. When the true model (JC) is a special case of the estimating model (GTR $+ \Gamma$), the overparameterized estimating model will converge on the special-case true model (i.e., base frequencies will be estimated to be equal). This situation will never occur in real data, for which all models are almost certainly wrong. Similarly, some nonnested models may be simpler than the most complex model available but may account for some important feature not addressed by the more highly parameterized model. In these cases, the simpler model may have better predictive ability.

Uniqueness of Phylogeny Estimation

The statistical nature of phylogeny estimation is very unusual. Standard statistical software packages, even ones as powerful as SAS or R, are unlikely to be of much use in phylogenetic analysis. The reason is that the fundamental parameter in phylogenetics is usually the tree topology, which is inherently discrete, whereas the wealth of statistical methodology and theory centers on continuously varying parametric models. Therefore, standard χ^2 goodness-of-fit tests are untrustworthy in the phylogenetic context, and methods such as parametric bootstrap or Bayesian posterior analysis (that do not rely on asymptotic theory) represent better statistical procedures for phylogeny estimation.

REVIEW OF MODELS

Reviews of models of nucleotide substitution have been provided by Swofford et al. (1996) and, more recently, by Felsenstein (2004). However, potentially important models are not presented in either of those publications and a brief review of models is therefore appropriate here.

GTR Family

Widely used models of nucleotide substitution are usually time reversible; an A→T transversion is treated as equivalent to a T→A transversion [i.e., $r_{(AT)} = r_{(TA)}$]. Thus, six possible substitution types exist among the four nucleotides. Each of these transformation types may be treated as equivalent (Jukes & Cantor 1969), transitions may be treated separately from transversions (e.g., Hasegawa et al. 1985, Kimura 1980), all six may be treated as unique (Tavaré 1986, Yang 1994), or any combination of the six types may be grouped. Thus, 203 transformation matrices are possible, each of which represents a special case of the GTR model. Furthermore, base frequencies may be assumed to be equal (i.e., Jukes & Cantor 1969, Kimura 1980) or allowed to vary.

Early models assumed that all sites in a collection of sequences evolve at a uniform rate. However, several methods have been developed to account for the observation that sites usually evolve at different rates (e.g., Uzzell & Corbin 1971). One may assume that some portion of the sites are invariable (e.g., Hasegawa et al. 1985), that rates across sites conform to a Γ-distribution (e.g., Uzzell & Corbin 1971, Yang 1993), or that rate heterogeneity is better described by a mixture of invariable sites and Γ-distributed rates, the I + Γ model, in which some sites are invariable (p_{inv}) and rates at variable sites conform to a Γ-distribution (Gu et al. 1995, Waddell & Penny 1996).

Swofford et al. (1996) reviewed the development and conceptual relationships among some of the commonly used equal-rates models; these relationships can be expanded to accommodate the heterogeneous-rates models mentioned above. Under this framework, the most general and parameter-rich model (GTR + I + Γ) has the following substitution parameters:

- Rate matrix parameters: $r_{(AC)}$, $r_{(AG)}$, $r_{(AT)}$, $r_{(CG)}$, and $r_{(CT)}$, with $r_{(GT)} = 1$
- Base frequencies: π_A, π_C, π_G, with $\pi_T = 1 - (\pi_A + \pi_C + \pi_G)$
- Rate heterogeneity parameters: gamma shape (α), proportion of sites that are invariable (p_{inv})

All other submodels within this family are special cases of GTR + I + Γ, with one or more of the parameters constrained.

Nonreversible Models

In many data sets, base frequencies change in different parts of the tree, and a few models have been proposed that accommodate this change. Base frequencies may be allowed to change on every branch, for $3(2n - 2)$ compositional parameters (because trees must now be rooted), or only on terminal branches, for $3n$ compositional parameters (Yang & Roberts 1995). Alternatively, nucleotide frequencies may be pooled, so that only GC content varies across a tree (Galtier & Guoy 1998). Foster (2004) has made important advances in modeling nonuniform base frequencies. In particular, he has made the number of base-frequency vectors a parameter that can be estimated and, for several real data sets, has demonstrated that even a single change in base frequencies on the tree is sufficient to provide an adequate improvement in model fit.

Other nonreversible models are based on the covarion hypothesis of Fitch & Markowitz (1970), in which rates of sites can change across the tree. Tuffley & Steele (1998) were the first to model this situation explicitly, and it has been incorporated into corrections for evolutionary distances and likelihood frameworks (Galtier 2001, Hueslenbeck 2002). These advances are likely to be important in phylogeny estimation across the tree of life and will almost certainly require application of Markov chain Monte Carlo approaches (Felsenstein 2001).

Nonindependence Across Sites

CODON-BASED MODELS   Because of the nature of the genetic code, one can expect nonindependence across sites within a codon. Codon-based models are particularly appealing for protein-coding genes because they account for the genetic code explicitly. Instead of a 4×4 rate matrix for transformations among nucleotides at a site, these models approximate a 61×61 matrix (with 3660 implied relative rates for the nonreversible version) to account for transformations among all possible (non–stop) codons for each triplet. The rate matrix is filled by use of the relevant genetic code, and rates of synonymous versus nonsynonymous codon substitution are optimized. Underlying nucleotide substitution models assume uniform rates [i.e., a single-nucleotide substitution type but with nonequal base frequencies (Muse & Gaut 1994)], a difference between transitions and transversions [i.e., two nucleotide substitution types (Goldman & Yang 1994)], or allow all six substitution types (Halpern & Bruno 1998).

Another approach to deal with nonindependence of sites is use of hidden Markov models (Felsenstein 2001) to permit the autocorrelation of rates regionally. For some reason, the hidden Markov models have not been utilized extensively.

rRNA MODELS For ribosomal RNA (rRNA) genes, the primary transcript is the functional product. These rRNAs fold into a secondary structure in which some regions form pair-bonded stems and others form single-stranded loops. Substitutions in stem regions are constrained by the complementary nucleotide and compensatory changes (substitutions that maintain pair bonding) are well known. Models specific to rRNA have been developed (e.g., Smith et al. 2004, Tillier & Collins 1995) in which loop regions are treated as distinct from stem regions and the latter treated as hydrogen-bonded pairs, although these models are yet to be implemented in many phylogeny estimation packages [with the exception of MrBayes (Huelsenbeck & Ronquist 2001)]. Kjer (2004) used this model, coupled with a mixed-distribution model of among-site rate variation (the Doublet $+ \mathrm{I} + \Gamma$ model) in analysis of 18S rRNA among insects. The parameters of the doublet model include 16 doublet frequencies (which sum to 1 for 15 free parameters), 3 free base frequencies, 5 free transformation rates for loops (from the reversible 4×4 nucleotide matrix), 119 free transformation rates for stems (from the reversible 16×16 doublet matrix), a separate p_{inv} for stems and loops (2 parameters), and a gamma across all variable sites. Clearly, this model is extremely parameter-rich.

Partitioned Models

If one has natural partitions in ones data sets (e.g., codon positions, multiple genes, etc.), an intuitively appealing option is to apply different models to the various partitions. The simplest of these approaches are the site-specific rate (SSR) models (although they really should be called partition-specific rate models), and these models apply a separate, equal-rates, GTR to each partition. Because partitions often have very different nucleotide frequencies, the simple SSR models often improve the likelihood score considerably. However, this improvement in fit may not equate to improved phylogeny estimates, because other simpler models (nonnested) may better account for rate variation among sites (Buckley et al. 2001).

Alternatively, one may apply a full GTR $+ \mathrm{I} + \Gamma$ model to each partition (e.g., Castoe et al. 2004), and any of the parameters may be linked (apply across partitions) or unlinked (be partition specific). If one had, for example a 10-gene data set, from two genomes (nuclear and organellar), one could imagine an enormous array of potential, plausible partitioning schemes. Some way of evaluating the partitioned models is necessary to guide the choice.

MODEL SELECTION CRITERIA IN PHYLOGENTICS

Given that model choice is critical in phylogeny estimation and the vast array of potential models from which to choose, one is faced with the decision of how to select from among these. Obviously, the requirement is to select a model or models from the set available that account for processes that impinge on phylogeny estimation sufficiently well to avoid the biases discussed above without sacrificing the predictive power of the chosen model. Posada & Buckley (2004) recently published an excellent overview of model choice in systematics and focus on a justification for model averaging by use of AIC weights (see below).

Likelihood-Ratio Tests

By far, the most widely used method of choosing a model objectively is through use of LRTs. This approach takes advantage of two issues. First, the likelihood score can be interpreted as measuring the fit between model and data that is comparable across models. Second, the commonly used models in phylogeny estimation from DNA sequences are members of the GTR $+ I + \Gamma$ family (i.e., are special cases or submodels). Thus, one may evaluate the effect of including one or more parameters by calculating the likelihood of a model in which the parameter of interest is optimized versus a model in which it is fixed and comparing the likelihoods of the two models by use of the classical test statistic

$$\delta = 2(\ln L_1 - \ln L_0),$$

where $\ln L_1$ is the likelihood score of the more complex model. The test statistic is then typically evaluated under the assumption of asymptotic convergence to a χ^2 distribution; the degrees of freedom are the difference in number of free parameters in the two models.

This approach was first used in a hierarchical fashion (the hLRT) by Frati et al. (1997) and Sullivan et al. (1997), who selected a model for phylogeny estimation from among a set of 16 models. It was suggested independently by Huelsenbeck & Crandall (1997). Posada & Crandall (1998) hard-coded this approach in the production of their program Modeltest and expanded the set of candidate models examined to include 56 members of the family. The release of Modeltest had an enormously important impact on phylogenetics because it permitted many systematists to select good models in a nonarbitrary fashion.

A potential weakness of LRTs (Sanderson & Kim 2000) is that an initial estimate of topology, usually from either a parsimony search or a neighbor-joining tree, is required to conduct hLRTs. However, although model parameters are not as invariant across tree topologies as initially postulated, analyses of real data have shown that extremely poor estimates of model parameters are typically only derived from very poor trees (e.g., Sullivan et al. 1996). Similarly, Posada & Crandall (2001) demonstrated that use of initial trees has little effect on the model chosen by hLRTs.

Nevertheless, serious weaknesses remain in the use of hLRTs for model se-lection. One of these weaknesses is the requirement to traverse model space via a series of pairwise comparisons without relevant theory to guide the traversal. Model space can be represented by a decision tree (Posada & Crandall 1998), and the first choice one must make in applying hLRTs is where to start on this tree. One may start with the most general and parameter-rich model (typically GTR + I + Γ) and simplify by fixing the values of certain parameters (e.g., setting the pro-portion of invariable sites equal to zero). Conversely, one may start with the simplest model (JC) and add parameters (e.g., base frequencies) that are then optimized. Once the decision has been made as to which direction to follow (top down or bottom up) in traversing model space, one must decide the order in which to subtract or add parameters. This traversal may either be hard-coded, as is the case with Modeltest, or be done interactively. Swofford & Sullivan (2003) and Sullivan (2005) demonstrate the interactive approach to hLRTs, by starting with the most general model and subtracting parameters that appear closest to their fixed values in the simpler model. Not surprisingly, this approach often leads to selection of models that would never be examined in current hard-coded approaches.

Similarly, several authors have demonstrated that the manner in which the model space is traversed influences model choice (Cunningham et al. 1998, Felsen-stein 2004, Pol 2004). In the most extensive examination, Pol (2004) examined 32 different traversals for 18 data sets and found that mode of traversal influenced model selection in 15 of the 18 data sets and that the selected models differed by as many a 6 parameters (for one data set). He further demonstrated for two data sets that the ML tree was different under models selected by use of different traversal schemes (however, in both cases, trees only differed very slightly, by one or two nearest-neighbor interchanges). These problems in how best to imple-ment hLRTs arise because no relevant theory exists to guide traversal of model space.

In addition to these issues of implementation (as well as others; for example, multiple testing), several authors have pointed out that LRTs were not intended to be used to select from a series of models (e.g., Posada & Buckley 2004). Similarly, the hypothesis-testing approach inherent in hLRTs is poorly suited to model se-lection, and LRTs typically favor the complex model (e.g., Burnham & Anderson 2002). Thus, despite the extremely widespread use of hLRTs to select models for phylogenetics, and the enormous improvement that this approach has made to model-based phylogenetics, time has probably come to move to other alternatives, including some that have been developed recently.

Akaike Information Criterion

The Akaike information criterion (AIC) (Akaike 1973) is a simple measure with a complex derivation. The AIC for model i (AIC$_i$) is calculated as follows:

$$\text{AIC}_i = -2 \ln L_i + 2k_i,$$

where ln L_i is the maximum log-likelihood of the model (i.e., with joint ML estimates across parameters) and k_i is the number of parameters in model i. In addition, a modification to correct for small sample sizes (where small is defined as $n/k_i \leq 40$, and n is typically the number of sites), the AIC_c (Burnham & Anderson 2002, 2004) is given by the following:

$$AIC_{Ci} = -2 \ln L_i + 2k_i + \frac{2k_i(k_i + 1)}{n - k_i - 1}.$$

The simple interpretation of the AIC is that it provides a measure of fit between model and data $(-2 \ln L_i)$ and includes a penalty for overparameterization. Its first application to phylogenetics was by Hasegawa (1990), and the model favored is that model with the lowest AIC (or AIC_c). Ideally, one would find the ML topology and parameters for each model, but usually, some initial tree is used across all models. In other words, as typically applied, the AIC shares the reliance on an initial tree with hLRTs. However, Posada & Crandall (2001) demonstrated that this reliance has virtually no effect on the model chosen by comparing the AIC rankings based on the true tree (in simulated data) with the rankings based on an initial (NJ) tree. A similar conclusion was reached by Abdo et al. (2005), who compared the models selected by AIC calculated on an initial tree with those chosen by optimizing the tree under each model examined (i.e., on the ML tree for each model).

An obvious advantage of AIC over LRTs in model selection is that the AIC is calculated for each model in isolation, which eliminates the need to traverse model space by a series of pairwise comparisons. The AIC can, therefore, be used to compare nonnested models. Another advantage of the AIC is that it can allow for generation of a plausible set of models by computation of the Δ_i for each model as follows:

$$\Delta_i = AIC_i - AIC_{min},$$

where AIC_{min} is the score of the preferred model. These Δ_i values provide for evaluating the support in the data for each of the models that is examined (i.e., quantifying uncertainty in model selection). Burnham & Anderson (2002, 2004) provide the following benchmarks for discerning the relative support for alternative models: $\Delta_i \leq 2$ indicates substantial support, $4 \leq \Delta_i \leq 10$ indicates weak support, and $\Delta_i \geq 10$ indicates no support. Furthermore, these Δ_i values can be used to erect AIC weights for multimodel inferences (see below).

Although the interpretation of the AIC given above is sufficient to understand the properties of the AIC, the approach has a formal derivation from information theory. Suppose we have a distribution that has been generated by some true but unknown process. The AIC represents the Kullback-Leibler (K-L) distance between that model and the model being examined. The K-L distance can be thought of as quantifying the information lost by approximation to the true model. More details are provided in the online Supplemental Material of this review;

follow the Supplemental Material link from the Annual Reviews home page at http://www.annualreviews.org.

Bayesian Model Selection

BAYES FACTORS   In Bayesian comparison of two models, the Bayes factor permits direct evaluation of the support in the data for one model versus another (Kass & Raftery 1995). This support is calculated as by $B_{12} = \text{pr}(D|M_1)/\text{pr}(D|M_2)$, and it can be multiplied by the ratio of the prior probabilities of each model to give the posterior odds that favor one model. Thus, if the priors are uniform (i.e., the ratio of priors equals 1), the posterior odds take a similar form as the LRT, with the important difference that $\text{pr}(D|M_i)$ is calculated by integrating across the parameters of M_i rather than by fixing parameter values at the ML point estimates. Bayes factors, therefore, account for uncertainty in parameter estimation, unlike hLRTs. As with the Δ_i under the AIC, benchmarks are provided by Raftery (1996) to interpret relative support on the basis of the magnitude of the Bayes factor. When $B_{12} > 20$, support for M_1 is strong; when $3 \leq B_{12} \leq 20$, M_1 is slightly favored; and when $1 \leq B_{ij} < 3$, the two models are supported roughly equally by the data. Suchard et al. (2002) used Bayes factors to examine a nested subset of the GTR $+$ I $+$ Γ family and rejected the K2P and HKY models in favor of the Tamura-Nei model (Tamura & Nei 1993). However, unlike in the case of LRTs, Bayes factors are not restricted to comparisons of nested models. For example, Nylander et al. (2004) used Bayes factors to select from an array of partitioned models that included nonnested variants. Interestingly, simpler models were preferred over more complex models only in comparisons of nonnested models. In this example, because no penalty was imposed for overparameterization, Bayes factors always favored the more general of two nested modes. They also noted symptoms of nonidentifiability (diffuse and highly skewed marginal posterior distributions) of p_{inv} and the Γ-shape parameter in the smallest partitions. Sullivan et al. (1999) have demonstrated the correlation of error in these two parameters, and this error impedes their estimation with limited data and likely explains the issues of nonidentifiability seen by Nylander et al. (2004).

BAYESIAN INFORMATION CRITERION   An approximation of full Bayesian model evaluation was devised by Schwarz (1978): the Bayesian information criterion (BIC). In calculation, this quantity is similar to the AIC,

$$\text{BIC}_i = -2 \ln Li + k_i \ln n,$$

where k_i is the number of parameters in model i, $\ln L_i$ is the ML score (i.e., with all parameters fixed to their ML point estimates), and n is the sample size. As above, sample size is typically taken to be the number of nucleotide sites, but its appropriate interpretation in phylogenetics is not entirely clear. Again, a superficial characterization of the BIC is that it assesses fit via the ML score and penalizes overparameterization (more heavily than is the case for the AIC, especially with large n). Moreover, the BIC resists the tendency for model selection to favor more complex models as n increases.

Again, as typically employed, BIC_i values are calculated on an initial tree, rather than the ML tree, under the model M_i. Just as for the AIC, Posada & Crandall (2001) demonstrated by use of simulations that this approximation is quite good, and Abdo et al. (2005) demonstrated the same by actually calculating the BIC_i on the ML tree for all M_i.

Just as the AIC has a more rigorous statistical justification than simply assessing fit plus a penalty for overparameterization (i.e., minimizing the K-L distance), the model with the minimum BIC will be the same as the model with the highest posterior probability, $pr(M_i|D)$, at least if one assumes uniform priors across models and certain approximations are valid. This derivation is discussed in more detail in the Supplemental Material available online.

Performance-Based Model Selection

Minin et al. (2003) developed a model-selection approach that ranks models on the basis of the weighted expected error in branch-length estimates, with the weights are derived from the BIC. This method focuses on the fact that both the tree topology and the branch lengths (the rate of evolution × the time between each node or speciation event in the tree) are critical. If we assume momentarily that topology is known, we can focus attention on accurate branch-length estimates; rather than worry about whether a model is correct, the accuracy of the branch lengths estimated under various models can be used to assess model quality.

Because the method of Minin et al. (2003) (available in the program DT-ModSel) relies on decision theory (DT), we defer explanation of the details of the method to the Supplemental Material available online; that material focuses on the decision-theoretic foundations of all the model-selection criteria. However, a few points are worth noting here. First, accuracy in branch-length estimation is justified as a performance measure by the observation that the reason ML estimation can be inconsistent under some topological conditions under strongly violated models is because of the underestimation of long branches discussed above. Thus, models that are expected to estimate branch lengths similarly are expected to perform similarly in phylogeny estimation. Abdo et al. (2005) validated the assumption by using data simulated under very complex conditions. Second, because the approach uses BIC weights, it typically selects simpler models than does either hLRTs (Minin et al. 2003) or AIC (Abdo et al. 2005). Nevertheless, these simpler models produce estimates of branch length with less error (both absolute error and relative error) and produce phylogeny estimates at least as accurate as the complex models selected by hLRTs, AIC, and BIC (Abdo et al. 2005). Third, inclusion of several poor models in the set examined has no effect on model choice, because the poor models receive extremely low BIC weights (Abdo et al. 2005). Fourth, although the method uses an initial estimate of topology (as do the other methods), this approximation does not compromise model choice (Abdo et al. 2005).

Finally, the loss function need not focus on branch-length estimates; any feature of the analysis can be used to erect a loss function.

Tests of Model Adequacy

Given the increasing uses of both Bayesian and frequentist tests of evolutionary hypotheses on model-based phylogenies, the adequacy of models should be assessed in an absolute sense. That is, all the methods described above permit us to choose objectively one or more models from a preselected set, but, although we certainly do not anticipate that the selected model or models will be true, any statistical tests conducted by use of the selected model or models may be compromised if the best available alternative is nevertheless insufficient. Thus, an absolute test of model adequacy is critical (Sanderson & Kim 2000), and two have been used in phylogenetics.

PARAMETRIC BOOTSTRAP   The first test of the absolute goodness-of-fit between model and data in phylogeoentics was proposed by Goldman (1993) and is described in detail in Whelan et al. (2001). This test is a simulation-based test, and it uses as a test statistic the difference between the multinomial likelihood, which sets an upper bound on the likelihood for the data set under examination, and the ML achievable under that model. This difference measures the deterioration in fit associated with forcing all the data to conform to a single (albeit potentially heterogeneous-rates) model and a single tree. Replicate data sets are then simulated on the ML tree under the model being examined, with parameters fixed to their ML estimates, and the difference between multinomial likelihood and ML under the model is examined for each data set. This difference represents the expected difference under the null hypothesis of a perfect fit between model and data (simply due to stochasticity) because the model was used to generate the data. The distribution of this difference across replicates then becomes the null distribution to which the observed difference is compared.

In the first application of this test, Goldman (1993) evaluated the absolute fit of the simple equal-rates models available at the time and could reject them for real data sets. Similarly, Whelan et al. (2001) rejected the GTR model (without rate variation) for primate mtDNA by use of this test, and these results have led to the perception that current modes are inadequate (e.g., Sanderson & Kim 2000). However, a number of studies have applied the multinomial test of model adequacy to heterogeneous rates models, and in many of these studies (e.g., Carstens et al. 2004, Demboski & Sullivan 2003, Sullivan et al. 2000), the model selected by one of the selection methods could not be rejected in terms of absolute goodness-of-fit. Thus, despite the early conclusions, many conditions exists in which models chosen from among a pool of candidates appear to be adequate, at least as judged by these tests.

However, one limitation of this test is that it relies on point estimates of topology, branch lengths, and model parameters to simulate null distributions. This limitation has the effect of underrepresenting uncertainty in the simulations and may compromise the power of those tests. An analysis of error rates by use of this approach is currently lacking, and the effect of its reliance on point estimates is not known.

POSTERIOR PREDICTIVE SIMULATIONS   Huelsenbeck et al. (2001) and Bollback (2002) have circumvented the weakness of Goldman's test by making use of posterior predictive simulations. This approach uses Bayesian estimation under the model being examined to provide posterior probability distributions of topologies, branch lengths, and substitution-model parameters. Simulations are then conducted under the model under examination, and each replicate samples the tree, branch lengths, and parameter values from the marginal posterior distributions. The idea is that future data should be predictable under a good model, but future data do not exist. Therefore, future data are simulated under conditions selected from the marginal posterior distributions derived from Bayesian analysis of real data, and replicates, therefore, account for uncertainty in parameter estimation. The multinomial likelihood from the real data is used as the test statistic and it is compared with the distributions of multinomial likelihoods derived from the posterior predictive simulations.

Interestingly, Bollback (2002) examined one of the same data sets that Goldman (1993) examined: the primate $\psi\eta$-globin data set. Whereas Goldman (1993) rejected the JC model for this data set by use of the parametric bootstrap test of absolute goodness-of-fit, Bollback could not (the P value was 0.123). Bollback attributes this outcome to the uncertainty in model parameters, topology, and branch lengths and the fact that the posterior predictive simulations account for this uncertainty explicitly. Comparison of the two methods on a diversity of real data sets would be extremely useful (e.g., Foster 2004). A second interesting result from Bollback's analysis of that data set is that the PPS test suggested that the HKY model (four parameters) is a better fit than the more general GTR model (eight parameters).

INCORPORATING UNCERTAINTY IN MODEL SELECTION

Classical parameter estimation involves choosing the appropriate statistical model and then estimating the parameter in the context of that model. Typically one only accounts for error in the estimate assuming the particular model chosen but does not account for the error associated with the model choice. This approach produces bias in the estimates, and the standard error of estimates calculated with a single model underrepresents the true error in the estimates. Model averaging is a way to overcome these problems (Burnham & Anderson 2002); this technique involves assigning each model a certain weight, estimating the parameter of interest under each model, and then producing an average estimate that is weighted across models. In the phylogeny context, Posada & Buckley (2004) have advocated AIC weights (w_i). These weights are a function of Δ_i as defined above (Burnham & Anderson 2002, 2004), and a few examples of model-averaged phylogenies that use AIC weights are in the literature (e.g., Posada & Buckley 2004).

However, model averaging requires that one accepts that models can be viewed as random variables, and one assigns a probability distribution to each of the models given the data. From the perspective of statistical philosophy, this approach requires the Bayesian view of statistical inference. Under the Bayesian view, the only logically coherent way to weight each model is to assign each model a weight according to the posterior probability of the model given the data. Thus, one could make the argument that if one is willing to use model averaging as a legitimate statistical procedure, only Bayesian approaches make sense, although Burnham & Anderson (2004) provide a Bayesian interpretation of AIC weights. In particular, the posterior probability of a model is equivalent to the AIC weight [pr($M_i \mid D$) = w_i], when the prior probabilities across models assume a particular form (for the derivation, see Burnham & Anderson 2004). Therefore, model averaging by use of AIC weights can be viewed as ad hoc; that is, to be consistent with Bayesian statistics, one is required to assume particular priors across models.

An alternative approach to model averaging by use of AIC weights in phylogenetics is reversible-jump Markov chain Monte Carlo (Huelsenbeck et al. 2004, Nylander et al. 2004, Suchard et al. 2002). This approach includes proposals to change models randomly in the Markov chain Monte Carlo proposal mechanism. Because this approach does not require any particular form of the priors across models, it seems to us to be a theoretically more justifiable approach to model averaging than is the use of AIC weights. Alternatively, many researchers seem to take a pragmatic approach to statistics and use methods that can be shown to work well under a variety of relevant conditions. AIC weights may prove to work sufficiently well in model averaging in phylogenetics.

CONCLUSIONS

Phylogenetics is beginning to grapple with model-selection issues, just as have other disciplines. Although what will ultimately be viewed as optimal model selection may depend on whether one is willing to adopt a Bayesian statistical philosophy, the fact that all current approaches to model selection can be formalized as a loss function within a DT framework facilitates direct comparison of the various approaches (Table 1). Minimizing loss in the DT interpretation of LRTs is equivalent to minimizing type II error (for a fixed type I error). The loss function for the AIC is the K-L distance, that is, the information lost by use of an assumed model rather than the true model. In Bayesian model selection, if we assume uniform priors across models, a binary loss function is proportional to the inverse of the posterior probability of a model, given the data. In performance-based methods, a nonbinary loss function can be erected on the basis of any feature of an analysis that one deems important to method performance (such as expected branch-length error). The derivations of these methods in the decision-theory framework is provided in the Supplementary Material available online at http://www.annualreviews.org/. Of the methods examined here, all but LRTs can easily be incorporated into model averaging, either manually (e.g., Posada & Buckley 2004) or through incorporation into reversible-jump Markov chain Monte Carlo (e.g., Huelsenbeck et al. 2004). Given the increasing numbers of taxa in phylogenetics data sets and the advantages of using partitioned models (e.g., Castoe et al. 2004, Nylander et al.

TABLE 1   Model-selection approaches interpretable from the perspective of decision theory

Approach[a]	Loss	Decision rule	Philosophy	Comments
hLRT	Binary	Minimize type II error rate	Non-Bayesian	Assume a fixed type I error rate
AIC	Nonbinary	Miminize Kullback-Leibler distance	Non-Bayesian	Assume candidate models are close to true model; Taylor expansion approximation[b]
BIC	Binary	Maximize posterior probability	Bayesian	Assume uniform priors across models; Taylor expansion approximation[b]
Performance based	Nonbinary	Minimize risk based on any feature of analysis (e.g., branch-length error)	Bayesian	Performance-measure dependence; Taylor expansion approximation[b]

[a]The derivations for interpreting these approaches in this framework are presented in the Supplemental Material online at http://www.annualreviews.org/.

[b]The Taylor expansion approximation permits priors across model parameters to be ignored and evaluation of a model at its joint maximum-likelihood estimates (Raftery 1995).

2004), simply choosing the most complex model available may result in loss of predictive ability and nonidentifiability of model parameters, both a function of too few degrees of freedom. Simulation studies with extremely complex models of sequence evolution to generate data (e.g., Minin et al. 2003) are likely to be very fruitful in evaluating alternative model-selection and model-averaging strategies.

ACKNOWLEDGMENTS

We thank Z. Abdo, D. Althoff, K. Segraves, B. Shaffer, and D. Vanderpool for critiquing the manuscript and for their many helpful comments. This work is part of the University of Idaho Initiative in Bioinformatics and Evolutionary Studies (IBEST). Funding was provided by NSF EPS-0080935 (IBEST), NSF Systematic Biology DEB-9974124 (J.S.), NSF Probability and Statistics DMS-0072198 (P.J.), NSF EPS-0132626 (P.J.), NSF Population Biology DEB-0089756 (P.J.), and NIH NCCR 1P20PR016448-01 (IBEST: PI, L.J. Forney). Long-term interactions with several excellent scientists outside of IBEST have contributed to our thinking about model selection. They include T. Buckley, K. Crandall, V. Minin, D. Posada, C. Simon, and D. Swofford.

The *Annual Review of Ecology, Evolution, and Systematics* **is online at**
http://ecolsys.annualreviews.org

LITERATURE CITED

Abdo Z, Minin V, Joyce P, Sullivan J. 2005. Accounting for uncertainty in the tree topology has little effect on the decision theoretic approach to model selection in phylogeny estimation. *Mol. Biol. Evol.* 22:691–703

Akaike H. 1973. Information theory and an extension of the maximum likelihood principle. In *Second International Symposium on Information Theory*, ed. PN Petrov, F Csaki. pp. 267–81. Budapest: Akad. Kiado

Anderson FE, Swofford DL. 2004. Should we be worried about long-branch attraction in real data sets? Investigations using metazoan 18S rDNA. *Mol. Phylogenet. Evol.* 33:440–51

Bollback JP. 2002. Bayesian model adequacy and choice in phylogenetics. *Mol. Biol. Evol.* 19:1171–80

Box GEP. 1976. Science and statistics. *J. Am. Stat. Assoc.* 71:791–99

Brown W, Prager EM, Wang A, Wilson AC. 1982. Mitochondrial DNA sequences of primates. *J. Mol. Evol.* 18:225–39

Bruno WJ, Halpern AL. 1999. Topological bias and inconsistency of maximum likelihood using wrong models. *Mol. Biol. Evol.* 16:564–66

Buckley TR. 2002. Model misspecification and probabilistic tests of topology: evidence from empirical data sets. *Syst. Biol.* 51:509–23

Buckley TR, Cunningham CW. 2002. The effects of nucleotide substitution model assumptions on estimates of non-parametric bootstrap support. *Mol. Biol. Evol.* 19:394–405

Buckley TR, Simon C, Chambers GC. 2001. Exploring among-site rate variation models in a maximum likelihood framework using empirical data: effects of model assumptions on estimates of topology, branch lengths, and bootstrap support. *Syst. Biol.* 50:67–86

Burnham KP, Anderson DA. 2002. *Model Selection and Multimodel Inference: A Practical Information-Theoretic Approach*. New York: Springer-Verlag. 488 pp. 2nd ed.

Burnham KP, Anderson DA. 2004. Multimodel inference: understanding AIC and BIC in model selection. *Sociol. Method Res.* 33:261–304

Carstens BC, Stevenson AL, Degenhardt JD, Sullivan J. 2004. Testing nested phylogenetic and phylogeographic hypotheses in the *Plethodon vandykei* species group. *Syst. Biol.* 53:781–92

Castoe TA, Doan TM, Parkinson CL. 2004. Data partitions and complex models in Bayesian analysis: the phylogeny of gymnophthalmid lizards. *Syst. Biol.* 53:448–59

Cicero C, Johnson N. 2001. Phylogeny and character evolution in the Empidonax group of tyrant flycatchers (Aves: Tyrannidae): a test of W.E. Lanyon's hypothesis using mtDNA sequences. *Mol. Phylogenet. Evol.* 22:289–302

Cunningham CW, Zhu H, Hillis DM. 1998. Best-fit maximum likelihood models for phylogenetic inference: empirical tests with known phylogenies. *Evolution* 52:978–87

Demboski JR, Sullivan J. 2003. Extensive mtDNA variation within the yellow-pine chipmunk, *Tamias amoenus* (Rodentia: Sciuridae), and phylogeographic inferences for northwestern North America. *Mol. Phylogenet. Evol.* 26:389–408

Erixon P, Svennblad B, Britton T, Oxelman B. 2003. Reliability of Bayesian posterior probabilities and bootstrap frequencies in phylogenetics. *Syst. Biol.* 52:665–73

Felsenstein J. 1978. Cases in which parsimony and compatibility methods will be positively misleading. *Syst. Zool.* 27:401–10

Felsenstein J. 2001. Taking variation of evolutionary rates between sites into account in inferring phylogenies. *J. Mol. Evol.* 53:447–55

Felsenstein J. 2004. *Inferring Phylogenies.* Sunderland, MA: Sinauer. 664 pp.

Fisher RA. 1958. *Statistical Methods for Research Workers.* New York: Hafner. 239 pp. 13th ed.

Fitch WM, Markowitz E. 1970. An improved method for determining codon variability in a gene and its application to the rate of fixation of mutations in evolution. *Biochem. Genet.* 4:579–93

Foster PG. 2004. Modeling compositional heterogeneity. *Syst. Biol.* 53:485–95

Frati F, Simon C, Sullivan J, Swofford DL. 1997. Evolution of the mitochondrial COII gene in Collembola. *J. Mol. Evol.* 44:145–58

Galtier N. 2001. Maximum-likelihood phylogenetic analysis under a covarion-like model. *Mol. Biol. Evol.* 18:866–73

Galtier N, Guoy M. 1998. Inferring pattern and process: maximum-likelihood implementation of a nonhomogeneous model of DNA sequence evolution for phylogenetic analysis. *Mol. Biol. Evol.* 15:871–79

Gaut BS, Lewis PO. 1995. Success of maximum likelihood phylogeny inference in the four-taxon case. *Mol. Biol. Evol.* 12:152–62

Gillespie JH. 1986. Rates of molecular evolution. *Annu. Rev. Ecol. Syst.* 17:636–65

Golding GB. 1983. Estimates of DNA and protein sequence divergence: an examination of some assumptions. *Mol. Biol. Evol.* 1:125–42

Goldman N. 1993. Statistical tests of models of DNA substitution. *J. Mol. Evol.* 36:182–98

Goldman N, Yang Z. 1994. A codon-based model of nucleotide substitution for protein-coding DNA sequences. *Mol. Biol. Evol.* 11:511–23

Gu X, Fu YX, Li WH. 1995. Maximum likelihood estimation of the heterogeneity of substitution rate among nucleotide sites. *Mol. Biol. Evol.* 12:546–57

Halpern A, Bruno WJ. 1998. Evolutionary distances for protein-coding sequences: modeling site-specific residue frequencies. *Mol. Biol. Evol.* 15:910–17

Han HY, Ro KE. 2005. Molecular phylogeny of the superfamily Tephritoidea (Insecta: Diptera): new evidence from the mitochondrial 12S, 16S, and COII genes. *Mol. Phylogenet. Evol.* 34:416–30

Hasegawa M. 1990. Phylogeny and molecular evolution of primates. *Jpn. J. Genet.* 65:243–65

Hasegawa M, Kishino H, Yano T. 1985. Dating the human-ape split by a molecular clock of mitochondrial DNA. *J. Mol. Evol.* 22:160–74

Hueslenbeck JP. 2002. Testing a covariotide model of DNA substitution. *Mol. Biol. Evol.* 19:698–707

Huelsenbeck JP, Crandall KA. 1997. Phylogeny estimation and hypothesis testing using maximum likelihood. *Annu. Rev. Ecol. Syst.* 28:437–66

Huelsenbeck JP, Hillis DM. 1993. Success of phylogenetic methods in the four-taxon case. *Syst. Biol.* 42:247–64

Huelsenbeck JP, Rannala B. 2004. Frequentist properties of Bayesian posterior probabilities of phylogenetic trees under simple and complex substitution models. *Syst. Biol.* 904–13

Huelsenbeck JP, Ronquist F. 2001. MR-BAYES: Bayesian inference of phylogeny. *Bioinformatics* 17:754–55

Huelsenbeck JP, Ronquist F, Nielsen R, Bollback JP. 2001. Bayesian inference of phylogeny and its impact on evolutionary biology. *Science* 294:2310–14

Huelsenbeck JP, Larget B, Alfaro M. 2004. Bayesian phylogenetic model selection using reversible jump Markov chain Monte Carlo. *Mol. Biol. Evol.* 21:1123–33

Johnson JP, Omland KS. 2004. Model selection in ecology and evolution. *Trends Ecol. Evol.* 19:101–08

Jukes TH, Cantor CR. 1969. Evolution of protein molecules. In *Mammalian Protein Metabolism*, ed. N Munro, pp. 21–132. New York: Academic

Kass RE, Raftery AE. 1995. Bayes factors. *J. Am. Stat. Assoc.* 90:773–95

Kimura M. 1980. A simple model for estimating evolutionary rates of base substitutions between homologous nucleotide sequences. *J. Mol. Evol.* 16:111–20

Kjer K. 2004. Aligned 18S and insect phylogeny. *Syst. Biol.* 53:506–14

Lemmon AR, Moriarty EC. 2004. The importance of proper model assumption in Bayesian phylogenetics. *Syst. Biol.* 53:265–77

López-Fernández H, Honeycutt RL, Winemiller KO. 2005. Molecular phylogeny and evidence for an adaptive radiation of geophagine cichlids from South America (Perciformes: Labroidei). *Mol. Phylogenet. Evol.* 34:227–44

Metzker ML, Mindell DP, Liu X, Ptak RG, Gibbs RA, Hillis DM. 2002. Molecular evidence of HIV-1 transmission in a criminal case. *Proc. Natl. Acad. Sci. USA* 99:14293–97

Minin V, Abdo Z, Joyce P, Sullivan J. 2003. Performance-based selection of likelihood models for phylogeny estimation. *Syst. Biol.* 52:674–83

Muse SV, Gaut BS. 1994. A likelihood approach for comparing synonymous and nonsynonymous nucleotide substitutions rates, with application to the chloroplast genome. *Mol. Biol. Evol.* 11:1139–51

Nylander JAA, Ronquist F, Huelsenbeck JPP, Nieves-Aldrey JL. 2004. Bayesian phylogenetic analysis of combined data. *Syst. Biol.* 53:47–67

Pol D. 2004. Empirical problems of the hierarchical likelihood ratio test for model selection. *Syst. Biol.* 53:949–62

Posada D, Buckley TR. 2004. Model selection and model averaging in phylogenetics: advantages of Akaike information criterion and Bayesian approaches over likelihood ratio tests. *Syst. Biol.* 53:793–808

Posada D, Crandall KA. 1998. Modeltest: testing the model of DNA substitution. *Bioinformatics* 14:817–18

Posada D, Crandall KA. 2001. Selecting the best-fit model of nucleotide substitution. *Syst. Biol.* 50:580–601

Raftery AE. 1995. Bayesian model selection in social research (with discussion by A Gelman, DB Rubin, and RM Hauser). In *Sociological Methodology*, ed. PV Marsden, pp. 111–96. Oxford, UK: Blackwell Sci.

Raftery AE. 1996. Hypothesis testing and model selection. In *Markov Chain Monte Carlo in Practice*, ed. WR Gilks, S Richardson, DJ Speigelhalter, pp. 163–87. New York: Chapman & Hall

Rannala B. 2002. Identifiability of parameters in MCMC Bayesian inference of phylogeny. *Syst. Biol.* 51:754–60

Sanderson MJ, Kim J. 2000. Parametric phylogenetics? *Syst. Biol.* 49:817–29

Schwarz G. 1978. Estimating the dimensions of a model. *Ann. Stat.* 6:461–64

Siddall ME. 1998. Success of parsimony in the four-taxon case: long-branch repulsion by likelihood in the Farris zone. *Cladistics* 14:209–20

Siddall ME, Kluge AG. 1997. Probabilism and phylogenetic inference. *Cladistics* 13:313–36

Smith AD, Lui TWH, Tillier ERM. 2004. Empirical models for substitution in ribosomal RNA. *Mol. Biol. Evol.* 21:419–27

Suchard MA, Weiss RE, Sinsheimer JS. 2002. Bayesian selection of continuous-time Markov chain evolutionary models. *Mol. Biol. Evol.* 18:1001–13

Sullivan J. 2005. Maximum-likelihood estimation of phylogeny from DNA sequence data. In *Molecular Evolution, Producing the Biochemical Data. Part B. Methods in Enzymology*, ed. E Zimmer, E Roalson. In press

Sullivan J, Swofford DL. 1997. Are guinea pigs rodents? The importance of adequate models in molecular phylogenetics. *J. Mammal. Evol.* 4:77–86

Sullivan J, Swofford DL. 2001. Should we use model-based methods for phylogenetic inference when we know assumptions about among-site rate variation and nucleotide substitution pattern are violated? *Syst. Biol.* 50:723–29

Sullivan J, Arellano EA, Rogers DS. 2000.

Comparative phylogeography of Mesoamerican highland rodents: concerted versus independent responses to past climatic fluctuations. *Am. Nat.* 155:755–68

Sullivan J, Holsinger KE, Simon C. 1996. The effect of topology on estimates of among-site rate variation. *J. Mol. Evol.* 42:308–12

Sullivan J, Markert JA, Kilpatrick CW. 1997. Phylogeography and molecular systematics of the *Peromyscus aztecus* species group (Rodentia: Muridae) inferred using parsimony and likelihood. *Syst. Biol.* 46:426–40

Sullivan J, Swofford DL, Naylor GJP. 1999. The effect of taxon sampling on estimating rate heterogeneity parameters of maximum-likelihood models. *Mol. Biol. Evol.* 16:1347–56

Swofford DL. 1998. *PAUP*: phylogenetic analysis using parsimony (*and other methods)*. Version 4.0b3a. Sunderland, MA: Sinauer Assoc. CD-ROM

Swofford DL, Sullivan J. 2003. Phylogenetic inference using parsimony and maximum likelihood using PAUP*. In *The Phylogenetic Handbook: A Practical Approach to DNA and Protein Phylogeny*, ed. M Salemi, AM Vandamme, pp. 160–96. Cambridge, UK: Cambridge Univ. Press

Swofford DL, Olsen GJ, Waddell PJ, Hillis DM. 1996. Phylogenetic inference. In *Molecular Systematics*, ed. DM Hillis, C Moritz, BK Mable, pp. 407–514. Sunderland, MA: Sinauer Assoc. 2nd ed

Swofford DL, Waddell PJ, Huelsenbeck JP, Foster PG, Lewis PO, Rogers JS. 2001. Bias in phylogenetic estimation and its relevance to the choice between parsimony and likelihood methods. *Syst. Biol.* 50:525–39

Tamura K, Nei M. 1993. Estimation of the number of nucleotides substitutions in the control region of mitochondrial DNA in humans and chimpanzees. *Mol. Biol. Evol.* 10:512–26

Tavaré S. 1986. Some probabilistic and statistical problems in the analysis of DNA sequences. *Lect. Math. Life Sci.* 17:57–86

Tillier ERM, Collins RA. 1995. Neighbor joining and maximum likelihood with RNA sequences: addressing the interdependence of sites. *Mol. Biol. Evol.* 12:7–15

Tuffley C, Steele M. 1998. Modeling the covarion hypothesis of nucleotide substitution. *Math. Biosci.* 147:63–91

Uzzell T, Corbin KW. 1971. Fitting discrete probability distributions to evolutionary events. *Science* 172:1089–96

Waddell P, Penny D. 1996. Evolutionary trees of apes and humans from DNA sequences. In *Handbook of Symbolic Evolution*, ed. AJ Lock, CR Peters, pp. 53–73. Oxford: Clarendon

Whelan S, Lio P, Goldman N. 2001. Molecular phylogenetics: state of the art methods for looking into the past. *Trends Genet.* 17:262–72

Worobey M, Santiago ML, Keele BF, Ndjango JBN, Joy JB, Labamall BL, et al. 2004. Origin of AIDS: contaminated polio vaccine theory refuted. *Nature* 428:820

Yang Z. 1993. Maximum-likelihood estimation of phylogeny from DNA sequences when substitution rates differ over sites. *Mol. Biol. Evol.* 10:1396–1401

Yang Z. 1994. Estimating the pattern of nucleotide substitution. *J. Mol. Evol.* 39:105–11

Yang Z. 1997. How often do wrong models produce better phylogenies? *Mol. Biol. Evol.* 14:105–08

Yang Z, Roberts D. 1995. On the use of nucleic acid sequences to infer early branchings in the tree of life. *Mol. Biol. Evol.* 12:451–58

Annu. Rev. Ecol. Evol. Syst. 2005. 36:467–97
doi: 10.1146/annurev.ecolsys.36.102403.115320
Copyright © 2005 by Annual Reviews. All rights reserved
First published online as a Review in Advance on August 17, 2005

POLLEN LIMITATION OF PLANT REPRODUCTION:
Pattern and Process

Tiffany M. Knight,[1] Janette A. Steets,[2] Jana C. Vamosi,[3]
Susan J. Mazer,[4] Martin Burd,[5] Diane R. Campbell,[6]
Michele R. Dudash,[7] Mark O. Johnston,[8]
Randall J. Mitchell,[9] and Tia-Lynn Ashman[2]

[1]National Center for Ecological Analysis and Synthesis, Santa Barbara,
California 93101; email: tknight@biology2.wustl.edu
[2]Department of Biological Sciences, University of Pittsburgh, Pittsburgh,
Pennsylvania 15260; email: jsteets@pitt.edu, tia1@pitt.edu
[3]Department of Biological Sciences, University of Calgary, Calgary, Alberta, Canada
T2N 1N4; email: jvamosi@ucalgary.ca
[4]Department of Ecology, Evolution and Marine Biology, University of California,
Santa Barbara, California 93106; email: mazer@lifesci.ucsb.edu
[5]School of Biological Sciences, Monash University, Melbourne, Victoria 3800, Australia;
email: martin.burd@sci.monash.edu.au
[6]Department of Ecology and Evolutionary Biology, University of California, Irvine,
California 92697; email: drcampbe@uci.edu
[7]Department of Biology, University of Maryland, College Park, Maryland 20742;
email: mdudash@umd.edu
[8]Department of Biology, Dalhousie University, Halifax, Nova Scotia, Canada B3H 4J1;
email: mark.johnston@dal.ca
[9]Department of Biology, University of Akron, Akron, Ohio 44325;
email: rjm2@uakron.edu

Key Words breeding system, floral traits, meta-analysis, phylogenetically independent contrasts, pollination, resource limitation

■ **Abstract** Quantifying the extent to which seed production is limited by the availability of pollen has been an area of intensive empirical study over the past few decades. Whereas theory predicts that pollen augmentation should not increase seed production, numerous empirical studies report significant and strong pollen limitation. Here, we use a variety of approaches to examine the correlates of pollen limitation in an effort to understand its occurrence and importance in plant evolutionary ecology. In particular, we examine the role of recent ecological perturbations in influencing pollen limitation and discuss the relation between pollen limitation and plant traits. We find that the magnitude of pollen limitation observed in natural populations depends on both historical constraints and contemporary ecological factors.

INTRODUCTION

Pollen limitation occurs when plants produce fewer fruits and/or seeds than they would with adequate pollen receipt. Inadequate pollination can in turn affect plant abundance and population viability and cause selection on plant mating system and floral traits (e.g., Ashman et al. 2004; Johnston 1991a,b; Lennartsson 2002; Lloyd & Schoen 1992). Thus pollen limitation has attracted considerable attention from both ecologists and evolutionary biologists (Ashman et al. 2004, Burd 1994, Larson & Barrett 2000). Given that floral phenotype can affect pollen receipt (e.g., aerodynamic morphology for abiotic pollination and attraction and rewards for biotic pollination), selection for floral traits may be particularly strong in pollen-limited populations, and thus pollen limitation may play an important role in the evolution of secondary sexual traits (Ashman & Morgan 2004). Furthermore, there is concern that pollinators are declining in many habitats, which could lead to widespread pollen limitation and a global pollination crisis (e.g., Buchmann & Nabhan 1996), affecting not only the sustainability of plant populations but also that of the organisms that either directly or indirectly rely on them. Even ecosystem services provided to humans by plants and pollinators, such as pollination of crops, may be at risk (Buchmann & Nabhan 1996, Kremen et al. 2002, Kremen & Ricketts 2000, Ricketts et al. 2004).

To test for pollen limitation in natural populations, researchers often conduct pollen supplementation experiments in which they compare the reproductive success of control plants with that of plants given supplemental pollen. If plants (or inflorescences or flowers) produce more fruits or seeds when supplemented, then it is usually concluded that reproduction is limited by pollen receipt (e.g., Ashman et al. 2004, Bierzychudek 1981).

Sexual selection theory assumes that female reproductive success is limited by resources rather than by access to mates (receipt of pollen) (Bateman 1948, Janzen 1977, Willson & Burley 1983, Wilson et al. 1994). In addition, one set of predictions, based on optimality theory, suggests that plants should evolve a level of attraction in which the benefits of attraction balance the costs of seed maturation (Haig & Westoby 1988). Both frameworks suggest that pollen addition should not increase fruit or seed set in populations at evolutionary equilibrium because resources should be unavailable for maturation of the additional fertilized ovules. However, empirical tests indicate the contrary: Pollen insufficiency often limits seed production, sometimes severely (reviewed in Ashman et al. 2004, Burd 1994, Larson & Barrett 2000).

There are several possible explanations for the apparent prevalence and strength of pollen limitation. First, recent ecological perturbations may disrupt coevolved interactions between plants and pollinators leading to pollen limitation. Second, pollen limitation may represent an evolutionary equilibrium in a stochastic environment. Finally, plant traits that promote outcrossing may evolve even when pollen limitation increases as a correlated response to selection on these traits.

We begin our consideration of these possible explanations with a brief introduction to the theoretical frameworks proposed to explain the causes and consequences of pollen limitation. Second, we discuss the evidence that ecological perturbations may contribute to pollen limitation. Third, we provide an overview of the extent and distribution of pollen limitation within and among plant species using quantitative meta-analysis (Gurevitch et al. 2001) on the most extensive data set on pollen limitation available. And finally, we use comparative approaches that account for phylogeny to explore relationships between plant traits and the magnitude of pollen limitation.

THEORETICAL BACKGROUND

An optimal plant in a constant environment should allocate resources perfectly between plant attraction and seed provisioning to ensure that enough pollen arrives to fertilize ovules that will mature to seed (Haig & Westoby 1988). As a result, pollen supplementation experiments should not, at least on average, enhance seed production. This is because plants exhibiting the optimal strategy do not have extra resources available for maturation of ovules fertilized by supplemental pollen. However, plants in recently degraded habitats may be pollen limited because they have not had enough time to evolve to their new optimal level of allocation. In a section below, we examine evidence for increased pollen limitation following ecological perturbations.

Even if plants are at an evolved equilibrium, pollen limitation might be expected if there is intraplant variation in pollen receipt. In a theoretical treatment, Burd (1995) considered whether plants should package more ovules within each flower than resources would allow the plant to mature into seeds should they all be fertilized. Such overproduction of ovules allows a plant to capitalize on particularly plentiful pollination of some flowers. Because most plants package ovules in multiple flowers or inflorescences, there is an economy of scale within packages, where per-ovule costs decrease as ovule number increases (Thomson 1989). The model assumes that individual plants can direct resources toward fertilized ovules and can, therefore, take advantage of occasional abundant pollen receipt. Higher flower-to-flower variance in pollen receipt (which increases the frequency of occasional, abundant pollination), greater non-ovule flower costs (e.g., pedicel, corolla), and cheaper ovules all favor excessive ovule production. We expect to see greater pollen limitation in plants that have high ovule numbers per flower and high interflower variation in pollen receipt.

The presence of pollen limitation reduces seed production and, therefore, may have a variety of demographic consequences. Reductions in seed production may have effects on the overall population size, particularly in short-lived species without a seed bank, in species that respond to disturbances by re-establishing from seeds (Ashman et al. 2004, Bond 1994). As landscapes become increasingly fragmented, the persistence of plant populations may depend on the abilities of their pollinators to move between patches (Amarasekare 2004). Pollen limitation may also play a role in the coexistence among plant species within a community (Feldman et al. 2004, Ishii & Higashi 2001). It is well know that trade-offs among plants in performance traits can facilitate coexistence; therefore, plants that trade-off in their ability to compete for resources and attract pollinators might be better able to coexist. However, to date, the role of pollen limitation on plant community composition has received little theoretical and empirical attention.

Persistent pollen limitation has evolutionary consequences; most notably, pollen limitation may favor the evolution of self-compatibility and/or increased selfing when selfing offers reproductive assurance. Pollen discounting and inbreeding depression are expected to counteract any selection pressures to increase the selfing rate; however, pollen discounting may be reduced when plants are pollen limited, which would promote the evolution of increased selfing (Porcher & Lande 2005). Evolution of self-compatibility and/or increased selfing has been suggested for plants in recently established habitats (as in Baker's law; Pannell & Barrett 1998), for plants occurring at low densities (Morgan et al. 2005), and for plants in variable pollination environments (Morgan & Wilson 2005). Furthermore, pollen limitation has been proposed as a mechanism by which androdioecy can evolve from dioecy; hermaphrodites suffer less pollen limitation than female plants (Liston et al. 1990, Maurice & Fleming 1995, Wolf & Takebayashi 2004). Also, it has been suggested that plants evolve to have only partial expression of self-incompatibility as an adaptation to avoid pollen limitation (Vallejo-Marín & Uyenoyama 2004). In all, plants that are hermaphroditic and self-compatible are expected to have lower levels of pollen limitation than those that are obligate outcrossers. In the evolutionary consequences of pollen limitation section below, we examine the evidence for this from pollen supplementation experiments.

Likewise, wind pollination should reduce the reliance of plants on pollinators and as such may evolve when plants are pollen limited; this possibility has been discussed in detail for dioecious species. Dioecious species are often sexually dimorphic (females typically inconspicuous relative to males), and during years of low pollinator abundance, this makes female reproductive success particularly prone to pollen limitation (Vamosi & Otto 2002). While the order of origin of dioecy and of wind pollination is a matter of debate, it may be that dioecious species that persist evolve mechanisms, such as wind pollination, that reduce reliance on pollinators. A related argument notes that separation of genders in dioecious species increases variance in reproductive success (i.e., pollen limitation) (Wilson & Harder 2003) and leads to selection for variance-reducing mechanisms such as high pollen or seed dispersal.

Pollen limitation may cause stronger selection on attractive traits that enhance reliability of pollinator visits (Ashman & Diefenderfer 2001; Johnston 1991a,b; Wilson et al. 1994). Similarly, pollen limitation may select for changes in floral shape; zygomorphic flowers attract more specialized pollinators than actinomorphic flowers, and these pollinators may provide more reliable pollination (Neal et al. 1998). Thus we expect that plants that make larger and more specialized flowers may have lower levels of pollen limitation than those that make smaller actinomorphic flowers.

ECOLOGICAL PERTURBATIONS AND POLLEN LIMITATION

Changes in the abiotic or biotic features of the habitat and changes in the range of plants and/or pollinators can disrupt plant-pollinator interactions (Bond 1994, Kearns et al. 1998, Wilcock & Neiland 2002). Because plants in recently perturbed environments will not have made evolutionary adjustments, these plants may show unusually high levels of pollen limitation.

Types of Ecological Perturbations

Here, we briefly discuss several types of ecological perturbations, the mechanisms by which they alter the magnitude of pollen limitation, and the evidence of their occurrence from pollen supplementation experiments (Table 1).

PRESENCE OF CO-FLOWERING PLANT SPECIES   Changes in abiotic or biotic conditions may change the identity or abundance of co-flowering plant species, and the presence of co-flowering species can reduce or increase pollination success of a focal species (Table 1). Facilitation may be more likely when plant species offering little reward are surrounded by rewarding plant species (Johnson et al. 2003, Laverty 1992), whereas competition for pollinators may be more likely between equally rewarding species.

DECREASES IN PLANT POPULATION SIZE/DENSITY   Many ecological perturbations such as habitat fragmentation (Cunningham 2000), human harvesting (Hackney & McGraw 2001), and increases in herbivore abundance (Vàzquez & Simberloff 2004) reduce plant abundance. Decreases in plant density are generally assumed to decrease pollination success and increase the magnitude of pollen limitation of both animal- and wind-pollinated plant species (Table 1). In addition, reductions in plant density may affect offspring quality. If pollinators visit more flowers on the same plant when density is low, this will increase the level of inbreeding for self-compatible plants (e.g., Karron et al. 1995). Alternatively, increases in the interplant distance traveled by pollinators may increase the outcrossing rate if plants that are further away are also less related (Lu 2000, Schaal 1978).

POLLINATOR LOSS   Ecological perturbations that displace pollinators may result in pollen limitation in plants that utilize those pollinators. However, there are few studies demonstrating that the loss of a pollinator results in increased pollen limitation (Table 1). It may be that such cases occur but have not been documented

TABLE 1 For each type of ecological perturbation, we provide the predicted consequences for pollen limitation (PL), an explanation of the mechanism, and supporting studies

Ecological perturbation	Predicted consequence for pollen limitation	Explanation	Empirical support
Presence of other plant species	$PL_{\text{coflowering species}} > PL_{\text{single flowering species}}$	Co-flowering results in pollinator competition, increased heterospecific pollen delivery, and/or stigma clogging by heterospecific pollen[a]	Campbell 1985, Gross 1996, Gross & Werner 1983
	$PL_{\text{coflowering species}} < PL_{\text{single flowering species}}$	Co-flowering results in increased pollinator attraction[b]	Moeller 2004
Plant population size/density	$PL_{\text{in small populations}} > PL_{\text{in large populations}}$	Small populations have reduced pollinator visitation, pollen deposition[c], ratio of conspecific to heterospecific pollen delivered[d] and more intraplant pollinator visits[e]	Ågren 1996, Davis et al. 2004, Forsyth 2003, Knight 2003, Kunin 1997, Moeller 2004, Sih & Balthus 1987, Waites & Ågren 2004, Ward & Johnson 2005
Pollinator loss	$PL_{\text{with fewer pollinators}} > PL_{\text{with more pollinators}}$	Pollinator visitation rate increases with pollinator abundance and diversity[f]	Liu & Koptur 2003
Resource additions	$PL_{\text{in resource rich habitat}} > PL_{\text{in resource poor habitat}}$	Seed production depends solely on pollen receipt when resources are unlimited	Galen et al. 1985
Habitat size and isolation	$PL_{\text{in fragmented habitat}} > PL_{\text{in continuous landscape}}$	Habitat fragmentation reduces the abundance of plants and/or pollinators, alters pollinator composition[g]	Cunningham 2000, Groom 2001, Johnson et al. 2004, Moody-Weis & Heywood 2001, Steffan-Dewenter & Tscharntke 1999, Wolf & Harrison 2001

(Continued)

TABLE 1   (*Continued*)

Ecological perturbation	Predicted consequence for pollen limitation	Explanation	Empirical support
Plant enemies (herbivores, pathogens)	$PL_{\text{with high enemy abundance}} >$ $PL_{\text{with low enemy abundance}}$	Enemies decrease pollinator attraction and pollinator visitation[h]	None
	$PL_{\text{with high enemy abundance}}$ $< PL_{\text{with low enemy abundance}}$	Enemies decrease plant resource status; plants become more limited by resources than pollen	Parker 1987
Plant mutualists (mycorrhizal fungi)	$PL_{\text{with high mutualist abundance}}$ $< PL_{\text{with low mutualist abundance}}$	The presence of soil mutualists facilitates plant resource acquisition, pollinator visitation and seed set[i]	None
Pollinator predators	$PL_{\text{with pollinator predators}} >$ $PL_{\text{without predators}}$	Predators reduce pollinator abundance and visitation rate[j]	Knight et al. 2005
Non-native plant species	$PL_{\text{non-native plants}} >$ $PL_{\text{native plants}}$	Non-native plants lack effective pollinators[k]	None
	$PL_{\text{non-native plants}} <$ $PL_{\text{nativeplants}}$	Non-native plants have a higher frequency of autogamy[l]	None
Non-native pollinators	$PL_{\text{with non-native pollinator}} >$ $PL_{\text{without non-native pollinator}}$	Non-native pollinators compete with native pollinators, and are less efficient pollinators of crops and wild plants[m]	None

[a]Campbell & Motten 1985; Caruso 1999, 2001; Galen & Gregory 1989; Waser 1983.

[b]Moeller 2005, Rathcke 1983.

[c]Fausto et al. 2001, Feinsinger et al. 1991, Regal 1982, Whitehead 1983.

[d]Caruso 2002, Kunin 1993.

[e]Franceschinelli & Bawa 2000, Iwaizumi & Sakai 2004, Klinkhamer & de Jong 1990, Mustajärvi et al. 2001.

[f]Buchmann & Nabhan 1996, Kearns et al. 1998, Thomson 2001.

[g]Jennersten 1988, Linhart & Feinsinger 1980.

[h]Mothershead & Marquis 2000; Steets & Ashman 2004; Strauss et al. 1996; Irwin, Brody & Waser 2001.

[i]Wolfe et al. 2005.

[j]Dukas 2001, Dukas & Morse 2003, Muñoz & Arroyo 2004, Suttle 2003.

[k]Parker 1997.

[l]Rambuda & Johnson 2004.

[m]O'Toole 1993, Paini 2004, Paton 1993, Sugden & Pyke 1991.

or that the majority of plants are buffered against the loss of any single pollinator species (Morris 2003) because of generalized pollination systems (Waser et al. 1996), the presence of other functionally redundant pollination species (Balvanera et al. 2005, Fenster et al. 2004), and/or autogamous pollination.

RESOURCE AVAILABILITY   Several perturbations may rapidly increase or decrease the resources available to plants (i.e., agricultural runoff, competition with non-native plant species). Based on theoretical predictions (Haig & Westoby 1988), increases in resource availability should cause plant reproductive output to become more limited by pollen receipt than by resources. Although several studies have factorially manipulated pollen and resources (i.e., water, nutrients, light), only one study reported a significant interaction between pollination and resource treatments in the expected direction (Table 1). This may reflect the abilities of plants to plastically respond to resource additions in their attraction traits or could suggest that factors in addition to resource availability and pollen receipt are influencing seed production (Campbell & Halama 1993).

HABITAT FRAGMENTATION   Habitat fragmentation simultaneously affects a variety of abiotic and biotic factors, which directly or indirectly change the abundance and composition of plants and pollinators, and may lead to increased pollen limitation (Table 1). The effect of fragmentation on the magnitude of pollen limitation will depend on the extent of habitat loss and isolation. Pollinators with limited movement may be lost from fragmented habitats unless corridors are present (Townsend & Levey 2005). Furthermore, self-incompatible plants with specialized pollinators may be particularly prone to pollen limitation following habitat fragmentation (Bond 1994), although there are not enough studies of pollination within fragmented landscapes to make generalizations at this time (Aizen et al. 2002).

OTHER INTERACTING SPECIES   Changes in the abundance of plant enemies (e.g., herbivores, pathogens, nectar robbers), plant mutualists (e.g., mycorrhizal fungi), and pollinator enemies (predators) may all affect interactions between plants and pollinators and thus the magnitude of pollen limitation (Table 1). Whereas many studies have factorially manipulated plant enemies and pollen receipt (Bertness & Shumway 1992, Garcia & Ehrlen 2002, Holland 2002, Juenger & Bergelson 1997, Krupnick & Weis 1999, Lehtila & Syrjanen 1995, Mizui & Kikuzawa 1991, Mothershead & Marquis 2000), only Parker (1987) found a significant interaction; enemies decreased the magnitude of pollen limitation. Other interacting species in the community have received much less attention and should be the target of future empirical work.

INTRODUCED PLANTS   To date it is unclear what role pollen limitation has in the establishment and spread of non-native plants (Richardson et al. 2000). Plant ecologists have argued that introduced plants may be more or less likely to suffer pollen limitation (Table 1). Studies comparing the breeding system and magnitude of pollen limitation of native and non-native plant species would be particularly informative, especially if phylogeny could be accounted for (i.e., by comparing related species). Once established, non-native plants may have consequences for the pollination of native species (Brown et al. 2002, Grabas & Laverty 1999, Moragues & Traveset 2005).

INTRODUCED POLLINATORS   Pollinators, and in particular honeybees (*Apis mellifera*), have been introduced widely outside of their native range (Hanley & Goulson 2003). The presence of non-native pollinators is expected to increase the magnitude of pollen limitation in native plants (Table 1). However, no study to date has demonstrated this.

Population-Level Effects

Although we have shown that a variety of ecological perturbations can alter the magnitude of pollen limitation among plant species, plant species will not be equally responsive. We expect that plant species that are pollinated by specialists, those that are non-rewarding, and those that are self-incompatible or dioecious should be particularly vulnerable to ecological perturbations leading to increased levels of pollen limitation (Bond 1994). In addition, even if pollen limitation does occur, this will not necessarily have large effects on the abundance and viability of plant populations (Ashman et al. 2004). Plants that are long-lived or capable of asexual reproduction are less vulnerable to pollination-driven extinction (Bond 1994, Johnson et al. 2004). Few studies have examined whether plant traits related to pollination biology can help to explain large-scale patterns of rarity in plants; those that have, however, provide intriguing results (Neiland & Wilcock 1998, Vamosi & Vamosi 2005a, Wilcock & Neiland 1998). For example, non-rewarding orchid species are rarer than their rewarding counterparts (Neiland & Wilcock 1998), and dioecious families were more often listed as threatened or endangered than their hermaphrodite sister taxa (Vamosi & Vamosi 2005a).

PATTERNS OF POLLEN LIMITATION

Here, we describe the pattern of pollen limitation documented from pollen supplementation experiments conducted over the past 25 years. In particular, we ask the following questions:

1. Which components of seed production (e.g., fruit set, seeds per plant) are most commonly studied, and do these components show similar patterns?

2. What is the prevalence and magnitude of pollen limitation?

3. How much of the variation in the magnitude of pollen limitation occurs across plant taxa, across populations within a species, and across individuals within a population?

To describe patterns in pollen limitation within and across plant species, we created a data set from published and unpublished pollen supplementation experiments. We searched ISI's *Web of Science* and *Biological Abstracts* for the publication years 1981–2003 using the key words pollen limit*, supplement* poll*, and hand poll*. We also solicited unpublished data, which were generously provided by Chris Ivey, Susan Kephart, Renate Wesselingh, Lorne Wolfe, and Helen Young. We only included studies that reported the sample size, mean, and some measure of variance among plants in pollen-supplemented and control treatments, and studies that measured one of the following five response variables: percent fruit set (percent of flowers setting fruit), percent seed set (percent of ovules setting seed), number of seeds per fruit, number of seeds per flower, and number of seeds per plant. When variance was not presented and data were binary (i.e., fruit set), we calculated variance from the mean and sample size of each treatment. Data published in graphical form were estimated using digitizing software (Grab It, version 1, 1998). We also recorded other information regarding plant traits, pollinator environment, habitat, and methodology; these are discussed in detail in the analyses below.

To describe general patterns of pollen limitation, we included data only from plants that were not manipulated in other ways (e.g., nutrient addition). When a study involved multiple populations, we considered each population to be a separate data record. Likewise, when a study considered multiple years, each year was considered a separate data record (as long as the treatments were not applied to the same plants each year). However, when studies included within-year variation (e.g., studies that applied treatments to some plants early in the season and others later in the season) and/or within-population variation (e.g., plants with different style morphs within a population), we averaged across these to produce a single value of pollen limitation for each response variable. For the few studies that applied pollination treatments in more than one year on the same plants, only the first year of the study was included. In all, we had 655 records from 263 studies, which were conducted on 306 plant species in 80 plant families.

To examine the magnitude of pollen limitation for each response variable, we calculated effect size from pollen supplementation experiments as the log response ratio (ln R),

$$\ln R = \ln(\bar{X}^E / \bar{X}^C),$$

where $\bar{X}$ is the mean of the response variable, and E and C denote the experimental (supplemented) and control treatments (Gurevitch et al. 2001, Hedges et al. 1999). A value of 0 reflects no difference in reproductive success between plants in the supplemented and control treatments, a positive value indicates higher reproductive success in the supplemented treatment, and a negative value indicates higher reproductive success in the control treatment. Because studies varied in sample size and therefore sampling variance, we weighted individual studies by their variances before calculating the average effect size (Gurevitch et al. 2001).

Pollen supplementation experiments may indicate higher magnitudes of pollen limitation when only a fraction of the plant's flowers receive the experimental treatment relative to whole-plant treatment (Zimmerman & Pyke 1988). Many plant species can reallocate resources among flowers within inflorescences, and many polycarpic species can reallocate unused resources to reproduction in future years (Calvo 1993, Campbell & Halama 1993, Dudash & Fenster 1997, Ehrlen & Eriksson 1995, Primack & Hall 1990, Primack & Stacy 1998, Stephenson 1981, Zimmerman & Pyke 1988). As a result, fruit set for the flowers hand-pollinated may exceed the controls even though a whole plant would be unable to respond with higher fruit set. Although these reallocation problems were pointed out in detail by Zimmerman & Pyke (1988), most pollen supplementation experiments are still conducted on only a portion of a plant's flowers, likely because of the difficulty in pollinating all flowers on long-lived plants. Because the level at which the treatments are applied influences the overall magnitude of pollen limitation (T.M. Knight, J.A. Steets, T.-L. Ashman, unpublished data), we distinguished between studies conducted on all or a portion of the flowers on a plant in our analyses.

Which Components of Seed Production Are Most Commonly Studied and Do These Components Show Similar Patterns?

Studies differed in the number and type of response variables (e.g., fruit set, seeds per plant) measured. Of the 655 records described above, 482 report measures of pollen limitation for percent fruit set, 170 for percent seed set, 182 for number of seeds per fruit, 94 for number of seeds per flower, and 87 for number of seeds per plant. Because percent fruit set was the most commonly measured response variable, we primarily use percent fruit set effect size for the analyses in this review. However, seeds per plant is the most appropriate response variable for most questions related to the study of pollen limitation, as it measures the effect on maternal fitness of an individual plant (Ashman et al. 2004, Dudash & Fenster 1997). To determine how well the magnitude of pollen limitation reported for fruit set (the most commonly measured response variable) correlated with the magnitude of pollen limitation reported for seeds per plant (the best response variable), we correlated these effect sizes for the 63 data records in which both response variables were reported. Most of these 63 data records were for herbaceous species, in which pollination treatments at the whole plant level were possible. We find a strong correlation between these effect sizes (Pearson's $r = 0.567, P < 0.001$) (Figure 1), suggesting that pollen limitation in percent fruit set is a good indicator of pollen limitation in the number of seeds produced by the entire plant. However, we note that there are two outlier points in which strong pollen limitation was present for percent fruit set but not in total seed production. This could result from plants in the supplement treatment making fewer flowers than plants in the control treatment.

What Is the Prevalence and Magnitude of Pollen Limitation?

Sixty-three percent of the 482 data records on percent fruit set showed significant pollen limitation (for each data record, significant pollen limitation was determined

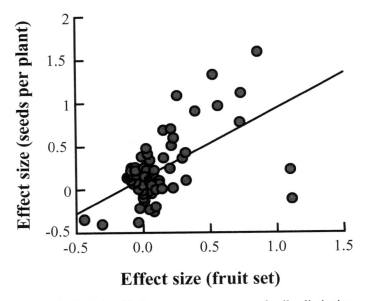

Figure 1   Relationship between two measures of pollen limitation effect size, percent fruit set and number of seeds per plant, for data records in which both were measured.

by a 1-tailed *t*-test). These results are similar to the prevalence calculated in slightly different ways in previous reviews of pollen limitation. For example, Burd (1994) reported that 62% of plant species showed pollen limitation in some populations or years (see also Ashman et al. 2004, Larson & Barrett 2000, Young & Young 1992).

The distribution of effect sizes for percent fruit set across all our data records suggests a high magnitude of pollen limitation in angiosperms. The weighted average effect size of 0.52 was significantly positive (i.e., 95% confidence intervals do not overlap with zero) and the distribution was leptokurtic distribution. The weighted average effect size corresponds to 75% higher fruit set in the supplemented compared with that in the control treatment (Figure 2).

How Much of the Variation in the Magnitude of Pollen Limitation Occurs Across Plant Taxa, Within Plant Species, and Within Populations?

Although we did not find differences in effect size for percent fruit set among Classes or Orders of plants in our full data set, the effect size differed significantly among plant families (variance component = 22% of the total variance within Class), among genera within families (variance component = 15%), and among

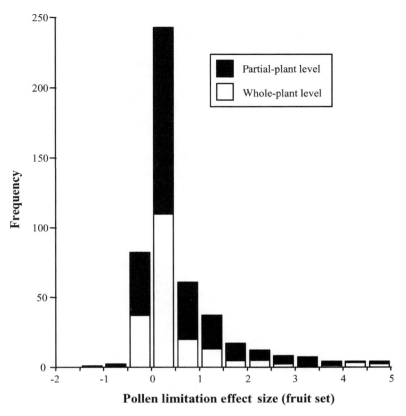

Figure 2   Histogram showing pollen limitation effect size for percent fruit set ($N =$ 482 data records). See text for calculation. Shading indicates the level of treatment (whole-plant versus partial-plant) used in the pollen supplementation experiment.

species within genera (33%)[1]. Variation among families still existed when we restricted the analysis to the studies conducted at the partial-plant level (variance component $= 37\%$), suggesting that differences in pollen limitation among taxa are not solely due to methodological differences in how these taxa are studied. For studies at the whole-plant level, which give the least biased estimates of pollen limitation, significant taxonomic variation was retained only for variation among species in fruit set ($P < 0.05$). These findings suggest that related taxa may share

---

[1]For these analyses we treated Class (Monocots vs. Eudicots) as a fixed factor and all other levels (Order, Family, Genus, Species) as random factors fully nested within the higher levels. Because our sampling at various taxonomic levels was unbalanced, we used the Mixed procedure in SAS software (version 8.2; SAS 2001) to find restricted maximum likelihood estimates of the variance components (Littell et al. 1996).

traits that affect the degree of pollen limitation and justify further analyses using phylogenetically corrected methods (see below).

Significant variation among populations of a species may signify that ecological features of the habitat, which vary in space and time, drive patterns of pollen limitation. In addition, high variation in pollen limitation may occur among individuals within a population as a result of microsite differences or phenotypic differences between individual plants. In our data set, three species were studied in more than 10 localities or years, and several studies examined within season and within individual variation in pollen limitation. In general, variation within these species, seasons, and across individuals within a population was small relative to the variation observed across taxa. Specifically, the range of pollen limitation observed for three well-studied species was 0 to 0.30 for *Trillium grandiflorum* (Kalisz et al. 1999, Knight 2003); −0.31 to 0.31 for *Silene dioica* (Carlsson-Graner et al. 1998); and −0.03 to 1.11 for *Narcissus assoanus* (Baker et al. 2000).

Pollen limitation may vary among individuals within a population owing to variation in plant traits. A trait that has received much attention is flowering time, and several studies have suggested that pollination success can select for changes in flowering time (e.g., Campbell 1985, Gross & Werner 1983, Ramsey 1995, Santrandreu & Lloret 1999, Widén 1991). For example, Santrandreu & Lloret (1999) found that *Erica multiflora* individuals flowering at the peak of the population bloom were less pollen limited than those flowering early or late, possibly because peak flowers are more likely to receive pollinator visits and outcross pollen. Others have suggested that temporal variation in pollen limitation may be caused by interspecific competition for pollinators (e.g., Campbell 1985, Gross & Werner 1983, Ramsey 1995) and temporal shifts in the sex ratio of the population (Jennersten et al. 1988, Le Corff et al. 1998).

MACROEVOLUTIONARY PATTERNS OF POLLEN LIMITATION

A number of traits may cause, or coevolve with, increased pollen limitation, and these can be divided into two main categories: (*a*) traits associated with sexual reproduction, including flower size, floral longevity, breeding system, floral shape, pollination syndrome, and ovule number per flower; and (*b*) life-history traits, including the number of reproductive episodes and the capability for asexual reproduction. Associations or correlations among species between a trait and pollen limitation can be from two distinct phenomena: (*a*) A direct causal relationship owing to the effects of the trait phenotype on pollinator behavior (e.g., flower size and pollen limitation; large-flowered species may attract more pollinators and are therefore less pollen limited) and (*b*) a coevolutionary adaptation (e.g., it may be adaptive for species with chronically unpredictable pollinator service to produce excess ovules per flower, which frequently make its flowers appear pollen limited). In the latter case, there is the joint evolution of two (or more) traits because the particular combination of traits represents the outcome of natural selection.

We conducted two types of analyses to detect associations between plant traits and pollen limitation. First, we performed traditional statistical analyses (in this case, meta-analysis), often referred to as TIPS analyses, in which each species represents an independent data point (Burd 1994, Sutherland 1986). When more than one data record was present for a species, we calculated a weighted average effect size for each species (see above and Verdu & Traveset 2004). Randomization tests were performed to determine whether qualitative variables such as self-compatibility predict levels of pollen limitation among species; linear regression was used to determine whether quantitative variables such as ovule number or flower size predict the magnitude of pollen limitation. TIPS analyses are vulnerable to two primary criticisms. First, closely related and phenotypically similar species may not represent statistically independent data (Felsenstein 1985). Second, unmeasured variables that are associated with taxonomic membership may confound the statistical relationships among the measured variables. Nevertheless, TIPS analyses of taxonomically highly diverse data sets can provide ecologists with tests of whether observed relationships among traits support predictions based on clear causal mechanisms.

Second, we used a more conservative approach—the analysis of phylogenetically independent contrasts (or PICs) that takes into account the evolutionary relationships among species—to detect the direction or magnitude of joint evolutionary change in focal traits (Felsenstein 1985, 1988; Garland et al. 1992; Harvey & Mace 1982; Harvey & Partridge 1989; Pagel & Harvey 1988). PICs statistically analyze contrasts, i.e., differences in trait values between each pair of sister taxa or clades that represent independent evolution of the trait from a common starting point (common ancestor). Wilcoxon signed-rank tests (Vamosi & Vamosi 2005b) and regression tests (Webb et al. 2002) were used to determine whether a unidirectional change in an independent trait is associated with a positive (or negative) directional change in a dependent trait more frequently than one would expect by chance for discrete and continuous traits, respectively.

Larson & Barrett (2000) conducted both TIPS and PICs analyses on 224 species of biotically pollinated plants to examine the association between plant traits and pollen limitation. Their TIPS analyses indicated that pollen limitation was significantly associated with numerous plant traits, whereas their PICs analyses of the same data detected limited significant results; pollen limitation was found to be lower in clades that are self-compatible and herbaceous. The TIPS analyses may have detected more significant associations than the PICs analyses because the focal traits exhibit strong phylogenetic signals and because pollen limitation was measured on a nonrandom sample of angiosperm taxa. Since this analysis was performed, however, not only has a more extensive data set been compiled, but there have been many refinements and adjustments in the phylogenetic reconstruction of angiosperms, in the use of meta-analytical approaches in phylogenetic analyses (Verdu & Traveset 2004) and the recognition of the impact of methodology on pollen limitation estimates (T.M. Knight, J.A. Steets, T.-L. Ashman, unpublished data), making a re-analysis warranted.

We used the online software utility, Phylomatic (Webb & Donoghue 2005), to provide a phylogenetic tree with taxonomic resolution to at least the family level. All branch lengths in the phylogeny were set to unity. When more than one data record was present for a species, we calculated a weighted average effect size for each species (see above and Verdu & Traveset 2004). We recorded the plant traits reported by authors of the studies in our data set (or through personal communication with those authors), or obtained from other published sources. We considered studies conducted on only portions of the plant (partial-plant level) and measuring percent fruit set as the greatest number of data were available for this level of treatment and response variable. We used the software program MetaWin (Rosenberg et al. 2000) for TIPS analyses and Phylocom (Webb et al. 2004) for PICs analyses.

Is There a Phylogenetic Signal for Pollen Limitation or Plant Traits of Interest?

We detected a phylogenetic signal (sensu Blomberg et al. 2003) for pollen limitation among species ($N = 166$, mean contrast $= 0.494$, $P = 0.002$) and among the plant traits we examined (Table 2A). This suggests that an examination of the relationship between pollen limitation and each plant trait is warranted.

Is There an Association Between Pollen Limitation and Plant Traits?

Below, for each trait evaluated, we discuss its predicted association with pollen limitation and present results from TIPS and PICs analyses. A potential shortcoming of these analyses is that sample size limitations precluded conducting multivariate analyses, i.e., our bivariate analyses did not control for potentially confounding third variables.

Traits Associated with Sexual Reproduction

FLOWER SIZE   Species with large flowers may attract more pollinators than small-flowered species (Momose 2004, Valido et al. 2002) and are therefore less pollen limited. However, flower size is likely confounded with mating system, as species that are capable of autogamous selfing often have smaller flowers than those that are primarily outcrossing (Armbruster et al. 2002, Jain 1976). We conducted analyses with and without the inclusion of autogamous species, predicting that a negative relationship between flower size and pollen limitation would be more likely when

TABLE 2 Results from TIPs and PICs analyses examining the relationship between plant traits and pollen limitation

Trait	Test for phylogenetic signal in plant traits[A]			TIPS[B]				PICs[C]		
	Sample size (number of species)[a]	Mean contrast[b]	P-value[c]	Sample size (number of species)[d]	P-value		Mean contrast or slope	Positive/ total contrasts		P-value
Flower size	115	16.41	0.03	105	0.04		0.27	43/76		0.24
Floral longevity	96	2.62	0.09	74	0.39		−0.31	20/49		0.40
Breeding system: self-compatible versus self-incompatible	182	0.29	<0.01	155	<0.01		0.30	11/19		0.08
Breeding system: perfect versus imperfect	206	0.17	<0.01	166	0.43		−0.03	9/12		0.47
Floral shape: zygomorphic versus actinomorphic	207	0.14	<0.01	166	0.46		−0.20	5/14		0.15
Pollination syndrome: abiotic versus biotic	189	0.03	<0.01	150	0.30		0.10	2/3		0.38
Number of pollinating species: one, few, many	141	0.38	<0.10	111	0.01		−0.32	29/69		0.048
Ovule number per flower	183	144.3	<0.01	148	<0.01		0.29	51/84		0.04

(Continued)

TABLE 2 (*Continued*)

Trait	Test for phylogenetic signal in plant traits[A]			TIPS[B]		PICs[C]		
	Sample size (number of species)[a]	Mean contrast[b]	P-value[c]	Sample size (number of species)[d]	P-value	Mean contrast or slope	Positive/total contrasts	P-value
Number of reproductive episodes: monocarpic versus polycarpic	194	0.10	<0.01	156	*0.09*	0.11	3/5	0.41
Growth form: woody versus herbaceous	207	0.32	<0.01	166	*0.10*	0.27	44/91	*0.07*
Capacity for asexual reproduction: clonal versus not clonal	160	0.24	<0.01	130	*0.08*	−0.03	5/16	0.22

For each plant trait, the results of a test for phylogenetic signal,[A] and the relationship between the trait and pollen limitation from TIPS[B], and PICs[C] analyses. TIPS and PICs analyses were restricted to those studies treating portions of the flowers on a plant and measuring percent fruit set. For TIPS, randomization tests were used to test for significant associations between pollen limitation and discrete plant traits, and linear regression was used to detect significant associations between pollen limitation and continuous plant traits. For PICs, the ratio of positive contrasts to total contrasts is shown. In the case of discrete traits, the P-value is derived from Wilcoxon-signed-rank tests to detect whether the mean contrast differs significantly from zero. In the case of continuous traits, the P-value is derived from a regression analysis to detect whether the slope differs from zero. All continuous traits were log-transformed before analysis. For both TIPS and PICs, the P-values are shown (bold: $P < 0.05$; italics $0.05 < P < 0.10$).

[a] Sample size depended on the availability of trait information for each species.

[b] The average phenotypic difference in the focal trait between sister taxa.

[c] P-values are based on the rank of the mean contrast compared with the 999 mean contrasts calculated in the randomized phylogenies. A drank below 50/1000 indicates a probability less than 5% of obtaining an observed mean contrast as low as it is by chance. This is interpreted as evidence of a significant phylogenetic signal.

[d] Sample size identical for TIPS and PICs analyses.

autogamous species are excluded. Flower size can be measured in many ways, so in each species we used the measurement that we thought would best indicate the floral dimension visible to prospective pollinators (generally, flower diameter). The TIPS analysis (Table 2B) revealed that pollen limitation decreased with increasing flower size; however, when we controlled for phylogenetic history or removed autogamous species from the analysis, this relationship was not statistically significant (Table 2B,C).

FLORAL LONGEVITY The longer flowers are open, the more likely they are to receive pollinator visits and pollen (Ashman & Schoen 1994; Campbell et al. 1994, 1996; Rathcke 2003). Consequently, we expect a negative relationship between floral longevity and the magnitude of pollen limitation. However, we did not find evidence for a relationship between floral longevity and pollen limitation in the TIPS or PICs analyses (Table 2B,C).

BREEDING SYSTEM Plants facing chronic pollen limitation may evolve self-compatibility for reproductive assurance (Baker 1955, 1967). In our data set, we classified plants as either self-compatible or self-incompatible using the categorizations stated by the authors of the pollen limitation study. In accord with other reviews (Burd 1994, Larson & Barrett 2000), we found that self-incompatible plants were more pollen limited than self-compatible plants in the TIPS and PICs analyses (but only marginally so in the PICs analysis; Table 2C).

Perfect-flowered species have male and female structures in the same flower and thus a single pollinator visit may be adequate to transfer pollen; however, species with imperfect (unisexual) flowers often require a pollinator to move between a male and a female flower for pollen transfer. Moreover, female flowers are often avoided by pollinators (e.g., Ashman 2000), indicating that the potential exists for species with unisexual flowers to be more pollen limited than hermaphroditic ones, especially when sexual-dimorphism between the sexes is high and pollinators are scarce (Ashman & Diefenderfer 2001, Vamosi & Otto 2002). On the other hand, unisexual species are more often pollinated by generalists (Charlesworth 1993), which may make them less prone to pollen limitation than perfect-flowered species (see below). We do not find differences in pollen limitation between perfect-flowered species and species with unisexual flowers or unisexual morphs (Table 2B,C). This result corroborates findings of a meta-analysis by Shykoff et al. (2003), who found no difference in pollen limitation between hermaphrodite and female morphs of gynodioecious species.

FLORAL SHAPE, POLLINATION SYNDROME, AND NUMBER OF POLLINATING SPECIES
Zygomorphic flowers encourage more precise placement of pollen and generally attract more specialized pollinators than actinomorphic flowers (Fenster et al. 2004, Neal et al. 1998, Sargent 2004). Similarly, plants pollinated by biotic pollinators are usually considered to be more specialized than those pollinated by abiotic

vectors (Culley et al. 2002). Such specialization may provide more reliable pollination, resulting in lower levels of pollen limitation. Alternatively, plants relying on specialized pollinators may experience greater variation in their pollination success (e.g., because abundance of a few specialized pollinators may fluctuate more in space or time than that of many generalized species), and hence be more prone to pollen limitation (Eckhart 1992; Fishbein & Venable 1996; Herrera 1988, 1996; Horvitz & Schemske 1990; Waser et al. 1996). We classified the species in our data set on the basis of their floral shape (actinomorphic versus zygomorphic) and pollination syndrome (biotic versus abiotic pollination). When the appropriate information was provided in a study, we classified biotic pollination into three categories on the basis of the number of pollinating taxa: one (specialist plants visited primarily by one pollinating species), few (2–5 species of pollinators), and many (>5 pollinating species). In both TIPS and PICs analyses, pollen limitation decreased with an increase in the number of pollinating taxa (Table 2B,C). We detected no difference between zygomorphic and actinomorphic species or between biotic and abiotic pollinated plants in their degree of pollen limitation (Table 2B, C). However, the low sample size of abiotically pollinated species ($N = 10$) may have limited our ability to detect a difference in pollen limitation between biotically and abiotically pollinated species.

OVULE NUMBER PER FLOWER   Burd (1995) proposed that whole-plant seed output is maximized at an ovule number per flower determined by the resource cost of ovule production and the flower-to-flower variance in pollen receipt and fertilizations. This theory predicts that selection may favor the production of more ovules per flower than the average number of fertilizations per flower obtained in a given environment because individual flowers with many ovules are able to capitalize on occasional but unpredictable receipt of large pollen loads. If selection acts in this way, flowers would often have unused capacity for seed production and frequently appear to be pollen limited. Both TIPS and PICs analyses detected a positive correlation between ovule number and pollen limitation (Table 2B,C).

We explore this result further by asking, is high ovule number favored in highly stochastic pollination environments [as proposed by Burd (1995)]? To address this question, we characterized the stochasticity in pollination and fertilization by the degree of intraplant variation in seeds per fruit under natural pollination. For those species in the data set with appropriate data, we calculated the coefficient of variation in seeds per fruit (CV = standard deviation/mean) obtained under natural pollination. When there were multiple studies or multiple population-years for a species, we used the mean CV among studies to obtain one CV value per species. There were 43 species for which we could obtain both the CV of seeds per fruit and the ovule number per flower. We performed a PICs analysis to examine the association of these two variables. The regression through the origin of contrasts in ovule number per flower versus contrasts in the CV of seeds per fruit had a significant positive slope ($t_{34} = 2.16, P = 0.038$) (Figure 3). A sign test indicated that there were 23 positive contrasts in ovule number and 11 negative contrasts

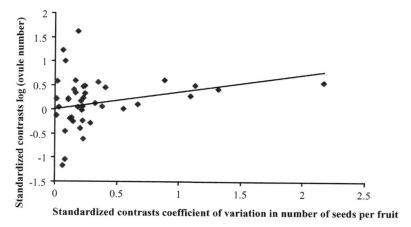

Figure 3   Ovule number per flower in relation to intraplant variation in floral fertilization. Individual points in the diagram represent standardized contrasts obtained from PICs analysis. The horizontal axis shows contrasts in the coefficient of variation (CV) of seed number per fruit (seed number is used as an easily measured surrogate for fertilization level); the vertical axis shows contrasts in ovule number per flower (logarithmically transformed). The least-squares regression line is shown [bivariate regressions of PICs must pass through the origin (Garland et al. 1992)]. The significantly positive slope ($t_{34} = 2.16, P = 0.038$) implies that increases in the variance of fertilization success (seed number) are, on average, accompanied by increases in ovule number per flower.

(2-tailed sign test, $P = 0.057$). This result is consistent with the idea that higher numbers of ovules per flower evolved in response to variable pollen receipt among flowers (Burd 1995). That is, high ovule number is a bet-hedging strategy to deal with stochastic pollen receipt.

Life-History Traits

NUMBER OF REPRODUCTIVE EPISODES   Monocarpic and short-lived species are expected to be less pollen limited than polycarpic and long-lived species for two reasons. First, pollination treatments are applied to all reproductive bouts for monocarpic plants. As a result, experimental measures of pollen limitation are not subject to between-year resource reallocation. Second, selection for traits that may reduce pollen limitation (e.g., larger flowers, self-compatibility) has probably been stronger and more effective in short-lived than in long-lived species. If pollinator abundance fluctuates, the lifetime fitness of a short-lived individual would be irrevocably reduced if it experiences pollen limitation for one or several successive years (Vamosi & Otto 2002) whereas a long-lived species could better endure a short pollinator drought (Calvo & Horvitz 1990, Primack & Hall 1990, Zhang

& Wang 1994). Similarly, because woody species live longer than most herbaceous species, trees and shrubs are likely to display more pollen limitation than herbs.

There was a trend for higher pollen limitation among polycarps than monocarps in the TIPS analysis (Table 2B). However, when we controlled for phylogenetic relationships, this pattern was not evident, perhaps owing to the reduced number of contrasts available for analysis (Table 2C). In both TIPS and PICs analyses there were trends for woody species to be more pollen limited than herbs (Table 2B,C). This result is in accord with Larson & Barrett (2000), who found higher pollen limitation in woody plants compared with that in herbs. However, their analyses did not control for the level at which the treatment was applied, and woody plants are more likely to have treatments applied at the partial-plant level than herbs.

CAPACITY FOR ASEXUAL REPRODUCTION   Pollen limitation may be more prevalent in species capable of asexual reproduction because fitness of these species depends less on pollination success and sexual reproduction. As a result, we expect species with the capacity for asexual reproduction to evolve lower levels of resource allocation to floral traits and pollinator attraction than their exclusively sexual counterparts (Charpentier 2001, Ronsheim & Bever 2000). However, we did not find higher pollen limitation among asexually relative to sexually reproducing species (Table 2B,C).

CONCLUSIONS AND RECOMMENDATIONS FOR FUTURE STUDIES

In this concluding section, we highlight our key results. Overall, our review and analyses suggest that the magnitude of pollen limitation found in pollen supplementation experiments is high (see also Burd 1994, Larson & Barrett 2000). The observed effect sizes, however, are quite variable among populations or treatments that differ in ecological factors and among taxa.

Because of the stochastic nature of pollination and resource availability, plants may be pollen limited at some times and resource limited at other times, but not severely limited by either over their lifetimes (e.g., Casper & Niesenbaum 1993). The magnitude of pollen limitation varies among flowers within a plant, among plants within a season, and among seasons. However, plants that are able to redistribute resources temporally may not be limited by pollen receipt or resource availability over longer time intervals. We found that evolutionary increases in the number of ovules per flower were correlated with increases in the magnitude of observed pollen limitation among species. Thus our analyses suggest that high ovule number may be a bet-hedging strategy that allows plants to tolerate intraplant stochastic pollen receipt.

Plant traits may evolve despite their association with strong pollen limitation. For example, self-incompatibility has evolved multiple times (Charlesworth 1985, Igic et al. 2003, Steinbachs & Holsinger 2002), and yet we find that this trait is associated with higher levels of pollen limitation. Thus it is likely that the benefits that these plants receive from outcrossing outweigh the costs of often being limited by pollen receipt. These benefits could include increased seed set through higher pollen quality, higher seed germination, and seedling survivorship probabilities. We suggest that pollination researchers will gain considerable insight into these benefits by increasing their scope of research beyond measuring seed production and including measurements of offspring quality.

Reduced seed production as a result of pollen limitation does not necessarily result in demographic consequences for the plant population. This is because for many plant populations, particularly long-lived species with multiple reproductive bouts, the sensitivity of the population to increased seed set is low relative to other vital rates such as adult survivorship (Ashman et al. 2004). To understand the role of pollination in the dynamics of species, such as the viability of rare species or the spread of invasive species, we suggest increasing the scope of the study to include other aspects of a plant's life-cycle.

Although these factors suggest that pollen limitation might not be as frequent or as severe as is often thought, or as demographically critical, pollen limitation may still be of great importance for evolutionary or ecological processes. In particular, in our rapidly changing world, variation in the pollination environment will probably alter the population dynamics and future evolutionary potential of many plant species. There is evidence that a variety of ecological perturbations increases the magnitude of pollen limitation. For example, the magnitude of pollen limitation often increases with habitat fragmentation. Currently, however, few studies are available that directly examine the effects of particular ecological conditions on the magnitude of pollen limitation. We hope that in the future, ecologists and conservation biologists will explore the role of pollen limitation in these and other contexts, so that we can begin to make such generalizations.

ACKNOWLEDGMENTS

We are grateful to Martin Morgan, Will Wilson, Jonathan Chase, and Doug Schemske for their comments on this manuscript; Abagail Matela and Erin Marnocha for their help with digitizing and entering of data; and Gregory Anderson, Helene Froburg, Leo Galetto, Damien Lanotte, Francisco Lloret, Ingrid Parker, Beverly Rathcke, and Charles Schick for providing supplementary plant trait information. This work was conducted as part of the Pollen Limitation Working Group supported by the National Center for Ecological Analysis and Synthesis, a Center funded by NSF (DEB-00,72909), the University of California, and the Santa Barbara campus. Additional support was also provided for the Post-doctoral Associate (TMK) in the group.

The *Annual Review of Ecology, Evolution, and Systematics* is online at http://ecolsys.annualreviews.org

LITERATURE CITED

Ågren J. 1996. Population size, pollinator limitation, and seed set in the self-incompatible herb *Lythrum salicaria*. *Ecology* 77:1779–90

Aizen MA, Ashworth L, Galetto L. 2002. Reproductive success in fragmented habitats: do compatibility systems and pollination specialization matter? *J. Veg. Sci.* 13:885–92

Amarasekare P. 2004. Spatial dynamics of mutualistic interactions. *J. Anim. Ecol.* 73:128–42

Armbruster WS, Mulder CPH, Baldwin BG, Kalisz S, Wessa B, et al. 2002. Comparative analysis of late floral development and mating system evolution in tribe Collinsieae (Scrophulariaceae S.L.). *Am. J. Bot.* 89:37–49

Ashman T-L. 2000. Pollinator selectivity and its implications for the evolution of dioecy and sexual dimorphism. *Ecology* 81:2577–91

Ashman T-L, Diefenderfer C. 2001. Sex ratio represents a unique context for selection on attractive traits: consequences for the evolution of sexual dimorphism. *Am. Nat.* 157:334–47

Ashman T-L, Knight TM, Steets JA, Amarasekare P, Burd M, et al. 2004. Pollen limitation of plant reproduction: ecological and evolutionary causes and consequences. *Ecology* 85:2408–21

Ashman T-L, Morgan MT. 2004. Explaining phenotypic selection on plant attractive characters: male function, gender balance or ecological context? *Proc. R. Soc. London Ser. B* 271:553–59

Ashman T-L, Schoen DJ. 1994. How long should flowers live? *Nature* 371:788–91

Ashman T-L, Schoen DJ. 1996. Floral longevity: fitness consequences and resource costs. In *Floral Biology*, ed. SCH Barrett, DG Lloyd, pp. 112–39. New York: Chapman & Hall

Baker AM, Barrett SCH, Thompson JD. 2000. Variation of pollen limitation in the early flowering Mediterranean geophyte *Narcissus assoanus* (Amaryllidaceae). *Oecologia* 124:529–35

Baker HG. 1955. Self-compatibility and establishment after "long-distance" dispersal. *Evolution* 9:347–49

Baker HG. 1967. Support for Baker's Law–as a rule. *Evolution* 21:853–56

Balvanera P, Kremen C, Martinez-Ramos M. 2005. Applying community structure analysis to ecosystem function: examples from pollination and carbon storage. *Ecol. App.* 15:360–75

Bateman AJ. 1948. Intra-sexual selection in *Drosophila*. *Heredity* 2:349–68

Bertness MD, Shumway SW. 1992. Consumer driven pollen limitation of seed production in marsh grasses. *Am. J. Bot.* 79:288–93

Bierzychudek P. 1981. Pollinator limitation of plant reproductive effort. *Am. Nat.* 117:838–40

Blomberg SP, Garland T, Ives AR. 2003. Testing for phylogenetic signal in comparative data: behavioral traits are more labile. *Evolution* 57:717–45

Bond WJ. 1994. Do mutualisms matter? Assessing the impact of pollinator and disperser disruption on plant extinction. *Philos. Trans. R. Soc. London Ser. B* 344:83–90

Brown BJ, Mitchell RJ, Graham SA. 2002. Competition for pollination between an invasive species (purple loosestrife) and a native congener. *Ecology* 83:2328–36

Buchmann SL, Nabhan GP. 1996 *The Forgotten Pollinators*. Washington DC: Island

Burd M. 1994. Bateman's principle and plant reproduction: the role of pollen limitation in fruit and seed set. *Bot. Rev.* 60:83–139

Burd M. 1995. Ovule packaging in stochastic

pollination and fertilization environments. *Evolution* 49:100–9

Calvo RN. 1993. Evolutionary demography of orchids: intensity and frequency of pollination and the cost of fruiting. *Ecology* 74:1033–42

Calvo RN, Horvitz CC. 1990. Pollinator limitation cost of reproduction, and fitness in plants: a transition-matrix demographic approach. *Am. Nat.* 136:499–516

Campbell DR. 1985. Pollinator sharing and seed set of *Stellaria pubera*: competition for pollination. *Ecology* 66:544–53

Campbell DR, Halama KJ. 1993. Resource and pollen limitations to lifetime seed production in a natural plant population. *Ecology* 74:1043–51

Campbell DR, Motten AF. 1985. The mechanism of competition for pollination between two forest herbs. *Ecology* 66:554–63

Campbell DR, Waser NM, Price MV. 1994. Indirect selection of stigma position in *Ipomopsis aggregata* via a genetically correlated trait. *Evolution* 48:55–68

Carlsson-Graner U, Elmqvist T, Agren J, Gardfjell H, Ingvarsson P. 1998. Floral sex ratios, disease and seed set in dioecious *Silene dioica*. *J. Ecol.* 86:79–91

Caruso CM. 1999. Pollination of *Ipomopsis aggregata* (Polemoniaceae): effects of intra- vs. interspecific competition. *Am. J. Bot.* 86:662–68

Caruso CM. 2001. Differential selection on floral traits of *Ipomopsis aggregata* growing in contrasting environments. *Oikos* 94:295–302

Caruso CM. 2002. Influence of plant abundance on pollination and selection on floral traits of *Ipomopsis aggregata*. *Ecology* 83:241–54

Casper BB, Niesenbaum RA. 1993. Pollen versus resource limitation of seed production: a reconsideration. *Curr. Sci.* 65:210–14

Charlesworth D. 1985. Distribution of dioecy and self-incompatibility in angiosperms. In *Evolution-Essays in Honour of John Maynard Smith*, ed. PJ Greenwood, M Slatkin, pp. 237–68. Cambridge, UK: Cambridge Univ. Press

Charlesworth D. 1993. Why are unisexual flowers associated with wind pollination and unspecialized pollinators? *Am. Nat.* 141:481–90

Charpentier A. 2001. Consequences of clonal growth for plant mating. *Evol. Ecol.* 15:521–30

Culley TM, Weller SG, Sakai AK. 2002. The evolution of wind pollination in angiosperms. *Trends Ecol. Evol.* 17:361–69

Cunningham SA. 2000. Depressed pollination in habitat fragments causes low fruit set. *Proc. R. Soc. London Ser. B* 267:1149–52

Davis HG, Taylor CM, Lambrinos JG, Strong DR. 2004. Pollen limitation causes an Allee effect in a wind-pollinated invasive grass (*Spartina alterniflora*). *Proc. Natl. Acad. Sci. USA* 101:13804–7

Dudash MR, Fenster CB. 1997. Multiyear study of pollen limitation and cost of reproduction in the iteroparous *Silene virginica*. *Ecology* 78:484–93

Dukas R. 2001. Effects of perceived danger on flower choice by bees. *Ecol. Lett.* 4:327–33

Dukas R, Morse DH. 2003. Crab spiders affect flower visitation by bees. *Oikos* 101:157–63

Eckhart VM. 1992. Spatio-temporal variation in abundance and variation in foraging behavior of the pollinators of gynodioecious *Phacelia linearis* (Hydrophyllaceae). *Oikos* 64:573–86

Ehrlén J, Eriksson O. 1995. Pollen limitation and population growth in a herbaceous perennial legume. *Ecology* 76:652–56

Fausto JA, Eckhart VM, Geber MA. 2001. Reproductive assurance and the evolutionary ecology of self-pollination in *Clarkia xantiana* (Onagraceae). *Am. J. Bot.* 88:1794–1800

Feinsinger P, Tiebout HM, Young BE. 1991. Do tropical bird-pollinated plants exhibit density-dependent interactions? Field experiments in a chaco dry forest, Argentina. *Ecology* 72:1953–63

Feldman TS, Morris WF, Wilson WG. 2004. When can two plant species facilitate each other's pollination? *Oikos* 105:197–207

Felsenstein J. 1985. Phylogenies and the comparative method. *Am. Nat.* 125:1–15

Felsenstein J. 1988. Phylogenies and quantitative traits. *Annu. Rev. Ecol. Syst.* 19:445–71

Fenster CB, Armbruster WS, Wilson P, Dudash MR, Thomson JD. 2004. Pollination syndromes and floral specialization. *Annu. Rev. Ecol. Syst.* 35:375–403

Fishbein M, Venable DL. 1996. Evolution of inflorescence design: theory and data. *Evolution* 50:2165–77

Forsyth SA. 2003. Density-dependent seed set in the Haleakala silversword: evidence for an Allee effect. *Oecologia* 136:551–57

Franceschinelli EV, Bawa KS. 2000. The effect of ecological factors on the mating system of a South American shrub species (*Helicteres brevispira*). *Heredity* 84:116–23

Galen C, Gregory T. 1989. Interspecific pollen transfer as a mechanism of competition: consequences of foreign pollen contamination for seed set in the alpine wildflower, *Polemonium viscosum*. *Oecologia* 81:120–23

Galen C, Plowright RC, Thompson JD. 1985. Floral biological and regulation of seed set and seed size in the lily, *Clintonia borealis*. *Am. J. Bot.* 72:1544–52

Garcia MB, Ehrlén J. 2002. Reproductive effort and herbivory timing in a perennial herb: fitness components at the individual and population levels. *Am. J. Bot.* 89:1295–302

Garland T, Harvey PH, Ives AR. 1992. Procedures for the analysis of comparative data using phylogenetically independent contrasts. *Syst. Biol.* 41:18–32

Goldingay RL. 2000. Further assessment of pollen limitation in the waratah (*Telopea speciosissima*). *Aust. J. Bot.* 48:209–14

Grab It, version 1. 1998. DataTrend Software

Grabas GP, Laverty TM. 1999. The effect of purple loosestrife (*Lythrum salicaria* L; Lythraceae) on the pollination and reproductive success of sympatric co-flowering wetland plants. *Ecoscience* 6:230–42

Groom MJ. 2001. Consequences of subpopulation isolation for pollination, herbivory, and population growth in *Clarkia concinna* (Onagraceae). *Biol. Cons.* 100:55–63

Gross CL. 1996. Is resource overlap disadvantageous to three sympatric legumes? *Aust. J. Ecol.* 21:133–43

Gross RS, Werner PA. 1983. Relationships among flowering phenology, insect visitors, and seed-set of individuals: experimental studies on four co-occurring species of goldenrod (*Solidago*: Compositae). *Ecol. Monogr.* 53:95–117

Gurevitch J, Curtis PS, Jones MH. 2001. Meta-analysis in ecology. *Adv. Ecol. Res.* 32:200–47

Hackney EE, McGraw JB. 2001. Experimental demonstration of an Allee effect in American ginseng. *Cons. Biol.* 15:129–36

Haig D, Westoby M. 1988. On limits to seed production. *Am. Nat.* 131:757–59

Hanley ME, Goulson D. 2003. Introduced weeds pollinated by introduced bees: cause or effect? *Weed Biol. Manage.* 3:204–12

Harvey PH, Mace GM. 1982. *Comparisons Between Taxa and Adaptive Trends: Problems of Methodology. King's.* New York: Cambridge Univ. Press

Harvey PH, Partridge L, eds. 1989. *Comparative Methods Using Phylogenetically Independent Contrasts.* Oxford, UK: Oxford Univ. Press

Hedges LV, Gurevitch J, Curtis PS. 1999. The meta-analysis of response ratios in experimental ecology. *Ecology* 80:1150–56

Herrera CM. 1988. Variation in mutualisms—the spatio-temporal mosaic of a pollinator assemblage. *Biol. J. Linn. Soc.* 35:95–125

Herrera CM. 1996. Floral traits and plant adaptation to insect pollinators: a devil's advocate approach. In *Floral Biology*, ed. D Llyod, SCH Barrett, pp. 65–87. New York: Chapman & Hall

Holland JN. 2002. Benefits and costs of mutualism: demographic consequences in a pollinating seed-consumer interaction. *Proc. R. Soc. London Ser. B* 269:1405–12

Horvitz CC, Schemske DW. 1990. Spatiotemporal variation in insect mutualists of a neotropical herb. *Ecology* 71:1085–97

Igic B, Bohs L, Kohn JR. 2003. Historical inferences from the self-incompatibility locus. *New Phytol.* 161:97–105

Irwin RE, Brody AK, Waser NM. 2001. The impact of floral larceny on individuals, populations, and communities. *Oecologia* 129:161–68

Ishii R, Higashi M. 2001. Coexistence induced by pollen limitation in floweringplant species. *Proc. R. Soc. London Ser. B* 268:579–85

Iwaizumi MG, Sakai S. 2004. Variation in flower biomass among nearby populations of *Impatiens textori* (Balsaminaceae): effects of population plant densities. *Can. J. Bot.* 82:563–72

Jain SK. 1976. The evolution of inbreeding in plant. *Annu. Rev. Ecol. Syst.* 7:469–95

Janzen DH. 1977. A note on optimal mate selection in plants. *Am. Nat.* 111:365–71

Jennersten O. 1988. Pollination of *Dianthus deltoides* (Caryophyllaceae): effects of habitat fragmentation on visitation and seed set. *Cons. Biol.* 2:359–66

Jennersten O, Berg L, Lehman C. 1988. Phenological differences in pollinator visitation, pollen deposition and seed set in the sticky catchfly, *Viscaria vulgaris. J. Ecol.* 76:1111–32

Johnson SD, Collin CL, Wissman HJ, Halvarsson E, Ågren J. 2004. Factors contributing to variation in seed production among remnant populations of the endangered daisy *Gerbera aurantiaca. Biotropica* 36:148–55

Johnson SD, Peter CI, Nilsson LA, Agren J. 2003. Pollination success in a deceptive orchid is enhanced by co-occurring rewarding magnet plants. *Ecology* 84:2919–27

Johnston MO. 1991a. Natural selection of floral traits in 2 species of *Lobelia* with different pollinators. *Evolution* 45:1468–79

Johnston MO. 1991b. Pollen limitation of femal reproduction in *Lobelia cardinalis* and *L. siphilitica. Ecology* 72:1500–3

Juenger T, Bergelson J. 1997. Pollen and resource limitation of compensation to herbivory in scarlet gilia, *Ipomopsis aggregata. Ecology* 78:1684–95

Kalisz S, Hanzawa FM, Tonsor SJ, Thiede DA, Voigt S. 1999. Ant-mediated seed dispersal alters pattern of relatedness in a population of *Trillium grandiflorum. Ecology* 80:2620–34

Karron JD, Thumser NN, Tucker R, Hessenauer AJ. 1995. The influence of population density on outcrossing rates in *Mimulus ringens. Heredity* 75:175–80

Kearns CA, Inouye DW, Waser NM. 1998. Endangered mutualisms: the conservation of plant-pollinator interactions. *Annu. Rev. Ecol. Syst.* 29:83–112

Klinkhamer PGL, de Jong TJ. 1990. Effects of plant size, plant-density and sex differential nectar reward on pollinator visitation in the Protandrous *Echium vulgare* (Boraginaceae). *Oikos* 57:399–405

Knight TM. 2003. Floral density, pollen limitation, and reproductive success in *Trillium grandiflorum. Oecologia* 137:557–63

Knight TM, McCoy MW, Chase JM, McCoy KA, Holt RD. 2005. Tropic cascades across ecosystems. *Nature.* In press

Kremen C, Ricketts TH. 2000. Global perspectives on pollination disruptions. *Cons. Biol.* 14:1226–28

Kremen C, Williams NM, Thorp RW. 2002. Crop pollination from native bees at risk from agricultural intensification. *Proc. Natl. Acad. Sci. USA* 99:16812–16

Krupnick GA, Weis AE. 1999. The effect of floral herbivory on male and female reproductive success in *Isomeris arborea. Ecology* 80:135–49

Kunin WE. 1993. Sex and the single mustard: Population-density and pollinator behavior effects on seed-set. *Ecology* 74:2145–60

Kunin WE. 1997. Population size and density effects in pollination: pollinator foraging and plant reproductive success in experimental arrays of *Brassica kaber. J. Ecol.* 85:225–34

Larson BMH, Barrett SCH. 2000. A comparative analysis of pollen limitation in flowering plants. *Biol. J. Linn. Soc.* 69:503–20

Laverty TM. 1992. Plant interactions for pollinator visits: a test of the magnet species effect. *Oecologia* 89:502–8

Le Corff J, Ågren J, Schemske DW. 1998.

Floral display, pollinator discrimination, and female reproductive success in two monoecious *Begonia* species. *Ecology* 79:1610–19

Lehtila K, Syrjanen K. 1995. Positive effects of pollination on subsequent size, reproduction, and survival of *Primula veris*. *Ecology* 76:1084–98

Lennartsson T. 2002. Extinction thresholds and disrupted plant pollinator interactions in fragmented plant populations. *Ecology* 83:3060–72

Linhart Y, Feinsinger P. 1980. Plant–hummingbird interactions: effects of island size and degree of specialization on pollination. *J. Ecol.* 68:745–55

Liston A, Rieseberg LH, Elias TS. 1990. Functional androdioecy in the flowering plant *Datisca glomerata*. *Nature* 343:641–42

Liu H, Koptur S. 2003. Breeding system and pollination of a narrowly endemic herb of the Lower Florida Keys: impacts of the urban-wildland interface. *Am. J. Bot.* 90:1180–87

Lloyd DG. 1979. Some reproductive factors affecting the selection of self-fertilization in plants. *Am. Nat.* 113:6779

Lloyd DG, Schoen DJ. 1992. Self-fertilization and cross fertilization in plants. 1. Functional dimensions. *Int. J. Plant Sci.* 153:358–69

Lu YQ. 2000. Effects of density on mixed mating systems and reproduction in natural populations of *Impatiens capensis*. *Int. J. Plant Sci.* 161:671–81

Maurice S, Fleming TH. 1995. The effect of pollen limitation on plant reproductive systems and the maintenance of sexual polymorphisms. *Oikos* 74:55–60

Mizui N, Kikuzawa K. 1991. Proximate limitations to fruit and seed set in *Phellodendron amurense* var. *sachalinense*. *Plant Species Biol.* 6:39–46

Moeller DA. 2004. Facilitative interactions among plants via shared pollinators. *Ecology* 85:3289–301

Moeller DA. 2005. Pollinator community structure and sources of spatial variation in plant-pollinator interactions in *Clarkia xantiana* ssp. *xantiana*. *Oecologia* 142:28–37

Momose K. 2004. Plant reproductive interval

and population density in aseasonal tropics. *Ecol. Res.* 19:245–53

Moody-Weis JM, Heywood JS. 2001. Pollination limitation to reproductive success in the Missouri evening primrose, *Oenothera macrocarpa* (Onagraceae). *Am. J. Bot.* 88:1615–22

Moragues E, Traveset A. 2005. Effect of *Carpobrotus* spp. on the pollination success of native plant species of the Balearic Islands. *Biol. Cons.* 122:611–19

Morgan MT, Wilson WG. 2005. Self-fertilization and the escape from pollen limitation in variable pollination environments. *Evolution* 59:1143–48

Morgan MT, Wilson WG, Knight TM. 2005. Plant population dynamics, pollinator foraging, and the selection of self-fertilization. *Am. Nat.* 166:169–183

Morris WF. 2003. Which mutualists are most essential?: buffering of plant reproduction against the extinction of pollinators. In *The Importance of Species: Perspectives on Expendability and Triage*, ed. P Karieva, S Levin, pp. 260–80. Princeton, NJ: Princeton Univ. Press

Mothershead K, Marquis RJ. 2000. Fitness impacts of herbivory through indirect effects on plant-pollinator interactions in *Oenothera macrocarpa*. *Ecology* 81:30–40

Muñoz AA, Arroyo MTK. 2004. Negative impacts of a vertebrate predator on insect pollinator visitation and seed output in *Chuquiraga oppositifolia*, a high Andean shrub. *Oecologia* 138:66–73

Mustajärvi K, Siikamaki P, Rytkonen S, Lammi A. 2001. Consequences of plant population size and density for plant-pollinator interactions and plant performance. *J. Ecol.* 89:80–87

Neal PR, Dafni A, Giurfa M. 1998. Floral symmetry and its role in plant-pollinator systems: terminology, distribution, and hypotheses. *Annu. Rev. Ecol. Syst.* 29:345–73

Neiland MRM, Wilcock CC. 1998. Fruit set, nectar reward, and rarity in the Orchidaceae. *Am. J. Bot.* 85:1657–71

Niesenbaum RA. 1993. Light or pollen:

seasonal limitations on female reproductive success in the understory shrub *Lindera benzoin. J. Ecol.* 81:315–23

O'Toole C. 1993. Diversity of native bees and agroecosystems. In *Hymenoptera and Biodiversity, 3rd Quadrennial Symp. Int. Soc. Hymenopterists, London,* ed. J LaSalle ID Gauld, pp. 69–106. London: Commonwealth Agricultural Bureau Int.

Pagel MD, Harvey PH. 1988. Recent developments in the analysis of comparative data. *Q. Rev. Biol.* 63:413–40

Paini DR. 2004. Impact of the introduced honey bee (*Apis mellifera*) (Hymenoptera: Apidae) on native bees: a review. *Aust. Ecol.* 29:399–407

Pannell JR, Barrett SCH. 1998. Baker's law revisited: reproductive assurance in a metapopulation. *Evolution* 52:657–68

Parker IM. 1997. Pollinator limitation of *Cytisus scoparius* (Scotch broom), an invasive exotic shrub. *Ecology* 78:1457–70

Parker MA. 1987. Pathogen impact on sexual vs. asexual reproductive success in *Arisaema triphyllum. Am. J. Bot.* 74:1758–63

Paton DC. 1993. Honeybees *Apis mellifera* in the Australian environment. Does *Apis mellifera* disrupt or benefit native biota? *BioScience* 43:95–103

Porcher E, Lande R. 2005. The evolution of self-fertilization and inbreeding depression under pollen discounting and pollen limitation. *J. Evol. Biol.* 18:497–508

Primack RB, Hall P. 1990. Costs of reproduction in the pink lady's slipper orchid: a 4-year experimental study. *Am. Nat.* 136:638–55

Primack RB, Stacy E. 1998. Cost of reproduction in the pink lady's slipper orchid, (*Cypripedium acaule*, Orchidaceae): an eleven-year experimental study of three populations. *Am. J. Bot.* 85:1672–79

Rambuda TD, Johnson SD. 2004. Breeding systems of invasive alien plants in South Africa: does Baker's rule apply? *Divers. Distrib.* 10:409–16

Ramsey M. 1995. Causes and consequences of seasonal variation in pollen limitation of seed production in *Blandfordia grandiflora* (Liliaceae). *Oikos* 73:49–58

Rathcke BJ. 1983. Competition and facilitation among plants for pollination. In *Pollination Biology,* ed. L Real, pp. 305–29. New York: Academic

Rathcke BJ. 2003. Floral longevity and reproductive assurance: seasonal patterns and an experimental test with *Kalmia latifolia* (Ericaceae). *Am. J. Bot.* 90:1328–32

Regal PJ. 1982. Pollination by wind and animals: ecology of geographic patterns. *Annu. Rev. Ecol. Syst.* 13:497–524

Richardson DM, Allsop N, D'Antonio CM, Milton SJ, Rejmánek M. 2000. Plant invasions: the role of mutualism. *Biol. Rev.* 75:65–93

Ricketts TH, Daily GC, Ehrlich PR, Michener CD. 2004. Economic value of tropical forest to coffee production. *Proc. Natl. Acad. Sci. USA* 101:12579–82

Ronsheim ML, Bever JD. 2000. Genetic variation and evolutionary trade-offs for sexual and asexual reproductive modes in *Allium vineale* (Liliaceae). *Am. J. Bot.* 87:1769–77

Rosenberg MS, Adams DC, Gurevitch J. 2000. MetaWin statistical software for meta-analysis version 2. Sunderland, MA: Sinauer Assoc. Inc.

Santrandreu M, Lloret F. 1999. Effect of flowering phenology and habitat on pollen limitation in *Erica multiflora. Can. J. Bot.* 77:734–43

Sargent RD. 2004. Floral symmetry affects speciation rates in angiosperms. *Proc. R. Soc. London Ser. B* 271:603–8

SAS Institute. 2001. *Version 8.2.* Cary, NC: SAS Inst. Inc.

Schaal BA. 1978. Density dependent foraging on *Liatris pycnostachya. Evolution* 32:452–54

Shykoff JA, Kolokotronis S-O, Collin CL, Lopez-Villavicencio M. 2003. Effects of male sterility on reproductive traits in gynodioecious plants: a meta-analysis. *Oecologia* 135:1–9

Sih A, Baltus MS. 1987. Patch size, pollinator

behavior, and pollinator limitation in Catnip. *Ecology* 68:1679–90

Steets JA, Ashman T-L. 2004. Herbivory alters the expression of a mixed-mating system. *Am. J. Bot.* 91:1046–51

Steffan-Dewenter I, Tscharntke T. 1999. Effects of habitat isolation on pollinator communities and seed set. *Oecologia* 121:432–40

Steinbachs JE, Holsinger KE. 2002. S-RNase-mediated gametophytic self-incompatibility is ancestral in eudicots. *Mol. Biol. Evol.* 19:825–29

Stephenson AG. 1981. Flower and fruit abortion: proximate causes and ultimate functions. *Annu. Rev. Ecol. Syst.* 12:253–79

Strauss SY, Conner JK, Rush SL. 1996. Foliar herbivory affects floral characters and plant attractiveness to pollinators: implications for male and female plant fitness. *Am. Nat.* 147:1098–1107

Sugden EA, Pyke GH. 1991. Effects of honey bees on colonies of *Exoneura asimillima*, an Australian native bee. *Aust. J. Ecol.* 16:171–81

Sutherland S. 1986. Patterns of fruit-set: what controls fruit-flower ratios in plants? *Evolution* 40:117–28

Suttle KB. 2003. Pollinators as mediators of top-down effects on plants. *Ecol. Lett.* 6:688–94

Thomson JD. 1989. Deployment of ovules and pollen among flowers within inflorescences. *Evol. Trends Plants* 3:65–68

Thomson JD. 2001. Using pollination deficits to infer pollinator declines: Can theory guide us? *Cons. Ecol.* 5(1):6. URL: http://www.consecol.org/vol5/iss1/art6/

Townsend PA, Levey DJ. 2005. An experimental test of whether habitat corridors affect pollen transfer. *Ecology* 86:466–75

Valido A, Dupond YL, Hansen DM. 2002. Native birds and insects, and introduced honey bees visiting *Echium wildpretii* (Boraginaceae) in the Canary Islands. *Acta Oecol.* 23:413–19

Vallejo-Marín M, Uyenoyama MK. 2004. On the evolutionary costs of self-incompatibility: incomplete reproductive compensation

due to pollen limitation. *Evolution* 58:1924–35

Vamosi JC, Otto SP. 2002. When looks can kill: the evolution of sexually dimorphic floral display and the extinction of dioecious plants. *Proc. R. Soc. London Ser. B* 269:1187–94

Vamosi JC, Vamosi SM. 2005a. Present day risk of extinction may exacerbate the lower species richness of dioecious clades. *Divers. Distrib.* 11:25–32

Vamosi JC, Vamosi SM. 2005b. Endless tests: guidelines for analyzing non-nested sister-group comparisons. *Evol. Ecol. Res.* 7:567–79

Vàzquez DP, Simberloff D. 2004. Indirect effects of an introduced ungulate on pollination and plant reproduction. *Ecol. Monogr.* 74:281–308

Verdu M, Traveset A. 2004. Bridging meta-analysis and the comparative method: a test of seed size on germination after frugivore's gut passage. *Oecologia* 138:414–18

Waites AR, Ågren J. 2004. Pollinator visitation, stigmatic pollen loads and among-population variation in seed set in *Lythrum salicaria*. *J. Ecol.* 92:512–26

Ward M, Johnson SD. 2005. Pollen limitation and demographic structure in small fragmented populations of *Brunsvigia radulosa* (Amaryllidaceae). *Oikos* 108:253–62

Waser NM. 1983. The adaptive nature of floral traits: ideas and evidence. In *Pollination Biology*, ed. L Real, pp. 242–85. Orlando, FL: Academic

Waser NM, Chittka L, Price MV, Williams NM, Ollerton J. 1996. Generalization in pollination systems and why it matters. *Ecology* 77:1043–60

Webb C, Ackerly D, Kembel S. 2004. *Phylocom: Software for the analysis of community phylogenetic structure and character evolution.* Version 3.22. http://www.phylodiversity.net/phylocom/

Webb CO, Ackerly DD, McPeek MA, Donoghue MJ. 2002. Phylogenies and community ecology. *Annu. Rev. Ecol. Syst.* 33:475–505

Webb CO, Donoghue MJ. 2005. Phylomatic:

tree assembly for applied phylogenetics. *Mol. Ecol. Notes* 5:181–83

Whitehead DR. 1983. Wind pollination: some ecological and evolutionary perspectives. In *Pollination Biology*, ed. L Real, pp. 97–108. New York: Academic

Widén B. 1991. Phenotypic selection on flowering phenology in *Senecio integrifolius*, a perennial herb. *Oikos* 61:205–15

Wilcock C, Neiland R. 2002. Pollination failure in plants: why it happens and when it matters. *Trends Plant Sci.* 7:270–77

Wilcock CC, Neiland MRM. 1998. Reproductive characters as priority indicators for rare plant conservation. In *Planta Europa: Proceedings of the Second European Conference on the Conservation of Wild Plants*, ed. H Synge, J Akeroyd, pp. 221–30. Uppsala, Sweden/London: UK: Swedish Threatened Plants Unit & Plantlife

Willson MF, Burley N. 1983. *Mate Choice in Plants*. Princeton, NJ: Princeton Univ. Press

Wilson P, Thomson JD, Stanton ML. 1994. Beyond floral Batemania: gender biases in selection for pollination success. *Am. Nat.* 143:283–96

Wilson WG, Harder LD. 2003. Reproductive uncertainty and the relative competitiveness of simultaneous hermaphroditism versus dioecy. *Am. Nat.* 162:220–41

Wolf AT, Harrison SP. 2001. Effects of habitat size and patch isolation on reproductive success of the serpentine morning glory. *Cons. Biol.* 15:111–21

Wolf DE, Takebayashi N. 2004. Pollen limitation and the evolution of androdioecy from dioecy. *Am. Nat.* 163:122–37

Wolfe BE, Husband BC, Klironomos JN. 2005. Effects of a belowground mutualism on an aboveground mutualism. *Ecol. Lett.* 8:218–23

Young HJ, Young TP. 1992. Alternative outcomes of natural and experimental high pollen loads. *Ecology* 73:639–47

Zhang D-Y, Wang G. 1994. Evolutionarily stable reproductive strategies in sexual organisms: an integrated approach to life-history evolution and sex allocation. *Am. Nat.* 144:65–75

Zimmerman M, Pyke GH. 1988. Reproduction in *Polemonium*: assessing the factors limiting seed set. *Am. Nat.* 131:723–38

Annu. Rev. Ecol. Evol. Syst. 2005. 36:499–518
doi: 10.1146/annurev.ecolsys.36.113004.083814
First published online as a Review in Advance on August 23, 2005

EVOLVING THE PSYCHOLOGICAL MECHANISMS FOR COOPERATION

Jeffrey R. Stevens, Fiery A. Cushman, and Marc D. Hauser

Primate Cognitive Neuroscience Laboratory, Department of Psychology, Harvard University, Cambridge, Massachusetts 02138; email: jstevens@wjh.harvard.edu, cushman@wjh.harvard.edu, mdh@wjh.harvard.edu

Key Words altruism, individual recognition, memory, mutualism, temporal discounting

■ **Abstract** Cooperation is common across nonhuman animal taxa, from the hunting of large game in lions to the harvesting of building materials in ants. Theorists have proposed a number of models to explain the evolution of cooperative behavior. These ultimate explanations, however, rarely consider the proximate constraints on the implementation of cooperative behavior. Here we review several types of cooperation and propose a suite of cognitive abilities required for each type to evolve. We propose that several types of cooperation, though theoretically possible and functionally adaptive, have not evolved in some animal species because of cognitive constraints. We argue, therefore, that future modeling efforts and experimental investigations into the adaptive function of cooperation in animals must be grounded in a realistic assessment of the psychological ingredients required for cooperation. Such an approach can account for the puzzling distribution of cooperative behaviors across taxa, especially the seemingly unique occurrence of cooperation observed in our own species.

1. INTRODUCTION

Vampire bats regurgitate blood to others despite the possibility of dying if three days elapse without consuming blood. Ground squirrels give alarm calls even though they alert predators to their own presence. Cleaner fish enter the mouths of their hosts to remove parasites even at risk of being eaten. Florida scrub jays often stay at home with their parents, foregoing the benefits of personal reproduction to help rear their younger siblings. These cases of cooperation have generated a substantial amount of theoretical and empirical interest over the past several decades, primarily focusing on adaptive accounts of cooperative behaviors. This adaptive perspective has been fruitful; indeed, the crowning glory of the sociobiological revolution beginning in the 1960s has been the overwhelming empirical support for its theoretical predictions targeted at adaptive accounts of social behavior (Alcock 2001, Hamilton 1964, Trivers 2002, Williams 1966, Wilson 1975). The adaptive view, however, fails to fully account for the empirical data on cooperative behavior. This weakness, in our opinion, results from a strictly ultimate perspective that ignores proximate mechanisms of cooperation. Here, we argue that evolutionary puzzles concerning the phylogenetic distribution of cooperative behaviors can be resolved by unraveling the psychological machinery upon which they depend.

We first discuss the adaptive challenges of cooperation and briefly review potential solutions. We then describe the empirical evidence for the different solutions, demonstrating that some types of cooperation occur much less frequently in non-human animals (hereafter animals) than theory predicts. Finally, we introduce the idea of cognitive constraints on cooperation to explain the limited taxonomic scope of certain kinds of cooperation. We conclude by discussing how a psychologically informed approach to cooperation opens a new set of questions, guides how we design our experiments, and helps resolve apparently contradictory findings concerning the uniqueness of human cooperation.

2. THE PROBLEM OF COOPERATION

We define cooperation as any behavior that provides a benefit to an individual other than the cooperator (where benefit is defined as an increase in reproductive success). Given the generally accepted Darwinian assumption that behavior evolves via natural selection, cooperation poses an apparent problem: What selective pressure favors individuals who provide benefits to other individuals? A number of models address this problem, of which we review four: mutualism, kin selection, reciprocity, and sanctioning (for more complete treatment, see Dugatkin 1997).

2.1. Mutualism

The simplest explanation for cooperative behavior is that it provides direct benefits to the cooperator, in addition to other individuals. This model of cooperative behavior is termed mutualism (Brown 1983, West Eberhard 1975). Any individual that defects (i.e., does not cooperate) in mutualistic situations will, by definition, do worse than a cooperator; therefore, in the absence of a temptation to defect, cooperation provides the best option. Importantly, mutualism does not depend on the identity of your partner and, therefore, can occur between any members of the same species and even members of different species (Boucher 1985, Herre et al. 1999).

2.2. Kin Selection

Cooperation that does not yield the direct benefits of mutualism poses an even deeper evolutionary paradox: altruism. Why would an individual help others, especially at a cost to itself? Kin selection provided the first clear theoretical solution to the paradox of altruism. Although introduced by Darwin (1859), Hamilton (1964) first mathematically formalized kin selection as a mechanism to maintain cooperation among genetic relatives. He suggested that individuals may bias cooperation toward their genetic relatives because it helps propagate their own genes. What looks altruistic from an individual's perspective actually serves self-interest from the gene's view (Dawkins 1976). Individuals share a certain proportion of their genes (r—the coefficient of relatedness) with relatives due to common descent. If the benefits to kin discounted by this coefficient of relatedness outweigh the costs of helping, altruism toward kin can evolve.

2.3. Reciprocity

Reciprocity, in which individuals pay a short-term cost of cooperation for the future benefit of a social partner's reciprocated cooperation, has probably been the most celebrated type of cooperation. Reciprocity aims to explain cooperative behavior in a unique type of social interaction termed the prisoner's dilemma (Flood 1958, Rapoport & Chammah 1965). The key aspects of the prisoner's dilemma are (a) cooperation maximizes the total payoff to everyone involved in the interaction (mutual cooperation provides more benefits than mutual defection); however, (b) any individual will receive a higher personal payoff by defecting, so a sizable temptation to cheat exists (Figure 1a). Pursuing unilateral cooperation in this game is not an evolutionarily stable strategy (Maynard Smith 1982, Maynard Smith & Price 1973).

Trivers (1971) suggested that reversing roles as donor and recipient of altruism may reduce the temptation to defect because individuals are investing in future cooperation. Reciprocity can stabilize cooperation if the following conditions are met: (a) the benefits to the recipient outweigh the costs to the donor, (b) individuals interact repeatedly, and (c) individuals recognize partners so they can detect cheaters. When the fitness payoffs sum over a series of interactions with the same partner, reciprocal strategists can reap the benefits of mutual cooperation (Figure 1b). The reciprocal strategy tit-for-tat (TFT), in which a player starts out cooperating and copies its opponent's behavior in previous interactions, can successfully invade and dominate simulated populations of social partners engaging in prisoner's dilemma games, winning out over many alternative behavioral strategies (Axelrod 1984, Axelrod & Hamilton 1981). If the probability of interacting again exceeds a critical level, a reciprocal strategy can maintain cooperation.

2.4. Sanctioning

Punishing defection can impose enough costs to offset the temptation to cheat and, like reciprocity, can elicit future cooperation (Boyd & Richerson 1992, Clutton-Brock & Parker 1995a). Punishment involves energetic costs and, when accomplished by aggression, also involves the cost of risked injury. Punishment can only be an adaptive behavior at the individual level of selection, therefore, when it successfully elicits cooperative behavior directed strictly at the punisher (Gardner & West 2004). In this manner punishment resembles reciprocity,

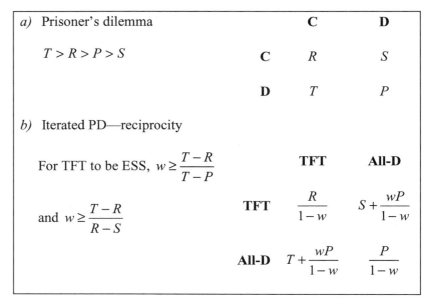

Figure 1 Economics of cooperation. Evolutionary biologists have used the prisoner's dilemma as the standard model of altruistic cooperation. (*a*) The payoffs are structured such that a defector playing against a cooperator receives the highest payoff (*T*), mutual cooperators receive the next highest payoff (*R*), mutual defectors receive the next highest payoff (*P*), and cooperators playing against defectors receive the lowest payoff (*S*). Because no matter what your opponent chooses, you do better by defecting (*T* > *R*, *P* > *S*), defection is the only stable strategy when playing a one-shot prisoner's dilemma. (*b*) When iterating the game over several interactions, however, cooperation can stabilize (Axelrod & Hamilton 1981). A population of reciprocal strategies playing tit-for-tat (TFT) can avoid invasion of all defectors (All-D) with a high probability of future interaction *w*.

which should occur when it elicits cooperative behavior directed strictly at the reciprocator.

Another form of sanctioning is harassment. Whereas punishment penalizes past behavior with the hope of future reward, harassment penalizes present behavior with the hope of present reward. For instance, if a defector has consumed 100% of a food resource, a punisher could punish the defector in the hope of obtaining 50% of the food resource in some future interaction; however, if a defector is in the process of consuming a food resource, a harasser could impose costs on the defector aimed at obtaining 50% of the food resource immediately. By harassing during defection, individuals may induce cooperation, thereby providing an immediate individual benefit for the harasser, rather than the delayed benefit of future cooperation (Stevens & Stephens 2002).

3. EVIDENCE FOR COOPERATION

Despite the abundance of theoretical interest in the different models of cooperation, over 30 years of research on a wide variety of species, under lab and field conditions, reveals that some occur much more frequently in animals than others. In particular, mutualism and kin-biased cooperation account for the vast majority of instances of cooperation. In stark contrast, only a handful of species have demonstrated reciprocity, and even within these species reciprocity occurs infrequently (Hammerstein 2003, Noë 1990, Stevens & Hauser 2004). Sanctioning is also inconsistently distributed across taxa; punishment appears to be less common than mutualism or kin selection, but harassment may occur frequently, perhaps especially among primates (Stevens & Gilby 2004). Here we describe only a fraction of the empirical evidence for the four models of cooperation. The goal is to look at a few selective cases and then attempt to explain the patterns in light of our argument for cognitive constraints.

3.1. Mutualism

Mutualism may be the most common form of cooperation; it occurs frequently across a wide variety of taxa. Cooperative hunting provides a prime example described in numerous species (Dugatkin 1997, Krause & Ruxton 2002). Cooperative hunting provides mutualistic benefits only when the per capita intake rate increases with group size. Therefore, a pair of hunters would have to capture more than twice as many prey items as a solitary hunter. For cooperative hunting to benefit hunters, the success rate of solitary hunters should be fairly low, making cooperation particularly successful (Packer & Ruttan 1988). In addition to increasing the probability of capture, cooperative hunting may also reduce the individual costs of hunting (Creel 1997). Combined, these two mechanisms can lead to direct, immediate, and simultaneous fitness benefits for cooperative hunters, particularly when hunting small or difficult prey.

Another example of mutualism is food recruitment in ravens (*Corvus corax*). Heinrich and colleagues (Heinrich 1989, Heinrich & Marzluff 1991) observed that when ravens discovered animal carcasses, they often gave food calls and returned to communal roosts to recruit others. After ruling out reciprocity and kin-biased cooperation (Heinrich 1988, Parker et al. 1994), Heinrich proposed that the callers recruited others to prevent territory owners from monopolizing the food. Therefore, though it appears altruistic, recruiting actually increased access to an otherwise unavailable food source.

Cooperative breeders may also accrue mutualistic benefits (Woolfenden & Fitzpatrick 1978). Individuals help raise their younger siblings because larger group sizes can yield benefits such as larger territories. These territories often have better access to food, reduced predation risk, and space to establish their own territories (Clutton-Brock 2002, Cockburn 1998, Stacey & Koenig 1990). Some species even adopt or kidnap unrelated offspring, presumably to reap the benefits of increasing their group size (Connor & Curry 1995, Heinsohn 1991, Zahavi 1990).

3.2. Kin Selection

Kin-biased cooperation commonly occurs in species ranging from arthropods to apes (Bourke 1997, Chapais & Berman 2004, Silk 2002). Darwin's original paradox of altruism was motivated by the kin-biased cooperation that occurs in eusocial insects such as ants, bees, and wasps (Bourke 1997). In eusocial species, individuals cooperate to care for the young, divide reproductive opportunities such that only a few individuals reproduce, and have overlapping generations (Hölldobler & Wilson 1990). Importantly, many individuals forego reproduction completely to aid in the care of their siblings. Some species even express haplodiploid genotypes with haploid males and diploid females. This results in sisters being highly related to each other ($r = 3/4$) but less closely related to their brothers ($r = 1/4$) and even their own daughters and sons ($r = 1/2$); therefore, females should bias more cooperation toward other sisters. In a now classic study, Trivers & Hare (1976) used this unique system to test kin selection in ants. They found that females matched their altruistic allocations to the degree of genetic relatedness: Investment in other females exceeds investment in males by three orders of magnitude (for an alternative perspective, see Reeve 1993).

Kin selection may be particularly powerful in cooperatively breeding species, although it by no means accounts for all cooperative behaviors in these systems (Clutton-Brock 2002, Cockburn 1998, Griffin & West 2002). Kin selection theory predicts that individuals should preferentially help relatives and should help in proportion to their coefficient of relatedness. Reviews of the literature support this prediction, demonstrating that the probability and amount of help correlate with the coefficient of relatedness (Emlen 1997, Griffin & West 2003). Griffin & West (2003) tested even stronger predictions in a meta-analysis by demonstrating that the effect of kin selection (i.e., the correlation between helping and relatedness) correlates highly with the benefit of helping. This meta-analysis extended the findings of Trivers & Hare (1976) to other species, suggesting that helping scales with not only the coefficient of relatedness (r) but also the benefit of helping.

3.3. Reciprocity

Many authors have reported reciprocity—the alternation of receiving costs and benefits—in numerous contexts including food sharing (de Waal & Berger 2000, Hauser et al. 2003, Watts & Mitani 2002), grooming (Barrett & Henzi 2001, Rowell et al. 1991), predator inspection (Dugatkin 1988, Milinski 1987), and coalitions associated with mating opportunities (Packer 1977). Unfortunately, most examples of reciprocity suffer from one of two problems: (*a*) they have not been replicated, and (*b*) alternative explanations, such as kin selection and mutualism, can account for the observed reciprocal pattern.

One of the first reported instances of reciprocity occurred in olive baboons (*Papio anubis*) (Packer 1977). Males formed coalitions in order to drive off rival males and gain access to reproductively active females. Packer's analyses suggested that males took turns reaping the benefits, implying evidence of reciprocity. A subsequent study of a different population of baboons did not find the same reciprocal patterns (Bercovitch 1988), and a study of a closely related species proposes an alternative explanation. Noë (1990) suggested that male savannah baboons (*Papio cyanocephalus*) implemented multiplayer market strategies during coalition formation. Coalition partners did not face a prisoner's dilemma; rather they faced a veto game in which the coalition initiator forces other males to "bid" on joining the coalition. If the potential helpers did not join, they would not receive mating opportunities. Therefore, the coalition initiator can choose the partner which will take the lowest share of mating opportunities.

Probably the best-known putative example of reciprocity is Wilkinson's (1984) study of blood sharing in vampire bats (*Desmodus rotundus*). In this study, vampire bats regurgitated blood to individuals that failed to obtain blood meals on their own. The bats shared mainly with individuals with whom they associated, leading Wilkinson to conclude that the bats reciprocated. Although commonly cited as evidence for reciprocity, only 5 of the 98 instances of sharing between individuals of known genetic relatedness involved individuals less related than grandparent to grandchild ($r < 0.25$); most occurred between mother and offspring. Therefore, direct fitness benefits and kin selection can account for the vast majority of sharing, leaving instances of potential reciprocal sharing quite rare and possibly attributable to recognition errors.

Another controversial example of reciprocity is predator inspection in stickleback fish (*Gasterosteus aculeatus*) and guppies (*Poecilia reticulata*). Milinksi (1987) and Dugatkin (1988) found that when inspecting a predator fish, individuals copied their partners' approach or retreat (the "partners" were actually mirrors that mimicked a fish either swimming with or away from the subject). These experiments elicited a flood of criticisms. In particular, the fitness payoffs of cooperation and defection were unclear (Connor 1996, Lazarus & Metcalfe 1990), and rather than reciprocating, the target fish may simply have preferred to stay in groups to reduce predation risk (Masters & Waite 1990, Stephens et al. 1997). Minimally, we consider the stickleback and guppy work on reciprocity as unresolved.

Finally, we describe several cases that appear to offer viable evidence of reciprocity. Hauser et al. (2003) conducted a series of experiments in which cotton-top tamarins (*Saguinus oedipus*) could altruistically pull a tool to give food to an unrelated recipient without getting any food for itself. Subjects alternated which partner had the opportunity to pull with a short time interval between trials. Tamarins pulled the tool most often for partners that always pulled and infrequently for partners that never pulled. The tamarins, however, cooperated less than 50% of the time, and as each game progressed, the amount of food given decreased. Tamarins, therefore, maintained a moderate level of cooperation when receiving food closely followed giving food.

Some researchers have proposed reciprocal egg swapping in simultaneously hermaphroditic fish (chalk bass—*Serrannus tortugarum*) and polychaete worms (*Ophryotrocha gracilis*) as examples of reciprocity (Fischer 1988, Sella et al. 1997). These species produce both male and female gametes and, therefore, can both give and receive fertilizations. Because eggs cost more to produce, a defector could fertilize a cooperator's eggs but offer none of its own, thereby avoiding costly egg production. Both the fish and worms repeatedly alternate depositing packets of eggs and fertilizing their opponents' packets of eggs. Depositing eggs depends on the partner's behavior because the interaction stops when the partner fails to deposit eggs. Reciprocal allogrooming in impala (*Aepyceros melampus*) follows a similar pattern: Individuals groom one another for short bouts and then receive grooming from their partner, repeatedly alternating who grooms (Hart & Hart 1992). In all of these examples, the small time delay between paying the cost of cooperation and receiving the benefit minimizes the chance of defection. In Section 4.2, we further discuss the significance of time in the evaluation of reciprocal possibilities.

3.4. Sanctioning

Despite the theoretical interest, punishment is not well documented in animals. Clutton-Brock & Parker (1995a) offered several examples of punishment enforcing cooperation in animals, but we would characterize these examples as harassment because the punisher usually receives immediate benefits from punishing. For instance, queen paper wasps (*Polistes fuscatus*) attack lazy workers; when Reeve & Gamboa (1987) removed the queen from the colony, the workers stopped working. Although reported as punishment, one should categorize this as harassment because the queen's aggression immediately increased the activity of the workers.

Animals often impose costs on others to influence their current behavior (Clutton-Brock & Parker 1995b, Stevens 2004, Wrangham 1975). For instance, Gilby (2004) studied the food-sharing patterns of wild chimpanzees (*Pan troglodytes*). After capturing prey, the chimpanzees frequently allowed other individuals to consume part of the meat. Gilby showed that harassment accounted for the pattern of food sharing, because harassment was costly for the food owner (food intake rate decreased as the number of beggars increased), owners shared more often when beggars harassed frequently and intensely, and when sharing occurred harassment levels decreased. Controlled experiments corroborated these findings with captive chimpanzees and extended them to other species such as squirrel monkeys (*Saimiri boliviensis*), a species that rarely cooperates (Stevens 2004).

Harassment may, of course, influence future as well as current cooperation, suggesting that it may lead to punishment strategies. For example, when rhesus monkeys (*Macaca mulatta*) discovered food, those individuals who announced their discovery by vocalizing faced fewer attacks than those individuals who made their discoveries in silence, apparently withholding information (Hauser 1992,

Hauser & Marler 1993). This sanctioning imposed costs on the discoverer not only in terms of potential for injury and wasted energy but also by reduced food intake relative to vocal discoverers. This sanctioning yielded an immediate benefit of accessing food, clearly qualifying it as a case of harassment. In addition, these data suggest that the sanctioning may have had a punishing effect on silence, eliciting future food calls. That is, sanctioning behavior resulted in both immediate and future benefits. To the authors' knowledge, no clear evidence in animals demonstrates that punishment influences future cooperation in the absence of harassment.

In summary, although ample evidence of cooperation exists in the animal kingdom, reciprocity and punishment rarely occur in animals, especially when contrasted with humans (Fehr & Gächter 2002, Ostrom & Walker 2003). We next turn to an explanation for this taxonomic distribution.

4. COGNITIVE CONSTRAINTS ON COOPERATION

Mutualism and kin selection are both theoretically well understood and empirically well documented (Dugatkin 1997). By contrast, reciprocity and punishment, although theoretically feasible, do not frequently occur in animals (although harassment may be more widespread). Therefore, despite models purporting the evolutionary stability of all of these types of cooperation, some types occur much more frequently than others. Unfortunately, a strictly adaptive perspective has limited power to explain the frequency of mutualism and kin-biased cooperation, and the rarity of reciprocity and punishment. A proximate perspective that keeps its eye on the ultimate problem can, however, reveal how psychological constraints limit or facilitate particular forms of cooperation.

The proximate approach emphasizes critical aspects of reciprocity and punishment that differ markedly from mutualism and kin-biased cooperation. In both reciprocity and punishment, the fitness benefits associated with cooperation depend on the partner's behavior: Cooperation should only occur when the partner responds by reciprocating or punishing. When this contingent response occurs in the future, the temporal delay introduces cognitive challenges that may constrain the emergence and stability of cooperation (Stevens & Gilby 2004). Animals can easily implement strategies that yield immediate benefits, such as mutualism and harassment, because individuals do not have to track benefits over time. With a time delay between cooperating and receiving return benefits, however, individuals must invest in an uncertain future. Delayed benefits impede learning the consequences of cooperation, require more memory capacity for previous interactions, and trade off short-term fitness gains for long-term gains. Here we provide a sketch of our proximate perspective on cooperation, highlighting several key cognitive constraints as a way to distinguish mutualism, kin-biased cooperation, reciprocity, and sanctioning.

4.1. Cognitive Constraints on Mutualism and Kin-Biased Cooperation

Because no temptation to cheat exists in mutualistic interactions, individuals should always cooperate. As a result, mutualism requires no special cognitive abilities above and beyond the challenges inherent in the cooperative behavior itself. Although Dugatkin & Alfieri (2002) contend that animals must recognize whether they are in a mutualistic situation, we argue that recognition is neither necessary for nor specific to mutualism. Kin-biased cooperation, on the other hand, does require additional cognitive capacities. At a minimum, it requires the capacity to direct cooperative actions to related individuals (Dugatkin & Alfieri 2002). Indeed, Hamilton's formulation of kin selection spawned a critical series of empirical studies showing that numerous species had the ability to make certain kin discriminations (Fletcher & Michener 1987, Hepper 1991). These studies were critical in the general acceptance of kin selection.

Mechanisms of kin recognition include recognition alleles, phenotype matching, and spatial and familiarity cues (Hepper 1991, Sherman et al. 1997, Wilson 1987). The recognition allele hypothesis predicts that individuals can compare a particular phenotypic cue (auditory, olfactory, visual, etc.) to an innately specified template (e.g., the "green beard effect"—Dawkins 1976, Hamilton 1964). Such a model requires few cognitive skills other than discriminating the cue associated with relatedness. Keller & Ross (1998) suggested that fire ants (*Solenopsis invicta*) may use recognition alleles to selectively kill queens that do not share their genotype. Phenotypic matching occurs when an individual compares a conspecific's phenotypic cues to a learned template. This requires specialized perceptual and computational systems that detect cues at an early stage to form a template, then test cues against the template to discriminate kin (Hauber & Sherman 2001). Peacocks (*Pavo cristatus*), golden hamsters (*Mesocricetus auratus*), and a number of other species have demonstrated phenotypic matching (Mateo & Johnston 2000, Petrie et al. 1999). Finally, a common alternative is to use a simple set of rules such as spatial and familiarity cues to discriminate kin. Often, animals may use rules such as "be nice to individuals near your home" or "help those that you grew up with" to direct the benefits of cooperation toward kin. These mechanisms occur regularly in a variety of animal taxa.

4.2. Cognitive Constraints on Reciprocity and Punishment

Trivers' (1971) classic formulation of reciprocity had three requirements for evolutionary stability: (*a*) the reciprocated benefit must outweigh the immediate cost, (*b*) individuals must interact repeatedly, and (*c*) individuals must recognize each other. We contend that these requirements do not capture the cognitive sophistication required for utilizing reciprocal strategies. In particular, the delay between the cost of a cooperative act and the benefit of reciprocated cooperation introduces a number of cognitive challenges. Like reciprocity, punishment can involve a delay between a costly act and a beneficial payoff, and in these cases it faces similar constraints. For this reason, we consider the constraints on reciprocity and punishment together.

4.2.1. INDIVIDUAL RECOGNITION   Trivers does propose individual recognition as a cognitive ability needed to avoid cheaters and stabilize cooperation via reciprocity (see also Dugatkin & Alfieri 2002). Punishment also requires individual recognition to ensure that individuals only cooperate with punishers, thereby preventing punishment from benefiting nonpunishers (Gardner & West 2004). Therefore, the delayed, contingent response required for both reciprocity and punishment necessitates that individuals can distinguish different partners. Numerous species across the animal kingdom possess the ability to recognize individuals, however, so the necessity for a mechanism of individual recognition cannot explain the paucity of cooperative behavior across most nonhuman taxa.

4.2.2. TEMPORAL DISCOUNTING   Temporal discounting is the devaluing of future rewards, which often results in a preference for smaller, immediate rewards over larger, delayed rewards. Many psychologists who study discounting consider the prisoner's dilemma to be analogous to the discounting problem (Green et al. 1995, Rachlin 2000). Individuals must choose between the immediate reward of defecting and the long-term reward of cooperating. Indeed, a number of researchers have predicted that temporal discounting can reduce the value of reciprocated benefits (Frank 1988, May 1981, Trivers 1971). Experimental data on variation in human discounting and cooperation validate the view that a preference for immediate rewards may inhibit reciprocity. Discounting correlates with cooperation such that individuals who prefer immediacy cooperate less frequently (Harris & Madden 2002). In parallel, blue jays (*Cyanocitta cristata*) show stable cooperation in the prisoner's dilemma only following a reduction in their preference for immediacy resulting from accumulating payoffs over several trials (Stephens et al. 2002). Therefore, if animals highly discount future rewards (Figure 2; Mazur 1987, Richards et al. 1997, Stevens et al. 2005), the immediate benefits of defections may outweigh the future reciprocated benefits.

4.2.3. MEMORY   Limitations in memory decay, interference, and capacity can also constrain the frequency of reciprocity and punishment. Models of forgetting predict exponential or power functions (Sikstrom 2002, White 2001, Wixted 2004), because memories decay rapidly over time (Figure 3). Therefore, longer time intervals between cooperative acts may make reciprocity and punishment more difficult. Even with short time delays between cooperative interactions and few distractions, every potential new partner increases the computational load of tracking debts owed, favors given, and costs imposed. Keeping score of reciprocal obligations and punishment with multiple individuals may place a computationally intensive burden on memory systems. Although few studies examine learning and memory constraints in animal cooperation, human studies suggest that these constraints can pose challenges for maintaining stable cooperative relationships (Milinski & Wedekind 1998).

Existing evidence demonstrates the importance of the time delay between paying the costs of cooperation and receiving the benefits. In the previously described examples of food sharing in tamarins, reciprocal egg swapping in fish and worms,

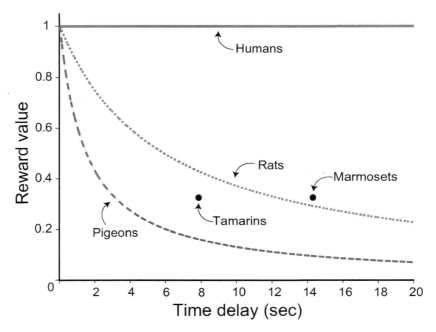

Figure 2 Species comparison of temporal discounting. Although the actual form of the discounting function is debated (Frederick et al. 2002, Kacelnik 2003), the value of a delayed reward decreases with the time to receiving the reward. Plotting estimated hyperbolic functions for pigeons and rats (Mazur 1987, Richards et al. 1997) and individual data points for tamarins and marmosets (Stevens et al. 2005) shows very high levels of impulsivity. The value of a reward decreases by 50% in the first 2–6 s. In contrast, humans show parallel reductions in value in months rather than seconds (Rachlin et al. 1991). Note that there are a number of important differences between the human and nonhuman studies including the reward currency (money versus food) and experimental techniques (hypothetical situations versus operant training).

and reciprocal allogrooming in impala, the time delay is minimal. This greatly reduces the cognitive demands for reciprocity: Individuals no longer need to recognize each other because they are always in proximity during the interactions, the benefits accrue immediately, and thus avoid discounting, and memory is not needed because individuals can leave when their partner cheats.

4.3. Cognitive Constraints on Harassment

Like mutualism and kin-biased cooperation, harassment does not suffer from the same cognitive limitations as reciprocity and punishment, primarily because of the brief time delay between imposed costs and potential cooperation. Harassment does not require individual recognition because individuals need not interact

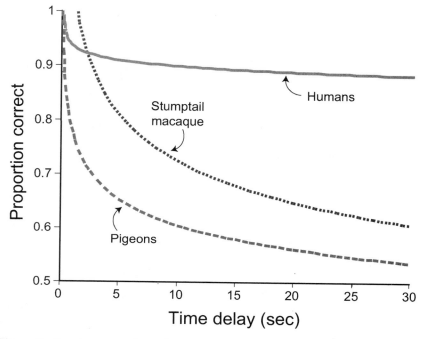

Figure 3 Species comparison of memory. Comparisons of memory across species is probably even more problematic than discounting, again because of methodological differences. Stumptail macaques (*Macaca arctoides*) and pigeons show fairly steep decreases in memory retention in binary delayed matching-to-sample tasks (Jarrad & Moise 1970, Wixted & Ebbesen 1991). Although not a perfectly analogous comparison, face recognition tasks in humans show a much longer retention interval (Wixted & Ebbesen 1991).

repeatedly; harassment can elicit cooperation between perfect strangers. Temporal discounting plays a minor role because harassers receive immediate benefits to offset their own costs of harassing. In addition, eliminating the time delay removes limitations associated with memory. Although harassment is less well studied than the other forms of cooperation, we predict that it occurs frequently in the animal kingdom given its weak demands on psychological capacities and its utility in reaping benefits at a small cost.

5. CONCLUSIONS

Several different models have solved the adaptive paradox posed by cooperative behavior; here, we have reviewed mutualism, kin selection, reciprocity, and sanctioning. Though empirical evidence for mutualism and kin-biased cooperation is widespread, reciprocity occurs relatively infrequently among nonhuman animals. Many of the current studies on sanctioning in animals are better explained by harassment models than by punishment models. We propose that a unique feature of reciprocity and punishment can explain this mismatch between the models and the data: Individuals must balance the costs and benefits of repeated interactions over periods of delay. Balancing costs and benefits over time poses several cognitive challenges, and therefore the emergence of reciprocity and punishment faces nontrivial psychological constraints. Here we focus on the possible role of individual recognition, temporal discounting, and memory as specific constraints on reciprocity and punishment. This is by no means an exhaustive list of abilities required to implement these strategies. Instead, this approach raises important questions about the nature of cooperation. What other abilities does cooperation require? Does incorporating these constraints lead to more predictive power in models of cooperation? How do psychological and evolutionary prerequisites interact to allow the implementation of reciprocal and punishment strategies? How do these and other factors influence not only animal cooperation but also the ontogeny of cooperation in our own species?

Knowing that a variety of psychological mechanisms facilitate cooperation allows us to design more appropriate experiments. Investigations of reciprocity and punishment must first evaluate the limitations of animals in the areas we have described, as well as others. What are the recognition abilities, discounting rates, memory features, numerical discrimination abilities, and learning rates of the species being investigated? For example, given evidence of limits on number discrimination (Dehaene 1997, Gallistel 1990, Hauser 2000), it makes little sense to set up payoff matrices that entail nondiscriminable alternatives. When testing reciprocity and punishment, researchers must first consider the cognitive constraints operating in the target animal and, based on this analysis, design an appropriate experiment.

We emphasize that the cognitive constraints discussed in this paper pose a challenging hurdle to the evolution of reciprocity and punishment but not an insurmountable barrier. Instances abound of extraordinary cognitive adaptations narrowly tailored to specific behavioral routines, overcoming the more general cognitive limitations. For instance, Clark's nutcrackers (*Nucifraga columbiana*) stash several thousand seeds each fall, foregoing the opportunity of immediate consumption in favor of future benefits. This behavior is grossly inconsistent with the typical rate of temporal discounting in nonhuman animals, implicating a specific cognitive adaptation bypassing a more general constraint. The nutcracker also depends on extraordinary spatial memory, which greatly facilitates its capacity to retrieve stashed food from thousands of hiding locations several months later (Balda & Kamil 1992, Kamil et al. 1994). These impressive cognitive adaptations enable a single, specific behavioral routine. By analogy, we should expect to find reciprocity and punishment in instances where adaptation has overcome the initial cognitive constraints—where narrowly tailored cognitive mechanisms have evolved to support specific behavioral routines.

An exception to this pattern, gratifyingly, seems to prove the rule. In humans, reciprocity and punishment commonly occur across a broad array of social interactions (Camerer 2003, Fehr & Gächter 2002, Fehr et al. 2002, Gurven 2004, Ostrom & Walker 2003, Ostrom et al. 1992). Far from being narrowly tailored to specific behavioral routines, reciprocity and punishment are broad and flexible strategies that can be applied to novel circumstances. Traditional, ultimate models fail to explain the ubiquity of cooperation among humans, where factors like population size, migration rates, frequency of interaction, and the cost-benefit structure of social interaction play the key roles in constraining or enabling reciprocity and punishment. By these measures, nothing about humans is very unique. Factoring in the role of cognitive constraints, however, helps explain the uniqueness of human cooperative behavior. In important ways, human cognition differs from nonhuman cognition and may pose fewer constraints on the emergence of reciprocity and sanctioning. For instance, rats, pigeons, and even nonhuman primates devalue rewards postponed by just a few seconds; to see this kind of discounting function in humans requires extending choices over months rather than seconds (Figure 2; Mazur 1987, Rachlin et al. 1991, Richards et al. 1997, Stevens et al. 2005). As a result, humans do not need to evolve narrowly tailored exceptions to their general rate of temporal discounting because it does not impose a severe constraint on reciprocity or punishment. Other features of human cognition that may enable reciprocity and punishment include face recognition and episodic memory, allowing for specific social interactions to be recalled; language, allowing for the negotiation of threats and promises and for facilitated bookkeeping by tagging cooperators and cheaters with symbols or labels; and theory of mind, allowing for inferences of intent and motivation in social exchange. Although some of these cognitive capacities are shared in part or in whole with nonhuman animals, others appear unique to humans. In his original formulation of reciprocity, Trivers (1971) emphasized the myriad cognitive abilities that humans may use to implement reciprocal strategies. We, however, disagree with Trivers' suggestion that these abilities evolved after reciprocity as regulating mechanisms. Instead, we concur with Darwin (1872) and Williams (1966) in that reciprocity requires the existence of these faculties before it can evolve.

Why have no other species evolved these mechanisms to allow reciprocity? The effort poured into the theoretical analyses of reciprocity may not reflect its frequency in the wild. Animals may not face altruistic situations in which they interact repeatedly with nonkin. Many instances of cooperation that appear altruistic may, instead, provide immediate mutualistic benefits such as the raven food-calling example. Similarly, individuals interact with genetic relatives so often that kinship may drive the majority of their social situations (e.g., vampire bat blood sharing). With few opportunities for reciprocity to provide benefits, selection may have been too weak to overcome cognitive constraints.

Mutualism, kin selection, reciprocity, and sanctioning elegantly explain how we can reconcile cooperative behavior against the Darwinian maxim that selection favors behavior that maximizes personal gains in terms of survival and reproduction.

Developed in response to an adaptive paradox, however, these models have long neglected the role of mechanistic constraints. Integrating animal psychology into current models of cooperative behavior can help explain the curious taxonomic distribution of reciprocity and punishment, which are rare among nonhuman animals but ubiquitous among humans. We have touched upon a few of the possible cognitive constraints on cooperation, and eagerly anticipate future research to expand and clarify the role of others.

ACKNOWLEDGMENTS

We gratefully acknowledge funding from the National Institutes of Health (NRSA for J.R.S.) and the National Science Foundation (ROLE grant for M.D.H.). We appreciate comments on the manuscript from Mike Ryan, David Stephens, and Robert Trivers.

The *Annual Review of Ecology, Evolution, and Systematics* is online at http://ecolsys.annualreviews.org

LITERATURE CITED

Alcock J. 2001. *The Triumph of Sociobiology.* Oxford: Oxford Univ. Press

Axelrod R. 1984. *The Evolution of Cooperation.* New York: Basic Books. 241 pp.

Axelrod R, Hamilton WD. 1981. The evolution of cooperation. *Science* 211:1390–96

Balda RP, Kamil AC. 1992. Long-term spatial memory in Clark's nutcracker, *Nucifraga columbiana. Anim. Behav.* 44:761–69

Barrett L, Henzi SP. 2001. The utility of grooming in baboon troops. In *Economics in Nature: Social Dilemmas, Mate Choice, and Biological Markets*, ed. P Hammerstein, pp. 119–45. Cambridge: Cambridge Univ. Press

Bercovitch F. 1988. Coalitions, cooperation and reproductive tactics among adult male baboons. *Anim. Behav.* 36:1198–209

Boucher DH. 1985. *The Biology of Mutualism: Ecology and Evolution.* New York: Oxford Univ. Press

Bourke AFG. 1997. Sociality and kin selection in insects. See Krebs & Davies 1997, pp. 203–27

Boyd R, Richerson PJ. 1992. Punishment allows the evolution of cooperation (or anything else) in sizable groups. *Ethol. Sociobiol.* 13:171–95

Brown JL. 1983. Cooperation—a biologist's dilemma. In *Advances in the Study of Behaviour*, ed. JS Rosenblatt, pp. 1–37. New York: Academic

Camerer C. 2003. *Behavioral Game Theory: Experiments in Strategic Interaction.* Princeton: Princeton Univ. Press

Chapais B, Berman CM. 2004. *Kinship and Behavior in Primates.* Oxford: Oxford Univ. Press

Clutton-Brock TH. 2002. Breeding together: kin selection and mutualism in cooperative vertebrates. *Science* 296:69–72

Clutton-Brock TH, Parker GA. 1995a. Punishment in animal societies. *Nature* 373:209–16

Clutton-Brock TH, Parker GA. 1995b. Sexual coercion in animal societies. *Anim. Behav.* 49:1345–65

Cockburn A. 1998. Evolution of helping behavior in cooperatively breeding birds. *Annu. Rev. Ecol. Syst.* 29:141–77

Connor RC. 1996. Partner preferences in by-product mutualisms and the case of predator inspection in fish. *Anim. Behav.* 51:451–54

Connor RC, Curry RL. 1995. Helping

non-relatives: a role for deceit? *Anim. Behav.* 49:389–93

Creel S. 1997. Cooperative hunting and group size: assumptions and currencies. *Anim. Behav.* 54:1319–24

Darwin C. 1859. *On the Origin of Species.* London: Murray

Darwin C. 1872. *The Descent of Man and Selection in Relation to Sex.* London: Murray

Davies NB, Krebs JR, ed. 1997. *Behavioural Ecology: An Evolutionary Approach.* Oxford: Blackwell Sci. 4th ed.

Dawkins R. 1976. *The Selfish Gene.* Oxford: Oxford Univ. Press

Dehaene S. 1997. *The Number Sense.* New York: Oxford Univ. Press

de Waal FBM, Berger ML. 2000. Payment for labour in monkeys. *Nature* 404:563

Dugatkin LA. 1988. Do guppies play Tit for Tat during predator inspection visits? *Behav. Ecol. Sociobiol.* 23:395–99

Dugatkin LA. 1997. *Cooperation Among Animals: An Evolutionary Perspective.* New York: Oxford Univ. Press

Dugatkin LA, Alfieri MS. 2002. A cognitive approach to the study of animal cooperation. In *The Cognitive Animal: Empirical and Theoretical Perspectives on Animal Cognition,* ed. M Bekoff, C Allen, GM Burghardt, pp. 413–19. Cambridge, MA: MIT Press

Emlen ST. 1997. Predicting family dynamics in social vertebrates. See Krebs & Davies 1997, pp. 228–53

Fehr E, Fischbacher U, Gächter S. 2002. Strong reciprocity, human cooperation and the enforcement of social norms. *Hum. Nat.* 13:1–25

Fehr E, Gächter S. 2002. Altruistic punishment in humans. *Nature* 415:137–40

Fischer EA. 1988. Simultaneous hermaphroditism, Tit-for-Tat, and the evolutionary stability of social systems. *Ethol. Sociobiol.* 9:119–36

Fletcher DJC, Michener CD. 1987. *Kin Recognition in Animals.* New York: John Wiley & Sons. 476 pp.

Flood MM. 1958. Some experimental games. *Manag. Sci.* 5:5–26

Frank RH. 1988. *Passions Within Reason: The Strategic Role of the Emotions.* New York: Norton

Frederick S, Loewenstein G, O'Donoghue T. 2002. Time discounting and time preference: a critical review. *J. Econ. Lit.* 40:351–401

Gallistel CR. 1990. *The Organization of Learning.* Cambridge, MA: MIT Press

Gardner A, West SA. 2004. Cooperation and punishment, especially in humans. *Am. Nat.* 164:753–64

Gilby IC. 2004. *Hunting and meat sharing among the chimpanzees of Gombe National Park, Tanzania.* PhD thesis. Univ. Minn., Minneapolis

Green L, Price PC, Hamburger ME. 1995. Prisoner's dilemma and the pigeon: control by immediate consequences. *J. Exp. Anal. Behav.* 64:1–17

Griffin AS, West SA. 2002. Kin selection: fact and fiction. *Trends Ecol. Evol.* 17:15–21

Griffin AS, West SA. 2003. Kin discrimination and the benefit of helping in cooperatively breeding vertebrates. *Science* 302:634–36

Gurven M. 2004. To give and to give not: the behavioral ecology of human food transfers. *Behav. Brain Sci.* 27:543–83

Hamilton WD. 1964. The genetical evolution of social behaviour. I, II. *J. Theor. Biol.* 7:1–52

Hammerstein P. 2003. Why is reciprocity so rare in social animals? A protestant appeal. In *Genetic and Cultural Evolution of Cooperation,* ed. P Hammerstein, pp. 83–94. Cambridge, MA: MIT Press

Harris AC, Madden GJ. 2002. Delay discounting and performance on the prisoner's dilemma game. *Psychol. Rec.* 52:429–40

Hart BL, Hart LA. 1992. Reciprocal allogrooming in impala, *Aepyceros melampus. Anim. Behav.* 44:1073–83

Hauber ME, Sherman PW. 2001. Self-referent phenotype matching: theoretical considerations and empirical evidence. *Trends Neurosci.* 24:609–16

Hauser MD. 1992. Costs of deception: cheaters are punished in rhesus monkeys (*Macaca mulatta*). *Proc. Natl. Acad. Sci. USA* 89:12137–39

Hauser MD. 2000. *Wild Minds: What Animals Really Think.* New York: Holt

Hauser MD, Chen MK, Chen F, Chuang E. 2003. Give unto others: genetically unrelated cotton-top tamarin monkeys preferentially give food to those who altruistically give food back. *Proc. R. Soc. London. Ser. B* 270:2363–70

Hauser MD, Marler P. 1993. Food-associated calls in rhesus macaques (*Macaca mulatta*): II. Costs and benefits of call production and suppression. *Behav. Ecol.* 4:206–12

Heinrich B. 1988. Food sharing in the raven, *Corvus corax.* In *The Ecology of Social Behavior,* ed. CN Slobodchikoff, pp. 285–311. San Diego: Academic

Heinrich B. 1989. *Ravens in Winter.* New York: Simon & Schuster

Heinrich B, Marzluff JM. 1991. Do common ravens yell because they want to attract others? *Behav. Ecol. Sociobiol.* 28:13–21

Heinsohn RG. 1991. Kidnapping and reciprocity in cooperatively breeding white-winged choughs. *Anim. Behav.* 41:1097–100

Hepper PG. 1991. *Kin Recognition.* Cambridge, UK: Cambridge Univ. Press. 469 pp.

Herre EA, Knowlton N, Mueller UG, Rehner SA. 1999. The evolution of mutualisms: exploring the paths between conflict and cooperation. *Trends Ecol. Evol.* 14:49–53

Hölldobler B, Wilson EO. 1990. *The Ants.* Cambridge, MA: Harvard Univ. Press

Jarrad LE, Moise SL. 1970. Short-term memory in the stumptail macaque: effect of physical restraint of behavior on performance. *Learn. Motiv.* 1:267–75

Kacelnik A. 2003. The evolution of patience. In *Time and Decision: Economics and Psychological Perspectives on Intertemporal Choice,* ed. R Baumeister, pp. 115–38. New York: Russell Sage Found.

Kamil AC, Balda RP, Olson DJ. 1994. Performance of four seed-caching corvid species in the radial-arm maze analog. *J. Comp. Psychol.* 108:385–93

Keller L, Ross KG. 1998. Selfish genes: a green beard in the red fire ant. *Nature* 394:573–75

Krause J, Ruxton GD. 2002. *Living in Groups.* Oxford: Oxford Univ. Press

Lazarus J, Metcalfe NB. 1990. Tit-for-tat cooperation in sticklebacks: a critique of Milinski. *Anim. Behav.* 39:987–88

Masters WM, Waite TA. 1990. Tit-for-tat during predator inspection or shoaling? *Anim. Behav.* 39:603–4

Mateo JM, Johnston RE. 2000. Kin recognition and the 'armpit effect': evidence of self-referent phenotype matching. *Proc. R. Soc. London Ser. B* 267:695–700

May RM. 1981. The evolution of cooperation. *Nature* 292:291–92

Maynard Smith J. 1982. *Evolution and the Theory of Games.* Cambridge, UK: Cambridge Univ. Press

Maynard Smith J, Price GR. 1973. The logic of animal conflict. *Nature* 246:15–18

Mazur JE. 1987. An adjusting procedure for studying delayed reinforcement. In *Quantitative Analyses of Behavior: The Effect of Delay and of Intervening Events on Reinforcement Value,* ed. H Rachlin, pp. 55–73. Hillsdale, NJ: Erlbaum

Milinski M. 1987. TIT FOR TAT in sticklebacks and the evolution of cooperation. *Nature* 325:433–35

Milinski M, Wedekind C. 1998. Working memory constrains human cooperation in the Prisoner's Dilemma. *Proc. Natl. Acad. Sci. USA* 95:13755–58

Noë R. 1990. A Veto game played by baboons: a challenge to the use of the Prisoner's Dilemma as a paradigm for reciprocity and cooperation. *Anim. Behav.* 39:78–90

Ostrom E, Walker J. 2003. *Trust and Reciprocity: Interdisciplinary Lessons from Experimental Research.* New York: Russell Sage Found.

Ostrom E, Walker J, Gardner R. 1992. Covenants with and without a sword: self-governance is possible. *Am. Polit. Sci. Rev.* 86:404–17

Packer C. 1977. Reciprocal altruism in *Papio anubis. Nature* 265:441–43

Packer C, Ruttan L. 1988. The evolution of cooperative hunting. *Am. Nat.* 132:159–98

Parker PG, Waite TA, Heinrich B, Marzluff JM. 1994. Do common ravens share ephemeral food resources with kin? DNA fingerprinting evidence. *Anim. Behav.* 48:1085–93

Petrie M, Krupa A, Burke T. 1999. Peacocks lek with relatives even in the absence of social and environmental cues. *Nature* 401:155–57

Rachlin H. 2000. *The Science of Self-Control.* Cambridge, MA: Harvard Univ. Press

Rachlin H, Raineri A, Cross D. 1991. Subjective probability and delay. *J. Exp. Anal. Behav.* 55:233–44

Rapoport A, Chammah AN. 1965. *Prisoner's Dilemma: A Study in Conflict and Cooperation.* Ann Arbor: Univ. Mich. Press

Reeve HK. 1993. Haplodiploidy, eusociality and absence of male parental and alloparental care in Hymenoptera: a unifying genetic hypothesis distinct from kin selection theory. *Philos. Trans. R. Soc. London Ser. B* 342:335–52

Reeve HK, Gamboa GJ. 1987. Queen regulation of worker foraging in paper wasps: a social feedback control system, *Polistes fuscatus* (Hymenoptera: Vespidae). *Behaviour* 102:147–67

Richards JB, Mitchell SH, de Wit H, Seiden LS. 1997. Determination of discount functions in rats with an adjusting-amount procedure. *J. Exp. Anal. Behav.* 67:353–66

Rowell TE, Wilson C, Cords M. 1991. Reciprocity and partner preference in grooming of female blue monkeys. *Int. J. Primatol.* 12:319–36

Sella G, Premoli MC, Turri F. 1997. Egg trading in the simultaneously hermaphroditic polychaete worm *Ophryotrocha gracilis* (Huth). *Behav. Ecol.* 8:83–86

Sherman PW, Reeve HK, Pfennig DW. 1997. Recognition systems. See Krebs & Davies 1997, pp. 69–96

Sikstrom S. 2002. Forgetting curves: implications for connectionist models. *Cogn. Psychol.* 45:95–152

Silk JB. 2002. Kin selection in primate groups. *Int. J. Primatol.* 23:849–75

Stacey PB, Koenig WD. 1990. *Cooperative Breeding in Birds: Long-Term Studies of Ecology and Behavior.* Cambridge, UK: Cambridge Univ. Press

Stephens DW, Anderson JP, Benson KE. 1997. On the spurious occurrence of Tit for Tat in pairs of predator-approaching fish. *Anim. Behav.* 53:113–31

Stephens DW, McLinn CM, Stevens JR. 2002. Discounting and reciprocity in an Iterated Prisoner's Dilemma. *Science* 298:2216–18

Stevens JR. 2004. The selfish nature of generosity: harassment and food sharing in primates. *Proc. R. Soc. London Ser. B* 271:451–56

Stevens JR, Gilby IC. 2004. A conceptual framework for non-kin food sharing: timing and currency of benefits. *Anim. Behav.* 67:603–14

Stevens JR, Hallinan EV, Hauser MD. 2005. The ecology and evolution of patience in two New World primates. *Biol. Lett.* 1:223–26

Stevens JR, Hauser MD. 2004. Why be nice? Psychological constraints on the evolution of cooperation. *Trends Cogn. Sci.* 8:60–65

Stevens JR, Stephens DW. 2002. Food sharing: a model of manipulation by harassment. *Behav. Ecol.* 13:393–400

Trivers RL. 1971. The evolution of reciprocal altruism. *Q. Rev. Biol.* 46:35–57

Trivers RL. 2002. *Natural Selection and Social Theory: Selected Papers of Robert Trivers.* Oxford: Oxford Univ. Press

Trivers RL, Hare H. 1976. Haplodiploidy and the evolution of the social insects. *Science* 191:249–63

Watts DP, Mitani JC. 2002. Hunting and meat sharing by chimpanzees at Ngogo, Kibale National Park, Uganda. In *Behavioural Diversity in Chimpanzees and Bonobos*, ed. LF Marchant, pp. 244–55. Cambridge, UK: Cambridge Univ. Press

West Eberhard MJ. 1975. The evolution of social behavior by kin selection. *Q. Rev. Biol.* 50:1–33

White KG. 2001. Forgetting functions. *Anim. Learn. Behav.* 29:193–207

Wilkinson GS. 1984. Reciprocal food sharing in the vampire bat. *Nature* 308:181–84

Williams GC. 1966. *Adaptation and Natural Selection.* Princeton: Princeton Univ. Press

Wilson EO. 1975. *Sociobiology: The New Synthesis*. Cambridge, MA: Harvard Univ. Press

Wilson EO. 1987. Kin recognition: an introductory synopsis. In *Kin Recognition in Animals*, ed. CD Michener, pp. 7–18. Chichester: Wiley

Wixted JT. 2004. The psychology and neuroscience of forgetting. *Annu. Rev. Psychol.* 55: 235–69

Wixted JT, Ebbesen EB. 1991. On the form of forgetting. *Psychol. Sci.* 2:409–15

Woolfenden GE, Fitzpatrick JW. 1978. The inheritance of territory in group breeding birds. *BioScience* 28:104–8

Wrangham RW. 1975. *The behavioural ecology of chimpanzees in Gombe National Park, Tanzania*. PhD thesis. Cambridge Univ.

Zahavi A. 1990. Arabian Babblers: the quest for social status in a cooperative breeder. In *Cooperative Breeding in Birds: Long-Term Studies of Ecology and Behavior*, ed. WD Koenig, pp. 103–30. Cambridge, UK: Cambridge Univ. Press

Annu. Rev. Ecol. Evol. Syst. 2005. 36:519–39
doi: 10.1146/annurev.ecolsys.36.102803.095431
Copyright © 2005 by Annual Reviews. All rights reserved
First published online as a Review in Advance on August 17, 2005

NICHE CONSERVATISM: Integrating Evolution, Ecology, and Conservation Biology

John J. Wiens and Catherine H. Graham

*Department of Ecology and Evolution, Stony Brook University, Stony Brook, New York
11794-5245; email: wiensj@life.bio.sunysb.edu, cgraham@life.bio.sunysb.edu*

Key Words   biogeography, climate, invasive species, speciation, species richness

■ **Abstract**   Niche conservatism is the tendency of species to retain ancestral ecological characteristics. In the recent literature, a debate has emerged as to whether
niches are conserved. We suggest that simply testing whether niches are conserved is
not by itself particularly helpful or interesting and that a more useful focus is on the
patterns that niche conservatism may (or may not) create. We focus specifically on how
niche conservatism in climatic tolerances may limit geographic range expansion and
how this one type of niche conservatism may be important in (*a*) allopatric speciation,
(*b*) historical biogeography, (*c*) patterns of species richness, (*d*) community structure,
(*e*) the spread of invasive, human-introduced species, (*f*) responses of species to global
climate change, and (*g*) human history, from 13,000 years ago to the present. We describe how these effects of niche conservatism can be examined with new tools for
ecological niche modeling.

INTRODUCTION

The niche is a central concept in ecology and evolution that dates back at least to
Grinnell (1917). Although many definitions of the niche have been proposed, the
definition introduced by Hutchinson (1957) is particularly widespread and useful:
The niche is the set of biotic and abiotic conditions in which a species is able
to persist and maintain stable population sizes. Hutchinson (1957) also made the
valuable distinction between the fundamental and realized niche. The fundamental
niche describes the abiotic conditions in which a species is able to persist, whereas
the realized niche describes the conditions in which a species persists given the
presence of other species (e.g., competitors and predators).

Many aspects of the fundamental niche can be conserved over long evolutionary
timescales. For example, tens of thousands of actinopterygian fish species are
confined to aquatic habitats, and many fish clades are confined to either saltwater
or freshwater. The tendency of species to retain aspects of their fundamental niche
over time is called niche conservatism. We refer to niche conservatism as a process,
although it may be caused by more than one factor at the population level (a feature
it shares with other evolutionary processes, such as speciation and anagenesis).

1543-592X/05/1215-0519$20.00

In this review, we describe the importance of niche conservatism to evolution, ecology, and conservation biology. We outline how answers to some long-standing questions in these fields may lie (at least in part) in the inability of species to adapt to novel abiotic conditions over a given timescale. These questions include the following: How does a single species split to create two new species? Why are there more species in tropical regions than in temperate regions? Which introduced species are likely to invade a given region, and how far will they spread? How will species respond to global warming? The importance of niche conservatism does not depend on ecological traits being maintained indefinitely. Instead, these diverse patterns may be explained by niche conservatism at different timescales.

This review has three objectives: (*a*) to address the controversy over whether or not niches are conserved, (*b*) to highlight the diverse areas that niche conservatism might help explain, a topic that has not been reviewed previously, and (*c*) to describe methodological tools that can be used to test the effects of niche conservatism (i.e., GIS-based ecological niche modeling).

An important theme of our review is that both evolution and ecology are important in explaining biogeographic patterns. At the same time, we show how biogeography is important to diverse topics in evolution, ecology, conservation biology, and even human history, areas in which biogeography may not be widely appreciated.

ARE NICHES CONSERVED OR AREN'T THEY?

Considerable debate has emerged in the recent literature as to whether or not niches are conserved. Much of this debate was sparked by the landmark paper by Peterson et al. (1999). These authors combined museum locality data, climatic data, and niche modeling to show that climatic niches were conserved (similar) between many sister-species pairs of mammals, birds, and butterflies in Mexico. Their general assertion that niches are conserved was countered by other studies, which claimed that niches are evolutionarily labile, including studies of microhabitat preferences in *Anolis* lizards (Losos et al. 2003), morphometric variation in warblers (*Sylvia*) (Böhning-Gaese et al. 2003), and environmental niche models in dendrobatid frogs (Graham et al. 2004b). Other studies have supported niche conservatism. For example, Ricklefs & Latham (1992) showed that many congeneric plant species in Europe and North America had similar geographic ranges on each continent, a pattern interpreted as evidence for niche conservatism. Prinzig et al. (2001) argued for niche conservatism in six environmental variables in a sample of more than 1000 species of higher plants in Europe.

We believe that the question of whether niches are conserved or labile is not in itself particularly fruitful. Species will always inhabit environments that bear some similarity to those of their close relatives (i.e., few tropical rainforest species have a sister species in undersea vents). Thus, to some extent, niches will always be conserved. Yet, few sister species may share identical niches; so niches may never be perfectly conserved either. The answer to the question "Are niches conserved?" may simply depend on exactly how similar niches must be among species to be considered conserved.

Instead, a more constructive way to think about this issue may be to focus on specific outcomes of niche conservatism. For example, does niche conservatism drive allopatric speciation, high tropical species richness, or responses of species to climate change? In the next section, we describe the many effects of niche conservatism and their predicted empirical signatures.

In this review, we focus on niche conservatism in a very restricted sense in that we emphasize how conservatism in climatic tolerances limits geographic ranges of species and clades. We argue that even this one limited aspect of niche conservatism has a plethora of important consequences for ecology, evolution, and conservation biology. This aspect of niche conservatism is also the most readily studied through ecological niche modeling. Before we begin, however, we strongly emphasize that even though we believe niche conservatism may play some role in all of these areas in some cases, we do not think that it does so in every single case.

WHAT DOES NICHE CONSERVATISM DO?

Allopatric Speciation

The importance of niche conservatism to speciation may not be immediately obvious. After all, speciation is typically equated with divergence (e.g., Coyne & Orr 2004, Futuyma 1998) and not maintenance of ecological similarity over time. The importance of any process to speciation depends upon our concept of what species are and what speciation is.

Several authors (e.g., de Queiroz 1998, Mayden 1997, Wiens 2004b) have argued that species are lineages and that the characteristics used to define species under most concepts are simply traits that evolve in lineages given enough time (e.g., postzygotic isolating mechanisms, diagnostic morphological traits, monophyletic gene genealogies). Given this view, speciation is the actual splitting of one lineage into two. For parapatric and sympatric speciation, lineage splitting is intimately related to divergence in traits associated with intrinsic reproductive isolation (Coyne & Orr 2004, Turelli et al. 2001). However, for allopatric speciation [often considered the most common geographic mode (Barraclough & Vogler 2000, Coyne & Orr 2004)] new lineages arise from the geographic separation of ancestral species into isolated sets of populations (Wiens 2004b). Even if one does not equate lineage splitting with speciation, the geographic separation of one lineage into two is considered an essential part of allopatric speciation under many concepts [e.g., biological species concept (Futuyma 1998)].

In many cases, this geographic separation may be associated with niche conservatism (Wiens 2004a). Allopatry is generally caused by a geographic barrier that consists of suboptimal environmental conditions for the species in question (e.g., deserts, mountains, or oceans). If a species can adapt to ecological conditions at this barrier, then gene flow will continue across it and allopatric speciation will not occur. Niche conservatism (in the broad sense) may be generally important in allopatric speciation because it will limit adaptation to ecological conditions at the geographic barrier. In some cases, this barrier may be associated with different microhabitat preferences (e.g., a river that separates terrestrial habitats), but in other cases, it may be associated with differences in climatic regimes (e.g., montane endemics separated by intervening lowlands or lowland endemics separated by intervening mountain ranges).

So far, no studies have adequately addressed the role of niche conservatism in allopatric speciation. Peterson et al. (1999) showed similarity in the climatic niche space of allopatric species pairs of birds, butterflies, and mammals on either side of the Isthmus of Tehuantepec in Mexico. However, they did not address whether the vicariance event that created these lineages was associated with conservatism in climatic tolerances. Graham et al. (2004b) showed differences in the climatic niches of allopatric species pairs of Andean frogs but did not address whether these niche differences were the cause of lineage splitting (e.g., as in parapatric speciation) or arose after speciation (e.g., species on different mesic mountaintops may have adapted to different climatic regimes, but their splitting into separate lineages may have been caused by a dry valley between them). Wiens (2004b) discussed how the role of niche conservatism in allopatric speciation might be tested.

Niche conservatism (and niche evolution) may also be important for species delimitation, the process of identifying and diagnosing species (Sites & Marshall 2003). When species are diagnosed, sample sizes are almost never large enough to infer with statistical confidence that a putative species is truly fixed for a trait or allele (Wiens & Servedio 2000) or that all alleles of a given locus in a given species form a monophyletic group. This situation makes species-level decisions based on traditional morphological and genetic data alone potentially problematic. However, if a set of populations of uncertain taxonomic status is geographically separated from closely related species by areas that are outside of the climatic niche envelope of all of these species (Figure 1), then gene flow within these species is unlikely because it would involve crossing unsuitable habitat. This pattern would support the hypothesis that the populations of uncertain status represent a distinct species. Differences in niche characteristics might also be important. If a set of populations occurs under climatic conditions that do not overlap those of closely related species (e.g., Graham et al. 2004b), then gene flow between these populations and other species may also be unlikely, and these populations may represent a distinct species. These similarities and differences in niche characteristics can be visualized and analyzed statistically with methods from ecological niche modeling (see below).

In a similar vein, Raxworthy et al. (2003) used niche modeling to infer areas in which species of chameleons (lizards) might be expected to occur given their known geographic distribution. They found that niche models for some species predicted their occurrence in disjunct areas of seemingly suitable habitat where

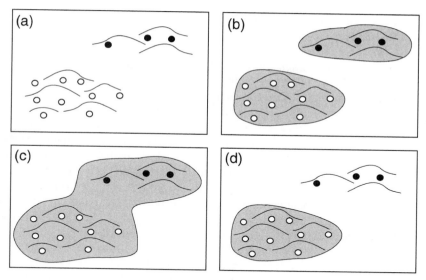

Figure 1 Hypothetical example illustrating niche conservatism, ecological niche modeling, and species delimitation. (*a*) Two sets of allopatric populations occur in two geographically separate montane regions. One set (*open circles*) is a previously described species, the other set (*closed circles*) is of unknown taxonomic status. (*b*) Ecological niche modeling shows that the two sets of populations share a similar climatic niche envelope (*shown in gray*). The intervening lowland areas between the montane regions are outside the envelope of acceptable environmental conditions, which suggests that niche conservatism may prevent gene flow between these two sets of populations and supports the hypothesis that they are distinct species. (*c*) The two sets of populations share a similar climatic niche envelope, but this niche envelope also includes the areas between them. This result suggests that niche conservatism is not important in isolating them and that there may be ongoing dispersal and gene flow between them (if no other barriers are present). This pattern would not add support to the hypothesis that they are distinct species. (*d*) The two sets of populations have dissimilar climatic niche envelopes (illustrated here by the restricted niche envelope of the known species). This result suggests that past niche evolution (and current niche conservatism) maintains the geographic separation of these populations and supports the hypothesis that they are distinct species.

these species were not presently recorded. These areas subsequently were found to contain several undescribed, closely related chameleon species.

Historical Biogeography

Historical biogeography uses phylogenies to help explain the geographic distribution of species and clades. In recent decades, historical biogeography has focused primarily on addressing specific hypotheses of geological connections among areas, rather than on the general ecological and evolutionary processes that explain the large-scale distribution of clades (reviewed by Wiens & Donoghue 2004).

Wiens & Donoghue (2004) argued that the interplay of niche conservatism and niche evolution may be critical in the biogeographic history of many groups. A major challenge in biogeography is to explain why clades have dispersed to some areas but not others. Climatic niche specialization seems to be important in many groups. For example, many clades, such as crocodiles, caecilians, and trogoniform birds, occur in tropical regions on two or more continents but are largely absent from geographically adjacent temperate regions (Wiens & Donoghue 2004). Some of these clades are extremely old [e.g., caecilians are more than 140 million years old (Zug et al. 2001)], which suggests that the general tropical niche can be conserved over very long time periods in some groups. Similarly, many clades (both old and young) are largely confined to temperate regions and occur on two or more continents; these clades include numerous groups of plants, insects, and fish (e.g., Donoghue & Smith 2004, Sanmartín et al. 2001). For example, with the exception of a derived clade of plethodontids (bolitoglossines), almost all salamanders occur in temperate climate regions of the Northern Hemisphere (Zug et al. 2001). This pattern suggests that the general temperate niche has been maintained in most salamander clades for more than 100 million years (Zug et al. 2001). Similarly, the boundaries between some of Earth's major biogeographic realms, recognized by both zoologists (Wallace 1876) and botanists (Good 1947), correspond to transitions between temperate and tropical climatic regimes within continents (i.e., Nearctic versus Neotropical; Palearctic versus Oriental) rather than the edges of continents.

Patterns of Species Richness

Species richness is a central topic in ecology and biogeography (Brown & Lomolino 1998, Rosenzweig 1995, Willig et al. 2003). In recent years, species richness has been studied primarily by ecologists seeking correlations between environmental variables and species numbers at specific locations throughout the globe. Although this research program has found strong correlations between climate and species richness (e.g., Francis & Currie 2003, Hawkins et al. 2003), it has generally not addressed how climate actually influences species numbers in a region or community (Wiens & Donoghue 2004). For example, there is no direct explanation for how high energy or productivity in a community or region can lead to a greater number of species (i.e., just because there are more individuals of a species at a given location does not mean that there will be more species). Instead, the processes that will actually change species numbers within a region involve evolution and biogeography, such as speciation, extinction, and dispersal of species into or out of a region (Wiens & Donoghue 2004).

Niche conservatism offers a mechanism that can help explain large-scale species-richness patterns in a way that reconciles both ecological and evolutionary perspectives. Several evolutionary ecologists have converged on a very similar explanation for the latitudinal gradient in species richness (i.e., Brown & Lomolino 1998, Farrell et al. 1992, Futuyma 1998, Ricklefs & Schluter 1993). This explanation, dubbed the "tropical conservatism hypothesis" by Wiens & Donoghue (2004), has three parts. First, many groups have more species in the tropics because they originated in the tropics and have had more time to speciate there. Second, species disperse from tropical regions to temperate regions rarely or not at all, because they lack adaptations to survive cold winter temperatures. Thus, niche conservatism helps to create and maintain a disparity in species richness between tropical and temperate regions. Third, tropical regions were much more extensive until relatively recently (roughly 30 to 40 mya), which explains why many extant groups originated in the tropics. Overall, the tropical conservatism hypothesis is consistent with the observation of ecological studies that show high species richness in tropical communities with high temperatures and rainfall (and energy and productivity) but links this pattern to processes that directly affect the number of species in each region (i.e., dispersal and speciation).

The tropical conservatism hypothesis has important implications for conservation. If most groups show the predicted pattern of many, older clades in tropical regions and fewer, younger clades in temperate regions, then there may be higher genetic and phylogenetic diversity for the same number of species in tropical regions than in temperate regions. If this pattern occurs, then loss of tropical habitats will have two important consequences (relative to temperate regions): the loss of more species per unit area and the loss of more genetic and phylogenetic diversity per species (for a similar argument for biodiversity hot spots, see Sechrest et al. 2002). Conversely, some authors have used the finding of higher genetic diversity in tropical faunas relative to temperate faunas as evidence to support the tropical conservatism hypothesis [in New World birds (Gaston & Blackburn 1996, Ricklefs & Schluter 1993)].

Niche conservatism may explain many other patterns of species richness as well, such as the low species richness of many clades in arid regions and the reverse latitudinal gradient (higher richness in temperate regions) seen in some groups (e.g., Brown & Lomolino 1998, Ricklefs & Schluter 1993). Any novel set of environmental conditions is potentially a long-term barrier to dispersal for some clades, and niche conservatism should tend to create a disparity in species richness between habitats or regions over time for many different groups of organisms.

We do not claim that the tropical conservatism hypothesis is the sole explanation for the latitudinal diversity gradient in all groups. For example, some groups appear to have higher rates of diversification in the tropics [birds and butterflies (Cardillo 1999)], a pattern which suggests that the tropical conservatism hypothesis is unnecessary to explain high tropical species richness in these groups. The causes of this higher diversification rate are unclear. One potential cause is a greater tendency for montane endemism in tropical regions (as opposed to more elevational generalists in temperate regions), which may be related to a greater zonation of climatic regimes at different elevations in tropical regions [associated with reduced seasonal temperature variation (Janzen 1967)]. In support of this hypothesis, the Andes

Mountains of South America contain higher species richness of birds than do the adjacent Amazonian rainforests (Rahbek & Graves 2001). Interestingly, if Janzen's hypothesis is correct, the extensive speciation in tropical montane regions may also involve the limited ability of climatic specialists to disperse between climatic regimes at different elevations (i.e., niche conservatism at a smaller spatial scale).

Community Structure

Community structure is used here as the guild composition of an assemblage (e.g., the number of sympatric microhabitat and diet specialists). In some cases, differences in community structure between regions may result from niche conservatism, which limits dispersal of lineages with different ecological traits between regions (e.g., Ackerly 2003, Webb et al. 2002). In these cases, the ecological structure of a given community cannot be understood simply by examination of ecological characteristics of the species present in that community or even the phylogeny of those species. Instead, community structure may result from constraints on the dispersal of lineages that are not presently represented in those communities (Wiens & Donoghue 2004).

For example, in emydid turtles in eastern North America, communities in the northeast are dominated by semiaquatic emydine lineages, whereas communities in the southeast are dominated by aquatic deirochelyine species (Stephens & Wiens 2004). Although some aquatic deirochelyines have invaded northeastern communities, most semiaquatic emydines have failed to invade southern communities.

Why have these lineages remained in the northeast? Two obvious explanations are competition and niche conservatism. Competition with other emydids seems unlikely, given that southeastern deirochelyines either occur far south of these emydine species or overlap their geographic ranges extensively. Ecological niche modeling (P.R. Stephens & J.J. Wiens, unpublished data) suggests that high summer temperatures may limit the spread of these emydine lineages into southeastern communities. Thus, the tendency of the northern semiaquatic lineages to retain their ancestral niche seemingly has created significant differences in emydid community structure across eastern North America.

In this example, competition seemingly is unimportant in creating these geographic patterns of community structure. In other cases, competition may determine which guilds and lineages can invade a community and which cannot. Furthermore, in many cases, differences in community structure between regions seem to result from biogeographic constraints unrelated to climatic niche conservatism [e.g., islands and different continents (Cadle & Green 1993)] or absent guilds simply evolve from within the local species pool [some Caribbean *Anolis* (Losos et al. 1998)].

Invasive Species

Invasive species are thought to be one of the major threats to biodiversity (e.g., Wilcove et al. 1998, Wilson 1992). Niche conservatism may determine which species can invade which regions and where they will spread within those regions (reviewed by Peterson 2003, Peterson & Vieglais 2001). If their fundamental niches are conserved, species will only be able to invade regions that have a climate similar to that of their native range. Peterson and collaborators have shown several examples in which ecological niche modeling of the climatic characteristics of the native range of a species can predict its introduced range.

The introduced reptile and amphibian fauna of North America exemplifies the idea that niche conservatism determines which exotic species can become established and where (Conant & Collins 1991, Stebbins 2003). Southern Florida, a region whose native vegetation includes moist subtropical forests, now contains numerous introduced species of reptiles and amphibians from tropical regions around the world, including species from the West Indies (e.g., the frogs *Osteopilus septentrionalis*, *Eleutherodactylus coqui*, and *E. planirostris*; many lizards, including species of *Anolis*, *Leiocephalus*, and *Sphaerodactylus*), Central and South America (the toad *Bufo marinus* and lizards *Cnemidophorus lemniscatus*, *Ameiva ameiva*, *Ctenosaura pectinata*, *Iguana iguana*, and *Basiliscus vittatus*), and Southeast Asia (the lizard *Gekko gecko* and the snake *Rhamphotyphlops braminus*). None of these species have successfully invaded more temperate regions north of Florida, even though many have been present and widespread in Florida for decades. The only exotic species that are well established in temperate eastern North America are two lizards from temperate Europe, which have populations in Kansas and New York (*Podarcis muralis*) and Ohio (*Podarcis sicula*). A species of gecko from Mediterranean Europe (*Hemidactylus turcicus*) is widespread in the southern United States, and one from the Middle East and Central and South Asia (*Cyrtopodion scabrum*) is established in coastal Texas. In the western United States, successful introductions have consisted mostly of eastern species from similar latitudes (the amphibians *Rana berlandieri*, *R. catesbiana*, and *Ambystoma tigrinum* and the turtles *Apalone spinifera*, *Chelydra serpentina*, and *Trachemys scripta*). The African clawed frog (*Xenopus laevis*) has also been introduced into Mediterranean climate regions of southern California, and the natural range of this species includes Mediterranean climate regions of southern Africa. Although two species common in eastern North America have been widely introduced into tropical regions throughout the world (the frog *R. catesbiana* and the turtle *Trachemys scripta*), the native ranges of these species extend into tropical Mexico.

The correspondence between the native climate of these exotic species and their introduced range seems striking. As a crude quantitative index of this association, regression (Figure 2) of the estimated northernmost latitudinal range limits of 35 of these species (those for which information on introduced ranges were available) in their native and introduced ranges shows a highly significant relationship ($r^2 = 0.792, P < 0.0001$). This pattern seems to reflect the effects of niche conservatism.

Not all groups of invaders may show this level of climate matching. For example, some introduced mammalian species seem to tolerate both temperate and tropical regions (e.g., rats). Unlike most other plants and animals, mammals are endotherms, which physiologically maintain similar body temperatures across

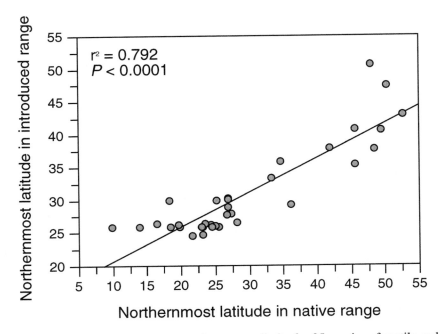

Figure 2 Relationship between northern range limits for 35 species of reptile and amphibian in their native and introduced ranges (in North America). The significant relationship implies that niche conservatism in climatic tolerances determines which exotic species become established in a region.

a range of environmental temperatures, and this characteristic may allow some species to tolerate a broad range of climatic conditions.

Invasive species are important as a threat to biodiversity, but they also offer an intriguing system by which to study the ecological and evolutionary causes of biogeographic patterns. Invasive species represent replicated biogeographic experiments that can be used to test for the impact of niche conservatism on range limits. If niche conservatism in climatic tolerance determines the range limits of species, then we should expect to see consistent parallels between their climatic distribution in their native and introduced ranges. Conversely, if introduced tropical species routinely spread into temperate regions or vice versa, then competition (or other biotic factors) may be more important in setting geographic range limits in their native ranges. Overall, we think that the application of data on invasive species to historical biogeography could be an exciting area for future research.

Responses to Global Climate Change

In many ways, niche conservatism is the underlying process that makes global climate change a danger to the world's biota. If species could simply adapt to changing climatic conditions, then we would have little cause for concern. On the other hand, given niche conservatism in climatic tolerances, species should shift their geographic ranges in the face of global warming to track their ancestral climatic regime, moving poleward in latitude and downward in elevation; species that cannot adapt and cannot shift their geographic ranges (e.g., due to habitat destruction or geographic constraints) may be at risk of extinction. Many studies have now addressed the effects of global warming on plant and animal distributions, and Parmesan & Yohe (2003) recently analyzed an extensive database that incorporates the results of hundreds of previous studies. Their analysis included more than 1700 species of plants, animals, and lichens from terrestrial, freshwater, and marine environments. For 99 species with quantitative data, they found that species (overall) moved an average of 6.1 km per decade poleward and 6.1 m per decade upward in elevation. For more than 1045 species with qualitative data, only 27% showed stable geographic ranges, and for the other 73%, most changes (75% to 81%) were in the direction predicted (poleward) for both northern and southern range limits. These results not only show a significant impact of global warming but also suggest that many species respond to climate change as predicted by niche conservatism, rather than with rapid evolution of climatic tolerances.

Recently, several authors have assumed niche conservatism to evaluate the potential large-scale impact of global warming on regional biotas. Peterson et al. (2002) used niche modeling to address how Mexican birds, butterflies, and mammals may be affected by global warming. Thomas et al. (2004) modeled the effects of climate change on many taxa and regions and concluded that climatic warming may be an important threat to global biodiversity. Although the effects of global warming may be very difficult to mitigate, some authors have suggested the intriguing possibility that species can be managed to minimize the effects of niche conservatism and maximize the potential for rapid adaptive evolution, including their response to changing climate (Rice & Emery 2003, Stockwell et al. 2003).

In addition to the present crisis, niche conservatism may also be associated with major historical mass extinctions. The history of life on earth has been marked by five major extinction events (Benton 1995). The ultimate cause of these events remains controversial, but relatively rapid climate change is considered to be among the most important proximate factors in many of them (Hallam & Wignall 1997). Mass extinctions associated with rapid climate change may be another manifestation of niche conservatism.

Human History and Agriculture

Niche conservatism also seems to have had a profound impact on human history. Diamond (1997) has proposed that the ultimate cause of the different histories and fates of human societies lies in the shape of the continents on which these societies developed. Although not discussed explicitly by Diamond, the reason that continental axis matters clearly is niche conservatism.

A major feature of recent human history was the conquest of much of Africa and the New World by Europeans, and its many consequences. But why were the peoples of Africa and the New World conquered by Europeans and not vice versa? Diamond argues for a chain of causality that explains the major features of human history from the end of the Ice Ages to the present. For continents that have their greatest length on their longitudinal axis (i.e., Eurasia), domesticated plant and animal species can be readily spread between regions, and a large "package" of domesticated species can accumulate in one place. More domesticated species lead to greater food production, food surpluses, and food storage. Greater food availability allows for high population densities in fixed locations, which permits many individuals to pursue other activities besides food production. These activities include development of technology (e.g., ocean-going ships, guns, and steel), writing, and government. Furthermore, a large number of domesticated mammal species may have led to a large pool of diseases ("germs") in European populations that eventually decimated the native peoples of Africa, Australia, and the New World. (Although Australia is shaped like Eurasia, it has a restricted pool of candidate species for domestication, seemingly because of biogeographic isolation and prehistoric extinction of the megafauna.)

Niche conservatism may explain the difficulty in spreading domesticated species (or any species) between different climatic regimes and the ease of spreading them across the same climatic regime. For example, food production is thought to have originated in southwest Asia (the Fertile Crescent) around 8000 B.C., and the same suite of domesticated species (e.g., pea, chickpea, wheat, barley, sheep, and goat) was then transported to Europe, North Africa, Central Asia, and the Indus Valley region (Pakistan). These species showed limited diffusion into adjacent tropical regions of Africa and Asia. In the New World, different sets of species were domesticated independently in North America, the Middle American highlands, and South America, or the same species (or close relatives) were domesticated independently in different regions (e.g., lima beans, cotton, chili peppers, and squashes). Relatively few species spread between regions, or diffusion was very slow. Diamond postulates that diffusion was limited because of differences in climate between regions (e.g., North America versus Mexico) or because areas of similar climate (e.g., montane Middle America and montane South America) were separated by areas of dissimilar climate, and these species could not tolerate radically different climatic regimes. Thus, the domesticated llamas, guinea pigs, and potatoes of the South American highlands never spread to montane Middle America, and the domestic turkeys of Mexico never spread to montane South America, presumably because of the intervening hot tropical lowlands that separate these cooler montane regions.

Diamond's analysis also has interesting implications for the study of niche conservatism. Domesticated species are raised under conditions in which competition with native biota presumably is limited. This observation supports the idea that physiological tolerances to climate may be sufficient to determine large-scale biogeographic patterns in the diverse plant and animal lineages that have been domesticated (although other biotic factors, such as disease, could be important in some cases). Furthermore, the spread of some domesticated species between climatic regions (e.g., corn from Mexico to temperate North America) suggests that the physiological tolerances that underlie climatic barriers can be overcome with artificial selection. In parallel to invasive species, the relationships between domesticated species and climate suggest that the agricultural sciences may offer a useful database for studies of niche conservatism, niche evolution, and biogeography.

TESTING FOR THE EFFECTS OF NICHE CONSERVATISM

We have discussed how niche conservatism can have a variety of consequences at different spatial and temporal scales. However, these diverse effects can all be reduced to a common cause: the evolutionary specialization of a species or clade to a particular climatic regime limits their dispersal. Ecological niche modeling can be used to help test whether climatic factors do limit the distribution of species and can set the stage for additional phylogenetic, ecological, and physiological studies.

Niche modeling applies powerful new computational tools to museum locality data assembled through decades of fieldwork (reviewed by Graham et al. 2004a). The general approach combines three elements: (*a*) georeferenced localities for the species in question (i.e., localities where a species has been collected and for which latitude and longitude coordinates are available), (*b*) data on climatic variables (e.g., yearly mean, variance, minimum, and maximum for temperature and precipitation at each site) at those sites and in surrounding areas, and (*c*) algorithms that estimate the climatic niche envelope of these species on the basis of the distribution of climatic variables where they occur and do not occur within a region.

Locality data for individual species are available from natural history museums. Many localities are georeferenced, and many databases of georeferenced localities are becoming available on the Internet, depending upon the organism and region (Graham et al. 2004a). Even if localities are not georeferenced, georeferencing can be quite straightforward (if time consuming) with Internet resources (e.g., Alexandria digital library gazetteer server, global directory of cities and towns, and Topozone).

Fine-scale climatic data sets are freely available that cover the entire planet (e.g., R. Hijman's WORLDCLIM at http://biogeo.berkeley.edu/worldclim/worldclim. htm). These data sets are based on information from a large number of weather stations, augmented by statistical extrapolations to locations without weather stations by use of digital elevation models. Many climatic variables are available, although many may be tightly correlated and largely redundant. Rather than analyze all of them, a better approach may be to choose a limited number that are not strongly correlated and that are considered (a priori) to be potentially important in limiting distributions within the group, such as coldest yearly temperatures or precipitation during the driest quarter.

A variety of methods are available to construct ecological niche models (reviewed by Guisan & Zimmerman 2000). These methods can be grouped into several categories, such as environmental envelopes [e.g., BIOCLIM (Nix 1986) and DOMAIN (Carpenter et al. 1993)], generalized regression methods (e.g., Lehman et al. 2002, Pearce & Ferrier 2000), ordination approaches (Austin 1985, Guisan et al. 1999, Hirzel et al. 2002), Bayesian methods (e.g., Gelfand et al. 2003), and genetic algorithms [e.g., GARP (Stockwell & Peters 1999)]. In general, a statistical model is used to establish a relationship between point-locality data (either sites where a species is recorded or localities for both presence and absence) and environmental layers (describing variation in a climatic variable over space). The model is then used to create a predicted map of a species' distribution, given these environmental variables.

When the niche envelope is projected onto a species' range map, one can visualize whether climatic variables predict (match) or overpredict the species range limits. Matching supports the hypothesis that the specialized climatic tolerances of a species may limit its geographic spread (but does not necessarily rule out other hypotheses; see below). In contrast, when the range is overpredicted, the climatic variables indicate that the species should have a more extensive geographic range than it actually does. This pattern of overprediction suggests that climate is not the primary factor that limits the geographic range of the species in that region, and that other factors may be responsible instead of niche conservatism (e.g., oceanic or riverine barriers to dispersal or competitors). In Table 1, we outline how the results of ecological niche modeling (and other types of evidence) might be used to determine the role of niche conservatism in each of the areas outlined in this paper.

The next challenge is to determine which climatic variables are most important in limiting the distribution of a species. Relatively few methods have been developed specifically for this purpose. However, most methods can be run with single variables to evaluate which variable most closely matches the geographic range of a species. Peterson & Cohoon (1999) have used bootstrapping to evaluate the performance of each variable. The DOMAIN and BIOCLIM methods in DIVA-GIS [http://www.diva-gis.org/ (Hijmans et al. 2002)] can identify the most-limiting variable for a given species for any point (pixel) on a map. Finally, logistic regression analyses of presence and (carefully selected) absence localities can also be used to identify the most important limiting variables.

Once the most-limiting variable (or combination of variables) is identified, its distribution among species can be mapped onto a phylogeny to determine how long this aspect of the niche has been conserved over the evolutionary history of the group (for a similar example, see Rice et al. 2003). Results from niche modeling can also set the stage for future observational and experimental studies to test how exactly this climatic variable interacts with the biology of the organism to set geographic range limits [e.g., physiological tolerances and interactions with other species (Gross & Price 2000, Kearney & Porter 2004)].

TABLE 1 Expected patterns resulting from niche conservatism

Topic	Pattern predicted from niche conservatism	Pattern that rejects niche conservatism
Allopatric speciation	Allopatric sister species have similar niche characteristics; geographic area that separates them is outside of their climatic niche envelope (see also Wiens 2004b)	Area that separates allopatric sister species is within their climatic niche envelope; nonclimatic barriers separate them (e.g., river or ocean)
Historical biogeography	Limited dispersal between different climatic regimes (e.g., temperate, tropical, mesic, or arid) relative to within-climatic regimes; species and clades fail to disperse into geographically adjacent regions with different climatic regimes	Dispersal between different climatic regimes equal to or greater than dispersal within-climatic regimes; dispersal within group is limited only by nonclimatic barriers (e.g., rivers or oceans for terrestrial organisms)
Species richness	(a) Group with high tropical species richness will originate in tropical regions (as shown by ancestral area reconstruction on a phylogeny), (b) significant relationship between amount of time the group has been present in each region and number of species in each region (e.g., Stephens & Wiens 2003); and (c) distribution of cool winter temperatures predicts poleward range limits of many or most tropical lineages in the group (Wiens & Donoghue 2004); expect similar patterns for ancestrally temperate groups or for groups in arid versus mesic environments	(a) Despite higher species richness in tropical region, group originated in temperate regions, (b) group dispersed to tropical regions relatively recently, despite higher species richness there, which suggests that latitudinal differences in species richness arose primarily from latitudinal differences in rates of diversification (rate of speciation—rate of extinction), and (c) even if group arose in tropical regions and dispersed to temperate regions recently, cool winter temperatures do not predict poleward range limits of tropical lineages, and poleward disperse is limited instead by traditional biogeographic barriers (e.g., water) or other climatic variables (e.g., limited precipitation)
Community structure	In a given region, the absence of a given guild is explained by specialized climatic tolerances that limit the large-scale dispersal of the clade representing that guild; climatic variables predict range limits of clade and do not overpredict into the region or community in question; geographic distribution of potential competitors do not abut the range of the clade but instead are either broadly disjunct or broadly overlapping	Environmental niche envelope for the guild/clade includes the community or region in which it is absent, which suggests that competition or other factors prevent clade from entering the region or community; if competition is important in setting range limits, then potential competitors are expected to geographically abut range of absent clade
Invasive species	Climatic conditions in invaded region similar to those of native range; niche modeling of native range predicts some or all of introduced range (Peterson 2003)	Significant differences between climatic conditions in native and introduced ranges; niche modeling of native range fails to predict introduced range
Climate change (global warming)	Species ranges will shift to track their ancestral (prewarming) climatic regime; ecological niche remains the same over time but geographic distributions do not (poleward shift)	Species adapt and shift environmental tolerances to cope with changing climate rather than changing geographic range as predicted; ecological niche changes and geographic distribution remains the same or changes in opposite direction than expected (i.e., toward equator)
Human history and agriculture	Similar to invasive species; for domesticated species, expect that ecological niche model of their native (nondomesticated) geographic distribution will predict into different regions with similar climates where these species are utilized today	Domesticated species thrive under climatic conditions that are outside the environmental niche envelope of the native (nondomesticated) populations

The observation that climatic variables predict the range limits of a species does not rule out a role for nonclimatic factors in limiting range expansion. For example, range limits might be set by the interactions of climate, resource availability, and competition (e.g., Case & Taper 2000, Darwin 1859, MacArthur 1972). There is an extensive literature that emphasizes the importance of competition in setting species geographic range limits (e.g., Connell 1961, Darwin 1859, MacArthur 1972). Although biotic interactions may be difficult to rule out, some patterns of distribution may favor climate as an explanation over competition (e.g., Anderson et al. 2002). For example, if the geographic range of a given species is predicted by climate and it shows only extensive geographic overlap or distant allopatry with those species that are most likely to be competitors, competition may be a less likely explanation than is climate alone. Again, local-scale studies that test the roles of biotic factors and physiological tolerances to abiotic conditions in setting geographic range limits are an important complement to ecological niche modeling at a biogeographic scale.

WHAT CAUSES NICHE CONSERVATISM?

At the population level, we see niche conservatism as the failure of adaptive evolution to allow range expansion into new climatic regimes. Empirical and theoretical work on species ranges suggests that four general factors may be important causes of niche conservatism: natural selection, gene flow, pleiotropy, and lack of variability (Wiens 2004a,b).

Natural (stabilizing) selection should be an important factor in the conservation of niches over time. If ecological conditions reduce fitness or population growth outside the niche, then natural selection should favor traits that keep individuals inside the niche (Holt 1996, Holt & Gaines 1992). An obvious example is behavioral habitat selection. For species that lack behavioral habitat selection (e.g., plants), natural selection will be biased toward those environmental conditions in which the largest number of individuals occurs (Holt & Gaines 1992).

Gene flow may also be an important force preventing niche expansion. Small populations at the edge of the geographic range may be flooded by individuals from the center, which may prevent these populations from adapting to environmental conditions outside the range (e.g., Haldane 1956, Holt 1996, Holt & Gaines 1992, Holt & Gomulkiewicz 1997, Kirkpatrick & Barton 1997, Stearns & Sage 1980).

Traits that would allow range expansion may be pleiotropically linked to traits that reduce fitness. In *Drosophila serrata* in Australia, range expansion into cooler temperate regions may be limited because evolution of increased cold resistance is associated with decreased fecundity (Jenkins & Hoffman 1999). Similarly, Etterson & Shaw (2001) have presented evidence that adaptation to warmer climatic regimes is slowed by genetic correlations among traits that are antagonistic to the direction of selection.

Finally, species may not evolve to expand their geographic range and niche because they lack genetic variation in the appropriate traits (e.g., Bradshaw 1991, Case & Taper 2000). Lack of variation may be very important in some cases (e.g., Hoffman et al. 2003), but several lines of evidence suggest that it may not be a universal explanation (Ackerly 2003). This evidence includes differences in climatic regimes among some closely related species (e.g., closely related montane and lowland endemics) and the evidence for genetic variation in most quantitative traits (e.g., Roff 1997), particularly those of ecological significance (e.g., Geber & Griffen 2003). Although the general causes niche conservatism at the population-genetic level have been discussed, the actual physiological traits that underlie niche conservatism (e.g., limited tolerance to heat, cold, or dessication) are poorly studied and may be relatively taxon specific.

CONCLUSIONS

In recent years, a controversy has developed over whether niches are evolutionarily conserved. Rather than debating whether niche conservatism exists, we suggest that a more useful focus for research would be to test the specifics of what niche conservatism may (or may not) do. In this review, we described the potential implications of one aspect of niche conservatism (the effects of climatic tolerances on dispersal) for many different areas of evolution, ecology, and conservation biology. Our review is not exhaustive, and this aspect of niche conservatism may be important in many other areas as well (e.g., intraspecific phylogeography). These diverse effects of niche conservatism may simply reflect the same process playing out over different temporal scales—whether for decades, or hundreds to thousands of years (i.e., invasive species, response to climate change, human history), thousands to millions of years (i.e., allopatric speciation), or tens or even hundreds of millions of years (i.e., historical biogeography, community structure, species richness). If this idea is true, then studies in diverse areas of ecology, evolution, and conservation biology may have unexpected relevance for each other. For example, invasive species offer many replicated "experiments" in large-scale biogeography, and studies of the role of niche conservatism in speciation and historical biogeography may offer insights both into how organisms have responded to climate change in the past and how future climate change may affect them. New tools from environmental bioinformatics should facilitate empirical tests of the role of niche conservatism across many different systems and questions, especially when coupled with phylogenetic analyses and with ecological and physiological studies at the local scale.

ACKNOWLEDGMENTS

Wiens' research on niche conservatism, speciation, and species richness has been supported by NSF grant 03–31,747. We thank H.B. Shaffer for helpful comments on the manuscript.

**The *Annual Review of Ecology, Evolution, and Systematics* is online at
http://ecolsys.annualreviews.org**

LITERATURE CITED

Ackerly DD. 2003. Community assembly, niche conservatism, and adaptive evolution in changing environments. *Int. J. Plant Sci.* 164(Suppl.):165–84

Anderson RP, Peterson AT, Gomez-Laverde M. 2002. Using niche-based GIS modeling to test geographic predictions of competitive exclusion and competitive release in South American pocket mice. *Oikos* 98:3–16

Austin MP. 1985. Continuum concept, ordination methods and niche theory. *Annu. Rev. Ecol. Syst.* 16:39–61

Barraclough TG, Vogler AP. 2000. Detecting the geographic pattern of speciation from species level phylogenies. *Am. Nat.* 155:419–34

Benton MJ. 1995. Diversification and extinction in the history of life. *Science* 268:52–58

Böhning-Gaese K, Schuda MD, Helbig AJ. 2003. Weak phylogenetic effects on ecological niches of *Sylvia* warblers. *J. Evol. Biol.* 16:956–65

Bradshaw AD. 1991. Genostasis and the limits to evolution. *Philos. Trans. R. Soc. London Ser. B* 333:289–305

Brown JH, Lomolino MV. 1998. *Biogeography*. Sunderland, MA: Sinauer Assoc. 2nd ed. 692 pp.

Cadle JE, Green HW. 1993. Phylogenetic patterns, biogeography, and the ecological structure of Neotropical snake assemblages. In *Species Diversity in Ecological Communities*, ed. RE Ricklefs, D Schluter, pp. 281–93. Chicago: Univ. Chicago Press

Cardillo M. 1999. Latitude and rates of diversification in birds and butterflies. *Proc. R. Soc. London Ser. B* 266:1221–25

Carpenter G, Gillison AN, Winter J. 1993. DOMAIN: a flexible modeling procedure for mapping potential distributions of plants and animals. *Biodivers. Conserv.* 2:667–80

Case TJ, Taper M. 2000. Interspecific competition, environmental gradients, gene flow, and the coevolution of species' borders. *Am. Nat.* 155:583–605

Conant R, Collins JT. 1991. *A Field Guide to Reptiles and Amphibians of Eastern and Central North America*. Boston: Houghton Mifflin. 3rd ed. 450 pp.

Connell JH. 1961. The influence of interspecific competition and other factors on the distribution of the barnacle *Chthamalus stellatus*. *Ecology* 42:710–23

Coyne JA, Orr HA. 2004. *Speciation*. Sunderland, MA: Sinauer Assoc. 545 pp.

Darwin C. 1859. *On the Origin of Species*. Cambridge, MA: Harvard Univ. Press. 436 pp.

de Queiroz K. 1998. The general lineage concept of species, species criteria, and the process of speciation: a conceptual unification and terminological recommendations. In *Endless Forms: Species and Speciation*, ed. DJ Howard, SH Berlocher, pp. 57–75. Oxford: Oxford Univ. Press

Diamond JM. 1997. *Guns, Germs, and Steel: the Fates of Human Societies*. New York: WW Norton. 480 pp.

Donoghue MJ, Smith SA. 2004. Patterns in the assembly of temperate forests around the Northern Hemisphere. *Philos. Trans. R. Soc. London Ser. B* 359:1633–44

Etterson JR, Shaw RG. 2001. Constraint to adaptive evolution in response to global warming. *Science* 294:151–54

Farrell BD, Mitter C, Futuyma DJ. 1992. Diversification at the insect-plant interface. *BioScience* 42:34–42

Francis AP, Currie DJ. 2003. A globally consistent richness-climate relationship for angiosperms. *Am. Nat.* 161:523–36

Futuyma DJ. 1998. *Evolutionary Biology*. Sunderland, MA: Sinauer Assoc. 3rd ed. 763 pp.

Gaston KJ, Blackburn TM. 1996. The tropics as

a museum of biological diversity: an analysis of the New World avifauna. *Proc. R. Soc. London Ser. B* 263:63–68

Geber MA, Griffen LR. 2003. Inheritance and natural selection on functional traits. *Int. J. Plant Sci.* 164(Suppl.):21–42

Gelfand AE, Silander JA Jr, Wu S, Latimer AM, Lewis PO, et al. 2003. Explaining species distribution patterns through hierarchical modeling. *Bayesian Analysis* 1:1–35

Good R. 1947. *The Geography of Flowering Plants.* New York: Longmans, Green & Co. 403 pp.

Graham CH, Ferrier S, Huettman F, Moritz C, Peterson AT. 2004a. New developments in museum-based informatics and application in biodiversity analysis. *Trends Ecol. Evol.* 19:497–503

Graham CH, Ron SR, Santos JC, Schneider CJ, Moritz C. 2004b. Integrating phylogenetics and environmental niche models to explore speciation mechanisms in dendrobatid frogs. *Evolution* 58:1781–93

Grinnell J. 1917. The niche-relationship of the California thrasher. *Auk* 34:427–33

Gross SJ, Price TD. 2000. Determinants of the northern and southern range limits of a warbler. *J. Biogeogr.* 27:869–78

Guisan A, Weiss SB, Weiss AD. 1999. GLM versus CCA spatial modeling of plant species distributions. *Plant Ecol.* 143:107–22

Guisan A, Zimmerman NE. 2000. Predictive habitat distributional models in ecology. *Ecol. Model.* 135:147–86

Haldane JBS. 1956. The relationship between density regulation and natural selection. *Proc. R. Soc. London Ser. B* 145:306–08

Hallam A, Wignall PB. 1997. *Mass Extinctions and Their Aftermath.* Oxford: Oxford Univ. Press. 320 pp.

Hawkins BA, Field R, Cornell HV, Currie DJ, Guégan JF, et al. 2003. Energy, water balance, and broad-scale geographic patterns of species richness. *Ecology* 84:3105–17

Hijmans RJ, Guarino L, Cruz M, Rojas E. 2002. Computer tools for spatial analysis of plant genetic resource data: 1. DIVA-GIS. *Plant Genet. Res. Newsl.* 127:15–19

Hirzel AH, Hausser J, Chessel D, Perrin N. 2002. Ecological-niche factor analysis: how to compute habitat-suitability maps without absence data? *Ecology* 83:2027–36

Hoffman AA, Hallas RJ, Dean JA, Schiffer M. 2003. Low potential for climatic stress adaptation in a rainforest *Drosophila* species. *Science* 301:100–2

Holt RD. 1996. Demographic constraints in evolution: towards unifying the evolutionary theories of senescence and niche conservatism. *Evol. Ecol.* 10:1–11

Holt RD, Gaines MS. 1992. Analysis of adaptation in heterogeneous landscapes: implications for the evolution of fundamental niches. *Evol. Ecol.* 6:433–47

Holt RD, Gomulkiewicz R. 1997. How does immigration influence local adaptation? A reexamination of a familiar paradigm. *Am. Nat.* 149:563–72

Hutchinson GE. 1957. *A Treatise on Limnology.* New York: Wiley & Sons. 1015 pp.

Janzen DH. 1967. Why mountain passes are higher in the tropics. *Am. Nat.* 101:233–49

Jenkins NL, Hoffman AA. 1999. Limits to the southern border of *Drosophila serrata:* cold resistance, heritability, and trade-offs. *Evolution* 53:1823–34

Kearney M, Porter WP. 2004. Mapping the fundamental niche: physiology, climate, and the distribution of a nocturnal lizard. *Ecology* 85:3119–31

Kirkpatrick M, Barton NH. 1997. Evolution of a species' range. *Am. Nat.* 150:1–23

Lehmann A, Overton JM, Austin MP. 2002. Regression models for spatial prediction: their role for biodiversity and conservation. *Biodivers. Conserv.* 11:2085–92

Losos JB, Jackman TR, Larson A, de Queiroz K, Rodríguez-Schettino L. 1998. Historical contingency and determinism in replicated adaptive radiations of island lizards. *Science* 279:2115–18

Losos JB, Leal M, Glor RE, de Queiroz K, Hertz PE, et al. 2003. Niche lability in the evolution of a Caribbean lizard community. *Nature* 424:542–50

MacArthur RH. 1972. *Geographical Ecology.* New York: Harper & Row. 269 pp.

Mayden RL. 1997. A hierarchy of species concepts: the denouement in the saga of the species problem. In *Species. The Units of Biodiversity,* ed. MF Claridge, HA Dawah, MR Wilson, pp. 381–424. London: Chapman & Hall

Nix HA. 1986. A biogeographic analysis of Australian elapid snakes. In *Atlas of Elapid Snakes of Australia,* ed. R Longmore, pp. 5–15. Canberra, Aust.: Aust. Gov. Publ. Serv.

Parmesan C, Yohe G. 2003. A globally coherent fingerprint of climate change impacts across natural systems. *Nature* 42137–42

Pearce J, Ferrier S. 2000. An evaluation of alternative algorithms for fitting species distribution models using logistic regression. *Ecol. Mod.* 128:127–47

Peterson AT. 2003. Predicting the geography of species' invasions via ecological niche modeling. *Q. Rev. Biol.* 78:419–33

Peterson AT, Cohoon KP. 1999. Sensitivity of distributional prediction algorithms to geographic data completeness. *Ecol. Mod.* 117: 159–64

Peterson AT, Ortega-Juerta M, Bartley J, Sanchez-Cordero V, Soberón J, et al. 2002. Future projections for Mexican faunas under global climate change scenarios. *Nature* 416: 626–29

Peterson AT, Soberón J, Sanchez-Cordero V. 1999. Conservatism of ecological niches in evolutionary time. *Science* 285:1265–67

Peterson AT, Vieglais DA. 2001. Predicting species invasions using ecological niche modeling: new approaches from bioinformatics attack a pressing problem. *BioScience* 51:363–71

Prinzig A, Durka W, Klotz S, Brandl F. 2001. The niche of higher plants: evidence for phylogenetic conservatism. *Proc. R. Soc. London Ser. B* 268:2383–89

Rahbek C, Graves GR. 2001. Multiscale assessment of patterns of avian species richness. *Proc. Natl. Acad. Sci. USA* 98:4534–39

Raxworthy CJ, Martinez-Meyer E, Horning N,

Nussbaum RA, Schneider GE, et al. 2003. Predicting distributions of known and unknown reptile species in Madagascar. *Nature* 426:837–41

Rice KJ, Emery NC. 2003. Managing microevolution: restoration in the face of global change. *Front. Ecol Environ.* 1:469–78

Rice NH, Martínez-Meyer E, Peterson AT. 2003. Ecological niche differentiation in the *Aphelocoma* jays: a phylogenetic perspective. *Biol. J. Linn. Soc.* 80:369–83

Ricklefs RE, Latham RE. 1992. Intercontinental correlation of geographical ranges suggests stasis in ecological traits of relict genera of temperate perennial herbs. *Am. Nat.* 139: 1305–21

Ricklefs RE, Schluter D. 1993. Species diversity: regional and historical influences. In *Species Diversity in Ecological Communities: Historical and Geographical Perspectives,* ed. RE Ricklefs, D Schluter, pp. 350–63. Chicago: Univ. Chicago Press

Roff DA. 1997. *Evolutionary Quantitative Genetics.* New York: Chapman and Hall. 493 pp.

Rosenzweig ML. 1995. *Species Diversity in Space and Time.* Cambridge: Cambridge Univ. Press. 436 pp.

Sanmartín I, Enghoff H, Ronquist F. 2001. Patterns of animal dispersal, vicariance and diversification in the holarctic. *Biol. J. Linn. Soc.* 73:345–90

Sechrest W, Brooks TM, da Fonseca GAB, Konstant WR, Mittermeier RA, et al. 2002. Hotspots and the conservation of evolutionary history. *Proc. Natl. Acad. Sci. USA* 99: 2067–71

Sites JW Jr, Marshall JC. 2003. Species delimitation: a renaissance issue in systematic biology. *Trends Ecol. Evol.* 18:462–70

Stearns SC, Sage RD. 1980. Maladaptation in a marginal population of the mosquitofish, *Gambusia affinis. Evolution* 34:65–75

Stebbins RC. 2003. *A Field Guide to Western Reptiles and Amphibians.* Boston: Houghton Mifflin. 3rd ed. 533 pp.

Stephens PR, Wiens JJ. 2003. Explaining species richness from continents to communities:

the time-for-speciation effect in emydid turtles. *Am. Nat.* 161:112–28

Stephens PR, Wiens JJ. 2004. Convergence, divergence, and homogenization in the ecological structure of emydid turtle communities: the effects of phylogeny and dispersal. *Am. Nat.* 164:244–54

Stockwell CA, Hendry AP, Kinnison MT 2003. Contemporary evolution meets conservation biology. *Trends Ecol. Evol.* 18:94–101

Stockwell DRB, Peters DB. 1999. The GARP modeling system: problems and solutions to automated spatial prediction. *Int. J. Geogr. Inf. Syst.* 13:143–58

Thomas CD, Cameron A, Green RE, Bakkenes M, Beaumont LJ, et al. 2004. Extinction risk from climate change. *Nature* 427:145–48

Turelli M, Barton NH, Coyne JA. 2001. Theory and speciation. *Trends Ecol. Evol.* 16:330–43

Wallace AR. 1876. *The Geographical Distribution of Animals.* London: MacMillan. 1110 pp.

Webb CO, Ackerly DD, McPeek MA, Donoghue MJ. 2002. Phylogenies and community ecology. *Annu. Rev. Ecol. Syst.* 33:475–505

Wiens JJ. 2004a. Speciation and ecology revisited: phylogenetic niche conservatism and the origin of species. *Evolution* 58:193–97

Wiens JJ. 2004b. What is speciation and how should we study it? *Am. Nat.* 163:914–23

Wiens JJ, Donoghue MJ. 2004. Historical biogeography, ecology, and species richness. *Trends Ecol. Evol.* 19:639–44

Wiens JJ, Servedio MR. 2000. Species delimitation in systematics: inferring diagnostic differences between species. *Proc. R. Soc. London Ser. B* 267:631–36

Wilcove DS, Rothstein D, Dubow J, Phillips A, Losos E. 1998. Quantifying threats to imperiled species in the United States. *BioScience* 48:607–15

Willig MR, Kaufman DM, Stevens RD. 2003. Latitudinal gradients of biodiversity: pattern, process, scale, and synthesis. *Annu. Rev. Ecol. Evol. Syst.* 34:273–309

Wilson EO. 1992. *The Diversity of Life.* New York: WW Norton. 424 pp.

Zug GR, Vitt LJ, Caldwell JP. 2001. Herpetology. *An Introductory Biology of Amphibians and Reptiles.* San Diego: Academic. 2nd ed. 630 pp.

Annu. Rev. Ecol. Evol. Syst. 2005. 36:541–62
doi: 10.1146/annurev.ecolsys.35.112202.130205
First published online as a Review in Advance on August 17, 2005

PHYLOGENOMICS

Hervé Philippe,[1] Frédéric Delsuc,[1] Henner Brinkmann,[1] and Nicolas Lartillot[2]

[1]*Canadian Institute for Advanced Research, Département de Biochimie, Université de Montréal, Montréal, Québec H3C3J7, Canada; email: herve.philippe@umontreal.ca, frederic.delsuc@umontreal.ca, henner.brinkmann@umontreal.ca*
[2]*Laboratoire d'Informatique, de Robotique et de Mathématiques de Montpellier, Centre National de la Recherche Scientifique, Université de Montpellier, 34392 Montpellier Cedex 5, France; email: nicolas.lartillot@lirmm.fr*

Key Words inconsistency, molecular phylogeny, systematic bias, taxon sampling, tree of life

■ **Abstract** The continuous flow of genomic data is creating unprecedented opportunities for the reconstruction of molecular phylogenies. Access to whole-genome data means that phylogenetic analysis can now be performed at different genomic levels, such as primary sequences and gene order, allowing for reciprocal corroboration of the results. We critically review the different kinds of phylogenomic methods currently available, paying particular attention to method reliability. Our emphasis is on methods for the analysis of primary sequences because these are the most advanced. We discuss the important issue of statistical inconsistency and show how failing to fully capture the process of sequence evolution in the underlying models leads to tree reconstruction artifacts. We suggest strategies for detecting and potentially overcoming these problems. These strategies involve the development of better models, the use of an improved taxon sampling, and the exclusion of phylogenetically misleading data.

INTRODUCTION

The newly arising discipline of phylogenomics owes its existence to the revolutionizing progress in DNA sequencing technology. The number of complete genome sequences is already high and increases at an ever-accelerating pace. The newly coined term "phylogenomics" (Eisen 1998, O'Brien & Stanyon 1999) comprises several areas of research at the interplay between molecular biology and evolution. The main issues are (*a*) using molecular data to infer species' relationships and (*b*) using information on species' evolutionary history to gain insights into the mechanisms of molecular evolution. The majority of publications on phylogenomics deal with the second aspect (see Sjolander 2004 for a review). However, our concern here is the use of data at the genomic scale to reconstruct the phylogeny of organisms.

A novel and interesting aspect of phylogenomics lies in the possibility of using molecular information above the primary sequence level. In particular, trees can be inferred from whole-genome features, such as gene content (Fitz-Gibbon & House 1999, Snel et al. 1999, Tekaia et al. 1999), gene order (Korbel et al. 2002, Sankoff et al.1992), intron positions (Roy & Gilbert 2005), or protein domain structure (Lin & Gerstein 2000, Yang et al. 2005). A potential advantage of these methods is that the complexity of some of these characters (e.g., gene order) renders the character-state space very large, reducing the risk of homoplasy by convergence and reversal, thus rendering the inferred phylogenies more reliable. However, these integrated approaches imply the use of a reduced number of characters relative to primary sequence-based approaches. There are about 300 times fewer genes than amino acid positions (assuming a mean protein length of about 300 amino acids), thus increasing the risk of stochastic error.

In fact, stochastic or sampling error constitutes one of the major limitations of standard phylogenetics based on single genes. Because the number of positions of a single gene is small, random noise influences the inference of numerous nodes, leading generally to poorly resolved phylogenetic trees. The idea of using large amounts of genomic data as a way to address this problem is not new. For example, to resolve the original tritomy between chimpanzee, human, and gorilla, about 10 kb were sequenced as early as the late 1980s (Miyamoto et al. 1988). However, technical and financial limitations have often confined molecular systematics to the use of a few markers (e.g., rRNA) for which a large diversity of organisms have been sequenced. Numerous important phylogenetic questions remained unsolved, and great hope was placed into the wealth of genomic data soon to be available.

Sampling (or stochastic) error should vanish as the number of genes added to the analysis gets large enough. In practice, this means that statistical support (e.g., bootstrap support) will eventually rise to 100% as more genes are considered. The use of tens of thousands, or millions, of aligned positions that provide a great deal of phylogenetic information should ultimately lead to fully resolved trees. Indeed, several empirical studies confirmed this premise (Bapteste et al. 2002, Madsen et al. 2001, Murphy et al. 2001, Qiu et al. 1999, Rokas et al. 2003, Soltis et al. 1999). This increased resolution leads to the optimistic view that phylogenomics would "end the incongruencies" observed in single-gene phylogenies (Gee 2003). However, whether the resulting, highly supported, phylogenetic trees are the true ones is not certain.

Systematic Error and Consistency

The most important challenge of phylogenomics is to verify that tree reconstruction methods are consistent, i.e., converge toward the correct answer as more and

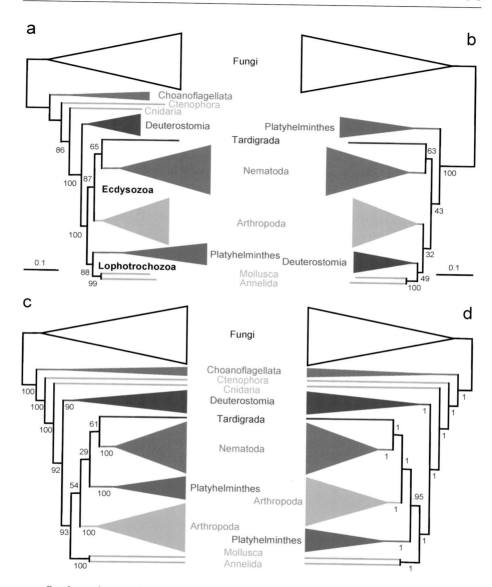

See legend on next page

Figure 1 Supermatrix, taxon sampling and supertree in animal phylogenomics. A dataset constituted of 71 slowly evolving nuclear proteins corresponding to 20,705 amino acid positions (Philippe et al. 2005) was used for phylogenetic analyses based on maximum likelihood (ML) with a separate WAG + F + Γ model. Panel (*a*) presents the ML tree obtained with 49 species (redrawn from Philippe et al. 2005). This tree strongly supports the new animal phylogeny (Aguinaldo et al. 1997), dividing Bilateria into Deuterostomia and Protostomia, which comprises Lophotrochozoa and Ecdysozoa. Note that the major division between Deuterostomia and Protostomia is supported by a 100% bootstrap value. In panel (*b*), the exclusion of only four close out-group sequences (two Choanoflagellata, Cnidaria, and Ctenophora) creates an LBA artifact via the distant fungal out-group leading to the successive early emergence of the fast-evolving Platyhelmintes and then nematoda plus Tardigrada. Note that the overall bootstrap support substantially decreases, requiring caution when the existence of a radiation is extrapolated from low statistical supports because limited resolution can also be due to poor taxon sampling or to unreliable tree reconstruction methods (not shown). Branch lengths are drawn proportionally to evolutionary rates, and the height of triangles represents the taxonomic diversity of the different groups. Panel (*c*) presents the supertree obtained by maximum parsimony analysis conducted with PAUP* (Swofford 2000) of the matrix representation of 71 source trees obtained from Bayesian analyses of individual genes with MrBayes (Ronquist & Huelsenbeck 2003) using a WAG + F + Γ model. This supertree recovers both the monophyly of Deuterostomia and Protostomia with strong bootstrap support, but fails to find Ecdysozoa and Lophotrochozoa. Panel (*d*) shows the supertree obtained on the same matrix from a Bayesian analysis using MrBayes and a simple two-state model. This supertree strongly supports the respective monophyly of Deuterostomia, Protostomia and Ecdysozoa, but not of Lophotrochozoa. Numbers on branches are bootstrap values (*a–c*) or posterior probabilities (*d*).

more characters are considered (Felsenstein 1978, 1988).[1] In principle, at least in a probabilistic framework, a lack of consistency can always be traced back to some violation of model assumptions by the data analyzed. Note that methods that are not explicitly model based, such as maximum parsimony, are equivalent to a statistical analysis under an implicit model (Steel & Penny 2000). The best understood causes of method inconsistency stem from models that do not properly account for (*a*) variable evolutionary rates across lineages, leading to the long-branch-attraction (LBA) artifact (Felsenstein 1978), (*b*) heterogeneous nucleotide/amino acid compositions, resulting in the artificial grouping of species that share the same bias (Lockhart et al. 1994), and (*c*) heterotachy, i.e., shift of position-specific evolutionary rates (Kolaczkowski & Thornton 2004, Lockhart et al. 1996, Philippe & Germot 2000). These systematic biases could be interpreted respectively as rate signal, compositional signal, and heterotachous signal, which we will collectively refer to as nonphylogenetic signals (Ho & Jermiin 2004). In other words, nonphylogenetic signals are due to substitutions that occurred along the true phylogeny but are misinterpreted by tree reconstruction methods as supporting an alternative topology.

Compared to single-gene studies, inconsistency is more pronounced in phylogenomic analyses. For example, a reanalysis of the large dataset of Rokas et al. (2003) demonstrates that, depending on the method used, mutually incongruent, yet 100% supported, trees could be obtained (Phillips et al. 2004). It is well accepted that the analysis of phylogenomic datasets will necessarily increase the resolution of the trees through the "increase of the signal-to-noise ratio" (Rokas et al. 2003). Indeed, the signal-to-random-noise ratio increases, but the phylogenetic-to-nonphylogenetic signal ratio remains constant whatever the number of genes considered (assuming that the gene sampling is not biased). In this review, we focus on best practices for enhancing the phylogenetic signal in genomic data, while reducing the impact of erroneous signals, in order to obtain accurate and robust trees.

ASSEMBLY OF PHYLOGENOMIC DATASETS

The reliability of a phylogenetic tree depends on the quality of the data and the accuracy of the reconstruction method. In 1988, Felsenstein noted that "molecular evolutionists who use methods for inferring phylogenies do not engage in much

---

[1]The concept of consistency is formally defined as a property of a statistical estimator. Bayesian statistics is not so much concerned with statistical estimators but with the posterior distribution of a parameter, leading some to question the relevancy of consistency to Bayesian analysis (Felsenstein 2004). However, this view is not shared by many Bayesian statisticians. Indeed, if the underlying model is misspecified, this may result in an asymptotically vanishing posterior probability for the true tree, which is exactly the problem that the concept of consistency tries to capture. This suggests a Bayesian analogue of consistency: A model is consistent if the posterior distribution tends to a point mass concentrated on the true value of the unknown parameter, as the number of observations tends to infinity (Diaconis 1986).

discussion of the properties of the methods they use since they focus on the difficult task of collecting the data" (Felsenstein 1988, p. 523). Almost 20 years later, molecular systematists still spend much of their time assembling larger and larger datasets, and the crucial discussion about inference methods remains neglected. In phylogenomics, the reliability of the inference is often simply justified by the large number of characters used. Nevertheless, the problem of data acquisition deserves further discussion, as it can heavily compromise the subsequent analysis.

Importance of a Rich Taxon Sampling

A long-standing debate in phylogenetics concerns the relative importance of improving taxon versus gene sampling (Graybeal 1998, Hillis et al. 2003, Rosenberg & Kumar 2003). In the genomic age, gene sampling would seem not to be an issue. However the limited resources devoted to systematics often prevent sequencing the genomes of all relevant species. Depending on the importance accorded to taxon sampling, two strategies can be used: (*a*) gathering complete genome sequences from a few key organisms or (2) gathering incomplete, yet large, genome sequences from a great diversity of organisms.

The first approach is supported by some computer simulation studies (Rosenberg & Kumar 2003) and is the most frequently used in phylogenomic analyzes (Blair et al. 2002, Goremykin et al. 2004, Misawa & Janke 2003, Philip et al. 2005, Rokas et al. 2003, Wolf et al. 2004). However, the design of computer simulations and the interpretation of their results make it difficult to draw firm conclusions from this approach (Hillis et al. 2003, Rosenberg & Kumar 2003). Empirical evidence seems nevertheless to argue against the taxon-poor approach, as illustrated by the phylogeny of metazoans. Two new clades (Ecdysozoa, the moulting animals, including among others arthropods and nematodes, and Lophotrochozoa, including among others annelids, molluscs and Platyhelminthes) were proposed from rRNA analyses (Aguinaldo et al. 1997). However, several phylogenomic studies strongly supported the paraphyly of Ecdysozoa when considering a few model organisms and using a distant out-group (Blair et al. 2002, Dopazo et al. 2004, Philip et al. 2005, Wolf et al. 2004). In contrast, when 49 species and 71 genes (20,705 positions) are used (Philippe et al. 2005), the monophyly of Ecdysozoa and Lophotrochozoa is recovered with strong support (Figure 1*a*, see color insert). However, the removal of close out-groups leads to drastic changes (Figure 1*b*): the fast-evolving lineages emerge paraphyletically at the base of the tree. Such an asymmetrical tree shape is expected to result from an LBA artifact when the out-group is distantly related (Philippe & Laurent 1998), as fungi are. Contrary to the situation with a few species discussed above, the statistical support for these incorrect placements is weak (bootstrap values between 32 and 63), demonstrating that an increased taxon sampling (from 4–10 to 45) has reduced, but not eliminated, the impact of LBA. In fact, the low bootstrap support (Figure 1*b*) demonstrates that the nonphylogenetic signal becomes equivalent to a phylogenetic signal when species sampling is impoverished.

A rich-taxon sampling is not, however, the panacea. First, computation time increases rapidly with the number of species, rendering exhaustive searches impossible with more than 20 species and most heuristic searches with probabilistic methods intractable with more than ~200 species. Second, adding taxa can sometimes degrade the phylogenetic inference (Kim 1996). Third, the number of extant species can be naturally sparse, forever preventing the assembly of a rich and balanced taxon sample. For example, *Amborella* is proposed to constitute the first, or one of the first, emerging angiosperm lineages (Qiu et al. 1999), but it is the only extant representative of an ancient group. Even if the heated debate about its placement (Goremykin et al. 2004, Soltis et al. 2004) could be solved by improved taxon sampling, the assumed basal position of *Amborella* might prove difficult to attest in the absence of closely related extant taxa (Stefanovic et al. 2004). In conclusion, an adequate taxon sampling, as balanced as possible, is important to increase the accuracy of phylogenomic trees.

Missing Data

Although genome sequencing has become ever easier, it seems unlikely that complete genomes will be soon available for a rich diversity of organisms. In addition, a bias in favor of sequencing small genomes leads to potential problems. Because small genomes are generally derived from larger genomes, whole-genome features, such as gene content or gene order, will evolve much faster, rendering tree reconstruction susceptible to artifacts such as LBA and compositional bias (Copley et al. 2004, House & Fitz-Gibbon 2002, Korbel et al. 2002, Lake & Rivera 2004, Wolf et al. 2001). Moreover, genome reduction is often associated with an accelerated rate of protein evolution (Brinkmann et al. 2005, Dufresne et al. 2005) or extremely biased nucleotide compositions (Herbeck et al. 2005). The sampling of a fraction of the genome from species with huge genomes is therefore a necessity to represent some key taxa and/or to include less biased representatives.

Two low-cost approaches can be used: (*a*) the selection of a limited set of genes potentially useful for the phylogenetic question of interest, followed by their targeted PCR amplification and sequencing (Murphy et al. 2001) and (*b*) the sequencing of thousands of expressed sequence tags (ESTs), which generally provides hundreds of relevant genes (Bapteste et al. 2002). The first method is more adapted to phylogeny at small evolutionary scales and has the advantage that genes can be a priori selected to obtain an optimal phylogenetic signal. The second might be preferable at larger evolutionary scales (e.g., among protists) and allows the discovery of many other genes, which can shed light on the evolution of important features such as metabolic pathways.

Phylogenomic reconstruction methods based on gene content/order cannot be applied to incomplete genomic sampling, but those based on DNA strings and on primary sequences can. In the latter case, missing data will occur even when complete genomes are used, especially at a large evolutionary scale because most, if not all, genes can be lost, duplicated, or horizontally transferred in some organisms.

To our knowledge, no theoretical reasons suggest that sequence-based approaches can not be used on incomplete alignments, i.e., containing cells coded as missing data. Nevertheless, "the problem of missing data is widely considered to be the most significant obstacle. . .in combining datasets. . . that do not include identical taxa," as suggested by empirical studies and computer simulations (Wiens 2003). Two problems need to be distinguished: (*a*) the potential lack of resolution induced by the presence of taxa with too many missing cells and (*b*) the possible interaction between missing entries and artifact-inducing model violations.

Recent computer simulations using a large number of characters (Philippe et al. 2004, Wiens 2003) suggest that the inaccurate placement of incomplete taxa is not the result of missing data but rather the result of an insufficient number of informative characters. As an extreme example, the tree reconstruction method remains accurate when positions have an average of 4 known and 32 unknown character states because each species is nevertheless represented by about 3000 amino acids (Philippe et al. 2004). However, the presence of missing cells unevenly distributed across the data matrix potentially affects estimates of model parameters. It is not yet clear how the induced model misspecifications in turn influence phylogenetic inference. Interestingly, it seems that the advantage of adding an incomplete taxon that breaks a long branch is greater than the disadvantage of the induced model misspecification (Wiens 2005).

Few attempts at assessing the effect of missing data have been made with empirical data (Bapteste et al. 2002, House & Fitz-Gibbon 2002, Philippe et al. 2004). For instance, a bipartition of the supermatrix (25% of missing data) into the most complete genes and the less complete genes appears to be indistinguishable from random bipartitions of the same size (Bapteste et al. 2002, Philippe et al. 2004). However, when the level of missing data is extreme (92%), the quality of the inference appears to be affected (e.g., strong support for the paraphyly of Glires and of Ecdysozoa) (Driskell et al. 2004), despite the large size of the dataset (70 taxa and 1131 genes). In summary, even if the problem generated by missing data has been overrated, additional work is needed to characterize its impact more precisely.

INFERENCE OF PHYLOGENOMIC TREES

The methods used in phylogenomic inference are either (*a*) primary sequence-based methods, which are very similar to the classical tree reconstruction, and for which several excellent reviews and textbooks are available (Felsenstein 2004, Holder & Lewis 2003, Swofford et al. 1996) or (*b*) methods above the sequence level (Wolf et al. 2002). The two approaches are fundamentally similar; the main difference is the characters used. In both cases, we believe that probabilistic methods are more powerful and more reliable. First, they have a more robust theoretical justification in that they rely on an explicit account of their assumptions by using stochastic models describing the pattern of molecular evolution (Felsenstein 2004). Second, not only do they allow estimates of the phylogeny and a confidence level, but they also provide general methods to evaluate the fit of the model used (Goldman 1993).

Approaches Based on Whole-Genome Features

Because genomes are the results of evolution, virtually any features comparable between organisms can be used to infer phylogenies, as evidenced by the plethora of new approaches recently published (Fitz-Gibbon & House 1999, Henz et al. 2005, House & Fitz-Gibbon 2002, House et al. 2003, Korbel et al. 2002, Lin & Gerstein 2000, Pride et al. 2003, Qi et al. 2004, Snel et al. 1999, Tekaia et al. 1999, Wolf et al. 2001, Yang et al. 2005). The justifications of whole-genome tree approaches are generally that the phylogeny of organisms can not be equated to the phylogeny of single genes (such as rRNA) and that this classical phylogeny is sensitive to hidden paralogy, horizontal gene transfer (HGT) or tree reconstruction artifacts (Fitz-Gibbon & House 1999, Lin & Gerstein 2000, Snel et al. 1999, Tekaia et al. 1999). Because tree reconstruction artifacts can affect any approach and are difficult to detect, we rather believe that these new approaches based on various character types are of paramount importance to corroborate phylogenetic inference (Miyamoto & Fitch 1995, Swofford 1991, Wolf et al. 2002).

DISTRIBUTION OF SEQUENCE STRINGS   The frequencies of small oligonucleotides (up to eight) or oligopeptides (up to six) observed in the genome or the proteome can be transformed into distances then used to construct phylogenies (Blaisdell 1986). Numerous well-accepted clades, including deep ones, were recovered using this approach, confirming that a phylogenetic signal is present in these characters (Edwards et al. 2002, Pride et al. 2003, Qi et al. 2004). However, several undisputed clades were significantly rejected, and the comparison of the branch lengths of the genome tree obtained using tetranucleotide frequencies with those of the tree obtained from the standard analysis of rRNA sequences (Pride et al. 2003) suggests that the phylogenetic signal contained in sequence strings saturates rapidly. However, the methods proposed so far are extremely crude. Oligonucleotide or oligopeptide frequencies are transformed into distances without any underlying model of evolution. It is nevertheless remarkable that something considered as a bias in standard sequence-based methods (Lockhart et al. 1994) contains a phylogenetic signal, but it is not yet clear whether accurate methods can be developed to extract it.

HOMOLOGY AND ORTHOLOGY ASSESSMENT   All other approaches require the establishment of homology, or more often, orthology of genes. By definition, the phylogenetic history displayed by orthologous genes is the organismal phylogeny (Fitch 1970), whereas paralogous or xenologous genes display a combination of organismal and gene-specific history. In practice, the identification of orthologous genes involves a certain amount of circularity because it requires an a priori knowledge of the organismal phylogeny. Indeed, finding orthologous genes is difficult because the organismal phylogeny is generally unknown (or at best partially known), and its reconstruction represents the goal. This problem is much akin to the alignment/phylogeny problem, in which alignment and phylogeny should be estimated simultaneously (Sankoff et al. 1973, Wheeler 2003), but the practical difficulties are such that the two steps are generally separated. Moreover, a careful phylogenetic reconstruction of all gene families is a Herculean labor most researchers want to avoid (but see Storm & Sonnhammer 2002). Therefore, an operational, yet approximate, definition of orthology is used. Schematically, all genomes are compared to each other at the amino acid level; only the pairs of sequences that are the best reciprocal hits are considered further. The clusters of orthologous groups are then constructed by a single-linkage analysis of all orthologous pairs (Tatusov et al. 1997) or by Markov cluster algorithms (Remm et al. 2001). These approaches, albeit reasonably effective in practice, do not guarantee the identification of orthologous genes because, if a gene has been transferred from a distant organism and replaced the original copy, it will fulfill all the requirements while being xenologous. Information from synteny will probably improve the accuracy of orthology assignment (Zheng et al. 2004), but much more work is needed before obtaining perfect assignment of orthologous genes.

GENE-ODER METHODS   The character space of gene-order data is huge, so it probably constitutes the most promising genome feature-based method. However, it is also technically and computationally the most difficult (see Moret et al. 2005 for a review). Briefly, distances can be computed by minimizing the number of inversions, transpositions, insertions, and deletions necessary to transform one unichromosomal genome into another (Sankoff et al. 1992), a method which is most often further simplified by considering only inversions (Bourque & Pevzner 2002). Alternatively, the "break-point distance" between two genomes, defined as the minimum number of pairs of genes next to each other in one of the two genomes, but not in the other (Nadeau & Taylor 1984), can be used to infer phylogeny (Blanchette et al. 1997). However, the evolution of gene order involves additional rearrangement mechanisms that are not easily accounted for using these methods, such as translocation (i.e., transposition between chromosomes) and fusion or fission events of chromosomes.

More recently, methods based on maximum-likelihood (ML) distances (Wang & Warnow 2005), or on Bayesian inference (Larget et al. 2005, York et al. 2002), have been proposed that take multiple changes in gene order into account. They revealed an important level of saturation (York et al. 2002) and a limited tree-resolving power (Larget et al. 2005). The small size of mitochondrial genomes might explain the predominance of stochastic noise in the latter case. Prokaryotic genomes contain much more information, but are too large for analysis by current software. The drastic simplifying assumption that gene order can be reduced to the presence/absence of gene pairs allows the inference of prokaryotic phylogenies that are similar to the ones based on gene content (Korbel et al. 2002, Wolf et al.

2001). Further methodological and computational developments are needed to realize the full potential of gene-order methods.

GENE-CONTENT METHODS Phylogenomic analyses based on gene content generally use orthologs, but a few variants exist that use homologous instead of orthologous genes (Fitz-Gibbon & House 1999), protein domain content (Yang et al. 2005), or fold occurrence (Lin & Gerstein 2000, Yang et al. 2005). Interestingly, an orthologous gene present in all organisms (the Holy Grail of the sequence-based approach) has the same character state and is not informative for gene-content methods. This is an important source of corroboration because patterns informative for sequence-based approaches are not informative for gene-content approaches. Otherwise, the distribution of orthologs will be informative in the cases of (a) gene loss, which have a more than negligible probability of being convergent, (b) gene genesis, which is potentially the most informative, and (c) horizontal gene transfers (i.e., the acquisition of a new gene from a distantly related organism at the base of a clade will create a synapomorphy for this clade and a homoplasy for locating this clade). Gene content, albeit more integrated than primary sequences, is thus far from providing an unambiguous phylogenetic signal.

The binary matrices of presence/absence of homologous or orthologous genes can be analyzed by distance methods (Lin & Gerstein 2000, Snel et al. 1999), parsimony (Fitz-Gibbon & House 1999), or Dollo parsimony (Wolf et al. 2001). However, big/small genome attraction (Lake & Rivera 2004) appears to affect all these methods, as demonstrated by the artificial grouping of unrelated species with small genomes (e.g., *Mycoplasma*, *Buchnera*, *Chlamydia*, or *Rickettsia*). For instance, Copley et al. (2004) demonstrated that phylogenies based on gene content and protein domain combinations support the paraphyly of Ecdysozoa, but these results are biased by a systematic high rate of character loss in nematodes. When this bias is accounted for by computing the number of losses expected randomly, a slight support for the monophyly of Ecdysozoa is recovered. Interestingly, this artifact yields the same inconsistency phenotype as the one displayed by sequence-based analyses (Philippe et al. 2005); in both cases, arthropods are the sister group of vertebrates to the exclusion of nematodes. Such a convergence between the two methods could be explained by the fact that the fast-evolving species are also those that have undergone the most extreme genome reduction. This observation weakens the strength of the corroboration between gene-content and primary sequence approaches.

Altogether, gene-content phylogenies are not in excellent agreement with previous knowledge, even if we ignore the problem caused by small genome attraction. The monophyly of the three domains (Archaea, Bacteria, and Eukaryota) is always recovered, but the phylogeny within domains is much more problematic. For instance, *Halobacterium* never clusters with *Methanosarcina*, probably because of many HGTs from Bacteria that attract it toward the base of Archaea (Korbel et al. 2002); except with threshold parsimony (House et al. 2003), the monophyly of Proteobacteria is never recovered (Dutilh et al. 2004, Gu & Zhang 2004, Henz et al. 2005, Wolf et al. 2001).

Several technical improvements have recently been proposed. Because big/small genome attraction is akin to the problem of compositional bias in sequence-based approaches, this nonphylogenetic signal can be reduced by the use of the LogDet/paralinear transformation (Lake & Rivera 2004). A simple model of gene genesis and gene loss allows ML estimates of evolutionary distances (Gu & Zhang 2004, Huson & Steel 2004), but simulations showed that their performance appears to be slightly poorer than the performance of Dollo parsimony (Huson & Steel 2004).

In summary, the methods for inferring trees based on whole-genome features are at an early stage of their development, which might be comparable to that of sequence-based methods in the early 1970s. In particular, they generally lack a global probabilistic modeling. Numerous works are ongoing, and it will be important to extensively evaluate the accuracy of present and future methods.

Approaches Based on Primary Sequences

SUPERMATRIX VERSUS SUPERTREE   The question of how to analyze multiple data-sets has been the subject of intense debate (for reviews see Bull et al. 1993, de Queiroz et al. 1995). In brief, three approaches are mainly used: (*a*) total evidence (Kluge 1989), in which all datasets are combined together, called hereafter the supermatrix approach; (*b*) separate analysis (Miyamoto & Fitch 1995), in which the datasets are analyzed individually and resulting topologies are combined using consensus or supertree methods, called hereafter the supertree approach; and (*c*) conditional combination (Bull et al. 1993, Lecointre & Deleporte 2005), in which only the datasets considered as congruent are combined, called hereafter the conditional supermatrix approach. Generally, their respective advantages are minimizing stochastic error, increasing the significance of corroboration, and minimizing conflicting signals. In the practice of phylogenomics, the supermatrix is by far preferred (Murphy et al. 2001, Philippe et al. 2005, Qiu et al. 1999, Rokas et al. 2003), followed by a few cases of the supertree method (Daubin et al. 2002, Philip et al. 2005).

Although Bull et al. (1993) state that "no rational systematist would suggest combining genes with different histories to produce a single reconstruction," the reliance in the power of the supermatrix approach is quite strong, and this methodology is generally applied. Against Bull et al., one may argue that discordant genes will each display a different discordant history, which will be averaged away through a combined analysis (but see Matte-Tailliez et al. 2002). In most of the large multigene studies, the possibility of incongruencies is voluntarily minimized by selecting genes having a priori the same evolutionary history [e.g., single-copy genes (Lerat et al. 2003, Murphy et al. 2001, Philip et al. 2005), organellar genes (Qiu et al. 1999, Soltis et al. 1999), or orthology assessment based on synteny (Rokas et al. 2003)]. But generally, the homogeneity of the datasets was not tested, indicating a strict application of the total evidence principle.

The problem of homogeneity is important because several recurrent processes (gene duplication, HGT, or lineage sorting) can lead to incongruent gene trees.

In particular, the very notion of a "tree of life" has been questioned because of rampant HGTs (Doolittle 1999). Several studies have nevertheless argued that a strong phylogenetic signal is present in the prokaryotic genomes (for reviews see Brown 2003, Philippe & Douady 2003) and therefore that HGTs do not wipe out the notion of organismal phylogeny. It is clear, however, that few, and probably none, of the genes have followed exactly the organismal phylogeny during the entire history of life on Earth. Thus, HGTs constitute a source of nuisance that should be addressed to improve the accuracy of phylogenetic reconstructions.

The infrequent use of the conditional supermatrix approach is likely not due to a philosophical rejection of its principle but rather to the difficulty of detecting homogeneous datasets in practice. The incongruence length difference (ILD) test (Farris et al. 1995) was initially designed for parsimony and has been recently adapted to distance methods (Zelwer & Daubin 2004). However, the interpretation of this test is complicated by the fact that stochastic noise can generate by itself significant results (Dolphin et al. 2000). Its efficiency in detecting incongruence and determining data combinability has been repeatedly questioned (Darlu & Lecointre 2002, Dowton & Austin 2002, Yoder et al. 2001). Parametric incongruence tests have also been proposed for ML methods (Huelsenbeck & Bull 1996), or in a Bayesian framework (Nylander et al. 2004), although they are not yet in widespread use. A promising model has been proposed in which each gene can, with a certain prior probability, choose between either conforming to the common global topology or relying on its own topology (Suchard et al. 2003).

Nevertheless, the conditional supermatrix approach is used, despite the fact that the criteria used to discard gene/sequence are not well validated (Brochier et al. 2002, Brown et al. 2001, Lecointre & Deleporte 2005, Matte-Tailliez et al. 2002). However, the nature of the test used to decide whether a gene significantly supports a different topology yields divergent interpretations of the same data with regards to the importance of HGTs (Bapteste et al. 2004, Lerat et al. 2003, Zhaxybayeva et al. 2004). An improvement of these incongruence tests (Goldman et al. 2000) therefore constitutes an important avenue of future research.

The supertree approach is most often used to combine trees from the literature that were obtained from diverse sources of data (Sanderson et al. 1998), the most popular method being matrix representation with parsimony (MRP) (Baum 1992, Ragan 1992). The comparative efficiency of supermatrix and supertree approaches is poorly studied, especially in a phylogenomic context (Gatesy et al. 2004, Philip et al. 2005). We have therefore analyzed the dataset of 71 genes (Figure 1a) using a supertree MRP approach. Interestingly, almost all nodes inferred from the supermatrix or using the supertree are identical, except the position of urochordates within deuterostomes (not shown) and the relationships among protostomes, indicating an excellent congruence of the supermatrix and supertree (only 3 differences for 46 bipartitions). When the supertree is reconstructed with maximum parsimony (Figure 1c), Platyhelminthes are grouped with tardigrads + nematods instead of with other lophotrochozoans (annelids and molluscs), disrupting the monophyly of both Ecdysozoa and Lophotrochozoa. When the supertree is reconstructed using

Bayesian inference (Figure 1*d*), the results were slightly more consistent because the monophyly of Ecdysozoa, but not Lophotrochozoa, was recovered. The MRP supertree approach appears to have difficulty in placing the fast-evolving lineages (e.g., Platyhelminthes). The synergy among all positions in the supermatrix might explain the ability of the method to better deal with LBA artifacts. However, refined studies are urgently required to evaluate the relative accuracy and efficiency of the supermatrix and supertree methods.

SUPERMATRIX: SCALING UP CURRENT METHODS   In contrast to the genome-feature approaches mentioned above, the main advantage of sequence-based methods is that their properties have been intensively explored, tested, and validated, so that many of their strengths and weaknesses are known. Indeed, in most sequence-based phylogenomic analyses published to date, almost the same protocols as for single-gene studies have been applied. The congruence among results obtained by different methods is high, with some notable exceptions (Canback et al. 2004, Goremykin et al. 2004, Soltis et al. 2002, Stefanovic et al. 2004). These incongruencies confirm the presence of a nonphylogenetic signal and suggest that the increase in resolution obtained by analyzing larger datasets is not in itself a guarantee of accuracy. Conversely, the agreement between the methods does not mean that the obtained tree is correct (see Brinkmann et al. 2005).

The dramatic change of scale of the data matrices implies the need for a corresponding increase in computational power, in particular for probabilistic methods. Two factors have to be considered. First, there is a simple scaling up of both the memory requirement and computational load. Second, the reliability of the heuristic search procedures underlying ML programs, or the Monte Carlo devices of the Bayesian samplers, is anything but guaranteed. A particular concern is that conflicting signals result in the presence of many secondary maxima separated by high potential barriers in the space of tree topologies. Standard procedures are likely to get trapped in these local optima (Salter 2001) and thus do not yield reliable phylogenetic estimates.

A number of algorithmic innovations have led to better heuristic searches in the space of topologies; these innovations include the following: genetic algorithms (Brauer et al. 2002, Lemmon & Milinkovitch 2002), disk-covering methods (Huson et al. 1999), and parallelized computing (Keane et al. 2005). In the case of the Bayesian methods, an interesting approach has been proposed consisting of using coupled "heated" Monte Carlo Markov Chains, which can be easily parallelized (Altekar et al. 2004, Feng et al. 2003). Thanks to these advances, ML and Bayesian phylogenetic reconstructions will soon be able to handle phylogenomic datasets. However, their overall reliability has been evaluated mainly on simulated data, which are probably much more "funnel shaped" toward the true phylogeny than are real sequences (Brinkmann et al. 2005, Stamatakis et al. 2005).

Another possible stance toward efficient tree space searches is to restrict the analysis by constraining nodes that have been found with high support when each gene of the concatenation was analyzed separately (Philippe et al. 2005). The number of trees compatible with these constraints is still large, but accessible to an exhaustive analysis. Such approaches may not be considered as a definitive method but could provide a good proxy.

TOWARD MORE COMPLEX MODELS   If the availability of large data matrices poses new computational challenges to probabilistic methods, it allows the development of more realistic, parameter-rich models. As long as the number of parameters increases more slowly than the number of sites, a model does not fall into the infinitely many parameter trap (Felsenstein 2004) and thus has good consistency properties. Of course, no probabilistic model will ever capture evolutionary patterns in their full complexity, but their most important aspects, at least the ones that cause inconsistency of the current methods, can be accounted for (Steel 2005). The main idea is an improved flexibility that accounts for the diverse kinds of heterogeneities and disparities of the substitution processes.

One possible research direction is to account for disparities in the evolutionary process across the genes that make up the concatenation. A simple solution is to constrain the model to have a global topology, but gene-specific branch lengths (Yang 1996). More generally, any parameter other than the topology can be considered as gene-specific in such "separate" (or partitioned) models. Another possible avenue of research is to account for site-specific patterns of substitution, using mixture models (Kolaczkowski & Thornton 2004, Lartillot & Philippe 2004, Pagel & Meade 2004). A mixture model combines several different classes to describe the substitution process, each of which is characterized by its own set of parameters (e.g., equilibrium frequencies or exchangeability probabilities).

Thus far, few studies have tried to address the relative performances of alternative probabilistic models on phylogenomic datasets. A recent study (Brinkmann et al. 2005) has confirmed that accounting for site-specific rates, or having a good empirical substitution rate matrix, is an important factor, resulting in a higher phylogenetic accuracy. In contrast, separate models, in spite of their overall better statistical fit, do not seem to fundamentally improve phylogenetic inference. Because separate models handle a substantial part of heterotachy, this suggests that heterotachy, despite recent interest, may not constitute a major source of systematic bias. In any case, a much wider analysis of the impact of model choice on the prevalence of artifacts in phylogenomic inference has to be performed.

Reducing Systematic Errors Through Data Exclusion

Phylogenomic datasets contain a large amount of genuine phylogenetic signal, but they also contain nonphylogenetic signals that current methods of tree reconstruction are not able to perfectly handle. To avoid the perils of inconsistency, one can take advantage of the fact that the quantity of phylogenetic signal is no longer a serious limiting factor. More precisely, the part of the datasets that contains mainly nonphylogenetic signals can be excluded, allowing the concentration of phylogenetic signal in the remaining dataset. This increase of the phylogenetic to nonphylogenetic signal ratio reduces the probability of inconsistency even without the use of improved tree reconstruction methods.

The rationale of most methods of data exclusion is straightforward: Tree reconstruction artifacts are due to multiple substitutions that are not correctly identified as convergences or reversions by inference methods (Olsen 1987). The simplest possibility consists in removing the fast-evolving species, which by definition accumulate multiple substitutions. This approach efficiently reduces the misleading effect of the rate signal (Philippe et al. 2005, Stefanovic et al. 2004). In many cases, the exclusion of odd taxa (Sanderson & Shaffer 2002) is implicit because investigators never envision using them in their analyses (e.g., microsporidia as a fungal representative).

When all of the available species representing a clade of interest are fast evolving, the specific removal of the fastest-evolving sequences of this clade from the supermatrix appears to be efficient. For example, when a complete supermatrix of 133 genes is used, all tree reconstruction methods strongly, albeit artifactually, locate microsporidia at the base of eukaryotes, but probabilistic methods with a complex model (WAG + F + Γ) avoid this LBA artifact when >70% of the microsporidial sequences are coded as missing data (Brinkmann et al. 2005). It is interesting to note that a highly incomplete taxon (70% of the positions are unknown) is more accurately located than a complete one. A similar approach, in which genes in their totality are discarded, was successful at avoiding the attraction between the fast-evolving nematodes and Platyhelminthes (Philippe et al. 2005) or between nematodes and the out-group (Dopazo & Dopazo 2005). Interestingly, in the first case, the statistical support for the monophyly of Ecdysozoa and Lophotrochozoa increases when more and more genes are discarded (as long as more than 40 genes are considered).

These approaches are rather crude because sequences from genes and/or species are discarded in their totality. Nevertheless, even if these sequences contain much nonphylogenetic signal, they likely also contain some phylogenetic signal. Refined methods have been proposed to selectively eliminate fast-evolving characters in part (Lopez et al. 1999) or completely (Brinkmann & Philippe 1999, Burleigh & Mathews 2004, Dutilh et al. 2004, Pisani 2004, Ruiz-Trillo et al. 1999). In several cases, taxa that emerged at the base of the tree when all the characters are used are relocated later in the tree when fast-evolving positions are removed, strongly suggesting that their observed basal position is due to an LBA artifact (Brochier & Philippe 2002, Philippe et al. 2000, Pisani 2004). However, the number of remaining slowly evolving positions is often too small to yield high statistical support for most of the clades in single-gene analyses but not in phylogenomic analyses (Burleigh & Mathews 2004, Delsuc et al. 2005).

The RY-coding strategy (Woese et al. 1991) discards all fast-evolving transitions and improves inference without drastically compromising the resolution (Phillips et al. 2004). Importantly, this coding not only addresses the problem of rate signal but also of compositional signal. Indeed, the G + C content can be extremely variable among homologous sequences from various organisms, whereas the frequency of purines is remarkably homogeneous (Woese et al. 1991). This constitutes a method of choice to avoid inconsistency resulting from compositional bias.

Finally, it is possible to remove characters whose evolutionary history violates most the assumptions of the underlying model of sequence evolution instead of the fastest-evolving characters/species. In fact, the RY coding eliminates transitions that are mainly responsible for the nonstationarity of the nucleotide composition (see Hrdy et al. 2004 for a similar approach in the case of proteins). Similarly, the constant sites violate the assumptions about the distribution of site rates (Lockhart et al. 1996), and their elimination constitutes an efficient way of improving inference (Hirt et al. 1999, Phillips et al. 2004). Heterotachous sites violate the assumption that the evolutionary rate of a position is constant through time, made by all current models except the covarion model (Fitch & Markowitz 1970). As expected, the elimination of these sites reduced LBA artifacts in the case of the eukaryotic phylogeny (Inagaki et al. 2004, Philippe & Germot 2000).

We believe that these data removal approaches are complementary to the improvement of tree reconstruction methods through the implementation of more realistic models of sequence evolution. So far, they are also more readily accessible in practice because they are less demanding in terms of the complexity of bioinformatic methods and computational time.

CONCLUDING REMARKS

In this review, we have emphasized that inconsistency of tree reconstruction methods constitutes the major limitation of phylogenomics. We have explored ways to reduce its impact, whether at the level of data assembly or at the level of tree building per se. Adequate taxon sampling, probabilistic methods, and the exclusion of phylogenetically misleading data constitute the three most important criteria required to obtain reliable phylogenomic trees. However, this might not be sufficient to ensure that the inferred tree is the correct one. We believe that an important test of the reliability of tree reconstruction methods is their robustness with respect to species sampling. In particular, a reliable method should be able to recover exactly the same topology with a taxon-poor and a taxon-rich sampling, which is far from being the case at present (Figure 1). This is particularly important to be confident in the phylogenetic location of poorly diversified clades (e.g., *Amborella*, or monotremes). The stability of phylogenies in face of variation of species sampling will constitute one of the best guarantees that the nonphylogenetic signal has been correctly handled.

The correctness of inferences should also be verified via corroboration from independent sources (Miyamoto & Fitch 1995, Swofford 1991). When complete genomes are used, this may appear hopeless (even if internal verifications of homogeneity are possible). However, genomes can be conveniently subdivided into various character types that can be considered as more or less independent, e.g., oligonucleotide composition, sequences of orthologous gene, gene content, and gene order. If inferences based on these "independent" sets of characters converge to the same results, an increased confidence can be placed in the corresponding phylogeny, although the same bias can theoretically affect several approaches. We therefore encourage the development of sophisticated probabilistic methods for all types of data and not just for primary sequences.

In the long term, one might envision that the tree of life, or at least its global scaffold, will be established within the next 10 years. Then, to numerous molecular systematists the important question will be: What next?

ACKNOWLEDGMENTS

We wish to thank David Bryant, Emmanuel Douzery, David Moreira, Davide Pisani, Nicolas Rodrigue, Naiara Rodríguez-Ezpeleta, Béatrice Roure, and Brad Shaffer for critical readings of the manuscript. This work was supported by operating funds from Genome Québec. H.P. is a member of the Program in Evolutionary Biology of the Canadian Institute for Advanced Research, which is acknowledged for salary and interaction support. H.P. is also grateful to the Canada Research Chairs Program and the Canadian Foundation for Innovation for salary and equipment support.

The *Annual Review of Ecology, Evolution, and Systematics* is online at http://ecolsys.annualreviews.org

LITERATURE CITED

Aguinaldo AM, Turbeville JM, Linford LS, Rivera MC, Garey JR, et al. 1997. Evidence for a clade of nematodes, arthropods and other moulting animals. *Nature* 387:489–93

Altekar G, Dwarkadas S, Huelsenbeck JP, Ronquist F. 2004. Parallel Metropolis coupled Markov chain Monte Carlo for Bayesian phylogenetic inference. *Bioinformatics* 20:407–15

Bapteste E, Boucher Y, Leigh J, Doolittle WF. 2004. Phylogenetic reconstruction and lateral gene transfer. *Trends Microbiol.* 12:406–11

Bapteste E, Brinkmann H, Lee JA, Moore DV, Sensen CW, et al. 2002. The analysis of 100 genes supports the grouping of three highly divergent amoebae: *Dictyostelium*, *Entamoeba*, and *Mastigamoeba*. *Proc. Natl. Acad. Sci. USA* 99:1414–19

Baum BR. 1992. Combining trees as a way of combining data sets for phylogenetic inference, and the desirability of combining gene trees. *Taxon* 41:3–10

Blair JE, Ikeo K, Gojobori T, Hedges SB. 2002. The evolutionary position of nematodes. *BMC Evol. Biol.* 2:7

Blaisdell BE. 1986. A measure of the similarity of sets of sequences not requiring sequence alignment. *Proc. Natl. Acad. Sci. USA* 83:5155–59

Blanchette M, Bourque G, Sankoff D. 1997. *Breakpoint phylogenies.* Presented at 8th Genome Inform. Conf. (GIW 1997), Tokyo

Bourque G, Pevzner PA. 2002. Genome-scale evolution: reconstructing gene orders in the ancestral species. *Genome Res.* 12:26–36

Brauer MJ, Holder MT, Dries LA, Zwickl DJ, Lewis PO, Hillis DM. 2002. Genetic algorithms and parallel processing in maximum-likelihood phylogeny inference. *Mol. Biol. Evol.* 19:1717–26

Brinkmann H, Philippe H. 1999. Archaea sister group of bacteria? Indications from tree reconstruction artifacts in ancient phylogenies. *Mol. Biol. Evol.* 16:817–25

Brinkmann H, van der Giezen M, Zhou Y, Poncelin de Raucourt G, Philippe H. 2005. An empirical assessment of long branch attraction artifacts in phylogenomics. *Syst. Biol.* In press

Brochier C, Bapteste E, Moreira D, Philippe H. 2002. Eubacterial phylogeny based on translational apparatus proteins. *Trends Genet.* 18:1–5

Brochier C, Philippe H. 2002. Phylogeny: a non-hyperthermophilic ancestor for bacteria. *Nature* 417:244

Brown JR. 2003. Ancient horizontal gene transfer. *Nat. Rev. Genet.* 4:121–32

Brown JR, Douady CJ, Italia MJ, Marshall WE, Stanhope MJ. 2001. Universal trees based on large combined protein sequence data sets. *Nat. Genet.* 28:281–85

Bull JJ, Huelsenbeck JP, Cunningham CW, Swofford DL, Wadell PJ. 1993. Partitioning and combining characters in phylogenetic analysis. *Syst. Biol.* 42:384–97

Burleigh JG, Mathews S. 2004. Phylogenetic signal in nucleotide data from seed plants: implications for resolving the seed plant tree of life. *Am. J. Bot.* 91:1599–613

Canback B, Tamas I, Andersson SG. 2004. A phylogenomic study of endosymbiotic bacteria. *Mol. Biol. Evol.* 21:1110–22

Copley RR, Aloy P, Russell RB, Telford MJ. 2004. Systematic searches for molecular synapomorphies in model metazoan genomes give some support for Ecdysozoa after accounting for the idiosyncrasies of *Caenorhabditis elegans*. *Evol. Dev.* 6:164–69

Darlu P, Lecointre G. 2002. When does the incongruence length difference test fail? *Mol. Biol. Evol.* 19:432–37

Daubin V, Gouy M, Perriere G. 2002. A phylogenomic approach to bacterial phylogeny: evidence of a core of genes sharing a common history. *Genome Res.* 12:1080–90

de Queiroz A, Donoghue MJ, Kim J. 1995. Separate versus combined analysis of phylogenetic evidence. *Annu. Rev. Ecol. Syst.* 26:657–81

Delsuc F, Brinkmann H, Philippe H. 2005. Phylogenomics and the reconstruction of the tree of life. *Nat. Rev. Genet.* 6:361–75

Dolphin K, Belshaw R, Orme CD, Quicke DL. 2000. Noise and incongruence: interpreting results of the incongruence length difference test. *Mol. Phylogenet. Evol.* 17:401–6

Doolittle WF. 1999. Phylogenetic classification and the universal tree. *Science* 284:2124–29

Dopazo H, Dopazo J. 2005. Genome-scale evidence of the nematode-arthropod clade. *Genome Biol.* 6:R41

Dopazo H, Santoyo J, Dopazo J. 2004. Phylogenomics and the number of characters required for obtaining an accurate phylogeny of eukaryote model species. *Bioinformatics* 20:i116–21

Dowton M, Austin AD. 2002. Increased congruence does not necessarily indicate increased phylogenetic accuracy—the behavior of the incongruence length difference test in mixed-model analyses. *Syst. Biol.* 51:19–31

Driskell AC, Ane C, Burleigh JG, McMahon MM, O'Meara BC, Sanderson MJ. 2004. Prospects for building the tree of life from large sequence databases. *Science* 306:1172–74

Dufresne A, Garczarek L, Partensky F. 2005. Accelerated evolution associated with genome reduction in a free-living prokaryote. *Genome Biol.* 6:R14

Dutilh BE, Huynen MA, Bruno WJ, Snel B. 2004. The consistent phylogenetic signal in genome trees revealed by reducing the impact of noise. *J. Mol. Evol.* 58:527–39

Edwards SV, Fertil B, Giron A, Deschavanne PJ. 2002. A genomic schism in birds revealed by phylogenetic analysis of DNA strings. *Syst. Biol.* 51:599–613

Eisen JA. 1998. Phylogenomics: improving functional predictions for uncharacterized genes by evolutionary analysis. *Genome Res.* 8:163–67

Farris JS, Källerjo M, Kluge AG, Bult C. 1995. Testing significance of incongruence. *Cladistics* 10:315–19

Felsenstein J. 1978. Cases in which parsimony or compatibility methods will be positively misleading. *Syst. Zool.* 27:401–10

Felsenstein J. 1988. Phylogenies from molecular sequences: inference and reliability. *Annu. Rev. Genet.* 22:521–65

Felsenstein J. 2004. *Inferring Phylogenies.* Sunderland, MA: Sinauer. 645 pp.

Feng XZ, Buell DA, Rose JR, Waddell PJ. 2003. Parallel algorithms for Bayesian phylogenetic inference. *J. Parallel Distrib. Comput.* 63:707–18

Fitch WM. 1970. Distinguishing homologous from analogous proteins. *Syst. Zool.* 19:99–113

Fitch WM, Markowitz E. 1970. An improved method for determining codon variability in a gene and its application to the rate of fixation of mutations in evolution. *Biochem. Genet.* 4:579–93

Fitz-Gibbon ST, House CH. 1999. Whole genome-based phylogenetic analysis of free-living microorganisms. *Nucleic Acids Res.* 27:4218–22

Gascuel O, ed. 2005. *Mathematics of Evolution and Phylogeny.* Oxford: Oxford Univ. Press

Gatesy J, Baker RH, Hayashi C. 2004. Inconsistencies in arguments for the supertree approach: supermatrices versus supertrees of Crocodylia. *Syst. Biol.* 53:342–55

Gee H. 2003. Evolution: ending incongruence. *Nature* 425:782

Goldman N. 1993. Statistical tests of models of DNA substitution. *J. Mol. Evol.* 36:182–98

Goldman N, Anderson JP, Rodrigo AG. 2000. Likelihood-based tests of topologies in phylogenetics. *Syst. Biol.* 49:652–70

Goremykin VV, Hirsch-Ernst KI, Wolfl S, Hellwig FH. 2004. The chloroplast genome of *Nymphaea alba*: whole-genome analyses and the problem of identifying the most basal angiosperm. *Mol. Biol. Evol.* 21:1445–54

Graybeal A. 1998. Is it better to add taxa or characters to a difficult phylogenetic problem? *Syst. Biol.* 47:9–17

Gu X, Zhang H. 2004. Genome phylogenetic analysis based on extended gene contents. *Mol. Biol. Evol.* 21:1401–8

Henz SR, Huson DH, Auch AF, Nieselt-Struwe K, Schuster SC. 2005. Whole-genome prokaryotic phylogeny. *Bioinformatics* 21:2329–35

Herbeck JT, Degnan PH, Wernegreen JJ. 2005. Non-homogeneous model of sequence evolution indicates independent origins of primary endosymbionts within the Enterobacteriales (γ-proteobacteria). *Mol. Biol. Evol.* 22:520–32

Hillis DM, Pollock DD, McGuire JA, Zwickl DJ. 2003. Is sparse taxon sampling a problem for phylogenetic inference? *Syst. Biol.* 52:124–26

Hirt RP, Logsdon JM Jr, Healy B, Dorey MW, Doolittle WF, Embley TM. 1999. Microsporidia are related to fungi: evidence from the largest subunit of RNA polymerase II and other proteins. *Proc. Natl. Acad. Sci. USA* 96:580–85

Ho SY, Jermiin L. 2004. Tracing the decay of the historical signal in biological sequence data. *Syst. Biol.* 53:623–37

Holder M, Lewis PO. 2003. Phylogeny estimation: traditional and Bayesian approaches. *Nat. Rev. Genet.* 4:275–84

House CH, Fitz-Gibbon ST. 2002. Using homolog groups to create a whole-genomic tree of free-living organisms: an update. *J. Mol. Evol.* 54:539–47

House CH, Runnegar B, Fitz-Gibbon ST. 2003. Geobiological analysis using whole genome-based tree building applied to the Bacteria, Archaea and Eukarya. *Geobiology* 1:15–26

Hrdy I, Hirt RP, Dolezal P, Bardonova L, Foster PG, et al. 2004. *Trichomonas* hydrogenosomes contain the NADH dehydrogenase module of mitochondrial complex I. *Nature* 432:618–22

Huelsenbeck JP, Bull JJ. 1996. A likelihood ratio test to detect conflicting phylogenetic signal. *Syst. Biol.* 45:92–98

Huson DH, Nettles SM, Warnow TJ. 1999. Disk-covering, a fast-converging method for phylogenetic tree reconstruction. *J. Comput. Biol.* 6:369–86

Huson DH, Steel M. 2004. Phylogenetic trees based on gene content. *Bioinformatics* 20: 2044–49

Inagaki Y, Susko E, Fast NM, Roger AJ. 2004. Covarion shifts cause a long-branch attraction artifact that unites microsporidia and archaebacteria in EF-1α phylogenies. *Mol. Biol. Evol.* 21:1340–49

Keane TM, Naughton TJ, Travers SA, McInerney JO, McCormack GP. 2005. DPRml: distributed phylogeny reconstruction by maximum likelihood. *Bioinformatics* 21:969–74

Kim J. 1996. General inconsistency conditions for maximum parsimony: effects of branch lengths and increasing numbers of taxa. *Syst. Biol.* 45:363–74

Kluge AG. 1989. A concern for evidence and a phylogenetic hypothesis of relationships among *Epicrates* (Boidae, Serpentes). *Syst. Zool.* 38:7–25

Kolaczkowski B, Thornton JW. 2004. Performance of maximum parsimony and likelihood phylogenetics when evolution is heterogeneous. *Nature* 431:980–84

Korbel JO, Snel B, Huynen MA, Bork P. 2002. SHOT: a web server for the construction of genome phylogenies. *Trends Genet.* 18:158–62

Lake JA, Rivera MC. 2004. Deriving the genomic tree of life in the presence of horizontal gene transfer: conditioned reconstruction. *Mol. Biol. Evol.* 21:681–90

Larget B, Simon DL, Kadane JB, Sweet D. 2005. A Bayesian analysis of metazoan mitochondrial genome arrangements. *Mol. Biol. Evol.* 22:486–95

Lartillot N, Philippe H. 2004. A Bayesian mixture model for across-site heterogeneities in the amino-acid replacement process. *Mol. Biol. Evol.* 21:1095–109

Lecointre G, Deleporte P. 2005. Total evidence requires exclusion of phylogenetically misleading data. *Zool. Scr.* 31:101–17

Lemmon AR, Milinkovitch MC. 2002. The metapopulation genetic algorithm: an efficient solution for the problem of large phylogeny estimation. *Proc. Natl. Acad. Sci. USA* 99:10516–21

Lerat E, Daubin V, Moran NA. 2003. From gene trees to organismal phylogeny in prokaryotes: the case of the γ-proteobacteria. *PLoS Biol.* 1:101–9

Lin J, Gerstein M. 2000. Whole-genome trees based on the occurrence of folds and orthologs: implications for comparing genomes on different levels. *Genome Res.* 10: 808–18

Lockhart PJ, Steel MA, Hendy MD, Penny D. 1994. Recovering evolutionary trees under a more realistic model of sequence evolution. *Mol. Biol. Evol.* 11:605–12

Lockhart PJ, Larkum AWD, Steel MA, Waddell PJ, Penny D. 1996. Evolution of chlorophyll and bacteriochlorophyll: the problem of invariant sites in sequence analysis. *Proc. Natl. Acad. Sci. USA* 93:1930–34

Lopez P, Forterre P, Philippe H. 1999. The root of the tree of life in the light of the covarion model. *J. Mol. Evol.* 49:496–508

Madsen O, Scally M, Douady CJ, Kao DJ, DeBry RW, et al. 2001. Parallel adaptive radiations in two major clades of placental mammals. *Nature* 409:610–14

Matte-Tailliez O, Brochier C, Forterre P, Philippe H. 2002. Archaeal phylogeny based on ribosomal proteins. *Mol. Biol. Evol.* 19: 631–39

Misawa K, Janke A. 2003. Revisiting the Glires concept–phylogenetic analysis of nuclear sequences. *Mol. Phylogenet. Evol.* 28:320–27

Miyamoto MM, Fitch WM. 1995. Testing species phylogenies and phylogenetic methods with congruence. *Syst. Biol.* 44:64–76

Miyamoto MM, Koop BF, Slightom JL, Goodman M, Tennant MR. 1988. Molecular systematics of higher primates: genealogical relations and classification. *Proc. Natl. Acad. Sci. USA* 85:7627–31

Moret BME, Tang J, Warnow T. 2005. Reconstructing phylogenies from gene-content and gene-order data. See Gascuel 2005, pp. 321–52

Murphy WJ, Eizirik E, Johnson WE, Zhang YP, Ryder OA, O'Brien SJ. 2001. Molecular phylogenetics and the origins of placental mammals. *Nature* 409:614–18

Nadeau JH, Taylor BA. 1984. Lengths of chromosomal segments conserved since divergence of man and mouse. *Proc. Natl. Acad. Sci. USA* 81:814–18

Nylander JA, Ronquist F, Huelsenbeck JP, Nieves-Aldrey JL. 2004. Bayesian phylogenetic analysis of combined data. *Syst. Biol.* 53:47–67

O'Brien SJ, Stanyon R. 1999. Phylogenomics. Ancestral primate viewed. *Nature* 402:365–66

Olsen G. 1987. Earliest phylogenetic branching: comparing rRNA-based evolutionary trees inferred with various techniques. *Cold Spring Harbor Symp. Quant. Biol.* 52:825–37

Pagel M, Meade A. 2004. A phylogenetic mixture model for detecting pattern-heterogeneity in gene sequence or character-state data. *Syst. Biol.* 53:571–81

Philip GK, Creevey CJ, McInerney JO. 2005. The Opisthokonta and the Ecdysozoa may not be clades: stronger support for the grouping of plant and animal than for animal and fungi and stronger support for the Coelomata than Ecdysozoa. *Mol. Biol. Evol.* 22:1175–84

Philippe H, Douady CJ. 2003. Horizontal gene transfer and phylogenetics. *Curr. Opin. Microbiol.* 6:498–505

Philippe H, Germot A. 2000. Phylogeny of eukaryotes based on ribosomal RNA: long-branch attraction and models of sequence evolution. *Mol. Biol. Evol.* 17:830–34

Philippe H, Lartillot N, Brinkmann H. 2005. Multigene analyses of bilaterian animals corroborate the monophyly of Ecdysozoa, Lophotrochozoa, and Protostomia. *Mol. Biol. Evol.* 22:1246–53

Philippe H, Laurent J. 1998. How good are deep phylogenetic trees? *Curr. Opin. Genet. Dev.* 8:616–23

Philippe H, Lopez P, Brinkmann H, Budin K, Germot A, et al. 2000. Early-branching or fast-evolving eukaryotes? An answer based on slowly evolving positions. *Proc. R. Soc. London Ser. B.* 267:1213–21

Philippe H, Snell EA, Bapteste E, Lopez P, Holland PW, Casane D. 2004. Phylogenomics of eukaryotes: impact of missing data on large alignments. *Mol. Biol. Evol.* 21:1740–52

Phillips MJ, Delsuc F, Penny D. 2004. Genome-scale phylogeny and the detection of systematic biases. *Mol. Biol. Evol.* 21:1455–58

Pisani D. 2004. Identifying and removing fast-evolving sites using compatibility analysis: an example from the arthropoda. *Syst. Biol.* 53:978–89

Pride DT, Meinersmann RJ, Wassenaar TM, Blaser MJ. 2003. Evolutionary implications of microbial genome tetranucleotide frequency biases. *Genome Res.* 13:145–58

Qi J, Wang B, Hao BI. 2004. Whole proteome prokaryote phylogeny without sequence alignment: a K-string composition approach. *J. Mol. Evol.* 58:1–11

Qiu YL, Lee J, Bernasconi-Quadroni F, Soltis DE, Soltis PS, et al. 1999. The earliest angiosperms: evidence from mitochondrial, plastid and nuclear genomes. *Nature* 402:404–7

Ragan MA. 1992. Matrix representation in reconstructing phylogenetic relationships among the eukaryotes. *Biosystems* 28:47–55

Remm M, Storm CE, Sonnhammer EL. 2001. Automatic clustering of orthologs and in-paralogs from pairwise species comparisons. *J. Mol. Biol.* 314:1041–52

Rokas A, Williams BL, King N, Carroll SB. 2003. Genome-scale approaches to resolving incongruence in molecular phylogenies. *Nature* 425:798–804

Ronquist F, Huelsenbeck JP. 2003. MrBayes 3: Bayesian phylogenetic inference under mixed models. *Bioinformatics* 19:1572–74

Rosenberg MS, Kumar S. 2003. Taxon sampling, bioinformatics, and phylogenomics. *Syst. Biol.* 52:119–24

Roy SW, Gilbert W. 2005. Resolution of a deep animal divergence by the pattern of intron conservation. *Proc. Natl. Acad. Sci. USA* 102:4403–8

Ruiz-Trillo I, Riutort M, Littlewood DT, Herniou EA, Baguna J. 1999. Acoel flatworms: earliest extant bilaterian Metazoans, not members of Platyhelminthes. *Science* 283:1919–23

Salter LA. 2001. Complexity of the likelihood surface for a large DNA dataset. *Syst. Biol.* 50:970–78

Sanderson MJ, Purvis A, Henze C. 1998. Phylogenetic supertrees: assembling the trees of life. *Trends Ecol. Evol.* 13:105–9

Sanderson MJ, Shaffer HB. 2002. Troubleshooting molecular phylogenetic analyses. *Ann. Rev. Ecol. Syst.* 33:49–72

Sankoff D, Leduc G, Antoine N, Paquin B, Lang BF, Cedergren R. 1992. Gene order comparisons for phylogenetic inference: evolution of the mitochondrial genome. *Proc. Natl. Acad. Sci. USA* 89:6575–79

Sankoff D, Morel D, Cedergren R. 1973. Evolution of 5S RNA and the nonrandomness of base replacement. *Nat. New Biol.* 245:232–34

Sjolander K. 2004. Phylogenomic inference of protein molecular function: advances and challenges. *Bioinformatics* 20:170–79

Snel B, Bork P, Huynen MA. 1999. Genome phylogeny based on gene content. *Nat. Genet.* 21:108–10

Soltis DE, Albert VA, Savolainen V, Hilu K, Qiu YL, et al. 2004. Genome-scale data, angiosperm relationships, and "ending incongruence": a cautionary tale in phylogenetics. *Trends Plant Sci.* 9:477–83

Soltis DE, Soltis PS, Zanis MJ. 2002. Phylogeny of seed plants based on evidence from eight genes. *Am. J. Bot.* 89:1670–81

Soltis PS, Soltis DE, Chase MW. 1999. Angiosperm phylogeny inferred from multiple genes as a tool for comparative biology. *Nature* 402:402–4

Stamatakis A, Ludwig T, Meier H. 2005. RAxML-III: a fast program for maximum likelihood-based inference of large phylogenetic trees. *Bioinformatics* 21:456–63

Steel M. 2005. Should phylogenetic models be trying to 'fit an elephant'? *Trends Genet.* 21:307–9

Steel M, Penny D. 2000. Parsimony, likelihood, and the role of models in molecular phylogenetics. *Mol. Biol. Evol.* 17:839–50

Stefanovic S, Rice DW, Palmer JD. 2004. Long branch attraction, taxon sampling, and the earliest angiosperms: *Amborella* or monocots? *BMC Evol. Biol.* 4:35

Storm CE, Sonnhammer EL. 2002. Automated ortholog inference from phylogenetic trees and calculation of orthology reliability. *Bioinformatics* 18:92–99

Suchard MA, Kitchen CM, Sinsheimer JS, Weiss RE. 2003. Hierarchical phylogenetic models for analyzing multipartite sequence data. *Syst. Biol.* 52:649–64

Swofford DL. 1991. When are phylogeny estimates from molecular and morphological data incongruent? In *Phylogenetic Analysis of DNA Sequences*, ed. MM Miyamoto, J Cracraft, pp. 295–333. New York: Oxford Univ. Press

Swofford DL. 2000. *PAUP*: Phylogenetic Analysis Using Parsimony and Other Methods.* Sunderland, MA: Sinauer

Swofford DL, Olsen GJ, Waddell PJ, Hillis DM. 1996. Phylogenetic inference. In *Molecular Systematics*, ed. DM Hillis, C Moritz, BK Mable, pp. 407–514. Sunderland, MA: Sinauer

Tatusov RL, Koonin EV, Lipman DJ. 1997. A genomic perspective on protein families. *Science* 278:631–37

Tekaia F, Lazcano A, Dujon B. 1999. The genomic tree as revealed from whole proteome comparisons. *Genome Res.* 9:550–57

Wang L-S, Warnow T. 2005. Distance-based genome rearrangement phylogeny. See Gascuel 2005, pp. 353–83

Wheeler WC. 2003. Iterative pass optimization of sequence data. *Cladistics* 19:254–60

Wiens JJ. 2003. Missing data, incomplete taxa, and phylogenetic accuracy. *Syst. Biol.* 52:528–38

Wiens JJ. 2005. Can incomplete taxa rescue phylogenetic analyses from long-branch attraction? *Syst. Biol.* In press

Woese CR, Achenbach L, Rouviere P, Mandelco L. 1991. Archaeal phylogeny:

reexamination of the phylogenetic position of *Archaeoglobus fulgidus* in light of certain composition-induced artifacts. *Syst. Appl. Microbiol.* 14:364–71

Wolf YI, Rogozin IB, Grishin NV, Koonin EV. 2002. Genome trees and the tree of life. *Trends Genet.* 18:472–79

Wolf YI, Rogozin IB, Grishin NV, Tatusov RL, Koonin EV. 2001. Genome trees constructed using five different approaches suggest new major bacterial clades. *BMC Evol. Biol.* 1:8

Wolf YI, Rogozin IB, Koonin EV. 2004. Coelomata and not Ecdysozoa: evidence from genome-wide phylogenetic analysis. *Genome Res.* 14:29–36

Yang S, Doolittle RF, Bourne PE. 2005. Phylogeny determined by protein domain content. *Proc. Natl. Acad. Sci. USA* 102:373–78

Yang Z. 1996. Maximum-likelihood models for combined analyses of multiple sequence data. *J. Mol. Evol.* 42:587–96

Yoder AD, Irwin JA, Payseur BA. 2001. Failure of the ILD to determine data combinability for slow loris phylogeny. *Syst. Biol.* 50:408–24

York TL, Durrett R, Nielsen R. 2002. Bayesian estimation of the number of inversions in the history of two chromosomes. *J. Comput. Biol.* 9:805–18

Zelwer M, Daubin V. 2004. Detecting phylogenetic incongruence using BIONJ: an improvement of the ILD test. *Mol. Phylogenet. Evol.* 33:687–93

Zhaxybayeva O, Lapierre P, Gogarten JP. 2004. Genome mosaicism and organismal lineages. *Trends Genet.* 20:254–60

Zheng XH, Lu F, Wang ZY, Zhong F, Hoover J, Mural R. 2004. Using shared genomic synteny and shared protein functions to enhance the identification of orthologous gene pairs. *Bioinformatics* 21:703–10

Annu. Rev. Ecol. Evol. Syst. 2005. 36:563–95
doi: 10.1146/annurev.ecolsys.36.102003.152626
Copyright © 2005 by Annual Reviews. All rights reserved
First published online as a Review in Advance on August 19, 2005

THE EVOLUTION OF AGRICULTURE IN INSECTS

Ulrich G. Mueller,[1,2] Nicole M. Gerardo,[1,2,3] Duur K. Aanen,[4] Diana L. Six,[5] and Ted R. Schultz[6]

[1] Section of Integrative Biology, University of Texas at Austin, Austin, Texas 78712;
email: umueller@mail.utexas.edu

[2] Smithsonian Tropical Research Institute, Apartado 2072, Balboa, Republic of Panama

[3] Department of Ecology and Evolutionary Biology, University of Arizona, Tucson,
Arizona 85721-0088; email: ngerardo@email.arizona.edu

[4] Department of Population Biology, Biological Institute, University of Copenhagen,
2100 Copenhagen, Denmark; email: dkaanen@bi.ku.dk

[5] Department of Ecosystem and Conservation Sciences, University of Montana,
Missoula, Montana 59812; email: diana.six@cfc.umt.edu

[6] Department of Entomology, National Museum of Natural History,
Smithsonian Institution, Washington, District of Columbia 20013-7012;
email: schultz@lab.si.edu

Key Words   Attini, Macrotermitinae, mutualism, symbiosis, Xyleborini

■ **Abstract**   Agriculture has evolved independently in three insect orders: once in ants, once in termites, and seven times in ambrosia beetles. Although these insect farmers are in some ways quite different from each other, in many more ways they are remarkably similar, suggesting convergent evolution. All propagate their cultivars as clonal monocultures within their nests and, in most cases, clonally across many farmer generations as well. Long-term clonal monoculture presents special problems for disease control, but insect farmers have evolved a combination of strategies to manage crop diseases: They (*a*) sequester their gardens from the environment; (*b*) monitor gardens intensively, controlling pathogens early in disease outbreaks; (*c*) occasionally access population-level reservoirs of genetically variable cultivars, even while propagating clonal monocultures across many farmer generations; and (*d*) manage, in addition to the primary cultivars, an array of "auxiliary" microbes providing disease suppression and other services. Rather than growing a single cultivar solely for nutrition, insect farmers appear to cultivate, and possibly "artificially select" for, integrated crop-microbe consortia. Indeed, crop domestication in the context of coevolving and codomesticated microbial consortia may explain the 50-million year old agricultural success of insect farmers.

1. INTRODUCTION

The cultivation of crops for nourishment has evolved only a few times in the animal kingdom. The most prominent and unambiguous examples include the fungus-growing ants, the fungus-growing termites, the ambrosia beetles and, of

1543-592X/05/1215-0563$20.00   **563** course, humans. For humans, who started the transition from an ancestral hunter-gatherer existence to farming only about 10,000 years ago (Diamond 1997, Smith 1998), sustainable, high-yield agriculture has become critical for survival in a global economy with projected food shortages, and diverse research programs are currently devoted to the optimization of agricultural productivity in the context of growing environmental challenges (Green et al. 2005). Agricultural progress has been achieved by humans through a combination of insight, creative planning, and a fair share of contingency and luck (Diamond 1997, Schultz et al. 2005, Smith 1998). However, humans have so far not examined nonhuman agricultural systems, such as the fungus-growing insects, for possible insights to improve agricultural strategies.

This lack of an applied interest in insect agriculture probably derives from a general perception that human agricultural systems (based largely on plant cultivation) function in a fundamentally different manner than insect systems (all based on fungus cultivation). However, humans have learned much of practical value through the close examination of adaptive features of other organisms (including insects), and comparable problems such as crop diseases affect all farmers regardless of their phylogenetic positions or those of their crops (plant, fungus, or otherwise). Because of the universality of crop diseases in both human and insect agriculture, it may be fruitful to examine the short-term and long-term solutions that have evolved convergently in insect agriculture for possible application to human agriculture (Denison et al. 2003). Such a synthesis is the goal of this review.

1.1. Behavioral and Nutritional Elements Defining Agriculture

Insect fungiculture and human farming share the defining features of agriculture (see Table 1): (*a*) *habitual planting* ("inoculation") of sessile (nonmobile) cultivars in particular habitats or on particular substrates, including the seeding of new gardens with crop propagules (seeds, cuttings, or inocula) that are selected by the farmers from mature ("ripe") gardens and transferred to novel gardens; (*b*) *cultivation* aimed at the improvement of growth conditions for the crop (e.g., manuring; regulation of temperature, moisture, or humidity), or protection of the crop against herbivores/fungivores, parasites, or diseases; (*c*) *harvesting* of the cultivar for food; and (*d*) obligate (in insects) or effectively obligate (in humans) *nutritional dependency* on the crop. Obligate dependencies of the insect farmers can be readily demonstrated by experimental removal of their cultivated crops, resulting in reduced reproductive output, increased mortality, or even the certain death of the cultivar-deprived insect (Francke-Grosmann 1967, Grassé 1959, Norris 1972, Sands 1956, Weber 1972). Our definition of agriculture does not require conscious intent in planting and harvesting. Conscious planning, learning, and teaching have clearly accelerated the development of complex agriculture in humans, but presumably not in insects (Schultz et al. 2005).

TABLE 1 Agricultural behaviors of farming ants, termites, beetles, and humans

Agricultural behavior	Agriculture in:			
	Attine ants	Macrotermitine termites	Xyleborine ambrosia beetles	Humans
Dependency on crop for food	Obligate	Obligate	Obligate	Facultative
Engineering of optimal growth conditions for crop (e.g., substrate preparation; moisture or humidity regulation)	Present	Present	Present	Present
Planting of crop on improved substrate	Present	Present	Present	Present
Intensive, continuous monitoring of growth and disease status of all crops	Present	Present	Present	Absent
Sustainable harvesting of crop for food	Present	Present	Present	Present
Protection of crop from diseases and consumers	Present	Present	Present	Present
Weeding of alien organisms invading the garden	Present	Present	Unknown	Present
Use of chemical herbicides to combat pests	Present	Unknown	Unknown	Present
Use of microbial symbionts for nutrient procurement for the crop	Present	Unknown	Unknown	Present
Use of disease-suppressant microbes for biological pest control	Present	Unknown	Unknown	Absent[1]
Sociality	Strictly eusocial	Strictly eusocial	Subsocial or communal[2]	Social
Task partitioning in agricultural processes	Present	Unknown	Present	Present
Application of artificial selection for crop improvement	Unknown, but ants exert symbiont choice	Unknown	Unknown	Present
Learning and cultural transmission of agricultural innovations	Absent	Absent	Absent	Present

[1] See Section 7 for some recently discovered microbes with potential disease-suppressant properties.
[2] One ambrosia beetle species is eusocial, all other species appear to be subsocial or communal (see text).

We restrict our review to ant, termite, and beetle fungiculturists. Cases analogous to human animal husbandry, such as the tending by ants of hemipteran insects (e.g., aphids, treehoppers; Hölldobler & Wilson 1990), are beyond the scope of this review. We also exclude cases that fail to meet all four of the requirements of agriculture as defined above, including, e.g., the ant *Lasius fuliginosus*, which promotes fungal growth in the walls of its nest, because the fungus is apparently not grown for food but instead for strengthening the walls (Maschwitz & Hölldobler 1970) or for antibiotic protection of the walls (Mueller 2002). On the same grounds, we exclude a number of possible cases of incipient agriculture. For example, *Littoria* snails may "protofarm" fungi by creating plant wounds that become infected with fungal growth that is part of the snails' diet, but the snails do not actively inoculate the plant wounds or otherwise garden the fungi (Silliman et al. 2003). Many more such protofarming species probably await discovery, particularly among invertebrates, and all of the known insect agriculturists (fungus-growing ants, termites, and beetles) probably originated from comparable protoagricultural ancestors (Mueller et al. 2001, Schultz et al. 2005). Comparison of these protofarming insects with "primitive" human agriculture exceeds the scope of this review.

1.2. A Coevolutionary Approach to Understanding Agriculture

We will analyze agriculture as a type of strong coevolutionary interaction, defined by the nutritional and behavioral criteria summarized above, in which natural selection acts upon both farmers and crops as reciprocally interdependent lineages (Futuyma & Slatkin 1983, Rindos 1984). Our coevolutionary approach to agriculture considers not only the interactions between a specific farming insect and a single cultivated crop, but also its interactions with other pathogenic and mutualistic microbes that have recently been discovered in insect gardens. Like the cultivars, some of these microbes are also managed by the insect farmers for specific purposes (Figure 1, see color insert). In other words, an insect garden is not a pure monoculture, but a sequestered and engineered ecological community consisting of several interacting microbes, some beneficial and others detrimental to the farmers. To gain a comprehensive understanding of the principles of insect agriculture, it will therefore be necessary to examine the nature of insect-microbe interactions in gardens, the evolutionary origins of these interactions, and the convergent and divergent evolutionary trajectories that culminated in the extant agricultural systems of insects.

2. THE THREE INSECT-AGRICULTURE SYSTEMS

Behaviorally complex systems of insect agriculture are known from only three groups of insects: ant, termites, and beetles.

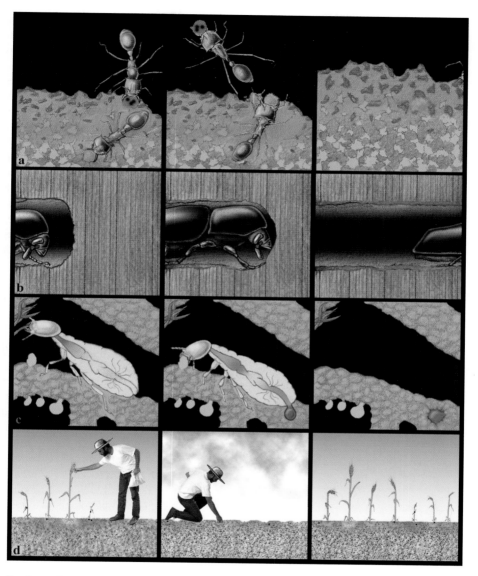

See legend on next page

Figure 1 Comparison of agriculture in attine ants, xyleborine beetles, macrotermitine termites, and humans. The time series (*left to right*) highlight the roles of beneficial auxiliary microbes (*blue shading*) that suppress diseases (*black dots*) or aid in buffering against contaminant microbes (*orange shading*). (*a*) **Ant agriculture**. Ants attempt to clean contaminant microbes from garden substrate (not shown) and remove garden diseases (*black dots*) through active weeding (*top ant*). The ants (*bottom ant*) then plant a crop-microbe consortium (*crop plus beneficial auxiliary blue microbes*) onto the prepared substrate, spreading beneficial microbes through the garden matrix. (*b*) **Beetle agriculture**. Primary fungus (crop) lining the tunnel grows intermixed with secondary microbes (*blue shading*) and occasional contaminant microbes (*orange dots*). No disease microbes (*black dots*) are indicated because they very rarely occur in young gardens near a tunnel head. The exact roles of the secondary microbes in beetle fungiculture are still unknown. (*c*) **Termite agriculture**. Hypothetical passage of a mixture of crop spores, auxiliary microbes, and substrate (ingested plant material) through the gut of a termite, followed by defecation of the substrate-crop-microbe consortium in fecal pellets that the termite adds to new garden. Other (external) modes of crop-microbe copropagation may exist in termite farmers, paralleling the planting of crop-microbe consortia in attine ants. Selective passage of microbes through the alimentary canals of attine ants and ambrosia beetles is unknown, but has never been investigated. No disease microbes (*black dots*) are indicated because no specialized pathogens have yet been identified in the fungus-growing termite system. (*d*) **Human agriculture** (wheat). A seed, fortuitously planted in soil enriched in antibiotic-secreting rhizosphere bacteria (*blue-shaded soil*), grows into a vigorous, disease-resilient crop plant (Weller et al. 2002). However, crops are often planted in microbially suboptimal soil (*orange-shaded soil*), leading to higher disease loads (*black dots*) on such plants. Traditional human planting schemes passage crops through a seed stage without copropagating disease-suppressant rhizosphere microbes. Illustrations by Barrett Klein.

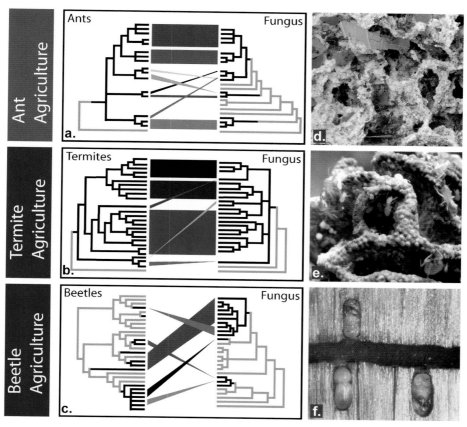

Figure 2 Evolutionary histories of insect agriculture. (*a–c*) Comparison of the patterns of evolutionary diversification in the insect farmers (*left cladograms*) and their cultivated fungi (*right cladograms*). In the left cladograms, farmer lineages are black and nonfarmer relatives are gray, whereas in the right cladograms, cultivated fungal lineages (cultivars) are black and noncultivated feral fungal lineages are gray. Independent origins of agricultural behavior are indicated for each farmer clade in the left cladograms, and independently domesticated fungal lineages appear as separate cultivar lineages in the right cladograms. (*d*) Garden of the fungus-growing ant *Atta texana* (photo by Greg Dimijian). The workers are cleaning and shredding leaf cuttings before expanding new gardens through the addition of leaf material. (*e*) Garden of the fungus-growing termite *Macrotermes bellicosus* (photo by Karen Machielsen). The fungus is grown on fecal pellets that are stacked into lamellar walls of the fungus garden (comb). (*f*) Gallery of the ambrosia beetle *Trypodendron lineatum* (photo by Susanne Kühnholz) with ambrosia fungus (*black*) lining the main gallery and beetle brood developing in niches adjacent to the gallery. Galleries are constantly patrolled by adult beetles (not shown). Figure adapted from Mueller & Gerardo 2002.

2.1. Ant Fungiculture

The fungus-growing ants are a monophyletic group of about 220 described and many more undescribed species in the tribe Attini (subfamily Myrmicinae) (Price et al. 2003, Schultz & Meier 1995). Attine ants occur only in the New World (Argentina to the southern United States) and attain their greatest diversity in the wet forests of equatorial South America, the region of their presumed evolutionary origin (Mueller et al. 2001). Attine ants are obligate agriculturists; their cultivated fungi are the sole source of food for the larvae and an important source of food for the adults. Although adults are able to supplement their diets by feeding on plant juices (Bass & Cherrett 1995, Murakami & Higashi 1997), the cultivated fungi are nutritionally sufficient to support the ants even in the absence of additional nutrients (Mueller 2002, Mueller et al. 2001). Garden fungi are transmitted vertically across generations when daughter queens transport small pellets of natal-nest mycelium within their infrabuccal pockets, pouches present in the mouthparts of all ants (Fernández-Marín et al. 2004, Huber 1905, Mueller 2002). In the derived leafcutter ants, the workers are divided into a remarkable range of differently sized morphological castes, each specialized on a different task (Hart et al. 2002, Weber 1972) (Figure 2*d*, see color insert).

Different attine ant lineages cultivate their fungi on different substrates. The ancestral gardening substrate, still used by the so-called lower attines, consists of flower parts, arthropod frass, seeds, wood fragments, or other similar plant debris, whereas the leafcutting genera *Atta* and *Acromyrmex* primarily use freshly cut leaves and flowers. Despite these distinct substrate specializations, all attine systems contain at least four symbionts: (*a*) the fungus-growing ants; (*b*) their fungal cultivars (basidiomycetes in the mushroom families Lepiotaceae and Pterulaceae; Mueller et al. 1998, Munkacsi et al. 2004); (*c*) mutualistic antibiotic-producing actinomycete bacteria (family Pseudonocardiaceae; Currie et al. 1999b); and (*d*) garden parasites in the ascomycete fungal genus *Escovopsis* (Currie et al. 1999a, Currie et al. 2003b). Additional bacteria and yeasts also occur in attine gardens and may function as mutualists, e.g., by secreting digestive enzymes or antibotics (Carreiro et al. 1997, Craven et al. 1970, Santos et al. 2004).

2.2. Termite Fungiculture

Of the more than 2600 described termite species, about 330 species in the subfamily Macrotermitinae cultivate a specialized fungus, genus *Termitomyces*, for food. Nests are generally founded by a single pair of reproductives, the future queen and king. They seal themselves permanently in a cell of hard clay (the so-called royal chamber) where they rear the first brood of sterile workers. In most termite species, a new colony acquires a fungal strain from wind-dispersed sexual *Termitomyces* spores shortly after nest founding and begins construction of the first gardens (De Fine Licht et al. 2005). These spores come from fruiting bodies (mushrooms) that arise from mature termite colonies. The fruiting of the fungus appears to be roughly synchronized to the period when the first foraging workers emerge from a new nest, a few months after the nest-founding stage. Termite gardens are grown on dead plant material that is only partially decomposed, such as leaf litter, dead grass, dead wood, or dry leaves.

Termite gardens are built from spore-containing fecal pellets in chambers that the termites construct either inside a mound or dispersed in the soil. Fecal pellets are added continuously to the top of the comb and fungal mycelium rapidly permeates the new substrate (Figure 2e). After a few weeks, the fungus starts to produce vegetative nodules that are consumed by the termites. These nodules are a rich source of nitrogen, sugars, and enzymes. The nodules are also covered with indigestible asexual spores (conidia), so that consumption serves the additional function of inoculating the feces with spores, which pass through the gut unharmed and are then planted in new comb with the deposition of feces (Leuthold et al. 1989). Mature comb is also consumed (Darlington 1994), but it is nutritionally inferior to the nodules.

2.3. Beetle Fungiculture

Ambrosia beetles make up around 3400 of the 7500 species in the weevil subfamily Scolytinae (the bark and ambrosia beetles, including the traditionally separate Platypodinae; Farrell et al. 2001, Harrington 2005, Wood 1982). Most ambrosia beetles construct tunnel systems (galleries; Figure 2f) in woody tissues of trees (typically in weakened or recently dead trees or, more rarely, in vigorous hosts), although some species are specialized to colonize pith, large seeds, fruits, and leaf petioles (Harrington 2005, Wood 1982). The term ambrosia refers to the fungi cultivated by the beetles on gallery walls, upon which they feed as an exclusive, or near exclusive, food source. The beetles are obligately dependent upon the fungi, from which they acquire essential vitamins, amino acids, and sterols (Beaver 1989, Kok et al. 1970).

The most advanced fungiculturists among the ambrosia beetles occur in the Xyleborini, a large monophyletic tribe of about 1300 species (Farrell et al. 2001, Jordal 2002). It is this group of ambrosia beetles that we primarily focus on in this review. Although life histories among the Xyleborini vary considerably, most share a number of fungicultural characteristics. There is a sexual division of labor in the Xyleborini; only females perform gardening tasks, whereas males are short-lived and flightless (Norris 1979). After mating, females disperse to new host substrate, carrying the fungi in specialized pockets termed mycangia. Once within a new host, founding females "plant" the fungi on the walls of the excavated tunnels, lay eggs, and tend the resulting garden and brood (Norris 1979). In ways not fully understood, they are able to control the growth of the fungal crop, as well as, to a degree, the composition of its multiple fungal species (Beaver 1989, French & Roeper 1972, Kingsolver & Norris 1977, Roeper et al. 1980). If the female dies, the garden is quickly overrun by contaminating fungi and bacteria, which ultimately results in the death of the brood (Borden 1988, Norris 1979).

The ambrosia gardens of xyleborine beetles are not pure monocultures as was once believed, but are typically composed of an assemblage of mycelial fungi, yeasts, and bacteria (Batra 1966, Haanstadt & Norris 1985). These assemblages were termed multi-species complexes by Norris (1965), who suggested that it is a complex as a whole, rather than any one individual microbe, that allows the beetles to exploit nutrient-poor substrates such as wood. However, most subsequent work has revealed that one "primary" fungus always dominates in beetle gardens (Baker 1963, Batra 1966, Gebhardt et al. 2004, Kinuura 1995). Furthermore, the beetles typically carry only the primary fungus in the mycangium (although secondary fungi are sometimes also isolated from mycangia), and the cultivation efforts of female beetles tend to favor the primary fungus, which imparts the greatest nutritional benefit (Francke-Grosmann 1967, Gebhardt et al. 2004, Morelet 1998, Norris 1979). Some auxiliary fungi also support beetle development, but survival on the auxiliary fungi alone is often greatly reduced (Norris 1979). These observations implicate the primary fungus as the intended crop, whereas the secondary fungi, yeasts, and bacteria may be contaminant "weeds" or may play additional auxiliary roles in the gardens, paralleling the hypothesized roles of the auxiliary bacteria and yeasts in attine gardens (see above).

3. EVOLUTIONARY ORIGINS OF INSECT AGRICULTURE

Phylogenetic analyses reveal nine independent origins of insect agriculture (Figure 2; Table 2). In ants, fungal cultivation arose only once, probably 45–65 Mya in the Amazon rainforest (Mueller et al. 2001; Schultz & Meier 1995). In termites, fungiculture likewise had a single origin, approximately 24–34 Mya in the African rainforest (Aanen et al. 2002; D.K. Aanen & P. Eggleton, submitted). In ambrosia beetles, however, agriculture arose independently seven times between 20–60 Mya, six times in various nonxyleborine lineages, and once in the ancestor of the Xyleborini about 30–40 Mya (Farrell et al. 2001). Whereas the common ancestors of the macrotermitines and of the xyleborines each domesticated a single, specific primary cultivar clade to which their descendants have adhered throughout subsequent evolution (Figure 2, Table 2), attine ants maintain associations with multiple independently domesticated cultivar lineages (which are for the most part very closely related; Mueller et al. 1998; Munkacsi et al. 2004; Table 2). Interestingly, there are no known cases of reversal from agricultural to nonagricultural life in any of the nine agricultural insect lineages (Figure 2a, 2b, and 2c), suggesting that the transition to fungiculture is a drastic and possibly irreversible change that greatly constrains subsequent evolution.

Two main models have been suggested for the independent evolutionary transitions to agriculture in insects, the "consumption-first" versus the "transmission-first" models (Mueller et al. 2001). In the consumption-first model (the likely model for the termites), an insect lineage initially begins to incorporate fungi into its more generalist diet, then becomes a specialized fungivore, and finally evolves adaptations for cultivating fungi. In the transmission-first model (the likely model for the beetles), the insect lineage begins its association with a fungus by serving

TABLE 2 Crop ecology and evolution in ant, termite, beetle, and human agriculture

Crop ecology & evolution	Agriculture in:			
	Attine ants	Macrotermitine termites	Xyleborine ambrosia beetles	Humans
Number of inferred evolutionary origins of agricultural behavior	Single origin of agriculture in the Attini No other ant practices true agriculture, but some ants allow fungi to grow in their nest walls for structural or antibiotic purposes.	Single origin of agriculture in the Macrotermitini No other termite practices true agriculture, but the distantly related termite *Sphaerotermes sphaerothorax* promotes the growth of bacteria in food stores.	Single origin of agriculture in the Xyleborini Agriculture originated independently in six additional ambrosia beetle lineages; these other beetle systems are less well understood (see text).	Multiple independent origins of agriculture
Estimated date and region of origin of agricultural behavior	45–65 Mya in Amazonian rainforests	24–34 Mya in African rainforests	30–40 Mya in xyleborine beetles; region of origin unknown 21–60 Mya in the other six ambrosia beetle lineages	10,000 years ago for the earliest known origins of human agriculture; multiple regions of origin
Number of crop clades cultivated	Multiple cultivar clades One cultivar clade belongs to the Pterulaceae; at least three additional polyphyletic cultivar clades belong to the Lepiotaceae (tribe Leucocoprineae).	Single cultivar clade All termite cultivars belong to the genus *Termitomyces*.	Multiple cultivar clades The ambrosia cultivars *Ambrosiella* and *Raffaellea* have a polyphyletic origin within the genera *Ophiostoma* and *Ceratocystis*.	Multiple crop clades from diverse lineages of plants and fungi

Crop transmission				
Vertical cultivar inheritance from parent to offspring (i.e., cultivar transfer between generations)	Present Vertical cultivar inheritance is the rule in all attine ant lineages.	Variable Vertical cultivar inheritance is typical for two independently derived macrotermite lineages; all other macrotermitines acquire cultivars horizontally via wind-dispersed spores from other colonies.	Present Vertical cultivar inheritance is the rule in all xyleborine lineages.	Present
Specialization on crops				
Specialization at higher phylogenetic levels	Clade-clade correspondence Defined clades of ant species are specialized in that they only grow fungi typical for their own ant-specific cultivar clade.	Clade-clade correspondence Defined clades of termite species are specialized in that they only grow fungi typical for their own termite-specific cultivar clade.	No strict clade-clade correspondence The fungal lineages associated with xyleborine ambrosia beetles are only distantly related to each other, and all are also associated with some bark beetles.	Not applicable
Specialization at level of farmer species	Present In all cases studied in detail, single species of ants grow only a single, phylogenetically narrow group of fungi (i.e., a single, phylogenetically defined "species" of fungus).	Present Single termite species generally associate with multiple fungal lineages, but within the limits of specific cultivar clades (see above); however, some termite species are very specialized on a particular fungus.	Present Single beetle species generally associate with multiple, distantly related fungal lineages; the primary cultivar is typically a single fungal species, but secondary cultivars vary and often come from different fungal lineages than the primary-cultivar lineage.	Not applicable Specialization on crops is historically manifested by different human lineages (reflecting separate domestication events), but no single human lineage is nearly as specialized on a single crop as are the insect farmers.

(Continued)

Table 2 (*Continued*)

	Agriculture in:			
Crop ecology & evolution	Attine ants	Macrotermitine termites	Xyleborine ambrosia beetles	Humans
Crop sharing and crop exchange				
Lateral exchange of cultivars within the same farmer species (i.e., exchange within same farmer generation)	Present Within a species of ant, cultivars are probably occasionally exchanged between colonies (e.g., after cultivar loss when cultivar-deprived colonies may acquire replacement cultivars from neighboring colonies).	Unknown	Unknown Lateral cultivar exchange is probably rare, but may occur when galleries containing brood of two or more foundress females, each carrying a different cultivar (e.g., different species of fungi), occur in close proximity in the same tree.	Present Lateral exchange of cultivars between farmer societies is a prominent feature of human agriculture.
Cultivar exchange between farmer species	Present Horizontal cultivar exchange occurs between sympatric ant species if they are specialized on the same cultivar lineage.	Present Horizontal cultivar exchange (through wind-dispersed spores) is typical for most termite lineages.	Rare Horizontal cultivar exchange occurs at least occasionally.	Not applicable
Genetic exchange between domesticated and wild populations (e.g., genetic interbreeding between	Variable Primitive attine ants probably import novel cultivar genotypes	Absent Wild populations are unknown for termite cultivars.	Absent Wild populations are unknown for ambrosia beetle cultivars.	Absent Some traditional cultivation regimes (e.g., potato farming in the

	Attine ants	Termites	Ambrosia beetles	Humans
domesticated and wild populations; or import of new domesticates from wild populations)	regularly from wild populations; cultivars of the derived higher attine ants do not appear to have free-living populations.			Andes) are based on regular genetic exchange between domesticated and wild populations, but most modern cultivars are highly derived and do not readily outcross with wild, ancestral populations.
Crop monoculture	Present — All available data indicate that only a single cultivar is grown within a nest.	Present — All available data indicate that only a single cultivar is grown within a nest.	Present — Most beetle species appear to be associated with a single primary cultivar and one or more secondary cultivars; it is unclear whether secondary cultivars are truly cultivars, weeds, or switch between these roles.	Variable — Monoculture predominates agriculture in many but not all regions.
Crop sexuality	Variable, but sexual recombination is rare — Predominant clonal cultivar propagation within nests and between nests is punctuated occasionally by genetic exchanges, outbreeding, or other events generating recombinants.	Variable, but most cultivars are propagated sexually — Cultivar sexuality is typical for those termite lineages that acquire cultivars from the environment every generation, but cultivar sexuality appears absent in the two agricultural systems with vertical cultivar transmission.	Variable, but sexual recombination is probably rare — Asexuality is typical for both primary and secondary cultivars; some secondary cultivars are sexual.	Variable, but most cultivars are propagated sexually — The great majority of human cultivars are sexually propagated, but some cultivars are largely clonally (e.g., potato) or strictly clonally propagated (e.g., banana).

as a vector of that fungus, then begins to derive nutrition from it, and finally becomes a fungus cultivator. In a third possibility, an insect-fungus association evolves because the insects originally use fungi as a source of antibiotics, as for example in the lower termite *Reticulotermes speratus* that derives antibiotic protection from fungal sclerotia mixed into egg piles (Matsuura et al. 2000). Lastly, insect-associated fungi may have undergone even more complicated evolutionary histories, originating from the exploitation by one insect lineage (e.g., the ancestor of attine ants) of a preexisting insect-fungus association (the fungi ancestrally associating with beetles) when it encounters these insect-adapted fungi in a shared nest environment (e.g., decaying wood; Sanchez-Peña 2005). This latter hypothesis, however, is not supported by the phylogenetic relationships between beetle and ant fungi and is inconsistent with the estimated dates of origin of these insect-fungal associations (i.e., attine agriculture probably arose well before ambrosia beetle agriculture; see Table 2).

For attine ants, it is unclear whether agriculture arose from a state of ancestral fungivory, antibiotic acquisition, or fungal vectoring (Mueller et al. 2001). Termite agriculture most likely originated via the consumption-first route, because many nonfarming termite species are attracted to and feed on fungus-infested wood, which suggests that the nonfarming ancestors of the farming termites may have fed on fungi as well (Batra & Batra 1979, Rouland-Lefevre 2000). The nonfarming ancestors of the fungus-growing beetles appear to have associated with fungi even before the origin of fungiculture, because many of the more primitive nongardening scolytines act as fungal vectors without apparent dependence on their fungal associates (Harrington 2005, Malloch & Blackwell 1993, Six 2003, Six & Klepzig 2004). This suggests non-nutritional dependencies on fungi that predate the origins of fungiculture in the various ambrosia beetle lineages (Six 2003). However, many nonambrosial scolytines carry fungi in mycangia and feed as larvae on ungardened mycelium that colonizes host plants and feed as new adults on spore layers lining pupal chambers (Ayres et al. 2000, Barras 1973, Six & Klepzig 2004, Six & Paine 1998; also A. Adams & D.L. Six, unpublished data), suggesting a stage of nutritional dependency predating the origin of fungiculture. Thus, some of the seven agricultural origins in beetles appear to have followed the transmission-first route, whereas others followed the consumption-first route.

Insect agriculture is restricted to the cultivation of fungi rather than plants, which predominate in human agriculture. Although it is true that some insects are specialized on host plants that they protect from other herbivores (e.g., *Pseudomyrmex* ants protect acacia trees in exchange for shelter and nutritional benefits; Janzen 1966, Hölldobler & Wilson 1990), none of these insect-plant mutualisms possesses all four of the components of agriculture listed above. One could therefore ask what factors have predisposed insects to evolve fungal rather than plant agriculture. Indeed, there are several advantages of fungal agriculture over plant agriculture, and several characteristics of plants may even preclude their easy cultivation. First, unlike fungi, plants typically have stringent light and space requirements, excluding them from cultivation in the subterranean or otherwise enclosed nests of insects. Such nesting habits may facilitate fungiculture by shielding fungal crops from unwanted consumers (i.e., other fungivores) and wind-dispersed diseases. Furthermore, unlike plants, which usually require regular pollination for long-term cultivation, fungi can be maintained indefinitely in a nonsexual mycelial state, yielding a more consistent food source. Thus, although seeds and plant material can be readily harvested, fungi are likely more cultivatable, explaining the predominance of fungal rather than plant agriculture among insects.

Ant, termite, and most beetle agriculturists are social. All ants and termites are eusocial (characterized by reproductive division of labor, cooperative brood care, and overlap of generations; Hölldobler & Wilson 1990). Only one ambrosia beetle (*Austroplatypus incompertus*) is known to be eusocial (Kent & Simpson 1992); the remainder are subsocial, in which a single female cares for her brood, or communal, in which several reproductive females cooperate in brood care and gardening (Kirkendall et al. 1997). Sociality may have facilitated the evolution of agriculture because of the inherent advantage to agriculture of division of labor, which enables the partitioning of agricultural tasks and augments agricultural efficiency (Hölldobler & Wilson 1990, Hart et al. 2002). In ant and termite farmers, for example, agricultural tasks are partitioned in a conveyor-belt-like series between different worker castes, each specialized on one main task: foraging; processing and cleaning of substrate before incorporation into the garden; planting of mycelium onto new substrate; monitoring and weeding of the garden; or disposal of diseased or senescent garden (Bot et al. 2001a, Hart et al. 2002, Traniello & Leuthold 2000). Task partitioning has so far not been investigated in the ambrosia beetles because of the logistical difficulties of studying beetle behavior in their concealed tunnels. Task partitioning likely facilitates great efficiency in defense against nest and garden robbers (Adams et al. 2000a,b; LaPolla et al. 2002), in monitoring gardens for diseases, and in modulating optimal environmental conditions for crop growth.

4. AGRICULTURAL EVOLUTION AND ECOLOGY

A series of convergent and divergent features of agricultural evolution emerge from a comparative analysis of ant, termite, and beetle fungiculture (summarized in Tables 1–3).

4.1. Cultivar Transmission Between Farmer Generations

In attine ant and xyleborine beetle agriculture, fungal cultivars are transmitted vertically by trophophoresy from parent to offspring generations (Fernández-Marín et al. 2004, Francke-Grosmann 1967, Haanstad & Norris 1985, Huber 1905). Female reproductive ants and beetles acquire inocula from their natal gardens, carry these inocula with them in specialized pockets during dispersal flights early in life, and use these inocula as starter cultures for their new gardens. Trophophoretic

TABLE 3 Disease ecology and evolution in ant, termite, beetle, and human agriculture

Disease ecology and evolution	Agriculture in:			
	Attine ants	Macrotermitine termites	Xyleborine ambrosia beetles	Humans
1. Disease and pest types	Crops are attacked by fungal and bacterial pests, including specialized (*Escovopsis*) and unspecialized (*Trichoderma*) fungal pathogens. Microbial "weeds" (e.g., *Xylaria* fungi) and fungivorous arthropods (e.g., mites) can also inhabit gardens.	Specialized crop parasites have not been documented, but some species of *Xylaria* are specialized as weeds of fungus gardens. Common soil fungi are present in gardens as well.	Most ambrosial gardens consist of a complex of mycelial fungi, yeasts, and bacteria. At least some of these are likely to cause disease or act as pests.	Crops are infected by fungal, bacterial, and viral diseases and are attacked by invertebrate and vertebrate herbivores. Weeds are an additional problem in human agriculture.
2. Disease prevalence	Infection of gardens by the crop parasite *Escovopsis* is frequent and sometimes devastating.	Unknown	Unknown	Ubiquitous and diverse diseases cause immense loss in crop harvests.
3. Defense mechanisms by farmers against crop disease				
3a. Protection from disease				
-Sequestration, sheltering	Present	Present	Present	Absent[a]
-Substrate sterilization or cleaning	Present	Present	Unknown	Uncommon
-Guarding or protecting against disease vectors	Present	Present	Present	Uncommon
-Partitioning of disease-removal tasks to minimize contact between contaminated workers and healthy crop	An uncontaminated worker-caste specializes on gardening; a contaminated worker-caste specializes on disease removal.	Foraging and gardening are performed by different worker castes in many species.	Unknown	Uncommon

3b. Eradication of disease				
-Continuous, intensive monitoring of all crop	Present	Present	Present	Very rare
-Physical weeding	Present Ant gardeners continuously "weed" and "groom" their gardens, excising infected garden fragments.	Present Termite gardeners groom gardens to eliminate some alien microbes (e.g., *Xylaria stromata*).	Unknown Female beetles continually tend crop gardens but whether directed weeding is part of this activity is unknown.	Present Physical weeding is more integral in primitive agricultural systems, but less so in modern monoculture systems.
-Application of chemical herbicides	Present Metapleural and mandibular gland secretions have general antibiotic properties; antibiotics derived from actinomycete bacteria have specific antibiotic effects against *Escovopsis* parasites.	Present Salivary gland secretions have antibiotic properties; defensive secretions of soldiers have antibiotic properties, but their sanitary function in the garden is unknown.	Unknown	Present Chemical control of diseases and herbivores is an essential element of most modern monoculture systems.
-Routine use of disease-suppressant microbes	Present	Unknown	Unknown	Absent[b]
4. Defense mechanisms of crop against crop diseases				
4a. Induced resistance of crop	Unknown	Unknown	Unknown	Present
4b. Constitutive resistance of crop	Present Cultivars produce antibiotics, which appear to mediate resistance of some cultivar genotypes to some *Escovopsis* genotypes.	Unknown	Unknown	Present

[a] Greenhouses provide a sequestered environment for some human crops, but greenhouse farming is costly and contributes only a small fraction (about 0.02%, Paulitz & Bélanger 2001) to the total productivity of human agriculture; the bulk of human agriculture operates in open landscapes that expose crops to environmental stresses (fluctuations in moisture and temperature; wind-borne pathogens; migratory herbivores; etc.).

[b] Some recently developed uses of disease-suppressant microbes in agricultural pest control are discussed in the text; application of disease-suppressant microbes has yet to become an integral part of mainstream human agriculture (but see Morrissey et al. 2004).

vertical transmission also occurs in two macrotermitine groups, except that in one of these two groups the fungal cultivar is transmitted via the king (the single species *Macrotermes bellicosus*), whereas in the other group (the genus *Microtermes*) the fungus is transmitted via the queen. In the few cases where fungal transmission has been studied in the remaining macrotermitines (Johnson 1981, Johnson et al. 1981, Korb & Aanen 2003, Sieber 1983), these termites rely on horizontal acquisition of fungal crops from the environment in each generation.

4.2. Higher-Level Specialization (Clade-Clade Congruence) Between Farmers and Crops

Vertical transmission of cultivars leads to the expectation of clade-clade correspondences and topological congruence between the phylogenies of insect farmers and those of their cultivars. Indeed, in all insect farming systems, major groups of farmers (large clades or paraphyletic grades, e.g., the lower attine ants) strictly specialize on major groups of corresponding fungal cultivars (Figure 2). The expected farmer-cultivar congruence therefore does occur at higher (i.e., broad) phylogenetic levels, possibly because of ancient evolutionary codependencies (e.g., physiological/nutritional requirements of the farmers, cultivation requirements of the fungi, etc.) that strictly preclude switches by farmers to cultivars outside of their specialized major cultivar groups.

Phylogenetic patterns (Figure 2) indicate, however, that within these strictly constrained major cultivar groups, insect-farmer species occasionally switch between fungal species or strains. This combination of lower-level, within-group switching and higher-level major-group specialization in insect farmers would be analogous, in humans, to defined clades of specialized wheat-farmers, rice-farmers, potato-farmers, bean-farmers, etc., each of which is able to switch between varieties within their area of specialization (e.g., between varieties of wheat and to closely related species such as barley), but which cannot switch across major groups (e.g., from wheat to beans). Among insect farmers, switches to novel major cultivar groups have been exceedingly rare evolutionary events (Villesen et al. 2004).

4.3. Lower-Level Specialization on Cultivars

Though low-level switching between cultivar species and strains within major cultivar groups occurs occasionally over evolutionary time, over ecological spans of time most insect farmer species associate with only a very narrow subgroup of cultivars (species or strains). For example, every attine ant species surveyed to date cultivates only a phylogenetically narrow set of cultivars (e.g., a single species of fungus), implicating species specificity between ants and cultivars at very recent levels of evolutionary diversification (Bot et al. 2001b, Green et al. 2002, Schultz et al. 2002). In ambrosia beetles, like ants, only one primary cultivar is associated with a particular beetle species within a particular geographic region (Gebhardt et al. 2004, Batra 1967). However, although most beetles are associated with a species-specific, primary fungus across their entire geographic ranges, some beetle species associate with different primary cultivars in different geographic regions (Baker 1963, von Arx & Hennebert 1965). Among macrotermitine species, there exists considerable variation in cultivar specialization: Some species are limited to a single, unique cultivar, whereas other species cultivate a great diversity of fungal cultivars, which they sometimes share with other, usually closely related, macrotermitine species (Katoh et al. 2002; D.K. Aanen & P. Eggleton, submitted). The factors underlying variation in termite specialization are unknown, but different cultivars may serve different primary functions, providing specific, termite-adapted enzymes in some cases (leading to termite-cultivar specialization), while providing generalized food in other cases (permitting exchange between termite species; D.K. Aanen, V.I. Ros, H.H. de Fine Licht, C. Roulant-LeFévre, J. Mitchel, et al., in review).

4.4. Cultivar Sharing and Exchange Between Farmer Species

Even though each attine ant species is specialized on a single cultivar species, a given cultivar species may be cultivated by several sympatric species of ants, and these sympatric ant species may not necessarily be closely related to one another (e.g., they may represent different ant genera) (Bot et al. 2001b, Green et al. 2002, Villesen et al. 2004). Cultivar transfer between ant species may occur via direct or indirect avenues. Direct avenues may include raiding of neighboring colonies (Adams et al. 2000a, Rissing et al. 1989) or, in polygynous species, cofounding of colonies by multiple queens that exchange cultivars or recombine them in the cofounded garden. Indirect avenues may include cultivar escapes from gardens, followed by a free-living (feral) existence and subsequent reincorporation into a symbiosis when a different attine colony imports the free-living strain into its nest (Mueller et al. 1998).

For ambrosia beetles, the available phylogenetic evidence points to cultivar sharing between different sympatric beetle species (Farrell et al. 2001, Gebhardt et al. 2004) but few investigations have addressed this question. Distantly related ambrosia beetle species are sometimes associated with the same cultivar (Gebhardt et al. 2004), implicating fungal exchange, either direct or indirect, as explained above for the ants. Cultivar exchange between and within beetle species may occur when different female beetles colonize the same tree and the fungal associates cross-contaminate adjacent galleries.

In contrast to attine ants and ambrosia beetles that all transmit their cultivars vertically between generations, most macrotermitine species acquire their fungi horizontally each generation. This implies that new termite-cultivar combinations arise each generation, which should facilitate cultivar exchange between species, as well as between lineages of the same species. Cultivar surveys of sympatric macrotermitine communities indeed indicate that cultivars are generally shared between closely related species via interspecific cultivar exchanges (Aanen et al. 2002; but see the exceptions mentioned above in Section 4.3). Intra-specific cultivar exchanges have so far not been investigated in macrotermitines.

4.5. Propagation of Sexual Versus Asexual Cultivars and Links to Free-Living Cultivar Populations

All vertically transmitted insect cultivars, including the cultivars of attine ants, ambrosia beetles, and termites in the genus *Microtermes* and the species *Macrotermes bellicosus*, seem to be asexually propagated by their insect farmers across multiple farmer generations. In contrast, the horizontally transmitted termite cultivars (propagated by all other macrotermitine genera) undergo regular meiosis and sexual recombination (see above).

Evidence for cultivar asexuality in attine ants comes from DNA fingerprinting studies that indicate that all gardens of a single leafcutter colony contain a single cultivar clone (monoculture) (Kweskin 2003, Poulsen & Boomsma 2005; J. Scott & U.G. Mueller, unpublished data); that identical cultivar clones occur in different colonies of the same geographically widespread attine ant species (Bot et al. 2001b, Green et al. 2002, Mueller et al. 1996); and that different sympatric ant species occasionally share genetically identical cultivar clones (see above; Bot et al. 2001b, Green et al. 2002, Mueller et al. 1998). Contrary to previous suggestions (Chapela et al. 1994), however, attine cultivar clones are not ancient. Although attine cultivars are clonally propagated across many ant generations (Mueller 2002), this clonality is punctuated by occasional recombination events, involving either sexual (meiosis, mating) or parasexual (e.g., mitotic recombination, exchange of haploid nuclei) processes. Evidence for occasional recombination includes: (*a*) fruiting structures (mushrooms), which are known for cultivars of nearly all genera of attine ants, contradicting the expectation of loss of fruiting ability under strict clonality spanning millions of years (Mueller 2002); (*b*) rates of allele sequence divergence in attine cultivars that are similar to those of closely related, sexually reproducing fungi (Mueller et al. 1998; S.A. Rehner, unpublished data); and (*c*) cultivars of the lower attine ants that have close genetic links to free-living fungal populations (Green et al. 2002, Mueller et al. 1998), suggesting that these fungi are capable of moving in and out of the symbiosis, that cultivar lineages may regularly interbreed with wild lineages, or both. Taken together, the genetic and natural-history information suggest predominantly asexual cultivar propagation within ant nests and across many ant generations, punctuated by occasional genetic recombination events.

As in attine ants, a single cultivar monoculture is grown in a single termite colony (Aanen et al. 2002, Katoh et al. 2002). Within termite nests, the *Termitomyces* cultivar is propagated asexually by inoculating fresh garden substrate with asexual spores (Leuthold et al. 1989), and probably also by transplanting mycelium from older to younger gardens. Although *Termitomyces* species have no known free-living populations existing entirely independent of the termite farmers, they have nonetheless retained the ancestral (presymbiotic) condition of regular sexual reproduction, and most *Termitomyces* cultivars are spread from one termite nest to another horizontally via sexual spores produced by fruiting bodies (mushrooms) growing on the external surfaces of mature nests. The *Termitomyces* cultivar of the termite *Macrotermes natalensis*, for example, has an outcrossing mating system (De Fine Licht et al. 2005). Asexual cultivar propagation spanning several termite generations only occurs in those species with vertical uniparental propagation. Phylogenetic patterns implicate occasional horizontal cultivar exchange between nests of the same and different termite species (Aanen et al. 2002), but whether such horizontal exchange is associated with cultivar sexual reproduction remains unknown.

In xyleborine beetles, the primary fungi are strictly asexual (Jones & Blackwell 1998, Rollins et al. 2001), whereas the less specific, auxiliary fungi are often sexual (Francke-Grosmann 1967). A preponderance of asexual reproduction in fungal cultivars also occurs in the primary fungi of all other non-xyleborine ambrosia beetles, whereas, again, the more incidental fungi are often sexual (Six 2003, Six & Paine 1999), suggesting that this may have been the ancestral condition at the origin of the xyleborine beetle-fungus symbiosis.

4.6. Coevolutionary Modifications

Farmer-cultivar specialization enhances the potential for coadaptation, in which evolutionary modification in one of the partners causes a reciprocal coevolutionary modification in the other partner (Futuyma & Slatkin 1983). It is relatively easy to identify evolutionary modifications in the farmer species, such as specialized morphological structures for the trophophoretic transport of cultivars by females during the dispersal flight (e.g., mycangia in the beetles, infrabuccal pocket in the ants), modifications of mandibles and guts of beetle and ant larvae for fungus-feeding (Browne 1961, Schultz & Meier 1995), or the suite of behavioral, glandular, or physiological modifications that form the basis of insect farming. Examples of evolutionary modifications in the cultivars have been more difficult to identify, however, because the cultivated fungi are inherently more difficult to study.

The clearest examples of cultivar modifications are the hyphal-tip swellings (gongylidia) produced by the cultivars of the higher attines and the analogous nodules produced by macrotermitine cultivars. Both gongylidia and nodules are nutrient-rich structures designed for easy harvesting by the farmers, ingesting, and feeding to the larvae or nymphs. Nutrient-rich structures are not known for beetle gardens, although the ambrosia morphology of the beetle cultivars suggests evolutionary modification designed specifically for efficient consumption and digestion by the beetle larvae. Ambrosial growth consists of tightly packed conidiophores with copious spores and is only formed in the presence of the beetles (French & Roeper 1972). Ambrosia formation has not been reported from nonsymbiotic fungus species. Interestingly, two of the major genera of fungi associated with ambrosia beetles (*Ambrosiella, Raffaelea*) are each polyphyletic, and the multiple lineages within each genus have converged on the same ambrosial morphology (Blackwell & Jones 1997, Jones & Blackwell 1998), suggesting evolutionary convergence due to selection. Other likely coevolutionary modifications that have yet to be investigated include predominantly asexual reproduction in the insect cultivars while under cultivation and the cultivars' capacity to survive storage in the dispersal pockets of the beetles and ants, or the passage through the alimentary canal of the termites.

4.7. Symbiont Choice and "Artificial Selection" of Cultivars

From an evolutionary perspective, insect agriculture represents a case of cooperative interaction between farmer and cultivar lineages, each exploiting the other for its own reproductive purposes (Herre et al. 1999, Mueller 2002). Such cooperative interactions are frequently unstable and can erode over evolutionary time, for example, when mutant overexploiters arise (so-called cheater cultivars) and invade a mutualism. A series of additional farmer-cultivar conflicts are predicted that could destablilize the mutualism (Mueller 2002, Aanen & Boomsma 2005, Schultz et al. 2005), but at least two evolutionary mechanisms preserve the cooperative nature of the farmer-cultivar association: First, partner feedback, inherent in vertical cultivar transmission, is an automatic feedback mechanism in which an uncooperative partner reduces the other partner's fitness to the extent that it reduces its own fitness as well; and, second, partner (symbiont) choice in which farmers favor associations with productive cultivars and discriminate against inferior cultivars in specific choice situations (e.g., a choice between cultivar strains that may coexist in a garden or in proximate gardens, exercised either by workers during the planting of new gardens or by reproductives when choosing a cultivar strain for dispersal). In cases where the evolutionary rates differ between two cooperating partners, partner choice is a particularly important mechanism (Sachs et al. 2004). The slower-evolving partner (e.g., the insect farmer) is expected to exert the choice between variants of the faster evolving partner (e.g., the fungal cultivar), and thus the slower-evolving farmer imposes selection favoring beneficial symbiont variants (e.g., productive cultivars) and prevents the spread of nonbeneficial cultivar mutants (e.g., degenerate or suboptimal cultivars; Mueller 2002, Sachs et al. 2004). Symbiont choice has yet to be investigated for termite and beetle farmers, but ant farmers are able to discern surprisingly fine genotypic differences between cultivars (Mueller et al. 2004; also N.K. Advani & U.G. Mueller, submitted), suggesting that cultivar diversity in ant gardens, arising, for example, through mutation in a garden or through the import of novel strains, may evolve under an analog of "artificial selection."

5. ROLE OF DISEASE IN INSECT AGRICULTURE

"Weedy" fungi frequently invade the gardens of ants, termites, and beetles, and may coexist at low or manageable levels along with the crop. If the gardening insects are removed or if they abandon their nests, the garden is quickly overrun by these weeds (Batra & Batra 1979, Norris 1979). One such group of weeds, wood-degrading fungi in the endophytic genus *Xylaria,* is found in most gardens of fungus-growing ants and termites, probably because it is introduced with garden substrate (Fisher et al. 1995; V.I. Ros, A.J. Debets, T. Læssøe, D.K. Aanen, submitted; N.M. Gerardo & U.G. Mueller, personal observation). Though weeds like *Xylaria* do not directly attack the cultivar, they compete with it for nutrients and thus decrease crop yield (V.I. Ros, A.J. Debets, T. Læssøe, D.K. Aanen, submitted). *Escovopsis* species, ascomycete fungi found in colonies of fungus-growing ants, are specialized parasites that subsist directly on the cultivars and reduce the nutrients available to the ants (Currie 2001a, Currie et al. 2003b). Weed fungi and bacteria are also known in termite and beetle agriculture, but have yet to be studied in detail (Six 2003; V.I. Ros, A.J. Debets, T. Læssøe, D.K. Aanen, submitted; D.K. Aanen, unpublished data).

Escovopsis infections reduce garden productivity, which in turn reduces ant colony growth and the likelihood of colony survival (Currie 2001a, 2001b, Currie et al. 1999a). *Escovopsis* is geographically widespread and taxonomically diverse. The parasite has been isolated from colonies of every attine genus throughout their geographic ranges, and particular *Escovopsis* lineages are specialized to parasite particular cultivar lineages. This high degree of host specificity suggests a long history of host-parasite coevolution in *Escovopsis*, in which the cultivars, the ants, and their mutualistic bacteria have likely coadapted to defend against *Escovopsis* attack and in which each *Escovopsis* species has become narrowly specialized to overcome the defenses of some hosts but not others (Gerardo et al. 2004).

6. DISEASE AND MICROBIAL MANAGEMENT STRATEGIES

In response to the persistent selection pressure imposed by weeds and pathogens, fungus-farming insects have evolved an arsenal of strategies for preventing and suppressing infection (Table 3).

6.1. Sequestration of Gardens

All insect agriculturists sequester and separate their gardens from the surrounding environment, e.g., by growing their gardens in underground chambers or galleries in wood, or by covering them with a protective mycelial veil in the case of some *Apterostigma* ants (Villesen et al. 2004). Although sequestered nests are the ancestral condition in fungus-growing ants, termites, and beetles, and although sequestered nests may serve other agricultural functions such as the regulation of temperature and humidity, sequestration no doubt also buffers the garden against fungivores, wind-borne pathogens, and arthropod vectors of diseases (e.g., mites, collembolans).

6.2. Maintenance of Cultivar Genetic Variability

Although attine ants clonally propagate their cultivars across generations over short evolutionary time spans, no attine cultivar has been found to be an ancient clone. Instead, the evidence indicates that the lower attines occasionally acquire new cultivars from wild (free-living), sexually reproducing fungal populations and that both lower and higher attines occasionally acquire new cultivars from the nests of other attines. The cultivars of higher attines, which are not known to have free-living populations (Mueller 2002), nonetheless retain the ability to fruit (Mueller 2002) and demonstrate patterns of DNA-sequence diversity that suggest occasional genetic recombination through self-mating (S.A. Rehner, personal communication) or through true intercrossing between different cultivar strains (M. Bacci, personal communication). Thus, whereas the crop employed by any attine ant colony at any given time is a clonally propagated monoculture, the genetic variability and resilience necessary for long-term disease management resides in the fungal populaton external to the nest. As already noted, sexual reproduction is the norm in the cultivars of those termites that reacquire their cultivars horizontally each generation, and sexual reproduction may also occur in the fungi of those termites that transmit their cultivars vertically across generations. Whether the primary cultivars of the ambrosia beetles occasionally reproduce sexually remains unknown. At least for the termites and ants, then, and possibly for the beetles as well, access to a population-level reservoir of cultivar genetic variability is a consistent feature of insect agriculture that may provide alternative crops for dealing with disease.

6.3. Intensive Monitoring of Gardens, Weeding, and Herbicide Application

All insect agriculturists constantly inspect their gardens, and no part of the garden is left untended for periods of time sufficient to allow the establishment and spread of diseases and fungivores (Batra & Batra 1979, Currie & Stuart 2001). Insect agriculturists, particularly the ants and termites, are able to invest in such intensive monitoring because their societies possess a nonreproducing worker caste, a large portion of which is dedicated to garden care. In the beetles, the relatively small size of the garden allows for intensive monitoring by a single female or by a small family of females. Intensive monitoring ensures that diseases are discovered and eradicated in the early stages of infection before they are able to spread and cause significant crop loss. Early detection is an effective defense against novel disease mutants that might evolve greater virulence if left untreated, because in the early stages of infection these strains can more readily be controlled with standard treatments.

Garden treatment in attine ants includes the use of secretions from their metapleural and mandibular glands to clean substrate as it is brought into the nest, probably removing some or most weeds and pathogens from the surface of the substrate before it is added to the fungus garden (Maschwitz et al. 1970, Ortius-Lechner et al. 2000). Although antibiotic-producing glands have not been studied in fungus-growing termites, some secretions of nonfungus-growing termites have antibiotic properties (Rosengaus et al. 1998, 2004). Antimicrobial glands in ambrosia beetles remain unknown and unstudied. In addition to glandular secretions, attine ants have another antimicrobial defense. Some or all of their integuments are covered with actinomycete bacteria. These bacteria are known to inhibit *Escovopsis* growth (Currie et al. 1999b), and experimental reduction of actinomycetes in colonies increases *Escovopsis* infection (Currie et al. 2003a). Garden bacteria in the genus *Burkholderia* (Santos et al. 2004) also provide antibiotics that provide protection against the garden parasite *Escovopsis* and against entomopathogenic diseases of the ants themselves. Termite gardens contain actinomycetes and other bacteria (Batra & Batra 1979), and beetle gardens contain a great diversity of bacterial secondary symbionts; however, the exact roles of these bacterial associates remain unknown.

6.4. Microbial Buffering

Beyond the known antibiotic-producing, disease-suppressing bacteria in attine colonies (Currie 2001a, Currie et al. 1999b, Santos et al. 2004), other secondary bacteria and fungi occur in insect gardens (Carreiro et al. 1997, Craven et al. 1970; C. Wang & U.G. Mueller, unpublished data), but their roles remain largely unknown. Although some of these secondary microbes may be neutral or detrimental to garden health and productivity, others may provide disease-modulating effects through competitive exclusion, antibiotic suppression of disease-causing microbes, resistance induction, or other mechanisms of microbial interaction. Competitive exclusion, disease suppression, and resistance induction have been demonstrated in both experimental and natural microbial systems (Hood 2003, Paulitz & Bélanger 2001, Wille et al. 2001). Some secondary microbes may even facultatively switch between beneficial and detrimental roles, depending on garden growth conditions, seasonal factors, or interactions with the insects or other microbes. For example, although the detrimental effects of *Escovopsis* are obvious in natural garden outbreaks and in interactions with cultivars in vitro (Currie 2001a,b; Currie et al. 1999a; Gerardo et al. 2004), it remains an untested possibility that *Escovopsis* may provide beneficial effects when present at low levels in the garden matrix.

Secondary microbes in termite gardens remain uninvestigated, but the possible significance of a secondary microbial flora in beetle gardens has been recognized for some time (Norris 1965). Norris (1965) suggested that it is the microbial complex as a whole (filamentous fungi, yeasts, and bacteria), rather than the dominant ambrosia fungus per se, that allows the beetles to exploit nutrient-poor substrates such as wood. Norris did not speculate on any additional roles for the secondary microbes, such as suppressing diseases, but such auxiliary roles deserve further study.

6.5. Management of Crop-Associated Microbial Consortia

If, as recent evidence indicates, secondary microbes serve ancillary functions in gardens by buffering against disease organisms or by producing antibiotics, enzymes, and metabolites, it is possible that insect farmers have evolved the capacity to manage these microbial consortia. Such microbial management strategies by insect farmers could include the following three methods.

6.5.1. STERILIZATION OF SUBSTRATE BEFORE INCORPORATION INTO THE GARDEN (ANTS, TERMITES) OR USE OF ESSENTIALLY STERILE SUBSTRATE FROM THE OUTSET (BEETLES)   In termites, the passage of substrate through the gut before incorporation into the garden probably eliminates many unwanted microbes and may increase the abundance of desired microbes (Figure 1). In attine ants, the considerable effort spent cleaning substrate surfaces appears to partially sterilize the substrate (Weber 1972). In the case of the ambrosia beetles, gardening occurs in a closed system because the beetles do not need to leave the nest to forage and because galleries are excavated in what is essentially a sterile medium, the sapwood or heartwood of living or recently killed trees, which are generally free of endophytic fungi and other microbes. This closed system greatly reduces the potential for accidental introduction of unwanted microbes and likely facilitates the management of desired microbes in beetle gardens.

6.5.2. SPATIALLY STRUCTURED GARDEN MATRIX   Structuring of gardens allows insect farmers to assess properties of particular, localized crop-microbe consortia. Any unwanted mutant genotypes, arising locally under particular microbe-microbe competitons, thus can be identified indirectly through the detection of their detrimental effects on the properties of the subgarden, and that subgarden piece can then be excised. Conversely, novel microbial mutants with beneficial effects can be identified indirectly by their beneficial effects on the subgarden, and preferentially subcultured and propagated across the rest of the garden (Figure 1). Such "symbiont-community choice" is possible only because of the fixed garden matrix, enabling farmers to assess properties of local consortia.

6.5.3. CONTINUOUS CO-OCCURRENCE OF GARDENS OF ALL AGES IN CLOSE PROXIMITY, RANGING FROM "UNPLANTED" TO MATURE GARDENS   Coexistence of gardens at different developmental stages is inherent in the vertical structuring of attine and termite gardens into younger subgardens at the top and older subgardens at the bottom. A range of differently-aged gardens allows farmers to efficiently practice one-way, selective transfer of only beneficial crop-microbe consortia from mature to younger gardens. Age-structuring also delays the spread of mutant microbes from older garden material to younger, more sterile gardens and thus prevents deterioration of the symbiont-community.

6.6. Multipartner Coevolution and Coevolving Antibiotic Defenses

One hypothetical advantage of secondary mutualistic microbes is that, unlike the insect farmers, microbes can potentially evolve at the same rate as the coevolving garden pests, enabling mutualistic insect-microbe systems to respond rapidly to the emergence of novel disease genotypes (Currie 2001a, Mueller & Gerardo 2002). Although such rapid microbial antibiotic defenses would obviously confer clear advantages, their evolutionary maintenance remains unclear. One possibility is that any single farmer society may have access to a diverse array of microbes from which it can select particular, desired types as needed. This scenario raises the question of how the insect farmers could maintain such a diverse array in their colonies in the face of both competition between microbes and the regular bottle-necking of the entire microbial "library" that presumably occurs at the founding of every new insect colony. Alternatively, the associated secondary microbes may be inherently fast-mutating, so that novel beneficial genotypes can rapidly arise to muster an appropriate defensive response. This scenario raises the question of how the most beneficial genotypes are recognized by the insect farmers and chosen for selective "amplification" against particular pathogens. The lack of clear evolutionary mechanisms for maintaining functional associations with coevolving, mutualistic microbes is not trivial, and future research needs to assess not only the diversity of microbial genotypes within single farmer colonies, but also to identify the mechanisms underlying adaptive symbiont-choice selection of beneficial, novel microbial genotypes. Future research also needs to address whether the coevolution of several, mutualistically-aligned partners (i.e., a "multidefense alliance" of ants, cultivar, and auxiliary microbes), each mustering its own defense, provides for a more evolutionarily stable disease-management strategy compared to a strategy in which the insect farmers act alone in a coevolutionary arms race against particular pathogens.

7. PRINCIPLES OF INSECT AGRICULTURE: LESSONS FOR HUMAN AGRICULTURE?

Perhaps the most striking feature of insect agriculture is the long-term cultivation of clonal monocultures. Monoculture increases agricultural efficiency through an economy of scale (Wolfe 1985), and clonality preserves the desirable properties of the crop by eliminating sexual recombination, but these advantages come at two costs: (*a*) increased vulnerability to the rapid spread of disease mutants (Barrett 1981, Shipton 1977, Mitchell et al. 2002, Mundt 2002, Peacock et al. 2001, Piper et al. 1996, Wolfe 1985), and (*b*) decreased resistance to fast-evolving diseases due to decreased genetic variability in the crop (Barrett 1981, Gustafson et al. 2003, Hamilton et al. 1990, Jaenike 1978, Zhu et al. 2000). These economic trade-offs (i.e., monoculture/clonality efficiency versus disease vulnerability) apply to both human and insect farmers.

The insect farmers' solution to the monoculture-disease problem appears to be not a single, "magic bullet" strategy (e.g., sole reliance on pesticides), but rather a combination of several strategies consisting of (*a*) crop sequestration, (*b*) intensive monitoring of crops for diseases, (*c*) access to a population-level reservoir of crop genetic variability, and (*d*) management of disease-suppressant microbes associated with the crop (Table 3). Of these strategies, large-scale crop sequestration is the least feasible in human agriculture because human crops need exposure to sunlight and because greenhouse cultivation is costly (Paulitz & Bélanger 2001). Intensive (e.g., daily) monitoring of every single crop plant for diseases may be feasible for some crops (e.g., in greenhouse environments); however, hourly monitoring of the kind implemented in insect agriculture seems cost-prohibitive for human agriculture at large.

A more novel approach is to design human agricultural systems that more efficiently take advantage of the microbial consortia that are known to play beneficial roles in crop nutrient uptake and disease resistance (Morrissey et al. 2004, Paulitz & Bélanger 2001, Wardle et al. 2004). Microbes of the rhizosphere (e.g., nitrogen-fixing bacteria and mycorrhizal fungi) have long been managed as critical associates of certain crops and trees (Finlay 2004, Johansson et al. 2004). More recently, disease-suppressant bacteria have been discovered that live on the root exudates of crops and produce antibiotics that protect the crop against pathogens (Haas & Keel 2003, Mazzola 2004, Morrissey et al. 2004, Weller et al. 2002, Whipps 2001). Disease-suppressant effects on crop plants have also been documented for phyllosphere microbes (Lindow & Brandl 2003) and endophytic microbes (Narisawa et al. 2002, Sturz et al. 2000). Agricultural research on rhizosphere, phyllosphere, and endophyte microbes of human crops is a very new field, however, and many beneficial microbes remain to be discovered and put to use (e.g., inoculation of crops with phyllosphere microbes to deter herbivores or to suppress airborne diseases).

Two problems commonly encountered in human agricultural experiments with beneficial microbial consortia are, first, that the composition of microbial species is difficult to manage and stabilize (Garbeva et al. 2004, Mazzola 2004), and, second, that beneficial microbes can rapidly evolve into detrimental ones (Alves et al. 2003, Morrissey et al. 2002). The farming insects' solution to these problems appears to consist of (*a*) selection on spatially limited microbial consortia (i.e., high-resolution, spatial separation of evolutionary processes, preventing the uncontrolled spread of microbes from inferior consortia); (*b*) propagation of crops with fast generational turnovers, thus minimizing the time for the evolution of any deleterious traits in the microbes; and (*c*) partial or complete sterilization of the substrate prior to planting, thus minimizing the influx of microbial contaminants into a largely closed agricultural system.

Perhaps it is from strategies such as these that humans have the most to learn from insect farmers, certainly if disease-suppressant microbes are ever to be managed in human agriculture (Morrissey et al. 2004). In developing these strategies, agriculturists would need to keep in mind that, during the domestication process, current human crops were not necessarily selected for capacities to interact with auxiliary microbes, i.e., the alleles in the wild ancestors optimally mediating such interactions may have been lost during the domestication process. Thus, a full evaluation of the potential uses of auxiliary microbes in human agriculture may require the study of the microbial consortia associated with the wild populations from which human-domesticated crops were originally derived. Such domestication within the context of coevolving and codomesticated microbial consortia may well be the key element explaining the 50-million year old agricultural success of the insect farmers.

ACKNOWLEDGMENTS

This review synthesizes work supported by the National Science Foundation (awards DEB-9707209 and IRCEB DEB-0110073 to U.G.M. & T.R.S.; CAREER DEB-9983879 to U.G.M.; DEB-0308757 to N.M.G.; INT-0434171 to D.L.S.), the Smithsonian Tropical Research Institute (to U.G.M., N.M.G. & D.K.A.), the Smithsonian Institution (to T.R.S.), and the Danish Natural Science Research Council (to D.K.A.). Some of the ideas presented here evolved out of discussions with P. Abbot, L. Ancel-Meyers, M. Bacci, K. Boomsma, J. Bull, Y. Carriere, C. Currie, A. Herre, A. Himler, N. Mehdiabadi, A. Mikheyev, T. Murakami, F. Pagnocca, S. Rehner, J. Sachs, R. Samuels, J. Scott, S. Solomon, L. Thomashow, and B. Wcislo. Special thanks to P. Abbot, R. Adams, K. Boomsma, S. Brady, S. Bruschi, E. Caldera, H. De Fine Licht, M. Dijkstra, A. Green, A. Himler, D. Kronauer, J. LaPolla, C. Marshall, A. Mikheyev, J. Miller, A. Mosegaard, D. Nash, J. Pedersen, M. Poulsen, C. Rabeling, J. Scott, B. Slippers, S. Solomon, J. Sosa-Calvo, and one anonymous reviewer for constructive comments on the manuscript; and to M. Bacci, A. Himler, C. Currie, and S. Rehner for unpublished information. We are most grateful to Greg Dimijian, Susanne Kühnholz, and Karen Machielsen for photos in Figure 2; and especially to Barrett Klein for producing Figure 1.

The *Annual Review of Ecology, Evolution, and Systematics* is online at
http://ecolsys.annualreviews.org

LITERATURE CITED

Aanen DK, Boomsma JJ. 2005. Evolutionary dynamics of the mutualistic symbiosis between fungus-growing termites and *Termitomyces* fungi. See Vega & Blackwell 2005, pp. 191–210

Aanen DK, Eggleton P, Rouland-Lefèvre C, Guldberg-Frøslev T, Rosendahl S, Boomsma JJ. 2002. The evolution of fungus-growing termites and their mutualistic fungal symbionts. *Proc. Natl. Acad. Sci. USA* 99:14887–92

Abe T, Bignell DE, Higashi M, eds. 2000. *Termites: Evolution, Sociality, Symbiosis, Ecology*. Dordrecht: Kluwer Acad. 488 pp.

Adams RMM, Mueller UG, Green AM, Narozniak JM. 2000a. Garden sharing and garden stealing in fungus-growing ants. *Naturwissenschaften* 87:491–93

Adams RMM, Mueller UG, Schultz TR, Norden B. 2000b. Agro-predation: usurpation of attine fungus gardens by *Megalomyrmex* ants. *Naturwissenschaften* 87:549–54

Alves BJR, Boddey RM, Urquiaga S. 2003. The success of BNF in soybean in Brazil. *Plant Soil* 252:1–9

Ayres MP, Wilkens RT, Ruel JJ, Lombardero MJ, Vallery E. 2000. Nitrogen budgets of phloem-feeding bark beetles with and without symbiotic fungi (Coleoptera: Scolytidae). *Ecology* 81:2198–10

Baker JM. 1963. Ambrosia beetles and their fungi, with particular reference to *Platypus cylindricus* Fab. *Symp. Soc. Gen. Microbiol.* 13:232–65

Barras SJ. 1973. Reduction of progeny and development in the southern pine beetles following removal of symbiotic fungi. *Can. Entomol.* 105:1295–99

Barrett JA. 1981. The evolutionary consequences of monocultures. In *Genetic Consequences of Man-Made Change*, ed. JA Bishop, LM Cook, pp. 209–48. London: Academic. 409 pp.

Bass M, Cherrett JM. 1995. Fungal hyphae as a source of nutrients for the leaf-cutting ant *Atta sexdens*. *Physiol. Entomol.* 20:1–6

Batra LR. 1966. Ambrosia fungi: extent of specificity to ambrosia beetles. *Science* 153:193–95

Batra LR. 1967. Ambrosia fungi: a taxonomic revision and nutritional studies of some species. *Mycologia* 59:976–1017

Batra LR. 1979. *Insect-Fungus Symbiosis.* Monclair: Allanheld, Osmun & Co. 288 pp.

Batra LR, Batra SWT. 1979. Termite-fungus mutalism. See Batra 1979, pp. 117–63

Beaver RA. 1989. Insect-fungus relationships in the bark and ambrosia beetles. In *Insect-fungus Interactions, 14th Symp. R. Entomol. Soc. London*, ed. N Wilding, NM Collins, PM Hammond, JF Webber, pp. 121–43. London: Academic

Blackwell M, Jones K. 1997. Taxonomic diversity and interactions of insect-associated ascomycetes. *Biodivers. Conserv.* 6:689–99

Borden JH. 1988. The striped ambrosia beetles. In *Dynamics of Forest Insect Populations*, ed. AA Berryman, pp. 579–96. New York: Plenum. 624 pp.

Bot ANM, Currie CR, Hart AG, Boomsma JJ. 2001a. Waste management in leaf-cutting ants. *Ethol. Ecol. Behav.* 13:225–37

Bot ANM, Rehner SA, Boomsma JJ. 2001b. Partial incompatibility between ants and symbiotic fungi in two sympatric species of *Acromyrmex* leaf-cutting ants. *Evolution* 55:1980–91

Browne F. 1961. The biology of Malayan Scolytidae and Platypodidae. *Malay. For. Rec.* 22:1–255

Carreiro SC, Pagnocca FC, Bueno OC, Bacci M, Hebling MJA, de Silva OA. 1997. Yeasts associated with the nests of the leaf-cutting ant *Atta sexdens rubropilosa* Forel, 1908. *Antonie van Leeuwenhoek* 71:243–48

Chapela IH, Rehner SA, Schultz TR, Mueller UG. 1994. Evolutionary history of the symbiosis between fungus-growing ants and their fungi. *Science* 266:1691–94

Craven SE, Dix MD, Michaels GE. 1970. Attine fungus gardens contain yeasts. *Science* 169:184–86

Currie CR. 2001a. A community of ants, fungi, and bacteria: a multilateral approach to studying symbiosis. *Annu. Rev. Microbiol.* 55:357–80

Currie CR. 2001b. Prevalence and impact of a virulent parasite on a tripartite mutualism. *Oecologia* 128:99–106

Currie CR, Bot ANM, Boomsma JJ. 2003a. Experimental evidence of a tripartite mutualism: bacteria protect ant fungus gardens from specialized parasites. *Oikos* 101:91–102

Currie CR, Mueller UG, Malloch D. 1999a. The agricultural pathology of ant fungus gardens. *Proc. Natl. Acad. Sci. USA* 96:7998–8002

Currie CR, Scott JA, Summerbell RC, Malloch D. 1999b. Fungus-growing ants use antibiotic-producing bacteria to control garden parasites. *Nature* 398:701–4

Currie CR, Stuart AE. 2001. Weeding and grooming of pathogens in agriculture by ants. *Proc. R. Soc. London Ser. B* 268:1033–39

Currie CR, Wong B, Stuart AE, Schultz TR, Rehner SA, et al. 2003b. Ancient tripartite coevolution in the attine ant-microbe symbiosis. *Science* 299:386–88

Darlington JECP. 1994. Nutrition and evolution in fungus-growing termites. In *Nourishment and Evolution in Insect Societies*, ed. JH Hunt, CA Nalepa, pp. 105–30. Boulder, CO: Westview. 449 pp.

de Fine Licht HH, Andersen A, Aanen DK. 2005. *Termitomyces* sp. associated with the termite *Macrotermes natalensis* has a heterothallic mating system and multinucleate cells. *Mycol. Res.* In press

Denison RF, Kiers ET, West SA. 2003. Darwinian agriculture: When can humans find solutions beyond the reach of natural selection? *Q. Rev. Biol.* 78:145–68

Diamond J. 1997. *Guns, Germs, and Steel: the Fates of Human Societies.* New York: Norton. 480 pp.

Farrell BD, Sesqueira AS, O'Meara BC, Normark BB, Chung JH, et al. 2001. The evolution of agriculture in beetles (Curculionidae: Scolytinae and Platypodinae). *Evolution* 55:2011–27

Fernández-Marín H, Zimmerman JK, Wcislo WT. 2004. Ecological traits and evolutionary sequences of nest establishment in fungus-growing ants (Hymenoptera, Formicidae, Attini). *Biol. J. Linn. Soc.* 81:39–48

Finlay RD. 2004. Mycorrhizal fungi and their multifunctional roles. *Mycologist* 18:91–96

Fisher PJ, Stradling DJ, Sutton BC, Petrini LE. 1995. Microfungi in the fungus gardens of the leaf-cutting ant *Atta cephalotes*: a preliminary study. *Mycol. Res.* 100:541–46

Francke-Grosmann H. 1967. Ectosymbiosis in wood-inhabiting insects. In *Symbiosis*, ed. SM Henry, 2:141–205. New York: Academic. 443 pp.

French JRJ, Roeper RA. 1972. Interactions of the ambrosia beetle *Xyleborus dispar* (Coleoptera: Scolytidae) with its symbiotic fungus, *Ambrosiella hartigii* (Fungi Imperfecti). *Can. Entomol.* 104:1635–41

Futuyma DJ, Slatkin M. 1983. *Coevolution.* Sunderland, MA: Sinauer. 555 pp.

Garbeva P, van Veen JA, van Elsas JD. 2004. Microbial diversity in soil: selection of microbial populations by plants and soil type and implications for disease suppressiveness. *Annu. Rev. Phytopathol.* 42:243–70

Gebhardt H, Bergerow D, Oberwinkler F. 2004. Identification of the ambrosia fungus of *Xyleborus monographus* and *X. dryographus* (Curculionidae, Scolytinae). *Mycol. Prog.* 3:95–102

Gerardo NM, Mueller UG, Price SL, Currie CR. 2004. Exploitation of a mutualism: special-ization of fungal parasites on cultivars in the attine ant symbiosis. *Proc. R. Soc. London Ser. B* 271:1791–98

Grassé PP. 1959. Une nouveau type de symbiose: La meule alimentaire des termites champignonnistes. *Nature* 3293:385–89

Green AM, Adams RM, Mueller UG. 2002. Extensive exchange of fungal cultivars between two sympatric species of fungus-growing ants. *Mol. Ecol.* 11:191–95

Green RE, Cornell SJ, Scharlemann JPW, Balmford A. 2005. Farming and the fate of wild nature. *Science* 307:550–55

Gustafson DM, Boe A, Jin Y. 2003. Genetic variation for *Puccinia emaculata* infection in switchgrass. *Crop Sci.* 43:755–59

Haanstad JO, Norris DM. 1985. Microbial symbionts of the ambrosia beetle *Xyloterinus politus*. *Microb. Ecol.* 11:267–76

Haas D, Keel C. 2003. Regulation of antibiotic production in root-colonizing *Pseudomonas* spp. and relevance for biological control of plant disease. *Annu. Rev. Phytopathol.* 41:117–53

Hamilton WD, Axelrod R, Tanese R. 1990. Sexual reproduction as an adaptation to resist parasites (a review). *Proc. Natl. Acad. Sci. USA* 87:3566–73

Harrington TC. 2005. Ecology and evolution of mycophagous bark beetles and their fungal partners. See Vega & Blackwell 2005, pp. 257–91

Hart AG, Anderson C, Ratnieks FL. 2002. Task partinioning in leafcutting ants. *Acta Ethol.* 5:1–11

Herre EA, Knowlton N, Mueller UG, Rehner SA. 1999. The evolution of mutualisms: exploring the paths between conflict and cooperation. *Trends Ecol. Evol.* 14:49–53

Hölldobler B, Wilson EO. 1990. *The Ants.* Cambridge: Harvard Univ. Press. 732 pp.

Hood ME. 2003. Dynamics of multiple infection and within-host competition by the anther-smut pathogen. *Am. Nat.* 162:122–33

Huber J. 1905. Über die Koloniegründung bei *Atta sexdens*. *Biol. Cent.* 25:606–19, 625–35

Jaenike J. 1978. A hypothesis to account for the

maintenance of sex within populations. *Evol. Theory* 3:191–94

Janzen DH. 1966. Coevolution of mutualism between ants and acacias in Central America. *Evolution* 20:249–75

Johansson JF, Paul LR, Finlay RD. 2004. Microbial interactions in the mycorrhizosphere and their significance for sustainable agriculture. *FEMS Microbiol. Ecol.* 48:1–13

Johnson RA. 1981. Colony development and establishment of the fungus comb in *Microtermes* sp. nr. *Usambaricus* (Isoptera, Macrotermitinae) from Nigeria. *Insect. Soc.* 28:3–12

Johnson RA, Thomas RJ, Wood TG, Swift MJ. 1981. The inoculation of the fungus comb in newly founded colonies of the Macrotermitinae (Isoptera). *J. Nat. Hist.* 15:751–56

Jones KG, Blackwell M. 1998. Phylogenetic analysis of ambrosial species in the genus *Raffaelea* based on 18S rDNA sequences. *Mycol. Res.* 102:661–65

Jordal BH. 2002. Elongation factor 1 resolves the monophyly of the haploid ambrosia beetles Xyleborini (Coleoptera: Curculionidae). *Insect Mol. Biol.* 11:453–65

Katoh H, Miura T, Maekawi K, Shinzato N, Matsumoto T. 2002. Genetic variation of symbiotic fungi cultivated by the macrotermitine termite *Odontotermes formosanus* (Isoptera: Termitidae) in the Ryukyu Archipelago. *Mol. Ecol.* 11:1565–72

Kent DS, Simpson JS. 1992. Eusociality in the beetle *Austroplatypus incompertus* (Coleoptera: Curculionidae). *Naturwissenschaften* 79:86–87

Kingsolver JG, Norris DM. 1977. External morphology of *Xyleborus ferrugineus* (Fabr.) (Coleoptera: Scolytidae) I. Head and prothorax of adult male and females. *J. Morphol.* 154:147–56

Kinuura H. 1995. Symbiotic fungi associated with ambrosia beetles. *Jpn. Agric. Res. Q.* 29:57–63

Kirkendall LR, Kent DS, Raffa KF. 1997. Interactions among males, females, and offspring in bark and ambrosia beetles: the significance of living in tunnels for the evolution of social behavior. In *Evolution of Social Behavior in Insects and Arachnids*, ed. JC Choe, BJ Crespi, pp. 181–215. Cambridge, UK: Cambridge Univ. Press. 561 pp.

Kok LT, Norris DM, Chu HM. 1970. Sterol metabolism as a basis for a mutualistic symbiosis. *Nature* 225:661–62

Korb J, Aanen DK. 2003. The evolution of uniparental transmission of fungal symbionts in fungus-growing termites (Macrotermitinae). *Behav. Ecol. Sociobiol.* 53:65–71

Kweskin M. 2003. *Molecular and behavioral ecology of fungus-growing ants and their fungi.* MA thesis. Univ. Tex., Austin. 79 pp.

LaPolla JS, Mueller UG, Seid M, Cover SP. 2002. Predation by the army ant *Neivamyrmex rugulosus* on the fungus-growing ant *Trachymyrmex arizonensis*. *Insect. Soc.* 49:251–56

Leuthold RH, Badertscher S, Imboden H. 1989. The inoculation of newly formed fungus comb with *Termitomyces* in *Macrotermes* colonies (Isoptera, Macrotermitinae). *Insect. Soc.* 36:328–38

Lindow SE, Brandl MT. 2003. Microbiology of the phyllosphere. *Appl. Environ. Microbiol.* 69:1875–83

Malloch D, Blackwell M. 1993. Dispersal biology of the ophiostomatoid fungi. In *Ceratocystis and Ophiostoma: Taxonomy, Ecology and Pathogenicity*, ed. MJ Wingfield, KA Seifert, J Webber, pp. 195–206. St. Paul: Am. Phytopathol. Soc. 304 pp.

Maschwitz U, Hölldobler B. 1970. Der Kartonbau bei *Lasius fuliginosus* Latr. (Hymenoptera: Formicidae). *Z. Vgl. Physiol.* 66:176–89

Maschwitz U, Koob K, Schildknecht H. 1970. Ein Beitrag zur Funktion der Metathorakaldrüse der Ameisen. *J. Insect Physiol.* 16:387–404

Matsuura K, Tanaka C, Nishida T. 2000. Symbiosis of a termite and a sclerotium-forming fungus: Sclerotia mimic termite eggs. *Ecol. Res.* 15:405–14

Mazzola M. 2004. Assessment and management of soil microbial community structure for disease suppression. *Annu. Rev. Phytopathol.* 42:35–59

Mitchell CE, Tilman D, Groth JV. 2002. Effects of grassland plant species diversity, abundance, and composition on foliar fungal disease. *Ecology* 83:1713–26

Morelet M. 1998. Une espece nouvelle de *Raffaelea*, isolee de *Platypus cylindrus*, coleoptere xylomycetophage des chenes. *Extr. Ann. Soc. Sci. Nat. Archeol. Toulon Var* 50:185–93

Morrissey JP, Dow JM, Mark GL, O'Gara F. 2004. Are microbes at the root of a solution to world food production? *EMBO Rep.* 10:922–26

Morrissey JP, Walsh UF, O'Donnell A, Moënne-Loccoz Y, O'Gara F. 2002. Exploitation of genetically modified inoculants for industrial ecology applications. *Antonie van Leeuwenhoek* 81:599–606

Mueller UG. 2002. Ant versus fungus versus mutualism: Ant-cultivar conflict and the deconstruction of the attine ant-fungus symbiosis. *Am. Nat.* 160(Suppl.):S67–98

Mueller UG, Gerardo N. 2002. Fungus-farming insects: Multiple origins and diverse evolutionary histories. *Proc. Natl. Acad. Sci. USA* 99:15247–49

Mueller UG, Lipari SE, Milgroom MG. 1996. Amplified Fragment Length Polymorphism (AFLP) fingerprinting of fungi cultured by the fungus-growing ant *Cyphomyrmex minutus*. *Mol. Ecol.* 5:119–22

Mueller UG, Poulin J, Adams RMM. 2004. Symbiont choice in a fungus-growing ant (Attini, Formicidae). *Behav. Ecol.* 15:357–64

Mueller UG, Rehner SA, Schultz TR. 1998. The evolution of agriculture in ants. *Science* 281:2034–38

Mueller UG, Schultz TR, Currie CR, Adams RMM, Malloch D. 2001. The origin of the attine ant-fungus mutualism. *Q. Rev. Biol.* 76:169–97

Mundt CC. 2002. Use of multiline cultivars and cultivar mixtures for disease management. *Annu. Rev. Phytopathol.* 40:381–410

Munkacsi AB, Pan JJ, Villesen P, Mueller UG, Blackwell M, et al. 2004 Convergent coevolution in the domestication of coral mushrooms by fungus-growing ants. *Proc. R. Soc. London Ser. B* 271:1777–82

Murakami T, Higashi S. 1997. Social organization in two primitive attine ants, *Cyphomyrmex rimosus* and *Myrmicocrypta ednaella*, with reference to their fungus substrates and food sources. *J. Ethol.* 15:17–25

Narisawa K, Kawamata H, Currah RS, Hashiba T. 2002. Suppression of *Verticillium* wilt in eggplant by some fungal root endophytes. *Eur. J. Plant Pathol.* 108:103–9

Norris DM. 1965. The complex of fungi essential to growth and development of *Xyleborus sharpi* in wood. *Mater. Org. Beih.* 1:523–29

Norris DM. 1972. Dependence of fertility and progeny development of *Xyleborus ferrugineus* upon chemicals from its symbiotes. In *Insect and Mite Nutrition*, ed. JC Rodriguez, pp. 299–310. North-Holland: Amsterdam. 702 pp.

Norris DM. 1979. The mutualistic fungi of Xyleborini beetles. See Batra 1979, pp. 53–65

Ortius-Lechner D, Maile R, Morgan ED, Boomsma JJ. 2000. Metapleural gland secretions of the leaf-cutter ant *Acromyrmex octospinosus*: New compounds and their functional significance. *J. Chem. Ecol.* 26:1667–83

Paulitz TC, Bélanger RR. 2001. Biological control in greenhouse systems. *Annu. Rev. Phytopathol.* 39:103–33

Peacock L, Hunter T, Turner H, Brain P. 2001. Does host genotype diversity affect the distribution of insect and disease damage in willow cropping systems? *J. Appl. Ecol.* 38:1070–81

Piper JK, Handley MK, Kulakow PA. 1996. Incidence and severity of viral disease symptoms on eastern gamagrass in monoculture and polycultures. *Agric. Ecosyst. Environ.* 59:139–47

Poulsen M, Boomsma JJ. 2005. Mutualistic fungi control crop-diversity in fungus-growing ants. *Science* 307:741–44

Price SL, Murakami T, Mueller UG, Schultz TR, Currie CR. 2003. Recent findings in fungus-growing ants: Evolution, ecology, and behavior of a complex microbial symbiosis. In *Genes, Behavior, and Evolution in Social Insects*, ed. M Kikuchi, S Higashi, pp. 255–80. Sapporo: Hokkaido Univ. Press. 314 pp.

Rindos D. 1984. *The Origins of Agriculture.* Orlando: Academic. 325 pp.

Rissing SW, Pollock GB, Higgins MR, Hagen RH, Smith DR. 1989. Foraging specialization without relatedness or dominance among co-founding ant queens. *Nature* 338: 420–22

Roeper RA, Treeful LM, O'Brien KM, Foote RA, Bunce MA. 1980. Life history of the ambrosia beetle *Xyleborus affinis* (Coleoptera: Scolytidae) from in vitro culture. *Great Lakes Entomol.* 13:141–44

Rollins F, Jones KG, Krokene P, Solheim H, Blackwell M. 2001. Phylogeny of asexual fungi associated with bark and ambrosia beetles. *Mycologia* 93:991–96

Rosengaus RB, Guldin MR, Traniello JFA. 1998. Inhibitory effect of termite fecal pellets on fungal spore germination. *J. Chem. Ecol.* 24:1697–706

Rosengaus RB, Traniello JFA, Lefebvre ML, Maxmen AB. 2004. Fungistatic activity of the sternal gland secretion of the dampwood termite *Zootermopsis angusticollis*. *Insect. Soc.* 51:259–64

Rouland-Lefevre C. 2000. Symbiosis with fungi. See Abe et al. 2000, pp. 289–306

Sachs J, Mueller UG, Wilcox TP, Bull JJ. 2004. The evolution of cooperation. *Q. Rev. Biol.* 79:135–60

Sanchez-Peña, SR. 2005. New view on origin of attine ant-fungus mutualism: exploitation of a preexisting insect-fungus symbiosis (Hymenoptera: Formicidae). *Ann. Entomol. Soc. Am.* 98:151–64

Sands WA. 1956. Some factors affecting the survival of *Odontotermes badius*. *Insect. Soc.* 3:531–36

Santos AV, Dillon RJ, Dillon VM, Reynolds SE, Samuels RI. 2004. Occurrence of the antibiotic producing bacterium *Burkholderia* sp. in colonies of the leaf-cutting ant *Atta*

sexdens rubropilosa. *FEMS Microbiol. Lett.* 239:319–23

Schultz TR, Meier R. 1995. A phylogenetic analysis of the fungus-growing ants (Formicidae: Attini) based on morphological characters of the larvae. *Syst. Entomol.* 20:337–70

Schultz TR, Mueller UG, Currie CR, Rehner SA. 2005. Reciprocal illumination: A comparison of agriculture in humans and ants. See Vega & Blackwell 2005, pp. 149–90

Schultz TR, Solomon SA, Mueller UG, Villesen P, Boomsma JJ, et al. 2002. Cryptic speciation in the fungus-growing ants *Cyphomyrmex longiscapus* Weber and *Cyphomyrmex muelleri* Schultz and Solomon, new species (Formicidae, Attini). *Insect. Soc.* 49: 331–43

Shipton PJ. 1977. Monoculture and soil borne pathogens. *Annu. Rev. Phytopathol.* 15:387–407

Sieber R. 1983. Establishment of fungus comb in laboratory colonies of *Macrotermes michaelseni* and *Odontotermes montanus* (Isoptera, Macrotermitinae). *Insect. Soc.* 30:204–9

Silliman BR, Newell SY. 2003. Fungus farming in snails. *Proc. Natl. Acad. Sci. USA* 100: 15643–48

Six DL. 2003. Bark beetle-fungus symbiosis. In *Insect Symbiosis*, ed. T Miller, K Kourtzis, pp. 99–116. Boca Raton, FL: CRC. 368 pp.

Six DL, Klepzig KD. 2004. *Dendroctonus* bark beetles as model systems for studies on symbiosis. *Symbiosis* 37:207–32

Six DL, Paine TD. 1998. Effects of mycangial fungi and host tree species on progeny survival and emergence of *Dendroctonus ponderosae* (Coleoptera: Scolytidae). *Environ. Entomol.* 27:1393–1401

Six DL, Paine TD. 1999. Allozyme diversity and gene flow in *Ophiostoma clavigerum* (Ophiostomatales: Ophiostomataceae), the mycangial fungus of the Jeffrey pine beetle, *Dendroctonus jeffreyi* (Coleoptera: Scolytidae). *Can. J. For. Res.* 29:324–31

Smith BD. 1998. *The Emergence of Agriculture.* New York: Sci. Am. Libr. 230 pp.

Sturz AV, Christie BR, Nowak J. 2000.

Bacterial endophytes: Potential role in developing sustainable systems of crop production. *Crit. Rev. Plant Sci.* 19:1–30

Traniello JFA, Leuthold RH. 2000. The behavioral ecology of foraging in termites. See Abe et al. 2000, pp. 141–68

Vega F, Blackwell M, eds. 2005. *Insect-Fungal Associations: Ecology and Evolution.* Oxford: Oxford Univ. Press. 333 pp.

Villesen P, Mueller UG, Schultz TR, Adams RMM, Bouck MC. 2004. Evolution of antcultivar specialization and cultivar switching in *Apterostigma* fungus-growing ants. *Evolution* 58:2252–65

von Arx JA, Hennebert GL. 1965. Deux champignons ambrosia. *Mycopathol. Mycol. Appl.* 25:309–15

Wardle DA, Bardgett RD, Klironomos JN, Setälä H, van der Putten WH, Wall DH. 2004. Ecological linkages between aboveground and belowground biota. *Science* 304:1629–33

Weber NA. 1972. *Gardening Ants: The Attines.* Philadelphia: Am. Philos. Soc. 146 pp.

Weller DM, Raaijmakers JM, Gardener BB, Thomashow LS. 2002. Microbial population responses for specific soil suppressiveness to plant pathogens. *Annu. Rev. Phytopathol.* 40: 309–48

Whipps JM. 2001. Microbial interactions and biocontrol in the rhizosphere. *J. Exp. Bot.* 52: 487–511

Wille P, Boller T, Kaltz O. 2001. Mixed inoculation alters infection success of strains of the endophyte *Epichole bromicola* on its grass host *Bromus erectus. Proc. R. Soc. London Ser. B* 269:397–402

Wolfe MS. 1985. The current status and prospects of multiline cultivars and variety mixtures for disease resistance. *Annu. Rev. Phytopathol.* 23:251–73

Wood SL. 1982. The bark and ambrosia beetles of North and Central America (Coleoptera: Scolytidae), a taxonomic monograph. *Great Basin Nat. Mem.* 6:1–1359

Zhu Z, Chen H, Fan J, Wang Y, Li Y, Chen J, et al. 2000. Genetic diversity and disease control in rice. *Nature* 406:718–22

Annu. Rev. Ecol. Evol. Syst. 2005. 36:597–620
doi: 10.1146/annurev.ecolsys.36.091704.175520
First published online as a Review in Advance on September 16, 2005

INSECTS ON PLANTS: Diversity of Herbivore Assemblages Revisited

Thomas M. Lewinsohn,[1] Vojtech Novotny,[2] and Yves Basset[3]

[1]*Laboratório de Interações Insetos-Plantas, Instituto de Biologia, Universidade Estadual de Campinas, Brazil; email: thomasl@unicamp.br*
[2]*Institute of Entomology, CAS and Biological Faculty, University of South Bohemia, Czech Republic; email: novotny@entu.cas.cz*
[3]*Smithsonian Tropical Research Institute, Panama; email: bassety@tivoli.si.edu*

Key Words herbivory, insect-plant interactions, tropical insects, local and regional richness, species diversity

■ **Abstract** The diversity and composition of herbivore assemblages was a favored theme for community ecology in the 1970s and culminated in 1984 with *Insects on Plants* by Strong, Lawton and Southwood. We scrutinize findings since then, considering analyses of country-wide insect-host catalogs, field studies of local herbivore communities, and comparative studies at different spatial scales. Studies in tropical forests have advanced significantly and offer new insights into stratification and host specialization of herbivores. Comparative and long-term data sets are still scarce, which limits assessment of general patterns in herbivore richness and assemblage structure. Methods of community phylogenetic analysis, complex networks, spatial and among-host diversity partitioning, and metacommunity models represent promising approaches for future work.

INTRODUCTION

Seldom in the history of science is an unmistakable cornerstone laid for a new subject, but Southwood (1961) was undoubtedly the first to consider insect herbivore richness and its variation among host-plant species a phenomenon worthy of explanation. Southwood's inaugural papers remained largely unappreciated until MacArthur & Wilson (1967) presented their theory of island biogeography and Janzen (1968) proposed that its theoretical framework could be applied to the diversity of herbivores on host plants. Within a decade or so, sufficient evidence had been gathered on the diversity of herbivorous insects associated with various host plants to allow inferences on the role of various causative processes, and in 1984 this became the leading theme of the book *Insects on Plants* (Strong et al. 1984).

Insect-plant interactions grew into a research domain in its own right, but its emphasis has shifted toward population-level processes and interactions, and to phylogenetic analyses of herbivore and plant lineages (e.g., Futuyma & Mitter 1996, Herrera & Pellmyr 2002, Schoonhoven et al. 1998). Thus, despite their importance in the 1970s and 1980s, the size of herbivore-host communities and their determinants has seemingly drifted from attention.

Twenty years after the publication of *Insects on Plants*, the time seemed opportune to consider further developments on its leading questions, which remain as relevant as before. We thus set out to evaluate the extent to which subsequent work has produced (*a*) major data sets that support earlier findings and hypotheses, (*b*) new empirical results that lead to novel insights, and (*c*) substantial advances in theory or explanatory models.

Our concern here is the size, structure, and composition of herbivore assemblages on particular plant species as well as the processes or factors that determine their variation. We do not consider the effects of herbivores on their hosts or on plant communities.

Three fairly distinct approaches can be recognized in studies of host-associated herbivore assemblages; these approaches often address different questions (see Denno et al. 1995, Strong et al. 1984). Accordingly, we focus first on analyses of data sets compiled from country-wide or regional catalogs of insect records. Next, we review observational and experimental field studies of local herbivore communities. We then consider studies conducted across several sites or at different spatial scales. In the final sections, we concentrate on evolutionary and dynamic aspects of herbivore assemblages, and then highlight certain lines of enquiry that promise new insights into patterns of herbivore diversity on plants and their determinants.

INSECT-HOST LISTS

Species Richness on Different Hosts

Insects recorded on British trees formed one of the data sets that inaugurated the entire subject (Southwood 1961), and since then it has been amended and repeatedly reanalyzed. Thus, Kennedy & Southwood (1984) estimated host area from their occurrence both in 10 km × 10 km and 2 km × 2 km grid units. This combined variable accounted for 58% of variation in insect species numbers. Further variables in their multiple-regression model were historical (length of time a host has been present in Britain), phenological (deciduousness), evolutionary (taxonomic relatedness, the size of the host's order in Britain), and structural (host average height and leaf length). All variables contributed significantly to the multiple-regression model, which explained 82% of variation, but their correlation structure was not investigated.

Kelly & Southwood (1999) incorporated phylogenetic relationships among British host trees into the foregoing model. The independent contrasts' analysis returned only a statistically nonsignificant improvement in the variance explained by host area, compared with species analyzed as independent entities. Host phylogeny, thus, evinced no effect on herbivore richness, although this finding may be partly due to the recent (i.e., postglacial) establishment of the British biota. They also reexamined the separate explanatory contributions of host frequency at the local (2-km units) and the regional scale and found that host frequency at the smaller scale was a far better predictor than was host regional range. This result concurs with Straw & Ludlow (1994), who showed that host local biomass is a better predictor of the regional number of herbivore species of British trees than either their smaller-scale or larger-scale frequency in the United Kingdom.

Few other data sets have lent themselves to equivalent exercises. In British Rosaceae, host geographic range, growth form (herbs to trees), and architecture (a function of plant height times leaf length) correlate with the numbers of herbivores (Leather 1986). For German tree genera, host area, size, and postglacial age of establishment explained 88% of variation in herbivore richness (Brändle & Brandl 2001), whereas taxonomic relatedness did not contribute to the multiple-regression model. Furthermore, the genera shared by Britain and Germany had highly correlated assemblage sizes in these countries; on average, more associated herbivores were listed in Germany. As found by Kelly & Southwood (1999), host phylogeny entailed no significant improvement to the model.

Frenzel & Brandl (1998) analyzed a data set of insects on Brassicaceae in Poland. Unlike other studies, this data set derives from a set of host records obtained in five custom "collection gardens," combined with occurrences in natural populations. Sampling effort was the main predictor for species richness of generalist but not of specialist insects; however, because effort was correlated with plant distribution and presumably with the number of sites where each host species was inspected, sampling effort may reflect differences in beta-diversity among these herbivore groups (see below). No other factor explained variation in herbivore richness. Thus, these studies mostly reinforce findings from early analyses of herbivore/host catalogs.

Taxonomic Composition and Guild Structure

Assemblages can be evaluated on their taxonomic or functional composition; the latter is often characterized by guilds. The proportions of different taxa or guilds can be estimated either for herbivore species or for individuals. Two simple descriptors used for herbivore assemblages are the specialist to generalist ratio, and the ratio of ectophages to endophages (feeding on versus within the plant), which is a minimalistic guild classification. These descriptors are correlated, because endophages tend to be more specialized than ectophages (Gaston et al. 1992).

A central concern here is whether the relative richness and abundance of guilds is determined by the available plant resources, competitively partitioned up by individual guilds or by relative sizes of regional species pools for individual guilds that serve as sources for plant colonization. The latter alternative seems more likely because no compensation between guilds (i.e., negative correlation between their species richness or abundance across plant species) has been found for tropical

(Basset & Novotny 1999) or temperate (Cornell & Kahn 1989) guilds. Likewise, variable guild composition on the same host across different geographic areas (Lawton et al. 1993) points to the importance of regional species pools for assemblage composition of local communities.

Cornell & Kahn (1989) explored the insects on British trees for regularities in size or proportion of different guilds. Except for a positive correlation of sapsuckers and chewers, guild sizes varied independently. They were also not influenced by the host traits used to predict total herbivore numbers (see above). The lack of predictable and general features in feeding guilds—especially relative size and taxonomic composition—across different hosts led Cornell & Kahn (1989) to propose that the British herbivore fauna reflects idiosyncratic histories of the herbivores on each host, despite the predictability of their total richness.

Constant taxonomic and guild composition among tree species may simply reflect regional or global differences in relative species richness of herbivore taxa. For example, the relative species richness of various hemipteran families in the tropics is virtually constant across continents (Hodkinson & Casson 1991). Guild structure may nonetheless differ among distinct plant lineages. For instance, nitrogen-fixing Fabaceae appear to be particularly important for xylem-suckers (Young 1984).

Differences in patterns conveyed by these simple descriptors suggest that the functional or taxonomic composition of herbivore assemblages is hardly explainable by immediate ecological factors, without consideration of the regional biotic history and phylogenetic constraints of the implicated plant and herbivore lineages.

COMPARATIVE AND EXPERIMENTAL FIELD STUDIES

Sampling Effort and Rarity: Pervasive Problems

A problem that besets all field studies is that observed species richness is correlated with sampling intensity. Within highly diverse assemblages that contain many rare species (Novotny & Basset 2000, Price et al. 1995), even very large samples fail to reach an asymptote. For example, Basset & Novotny (1999), with aggregate samples collected from 15 *Ficus* species in Papua New Guinea, were unable to establish clear asymptotes from circa 13,000 leaf-chewer and 45,000 sap-sucker individuals.

To overcome this predicament, richness in different-sized samples can be statistically standardized, usually through a rarefaction procedure. Otherwise, total richness can be estimated for each herbivore assemblage through various parametric and nonparametric estimators (Gotelli & Colwell 2001). Sampling effort can be factored out before the statistical effects of proposed causative factors are assessed (e.g., Frenzel & Brandl 1998). However, in spatially extensive samples, sampling intensity often correlates with geographical range so that, by factoring out sample size or number, the effect of geographic range cannot be fully assessed. At the local level, a rarefaction procedure yields standardized richness estimates independent of local sampling intensity (Lewinsohn 1991). At the regional scale, the entire set of recorded herbivores is perforce the basis for assessment of regional richness and the contribution of beta-diversity to it, which can be teased out, for instance, through path analysis (Lewinsohn 1991).

Many rare species are not trophically associated with the plants on which they were collected, especially in mass samples. These species have to be excluded from community analyses by in situ feeding observations or laboratory feeding experiments. However, many rare species prove to be associated with studied hosts; hence, infrequent interactions form a substantial part of plant-herbivore assemblages (Novotny & Basset 2000, Price et al. 1995). Quantitative insect-host records enable analyses that deemphasize rare interactions and correct potential bias (e.g., Godfray et al. 1999).

Plant Traits and Local Assemblages

Plant size and architecture affect the number of associated herbivore species (Strong et al. 1984). In some local studies, larger plant individuals (Cytrynowicz 1991) or species (Marquis 1991) were shown to support more herbivore species, whereas in other studies, no such effect was found (Basset 1996). In general, plant size effects per se are hard to evaluate because other factors are correlated with size, such as density, life stage, phenology, and architecture. Recent studies have focused on intraspecific differences between host growth stages (see below).

Architecture itself has no agreed-upon definition or measure. It can signify life stage (Fowler 1985), growth form of different species (Strong et al. 1984), size or number of structures (Haysom & Coulson 1998), or some combination of these (Strong et al. 1984). Thus, its effects on herbivore assemblages, whether significant or not, are difficult to compare among different studies.

Few studies have examined effects of chemical differences among host species on local herbivore assemblages. In Florida, each one of six oak species supported a distinctive cynipid gall-former assemblage whose composition was strongly related to certain chemical constituents (Abrahamson et al. 2003); this set of highly specialized herbivores thus differentiates its hosts at the species level. Likewise, many studies have investigated adaptive responses of certain insects to particular physical defenses and anatomical traits, but very few have extended to herbivore assemblages. For instance, Ezcurra et al. (1987) showed that glabrous plants supported higher densities of leaf-chewers and gallers than did their pilose conspecifics, on which, however, sap-suckers were more abundant. Peeters (2002) found that herbivore guild composition was more influenced by leaf structural traits than by leaf nitrogen, fiber, or water content.

Succession

Herbivore assemblages on different successional stages have been mostly studied by mass sampling in vegetation plots (e.g., sweep-netting or vacuuming). Overall insect species richness or guild structure can then be compared against total plant richness or vegetation structure. Earlier studies (see Strong et al. 1984) had shown that total richness increases with succession, in accord with plant richness. Higher total herbivore richness can be driven by (*a*) increased richness per host (herbivore species density), (*b*) higher average host specificity, (*c*) higher host diversity, or a combination of these. Aggregate insect samples are unsuitable for discriminating among these effects, however.

Though pioneer hosts support higher densities of herbivores and experience more damage compared with late successional plant species, herbivore richness per host does not seem to differ between pioneer and climax plant species, either in temperate or tropical vegetation (Basset 1996, Leps et al. 2001, Marquis 1991, Novotny 1994). An overall increase in herbivore species richness from early successional to mature vegetation would, therefore, be caused either by insect richness tracking the changes in plant-species richness or by higher average host specialization.

In tropical forests, the limited data available suggest that herbivore host specificity does not change noticeably during succession (Basset 1996, Leps et al. 2001, Marquis 1991), probably because, in contrast to temperate vegetation, tropical succession is often dominated by woody plants from the onset.

Tropical Forests

Early studies of tropical forests relied on mass trapping methods (reviewed in Basset 2001b), especially fogging, which produced large series of insects from canopies (Stork et al. 1997). However, particularly in tropical forests, the spatial distribution of individual insects is a poor indicator of herbivore niches and host ranges, because highly mobile insects circulate freely within a botanically diverse forest canopy [including lianas (Ødegaard 2000)] rather than being limited to their hosts. Canopy fogging studies are, therefore, being superseded by studies that combine in situ feeding observations, experimental feeding tests, and rearing of immatures (Basset et al. 2003, Marquis 1991).

Tropical studies are mostly local, directed at particular insect communities, whose comparison or analysis at the regional level is hampered by the many unnamed species they usually contain, so that specimens rather than names must be cross-checked (Kitching 1993). Furthermore, the asymptotic species richness of host-associated herbivore communities is difficult to estimate because of the large number of rare species. Hence, there are more studies on herbivore host specificity than on their species richness (Novotny & Basset 2005). The few comprehensive studies to date suggest that local factors, especially the availability of resources such as young foliage or overall plant biomass, or the pressure of ants and other enemies, may be more important in the determination of local herbivore richness than are historical or regional plant traits (Basset 1996, Basset & Novotny 1999, Marquis 1991), although these factors require further investigation.

Studies of arthropod stratification in rainforests represent a relatively recent field, facilitated by improvements in canopy access. Stratification studies either compare mass samples among strata without regard to hosts (see Basset et al. 2003) or compare the fauna of mature trees with that of conspecific seedlings or saplings (e.g., Barrios 2003, Basset 2001a). Both approaches indicate that the abundance and diversity of herbivorous taxa tend to be higher in the upper canopy than in the understory [with the possible exception of gallers (Cuevas-Reyes et al. 2004, Price et al. 1998)] and that the faunal similarity between the understory and the upper canopy is low. Differences among conspecific plants from different strata include nutritional quality, complexity, and resource quantity, compounded by environmental differences in temperature, light, and exposure to rain and wind.

The higher and less variable supply of young foliage in the upper canopy may be a significant determinant of herbivore stratification. Lowman et al. (1993) suggested that the upper canopy of temperate forests has proportionally fewer niches than does the tropical forest canopy. This proposition may help explain the pronounced vertical stratification of tropical forest herbivores compared with temperate forest herbivores.

LARGER-SCALE PATTERNS

Diversity Across Spatial Scales

The straightforward comparison of herbivore assemblages that feed on the same plant species at different sites and under contrasting ecological conditions (Lawton et al. 1993) has been used surprisingly rarely to test the effect of environmental variables such as altitude (Novotny et al. 2005) or climate (Andrew & Hughes 2004). Forestry and agriculture are potentially rich sources for data on spatial variability of herbivore communities on economically important hosts, but these sources have been scarcely looked into by ecologists (but see Lill et al. 2002, Strong et al. 1977).

The limits to extraction of general inferences on diversity patterns from studies that pertain exclusively to the local or to the regional scale are obvious. In studies that combine local and regional sets, regional diversity can be partitioned into spatial components, such as alpha-diversity and beta-diversity; these components can be assessed in an additive ANOVA model (Lande 1996).

The relationship between local-community diversity and the size of the regional species pool can elucidate the process of community assembly from the regional set of species (Ricklefs 2004). Most studies (e.g., Cornell & Lawton 1992) report a linear increase of local diversity with regional diversity, which suggests that local communities are proportional samples of regional species pools, from which, presumably, their component species are drawn independently (although not necessarily randomly, as they may still be subject to assembly rules). These results were viewed as evidence for the nonsaturation of local herbivore assemblages; that is, their resource space is not fully occupied by existing species (Cornell 1985, Cornell & Lawton 1992, but see Loreau 2000).

Local communities may depend on regional pools, but the converse causal direction is also plausible: Communities are assembled through local dynamics and amalgamated into regional pools rather than determined by them. Zwölfer (1987) took this approach with thistle flowerheads in Europe, whereas Straw & Ludlow (1994) proposed a model that derives regional species-area relations from local dynamics, in which local host biomass is the key driver, and applied it successfully to British tree and European thistle data. Lewinsohn (1991) examined both the regional-to-local causal model and its reverse in herbivore assemblages on flower heads of Brazilian Asteraceae. Both studies found a strong correlation of local and regional herbivore richness, as well as substantial turnover among sites. Thus, the higher herbivore richness of widespread plants was caused both by higher alpha-diversity and by higher beta-diversity.

Fernandes & Price (1991) found no relationship between gall richness and host distribution, whereas Blanche & Westoby (1996) showed that host-plant range influenced regional, but not local, richness of gallers on *Eucalyptus* species in Australia. Blanche & Westoby (1996) also demonstrated consistent differences among host subgenera, which highlighted that some host taxa or lineages bear a greater galling diversity than do others. In fact, regional diversity patterns may be essentially driven by the spatial turnover of certain host taxa that support high gall richness (Fernandes & Price 1991, Price et al. 1998).

Geographical Variation of Host Associations

Herbivorous insects are often oligophages, whose hosts belong to a species group (e.g., Becerra 1997), a genus, or subtribe (Prado & Lewinsohn 2004). However, host affiliations can vary geographically (Fox & Morrow 1981), both in specialized (Thompson 1999) and in generalist (Sword & Dopman 1999) insects. Insect ranges can either be smaller than that of host plants (Strong et al. 1984) or extend beyond limits of their individual host species (Scriber 1988). Thus, spatial turnover among herbivore assemblages may be produced either by plant or by insect change, or by shifting interactions among different sites. Sword & Dopman (1999) demonstrated geographical shifts in food plants among *Schistocerca emarginata* populations, together with ontogenetic shifts: nymphs were more specialized than adults and on different plants as well.

Historical processes can explain geographical differentiation of local assemblages. Sobhian & Zwölfer (1985) found that herbivorous assemblages on *Centaurea solstitialis* L. (Asteraceae) decreased from the Balkans, the plant's center of origin, towards Spain, mostly through loss of specialists, so that smaller Western assemblages consisted largely of polyphages. A similar pattern was found in pine-feeding Hemiptera in Central Europe (Brändle & Rieger 1999). However, on *Onopordum* thistles, specialists tracked their hosts across the Mediterranean, and polyphages were responsible for most of the spatial turnover in their herbivore assemblages (Briese et al. 1994).

Even though additional studies are needed, evidence seems to indicate that the "regional species pool" is, in fact, a variable assortment of specialists and generalists that have different probabilities of pertaining to local communities, according to conditions such as host abundance and predictability.

Plant introductions can be viewed as large-scale manipulative experiments with controls, represented by the alien's herbivore assemblage in its own native range (Memmott et al. 2000, Strong et al. 1977, Zwölfer 1988) or by assemblages on native plants in the alien's area of introduction (Leather 1986). Early studies showed that ectophagous herbivores can assemble rapidly on novel hosts (Kennedy & Southwood 1984, Strong et al. 1977), whereas endophages may be slower colonizers (Strong et al. 1984), and specialized guilds, such as seed-feeders, may be missing on recently arrived plant species (Memmott et al. 2000, Zwölfer 1988). Herbivores are quicker to colonize alien plants with native close relatives (Burki & Nentwig 1997).

Geographical Trends in Herbivore Richness

Regional diversity of almost all taxa of insect herbivores is known to be highest in the tropics; aphids are a notable exception (Dixon et al. 1987). However, the diversity of herbivore communities that feed on particular plant species appears to be similar between tropical and temperate forests (Basset & Novotny 1999, Janzen 1988). The ratio of butterfly to plant species also shows no major difference among regions (Gaston 1992). Flowerhead-feeders on Asteraceae do seem to have a higher local richness per host species in the neotropics than in Europe, after richness is adjusted for sampling effort; however, different plant tribes were studied in each region (Lewinsohn 1991, Zwölfer 1987). Thus, the considerable increase in regional diversity of insect herbivores from temperate to tropical areas appears driven largely by increasing plant diversity, but higher turnover among sites (beta-diversity) or among hosts (i.e., higher specialization) can contribute to this effect as well. However, in California, the simple correlation of butterfly richness with plant richness disappeared from more comprehensive models that included elevational range and actual evapotranspiration as explanatory variables; therefore, in this case, herbivore richness cannot be said to respond to host diversity per se (Hawkins & Porter 2003).

Rapid local censuses of gall-maker diversity, based on their morphologically distinct galls, produced for numerous sites around the world, suggest a richness gradient that peaks around 25° to 38° latitude (Price et al. 1998), whose interpretation is hindered by unequal distribution of samples and interaction of local factors (e.g., climate and soil). Within regions, the positive relationship between host and gall richness was confirmed in the South African Fynbos (Wright & Samways 1998), in Mexican dry forest (Cuevas-Reyes et al. 2004), and in the western United States and southeast Brazil (Fernandes & Price 1991, who mistakenly dismissed it as a spurious effect of other variables).

The altitudinal trend in herbivore richness is controversial, particularly in the tropics. Maximum species richness has been reported at lowland or at midmontane elevations (Fernandes & Price 1991, Lees et al. 1999, McCoy 1990). Furthermore, at least one study found species richness of the entire moth community constant with elevation, in seeming contrast to host diversity (Brehm et al. 2003). Another study reported constant species richness of moths per plant species (Novotny et al. 2005).

Steep altitudinal gradients of climate and biotic factors, such as vegetation structure or ant predation, foster high herbivore beta-diversity. Altitudinal gradients are also subject to spatial constraints that, on purely geometrical grounds, can establish either a monotonic decrease in species richness with elevation because of diminishing area or a peak in species richness at mid elevations (Lees et al. 1999). Furthermore, few studies (e.g., Hawkins & Porter 2003) verify whether variables are spatially autocorrelated.

The extraordinarily high local diversity of insects in tropical forests was extrapolated by Erwin (1982) to a global diversity estimate of 30 million insect species. This estimate assumed that herbivores are extremely host specific and, therefore, exhibit high turnover among tree species. Further studies (Basset et al. 1996, Novotny et al. 2002a, Ødegaard et al. 2000, Thomas 1990) have not confirmed this key assumption and revised insect diversity estimates to approximately 5 to 7 million species. Other studies (Orr & Häuser 1996) indicate that local assemblages may represent a large proportion of the regional species pool in tropical forests; hence, beta-diversity would also be lower than supposed in the initial estimates. However, this suggestion needs substantiation.

HERBIVORE ASSEMBLAGES AS EVOLUTIONARY DYNAMIC SYSTEMS

Herbivores and Host Phylogeny

Only a small, although growing, number of studies consider the phylogeny of host plants or herbivores when analyzing their assemblage size or structure. Methods for analyzing phylogenetic effects in communities are in the process of development, and some issues are still contentious (Losos 1996, Ricklefs 1996). Thus, despite the manifest importance of phylogeny, to what extent new analyses of assemblages in phylogenetic context change our understanding of community structure remains to be seen. For instance, reanalyses of regional species richness on British trees by use of phylogenetically independent contrasts still identified plant local frequency and distribution as their main determinants (Brändle & Brandl 2001, Kelly & Southwood 1999).

The lack of adequate species-level phylogenies is often limiting, especially in highly diverse tropical plant groups. Analyses of host specificity have mostly approximated phylogenetic relationships between host-plant species by their supraspecific taxonomic ranks that, however, are not commensurate across plant lineages (Losos 1996). New phylogenetic measures of host specificity and breadth (Symons & Beccaloni 1999, Webb et al. 2002) have not yet been widely applied (but see Weiblen et al. 2005). Phylogenetic constraints on host-plant selection may be also examined as a relationship of species turnover between herbivore communities and the phylogenetic distance of their host-plant species (Novotny et al. 2002a).

The interpretation of community composition requires understanding the evolutionary dynamics of host affiliation by herbivores. Increasingly powerful phylogenetic analyses enable tests for congruence between plant and herbivore phylogenies (Lopez-Vaamonde et al. 2003, Weiblen & Bush 2002) as well as for the effect of plant traits, such as secondary metabolites, on plant colonization by herbivores (Becerra 1997). These studies indicate that strict cospeciation between herbivores and plants is rare, although it is found in insect herbivores that also serve as specialized pollinators (Kato et al. 2003, Weiblen & Bush 2002). In other herbivores, even those intimately associated with host plants, such as leaf-miners, multiple colonizations of host lineages are common (Lopez-Vaamonde et al. 2003, Jermy & Szentesi 2003). This finding does not mean that the pattern of host use is random, as closely related herbivore species often feed on closely related plants (Futuyma & Mitter 1996).

Given the importance of plant chemistry in mediating plant–insect associations and their evolution (Becerra 1997, Berenbaum 2001), surprisingly few studies have attempted to investigate how it affects insect diversity. In British umbellifers, chemical uniqueness was unrelated to insect richness, whereas chemical diversity had a positive but slight effect (Jones & Lawton 1991). This observation lends only modest support to the hypothesis that chemically more diverse plants share more chemicals with other species, which, therefore, would facilitate host switches and, thus, increase herbivore richness.

Are more diverse communities especially rich in specialists or generalists, or is species richness independent from host specificity? So far, only a few cases have been studied. For instance, higher species richness of herbivorous chalcid wasps in Germany compared with wasps in Britain was solely caused by a higher number of generalists (Tscharntke et al. 2001). Herbivore communities on alien plants often reach the same species richness as those on native plants, but these species include higher proportions of generalists (Novotny et al. 2003, Zwölfer 1988).

Herbivory appears to be conducive to speciation, given that herbivore lineages tend to be more diverse than other modes of life. Most phytophagous beetles belong to radiations provoked by new angiosperm lineages (Farrell 1998).

Are assemblages more similar on phylogenetically closer hosts? The general answer to this question, as expected, is yes. Herbivore similarity may decrease gradually at increasing taxonomic levels of their host plants (comparisons among host genera, families, etc.) or may decrease sharply at one particular level, which in turn signals two attributes: (*a*) the taxonomic level at which insects recognize plants as equivalent or distinct hosts and (*b*) an upper threshold within which insects can shift more easily among host species (Futuyma & Mitter 1996, Jermy & Szentesi 2003). Thus, many herbivores respond similarly to congeneric plants but discriminate allogeneric plants, whereas their responses to confamilial compared with allofamilial plants are less distinct (Novotny & Basset 2005).

The degree of similarity also differs among guilds, depending on their host specialization. It should be lower for specialists than for generalists and, therefore, lower for endophages than for ectophages. Frenzel & Brandl (2001) found that endophages were consistently less similar than ectophagous assemblages among different Brassicaceae but not among Cynaroideae hosts. Furthermore, overall herbivore assemblage similarity was substantially lower among host species in the Cynaroideae than in Brassicaceae. This example indicates the vast potential for studies of lineage-specific differences in herbivore assemblages across plant phylogenies (Futuyma & Mitter 1996).

Interactions: Competition and Facilitation

A number of early observational studies searched herbivore assemblages on host plants for evidence of resource partitioning, as expected from theory (Denno et al. 1995). Failure to find such evidence established herbivorous insects as prime evidence against the proposition of interspecific competition as a major organizing force in communities (Denno et al. 1995, Lawton & Strong 1981, Strong et al. 1984).

A reassessment of studies on interspecific competition among herbivorous insects (Denno et al. 1995) revealed that it was detected in 76% of the investigated pairwise interactions. It was less common only among external leaf-chewers— precisely the guild on which previous assessments and generalizations had been based (Lawton & Strong 1981).

The commonness of interspecific competition is not necessarily commensurate with its intensity or net effect (Denno et al. 1995). Thus, the effects of interspecific interactions on species richness and assemblage structure remain to be assessed. Indeed, analyses of co-occurrences and experimental studies both show a number of instances in which some herbivore species facilitate the presence of others by providing entry points, shelter, or otherwise modifying the host (Lill & Marquis 2003, Martinsen et al. 2000, Waltz & Whitham 1997).

Herbivore community studies can be further complicated by numerous indirect and diffuse effects (Strauss & Irwin 2004) mediated by their predators (Romero & Vasconcellos-Neto 2004) or shared parasitoids [apparent competition (e.g., Morris et al. 2004)]. Particularly in the tropics, ant-tending promotes the occurrence of certain sap-sucking and folivorous groups but reduces the presence of other external feeders (Dyer & Letourneau 1999). The herbivore guild composition is, thus, affected by ants, but we are not aware of any evaluation of their effect on herbivore species richness. Most studies of indirect interactions assess demographic responses of particular species (Strauss & Irwin 2004) rather than effects on community attributes.

Insect-Plant Arrays

The diversity of entire local plant–herbivore assemblages can be partitioned into within-host and among-host diversity (Lewinsohn et al. 2001, Summerville et al. 2003). Turnover among hosts in this case is an alternative sense of beta-diversity and an inverse measure of host specialization. Partitions of diversity within hosts, compared with diversity among host species, can be evaluated with ANOVA-like models (Lande 1996) or randomization procedures (Summerville et al. 2003). For caterpillars that feed on four temperate-forest tree species (three of them confamilial), beta-diversity among hosts was only occasionally different from chance expectations, because of low host specialization (Summerville et al. 2003), although differentiation among hosts increased seasonally.

Alternatively, comprehensive host–insect arrays can be analyzed as interaction matrices to ascertain properties such as nestedness or compartmentation, two ways in which such matrices can show nonrandom structure (Lewinsohn et al. 2005). Nestedness has long been probed in biogeographic studies and recently has been found to characterize many mutualistic assemblages (Jordano et al. 2005); however, it has scarcely been sought in plant-herbivore arrays. A nested structure means that specialists should accumulate on hosts with the most diverse assemblages, whereas hosts with poorer assemblages should only be associated with generalist herbivores (Lewinsohn et al. 2005); this situation could result, for instance, from source-sink dynamics among co-occurring host species.

Compartments are sets of densely linked plants and insects whose outer boundaries are set by evolutionary processes, whereas their inner patterns can reflect more immediate ecological conditions. Compartmentation has been predicted to be uncommon in food webs on theoretical grounds but was demonstrated through multivariate and randomization procedures in an assemblage of specialized endophages on closely related hosts (Prado & Lewinsohn 2004). In a tropical secondary forest, Novotny et al. (2004) showed that most Lepidoptera, although not strictly monophagous, concentrate locally on a single host, and that the entire assemblage is highly compartmented.

Assembly Rules for Herbivore Communities

Herbivore assemblages that feed on bracken fern on different continents (Lawton et al. 1993) are quite dissimilar in guild, niche, and taxonomic composition. Although comparable studies on other plants are required, this finding demonstrates the importance of regional herbivore species pools for the composition of local assemblages.

The predictability of the process of community assembly from a particular regional species pool remains controversial. In tropical forests, spatially and temporally replicated fogging samples of herbivorous assemblages, particularly of adult beetles, differ widely and unpredictably in composition on the same host-plant species or even individual (Floren & Linsenmair 1998, Mawdsley & Stork 1997). These assemblages may represent nonequilibrium stochastic communities; on the other hand, unpredictability can be a sampling artifact, derived from tourist species that do not feed on the sampled hosts and from the large numbers of rare species characteristic of tropical assemblages (Novotny & Basset 2000, Price et al. 1995). The latter alternative is supported by higher constancy in tropical vegetation of the species composition and abundance of locally common, feeding herbivores (Novotny et al. 2002b). Community assembly rules are amenable to experimental study in artificially defaunated vegetation (Floren & Linsenmair 1998).

PERSPECTIVES

New Theoretical Approaches

Various theoretical avenues may provide further understanding of the determinants of diversity and organization of herbivore assemblages. We deem the following especially worthy of attention.

MECHANISTIC MODELS   The distribution of plant species within extensive vegetation, such as lowland tropical forest, has been tested against predictions under the assumptions of uniform distribution, random but spatially autocorrelated distribution caused by dispersal limitation, and patchy, environmentally determined distribution (reviewed in Chave 2004). Such neutral models can be adapted to test spatial changes in herbivorous communities on particular plant species.

Straw & Ludlow (1994) proposed a mechanistic model in which herbivore richness can be derived from host abundance and availability, combined with insect-resource appropriation efficiencies. This model deserves further empirical tests, although the necessary data are laborious to obtain.

PARTITIONING HERBIVORE DIVERSITY   Partitioning of diversity into either spatial components or within-host and among-host components has been discussed before. Further advances are possible if both of these partitioning modes are combined into an integrated framework, outlined by Lewinsohn et al. (2001), that extends earlier hierarchical schemes (Routledge 1984; also see Couteron & Pélissier 2004). This framework can be envisioned as a three-way table whose entries are the herbivores found on a given host in a site, which then are aggregated by locality and by host; two turnover components (among-site and among-host beta-diversity), plus their potential interaction, are thus factored into total herbivore richness.

Diversity-partitioning models provide promising means of separating herbivore turnover among hosts from spatial turnover and, therefore, can be useful in clarifying the apportionment of total diversity. A comparison of this apportionment in different geographical regions or biomes would be of high interest. Components of diversity are also potentially useful in monitoring changes over time in systems of particular concern.

HERBIVORE ASSEMBLAGES AS METACOMMUNITIES   Extension of metapopulation theory to metacommunities is fairly incipient, but Hugueny & Cornell (2000) proposed a patch-occupancy model to predict local numbers of species drawn independently from a regional pool; this model agreed well with cynipid gallers on oaks. The herbivore assemblage on ragwort, *Senecio jacobaea* L., was studied experimentally in a metapopulation framework by Harrison et al. (1995), who found, however, that the theoretical assumptions were met only in part by the herbivores. Herbivore assemblages offer useful testing grounds for metacommunity theory and also for models of communities in spatially structured landscapes (see Tscharntke & Brandl 2004).

COMPLEX-NETWORK THEORY   The bloom of complex-network theory has had a strong impact on food-web studies. It has been fruitfully applied to the investigation of structural and dynamic properties of mutualistic assemblages, especially of plants with pollinators or frugivores (Jordano et al. 2005). Plant–herbivore assemblages can be similarly represented as bipartite networks; that is, as an array of interactions between elements of two distinct sets (Lewinsohn et al. 2005). Among properties of interest that can be explored through this approach are, for instance, the asymmetry of trophic-link distribution among species, and the effects of random or directed species loss on the structure and dynamics of the entire plant–herbivore assemblage.

PHYLOGENY AND COMMUNITY STRUCTURE   Recent exponential increase in molecular data and improved cladistic methods for their analysis provide rapidly expanding opportunities for community-wide studies that incorporate phylogeny (Webb et al. 2002). The introduction of a phylogenetic perspective is one of the most significant recent advances in community ecology, although it still needs specific models and agreement on procedures (Ricklefs 1996).

The relatedness of coexisting plant species and the distribution of their life-history traits have been compared through resampling procedures and yielded a null expectation based on random draws from their regional species pool (Webb et al. 2002). This approach holds promise for the study of herbivore species coexisting on a particular host species.

Long-Term and Spatially Extensive Studies

Some of the open questions on herbivore assemblages can only be answered with an increase of temporally and spatially extensive studies. Far too few geographically extensive studies of assemblages have been sampled in a large number of localities, and even fewer studies (e.g., Barbosa et al. 2000, Root & Cappuccino 1992) have followed local communities over several years. Both kinds of studies are essential to elucidate the organization of local communities and their relations to regional assemblages.

A potential opportunity for long-term studies is offered by permanent study plots established primarily to study vegetation structure and dynamics. The intercontinental network of 50-ha forest plots, whose vegetation is regularly censused (Losos & Leigh 2004), and extensive networks of replicated experiments on plant communities (e.g., Van der Putten et al. 2000), exemplify systems in which herbivore assemblages and their host associations could be recorded regularly to assess variation across seasons and years, track longer-term changes, or compare different geographic regions.

The major difficulty for sound cross-latitudinal comparisons is the lack of rigorously comparable data. Temperate field studies can resort to catalogs and atlases for complementary information on regional lists, host distribution, and herbivore-host ranges. Such information is usualy lacking for tropical areas. Conversely, extensive surveys of tropical communities (Janzen 1988, Novotny et al. 2002a) often do not have matching counterparts in temperate communities. Paradoxically, the best short-term strategy for broader comparisons, thus, seems to be the production of inventories in temperate regions with the same procedures and under the same restrictions that apply in tropical settings, rather than the reverse (Novotny & Basset 2005).

Response to Global Changes

Given that herbivorous insects are the largest single contingent of terrestrial biodiversity (Schoonhoven et al. 1998) and have a variety of effects on food plants, their response to global warming and other large-scale changes are of high interest (Wilf & Labandeira 1999). Paleontological evidence indicates that the intensity of herbivory as well as average per-host diversity of herbivores increased during the Cenozoic warming that peaked 53 mya (Wilf & Labandeira 1999). This finding raises the possibility that global warming may promote long-term increases in herbivore assemblages and in herbivory levels in temperate ecosystems. In tropical and subtropical regions, reverse effects are conceivable; more insects will be forced past their tolerance limits. In either case, effects will also depend on changes in humidity and precipitation. Other consequences of climate change, already detectable by now (Walther et al. 2002), include shifts in geographical distribution and in timing and synchronization of phenological events; all of these changes will necessarily affect herbivores both directly and by way of their host plants. Changes in local and regional assemblage size and structure are inescapable but harder to predict than those of single species or pairwise interactions, although at least as important. Few experiments (e.g., Hartley & Jones 2003) as yet attempt to assess effects of global changes on herbivore assemblages.

Labandeira et al. (2002) show that specialized herbivore-plant interactions were at greater risk and had lower rates of recovery than did generalist interactions from past catastrophic events. An overall loss of specialists and specialized links would simplify interaction webs, increase overlap among host assemblages, and diminish compartmentation. However, stochastic assortment and idiosyncratic local conditions could still enlarge differences among local assemblages and, therefore, increase the relative weight of beta-diversity in total diversity.

Global climate changes do not occur independently from other worldwide alterations, especially species introductions and landscape changes, such as fragmentation and habitat loss. Introductions are expected to homogenize biotas among world regions, so that total diversity would be reduced while local and regional diversity would increase (Rosenzweig 2001). However, habitat loss, especially in megadiverse tropical areas, is by far the major and most immediate menace to biotic diversity (Myers et al. 2000, Rosenzweig 2001).

CONCLUSIONS

In this review we have endeavored to highlight several themes on which the study of insect assemblages on plants made progress over the past two decades.

The description and understanding of herbivore assemblages at different spatial scales has seen substantial development. Beta-diversity is clearly a key component in the understanding of the spatial organization of such assemblages and is instrumental in the refinement of specific explanatory hypotheses and predictions.

Phylogenetic analyses have emerged as an important perspective for separation of the imprint of plant phylogenetic relationships from contemporaneous ecological effects and are successful in explaining the evolution of host affiliation in herbivorous insects. Similar progress may be expected in finding explanations for the assembly of herbivore communities on plants.

Finally, studies of tropical communities, particularly their host specificity and stratification in forests, have emerged as an active area of research.

To return then to our initial questions: Twenty years after the seminal book by Strong et al. (1984), are more or better data available for the same explanatory factors? Are models improved, and have noticeable theoretical advances occurred? What is the current contribution of plant–herbivore assemblages to the general understanding of terrestrial species diversity?

Most early studies made use of data obtained for purposes other than the questions they posed, so that testing the factors of interest with such data demanded substantial ingeniousness. Given the gradual shift to studies actually designed to investigate patterns of insect richness on plants, the same factors could be more rigorously evaluated; plant structural, historical, or distributional variables, although nominally the same, are in general more strictly defined and can thus be better assessed. Hence, with regard to our first question, there are now better data for the same models, and these data span a more diverse set of taxa, geographical regions, and biomes.

As for analytical models, prevailing practice has incorporated some improvements but made no major advances; thus far, few studies recognize or deal with structural or spatial correlation among explanatory variables. We have also indicated some theoretical directions, such as phylogenetic analysis or complex network theory, that can be expected to offer new insights but whose expanded application is still in process.

The current contribution of insect–herbivore assemblages to a general perspective of terrestrial diversity is not commensurate with their diversity or importance. We have seen fewer advances in the production of broader evaluations of proposed patterns, and few generalizations can be safely established. Although underlying theoretical impediments may exist, at this stage, generalizations are unquestionably limited by the extent and nature of the data at hand. In natural habitats, especially in the tropics, with few exceptions, only local data are available, and these data sets cover as yet far too few taxa, habitats, and localities, mostly in forest settings. For temperate areas, many studies are on smaller plants in open and modified habitats, combined with further explorations of extensive host catalogs and atlases. Our current version of comparing apples and oranges is comparing herbivore assemblages on collards or oaks in temperate settings with assemblages on fig trees in tropical rainforests.

Advances in the immediate future can be fostered in two ways: First, we need to increase the number of comprehensive studies of plant–insect arrays in which individual trophic interactions are tested and quantified. These studies will produce improved descriptions of the structure of herbivore assemblages and better tests of hypotheses on their spatial and functional organization. Second, we must have wide-ranging comparative studies. The broader questions on which this field of enquiry was inaugurated cannot be resolved without large-scale comparative and collaborative work.

ACKNOWLEDGMENTS

We thank Pedro Jordano, Donald Strong, and Howard Cornell for critically reviewing the manuscript. This review benefited from discussions with Martin Konvicka, John Lawton, Jan Leps, Scott Miller, Paulo Inácio Prado, George Weiblen, and many others. We acknowledge research funding by the Fundação de Amparo à Pesquisa do Estado de São Paulo (Biota/Fapesp 98/05085-2), Conselho Nacional de Densenvolvimento Científico e Tecnológico-Brasil (306049/2004-0), National Geographic Society (5398-94, 7659-04), U.S. National Science Foundation (DEB-02-11591), Czech Academy of Sciences (A6007106, Z5007907), Czech Ministry of Education (ME646), Czech Grant Agency (206/04/0725, 206/03/H034), and Darwin Initiative for the Survival of Species (162/10/030).

The *Annual Review of Ecology, Evolution, and Systematics* is online at
http://ecolsys.annualreviews.org

LITERATURE CITED

Abrahamson WG, Hunter MD, Melika G, Price PW. 2003. Cynipid gall-wasp communities correlate with oak chemistry. *J. Chem. Ecol.* 29:209–23

Andrew NR, Hughes L. 2004. Species diversity and structure of phytophagous beetle assemblages along a latitudinal gradient: predicting the potential impacts of climate change. *Ecol. Entomol.* 29:527–42

Barbosa P, Segarra A, Gross P. 2000. Structure of two macrolepidopteran assemblages on *Salix nigra* (Marsh) and *Acer negundo* L.: abundance, diversity, richness, and persistence of scarce species. *Ecol. Entomol.* 25:374–79

Barrios H. 2003. Insect herbivores feeding on conspecific seedlings and trees. See Basset et al. 2003, pp. 282–90

Basset Y. 1996. Local communities of arboreal herbivores in Papua New Guinea: predictors of insect variables. *Ecology* 77:1906–19

Basset Y. 2001a. Communities of insect herbivores foraging on saplings versus mature trees of *Pourouma bicolor* (Cecropiaceae) in Panama. *Oecologia* 129:253–60

Basset Y. 2001b. Invertebrates in the canopy of tropical rain forests: How much do we really know? *Plant Ecol.* 153:87–107

Basset Y, Novotny V. 1999. Species richness of insect herbivore communities on *Ficus* in Papua New Guinea. *Biol. J. Linn. Soc.* 67: 477–99

Basset Y, Novotny V, Miller SE, Kitching RL, eds. 2003. *Arthropods of Tropical Forests: Spatio-Temporal Dynamics and Resource Use in the Canopy.* Cambridge, UK: Cambridge Univ. Press. 474 pp.

Basset Y, Samuelson GA, Allison A, Miller SE. 1996. How many species of host-specific insects feed on a species of tropical tree?*Biol. J. Linn. Soc.* 59:201–16

Becerra JX. 1997. Insects on plants: macroevolutionary chemical trends in host use. *Science* 276:253–56

Berenbaum MR. 2001. Chemical mediation of coevolution: phylogenetic evidence for Apiaceae and associates. *Ann. Mo. Bot. Gard.* 88: 45–59

Blanche KR, Westoby M. 1996. The effect of the taxon and geographic range size of host eucalypt species on the species richness of gall-forming insects. *Aust. J. Ecol.* 21:332–35

Brändle M, Brandl R. 2001. Species richness of insects and mites on trees: expanding southwood. *J. Anim. Ecol.* 70:491–504

Brändle M, Rieger C. 1999. Die Wanzenfauna (Insecta, Heteroptera) von Kiefernstandorten (*Pinus sylvestris* L.) in Mitteleuropa. *Faun. Abh. Mus. Tierkde. Dresden* 21:239–58

Brehm G, Suessenbach D, Fiedler K. 2003. Unique elevational diversity patterns of geometrid moths in an Andean montane rainforest. *Ecography* 26:456–66

Briese DT, Sheppard AW, Zwölfer H, Boldt PE. 1994. Structure of the phytophagous insect fauna of *Onopordum* thistles in the northern Mediterranean basin. *Biol. J. Linn. Soc.* 53:231–53

Burki C, Nentwig W. 1997. Comparison of herbivore insect communities of *Heracleum sphondylium* and *H. mantegazzianum* in Switzerland (Spermatophyta: Apiaceae). *Entomol. Gen.* 22:147–55

Chave J. 2004. Neutral theory and community ecology. *Ecol. Lett.* 7:241–53

Cornell HV. 1985. Local and regional species richness of cynipine gall wasps on California oaks. *Ecology* 66:1247–60

Cornell HV, Kahn DM. 1989. Guild structure in the British arboreal arthropods: Is it stable and predictable?*J. Anim. Ecol.* 58:1003–20

Cornell HV, Karlson RH. 1996. Local and regional processes as controls on species richness. In *Spatial Ecology: The Role of Space in Population Dynamics and Species Interactions*, ed. D Tilman, P Kareiva, pp. 250–68. Princeton: Princeton Univ. Press

Cornell HV, Lawton JH. 1992. Species interactions, local and regional processes, and limits to the richness of ecological communities—a theoretical perspective. *J. Anim. Ecol.* 61:1–12

Couteron P, Pélissier R. 2004. Additive apportioning of species diversity: towards more sophisticated models and analyses. *Oikos* 107: 215–21

Cuevas-Reyes P, Quesada M, Hanson P, Dirzo R, Oyama K. 2004. Diversity of gall-inducing insects in a Mexican tropical dry forest: the importance of plant species richness, life-forms, host plant age and plant density. *J. Ecol.* 92:707–16

Cytrynowicz M. 1991. Resource size and predictability, and local herbivore richness in a subtropical Brazilian cerrado. See Price et al. 1991, pp. 561–89

Denno RF, McClure MS, Ott JR. 1995. Interspecific interactions in phytophagous insects: competition reexamined and resurrected. *Annu. Rev. Entomol.* 40:297–331

Dixon AFG, Kindlmann P, Leps J, Holman J. 1987. Why there are so few species of aphids, especially in the tropics? *Am. Nat.* 129:580–92

Dyer LA, Letourneau DK. 1999. Trophic cascades in a complex terrestrial community. *Proc. Natl. Acad. Sci. USA* 96:5072–76

Erwin TL. 1982. Tropical forests: their richness in Coleoptera and other arthropod species. *Coleopt. Bull.* 36:74–75

Ezcurra E, Gomez JC, Becerra J. 1987. Diverging patterns of host use by phytophagous insects in relation to leaf pubescence in *Arbutus xalapensis* (Ericaceae). *Oecologia* 72:479–80

Farrell BD. 1998. "Inordinate fondness" explained: Why are there so many beetles? *Science* 281:555–59

Fernandes GW, Price PW. 1991. Comparison of tropical and temperate galling species richness: the roles of environmental harshness and plant nutrient status. See Price et al. 1991, pp. 91–115

Floren A, Linsenmair KE. 1998. Non-equilibrium communities of Coleoptera in trees in a lowland rain forest of Borneo. *Ecotropica* 4:55–67

Fowler SV. 1985. Differences in insect species richness and faunal composition of birch seedlings, saplings and trees: the importance of plant architecture. *Ecol. Entomol.* 10:159–69

Fox LR, Morrow PA. 1981. Specialization: species property or local phenomenon? *Science* 211:887–93

Frenzel M, Brandl R. 1998. Diversity and composition of phytophagous insect guilds on Brassicaceae. *Oecologia* 113:391–99

Frenzel M, Brandl R. 2001. Hosts as habitats: faunal similarity of phytophagous insects between host plants. *Ecol. Entomol.* 26:594–601

Futuyma DJ, Mitter C. 1996. Insect-plant interactions: the evolution of component communities. *Philos. Trans. R. Soc. London Ser. B* 351:1361–66

Gaston KJ. 1992. Regional number of insects and plant species. *Funct. Ecol.* 6:243–47

Gaston KJ, Reavey D, Valladares GR. 1992. Intimacy and fidelity—internal and external feeding by the British microlepidoptera. *Ecol. Entomol.* 17:86–88

Godfray HCJ, Lewis OT, Memmott J. 1999. Studying insect diversity in the tropics. *Philos. Trans. R. Soc. London Ser. B* 354:1811–24

Gotelli NJ, Colwell RK. 2001. Quantifying biodiversity: procedures and pitfalls in the measurement and comparison of species richness. *Ecol. Lett.* 4:379–91

Harrison S, Thomas CD, Lewinsohn TM. 1995. Testing a metapopulation model of coexistence in the insect community on ragwort *Senecio jacobaea*. *Am. Nat.* 145:546–62

Hartley SE, Jones TH. 2003. Plant diversity and insect herbivores: effects of environmental change in contrasting model systems. *Oikos* 101:6–17

Hawkins BA, Porter EE. 2003. Does herbivore diversity depend on plant diversity? The case of California butterflies. *Am. Nat.* 161:40–49

Haysom KA, Coulson JC. 1998. The Lepidoptera fauna associated with *Calluna vulgaris*: effects of plant architecture on abundance and diversity. *Ecol. Entomol.* 23:377–85

Herrera CM, Pellmyr O, eds. 2002. *Plant-Animal Interactions: An Evolutionary Approach*. Oxford: Blackwell Sci. 313 pp.

Hodkinson ID, Casson D. 1991. A lesser predilection for bugs: Hemiptera (Insecta) diversity in tropical rain forests. *Biol. J. Linn. Soc.* 43:101–9

Hugueny B, Cornell HV. 2000. Predicting the relationship between local and regional species richness from a patch occupancy dynamics model. *J. Anim. Ecol.* 69:194–200

Janzen DH. 1968. Host plants as islands in evolutionary and contemporary time. *Am. Nat.* 102:592–95

Janzen DH. 1988. Ecological characterization of a Costa Rican dry forest caterpillar fauna. *Biotropica* 20:120–35

Jermy T, Szentesi A. 2003. Evolutionary aspects of host plant specialisation—a study on bruchids (Coleoptera: Bruchidae). *Oikos* 101:196–204

Jones CG, Lawton JH. 1991. Plant chemistry and insect species richness of British umbellifers. *J. Anim. Ecol.* 60:767–77

Jordano P, Bascompte J, Olesen JM. 2006. The ecological consequences of complex topology and nested structure in pollination webs. In *Specialization and Generalization in Plant-Pollinator Interactions*, ed. N Waser, J Ollerton. Chicago: Univ. Chicago Press. In press

Kato M, Takimura A, Kawakita A. 2003. An obligate pollination mutualism and reciprocal diversification in the tree genus Glochidion (Euphorbiaceae). *Proc. Natl. Acad. Sci. USA* 100:5264–67

Kelly CK, Southwood TRE. 1999. Species richness and resource availability: a phylogenetic analysis of insects associated with trees. *Proc. Natl. Acad. Sci. USA* 96:8013–16

Kennedy CEJ, Southwood TRE. 1984. The number of species of insects associated with British trees: a re-analysis. *J. Anim. Ecol.* 53:455–78

Kitching RL. 1993. Biodiversity and taxonomy: impediment or opportunity? In *Conservation Biology in Australia and Oceania*, ed. C Moritz, J Kikkawa, pp. 253–68. Chipping Norton: Surrey Beatty & Sons

Labandeira CC, Johnson KR, Wilf P. 2002. Impact of the terminal Cretaceous event on plant-insect associations. *Proc. Natl. Acad. Sci. USA* 99:2061–66

Lande R. 1996. Statistics and partitioning of species diversity, and similarity among multiple communities. *Oikos* 76:5–13

Lawton JH, Lewinsohn TM, Compton SG. 1993. Patterns of diversity for the insect herbivores on bracken. In *Species Diversity in Ecological Communities: Historical and Geographical Perspectives*, ed. RE Ricklefs, D Schluter, pp. 178–84. Chicago: Univ. Chicago Press

Lawton JH, Strong DR Jr. 1981. Community patterns and competition in folivorous insects. *Am. Nat.* 118:317–38

Leather SR. 1986. Insect species richness of the British Rosaceae: the importance of host range, plant architecture, age of establishment, taxonomic isolation and species-area relationships. *J. Anim. Ecol.* 55:841–60

Lees DC, Kremen C, Andriamampianina L. 1999. A null model for species richness gradients: bounded range overlap of butterflies and other rainforest endemics in Madagascar. *Biol. J. Linn. Soc.* 67:529–84

Leps J, Novotny V, Basset Y. 2001. Habitat and successional status of plants in relation to the communities of their leaf-chewing herbivores in Papua New Guinea. *J. Ecol.* 89:186–99

Lewinsohn TM. 1991. Insects in flower heads of Asteraceae in Southeast Brazil: a case study on tropical species richness. See Price et al. 1991, pp. 525–59

Lewinsohn TM, Jordano P, Prado PI, Olesen JM, Bascompte J. 2005. Structure in plant-animal assemblages. *Oikos* In press

Lewinsohn TM, Prado PIKL, Almeida AM. 2001. Inventários bióticos centrados em recursos: insetos fitófagos e plantas hospedeiras. In *Conservação Da Biodiversidade Em Ecossistemas Tropicais*, ed. I Garay, BFS Dias, pp. 174–89. Petrópolis: Ed. Vozes

Lill JT, Marquis RJ. 2003. Ecosystem engineering by caterpillars increases insect herbivore diversity on white oak. *Ecology* 84:682–90

Lill JT, Marquis RJ, Ricklefs RE. 2002. Host plants influence parasitism of forest caterpillars. *Nature* 417:170–73

Lopez-Vaamonde C, Godfray HCJ, Cook JM. 2003. Evolutionary dynamics of host-plant use in a genus of leaf-mining moths. *Evolution* 57:1804–21

Loreau M. 2000. Are communities saturated? On the relationship between α, β and γ diversity. *Ecol. Lett.* 3:73–76

Losos EC, Leigh EG Jr, eds. 2004. *Tropical Forest Diversity and Dynamism: Findings from A Large-Scale Plot Network.* Chicago: Univ. Chicago Press. 645 pp.

Losos JB. 1996. Phylogenetic perspectives on community ecology. *Ecology* 77:1344–54

Lowman MD, Taylor P, Block N. 1993. Vertical stratification of small mammals and insects in the canopy of a temperate deciduous forest: a reversal of tropical forest distribution? *Selbyana* 14:25

MacArthur RH, Wilson EO. 1967. *The Theory of Island Biogeography.* Princeton: Princeton Univ. Press. 203 pp.

Marquis RJ. 1991. Herbivore fauna of *Piper* (Piperaceae) in a Costa Rican wet forest: diversity, specificity and impact. See Price et al. 1991, pp. 179–208

Martinsen GD, Floate KD, Waltz AM, Wimp GM, Whitham TG. 2000. Positive interactions between leafrollers and other arthropods enhance biodiversity on hybrid cottonwoods. *Oecologia* 123:82–89

Mawdsley NA, Stork NE. 1997. Host-specificity and the effective specialization of tropical canopy beetles. See Stork et al. 1997, pp. 104–30

McCoy ED. 1990. The distribution of insects along elevational gradients. *Oikos* 58:313–22

Memmott J, Fowler SV, Paynter Q, Sheppard AW, Syrett P. 2000. The invertebrate fauna on broom, *Cytisus scoparius*, in two native and two exotic habitats. *Acta Oecol.* 21:213–22

Morris RJ, Lewis OT, Godfray HCJ. 2004. Experimental evidence for apparent competition in a tropical forest food web. *Nature* 428:310–13

Myers N, Mittermeier RA, Mittermeier CG, Fonseca GABD, Kent J. 2000. Biodiversity hotspots for conservation priorities. *Nature* 403:853–58

Novotny V. 1994. Association of polyphagy in leafhoppers (Auchenorrhyncha, Hemiptera) with unpredictable environments. *Oikos* 70:223–32

Novotny V, Basset Y. 2000. Rare species in communities of tropical insect herbivores: pondering the mystery of singletons. *Oikos* 89:564–72

Novotny V, Basset Y. 2005. Host specificity of insect herbivores in tropical forests. *Proc. R. Soc. London Ser. B* 272:1083–90

Novotny V, Basset Y, Miller SE, Weiblen GD, Bremer B, et al. 2002a. Low host specificity

of herbivorous insects in a tropical forest. *Nature* 416:841–44

Novotny V, Miller S, Leps J, Basset Y, Bito D, et al. 2004. No tree an island: the plant-caterpillar food web of a secondary rain forest in New Guinea. *Ecol. Lett.* 7:1090–100

Novotny V, Miller SE, Basset Y, Cizek L, Darrow K, et al. 2005. An altitudinal comparison of caterpillar (Lepidoptera) assemblages on *Ficus* trees in Papua New Guinea. *J. Biogeogr.* 32:1303–14

Novotny V, Miller SE, Basset Y, Cizek L, Drozd P, et al. 2002b. Predictably simple: assemblages of caterpillars (Lepidoptera) feeding on rainforest trees in Papua New Guinea. *Proc. R. Soc. London Ser. B* 269:2337–44

Novotny V, Miller SE, Cizek L, Leps J, Janda M, et al. 2003. Colonising aliens: caterpillars (Lepidoptera) feeding on *Piper aduncum* and *P. umbellatum* in rainforests of Papua New Guinea. *Ecol. Entomol.* 28:704–16

Ødegaard F. 2000. The relative importance of trees versus lianas as hosts for phytophagous beetles (Coleoptera) in tropical forests. *J. Biogeogr.* 27:283–96

Ødegaard F, Diserud OH, Engen S, Aagaard K. 2000. The magnitude of local host specificity for phytophagous insects and its implications for estimates of global species richness. *Conserv. Biol.* 14:1182–86

Orr AG, Häuser CL. 1996. Temporal and spatial patterns of butterfly diversity in a lowland tropical rainforest. In *Tropical Rainforest Research—Current Issues. Proc. Conf. Bandar Seri Begawan, 1993*, ed. DS Edwards, WE Booth, SC Choy, pp. 125–38. Dordrecht: Kluwer Acad.

Peeters PJ. 2002. Correlations between leaf structural traits and the densities of herbivorous insect guilds. *Biol. J. Linn. Soc.* 77:43–65

Prado PI, Lewinsohn TM. 2004. Compartments in insect-plant associations and their consequences for community structure. *J. Anim. Ecol.* 73:1168–78

Price PW, Diniz IR, Morais HC, Marques ESA. 1995. The abundance of insect herbivore species in the tropics: the high local richness of rare species. *Biotropica* 27:468–78

Price PW, Fernandes GW, Lara ACF, Brawn J, Barrios H, et al. 1998. Global patterns in local number of insect galling species. *J. Biogeogr.* 25:581–91

Price PW, Lewinsohn TM, Fernandes GW, Benson WW, eds. 1991. *Plant-Animal Interactions: Evolutionary Ecology in Tropical and Temperate Regions.* New York: John Wiley. 639 pp.

Ricklefs RE. 1996. Phylogeny and ecology. *Trends Ecol. Evol.* 11:229–30

Ricklefs RE. 2004. A comprehensive framework for global patterns in biodiversity. *Ecol. Lett.* 7:1–15

Romero GQ, Vasconcellos-Neto J. 2004. Beneficial effects of flower-dwelling predators on their host plant. *Ecology* 85:446–57

Root RB, Cappuccino N. 1992. Patterns in population change and the organization of the insect community associated with goldenrod. *Ecol. Monogr.* 62:393–420

Rosenzweig ML. 2001. The four questions: What does the introduction of exotic species do to diversity? *Evol. Ecol. Res.* 3:361–67

Routledge RD. 1984. Estimating ecological components of diversity. *Oikos* 42:23–29

Schoonhoven LM, Jermy T, Van Loon JJA. 1998. *Insect-Plant Biology: From Physiology to Evolution.* London: Chapman & Hall. 409 pp.

Scriber JM. 1988. Tale of the tiger: Beringial biogeography, binomial classification, and breakfast choices in the *Papilio glaucus* complex of butterflies. In *Chemical Mediation of Coevolution*, ed. KC Spencer, pp. 241–301. New York: Academic

Sobhian R, Zwölfer H. 1985. Phytophagous insect species associated with flower heads of yellow starthistle (*Centaurea solstitialis* L.). *Z. Angew. Entomol.* 99:301–21

Southwood TRE. 1961. The number of species of insect associated with various trees. *J. Anim. Ecol.* 30:1–8

Stork NE, Adis J, Didham RK, eds. 1997. *Canopy Arthropods.* London: Chapman & Hall. 567 pp.

Strauss SY, Irwin RE. 2004. Ecological and evolutionary consequences of multispecies plant-animal interactions. *Annu. Rev. Ecol. Evol. Syst.* 35:435–66

Straw NA, Ludlow AR. 1994. Small-scale dynamics and insect diversity on plants. *Oikos* 71:188–92

Strong DR Jr, Lawton JH, Southwood TRE. 1984. *Insects on Plants: Community Patterns and Mechanisms.* Oxford: Blackwell Sci. 313 pp.

Strong DR Jr, McCoy ED, Rey JR. 1977. Time and the number of herbivore species: the pests of sugarcane. *Ecology* 58:167–75

Summerville KS, Crist TO, Kahn JK, Gering JC. 2003. Community structure of arboreal caterpillars within and among four tree species of the eastern deciduous forest. *Ecol. Entomol.* 28:747–57

Sword GA, Dopman EB. 1999. Developmental specialization and geographic structure of host plant use in a polyphagous grasshopper, *Schistocerca emarginata* (=*lineata*) (Orthoptera: Acrididae). *Oecologia* 120:437–45

Symons FB, Beccaloni GW. 1999. Phylogenetic indices for measuring the diet breadths of phytophagous insects. *Oecologia* 119:427–34

Thomas CD. 1990. Fewer species. *Nature* 347:237

Thompson JN. 1999. The evolution of species interactions. *Science* 284:2116–18

Tscharntke T, Brandl R. 2004. Plant-insect interactions in fragmented landscapes. *Annu. Rev. Entomol.* 49:405–30

Tscharntke T, Vidal S, Hawkins BA. 2001. Parasitoids of grass-feeding chalcid wasps: a comparison of German and British communities. *Oecologia* 129:445–51

Van der Putten WH, Mortimer SR, Hedlund K, Van Dijk C, Brown VK, et al. 2000. Plant species diversity as a driver of early succession in abandoned fields: a multi-size approach. *Oecologia* 124:91–99

Walther GR, Post E, Convey P, Menzel A, Parmesan C, et al. 2002. Ecological responses to recent climate change. *Nature* 416:389–95

Waltz AM, Whitham TG. 1997. Plant development affects arthropod communities: opposing impacts of species removal. *Ecology* 78:2133–44

Webb CO, Ackerly DD, McPeek MA, Donoghue MJ. 2002. Phylogenies and community ecology. *Annu. Rev. Ecol. Syst.* 33:475–505

Weiblen GD, Bush GL. 2002. Speciation in fig pollinators and parasites. *Mol. Ecol.* 11:1573–78

Weiblen GD, Webb CO, Novotny V, Basset Y, Miller SE. 2005. Phylogenetic dispersion of host use in a tropical insect herbivore community. *Ecology*. In Press

Wilf P, Labandeira CC. 1999. Response of plant-insect associations to Paleocene-Eocene warming. *Science* 284:2153–56

Wright MG, Samways MJ. 1998. Insect species richness tracking plant species richness in a diverse flora: gall-insects in the Cape Floristic Region, South Africa. *Oecologia* 115:427–33

Young AM. 1984. On the evolution of Cicada x host-tree associations in Central America. *Acta Biotheor.* 33:163–98

Zwölfer H. 1987. Species richness, species packing, and evolution in insect-plant systems. *Ecol. Stud.* 61:301–19

Zwölfer H. 1988. Evolutionary and ecological relationships of the insect fauna of thistles. *Annu. Rev. Entomol.* 33:103–22

Annu. Rev. Ecol. Evol. Syst. 2005. 36:621–42
doi: 10.1146/annurev.ecolsys.36.091704.175513
Copyright © 2005 by Annual Reviews. All rights reserved
First published online as a Review in Advance on August 19, 2005

THE POPULATION BIOLOGY OF MITOCHONDRIAL DNA AND ITS PHYLOGENETIC IMPLICATIONS

J. William O. Ballard[1] and David M. Rand[2]

[1] Ramaciotti Centre for Gene Function Analysis, School of Biotechnology and
Biomolecular Sciences, University of New South Wales, Sydney 2052, Australia;
email: w.ballard@unsw.edu.au
[2] Department of Ecology and Evolutionary Biology, Brown University, Providence,
Rhode Island 02912; email: david.rand@brown.edu

Key Words   fitness, neutrality, phylogeny, selection

■ **Abstract**   The reconstruction of evolutionary trees from mitochondrial DNA (mtDNA) data is a common tool with which to infer the relationships of living organisms. The wide use of mtDNA stems from the ease of getting new sequence data for a set of orthologus genes and from the availability of many existing mtDNA sequences for a wide array of species. In this review we argue that developing a fuller understanding of the biology of mitochondria is essential for the rigorous application of mtDNA to inferences about the evolutionary history of species or populations. Though much progress has been made in understanding the parameters that shape the evolution of mitochondria and mtDNA, many questions still remain, and a better understanding of the role this organelle plays in regulating organismal fitness is becoming increasingly critical for accurate phylogeny reconstruction. In population biology, the limited information content of one nonrecombining genetic marker can compromise evolutionary inference, and the effects of nuclear genetic variation—and environmental factors—in mtDNA fitness differences can compound these problems. In systematics, the limited gene set, biased amino acid composition, and problems of compensatory substitutions can cloud phylogenetic signal. Dissecting the functional bases of these biases offers both challenges and opportunities in comparative biology.

INTRODUCTION

Mitochondria are thought to have originated from a free-living, aerobic, and motile α-proteobacteria containing 3000–5000 genes (Boussau et al. 2004). In animals, mitochondria are controlled by a dual genome system, with cooperation between endogenous mitochondrial genes and nuclear-encoded genes with two origins, 1) mitochondrial genes translocated to the nucleus over the course of evolution, and 2) nuclear genes that have acquired targeting signals and been recruited to derived functions in the mitochondrion (Rand et al. 2004). Mitochondrial genomes show varying degrees of reduction, ranging from 97 protein coding genes of the protozoan *Reclinomas americana* (Berg & Kurland 2000) to a mere 3 protein coding genes in the malarial parasite *Plasmodium falciparum*. Despite the enormous variations in size, the coding function of the mitochondrial genome remains relatively stable in animals. In general, mitochondrial DNAs (mtDNAs) code only for genes involved in the mitochondrial translation apparatus, electron transport, and oxidative phosphorylation.

Mitochondria oxidize metabolic substrates, including carbohydrates and fats, to generate water and ATP, with O_2 acting as the terminal electron acceptor for the electron transport chains that generate the proton gradient across the inner mitochondrial membrane. Reducing equivalents in the form of electron donors are recovered from carbohydrates in the tricarboxylic acid cycle, whereas those recovered from fats are obtained through β-oxidation. The resulting electrons are transferred to the mitochondrial electron transport chains via complex I (about 43 nuclear-encoded and 7 mtDNA-encoded loci) or complex II (4 nuclear-encoded loci) and then flow to ubiquinone. Ubiquinone transfers electrons to cytochrome c and eventually to complex IV (9 to 10 nuclear- and 3 mitochondrial-encoded loci), where 4 single-electron transfers to oxygen result in the formation of water. The energy released by the electron transport chain is used to pump protons out of the inner mitochondrial membrane, creating the transmembrane electron gradient. The potential energy stored in the gradient is used to condense ADP and Pi to make ATP via complex V (14 nuclear- and 2 mitochondrial-encoded loci). Generation of reactive oxygen species (ROS), such as superoxide, hydrogen peroxide, and organic hydroperoxides, at complexes I and III is believed to be important in aging with up to 1% to 2% of the oxygen consumed being converted to ROS. These ROS can damage DNA, lipids, and proteins, causing mitochondrial ROS production to increase with age.

In 1987, John Avise et al. illustrated how mtDNA can be used to bridge the gap between population genetics and systematics (Avise et al. 1987). This classic review considered many issues, including the impact of selection versus neutrality, heteroplasmy (the situation in which, within a single cell, there is a mixture of different mtDNA haplotypes), homoplasmy (resemblance due to parallelism or convergent evolution rather than to common ancestry), scale, and lineage sampling bias on phylogenetic hypotheses generated from mtDNA. Over the past two decades, a number of new issues have arisen while some debates have continued. One of the most important issues to have arisen is the influence of nuclear mitochondrial pseudogenes on inferences of population history (Bensasson et al. 2001, Thalmann et al. 2004). One dispute that continues is the selection versus neutrality debate. The selection versus neutrality debate is a focus of this review because it concerns the basic biology of mitochondria and mtDNA, and it has phylogenetic implications. We agree with Avise et al. that "the phylogenetic value of mtDNA does not ... completely hinge on the outcome [of this debate]." We do, however, propose that it can no longer be assumed that mtDNA is evolving in a manner consistent with a strictly neutral equilibrium model. In this review, we present examples showing that mtDNA variation influences organismal fitness. We then consider factors that may bias phylogenetic inference. Finally, we propose some statistical tests that should be routinely included in phylogenetic studies, which would help researchers determine the robustness of the genealogical hypotheses they have generated from mtDNA.

mtDNA AS A NEUTRAL MARKER

For practical reasons, mtDNA is widely used because it provides easy access to an orthologous set of genes with little or no recombination and rapid evolution. From a theoretical perspective, mtDNA is widely used in population genetic, phylogeographical, and phylogenetic studies owing to the belief that haplotype frequencies are governed primarily by migration and genetic drift and that most of the variation within a species is selectively neutral. Indeed, the majority of studies have assumed that mtDNA is evolving as a strictly neutral marker, and researchers have employed it to estimate a range of evolutionary and ecological parameters of interest such as divergence times and phylogeographic patterns. Ballard & Kreitman (1995) reviewed the literature and concluded that the widespread acceptance of the selective neutrality of mtDNA follows from a series of plausibility arguments connecting features of mtDNA evolution with (mis)conceptions of neutral theory. Deviations from a strictly neutral model of evolution have been found in a variety of organisms (Nachman 1998, Rand 2001, Rand & Kann 1998).

A priori, we suggest it is reasonable to predict that mtDNA variation may be under strong selection for three main reasons. First, the mitochondrion is the powerhouse of the cell and, in most organisms, a reduction in ATP production is expected to reduce fecundity. In humans, a reduction in the efficiency of ATP production is known to be highly deleterious and lethal in the extreme case. Second, proteins from mtDNA interact with those imported from the nuclear genome to form four of the five complexes of the electron transport chain. Third, the lack of normal recombination in mtDNA means that each genome has a single genealogical history and all genes will share that history. Any evolutionary force acting at any one site will equally affect the history of the whole molecule. Thus, the fixation of an advantageous mutation by selection, for example, will cause the fixation of all other polymorphisms by a process known as genetic hitchhiking (Maynard Smith & Haigh 1974). Even the quickly evolving noncoding origin-of-replication region cannot be assumed to have neutral allele frequencies: It is linked to the rest of the genome where selection has been documented, and conserved motifs within this region exhibit variation that affects mitochondrial transcription and replication in significant ways (Coskun et al. 2004). Alternatively, polymorphism within a mitochondrial genome may be depressed through selection against linked deleterious mutations, a process known as background selection (Charlesworth 1994, Charlesworth et al. 1993, 1995). We suggest that the hypothesis that mtDNA is under strong selection is rarely explored, and selective neutrality is assumed because there is a perceived lack of evidence suggesting that different mtDNA types within a species have unequal fitness (Ballard & Whitlock 2004). In the following paragraphs, we review the evidence and conclude that different mtDNA haplotypes (sometimes called mitotypes) can have a significant influence on organismal fitness in a wide variety of species including humans.

The direct impact of mtDNA variation on fitness has been measured in humans (Ruiz-Pesini et al. 2000), mice (Roubertoux et al. 2003, Takeda et al. 2000), *Drosophila* (Ballard 2004; de Stordeur 1997; Fos et al. 1990; Hutter & Rand 1995; James & Ballard 2003; Kilpatrick & Rand 1995; Nigro 1991, 1994; Rand et al. 2001), and copepods (Schizas et al. 2001). In humans, there is increasing evidence showing that human sperm motility is strongly dependent on the ATP supplied by oxidative phosphorylation (Ruiz-Pesini et al. 1998). The frequency of the known pathological mtDNA mutations within humans is insufficient to explain a significant proportion of the asthenozoospermic (reduced sperm motility) patients. As a consequence, it was proposed that mtDNA mutations affecting mitochondrial ATP production could cause reduced sperm motility. Ruiz-Pesini et al. (2000) analyzed the distribution of mtDNA in Caucasian men having fertility problems and found that the sperm-motility phenotype was indeed conditioned by the mtDNA type. mtDNA is maternally inherited and is not expected to suffer strong selective pressure in males as it is an evolutionary dead end (Frank & Hurst 1996). We suggest that this striking result illustrates an underappreciated feature of mtDNA evolution. If sperm dysfunction (or any male specific effect) is the main, or the only, phenotypic consequence of a mtDNA mutation, specific mutation(s) could accumulate within a population and reach high frequency.

In a somewhat controversial and highly publicized study Roubertoux et al. (2003) showed that mtDNA influenced learning, exploration, sensory development, and the anatomy of the brain in mice. The effects of the mtDNA type persisted with age, increasing in magnitude as the mice got older. To complete this study, the authors were very careful to develop a well-controlled set of strains to effectively determine true mtDNA contributions as well as interactions with nuclear DNA. The experimental design excludes both the effects of genomic imprinting and the influences of the cytoplasmic and maternal environment of a parental strain on its mtDNA congenic strain.

A body of research shows that the three distinct and geographically subdivided mtDNA types of the fly *Drosophila simulans* (*si*I, -II, and -III) have significantly different effects on organismal fitness (Ballard 2004, Solignac et al. 1986). de Stordeur (1997) conducted microinjection transfection studies between eggs carrying the mtDNA types and assayed the frequencies of the foreign injected mtDNA in heteroplasmic strains. He observed that the mtDNA types have unequal fitness within cells during transmission between generations. James & Ballard (2003) controlled the nuclear genome of flies by backcrossing and observed that the mtDNA haplotype influenced physical activity, development time, and longevity. Ballard & James (2004) observed that the relative fitness of the three mtDNA types in perturbation cages was positively correlated with the observed worldwide distribution of the mitotypes (*si*II > -III > -I). However, it was also clear that mitochondrial-nuclear (hereafter mitonuclear) interactions also influenced the fitness of flies in population cages (Ballard & James 2004). Mitonuclear interactions have also been shown to influence the frequencies of flies in cages and in biochemical assays of mitochondrial metabolism. Nigro (1991) employed wild-caught and microinjected lines to show that both the mtDNA and nuclear background influenced the competitive ability of flies in population cages. Mitonuclear interactions have been shown to influence cytochrome c oxidase (complex IV) activity in wild-type and introgressed *D. simulans* (Sackton et al. 2003).

Mitonuclear Interactions

Mitochondrial fitness effects may be conferred directly by the mitochondrial genotype, nuclear-encoded loci-producing proteins imported into the mitochondrion, and/or coadapted mitonuclear gene complexes. It has been suggested that this is a potential explanation for the basic conundrum, "Why are mitochondria maternally inherited?" Ross (2004) proposes that the specific degradation of the paternal mitochondrial genome could have been selected to prevent competition between mtDNA and nuclear DNA gene products. Minor mutations in either mtDNA or nuclear DNA coding for proteins essential for oxidative phosphorylation are known to lead to major and catastrophic diseases of humans, suggesting that very tight and precise interactions are required. Most often, paternal mtDNA within a species is quickly degraded following recognition of a ubiquitin tag (Sutovsky et al. 1999), but recognition of paternal mtDNA apparently depends on phylogenetic relatedness. It seems that when the paternal mtDNA is more than about 2.5% divergent from the maternal mtDNA the paternal mtDNA is not recognized and excluded. Under such conditions high rates of heteroplasmy are observed (Kaneda et al. 1995, Kondo et al. 1990, Satta et al. 1988). It remains to be determined whether diverged mitochondria simply escape tagging or the hybrid nuclear genomes from crosses between divergent species are compromised in the tagging process.

The evolutionary forces responsible for the movement of genes from the mitochondrion to the nucleus is better understood in plants than in animals (Adams et al. 2002, Bergthorsson et al. 2003). However, the physical proximity of mtDNA to sites where ROS are produced may be a major reason for the selective pressure reinforcing transfer of genes to the nuclear genome (ROS quickly damages mtDNA and this may cause a rapid decline in metabolic efficiency and fecundity). Certainly, the movement of genes from the mitochondrial to the nuclear genome places the gene in a very different chromosomal context, including differences in the genetic code and codon bias. As a consequence of this export, it may be expected that genes may show a burst of evolution or acceleration in the rate of evolution following transfer. Unfortunately, this may be very difficult to detect because there is likely to be overlap in the functioning of the mitochondrial- and the nuclear-encoded genes. From a phylogenetic perspective, researchers must be careful in the assigning of homology in such cases.

The movement of genes from the mitochondrial to the nuclear genome is documented both between and within species. Different species of *Rickettsia* show varying degrees of "mutational meltdown," suggesting that movement of genes from the mitochondrion to the nucleus is an ongoing process (Andersson & Andersson 1999, Andersson et al. 1998). A result of the export of DNA from the mitochondrion is that nuclear-encoded proteins must be reimported into the mitochondrion to ensure successful mitochondrial metabolism. Zeviani et. al. (1999) grouped nuclear-encoded proteins, which influence the structure and function of mitochondria, into three categories: structural components of the electron transport chain, factors influencing the structural integrity or copy number of mtDNA, and proteins that control the formation, assembly, and turnover of respiratory complexes.

Mitonuclear interactions influence complexes I, III, IV, and V of the electron transport chain. Schmidt et al. (2001) studied the functional interactions between mitochondrial and nuclear-encoded proteins in the multisubunit respiratory complex cytochrome c oxidase (complex IV) in six species of mammals, using chickens as an outgroup. In this complex, mtDNA-encoded residues in physical proximity to nuclear DNA-encoded residues evolved more rapidly than the other mtDNA-encoded residues, indicating that mitonuclear interactions can alter rates of amino acid substitutions. The complexity of these mitonuclear gene interactions is compounded by the fact that genes from both genomes show significant sequence polymorphism. On average, two unrelated humans differ at more than 50 nucleotide polymorphisms in their mtDNAs, about 20–30 of which lead to amino acid changes (Ingman et al. 2000, Rand & Kann 1996). Two humans also differ at ∼1 million polymorphic nucleotide sites across the nuclear genome, several thousand of which will alter a protein sequence (Sachidanandam et al. 2001).

Mitonuclear interactions influence a variety of biochemical and physiological processes and these may influence the pattern of evolution shown by specific genes. Kern & Kondrashov (2004) compared 86 pathogenic mutations in human tRNAs, encoded by mitochondrial genes, to the sequences of their mammalian orthologs and noted that 52 pathogenic mutations were present in normal tRNAs of one or several nonhuman mammals. The authors proposed that a pathogenic mutation and its compensating substitution are fixed in a lineage in rapid succession. At least 10%, and perhaps as many as 50%, of all nucleotide substitutions in evolving mammalian tRNAs participate in such interactions, indicating that the evolution of tRNAs proceeds along highly epistatic fitness ridges. Because mitochondrial translation requires many proteins encoded in the nucleus, a greater understanding of these processes will enhance our ability to interpret selection on tRNAs. This, in turn, will facilitate accurate determination of the models that are most appropriate for phylogenetic analyses.

Mitonuclear interactions have also been shown to influence organismal fitness, and this influences our interpretation of the selective neutrality of mtDNA. The first experiments considering the coevolution of nuclear and mitochondrial genomes were reported in 1971 (Clayton et al. 1971). In 1997, elongated cells from hominoid apes (chimpanzee, pigmy chimpanzee, gorilla, and orangutan) were fused with mtDNA-less human cells (Kenyon & Moraes 1997), thereby creating cells with ape mtDNA and human nuclear DNA. Only the combinations of human nuclear DNA with mitochondria from the most closely related species (chimpanzee, gorilla) yielded cells even capable of oxidative phosphorylation.

More recently, it was shown that mice with introgressed interspecific and intersubspecific mtDNA exhibit reduced physical performance (Nagao et al. 1998). Nagao and colleagues backcrossed *Mus spretus* and *M. musculus* mtDNA into *M. domesticus*. These represent the conditions of interspecific and intersubspecific mitonuclear mismatch, respectively. Using these backcross mice, they examined physical performance by measuring running time on a treadmill until exhaustion. The result clearly showed that mtDNA backcross lines manifested a significant decrease in their physical performance compared to their progenitor lines. In the marine copepod, *Tigriopus californicus*, Rawson & Burton (2002) found that the cytochrome c variants isolated from two different populations each had significantly higher activity with the cytochrome c oxidase derived from their respective source population. Three amino acid substitutions in the cytochrome c protein appear to be sufficient to confer population specificity. These results suggest that electron transport chain proteins form coadapted sets of alleles within populations and that disruption of the coadapted gene complex leads to functional incompatibilities that may lower hybrid fitness. Mitonuclear coadaptation not only is an interesting feature of many species in nature, but can also be observed to evolve within only 2000 generations under replicated laboratory conditions. Experimental evolution in yeast populations has shown that the competitive ability of evolved strains relative to ancestral strains is governed by mitonuclear epistatic interactions (Zeyl et al. 2005).

In humans, the same mtDNA mutations may induce longevity or diseases, depending on their interactions with nuclear loci. De Benedictis et al. (2000) observed that Italian male centenarians had a significantly higher frequency of the European mitotype J than sex-matched younger subjects having the same ethnic and geographic origin. This result suggests mtDNA-specific effects on the rate and quality of aging. Somewhat surprisingly, however, complete mtDNA sequencing demonstrated that the J haplogroup is characterized by a particular suite of six mutations often associated with disease (Rose et al. 2001). From these data it has been argued that key mtDNA mutations may induce death or extend life span depending on other mtDNA mutations and on stoichiometric mismatches with nuclear-encoded proteins in each oxidative phosphorylation subunit.

SYSTEMATIC BIASES IN EMPLOYING mtDNA AS AN EVOLUTIONARY MARKER

In this section, we consider how phylogenetic hypotheses based on mtDNA may be systematically biased. Most obviously, this could occur when limited sampling occurs in a species harboring multiple mtDNA types (Ballard 2000c). This may be more widespread and difficult to detect than previously believed. In the copepod

T. californicus there are at least nine haplogroups with up to 23% nucleotide divergences and up to 3% amino acid divergences (Edmands 2001). It has recently been shown that slightly deleterious mutations segregating within haplogroups may be removed by selection prior to their fixation among haplogroups (Dean & Ballard 2005). This results in distinct intraspecific mtDNA lineages behaving more like lineages between species.

One strong candidate for a type of selection that may cause population subdivision is thermal adaptation (Ballard & Whitlock 2004). This may result in the clustering of species that exist, or existed, in similar climates. Given the potential for temperature variation across species' ranges in nature and the sympatric distributions of closely related taxa, temperature may also play a strong role in selecting for introgression of alien mtDNA from locally better-adapted species (Ballard & Whitlock 2004). A second candidate that may cause population subdivision is infection with a maternally inherited symbiont. In this review, we focus on *Wolbachia,* as it has been shown to influence mitochondrial evolution in a variety of arthropods (Karr & Ballard 2005).

Thermal Adaptation

The potential for temperature to influence the evolution of mtDNA was the focus of much attention a decade ago (Martin & Palumbi 1993, Martin et al. 1992, Rand 1993, 1994). Martin et al. (1992) reported that the nucleotide substitution rates in the cytochrome b and cytochrome oxidase I genes in sharks are seven- to eightfold slower than in primates or ungulates. In the following year, Martin & Palumbi (1993) show that, in general, exothermic vertebrates have slower mtDNA substitution rates overall than do endotherms of similar size. Such differences in mtDNA substitution rates suggest that the thermal environment may influence rates of evolution and indicate that it is inappropriate to use a calibration for one group to estimate divergence times or demographic parameters for another group.

Temperature adaptation has been shown to be important in arctic fishes and other species (Somero 2002, Sommer & Portner 2002), but the role of mitochondrial variation in this adaptation is little investigated. In *Drosophila*, the fitness of the mtDNA haplogroups appears to be temperature dependent. In a series of papers Matsuura and colleagues (Matsuura et al. 1993, Nagata & Matsuura 1991) have systematically examined the transmission rates of *Drosophila* mtDNA haplogroups in flies made heteroplasmic by microinjection. The most recent paper showed that the nuclear genome is involved in determining the temperature dependency of mtDNA transmission (Doi et al. 1999).

It has also been argued that thermal adaptation may occur within humans. Ruiz-Pesini et al. (2004) conducted a phylogenetic study including 1125 globally distributed human mtDNA sequences and observed that the relative frequency and amino acid conservation of internal branch amino acid mutations increased from tropical Africa to temperate Europe to arctic northeastern Siberia. Highly conserved amino acid changes were found at the roots of multiple mtDNA lineages from higher latitudes prompting the authors to suggest that specific mtDNA non-synonymous mutations permitted our ancestors to adapt to more northern climates because their mitochondria produced more heat. However, Elson et al. (2004) analyzed complete mtDNA coding-region sequences for 560 maternally unrelated individuals of European, African, and Asian descent and were not able to replicate the results of Ruiz-Pesini et al. (2004). Elson et al. concluded that appropriate methodology with which to study climatic adaptation and the development of alternative methods is a goal for ongoing research.

A question for future research is, "How frequently has thermal adaptation within and among species affected mtDNA variation?" We suggest that selection is a viable alternative in two classic cases where mtDNA has been used as a phylogeographic marker. Populations of the American oyster from the Gulf of Mexico and the Atlantic coasts of the southern United States have a dramatic mtDNA discontinuity on the Florida panhandle. However, surveys of polymorphic allozymes reveal near uniformity of allele frequencies throughout the range (Reeb & Avise 1990). Subsequent surveys of four nuclear restriction fragment length polymorphisms (RFLPs) tended to support the Atlantic/Gulf mtDNA dichotomy (Karl & Avise 1992), although the pattern of variation of only one nuclear locus occurred in the same region of the coastline as the mtDNA discontinuity. Karl & Avise (1992) considered a variety of alternatives to reconcile these data and suggested that it is most likely that the allozyme loci are under balancing selection. One clear alternative is that the mtDNA is under strong thermal selection associated with the different temperature between Atlantic and Gulf waters.

A second example comes from the killifish. González-Villaseñor & Powers (1990) demonstrated that the distribution of mtDNA RFLP polymorphism among populations of the killifish showed a marked disjunction between 39.7°N to 40.7°N in northeastern North America. One mtDNA type was fixed in the north and another in the south. To investigate this result Ropson et al. (1990) examined the geographical variation in 15 nuclear loci. Four showed a pattern of variation similar to the mtDNA. However, in no case was the disjunction in the same geographic region as that observed for the mtDNA. Ropson et al. consider a variety of alternatives "in the absence of selection" but never fully explore the possibility that selection may operate on the mtDNA itself.

Wolbachia

In arthropods, infectious microorganisms like the bacterium *Wolbachia* are widespread (Werren et al. 1995). *Wolbachia*-induced incompatibility has been shown to cause the symbiont and the linked, maternally inherited mitochondrial genotype to rise in frequency in theory (Caspari & Watson 1959), in population cages (Kambhampati et al. 1992, Nigro & Prout 1990), and in nature (Turelli & Hoffmann 1991). For mtDNA, the key theoretical result is that, as the maternally inherited infection spreads, it carries along whatever mtDNA genotype was initially associated with it (Turelli et al. 1992). This horizontal spread of an infectious organism throughout a group is a source of discrepancy between mtDNA-based and nuclear gene-based phylogenies. In *D. simulans*, the genealogy of mtDNA reflects the spread of *Wolbachia* rather than the relatedness of populations as assessed by nuclear DNA (Ballard et al. 2002, Dean & Ballard 2004).

Wolbachia symbionts may influence mtDNA diversity in at least three other ways. First, the resident strain of *Wolbachia* may have lost the ability to cause incompatibility. In this case, mutations may be accumulating in the mtDNA but still be depressed below the neutral equilibrium value [this may be the case in *D. melanogaster* (Ballard et al. 1996)]. Second, *Wolbachia* itself may directly affect the fitness of infected individuals. Males of *Sphyracephala beccarii* (a Stalk-Eyed Fly) infected with *Wolbachia* may have higher fertility than uninfected males (Hariri et al. 1998). *Wolbachia* can extend longevity and increase fecundity of some strains of *D. melanogaster* (Fry & Rand 2002, Fry et al. 2004). Conversely, a significant reduction in sperm production and fertility has been found in a *D. simulans* line, suggesting a negative fitness effect (Hariri et al. 1998, Snook et al. 2000). Third, *Wolbachia* may protect a less fit mtDNA type from going to extinction. One example of this protection may occur in the *D. simulans si*I type, which is an island endemic. In microinjection (de Stordeur 1997) and cage experiments (Ballard & James 2004), the *si*I mtDNA type is out-competed by flies harboring *si*II and *si*III mtDNA. However, at least 99.9% of flies harboring *si*I mtDNA are infected with the *w*Ha *Wolbachia* strain that induces high levels of incompatibility. In this case the microorganism-induced incompatibility may effectively limit the colonization potential of the fitter mtDNA types.

SAMPLING METHODS AND STATISTICAL TESTS THAT SHOULD BE ROUTINELY INCLUDED IN PHYLOGENETIC STUDIES

Reconstructing phylogenies involves inferring the branching sequence, but also describing the rate and the pattern of character-state change along each branch. In this way we can test alternative processes to explain the patterns of variation we observe. In this review, we argue that developing a more complete understanding of the biology of mitochondria is essential for interpreting evolutionary processes that influence mtDNA. However, mtDNA in an individual is completely linked with little recombination, and information from different mtDNA genes does not give us statistical independence. If the goal is to reconstruct the species-level phylogeny, we advocate the inclusion of nuclear DNA. In the following sections, we discuss some issues to resolving the mtDNA genealogy.

Phylogenetic Tests

From a sequence alignment, a phylogenetic tree can be reconstructed by several means. Each method, however, uses different assumptions (stated either explicitly or implicitly), different optimality criteria, and different tree search algorithms. Evolutionary rate heterogeneity can have important phylogenetic implications. Here, we consider heterogeneity in rate along length of the sequence and then rate heterogeneity in different regions of the tree.

Heterogeneity in substitution rates can occur in regions of mtDNA (Ballard 2000a,b) and can influence phylogenetic hypotheses. Steinbach et al. (2001) compared the robustness of distinct genes and of different tree-building methods in resolving a well-corroborated *Drosophila* mitochondrial genealogy. Somewhat surprisingly, ND5, the longest gene (1713 bp), recovered the correct topology for fewer than 30% of the methods/models. One explanation for this result is that two regions of this gene have significant rate heterogeneity (Ballard 2000b), possibly causing its inconsistent performance. Rate heterogeneity can be investigated with the sliding-window maximum likelihood method, implemented in PLATO (partial likelihoods assessed through optimization) (Grassly & Holmes 1997). This method aims to detect regions that conflict with a single phylogenetic topology and nucleotide substitution process derived from the entire sequence. Such deviation along sequences, called spatial phylogenetic variation by Grassly & Holmes (1997), may reflect recombination or varying selective forces along the sequence. This approach calculates the likelihoods for each site independently. It then generates a measure of the average likelihood of a given window with respect to the rest of the sequence. Maximum values of this method are associated with regions showing low likelihoods given the maximum likelihood model derived from the complete sequence. If rate heterogeneity is detected, the phylogenetic and population inferences should be independently assessed (Dean & Ballard 2004).

Varying substitution patterns in different parts of a tree may affect the estimation of branch lengths and also possibly the branching pattern of a reconstructed tree. Cummings et al. (1995) analyzed complete mitochondrial genomes of 10 vertebrates and found that individual genes—or contiguous nucleotide sites—provided poor estimates of the tree taken from whole mtDNA sequences, reflecting biases in the information of individual mtDNA genes. Devauchelle et al. (2001) investigated proteins encoded by the mitochondrial genome of 26 multicellular animals, which include vertebrates, arthropods, echinoderms, mollusks, and nematodes, and showed that systematic deviations from a single-rate model are unmistakable and related to the evolutionary history of the species under consideration. Weiss & von Haeseler (2003) tested the assumption that there was homogeneity of the substitution process using simulation studies and analyzed two real data sets, one of which was complete hominid mtDNA data. Statistical analysis and the development of a new test showed that the substitution process was not homogeneous within the hominid mtDNA, suggesting a change in evolutionary pressures in different parts of the tree. If statistical tests suggest a rejection of the homogeneity assumption, results of a phylogenetic reconstruction should be interpreted with care. Varying substitution patterns affect the estimation of branch lengths and possibly also the branching pattern of the reconstructed tree. For example, it is well known that compositional changes in nucleotide frequencies can produce misleading trees (Galtier & Gouy 1995, Lockhart et al. 1994). Two examples from earlier literature on mtDNA illustrate this point. In mammals and primates, the directional nucleotide substitution in favor of A + T nucleotides increases with weight-specific metabolic rate (Martin 1995). In addition, mitochondrially encoded proteins have highly hydrophobic amino acid compositions, limiting the information content in these sequences for phylogenetic analysis (Naylor & Brown 1997). One approach to minimize the potential for obtaining incorrect trees has been to develop more sophisticated models of nucleotide substitution with the hopes that the problems of analysis will be eliminated. Indeed, model-based approaches to phylogenetic reconstruction can facilitate determination of the correct tree if the correct model can be estimated (Sanderson & Shaffer 2002). Here we argue that testing the constraints on mtDNA sequence evolution, which stem from the biology of mitochondria, is in need of additional scrutiny. For example, empirically derived mtDNA substitution matrices of either nucleotides or amino acids could be used as a set of priors in a Bayesian analysis to improve accuracy. Branch length priors may also influence phylogenetic analyses and the effect of this could be considerable for species where mtDNA evolution is particularly rapid.

mtDNA and Deep Phylogenies

mtDNA has been used to infer deep phylogenies. A recent review of the current state of understanding of the evolution of the mitochondrial genome by Bullerwell & Gray (2004) comprehensively discusses the use of mtDNA to infer deep evolutionary relationships. Knoop (2004) summarizes the unique aspects of land plant mitochondrial evolution from a phylogenetic perspective. However, neither of these reviews presents a robust mechanism by which the phylogeny can be corroborated, and a concern exists over the level of homoplasy in such data sets. One goal of Engstrom et al. (2004) was to examine strategies for analyzing highly homoplasious mtDNA data in deep phylogenetic problems where increased taxon sampling is not an option. The analyses of the combined data set from two mitochondrial protein-coding genes and an approximately 1-kb nuclear intron, and 59 morphological characters converged on a set of well-supported relationships. In this case, weeded and weighted parsimony, and model-based techniques, generally improved the phylogenetic performance of highly homoplasious mtDNA sequences, but no single strategy completely mitigated the problems associated with these highly homoplasious data. Indeed, many deep nodes in the softshell turtle phylogeny were confidently recovered only after the addition of largely nonhomoplasious data from the nuclear intron.

In an attempt to resolve deep, ordinal-level phylogenies with mtDNA a number of researchers have proposed using mtDNA gene order as a character set (Boore et al. 1995). If rearrangements in mtDNAs are relatively rare in evolutionary history, these events can be useful synapomorphies (derived character states that are shared by two or more taxa). Under such scenarios, convergent or parallel translocations and reversions in nonrelated lineages would be expected to be uncommon, though convergent rearrangement of the mitochondrial genome has been observed in the reptile group Amphisbaenia (Macey et al. 2004). One potential source of concern is the lack of understanding of the mechanism of mtDNA gene order rearrangement (Dowton & Campbell 2001, Sun et al. 2005). This is particularly the case where large-scale rearrangements have occurred, such as in *Tigriopus japonicus* (Machida et al. 2002). The difficulty in identifying homologs in short tRNA genes is an area of particular concern. Rawlings et al. (2003) suggests that through a process of tRNA duplication and mutation in the anticodon triplet, remolded leucine (L_{UUR}) tRNA genes have repeatedly taken over the role of isoaccepting L_{CUN} leucine tRNAs within metazoan mtDNA. They further suggest that tRNA leucine duplication and remolding events have occurred independently at least seven times within three major animal lineages. One clear area of future research is to examine the mechanism of mtDNA rearrangements in closely related groups. One such group is the insect order Hymenoptera (Dowton et al. 2003). Although different partitions or approaches to mtDNA sequence analysis are effective at different depths of evolutionary history, phylogeneticists need to accept the possibility that mtDNA may not be capable of resolving deep phylogenies. Phylogenetic analysis of whole nuclear genome sequences in yeasts revealed that at least 20 independent genes were needed to recover the whole-genome tree (Rokas et al. 2003).

mtDNA and Speciation

Many researchers have assumed explicitly or implicitly that differentiation within a character system is indicative of organismal differentiation or history. From the standpoint of mtDNA, attention has been focused on whether mtDNA differentiation is indicative of species trees or gene trees. A genealogical species is defined as a basal group of organisms whose members are all more closely related to each other than they are to any organisms outside the group and which contains no exclusive group within it (Baum & Shaw 1995, Cracraft 1983). In practice, a pair of species is so defined when phylogenies of alleles from a sample of loci show them to be reciprocally monophyletic at all, or monyphyletic for some specified fraction of the loci (though not all genealogical species concepts require reciprocal monophyly). Hudson & Coyne (2002) investigated the length of time it takes to attain reciprocal monophyly when an ancestral population divides into two descendant populations of equal size with no gene exchange and when genetic drift and mutation are the only operating evolutionary forces. A clear lesson from their results is that one should be cautious about recognizing genealogical species using only mtDNA. Such DNA becomes monophyletic more rapidly than does a single nuclear gene, and it does so far more rapidly than does a sample of several nuclear genes. This may make inferences of species-level monophyly erroneous. Funk & Omland (2003) also argue that data from mtDNA may not provide an accurate measure of species status. To evaluate the importance of species-level polyphyly, Funk & Omland (2003) conducted an intense survey of studies that evaluate mtDNA variation in a phylogenetic context and observed that the mtDNA monophyly of many biological species is not well supported. Funk & Omland (2003) detected species-level mtDNA paraphyly or polyphyly in 23% of 2319 assayed species, demonstrating that this problem is statistically supported, taxonomically widespread, and far more common than previously recognized. These patterns could be due to (*a*) sporadic hybridization among divergent lineages, (*b*) incomplete lineage sorting of mtDNA gene trees relative to organismal lineages, and/or (*c*) selection on mtDNA that might retain certain haplotypes within diverging lineages.

Moreover, mtDNA has great potential for becoming monophyletic by selective sweeps. This can decrease the time to monophyly of a clade and not be reflective of the genealogical processes in the nuclear genome. Advantageous mutations occurring on mtDNA will cause the entire organelle genome to become monophyletic because such genomes have little or no recombination. Although selective sweeps will also occur in nuclear DNA, causing monophyly for regions linked to the selected locus, recombination will whittle away the section of genome that becomes monophyletic through linkage.

Finally, when there is gene flow between diverging populations, one may encounter the opposite problem: mtDNA may be homogenized between the populations more readily than is nuclear DNA, so that mtDNA may appear paraphyletic when nuclear genes may be monophyletic. In fish, mice, and crickets, for example, mtDNA flows between taxa much more readily than does nuclear DNA (e.g., Bernatchez et al. 1995, Ferris et al. 1983, Shaw 2002, Taylor & McPhail 2000). In some cases, the mtDNA from one taxon completely replaces that in another, without any evidence of nuclear introgression or morphological signal. For example, the mtDNA in an allopatric population of brook trout in Lake Alain in Québec is identical to the Québec arctic char genotype, yet the brook trout are morphologically indistinguishable from normal brook trout and have diagnostic alleles at nuclear loci (Bernatchez et al. 1995). Ballard & Whitlock (2004) discussed the possible explanations of selection and drift on introgression.

Sampling and Neutrality Testing

When conducting phylogenetic analyses, we advocate including multiple individuals within all species possible. This will enable the researcher to test the basic assumptions and predictions of the neutral model: a constant mutation rate, a stationary allele frequency distribution, and a correlation between polymorphism levels and divergence. Specifically, we suggest researchers sample multiple individuals from a broadly distributed species or a species that occurs in a variety of niches. Consideration of the species to be sampled should also include contemplation of the outgroup of each species. In many statistical tests of neutrality, it is also important to include a closely related outgroup (Table 1).

Deviations from a strictly neutral model of evolution are found in a variety of organisms (Nachman 1998, Rand 2001, Rand & Kann 1998), and reviews have

TABLE 1   Approaches that can be employed to test the evolutionary dynamics of mitochondrial DNA (mtDNA) (modified from Ballard & Kreitman 1995)

Approach	Test	Prediction under neutrality	References
Direct	Competition between mitochondrial haplotypes	Haplotype frequency will not change in any predictable or repeatable way	Ballard & James 2004, Fos et al. 1990, Hutter & Rand 1995, MacRae & Anderson 1990, Nigro & Prout 1990, Singh & Hale 1990
Phylogenetic	Rates of evolution along different lineages	The variance should equal the mean after taking into account possible "lineage" effects, such as differences in mutation rate	Gillespie 1991, Tajima 1993
	Types of mutational changes along different branches of a tree	The type or pattern of substitution should be the same on all branches	Akashi 1995, Ballard 2000b, Ballard & Kreitman 1994
Statistical	Frequency spectrum of haplotypes	With no recombination and no selection, haplotype frequencies should conform to the neutral infinite alleles distribution	Ewens 1973, 1975; Fu 1997; Fu & Li 1993; Tajima 1989; Watterson 1977
	Distribution of polymorphism within species and divergence between species for two loci	The level of polymorphism and divergence, governed only by genetic drift and selective constraints, will be positively correlated	Hudson et al. 1987
	Distribution of synonymous and replacement changes within and between species. Easily extended to preferred or unpreferred synonymous sites, or conservative versus radical amino acids	If the synonymous and replacement variation is neutral, the divergence of the ratio of polymorphism will be the same when species compared are similar	Ballard & Kreitman 1994, McDonald & Kreitman 1991, Rand et al. 2000
	Distribution of substitutions on lineages	The substitution rate matrices do not differ for all lineages	Weiss & von Haeseler 2003
	Distribution of substitutions along the sequence	The substitution rate matrices do not vary along the length of the sequence	Grassly & Holmes 1997

collated the battery of statistical tests that can be applied to mtDNA (Ballard & Kreitman 1994, Gerber et al. 2001) (Table 1). These data typically show an excess of amino substitutions within species, suggesting the accumulation of slightly deleterious intraspecific changes. Most statistical tests are implemented in shareware computer programs, including Arlequin (http://lgb.unige.ch/arlequin/) written by Laurent Excoffier and DNASP (http://www.ub.es/dnasp/) written by Julio Rozas & Ricardo Rozas (1997). Rejection of the null hypothesis likely means that selection- and/or population-level processes (expansion, contraction, subdivision, etc.) are operating on the region of interest. In these cases it is unwise to overinterpret phylogeographic and phylogenetic hypotheses. Alternatively, rejection of the null hypothesis often opens up new and exciting areas of study comparing different properties of selection on mitochondrial genes and proteins (Rand & Kann 1996, Rand et al. 2000, Weinreich & Rand 2000). There are far more data available for population genetic analysis of mtDNA than have been used, and this represents a great opportunity for future study to gain a greater understanding of the biology of mtDNA.

CONCLUSIONS

The vast majority of studies employing mtDNA as an evolutionary marker have not attempted to expand our knowledge of the basic biology of mitochondria. We believe that this can and should be done in the forthcoming years so that hypotheses generated from mtDNA data are robust. We have identified a number of research areas in this review that would benefit from additional work. These include mechanisms of mtDNA rearrangement and research focusing on thermal adaptation as a mechanism for population subdivision. We have also identified specific tests that should be included with studies using mtDNA as an evolutionary marker. These include heterogeneity rate tests both along the length of the sequence, heterogeneity rate tests in different branches of the tree, and tests of the basic assumptions and predictions of the neutral model. The omission of these tests limits our ability to interpret the results of these analyses, but perhaps more importantly it misses an opportunity to understand the nature of the basic biology of mitochondria. We conclude that we should not throw the organelle baby out with the organismal bathwater. Rather, we should develop a greater understanding of the basic biology of the molecule so that evolutionary hypotheses are robust.

ACKNOWLEDGMENTS

We thank Brad Shaffer for his constructive comments. This work was supported by a grant from the National Institute of Health, Grant GM067862, to David Rand and Bill Ballard and NSF grants NSF DEB-0444766 to Bill Ballard and DEB-0343464 to David Rand.

**The *Annual Review of Ecology, Evolution, and Systematics* is online at
http://ecolsys.annualreviews.org**

LITERATURE CITED

Adams KL, Qiu YL, Stoutemyer M, Palmer JD. 2002. Punctuated evolution of mitochondrial gene content: high and variable rates of mitochondrial gene loss and transfer to the nucleus during angiosperm evolution. *Proc. Natl. Acad. Sci. USA* 99:9905–12

Akashi H. 1995. Inferring weak selection from patterns of polymorphism and divergence at "silent" sites in Drosophila DNA. *Genetics* 139:1067–76

Andersson JO, Andersson SG. 1999. Genome degradation is an ongoing process in Rickettsia. *Mol. Biol. Evol.* 16:1178–91

Andersson SG, Zomorodipour A, Andersson JO, Sicheritz-Ponten T, Alsmark UC, et al. 1998. The genome sequence of Rickettsia prowazekii and the origin of mitochondria. *Nature* 396:133–40

Avise JC, Arnold RM, Ball RM, Bermingham E, Lamb T, et al. 1987. Intraspecific phylogeography: the mitochondrial DNA bridge between population genetics and systematics. *Annu. Rev. Ecol. Syst.* 18:489–522

Ballard JWO. 2000a. Comparative genomics of mitochondrial DNA in Drosophila simulans. *J. Mol. Evol.* 51:64–75

Ballard JWO. 2000b. Comparative genomics of mitochondrial DNA in members of the Drosophila melanogaster subgroup. *J. Mol. Evol.* 51:48–63

Ballard JWO. 2000c. When one is not enough: introgression of mitochondrial DNA in Drosophila. *Mol. Biol. Evol.* 17:1126–30

Ballard JWO. 2004. Sequential evolution of a symbiont inferred from the host: Wolbachia and Drosophila simulans. *Mol. Biol. Evol.* 21:428–42

Ballard JWO, Chernoff B, James AC. 2002. Divergence of mitochondrial DNA is not corroborated by nuclear DNA, morphology, or behavior in Drosophila simulans. *Evolution* 56:527–45

Ballard JWO, Hatzidakis J, Karr TL, Kreitman M. 1996. Reduced variation in Drosophila simulans mitochondrial DNA. *Genetics* 144:1519–28

Ballard JWO, James AC. 2004. Differential fitness of mitochondrial DNA in perturbation cage studies correlates with global abundance and population history in Drosophila simulans. *Proc. R. Soc. London Ser. B* 271:1197–201

Ballard JWO, Kreitman M. 1994. Unraveling selection in the mitochondrial genome of Drosophila. *Genetics* 138:757–72

Ballard JWO, Kreitman M. 1995. Is mitochondrial DNA a strictly neutral marker? *Trends Ecol. Evol.* 10:485–88

Ballard JWO, Whitlock MC. 2004. The incomplete natural history of mitochondria. *J. Mol. Ecol.* 13:729–44

Baum DA, Shaw KL. 1995. Genealogical perspectives on the species problem. In *Experimental and Molecular Approaches to Plant Biosystematics*, ed. PC Hoch, AG Stephenson, pp. 289–303. St. Louis: Mo. Bot. Gard.

Bensasson D, Zhang D-X, Hartl DL, Hewitt GM. 2001. Mitochondrial pseudogenes: evolution's misplaced witness. *Trends Ecol. Evol.* 16:314–21

Berg OG, Kurland CG. 2000. Why mitochondrial genes are most often found in nuclei. *Mol. Biol. Evol.* 17:951–61

Bergthorsson U, Adams KL, Thomason B, Palmer JD. 2003. Widespread horizontal transfer of mitochondrial genes in flowering plants. *Nature* 424:197–201

Bernatchez L, Glémet H, Wilson CC, Danzmann RG. 1995. Introgression and fixation of Arctic char (Salvelinus alpinus) mitochondrial genome in an allopatric population of brook trout (Salvelinus fontinalis). *Can. J. Fish. Aquat. Sci.* 52:179–85

Boore JL, Collins TM, Stanton D, Daehler LL, Brown WM. 1995. Deducing the pattern of arthropod phylogeny from mitochondrial DNA rearrangements. *Nature* 376:163–65

Boussau B, Karlberg EO, Frank AC, Legault BA, Andersson SG. 2004. Computational inference of scenarios for alpha-proteobacterial genome evolution. *Proc. Natl. Acad. Sci. USA* 101:9722–27

Bullerwell CE, Gray MW. 2004. Evolution of the mitochondrial genome: protist connections to animals, fungi and plants. *Curr. Opin. Microbiol.* 7:528–34

Caspari E, Watson GS. 1959. On the evolutionary importance of cytoplasmic sterility in mosquitoes. *Evolution* 13:568–70

Charlesworth B. 1994. The effect of background selection against deleterious mutations on weakly selected, linked variants. *Genet. Res.* 63:213–27

Charlesworth B, Morgan MT, Charlesworth D. 1993. The effect of deleterious mutations on neutral molecular variation. *Genetics* 134:1289–303

Charlesworth D, Charlesworth B, Morgan MT. 1995. The pattern of neutral molecular variation under the background selection model. *Genetics* 141:1619–32

Clayton DA, Teplitz RL, Nabholz M, Dovey H, Bodmer W. 1971. Mitochondrial DNA of human-mouse cell hybrids. *Nature* 234:560–62

Coskun PE, Beal MF, Wallace DC. 2004. Alzheimer's brains harbor somatic mtDNA control-region mutations that suppress mitochondrial transcription and replication. *Proc. Natl. Acad. Sci. USA* 101:10726–31

Cracraft J. 1983. Species concepts and speciation analysis. *Curr. Ornithol.* 1:159–87

Cummings MP, Otto SP, Wakeley J. 1995. Sampling properties of DNA sequence data in phylogenetic analysis. *Mol. Biol. Evol.* 12:814–22

Dean MD, Ballard JWO. 2004. Linking phylogenetics with population genetics to reconstruct the geographic origin of a species. *Mol. Phylogenet. Evol.* 32:998–1009

Dean MD, Ballard JWO. 2005. High divergence among Drosophila simulans mitochondrial haplogroups arose in midst of long term purifying selection. *Mol. Phylogenet. Evol.* In press

de Benedictis G, Carrieri G, Varcasia O, Bonafé M, Franceschi C. 2000. Inherited variability of the mitochondrial genome and successful aging in humans. *Ann. NY Acad. Sci.* 908:208–18

de Stordeur E. 1997. Nonrandom partition of mitochondria in heteroplasmic Drosophila. *Heredity* 79:615–23

Devauchelle C, Grossmann A, Henaut A, Holschneider M, Monnerot M, et. al 2001. Rate matrices for analyzing large families of protein sequences. *J. Comput. Biol.* 8:381–99

Doi A, Suzuki H, Matsuura ET. 1999. Genetic anlaysis of temperature-dependent transmission of mitochondrial DNA in Drosophila. *Heredity* 82:555–60

Dowton M, Campbell NJH. 2001. Intramitochondrial recombination—is it why some mitochondrial genomes sleep around? *Trends Ecol. Evol.* 16:269–71

Dowton M, Castro LR, Campbell SL, Bargon SD, Austin AD. 2003. Frequent mitochondrial gene rearrangements at the Hymenopteran nad3-nad5 junction. *J. Mol. Evol.* 56:517–26

Edmands S. 2001. Phylogeography of the intertidal copepod Tigriopus californicus reveals substantially reduced population differentiation at northern latitudes. *Mol. Ecol.* 10:1743–50

Elson JL, Turnbull DM, Howell N. 2004. Comparative genomics and the evolution of human mitochondrial DNA: assessing the effects of selection. *Am. J. Hum. Genet.* 74:229–38

Engstrom TN, Shaffer HB, McCord WP. 2004. Multiple data sets, high homoplasy, and the phylogeny of softshell turtles (Testudines: Trionychidae). *Syst. Biol.* 53:693–710

Ewens WJ. 1973. Testing for increased mutation rate for neutral alleles. *Theor. Popul. Biol.* 4:251–58

Ewens WJ. 1975. A note on the variance of the number of loci having a given gene frequency. *Genetics* 80:221–22

Ferris SD, Sage RD, Huang C-M, Neilsen JT, Ritte U, Wilson AC. 1983. Flow of mitochondrial DNA across a species boundary. *Proc. Natl. Acad. Sci. USA* 80:2290–94

Fos M, Domínguez MA, Latorre A, Moya A. 1990. Mitochondrial DNA evolution in experimental populations of Drosophila subobscura. *Proc. Natl. Acad. Sci. USA* 87:4198–201

Frank SA, Hurst LD. 1996. Mitochondria and male disease. *Nature* 383:224

Fry AJ, Palmer MR, Rand DM. 2004. Variable fitness effects of Wolbachia infection in Drosophila melanogaster. *Heredity* 93:379–89

Fry AJ, Rand DM. 2002. Wolbachia interactions that determine Drosophila melanogaster survival. *Evolution* 56:1976–81

Fu YX. 1997. Statistical tests of neutrality of mutations against population growth, hitchhiking and background selection. *Genetics* 147:915–25

Fu YX, Li WH. 1993. Statistical tests of neutrality of mutations. *Genetics* 133:693–709

Funk DJ, Omland KE. 2003. Species-level paraphyly and polyphyly: Frequency, causes, consequences, with insights from animal mitochondrial DNA. *Annu. Rev. Ecol. Syst.* 34:397–423

Galtier N, Gouy M. 1995. Inferring phylogenies from DNA sequences of unequal base compositions. *Proc. Natl. Acad. Sci. USA* 92:11317–21

Gerber AS, Loggins R, Kumar S, Dowling TE. 2001. Does nonneutral evolution shape observed patterns of DNA variation in animal mitochondrial genomes? *Annu. Rev. Genet.* 35:539–66

Gillespie JH. 1991. *The Causes of Molecular Evolution*. Oxford: Oxford Univ. Press

González-Villaseñor LI, Powers DA. 1990. Mitochondrial-DNA restriction site polymorphisms in the teleost Fundulus heteroclitus support secondary intergration. *Evolution* 44:27–37

Grassly NC, Holmes EC. 1997. A likelihood method for the detection of selection and recombination using nucleotide sequences. *Mol. Biol. Evol.* 14:239–47

Hariri AR, Werren JH, Wilkinson GS. 1998. Distribution and reproductive effects of Wolbachia in stalk-eyed flies (Diptera: Diopsidae). *Heredity* 81:254–60

Hudson RR, Coyne JA. 2002. Mathematical consequences of the genealogical species concept. *Evolution* 56:1557–65

Hudson RR, Kreitman M, Aguade M. 1987. A test of neutral molecular evolution based on nucleotide data. *Genetics* 116:153–59

Hutter CM, Rand DM. 1995. Competition between mitochondrial haplotypes in distinct nuclear genetic environments: Drosophila pseudoobscura vs. D. persimilis. *Genetics* 140:537–48

Ingmann M, Kaessmann H, Paabo S, Gyllenstein U. 2000. Mitochondrial genome variation and the origin of modern humans. *Nature* 408:708–13

James AC, Ballard JWO. 2003. Mitochondrial genotype affects fitness in Drosophila simulans. *Genetics* 164:173–86

Kambhampati S, Rai KS, Verleye DM. 1992. Frequencies of mitochondrial DNA haplotypes in laboratory cage populations of the mosquito, Aedes albopictus. *Genetics* 132:205–9

Kaneda H, Hayashi J, Takahama S, Taya C, Lindahl KF, Yonekawa H. 1995. Elimination of paternal mitochondrial DNA in intraspecific crosses during early mouse embryogenesis. *Proc. Natl. Acad. Sci. USA* 92:4542–46

Karl SA, Avise JC. 1992. Balancing selection at allozyme loci in oysters: implications from nuclear RFLPs. *Science* 256:100–2

Karr TL, Ballard JWO. 2005. Biology of *Wolbachia*. *Nat. Encyclop. Sci.* In press

Kenyon L, Moraes CT. 1997. Expanding the functional human mitochondrial DNA database by the establishment of primate xenomitochondrial cybrids. *Proc. Natl. Acad. Sci. USA* 94:9131–35

Kern AD, Kondrashov FA. 2004. Mechanisms and convergence of compensatory evolution in mammalian mitochondrial tRNAs. *Nat. Genet.* 36:1207–12

Kilpatrick ST, Rand DM. 1995. Conditional hitchhiking of mitochondrial DNA: Frequency shifts of Drosophila melanogaster mtDNA variants depend on nuclear genetic background. *Genetics* 141:1113–24

Knoop V. 2004. The mitochondrial DNA of land plants: peculiarities in phylogenetic perspective. *Curr. Genet.* 46:123–39

Kondo R, Satta Y, Matsuura ET, Ishikawa H, Takahata N, Chigusa SI. 1990. Incomplete maternal transmission of mitochondrial DNA in Drosophila. *Genetics* 126:657–63

Lockhart PJ, Steel MA, Hendy MD, Penny D. 1994. Recovering evolutionary trees under a more realistic model of sequence evolution. *Mol. Biol. Evol.* 11:605–12

Macey JR, Papenfuss TJ, Kuehl JV, Fourcade HM, Boore JL. 2004. Phylogenetic relationships among amphisbaenian reptiles based on complete mitochondrial genomic sequences. *Mol. Phylogenet. Evol.* 33:22–31

Machida RJ, Miya MU, Nishida M, Nishida S. 2002. Complete mitochondrial DNA sequence of Tigriopus japonicus (Crustacea: Copepoda). *Mar. Biotechnol.* 4:406–17

MacRae AF, Anderson WW. 1990. Can mating preferences explain changes in mtDNA haplotype frequencies? *Genetics* 124:999–1001

Martin AP. 1995. Metabolic rate and directional nucleotide substitution in animal mitochondrial DNA. *Mol. Biol. Evol.* 12:1124–31

Martin AP, Naylor GJ, Palumbi SR. 1992. Rates of mitochondrial DNA evolution in sharks are slow compared with mammals. *Nature* 357:153–55

Martin AP, Palumbi SR. 1993. Body size, metabolic rate, generation time, and the molecular clock. *Proc. Natl. Acad. Sci. USA* 90: 4087–91

Matsuura ET, Niki Y, Chigusa SI. 1993. Temperature-dependent selection in the transmission of mitochondrial DNA in Drosophila. *Jpn. J. Genet.* 68:127–35

Maynard Smith J, Haigh J. 1974. The hitchhiking effect of a favourable gene. *Genet. Res.* 23:23–35

McDonald JH, Kreitman M. 1991. Adaptive protein evolution at the Adh locus in Drosophila. *Nature* 351:652–54

Nachman MW. 1998. Deleterious mutations in animal mitochondrial DNA. *Genetica* 103: 61–69

Nagao Y, Totsuka Y, Atomi Y, Kaneda H, Lindahl KF, et al. 1998. Decreased physical performance of congenic mice with mismatch between the nuclear and the mitochondrial genome. *Genes Genet. Syst.* 73:21–27

Nagata Y, Matsuura ET. 1991. Temperature-dependency of electron-transport activity in mitochondria with exogenous mitochondrial DNA in Drosophila. *Jpn. J. Genet.* 66:255–61

Naylor GJ, Brown WM. 1997. Structural biology and phylogenetic estimation. *Nature* 388:527–28

Nigro L. 1991. The effect of heteroplasmy on cytoplasmic incompatibility in transplasmic lines of Drosophila simulans showing a complete replacement of the mitochondrial DNA. *Heredity* 66:41–45

Nigro L. 1994. Nuclear background affects frequency dynamics of mitochondrial DNA variants in Drosophila simulans. *Heredity* 72: 582–86

Nigro L, Prout T. 1990. Is there selection on RFLP differences in mitochondrial DNA? *Genetics* 125:551–55

Rand DM. 1993. Endotherms, ectotherms, and mitochondrial genome-size variation. *J. Mol. Evol.* 37:281–95

Rand DM. 1994. Thermal habit, metabolic rate and the evolution of mitochondrial DNA. *Trends Ecol. Evol.* 9:125–31

Rand DM. 2001. The units of selection on mitochondrial DNA. *Annu. Rev. Ecol. Syst.* 32:415–48

Rand DM, Clark AG, Kann LM. 2001. Sexually antagonistic cytonuclear fitness interactions in Drosophila melanogaster. *Genetics* 159:173–87

Rand DM, Haney RA, Fry AJ. 2004. Cytonuclear coevolution: the genomics of cooperation. *Trends Ecol. Evol.* 19:645–53

Rand DM, Kann LM. 1996. Excess amino acid

polymorphism in mitochondrial DNA: contrasts among genes from Drosophila, mice, and humans. *Mol. Biol. Evol.* 13:735–48

Rand DM, Kann LM. 1998. Mutation and selection at silent and replacement sites in the evolution of animal mitochondrial DNA. *Genetica* 103:393–407

Rand DM, Weinreich DM, Cezairliyan BO. 2000. Neutrality tests of conservative-radical amino acid changes in nuclear- and mitochondrially-encoded proteins. *Gene* 261:115–25

Rawlings TA, Collins TM, Bieler R. 2003. Changing identities: tRNA duplication and remolding within animal mitochondrial genomes. *Proc. Natl. Acad. Sci. USA* 100:15700–5

Rawson PD, Burton RS. 2002. Functional coadaptation between cytochrome c and cytochrome c oxidase within allopatric populations of a marine copepod. *Proc. Natl. Acad. Sci. USA* 99:12955–58

Reeb CA, Avise JC. 1990. A genetic discontinuity in a continuously distributed species: mitochondrial DNA in the American oyster, Crassostrea virginica. *Genetics* 124:397–406

Rokas A, Williams BL, King N, Carroll SB. 2003. Genome-scale approaches to resolving incongruence in molecular phylogenies. *Nature* 425:798–804

Ropson IJ, Brown DC, Powers DA. 1990. Biochemical genetics of Fundulus heteroclitus (L.). VI. Geographical variation in the gene frequencies of 15 loci. *Evolution* 44:16–26

Rose G, Passarino G, Carrieri G, Altomare K, Greco V, et al. 2001. Paradoxes in longevity: sequence analysis of mtDNA haplogroup J in centenarians. *Eur. J. Hum. Genet.* 9:701–7

Ross IK. 2004. Mitochondria, sex, and mortality. *Ann. NY Acad. Sci.* 1019:581–84

Roubertoux PL, Sluyter F, Carlier M, Marcet B, Maarouf-Veray F, et al. 2003. Mitochondrial DNA modifies cognition in interaction with the nuclear genome and age in mice. *Nat. Genet.* 35:65–69

Rozas J, Rozas R. 1997. DnaSP version 3.0: a novel software package for extensive molecular population genetics analysis. *Bioinformatics* 15:174–75

Ruiz-Pesini E, Diez C, Lapena AC, Perez-Martos A, Montoya J, et al. 1998. Correlation of sperm motility with mitochondrial enzymatic activities. *Clin. Chem.* 44:1616–20

Ruiz-Pesini E, Lapena AC, Diez-Sanchez C, Perez-Martos A, Montoya J, et al. 2000. Human mtDNA haplogroups associated with high or reduced spermatozoa motility. *Am. J. Hum. Genet.* 67:682–96

Ruiz-Pesini E, Mishmar D, Brandon M, Procaccio V, Wallace DC. 2004. Effects of purifying and adaptive selection on regional variation in human mtDNA. *Science* 303:223–26

Sachidanandam R, Weissman D, Schmidt SC, Kakol JM, Stein LD, et al. 2001. A map of human genome sequence variation containing 1.42 million single nucleotide polymorphisms. *Nature* 409:928–33

Sackton TB, Haney RA, Rand DM. 2003. Cytonuclear coadaptation in Drosophila: disruption of cytochrome c oxidase activity in backcross genotypes. *Evolution* 57:2315–25

Sanderson MJ, Shaffer HB. 2002. Troubleshooting molecular phylogenetic analyses. *Annu. Rev. Ecol. Syst.* 33:49–72

Satta Y, Toyohara N, Ohtaka C, Tatsuno Y, Watanabe K, et al. 1988. Dubious maternal inheritance of mitochondrial DNA in D. simulans and evolution of D. mauritiana. *Genet. Res.* 52:1–6

Schizas NV, Chandler GT, Coull BC, Klosterhaus SL, Quattro JM. 2001. Differential survival of three mitochondrial lineages of a marine benthic copepod exposed to a pesticide mixture. *Environ. Sci. Technol.* 35:535–38

Schmidt TR, Wu W, Goodman M, Grossman LI. 2001. Evolution of nuclear- and mitochondrial-encoded subunit interaction in cytochrome c oxidase. *Mol. Biol. Evol.* 18:563–69

Shaw KL. 2002. Conflict between nuclear and mitochondrial DNA phylogenies of a recent species radiation: what mtDNA reveals and conceals about modes of speciation in Hawaiian crickets. *Proc. Natl. Acad. Sci. USA* 99:16122–27

Singh RS, Hale LR. 1990. Are mitochondrial DNA variants selectively non-neutral? *Genetics* 124:995–97

Snook RR, Cleland SY, Wolfner MF, Karr TL. 2000. Offsetting effects of Wolbachia infection and heat shock on sperm production in Drosophila simulans: analyses of fecundity, fertility and accessory gland proteins. *Genetics* 155:167–78

Solignac M, Monnerot M, Mounolou JC. 1986. Mitochondrial DNA evolution in the melanogaster species subgroup of Drosophila. *J. Mol. Evol.* 23:31–40

Somero GN. 2002. Thermal physiology and vertical zonation of intertidal animals: optima, limits, and costs of living. *Integr. Comp. Biol.* 42:780–89

Sommer AM, Portner HO. 2002. Metabolic cold adaptation in the lugworm Arenicola marina: comparison of a North Sea and a White Sea population. *Mar. Ecol.* 240:171–82

Steinbachs JC, Schizas NV, Ballard JWO. 2001. Efficiency of different genes and accuracy of different methods in recovering a known Drosophila genealogy. In *Proc. Pac. Symp. Biocomput.*, ed. RB Altman, AK Dunker, LA Hunter, K Lauderdale, TE Klein, pp. 606–17. Singapore: World Sci.

Sun HY, Zhou KY, Song DX. 2005. Mitochondrial genome of the Chinese mitten crab Eriocheir japonica sinenesis (Brachyura: Thoracotremata: Grapsoidea) reveals a novel gene order and two target regions of gene rearrangements. *Gene* 349:207–17

Sutovsky P, Moreno RD, Ramalho-Santos J, Dominko T, Simerly C, Schatten G. 1999. Ubiquitin tag for sperm mitochondria. *Nature* 402:371–72

Tajima F. 1989. Statistical method for testing the neutral mutation hypothesis by DNA polymorphism. *Genetics* 123:585–95

Tajima F. 1993. Statistical analysis of DNA polymorphism. *Jpn. J. Genet.* 68:567–95

Takeda K, Takahashi S, Onishi A, Hanada H, Imai H. 2000. Replicative advantage and tissue-specific segregation of RR mitochondrial DNA between C57BL/6 and RR heteroplasmic mice. *Genetics* 155:777–83

Taylor EB, McPhail JD. 2000. Historical contingency and ecological determinism interact to prime speciation in sticklebacks, Gasterosteus. *Proc. R. Soc. London Ser. B* 267:2375–84

Thalmann O, Hebler J, Poinar HN, Pääbo S, Vigilant L. 2004. Unreliable mtDNA data due to nuclear insertions: a cautionary tale from analysis of humans and other great apes. *Mol. Ecol.* 13:321–35

Turelli M, Hoffmann AA. 1991. Rapid spread of an inherited incompatibility factor in California Drosophila. *Nature* 353:440–42

Turelli M, Hoffmann AA, McKechnie SW. 1992. Dynamics of cytoplasmic incompatibility and mtDNA variation in natural Drosophila simulans populations. *Genetics* 132:713–23

Watterson GA. 1977. Heterosis or neutrality? *Genetics* 85:789–814

Weinreich DM, Rand DM. 2000. Contrasting patterns of nonneutral evolution in proteins encoded in nuclear and mitochondrial genomes. *Genetics* 156:385–99

Weiss G, von Haeseler A. 2003. Testing substitution models within a phylogenetic tree. *Mol. Biol. Evol.* 20:572–78

Werren JH, Windsor D, Guo LR. 1995. Distribution of Wolbachia among neotropical arthropods. *Proc. R. Soc. London Ser. B* 262:197–204

Zeviani M, Corona P, Nijtmans L, Tiranti V. 1999. Nuclear gene defects in mitochondrial disorders. *Ital. J. Neurol. Sci.* 20:401–8

Zeyl C, Andreson B, Weninck E. 2005. Nuclear-mitochondrial epistatsis for fitness in Saccharomyces cerevisiae. *Evolution* 59:910–14

Annu. Rev. Ecol. Evol. Syst. 2005. 36:643–89
doi: 10.1146/annurev.ecolsys.36.102003.152638
Copyright © 2005 by Annual Reviews. All rights reserved
First published online as a Review in Advance on September 9, 2005

INTRODUCTION OF NON-NATIVE OYSTERS:
Ecosystem Effects and Restoration Implications

Jennifer L. Ruesink,[1] Hunter S. Lenihan,[2] Alan C. Trimble,[1] Kimberly W. Heiman,[3] Fiorenza Micheli,[3] James E. Byers,[4] and Matthew C. Kay[2]

[1]*Department of Biology, University of Washington, Seattle, Washington 98195-1800; email: ruesink@u.washington.edu, trimblea@u.washington.edu*
[2]*Bren School of Environmental Science and Management, University of California, Santa Barbara, California 93106-5131; email: lenihan@bren.ucsb.edu, kay@lifesci.ucsb.edu*
[3]*Hopkins Marine Station, Stanford University, Pacific Grove, California 93950-3094; email: micheli@stanford.edu, heiman@stanford.edu*
[4]*Department of Zoology, University of New Hampshire, Durham, New Hampshire 03824-2617; email: jebyers@unh.edu*

■ **Abstract** Oysters have been introduced worldwide to 73 countries, but the ecological consequences of the introductions are not fully understood. Economically, introduced oysters compose a majority of oyster harvests in many areas. Oysters are ecosystem engineers that influence many ecological processes, such as maintenance of biodiversity, population and food web dynamics, and nutrient cycling. Consequently, both their loss, through interaction of overharvest, habitat degradation, disease, poor water quality, and detrimental species interactions, and their gain, through introductions, can cause complex changes in coastal ecosystems. Introductions can greatly enhance oyster population abundance and production, as well as populations of associated native species. However, introduced oysters are also vectors for non-native species, including disease-causing organisms. Thus, substantial population, community, and habitat changes have accompanied new oysters. In contrast, ecosystem-level consequences of oyster introductions, such as impacts on flow patterns, sediment and nutrient dynamics, and native bioengineering species, are not well understood. Ecological risk assessments for future introductions must emphasize probabilities of establishment, spread, and impacts on vulnerable species, communities, and ecosystem properties. Many characteristics of oysters lead to predictions that they would be successful, high-impact members of recipient ecosystems. This conclusion leaves open the discussion of whether such impacts are desirable in terms of restoration of coastal ecosystems, especially where restoration of native oysters is possible.

INTRODUCTION

Oysters (Family Ostreidae) occupy nearshore marine and estuarine habitats at temperate to tropical latitudes worldwide. The hundred or so living Ostreidae species include at least 18 species consumed by humans (Carriker & Gaffney 1996).

Their good flavor and relative accessibility have contributed to the overexploitation of many native populations (Menzel 1991). By the mid-1800s, *Ostrea edulis* in Germany, England, and France had experienced 10- to 30-year boom-and-bust cycles of yield (Mobius 1877). By the late 1800s, reefs of *Crassostrea virginica* in Chesapeake Bay contained low densities of adult oysters and evidence of poor recruitment (Brooks 1891). In western North America, *Ostreola conchaphila* declined severely in yield by two orders of magnitude between 1880 and 1915 (Ruesink et al. 2005). We do not know if oyster populations were overexploited in China, Japan, or Korea because aquaculture began in those countries at least 500 years ago without record of whether it replaced a failed wild-stock fishery (Kusuki 1991).

Oyster fisheries, in which fishers exploit a common resource that is repopulated by natural recruitment, have poor records of sustainability (Kirby 2004). Most native populations of oysters have not been successfully restored after overexploitation, but instead remain at low population abundance for extended periods of time (Grizel & Heral 1991, Utting & Spencer 1992, Rothschild et al. 1994, Drinkwaard 1998, Ruesink et al. 2005). Explanations for failure to recover are myriad and include continued exploitation, habitat degradation through destructive fishing practices, disease, reduced water quality, and detrimental species interactions (Lenihan & Peterson 1998, Lenihan et al. 1999, Jackson et al. 2001). Aquaculture, on the other hand, can provide long-term productivity by allowing growers to "reap what they sow" with seed (newly settled) oysters from hatcheries or wild populations. Sometimes aquaculture is focused on native species, for instance, in East Asia (Kusuki 1991, Nie 1991), New Zealand (Dinamani 1991a), India (Nagabhushanam & Mane 1991), the Caribbean (Baqueiro 1991), and Central (Nascimento 1991) and South America (Velez 1991), and thereby provides a form of conservation. This review documents the worldwide changes in oyster populations during recent history. It focuses primarily on the consequences of introductions intended to replace and augment native species that have declined through overexploitation or other causes.

Decline of these conspicuous members of the nearshore community has been accompanied by economic losses and ecological change. Oysters are ecosystem engineers that provide many ecosystem goods and services. As such, they can have strong ecosystem-level impacts that must be adequately considered prior to their introduction into estuarine, lagoon, and rocky shore coastal ecosystems. Major questions concerning future introductions include the following: Do introduced oyster species provide the same ecological goods and services provided by native species? Can the loss of natural populations be compensated through introductions of new oyster species? What are the potential ecological impacts associated with both purposeful and unintentional introductions?

Oyster Introductions

Oysters have proved highly amenable to aquaculture, and today, exploitation of wild populations contributes little to worldwide oyster production (FAO 2002).

Ecological impacts of aquaculture techniques may be substantial in terms of biodeposits, altered flow regimes, and disturbance of the substrate (Everett et al. 1995); other reports indicate low environmental effects (Buschmann et al. 1996, Crawford et al. 2003). An assessment of aquaculture impacts is beyond the scope of this review. Our focus is on the ecological roles of oysters themselves. Many oyster species have been introduced to new ecosystems through aquaculture. One of the first ecologists to sound an alarm about species introductions, Charles Elton, paid particular attention to oysters among marine species (Elton 1958): "The greatest agency of all that spreads marine animals to new quarters of the world must be the business of oyster culture." Introductions of oysters for aquaculture were already widespread by the 1950s, when Elton's book was published, often to replace ailing populations of native oyster species and sometimes in attempts to develop new exportable commodities.

Rising concern about harmful impacts of non-native species has prompted a substantial literature that evaluates risks of oyster introductions. Of course, ecological concerns must be balanced against human need. Introductions of oysters, and advances in oyster aquaculture, could provide an important source of protein and revenue, particularly in developing countries.

The volume edited by Mann (1979) covers successful introductions in western North America, the United Kingdom, and France, in addition to legislation and risk assessment for eastern North America. Mann et al. (1991) and Gottlieb & Schweighofer (1996) argued strongly for the introduction of new oysters to the eastern United States to replace lost ecosystem functions of *C. virginica* [now at less than 1% of historic densities (NRC 2004)]. Chew (1982) compiled overviews of North American oyster practices, and Menzel (1991) provided a more global perspective. Shatkin et al. (1997) reviewed the consequences of oyster introductions in the western United States, France, Australia, and New Zealand in their risk assessment for the introduction of *Crassostrea gigas* to Maine, and the Maryland Sea Grant (MDSG 1991) and the National Research Council (NRC 2004) presented similar assessments relevant to the possible introduction of *C. gigas* or *Crassostrea ariakensis* to Chesapeake Bay. Finally, 73 oyster introductions are on record in a database maintained by the Food and Agriculture Organization (FAO/FIGIS) and based on published literature and questionnaires (http://www.fao.org/figis/servlet/static?dom=collection&xml=dias.xml).

These reviews provided a launching point for our analyses, but we have pursued a substantially different strategy. Most importantly, we expanded our scope to include all oyster introductions, rather than the four or five examples that have received most attention. Rather than present information as a series of case studies organized by country, we have instead organized by impact and applied data from several different areas to each possible introduction outcome.

In our view, the ecological consequences of oyster introductions have not received sufficient critical scrutiny. This conclusion is the only way we can reconcile the following disparate statements: "Examples of serious alterations of biotic communities by importations of exotic oysters with their associated faunas are found on the maritime coasts of western Europe and western North America" (Andrews

1980) and "With regard to deliberate introduction of mollusks, none has led to significant ecological disruption" (Grizel 1996). Our objectives are to provide a detailed examination of the potential ecosystem impacts of oyster introductions, thereby extending incomplete ecological assessments made by prior reviews (e.g., NRC 2004), and to identify key research priorities. Oyster introductions may, in fact, be highly desirable in terms of the ecological goods and services they can provide. However, as ecosystem engineers, oysters can have disproportionately high impacts, many of which are potentially undesirable (Davis et al. 2000, Shea & Chesson 2002, Cuddington & Hastings 2004). This review addresses the general ecological role of oysters, then focuses on the ecological impacts of introduced oysters, with respect to novel ecosystem impacts. In the final section, we consider implications for restoration of nearshore systems where formerly abundant oysters have declined.

OYSTERS AS ECOSYSTEM ENGINEERS

Understanding the broader ecosystem impacts of oysters and how they vary among species is crucial for assessing the realized and potential ecological impacts of non-native oyster introductions. As ecosystem engineers (Margalef 1968), oysters have major impacts in coastal ecosystems: They create habitat used by other species and modify the physical and chemical environment with major consequences on estuarine populations, communities, and food webs. A critical service provided by oysters is the creation of hard-substrate biogenic reefs that form conspicuous habitat in otherwise large expanses of soft-sediment estuarine and lagoonal seascapes. *Crassostrea virginica* forms more extensive reefs than do other oyster species (e.g., Rothschild et al. 1994). Available evidence suggests that reefs created by *C. gigas* (mostly in the intertidal) and *C. ariakensis* (mostly subtidal) are much smaller in size, occupy less area in estuaries, and are a more heterogenous mix of shell and sediment compared with *C. virginica* reefs (Ruesink et al. 2003; M. Luckenbach, personal communication). Most descriptions of *O. edulis*, *O. conchaphila*, and *Tiostrea chilensis* assemblages emphasize mainly loose accumulations of shell in the subtidal and intertidal (Mobius 1877, Hopkins 1937, Yonge 1960, Miller & Morrison 1988, Chanley & Chanley 1991, Baker 1995).

Large *C. virginica* reefs occupy water depths from the high intertidal to deep subtidal (>5 m depth) in estuaries on the Atlantic Coast of the United States. Before being degraded and reduced in size by destructive harvesting practices, single reefs covered areas more than 1 ha and stood over 3 m tall in many subtidal areas (Rothschild et al. 1994, Lenihan & Peterson 1998). These reefs are habitat for sessile, mobile, and even infaunal invertebrates, such as sponges, bryozoans, hydroids, corals, anemones, tunicates, crabs, shrimp, amphipods, isopods, cumaceans, polychaete, oligochaete, and flat worms (Wells 1961, Bahr & Lanier 1981, Coen et al. 1999b, Meyer & Townsend 2000). On the West Coast of the United States, reefs created by the native *O. conchaphila* and introduced *C. gigas* also harbor many

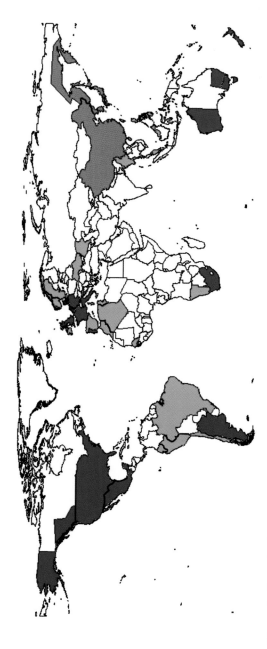

Figure 1 Countries that have received documented introductions of *Crassostrea gigas*. Blue represents countries where *C. Gigas* is known or highly likely to be established; green represents countries where the oyster is not established (in self-sustaining populations) or its status is unknown. Note that the map indicates where introductions have occurred and does not necessarily imply current importations. Also, *C. gigas* was not necessarily planted throughout the whole coastline of marked countries. The map probably underestimates certain areas of distribution where natural spreading or rogue (undocumented) plantings have occurred. The native range of *C. gigas* is colored red. It is worth noting that introductions of *C. gigas* have occurred back to the native range (Table 1), which may influence processes such as gene flow and the introduction of non-native hitchhiking species. Because of their extraordinary size, including coastline that borders several oceans, Canada, Russia, and Australia were broken up into regions/provinces. The United States meets these criteria as well; however, *C. gigas* has been planted along all of its coastlines. The Pacific Northwest is the only area where it is certain that populations of the species have been established (Table 1).

invertebrate species (Armstrong & Gunderson 1985, Miller & Morrison 1988), as do subtidal reefs in New Zealand created by *T. chilensis*, tunicates, bryozoans, and mussels (Cranfield et al. 1998, 2004). *Crassostrea* spp. and *O. conchaphila* reefs also support other bivalves, including mussels such as *Geukensia* spp. and clams such as *Macoma* spp., *Ensis* spp., *Mya arenaria*, and *Mercenaria mercenaria* (Miller & Morrison 1988, Micheli & Peterson 1999). Invertebrates occupy reefs because they provide refuge from predators and environmental stress, attachment surfaces, and populations of prey (Fernandez et al. 1993, Bartol & Mann 1999, Posey et al. 1999, Dumbauld et al. 2000). Many fishes utilize reefs as recruitment substrate (e.g., gobies, blennies, clingerfish, and oyster toadfish) (Hardy 1978a,b, Breitburg 1999, Lenihan et al. 2001, Grabowski 2004), nursery habitat (e.g., red drum, silver perch, pinfish, pigfish, and flounder) (Lenihan et al. 2001), and foraging ground (weakfish, bluefish, Atlantic croaker, pinfish, striped bass, mummichog, flounder, pigfish, toadfish, silver perch, and pompano) (Harding & Mann 2001a,b, 2003, Lenihan et al. 2001, Carbines et al. 2004).

The contribution of oysters as food for fish and invertebrates varies among species and locations. Bishop & Peterson (2005) found that blue crabs (*Callinectes sapidus*) in North Carolina had higher predation rates on non-native *C. ariakensis* than on native *C. virginica* because the shells of *C. ariakensis* are thinner than the native species, which makes them easier for crabs to crush. Relatively thin shells allow *C. ariakensis* faster growth rates than *C. virginica* (Grabowski et al. 2004). In addition, the physical structure of reef habitat is an important determinant of the foraging efficiency of consumers and other associated bivalves. Predation rates by the mud crab *Panopeus herbstii* are greatest in dense, structurally complex oyster beds because physical complexity likely decreases competitive interference among predators (Grabowski & Powers 2004). Similarly, blue crab predation on *C. virginica* is density-dependent, and foraging efficiency increases linearly with prey density (Eggleston 1990). Thus, variation in the shell morphologies, densities, and reef structural characteristics among different oyster species are important factors in the value of oysters as food resources for estuarine species.

Oyster populations and reef habitat also serve important ecosystem functions that extend beyond reef structures. Reefs influence the flow of water within estuaries and, in doing so, modify patterns of sediment deposition, consolidation, and stabilization (Dame & Patten 1981). Reefs disrupt flow on open bottoms or within tidal channels, and thereby create depositional zones, usually downstream of the reef structure, that accumulate sediment and organic material (Lenihan 1999). The alteration of flow and the physical barrier imposed by reefs influences the distribution and abundance of other biogenic habitats, such as seagrass beds, salt marshes, and algal beds, by preventing the erosion of channel banks, stabilizing and protecting the edges of salt marshes (Coen et al. 1999a), and providing attachment substrate for algae (Everett et al. 1995). Alteration of flow by reefs also influences biotic processes. Deposition of particles is enhanced downstream of reefs because of eddy formation, which thereby enhances settlement of fish (Breitburg et al. 1995) and invertebrate (Lenihan 1999) larvae. Acceleration of flow over reefs and the associated increase in the delivery rate of suspended food particles increases oyster growth, condition, and survivorship (Lenihan et al. 1995, Lenihan 1999) and influences in complex ways oyster disease dynamics (Lenihan et al. 1999). Enhanced flow probably has similar positive effects on other suspension feeders that inhabit reefs, such as tunicates, sponges, and bivalves.

Oyster populations influence energy flow and geochemical and ecological processes at the spatial scale of estuaries because they can filter large volumes of water through active suspension feeding. Oysters remove particles from the water column during suspension feeding and convert them to benthic sediments (feces and pseudofeces) and production (growth). Filtration rates are generally size related (Powell et al. 1992), and the relatively large size and high densities reached by oysters allow them to influence water properties and nutrient cycling. Research on *C. virginica* indicates that suspension feeding by oysters can reduce local concentrations of suspended solids, carbon, and chlorophyll a but increase ammonia and local deposition of fine-grained sediment and detritus (Dame 1976; Dame et al. 1984, 1986, 1992; Nelson et al. 2004). The removal of particulate matter through suspension feeding increases water clarity, which probably has a positive influence on the growth and abundance of seagrass and other benthic primary producers (Peterson & Heck 1999, Newell 2004, Newell & Koch 2004). Newell (1988) calculated that oyster abundance in Chesapeake Bay before 1870 was high enough that oysters could filter the entire volume of the bay in about 3 days, but after nearly a century of exploitation and habitat destruction, the reduced populations require 325 days to perform the same activity (see also Coen & Luckenbach 2000). Along with increased nutrient loading, loss of massive suspension-feeding capacity in Chesapeake Bay and other systems is thought to have caused shifts from primarily benthic to pelagic primary production, increased blooms of nuisance algae, and shifts in community dominance from macrophytes and nekton to bacteria and jellyfish (Jackson et al. 2001). Different oyster species exhibit significant variation in filtration rates. Filtration rates increase with size and result, for example, in higher filtration rates for larger (frequently cultivated) *Crassostrea* species as compared with the small-sized *Ostreola* species (Powell et al. 1992).

The influences of oyster habitat on associated populations, assemblages, and ecological processes can extend beyond the oyster reefs into adjacent habitats. Spatial configuration of estuarine habitats, such as salt marshes, seagrass beds, and oyster reefs, affects their use by fish and crustaceans, predator-prey interactions within each habitat type, and resulting diversity and structure of resident assemblages (Irlandi & Crawford 1997, Micheli & Peterson 1999). The specific locations, sizes, and relative proximity of introduced oyster reefs to native habitat patches is expected to influence their function as habitat and food for invertebrates and fish and possibly their influences on water quality, sediment erosion rates, and hydrodynamic patterns within estuaries. Thus, the ecological role and the effects of introduced oysters in estuaries and bays are likely to depend on context.

CONSEQUENCES OF OYSTER INTRODUCTIONS

We compiled published records of both introductions and transplantations of oysters on a country-by-country basis (Table 1). In total, we collected 182 records (168 introductions and 14 transplants) of 18 oyster species moved to 73 countries (or smaller regions). Almost all oyster introductions have occurred through oyster aquaculture; however, the introduction of the mangrove oyster (*C. rhizophorae*) from Brazil to the United Kingdom for research purposes and its subsequent eradication provides a notable exception (Spencer 2002).

Oyster introductions probably occurred as early as the seventeenth century, when the so-called Portuguese oyster (*Crassostrea angulata*) arrived in Europe from Asia (Carlton 1999). Overall, oysters have been introduced and established permanently in at least 24 countries outside their native ranges and have been introduced without successful establishment in 55 countries. Status of the remaining introductions is undocumented (Table 1). Most introductions (66) were of *C. gigas*, of which 17 established and 23 did not. *C. gigas* has been imported to most of the temperate zone (and some tropical areas) worldwide (Figure 1, see color insert). It is one of the most cosmopolitan macroscropic marine invertebrates. Other widely introduced species include *C. virginica* (14 cases), *O. edulis* (11 cases), and *Saccostrea commercialis* (6 cases); these species had slightly lower rates of establishment. France has been the recipient of the most introduced species; eight species were brought in for aquaculture or research in the past 150 years. The United Kingdom, Fiji, Tonga, and the US (West Coast), each received six introductions (Table 1). Only a few instances exist of an oyster arriving in a new location without deliberate introduction. *C. gigas* appeared on the northwest coast of New Zealand through an unknown pathway, potentially hull fouling from Asian boats or larval transport from Australia (Dinamani 1991a), and this species has also spread through the Mediterranean Sea after deliberate introduction to France and Italy (Galil 2000).

Failed introductions of *C. gigas* were the result mostly of transport to locations that are too warm and oligotrophic for survival of the species [Pacific Oceania (Eldredge 1994)] or too cold for successful reproduction (Alaska). For example, on Madeira Island in the subtropical Atlantic, *C. gigas* introduced at about half market size grew in shell dimensions but lost glycogen, and more than 70% died within 5 months (Kaufmann et al. 1994). However, even in "successful" introductions, particularly on western continental shores, spatfall occurs only in restricted locations that retain larvae and exceed critical temperatures (e.g., 18°C to 20°C for spawning and higher than 16°C for larval development in *C. gigas*) for several weeks. So, for instance, natural recruitment in western North America occurs regularly in perhaps only three locations in British Columbia, Canada, and in Hood Canal and Willapa Bay in Washington state (Kincaid 1951, Quayle 1969). The crash in summer 2004 of *C. gigas* populations introduced to France are causing concern that another case of a failed introduction is developing (P. Garcia Meunier, personal communication).

TABLE 1 Oyster introductions from one country (or smaller region) to another country (or smaller region) outside and inside the native range of the species

Introduced to	Species	Introduced from	Date	Established (yes/no, date if known)	Current aquaculture?	References
Algeria	*Crassostrea gigas*		<1984		Yes	FAO 2002a, Zibrowius 1992
Argentina	*Crassostrea gigas*	Chile	1982	1987	Yes	Orensanz et al. 2002
Australia (New South Wales)	*Crassostrea gigas*	Australia (Victoria, Tasmania)	1967[a]	1985	Yes	Chew 1990, Pollard & Hutchings 1990, Ayres 1991
Australia (Tasmania)	*Tiostrea chilensis*	New Zealand	1969			Pollard & Hutchings 1990
Australia (Victoria)	*Crassostrea gigas*	Australia (Tasmania)	1955	Yes	Yes	Thomson 1959
Australia (Western Australia, Tasmania)	*Crassostrea gigas*	Japan	1947–1970	Yes	Yes	Thomson 1952, 1959, Chew 1990, Pollard & Hutchings 1990, FAO/FIGIS
Bahamas	*Crassostrea virginica*			No		Glude 1981, Mann 1983
Belgium	*Crassostrea gigas*		1990[a]	Yes		Coutteau et al. 1997, FAO/FIGIS
Belize	*Crassostrea gigas*	United States (USA) (west)	1980			Chew 1990, FAO/FIGIS
Brazil	*Crassostrea gigas*	Chile	<1989	Unlikely	Yes	Nascimento 1991, Tavares 2003

Canada (east)	Ostrea edulis	UK	1957–1959	No		Mann 1983, Chew 1990, Hidu & Lavoie 1991, FAO/FIGIS
Canada (west)	Crassostrea gigas	Japan, USA (west)	1912–1977	1925	Yes	Bourne 1979, Chew 1990, FAO/FIGIS
	Crassostrea virginica	USA (east), Canada (east)	1883–1940	1917		Stafford 1913, Bourne 1979, Carlton & Mann 1996
Channel Islands [United Kingdom (UK)]	Crassostrea gigas		<1986		Yes	FAO 2002a
Chile	Crassostrea gigas	USA (west)	1983		Yes	Chew 1990, Buschmann et al. 1996, FAO/FIGIS
China	Crassostrea gigas[c]	Japan	1979			Tan & Tong 1989, FAO/FIGIS
Costa Rica	Crassostrea gigas	USA (west)	1979			Mann 1983, Chew 1990, FAO/FIGIS
Croatia	Crassostrea gigas		1980[a]		Unlikely	Zibrowius 1992, Galil 2000
Denmark	Crassostrea gigas	USA (west), Germany	1980	Yes	Unlikely	Mann 1983, Chew 1990, FAO/FIGIS
	Crassostrea virginica	Canada (east)	1880–1930	No	Unlikely	Carlton & Mann 1996
Ecuador	Crassostrea gigas	USA (west), Chile	1980	No	Yes	Chew 1990, FAO/FIGIS

(Continued)

TABLE 1 (*Continued*)

Introduced to	Species	Introduced from	Date	Established (yes/no, date if known)	Current aquaculture?	References
Fiji	*Crassostrea echinata*	Australia, Tahiti	1910, 1981	No	No	Eldredge 1994
	Crassostrea gigas	Japan, USA (west), Australia, Philippines	1968–1977	Unlikely		Bourne 1979, Eldredge 1994, FAO/FIGIS
	Crassostrea iredalei	Philippines	1975–1976	Unlikely		Eldredge 1994, FAO/FIGIS
	Crassostrea virginica	Hawaii	1970			Eldredge 1994, FAO/FIGIS
	Ostrea edulis	Japan	1977	Unlikely		Eldredge 1994, FAO/FIGIS
	Saccostrea commercialis	USA (west), Australia	1880, 1970–1973	No		Bourne 1979, Eldredge 1994
France	*Crassostrea angulata*	Portugal	1868	Unlikely	No longer	Andrews 1980
	Crassostrea ariakensis	USA (west)		No	No	NRC 2004
	Crassostrea densalamellosa	Korea	1982		Unlikely	Mann 1983
	Crassostrea gigas	Japan, Canada (west)	1966–1977	1975	Yes	Andrews 1980, Mann 1983, Chew 1990, Grizel & Heral 1991, Heral & Deslous-Paoli 1991, FAO/FIGIS

Country	Species	Location	Date			Reference
	Crassostrea rhizophorae	French Guyana	1976–1978	Unlikely	Unlikely	Maurin & Gras 1979
	Crassostrea virginica	USA (east)	1861–1875	No	No	Carlton & Mann 1996
	Ostrea puelchana	Argentina	1990		Unlikely	Pascual et al. 1991
	Tiostrea chilensis	Chile	1981	No		Mann 1983
	Ostrea edulis[c]	USA (west)	1970			Chew 1990, FAO/FIGIS
French Polynesia	*Crassostrea echinata*	New Caledonia	1972–1983	Unlikely	Unlikely	Eldredge 1994
	Crassostrea gigas	USA (west)	1972–1976	Unlikely	Unlikely	Eldredge 1994, FAO/FIGIS
Germany	*Crassostrea angulata*		1961	Unlikely		Drinkwaard 1999
	Crassostrea gigas	Scotland	1971	1991	Yes	Gollasch & Rosenthal 1994, Drinkwaard 1999
	Crassostrea virginica	UK	1913	No		Carlton & Mann 1996, Drinkwaard 1999, Wolff & Reise 2002
	Ostrea edulis[c]					Drinkwaard 1999
Greece	*Crassostrea gigas*				Unlikely	Zibrowius 1992
Guam	*Crassostrea echinata*	Palau	1979		Unlikely	Eldredge 1994
	Crassostrea gigas	Taiwan	1975	No	Unlikely	Eldredge 1994, FAO/FIGIS
	Saccostrea cucullata	Solomon Islands	1978		Unlikely	Eldredge 1994

(*Continued*)

TABLE 1 (*Continued*)

Introduced to	Species	Introduced from	Date	Established (yes/no, date if known)	Current aquaculture?	References
Ireland	*Crassostrea gigas*	France, UK	<1993		Yes	FAO/FIGIS
	Crassostrea virginica			No	Unlikely	Went 1962, Carlton & Mann 1996
Israel	*Crassostrea gigas*	UK	1976	Unlikely	Yes	Hughes-Games 1977, Chew 1990
	Ostrea edulis	UK	1976			Shpigel 1989
Italy	*Crassostrea angulata*	Portugal	1850		Unlikely	Zibrowius 1992, Galil 2000
	Crassostrea gigas	France	1972	Likely		Galil 2000, FAO/FIGIS
	Saccostrea commercialis	Australia	1985	Likely		Zibrowius 1992, Galil 2000, 2003
Japan	*Crassostrea virginica*	USA	1968			Chiba et al. 1989, FAO/FIGIS
	Ostrea edulis	France	1952			FAO/FIGIS
	Ostreola conchaphila	USA (west)	1948			FAO/FIGIS
	Crassostrea gigas[c]	USA (west)	1980			Chew 1990, FAO/FIGIS
Korea Republic	*Crassostrea gigas*[c]	USA (west)	1980			Chew 1990, FAO/FIGIS
Madeira Island (subtropical Atlantic)	*Crassostrea gigas*	UK	1991	No		Kaufmann et al. 1994
Malaysia	*Crassostrea gigas*	USA (west)	1980			Chew 1990, FAO/FIGIS

Location	Species	Source	Date			References
Mauritius (Indian Ocean)	Crassostrea gigas	USA (west)	1971	Unlikely	Likely	Bourne 1979, Macdonald et al. 2003
	Crassostrea virginica	USA (west)	1972		Likely	Macdonald et al. 2003
	Ostrea edulis	USA (west)	1972		Likely	Macdonald et al. 2003
	Saccostrea commercialis	Australia	1967		Likely	Macdonald et al. 2003
Mexico (east)	Crassostrea gigas					FAO 2002a
Mexico (west)	Crassostrea gigas	USA (west)	1973	Yes	Yes	Islas 1975, Chew 1990, FAO/FIGIS
	Crassostrea virginica					Carlton & Mann 1996
Morocco	Crassostrea gigas	France	<1966		Yes	Shafee & Sabatie 1986, Chew 1990
Myanmar	Crassostrea gigas				Unlikely	www.fao.org/documents/show_cdr.asp?url_file = /docrep/004/ad497e/ ad497e05.htm
Namibia	Crassostrea gigas	Chile	1990	No	Yes	FAO/FIGIS; P. Schneider, personal communication
Netherlands	Ostrea edulis		1990			FAO/FIGIS
	Crassostrea angulata	Portugal	1800s			Wolff & Reise 2002
	Crassostrea gigas	Canada (west), Belgium, France, USA (west)	1964–1981	1976	Yes	Chew 1990, Drinkwaard 1999, FAO/FIGIS

(*Continued*)

TABLE 1 (Continued)

Introduced to	Species	Introduced from	Date	Established (yes/no, date if known)	Current aquaculture?	References
	Crassostrea virginica	USA (east), UK	1939–1940	No	Unlikely	Carlton & Mann 1996, Wolff & Reise 2002
	Crassostrea sikamea		1964			Drinkwaard 1999
	Ostrea edulis[c]	France, Greece, Ireland, Italy, UK, Norway	1963–1977			Drinkwaard 1999
New Caledonia	Crassostrea echinata	Tahiti	1979–1980		Unlikely	Eldredge 1994
	Crassostrea gigas	Japan, USA (west), Australia, Tahiti	1967–1977	Unlikely	Yes	Bourne 1979, Eldredge 1994, FAO/FIGIS
	Saccostrea commercialis	Australia	1971	Unlikely	Unlikely	Eldredge 1994
New Hebrides	Crassostrea gigas	USA (west)	1972–1973	No		Bourne 1979
New Zealand	Crassostrea gigas	Japan or Australia (Victoria, Tasmania)	1958[a]	Yes	Yes	Chew 1990, Pollard & Hutchings 1990, FAO/FIGIS
	Ostrea edulis		1869	No		Cranfield et al. 1998
Norway	Crassostrea gigas	USA (west)	1985		Yes	Chew 1990, FAO/FIGIS
Palau	Crassostrea gigas	USA (west)	1972–1973	Unlikely		Bourne 1979, Eldredge 1994, FAO/FIGIS

Country	Species	Origin	Date			Reference
Peru	*Crassostrea gigas*		<1997		Likely	FAO 2002a
Philippines	*Crassostrea gigas*	Japan				Juliano et al. 1989, FAO/FIGIS
Portugal	*Crassostrea angulata*			Yes		Andrews 1980
	Crassostrea gigas	France, USA (west)	1977	Likely		Chew 1990, FAO/FIGIS
Puerto Rico	*Crassostrea gigas*	USA (west)	1980	No		Chew 1990, FAO/FIGIS
	Crassostrea virginica[d]			No		Walters & Prinslow 1975, Mann 1983
Russia (Black Sea)	*Crassostrea gigas*		1976	Unlikely		FAO/FIGIS
Samoa	*Crassostrea gigas*	USA (west)	1980			Chew 1990, FAO/FIGIS
Senegal	*Crassostrea gigas*		<2001			FAO 2002a
Serbia and Montenegro	*Crassostrea gigas*				Unlikely	Zibrowius 1992
Seychelles	*Crassostrea gigas*	Japan	1974	Unlikely		FAO/FIGIS
Singapore	*Crassostrea gigas*		2003	Unlikely	Yes	Quek 2004
Slovenia	*Crassostrea gigas*				Unlikely	Zibrowius 1992
South Africa	*Crassostrea gigas*	USA (west), Chile, France, UK	1950	2001	Yes	Chew 1990, Robinson et al. 2005, FAO/FIGIS
Spain	*Ostrea edulis*		<1992		Yes	FAO 2002a
	Crassostrea angulata			Yes	Likely	Andrews 1980
	Crassostrea gigas	France	1980		Yes	FAO/FIGIS

(Continued)

TABLE 1 *(Continued)*

Introduced to	Species	Introduced from	Date	Established (yes/no, date if known)	Current aquaculture?	References
Sweden	*Crassostrea gigas*		1980	No	Unlikely	Mann 1983
Tahiti	*Crassostrea gigas*	USA (west)	1972–1976	No		Bourne 1979
	Saccostrea echinata	New Caledonia	1978			Mann 1983
Tanzania	*Saccostrea cucullata*[d]					Macdonald et al. 2003
Tonga	*Crassostrea belcheri*	Malaysia (Sabah)	1977–1978	No	No	Bourne 1979, Eldredge 1994, FAO/FIGIS
	Crassostrea gigas	Japan, Australia (Tasmania)	1975	Unlikely		Bourne 1979, Eldredge 1994, FAO/FIGIS
	Crassostrea iredalei		1976			Eldredge 1994, FAO/FIGIS
	Crassostrea virginica	USA (west)	1973			Eldredge 1994, FAO/FIGIS
	Ostrea edulis	Japan, USA	1975	Unlikely		Eldredge 1994, FAO/FIGIS
	Saccostrea commercialis	New Zealand, USA (west)	1973	Unlikely		Eldredge 1994
Tunisia	*Crassostrea gigas*	France	<1984		Yes	Galil 2000, FAO 2002a
Turkey	*Crassostrea gigas*				Unlikely	Zibrowius 1992
	Saccostrea commercialis		2000[a]			Galil 2003
UK	*Crassostrea angulata*	Portugal		No		Andrews 1980

				Dis-agreement		
	Crassostrea gigas	Canada (west), USA (west), Hong Kong, Israel	1926, 1965–1979		Yes	Walne & Helm 1979, Mann 1983, Chew 1990, Drinkwaard 1999, FAO/FIGIS
	Crassostrea rhizophorae	Brazil	1980	No	No	Utting & Spencer 1992, Mann 1983, FAO/FIGIS
	Crassostrea virginica	Canada (east), USA (east)	1870–1939, 1984	Unlikely		Utting & Spencer 1992, Carlton & Mann 1996, FAO/FIGIS
	Saccostrea cucullata	Israel	1979	No	No	Mann 1983
	Tiostrea chilensis	Chile, New Zealand	1962–1963	Yes		Utting & Spencer 1992, Richardson et al. 1993, FAO/FIGIS
Ukraine (Black Sea)	Ostrea edulis[c]	Norway	1972			Askew 1972
U.S. Virgin Islands	Crassostrea gigas	USA (west)	1976	Unlikely		FAO/FIGIS
USA (Alaska)	Crassostrea gigas	USA (west)	1980		Yes	Chew 1990, FAO/FIGIS
	Crassostrea gigas	USA (west)	1980	Unlikely		Chew 1990, FAO/FIGIS
USA (east)	Crassostrea ariakensis	China, USA (west)	<2001	Unlikely		NRC 2004
USA (west)	Crassostrea gigas	USA (west)	1930–1990	Unlikely	No	Hickey 1979, Chew 1990, NRC 2004

(Continued)

TABLE 1 (*Continued*)

Introduced to	Species	Introduced from	Date	Established (yes/no, date if known)	Current aquaculture?	References
	Ostrea edulis	Netherlands	1949–1961	Likely	Yes	Mann 1983, Chew 1990, Hidu & Lavoie 1991, FAO/FIGIS
	Crassostrea virginica[c]	USA (east), USA (Gulf)	1808–1960			Carlton & Mann 1996
USA (Gulf)	*Crassostrea cortezensis*		1980		No	Mann 1983
	Crassostrea gigas		1930	No	No	NRC 2004
	Crassostrea rhizophorae		1980		No	Mann 1983
USA (Hawaii)	*Crassostrea gigas*	Japan, USA (west)	1926, 1980	1960? (Pearl Harbor)	Yes	Chew 1990, Eldredge 1994, FAO/FIGIS
	Crassostrea sikamea	Japan	1947	No		Woelke 1955
	Crassostrea virginica		1866–1949	1895	Likely	Carlton & Mann 1996

Location	Species	Origin	Date			References
USA (west)	*Crassostrea ariakensis*	Japan	1977[b]	No	Likely	Perdue & Erickson 1984, Langdon & Robinson 1996
	Crassostrea gigas	Japan, Korea	1902	Likely	Yes	Kincaid 1968, Andrews 1980, Chew 1990, FAO/FIGIS
	Crassostrea sikamea	Japan	1947	No	Yes	Woelke 1955
	Crassostrea virginica	USA (east)	1867–1935	Unlikely	Yes	Andrews 1980, Chew 1990, Carlton & Mann 1996
	Ostrea edulis	USA (east)			Yes	Chew 1990
Vanuatu	*Crassostrea gigas*	USA (west)	1972	Unlikely		Eldredge 1994, FAO/FIGIS
Yugoslavia	*Crassostrea gigas*				Unlikely	Zibrowius 1992

[a] Range expansion.
[b] Hitchhiker with other oysters.
[c] Transplantation in native range.
[d] Possible transplantation within native range, but taxonomy uncertain.

Oyster Production

One major consequence of introductions has been a shift in production from native to non-native oysters, largely in places where oysters have successfully established (e.g., *C. gigas* in the western United States, Europe, Australia, New Zealand, and South Africa) but also in places where they have not established and artificial reproduction is practiced (*C. gigas* in Namibia and *C. sikamea* in the western United States). The FAO compiles fishery statistics by species and country worldwide (FAO 2002). We used their recent data (1993–2002) to assess the contributions of non-native and native species to global oyster production (Table 2). These values differ substantially by region. In Asia, most production is based on native *Crassostrea* species [China: *C. plicatula* = *Saccostrea cucullata* (Nie 1991); Japan and Korea: *C. gigas* (Kusuki 1991)]; no records of cultured non-native species have emerged. *C. gigas* also contributes substantially to oyster production outside of Asia where it is not native. *C. gigas* constitutes 95% of European oyster production and 37% of African oyster production. On the western coast of North America, 99.8% of oyster production comes from non-native species, primarily *C. gigas*. However, only 20% of total U.S. production derives from introduced oysters, as much of the production still relies on the native *C. virginica* in Atlantic and Gulf Coast states. In the 26 countries where the FAO reports production from introduced oysters, 48% of production comes from introduced species (Table 2).

In most cases, historical yields of oysters are poorly known, so we cannot compare former productivity, on the basis of native species, with current productivity in which non-native species have replaced native species. However, isolated records do exist. In Willapa Bay, Washington, *C. gigas* yields about four times more shucked meat weight annually than at the peak of native oyster production in the late 1800s (Ruesink et al. 2005). The shift does not reflect an increase in area occupied by oysters (Townsend 1896, Hedgpeth & Obrebski 1981). In France, production of more recently introduced *C. gigas* outpaces the peak in *C. angulata* production by 30% (Goulletquer & Heral 1991, Heral & Deslous-Paoli 1991). Peak yields of the native *O. edulis* occurred more than 150 years ago, and data are not available for comparison. In New Zealand, aquaculture of the native *S. commercialis* yielded 500 metric tons a year until the 1970s (Dinamani 1991a), and its replacement by *C. gigas*, which reportedly grows twice as fast locally (Dinamani 1991a,b, Honkoop et al. 2003), has yielded 5,000 metric tons a year over the past decade (FAO 2002). Intrinsic differences between native and introduced oysters are difficult to distinguish from advances in hatchery techniques, more intensive aquaculture, and increased consumer demand. For comparison, China's oyster production, based exclusively on native species, is reported to have increased by a factor of 180 over the past 20 years (FAO 2002).

Habitat Impacts

Oysters have potentially high impact when introduced into ecosystems because of their influence on habitat quantity and quality (Crooks 2002). Their role as

TABLE 2   Production of native and non-native species of oysters by country

Country and region	Introduced*	Native*	Uncertain*	Introduced/ Total
Africa				
Algeria	5			1.00
Kenya		108		0.00
Mauritius		68		0.00
Morocco	1741	18		0.99
Namibia	310			1.00
Senegal	13	1381		0.01
South Africa	4513	1		1.00
Tunisia	13	9642		0.00
Regional total	6595	11218		0.37
Americas				
Argentina	82			1.00
Brazil	15313			1.00
Canada	55038	55553		0.50
Chile	33822	3355		0.91
Columbia		28		0.00
Cuba		16735		0.00
Dominican Republic		275		0.00
Ecuador	46			1.00
Mexico	16243	374194		0.04
Peru	90			1.00
USA	408831	1679965		0.20
Venezuela		24559		0.00
Regional total	514152	2154664	15313	0.19
Asia				
Australia	43478.5	53595	319	0.45
China		26067607		0.00
China, Hong Kong		4805		0.00
India		82		0.00
Indonesia		14717		0.00
Japan		2206168		0.00
Korea Republic		2049443		0.00
Malaysia			1335	0.00
New Caledonia	554			1.00
New Zealand	54638.5	11044		0.83
Philippines		143244		0.00
Taiwan		224856		0.00
Thailand		196024		0.00
Regional total	98671	30971585	1654	0.00
Europe				
Bosnia and Herzogovina		15		0.00

(Continued)

TABLE 2   *(Continued)*

Country and Region	Introduced*	Native*	Uncertain*	Introduced/ Total
Croatia		539		0.00
Channel Islands	2217	4		1.00
Denmark		698		0.00
France	1353313	20793.5		0.98
Germany	806			1.00
Greece		5842		0.00
Ireland	41306	7947		0.84
Italy	302			1.00
Netherlands	21486	1154		0.95
Norway	10.5	54		0.16
Portugal	6390	10.5	457	0.93
Russian Federation		38		0.00
Serbia andMontenegro		6		0.00
Slovenia		9		0.00
Spain	8312.5	30057		0.22
Sweden		27		0.00
United Kingdom	9425.5	7098		0.57
Regional total	1443568.5	74292	457	0.95
World total	2062986.5	33211759	17424	0.06
World total without China, Hong Kong, Japan, Korea and Taiwan	2062986.5	2658880	17424	0.44
Countries (n = 26) that report introduced oysters	2062986.5	2255866	776	0.48

*Numbers reported are shucked weights in metric tons/10 yr.

ecosystem engineers is particularly pronounced in soft-sediment environments, where hard substrate is rare except for shell deposits of oysters. Introduced ecosystem engineers are expected to improve conditions for some species and exclude others. Ideally, experiments would be conducted in which oyster reefs are created or removed, and associated communities are compared with those in unmanipulated areas. Lenihan et al. (2001) used the native oyster *C. virginica* to compare fish and epibenthic invertebrate (blue crab, mud crabs, grass shrimp, and amphipods) assemblages on experimentally constructed reefs with assemblages on soft-sediment bottom in Pamlico Sound, North Carolina. Fish abundance was 325% greater, and epibenthic invertebrate abundance was 213% greater per trap placed on reefs than on the unstructured sand/mud bottom, a finding consistent with observational studies (Kennedy 1996). However, few such manipulative experiments exist for introduced oysters (but see Escapa et al. 2004). Instead, most studies involve mensurative experiments that compare assemblages on existing habitat types.

Two soft-sediment systems have been examined in detail by use of this mensurative experimental approach. In Willapa Bay, infaunal, epifaunal, and nekton communities have been compared across habitats, including cultured oyster (introduced *C. gigas*) habitats and unstructured tideflat. Consistently, oysters harbor a higher diversity of epifauna (Hosack 2003) and higher densities of mussels, scaleworms, and tube-building amphipods (Dumbauld et al. 2001). Infaunal assemblages were unaffected (Dumbauld et al. 2001), as were small fish and year-old Dungeness crab (*Cancer magister*) (Hosack 2003). Nekton communities differed among regions of the bay, however, which suggests that small fish and crabs species may respond to habitat on scales larger than individual parcels of several hectares (Hosack 2003). Nevertheless, shells of *C. gigas* placed at high density in the intertidal zone provided excellent habitat for newly recruited crab (*C. magister*) in nearby Grays Harbor, Washington: crabs recruited preferentially to shell, and survival of tethered crabs was 70% higher on shell than over open bottom (Fernandez et al. 1993).

In Arcachon Bay, France, both seagrass (*Zostera noltii*) and oyster (*C. gigas*) culture contained higher densities of meiofauna (<0.5 mm) than did nearby sandflats; macrofauna reached highest densities in seagrass (Castel et al. 1989). The authors speculated that biodeposits of oysters provided a food resource for meiofauna, whereas macrofauna associated with oysters were negatively affected by hypoxic conditions. Alternatively, macrofauna could be depressed by effective predators foraging on oysters, in which case oyster habitats might support higher trophic levels (Lenihan et al. 2001, Leguerrier et al. 2004).

Clearly, the provision of hard surface in soft sediments influences many associated species, but few data exist on the rate of conversion of native habitats, such as unvegetated tideflat or eelgrass, into introduced oyster reefs. In many cases, these transitions are mediated by aquaculture practices (Simenstad & Fresh 1995). However, some evidence exists that oyster reefs can reduce eelgrass cover directly. In western Canada, eelgrass (*Zostera marina*) was relatively rare downslope from dense *C. gigas*, and transplanted shoots survived poorly relative to transplants within natural eelgrass beds located away from reefs (J. Kelly, unpublished data).

We found little published evidence of major impacts of introduced oysters on communities located on hard substrate. Natural recruitment of introduced *C. gigas* in British Columbia, Canada, occurs primarily in the rocky intertidal zone (Bourne 1979; J. Ruesink, unpublished data), which entails much less modification of substrate than in cases of reefs forming on soft sediment. In the Strait of Georgia, *C. gigas* are dominant in high (1.3 to 2.4 m) intertidal areas. This area is partially in the barnacle zone, and oysters may actually provide greater surface area for barnacles (Bourne 1979, p. 22). Introduced oysters inhabit a niche that was largely vacant and not dominated by any organism at the time of introduction. A more quantitative analysis has recently been published for Argentina, where *C. gigas* was introduced in 1982 and now occurs exclusively on rock outcrops (Escapa et al. 2004). Among eight epifaunal species, three occurred at higher densities inside oyster beds, and three occurred at higher densities outside. Shorebirds also spent a disproportionate amount of time associated with oysters, where foraging rate was often higher (Escapa et al. 2004).

Impacts on Species Interactions

Introduced oysters provide a new resource for native predators. Rocky intertidal predators such as seastars and crabs reduced monthly survival rates of *C. gigas*, introduced in western Canada, by 25% relative to caged oysters (J. Ruesink, unpublished data). Indeed, predator control is widely practiced to achieve higher aquaculture yields (see Menzel 1991). Some introduced oysters appear to be an easier resource than native species to handle or consume (Yamada 1993, Richardson et al. 1993), whereas other introduced oysters tend to be avoided (Richardson et al. 1993). In theory, then, introduced oysters may enhance the resource base for higher trophic levels of bivalve predators. Species interactions may also be modified by the shell habitat provided by oysters. In Grays Harbor, higher densities of crabs (*C. magister*) in oyster-shell habitats led to enhanced predation on and lower densities of native clams in these habitats, even though clam recruitment was not directly affected by shell (Iribarne et al. 1995). Grabowski (2004) demonstrated that the structural complexity of native-oyster reef habitat strongly controlled the strength of predation by oyster toadfish (*Opsanus tau*) on resident mud-crab populations (*P. herbstii*).

Competition between native and introduced oysters is expected to be most intense if they share similar habitat. Temperature, salinity, and desiccation are three primary physical factors that determine each species' fundamental niche. In many cases, native and introduced oysters differ in their environmental tolerances, which suggests the potential for few competitive interactions. On the western coast of North America, the native *O. conchaphila* tends to occur at lower depths with less temperature stress than does the introduced *C. gigas* (Stafford 1913). In contrast, in Australia, the native Sydney rock oyster *S. commercialis* actually survives longer out of water than does *C. gigas* (Pollard & Hutchings 1990). This difference in desiccation tolerance has been exploited to control *C. gigas* in places where it has been classified as noxious (e.g., in New South Wales, Australia). When both species settle on common substrate, *C. gigas* can be killed by holding the substrate out of water for sufficiently long time. Several examples exist in which native and introduced species do not overlap in their spatial distributions (Walne & Helm 1979, Andrews 1980).

Despite different habitats of many native and introduced oysters, they often overlap in some part of their range. When overlap occurs, introduced oysters consistently outgrow natives, presumably because higher-yielding species were specifically introduced for that characteristic. *C. gigas* grows five times faster than *O. conchaphila* in western North America (Baker 1995), possibly because of its higher per-area filtration rate (Galtsoff 1932). In the United Kingdom, *T. chilensis* (introduced from New Zealand) outgrows *C. gigas* (also introduced), which outgrows the native *O. edulis*, at least under some conditions (Askew 1972, Richardson et al. 1993). Ironically, in Chile, where *T. chilensis* is native, the relative growth rates are reversed; introduced *C. gigas* reaches market size "much more rapidly" than the 4 to 5 years required for the native species (Chanley & Chanley 1991). On the East Coast of the United States, *C. gigas* (introduced but not established) outgrows *O. edulis*, which outgrows the native *C. virginica* (Dean 1979). Indeed, *C. gigas* has been selected for worldwide introduction in part because of its rapid growth rate, which yields high biomass for growers.

Direct tests of competition between native and non-native oysters require comparisons of growth and survival in monocultures and mixed cultures, but few examples exist in the literature. In North Carolina, introduced *Crassostrea ariakensis* outgrows native *C. virginica* and introduced *C. gigas*, probably because *C. ariakensis* is better at assimilating food and has lower energy requirements to produce a relatively thin shell (Grabowski et al. 2004). Anecdotally, the arrival of *C. gigas* in New Zealand rapidly reduced native *S. commercialis*. On spat collectors, the ratio of *S. commercialis* to *C. gigas* in 1972 strongly favored the native (1000:1); they were evenly represented in 1977, and by 1978, the non-native outrecruited the native 4:1 (Dinamani 1991c). Whether this recruitment differential emerged from higher fecundity of *C. gigas*, better larval survival, or simply the introduced species' higher individual growth rate (Dinamani 1991c) is not clear.

A historical example in which an introduced species likely outcompeted a native oyster occurred in France after the introduction of *Crassostrea angulata* around 1868. Afterwards, native *O. edulis* began a steady decline until, by 1870, it was completely gone from certain sections of the French coast and fully replaced by *C. angulata*. French government figures on oyster production document this inverse relationship of the species' abundances. By 1925, 300 million *C. angulata* were produced; the figure climbed to 914 million by 1929. In sharp contrast, only 2.4 million *O. edulis* were harvested in 1925 and declined to 668,000 by 1929 (Galtsoff 1932). Mechanistic studies of these oysters' filtration rates by Viallanes (1892) demonstrated that *C. angulata* filtered water 5.5 times faster than did *O. edulis* and, thus, would be a superior competitor for seston resources. Furthermore, Danton (1914) observed that because *C. angulata* grows more quickly, it is superior at pre-empting settlement space. The possibility certainly exists that a disease helped mediate the rapid replacement by *C. angulata* because oyster diseases were not well known at the time. Nonetheless, the competitive advantages of *C. angulata* were pronounced, well documented, and certainly played some if not the central role in its dominance (Ranson 1926).

Competition between oyster species also occurs indirectly through habitat modification. The introduced *C. gigas* in Willapa Bay inhabits both feral oyster reefs and planted aquaculture beds, mostly in the intertidal zone (Kincaid 1951, Feldman et al. 2000). Neither of these habitat types likely provides a functional replacement for the largely subtidal accumulations of shell where the native *O. conchaphila* previously occurred (Townsend 1896). The native oyster has remained rare, although many observations over the past century suggest it is not recruitment limited (Kincaid 1968). Recent evidence suggests that native-oyster larvae disproportionately settle in areas with large accumulations of shell. Because intertidal *C. gigas* comprises most shell habitat in the bay, the native oysters only have the option of recruiting to zones where immersion times are too short for survival (A. Trimble, unpublished data). Thus, the introduced oyster has developed into a recruitment sink for natives, particularly in the absence of remnant subtidal native-oyster reefs.

Competition may also occur with species other than oysters. *C. gigas* introduced to Argentina recruits on native mussels that normally dominate intertidal rocky shores (Orensanz et al. 2002), and it similarly recruits to mussel beds that occupy tideflats of the Wadden Sea (Reise 1998). Oyster densities in these locations appear to be too low to achieve population-level impacts on mussels, but oysters can kill individual mussels (Reise 1998). In other locations, mussels are probably less vulnerable to novel oysters. On wave-exposed western North American shores, mussels are known to be dominant competitors (Paine 1966), and they reduce growth rates of *C. gigas* by more than 30% (J. Ruesink, unpublished data).

Many prior evaluations of oyster introductions suggest that introduced species had little impact on native populations in part because the native species was already at such low densities (Goulletquer & Heral 1991, NRC 2004). This suggestion begs the question of whether the new species has any impact on the ability of the native species to recover—certainly, competition can occur even when one species is rare. Native oysters have failed to recover in places where new species have been introduced (western North America and Europe), but they have also failed to recover where non-native species are not abundant (eastern North America). These comparisons are confounded by disease—the introduction of an oyster may not in itself prevent recovery, but rather the introduction of a disease carried by that oyster (reviewed by NRC 2004). The role of disease is explored more fully below (see Impacts of Hitchhiking Species).

Ecosystem Impacts

Oysters in high-density aquaculture experience reduced growth rates as their production increases and populations presumably approach carrying capacity (Kincaid 1968, Heral & Deslous-Paoli 1991, Kusuki 1991, Crawford 2003, Robinson et al. 2005). Such density dependence suggests that oysters can reach sufficiently high density, particularly via aquaculture, to reduce food availability to conspecifics as well as other species dependent on suspended particulate food. Filtration by large populations of introduced species (or restored native populations), therefore, has the potential to influence trophic dynamics and water quality (Newell 1988, Ulanowicz & Tuttle 1992, Coen & Luckenbach 2000, NRC 2004). For example, many investigators have hypothesized that overproduction of phytoplankton in Chesapeake Bay, generated by anthropogenic nutrient loading, could be reduced by increase of biofiltration rates through restoration of native populations of *C. virginica* or the introduction of *C. gigas* and *C. ariakensis* (Tuttle et al. 1987, Newell 1988, NRC 2004). Recent experimental results indicate that transplants of native oysters can significantly increase water quality in small bodies of water, such as tidal creeks (Nelson et al. 2004). Therefore, the probability is high that introductions of oysters that survive at high densities could improve water quality.

Oyster introductions may also enhance estuarine-wide production of other economically valuable species, such as finfish and crabs. Peterson et al. (2003) calculated that over a 20-year to 30-year period, a restored oyster reef could enhance the cumulative amount of fish and large decapod biomass by 38 to 50 kg per 10 m^{-2} of bottom area, discounted for present-day value. This positive effect would occur only where the introduction involved a reef builder and local species of fishes responded positively to that habitat through enhanced recruitment anduse of the substrate as refuge and as foraging ground.

Fecal pellets of suspension feeders on tidal flats tend to be organically rich relative to sediment and to provide sites for nutrient exchange, including nitrification and especially denitrification (Reise 1985). For introduced oysters, in particular, few data on biogeochemical impacts are available, and most come from aquaculture and should be applied tentatively to impacts of naturalized populations. At high densities, *C. gigas* generates biodeposits, which leads to reduced particle size and increased organic content in sediment (Castel et al. 1989), impacts that are avoided at lower oyster densities or higher flow rates (Crawford et al. 2003). The ability of suspension feeders, particularly oysters, to couple pelagic production to the benthos is well accepted (Dame et al. 1984), and researchers also hypothesized that release of inorganic nutrients into the water column by oysters may accelerate phytoplankton productivity (Leguerrier et al. 2004).

Impacts of Hitchhiking Species

The oyster industry has been one of the largest vectors of introduced marine invaders, despite early recognition that movement of oysters could also transport pests of aquaculture (Carlton 1992a,b). Early screening of imported oysters was driven entirely by the desire to prevent incidental importation of oyster pests such as drilling snails (*Urosalpinx cinerea* and *Ocinebrellus inornatus*) (Galtsoff 1932, McMillin & Bonnot 1932). For example, entire contents of infected shipments were often sacrificed to prevent the importation of oyster pest species; however, nonpest exotic species were not considered (Bonnot 1935). Nevertheless, few of the hitchhiker species of concern were ultimately prevented from introduction (Garcia-Meunier et al. 2002, Martel et al. 2004).

To explore the contribution of oyster culture to species invasions, we compiled data from the literature on the number of marine species introduced to nine regions of the world, where expert opinions had been expressed about the vectors of species introductions (Figure 2). A total of 78 established invasive marine algae,

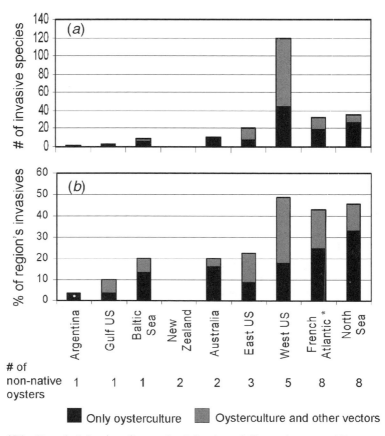

*The French Atlantic refers to the Atlantic and Channel coasts of France, Spain, and Portugal.

Figure 2   The number (*a*) and percentage (*b*) of known introduced species brought into different global regions exclusively through the culturing of oysters (black) or via oyster culture and some other vector such as shipping (gray). Established non-native oysters are included in these data. The regions are ordered by the number of non-native oyster species cultured in that region, from least to most. (Cranfield et al. 1998, Goulletquer et al. 2002, Olenin et al. 1997, Orensanz et al. 2002, Pollard & Hutchings 1990, Reise et al. 1999, Ruiz et al. 2000.)

invertebrates, and protozoa were introduced to the nine regions solely through the culturing of non-native oysters. If we include species with multiple vectors of introduction (oyster imports and some other vector such as shipping), then 46% of the introduced species in northern Europe and 20% in Australia likely entered with oyster aquaculture. The contribution of oyster aquaculture to invasion in coastal systems of the United States varies by region: 10% on the Gulf Coast, 20% on the East Coast, and 49% on the West Coast. Not unexpectedly, regions where a wider variety of oyster species have been cultured tend to have a greater number (Figure 2a) and percentage (Figure 2b) of hitchhiking non-native species.

Many of the species brought in with aquaculture present problems for the continued production of oysters in addition to potentially interacting with native species and altering the structure and function of surrounding communities and ecosystems (White et al. 1985, Wilson et al. 1988). Some invasives outcompete and ultimately displace native species. *Batillaria attramentaria*, an Asian snail introduced to the U.S. West Coast with *C. gigas*, outcompetes the mud snail *Cerithidea californica*, which has caused local extinction of the native snail in a number of estuaries (Byers 2000). Other hitchhikers alter the community structure in surrounding areas. In Great Britain, *Crepidula fornicata*, introduced with *C. virginica*, is found in densities greater than 4,000 individuals per m^2 and has positive effects on abundance, biomass, and species richness of the macrozoobenthos (de Montaudouin & Sauriau 1999). In Elkhorn Slough, a central California estuary, 38 of 58 known marine invasives were likely introduced through oyster culture (Wasson et al. 2001). In addition to free-living hitchhikers, parasites of introduced oysters can infest other native species. For example, the shell-boring sabellid polychaete, *Terebrasabella heterouncinata*, introduced with *C. gigas* in California, infested cultured red abalone, *Haliotis rufescens*, with great economic consequences to growers before it was successfully eradicated (Kuris & Culver 1999). Additionally, some hitchhikers provide structural habitats that can host a variety of other species. *Caulacanthus ustulatus*, an Asian turf-forming red alga also introduced with *C. gigas*, forms monospecific stands in the intertidal of Sao Miguel Island, Azores, and Elkhorn Slough, California, that are inhabited by both native and introduced invertebrates (Neto 2000, K. Heiman, unpublished data). With nearly 50% of the species invading some geographic regions attributed, at least in part, to the culturing of oysters (Figure 2), hitchhiking species must factor into assessments of further movement of oysters around the globe.

We have discussed oyster introductions to replace native species and, until now, paid little attention to the role of disease. However, disease is clearly a key factor in understanding both causes and consequences of oyster introductions. Introductions of oyster diseases via imported oysters have caused major ecological changes and economic loss in many estuaries worldwide. Aquaculture of native species may have been able to support high yields, but for high mortality caused by diseases in two high-profile examples: diseases that devastated *O. edulis* in Europe and separate diseases that affect *C. virginica* in North America. These diseases contributed to the decision to introduce *C. gigas* to Europe and to the intense discussion about whether to introduce new oysters to Chesapeake Bay and other eastern U.S. estuaries (Shatkin et al. 1997, MDSG 1991, NRC 2004).

We compiled information on the locations and impacts of 18 oyster diseases. We began with nine taxa recognized by the World Organization for Animal Health and added others reported in published studies (Table 3). Several additional bacterial

TABLE 3 Disease organisms affecting production of cultivated oysters

Disease agent	Disease name	Country	Species infected	Effect	Probable origin	Date	References
Protists							
Bonamia ostreae[a]		United States (USA) (west and east), Europe (e.g., Spain, France)	*O. edulis (T. chilensis, O. angasi, C. sikamea)*	70%–80% mortality	USA (east?)	1977, 1979	Chew 1990, Friedman & Perkins 1994, Cochennec et al. 1998, Cigarria 1997
Bonamia sp.		USA (east)	*C. ariakensis* (introduced)	Mortality	Native?	2003	Burreson et al. 2004
Bonamia exitiosa		Australia, Chile, New Zealand	*T. chilensis*	Mortality	Native and introduced?	Periodic (20–30 y)	Hine 1996, Hine et al. 2001
Haplosporidium nelsoni[a]	MSX	USA (west and east)	*C. virginica*	Mortality (S^b-specific)	Asia	West: 1900s East: 1957	Friedman 1996, Burreson et al. 2000
Haplosporidium costale[a]	SSO	USA (east)	*C. virginica*	Mortality	Native		Sunila et al. 2002
Marteilia refringens		Spain, France, Netherlands	*O. edulis, O. angasi, T. chilensis*	75%–100% mortality (T^c-specific)	Likely native that spread	1968	Grizel 1979, Berthe et al. 2004
Marteilia sydneyi[a]	QX	Australia	*S. commercialis*	>90% mortality	Likely native that spread	1990s	Nell 2002
Martilioides chungmuensis		Korean Republic, Japan	*C. gigas*	Reduced fecundity	Native	1990s	Park et al. 2003
Mikrocytos mackini[a]	Denman Island Disease	Canada and USA (west)	*C. gigas, O. conchaphila (C. virginica, O. edulis)*	Periodic high mortality (T-specific)	Native	1960	Bower et al. 1997
Mikrocytos roughleyi[a]		Australia, New Zealand	*S. commercialis, T. chilensis, O. angasi*	Mortality (T-specific)	Native?	1926	Nell 2002, Cochennec-Laureau et al. 2003

Perkinsus marinus[a]	Dermo	USA (east)	*C. virginica, (C. gigas* infected but no losses), *C. ariakensis*	Native, but spread	Mortality (T-specific) (S-specific)	East: 1948 Spread: 1985	Ford 1996, Villalba et al. 2004
Fungus							
Ostracoblabe implexa	Shell disease, maladie du pied	Europe, India, Canada (west, east)	*O. edulis, C. gigas, S. cucullata, C. angulata*		Shell blemish (T-specific)	<1950	Bower 2001
Bacteria							
Proteobacteria	Juvenile oyster disease	USA (east), France	*C. virginica*	Native in USA	Seed mortality (T-specific)	France: 1993	Renault et al. 2002, Paillard et al. 2004
Nocardia crassostreae		USA (west)	*C. gigas, O. edulis*	Asia?	Heat sensitivity	Native: 1945 Introduced: 1956	Friedman et al. 1998, Paillard et al. 2004
Vibrio splendidus		France	*C. gigas* larvae	Native	Summer mortality of spat	1991	Paillard et al. 2004
Viruses							
Iridolike viruses	Virus	France	*C. angulata* *C. gigas*	Asia?	Mortality	1967	Andrews 1980, Renault & Novoa 2004
Oyster virus velar disease	Virus	USA (west)	*C. gigas*	Asia?	Larval mortality		Elston 1985
Oyster herpesvirus		USA (west), New Zealand, France	*C. gigas, O. edulis* (*C. sikamea* infected but no losses)	Asia?	Larval, seed mortality (T-specific)	1990s	Renault & Novoa 2004

[a]Recognized as one of nine major oyster diseases by the World Organization for Animal Health (OIE), Aquatic Animal Health Code (www.oie.int).

[b]Salinity influence effect.

[c]Temperature influence effect.

and viral diseases not shown in Table 3 occur in larvae. In most cases, the diseases appeared in native oysters, but occasionally, introduced oysters contracted endemic diseases [e.g., *C. ariakensis* affected by *Bonamia* sp. in France and the eastern United States (Cochennec et al. 1998, Burreson et al. 2004)].

Disease theory suggests that pathogens and their hosts evolve toward coexistence, and impacts of native pathogens on native hosts are necessarily low (or the pathogen eliminates its host and goes extinct) (Price 1980). When a new combination of host and pathogen arises, the host may have innate resistance through physiological traits never encountered by the pathogen, or it may be highly susceptible to pathogen attack because selection for resistance has never occurred. Oysters appear to show a pattern similar to many marine species, namely, increased incidence of disease outbreaks and some entirely new (emerging) diseases (Harvell et al. 1999).

Our review of oyster diseases reveals the distressing pattern that oyster introductions or transplants of native species have been a major cause of emerging disease (Table 3). Among the 18 examples, two were definitely associated with introduced oysters (*Bonamia ostreae* and *Haplosporidium nelsoni*), and another five may have been. Three additional disease agents (*Marteilia refringens*, *Marteilia sydneyi*, and *Perkinsus marinus*) were moved via native-oyster transplants, and the pathogens infected naïve subpopulations. *B. ostreae*, a haplosporidian protist that kills three- to four-year-old *O. edulis*, appears to have infected this oyster species when the oyster was introduced to the United States and subsequently infected native-oyster populations when *O. edulis* was transplanted back to Europe (Chew 1990, Wood & Fraser 1996).

Diseases caused by two parasites, *H. nelsoni* (MSX) and *P. marinus* (dermo), are considered major factors in the decline of native *C. virginica* in the eastern United States. Molecular evidence indicated an Asian origin for *H. nelsoni*, which caused high mortality in *C. virginica* in the 1990s [although it was probably introduced with transfers of *C. gigas* much earlier (Burreson et al. 2000)]. In contrast, *P. marinus* probably originated in *C. virginica* along the southwest and Gulf Coast of the United States, but transplants of oysters within the native range spread it to locations where environmental conditions allowed the protist to become much more virulent (Table 1) (Reece et al. 2001). Substantial uncertainty remains in most cases about the origin of disease agents in oysters (Table 3).

CONSERVATION AND RESTORATION

Because oysters are often strong interactors in their native ecosystems, they pose several challenges for conservation. First, they require protection as key species that influence the structure and function of ecosystems. Yet, they are also directly exploited, which partly explains the genuine need for restoration in some places. In the past, oyster productivity has been restored through aquaculture and the introduction of novel species, and these activities can alter the species composition and ecological processes of coastal ecosystems. Decision makers are, thus, faced with the task of evaluating the costs and benefits of a potential introduction. The NRC (2004) reports an in-depth example of the difficulty of determining the consequences of different introduction decisions, ecologically, economically, and socially, in a book that evaluates the introduction of *C. ariakensis* as a means to recover oyster production in Chesapeake Bay. The approach hinges on ecological risk assessment.

Ecological Risk Assessment for Oyster Introductions

Ecological theory suggests that invasion success is a function of species traits, the recipient environment, the match between the species and the new environment, and effort applied to the introduction (number and size of introductions, often termed propagule pressure). Testing this theory requires data on both successful and failed introductions. Relevant data come from biocontrol releases (Beirne 1975), horticultural plants (Rejmanek & Richardson 1996, Reichard & Hamilton 1997), and imports of birds, mammals, and fishes (Veltman et al. 1996, Blackburn & Duncan 2001, Forsyth et al. 2004, Ruesink 2005). However, answers that emerge from these analyses tend to be idiosyncratic; different factors explain invasion in different taxa and at different scales. Factors also often differ in their predictive value for establishment and impact (Kolar & Lodge 2001).

Emerging ecological risk assessments for introductions embody ecological principles and include reproductive rates, species interactions, and propagule pressure, among others, in their guidelines (Ruesink et al. 1995). One widely accepted protocol for assessment of the risk of marine introductions was developed through the International Council for the Exploration of the Sea (ICES 2003). This protocol emphasizes four points:

1. Probability of colonization and establishment in the area of introduction, which depends on the match between the environment and the species' needs for food, reproduction, and habitat. This section also requires information on resistance to invasion from biotic or abiotic factors in the environment.

2. Probability of spread from the point of introduction, which includes the species' ability to disperse and the extent of suitable environmental conditions.

3. Magnitude of impact on native (especially natural) ecosystems, which includes trophic interactions, habitat transformation, and interactions with native species of concern (threatened or declining).

4. Probability of transport of a harmful pathogen or parasite. This final risk can be mitigated by a variety of methods to inspect and quarantine incoming organisms and release of only their progeny.

The ICES code recognizes two types of risks from introductions, namely, the possible negative impacts of the species itself and the undesirability of bringing in more hitchhiking species. Methods exist to reduce both of these risks: quarantine to reduce hitchhikers and triploidy to reduce establishment of oysters. The most feared organisms to import with oyster shipments are diseases (versus historical concern about predators) because of potential negative impacts on aquaculture and fisheries. Methods for disease reduction incidentally remove oyster predators and other hitchhikers as well (Barrett 1963, Mann 1983, Utting & Spencer 1992, Spencer 2002, NRC 2004). Non-native oysters are often planted as sterile triploids to prevent escape from cultivation and establishment of self-sustaining populations. However, a small percentage of triploid oysters typically revert toward diploidy with age (Guo & Allen 1994). Even triploid oysters are not completely sterile, although their fecundity relative to diploids is small. Nevertheless, the average triploid female still produces thousands of fertilization-capable eggs every year. A second problem with introductions of triploids is that a small percentage of nontriploids may be inadvertently stocked because of a failure in the screening (Dew et al. 2003).

The ICES (2003) also recommends that the risk assessment generates a hypothesis about the outcome of an introduction, which must be tested through postintroduction monitoring and experiments. We examine the history of introduction of *C. gigas* into western North America as a means of conducting an after-the-fact risk assessment. This species was introduced to Washington state in 1902, and regular imports began about 2 decades later and lasted until the 1970s. Imports of spat were initiated without any risk assessment and before another century of accumulated information on other oyster introductions. What would a risk assessment indicate if the species were only now considered for introduction? Here, we briefly consider each of the four points in the ICES protocol:

1. Because *C. gigas* has successfully established in warm bays on western continental shores (e.g., Europe and South Africa), it also would have a high probability of establishment in western North America. It has successfully colonized both rocky and soft-sediment habitats. However, resistance to invasion would be highly uncertain, because it has not been well studied anywhere.

2. *C. gigas* has planktonic larvae that increase the likelihood of long-distance spread from the point of introduction.

3. Impacts on natural ecosystems seem likely. Established populations in Germany occur at low density (Reise 1998), but high-density populations exist in New Zealand and South Africa (Robinson et al. 2005). Recent work in Argentina indicates community-level changes associated with high-density introduced oysters (up to 250 per m^2) (Escapa et al. 2004). However, the prediction is reasonable that *C. gigas* would occupy a higher tidal elevation than does the native species, *O. conchaphila*, and that, in places where it reached high density, it would transform habitat and increase epifaunal diversity. Thus, it would perform a novel ecosystem role in western North American estuaries. Evidence from other countries suggests that *C. gigas*

could be used to replace the economic production value of the native oyster, but it would not provide a functional replacement.

4. The probability of transporting harmful pathogens or parasites could be reduced by release of second-generation individuals, rather than by direct importation of spat. If this risk assessment had been applied, fewer byproduct introductions would have occurred (Figure 2). The high probability of establishment and uncertain impacts might have prompted greater efforts to protect and restore the native oyster, despite its slower growth and small size for aquaculture.

The ICES protocol can also be used to evaluate the potential ecological consequences of introducing *C. gigas* and *C. ariakensis* as replacement for diminished populations of native *C. virginica* in eastern North America:

1. Both introduced species have a high probability of establishing in bays occupied by *C. virginica*. The introduced species could occupy much of the same areas because of their high tolerance of temperature and salinity variation and because they could colonize remnant reefs created by the native species. However, any oyster introduced into the system will sustain high levels of predation from blue crabs, which will severely limit their recovery or establishment (C.H. Peterson, personal communication). Preliminary results from a multi-million-dollar research project recently initiated by the NOAA-Chesapeake Bay Program indicate that *C. ariakensis* has a thin shell compared with *C. virginica*, so is more vulnerable to crab predation (NRC 2004).

2. Both introduced species have long-lived larvae that would likely invade areas not intended for introduction.

3. Both species would have significant impacts on ecosystem functions. *C. ariakensis* and *C. gigas* filter large volumes of water and, therefore, could replace the biofiltration capacity lost with *C. virginica*, as well as fulfill some of the same functions regarding nutrient cycling. However, neither introduced species creates large subtidal reefs like *C. virginica* does. Therefore, the non-natives would not provide this critical ecological function.

4. Introduction of a harmful pathogen (e.g., *Bonamia* sp., via *C. ariakensis*) is possible.

To summarize, this risk assessment indicates that introductions of the two species into estuaries of the eastern United States are likely to have substantial ecological impacts, that introductions would possibly fail because of deleterious biotic interactions and disease, and that effort at restoration of native species should be increased. Powers et al. (2005), who evaluated the restoration success of 103 *C. virginica* reefs from 12 reef sanctuaries in North Carolina, found that restoration of native oysters has been largely successful from both an ecological and fisheries-productivity standpoint, which highlights the possibility that reintroductions of native oysters are a better option for ecosystem restoration than introduction of non-natives.

RESEARCH PRIORITIES

Ecological risk assessments associated with oyster introductions should place greater emphasis on ecosystem-level effects. Oyster introductions require that we advance our understanding of the functions and services provided by different marine species and assemblages. Major gaps in knowledge include how native and introduced species influence nutrient cycling, hydrodynamics, and sediment budgets; whether other native species use them as habitat and food; and the spatial and temporal extent of direct and indirect ecological effects within invaded and adjacent communities and ecosystems. Lack of information on community-level and ecosystem-level consequences of oyster introductions is surprising (but see Escapa et al. 2004), given that these introductions have occurred worldwide for more than a century. Studies that compare the ecosystem functions and services provided by native and introduced oysters are important research priorities, and they provide the framework for recent research projects, such as that supported by the NOAA-Chespaeake Bay Program to examine *C. ariakensis* and *C. gigas* introductions. Comparisons between introduced and native species must emphasize naturalized populations, rather than oysters in aquaculture, although impacts of aquaculture also warrant examination.

An important area of research is the possible context dependency of the impacts of oyster introductions. Introduction of the same species could have dramatically different consequences, depending on local environmental conditions, biological composition, and additional stressors at different sites. The broad geographic distribution of introductions of some oyster species, such as *C. gigas*, provides an opportunity for such spatial comparisons, both within (e.g., among estuaries along the western coast of the United States) and across regions (e.g., western versus eastern United States).

Another critical research area is the role of introduced oysters as vectors, refuges, and resources for other introduced species and diseases (Figure 2, Table 3). Widespread and unanticipated introductions of nonindigenous species and novel diseases through oyster introductions raise major concerns about the ecological and economic consequences of these introductions and call for careful screening of larvae, juveniles, and adults before introduction. Even introduced reef habitat could facilitate establishment and persistence of invasives and pathogens. Facilitation of invaders by species that provide biogenic habitat or other resources that enhance the recruitment, growth, or survival of the invaders has been proposed as a mechanism for "invasion meltdowns" in natural ecosystems (Simberloff & von Holle 1999, Ricciardi 2001). Evidence of invasion facilitation by habitat-creating invasive species exists for estuarine species, such as the cordgrass *Spartina alterniflora* in northern California (Brusati & Grosholz 2005, Neira et al. 2005), the reef-forming tubeworm *Ficopomatus enigmaticus* in central California (K. Heiman, unpublished data), and the bryozoan *Wateresipora subtorquata* in Queensland, Australia (Floerl et al. 2004). We found no similar evidence for oysters because such research has yet to be conducted.

Considering the large uncertainty about the functional equivalence of different oyster species and possible impacts of oyster introductions on native populations and assemblages (focus of this review), introductions should be considered with caution until further, well-directed, and designed research is conducted. The high potential for unintended consequences of oyster introductions suggests that the deliberate introduction of oysters, although often effective in providing the economic benefits of increased aquaculture production, is unlikely to provide an effective tool for the restoration of ecological functions lost from native oyster decline and habitat degradation.

ACKNOWLEDGMENTS

We thank Charles Griffiths and Tammy Robinson (South Africa); Peter Schneider (Namibia); Lobo Orensanz and Marcela Pascual (Argentina); Jenn Kelly and John Volpe (Canada); and Carolyn Friedman, John Grabowski, Terrie Klinger, Charles H. Peterson, and Sean Powers (United States) for ideas and information that improved our presentation. We extend special thanks to our editor Dan Simberloff for his thoughtful insights and guidance throughout this project.

The *Annual Review of Ecology, Evolution, and Systematics* is online at
http://ecolsys.annualreviews.org

LITERATURE CITED

Andrews J. 1980. A review of introductions of exotic oysters and biological planning for new importations. *Mar. Fish. Rev.* 42:1–11

Armstrong DA, Gunderson DR. 1985. The role of estuaries in Dungeness crab early life history: a case study in Grays Harbor, Washington. *Proc. Symp. Dungeness Crab Biol. and Manag.*, symp. coord. B.R. Melteff, pp. 145–70. Fairbanks, Alaska: Univ. Alaska

Askew CG. 1972. The growth of oysters *Ostrea edulis* and *Crassostrea gigas* in Emsworth Harbour. *Aquaculture* 1:237–59

Ayres P. 1991. Introduced Pacific Oysters in Australia. In *The Ecology of* Crassostrea gigas *in Australia, New Zealand, France and Washington State*, ed. MC Greer, JC Leffler, pp. 3–8. College Park, MD: Md. Sea Grant Coll. Prog.

Bahr LM, Lanier WP. 1981. *The Ecology of Intertidal Oyster Reefs of the South Atlantic Coast: A Community Profile*. Washington, DC: US Fish Wildl. FWS/OBS 81.15

Baker P. 1995. Review of ecology and fishery of the Olympia oyster, *Ostrea lurida* with annotated bibliography. *J. Shellfish Res.* 14:501–18

Baqueiro E. 1991. Culture of *Crassostrea corteziensis* in Mexico. See Menzel 1991, pp. 113–16

Barrett EM. 1963. The California oyster industry. *Fish. Bull.* 123:2–103

Bartol I, Mann R. 1999. Small-scale patterns of recruitment on a constructed intertidal reef: the role of spatial refugia. See Luckenbach et al., pp. 159–70

Beirne B. 1975. Biological control attempts by

introductions against pest insects in the field in Canada. *Can. Entomol.* 107:225–36

Berthe FCJ, Le Roux F, Adlard RD, Figueras A. 2004. Marteiliosis in molluscs: a review. *Aquat. Liv. Resour.* 17:433–48

Bishop MS, Peterson CH. 2005. Competitive interactions of a predator (*Callinectes sapidus*) and two oyster prey: *Crassostrea virginica* and *C. ariakensis*. *Mar. Ecol. Prog. Ser.* Submitted

Blackburn TM, Duncan RP. 2001. Determinants of establishment success in introduced birds. *Nature* 414:195–97

Bonnot P. 1935. A recent introduction of exotic species of mollucs in California waters from Japan. *Nautilus* 49:1–2

Bourne N. 1979. Pacific oysters, *Crassostrea gigas* Thunberg, in British Columbia and the South Pacific Islands. See Mann 1979, pp. 1–51

Bower SM, Hervio D, Meyer GR. 1997. Infectivity of *Mikrocytos mackini*, the causative agent of Denman Island disease in Pacific oysters *Crassostrea gigas*, to various species of oysters. *Dis. Aquat. Org.* 29:111–16

Breitburg DL. 1992. Episodic hypoxia in Chesapeake Bay—interacting effects of recruitment, behavior, and physical disturbance. *Ecol. Monogr.* 62:525–46

Breitburg DL. 1999. Are three-dimensional structure and healthy oyster populations the keys to an ecologically interesting and important fish community? See Luckenbach et al. 1999, pp. 239–50

Breitburg DL, Loher T, Pacey CA, Gerstein A. 1997. Varying effects of low dissolved oxygen on trophic interactions in an estuarine food web. *Ecol. Monogr.* 67:489–507

Breitburg DL, Palmer MA, Loher T. 1995. Larval distributions and the spatial patterns of settlement of an oyster reef fish—responses to flow and structure. *Mar. Ecol. Prog. Ser.* 125:45–60

Brooks WK. 1891. *The Oyster: A Popular Summary of A Scientific Study*. Baltimore, MD: Johns Hopkins Univ. Press. 230 pp.

Brusati ED, Grosholz ED. 2005. Effect of native and invasive cordgrass on *Macoma petalum*

density, growth, and isotopic signatures. *Mar. Ecol. Prog. Ser.* In press

Burreson EM, Stokes NA, Carnegie RB, Bishop MJ. 2004. *Bonamia* sp. (Haplosporidia) found in nonnative oysters *Crassostrea ariakensis* in Bogue Sound, North Carolina. *J. Aquat. Anim. Health* 16:1–9

Burreson EM, Stokes NA, Friedman CS. 2000. Increased virulence in an introduced pathogen: *Haplosporidium nelsoni* (MSX) in the Eastern oyster *Crassostrea virginica*. *J. Aquat. Anim. Health* 12:1–8

Buschmann A, Lopez DA, Medina A. 1996. A review of environmental effects and alternative production strategies of marine aquaculture in Chile. *Aquacul. Eng.* 15:397–421

Byers JE. 2000. Competition between two estuarine snails: implications for invasions of exotic species. *Ecology* 81:1225–39

Carbines G, Jiang WM, Beentjes MP. 2004. The impact of oyster dredging on the growth of blue cod, *Parapercis colias*, in Foveaux Strait, New Zealand. *Aquat. Cons. Mar. Freshw. Ecosyst.* 14:491–504

Carlton JT. 1992a. Dispersal of living organisms into aquatic ecosystems as mediated by aquaculture and fisheries activities. In *Dispersal of Living Organisms into Aquatic Ecosystems*, ed. A Rosenfield, R Mann, pp. 13–45. College Park, MD: Md. Sea Grant Prog.

Carlton JT. 1992b. Introduced marine and estuarine mollusks of North America: an end of the-20th-century perspective. *J. Shellfish Res.* 11:489–505

Carlton JT. 1999. Molluscan invasions in marine and estuarine communities. *Malacologia* 41:439–54

Carlton JR, Mann R. 1996. Transfers and worldwide introductions. In *The Eastern Oyster: Crassostrea virginica*, ed. VS Kennedy, RIE Newell, AF Eble, pp. 691–706. College Park, MD: Md. Sea Grant Coll. Prog.

Carriker MR, Gaffney PM. 1996. A catalogue of selected species of living oysters (Ostreacea) of the World. In *The Eastern Oyster: Crassostrea virginica*, ed. VS Kennedy, RIE

Newell, AF Eble, pp. 1–18. College Park, MD: Md. Sea Grant Coll. Prog.

Castel J, Labourg PJ, Escaravage V, Auby I, Garcia ME. 1989. Influence of seagrass beds and oyster parks on the abundance and biomass patterns of meiobenthos and macrobenthos in tidal flats. *Est. Coast. Shellfish S.* 28:71–85

Chanley MH, Chanley P. 1991. Cultivation of the Chilean oyster, *Tiostrea chilensis* (Philippi, 1845). See Menzel 1991, pp. 145–51

Chew KK, ed. 1982. *Proceedings of the North American Oyster Workshop: March 6–8, 1981, Seattle.* Baton Rouge, LA: La. State Univ. Div. Contin. Educ. 300 pp.

Chew K. 1990. Global bivalve shellfish introductions. *World Aquacult.* 21:9–22

Chiba K, Taki Y, Sakai K, Oozeki Y. 1989. Present status of aquatic organisms introduced into Japan. In *Exotic Aquatic Organisms in Asia, Proceedings of the Workshop on Introduction of Exotic Aquatic Organisms in Asia,* ed. SS De Silva, 3:63–70. Manila, Philippines: Asian Fisheries Society

Cigarria J, Elston R. 1997. Independent introduction of *Bonamia ostreae,* a parasite of *Ostrea edulis,* to Spain. *Dis. Aquat. Org.* 29:157–58

Cochennec N, Renault T, Boudry P, Chollet B, Gerard A. 1998. *Bonamia*-like parasite found in the Suminoe oyster *Crassostrea rivularis* reared in France. *Dis. Aquat. Org.* 34:193–97

Cochennec-Laureau N, Reece KS, Berthe FCJ, Hine PM. 2003. *Mikrocytos roughleyi* taxonomic affiliation leads to the genus *Bonamia* (Haplosporidia) *Dis. Aquat. Org.* 54:209–17

Coen LD, Knott DM, Wenner EL, Hadley NH, Ringwood AH. 1999a. Intertidal oyster reef studies in South Carolina: design, sampling and experimental focus for evaluating habitat value and function. See Luckenbach et al. 1999, pp. 133–58

Coen LD, Luckenbach MW. 2000. Developing success criteria and goals for evaluating oyster reef restoration: ecological function or resource exploitation? *Ecol. Eng.* 15:323–43

Coen LD, Luckenback MW, Breitburg DL.

1999b. The role of oyster reefs as essential habitat: a review of current knowledge and some new perspectives. In *Fish Habitat: Essential Fish Habitat and Rehabilitation,* ed. LR Benaka, 22:438–54. Bethesda MD: Am. Fish. Soc.

Coutteau P, Coolsaet N, Caers M, Bogaert P, De Clerck R. 1997. Re-introduction of oyster cultivation in the sluice-dock in Ostend, Belgium. *J. Shellfish Res.* 16:262

Cranfield HJ, Gordon DP, Willan RC, Marshall BA, Battershill CN, et al. 1998. Adventive marine species in New Zealand. *NIWA Tech. Rep. 34.* 48 pp

Cranfield HJ, Rowden AA, Smith DP, Gordon KP, Michael KP. 2004. Macrofaunal assemblages of benthic habitat of different complexity and the proposition of a model of biogenic reef habitat regeneration in Foveaux Strait, New Zealand. *J. Sea Res.* 52:109–25

Crawford C. 2003. Environmental management of marine aquaculture in Tasmania, Australia. *Aquaculture* 226:129–38

Crawford CM, Macleod CKA, Mitchell IM. 2003. Effects of shellfish farming on the benthic environment. *Aquaculture* 224:117–40

Crooks JA. 2002. Characterizing ecosystem-level consequences of biological invasions: the role of ecosystem engineers. *Oikos* 97:153–66

Cuddington K, Hastings A. 2004. Invasive engineers. *Ecol. Model.* 178:335–47

Dame RF. 1976. Energy flow in an intertidal oyster population. *Est. Coast. Mar. Sci.* 4:243–53

Dame RF, Chrzanowski T, Bildstein K, Kjerfve B, McKeller H, et al. 1986. The outwelling hypothesis and North Inlet, South Carolina. *Mar. Ecol. Prog. Ser.* 33:217–29

Dame RF, Patten BC. 1981. Analysis of energy flows in an intertidal oyster reef. *Mar. Ecol. Prog. Ser.* 5:115–24

Dame RF, Spurrier JD, Zingmark RG. 1992. In situ metabolism of an oyster reef. *J. Exp. Mar. Biol. Ecol.* 164:147–59

Dame RF, Zingmark RG, Haskin E. 1984. Oyster reefs as processors of estuarine materials. *J. Exp. Mar. Biol. Ecol.* 83:239–47

Danton M. 1914. L'Huitre Portugaise tend-elle à remplacer l'Ostrea edulis? *Comptes Rendus Acad. Sci.* 158:360–62

Davis MA, Grime JP, Thompson K. 2000. Fluctuating resources in plant communities: a general theory of invasibility. *J. Ecol.* 88: 528–34

Dean D. 1979. Introduced species and the Maine situation. See Mann 1979, pp. 149–61

De Montaudouin X, Sauriau PG. 1999. The proliferating Gastropoda *Crepidula fornicata* may stimulate macrozoobenthic diversity. *J. Mar. Biol. Assoc. UK* 79:1069–77

Dew JR, Berkson J, Hallerman EM, Allen SK. 2003. A model for assessing the likelihood of self-sustaining populations resulting from commercial production of triploid Suminoe oysters (*Crassostrea ariakensis*) in Chesapeake Bay. *Fish. Bull.* 101:758–68

Dinamani P. 1991a. The northern rock oyster, *Saccostrea glomerata* (Gould, 1850), in New Zealand. See Menzel 1991, pp. 335–41

Dinamani P. 1991b. The Pacific oyster, *Crassostrea gigas* (Thunberg, 1793), in New Zealand. See Menzel 1991, pp. 343–52

Dinamani P. 1991c. Introduced Pacific oysters in New Zealand. In *The Ecology of* Crassostrea gigas *in Australia, New Zealand, France and Washington State*, ed. MC Greer, JC Leffler, pp. 9–12. College Park, MD: Md. Sea Grant Coll. Prog.

Drinkwaard AC. 1998. Introductions and developments of oysters in the North Sea area: a review. *Helgol. Meeresunt.* 52:301–08

Dumbauld BR, Brooks KM, Posey MH. 2001. Response of an estuarine benthic community to application of the pesticide carbaryl and culture of Pacific oysters (*Crassostrea gigas*) in Willapa Bay, Washington. *Mar. Poll. Bull.* 42:826–44

Dumbauld BR, Visser EP, Armstrong DA, Cole-Warner L, Feldman KL, Kauffman BE. 2000. Use of oyster shell to create habitat for juvenile Dungeness crab in Washington coastal estuaries: status and prospects. *J. Shellfish Res.* 19:379–86

Eggleston DB. 1990. Foraging behavior of the blue crab, *Callinectes sapidus*, on juvenile oysters, *Crassostrea virginica*: effects of prey density and size. *Bull. Mar. Sci.* 46:62–82

Eldredge LG. 1994. *Introductions of Commercially Significant Aquatic Organisms to the Pacific Islands*. Noumea, New Caledonia: South Pac. Comm. 127 pp.

Elston RA, Wilkinson MT. 1985. Pathology, management and diagnosis of oyster velar virus-disease (OVVD). *Aquaculture* 48:189–210

Elton CS. 1958. *The Ecology of Invasions by Animals and Plants*. Chicago: Univ. Chicago Press. 181 pp.

Escapa M, Isacch JP, Daleo P, Alberti J, Iribarne O, et al. 2004. The distribution and ecological effects of the introduced Pacific oyster *Crassostrea gigas* (Thunberg, 1793) in northern Patagonia. *J. Shellfish Res.* 23:765–72

Everett RA, Ruiz GM, Carlton JT. 1995. Effect of oyster mariculture on submerged aquatic vegetation: an experimental test in a Pacific Northwest estuary. *Mar. Ecol. Prog. Ser.* 125:205–17

FAO/FIGIS. *Database on introduction of aquatic species, Fisheries Global Information System*. http://www.fao.org/figis/servlet/static?dom=collection&xml=dias.xml

FAO. 2002a. *FAO Aquaculture Production: 1950–2002: FAO Yearbook. Fishery Statistics.* Vol 94/2. New York: Food Agric. Organ., U.N.

FAO. 2002b. *FAO Capture Production: 1950–2002: FAO Yearbook. Fishery Statistics.* Vol. 94/1. New York: Food Agric. Organ., U.N.

Feldman KL, Armstrong DA, Dumbauld BR, DeWitt TH, Doty DC. 2000. Oysters, crabs, and burrowing shrimp: Review of an environmental conflict over aquatic resources and pesticide use in Washington state's (USA) coastal estuaries. *Estuaries* 23:141–76

Fernandez M, Iribarne O, Armstrong D. 1993. Habitat selection by young-of-the-year Dungeness crab *Cancer magister* and predation risk in intertidal habitats. *Mar. Ecol. Prog. Ser.* 92:171–77

Floerl O, Pool TK, Inglis GJ. 2004. Positive interactions between nonindigenous species facilitate transport by human vectors. *Ecol. Appl.* 14:1724–36

Ford SE. 1996. Range extension by the oyster parasite *Perkinsus marinus* into the northeastern United States: response to climate change? *J. Shellfish Res.* 15:45–56

Forsyth DM, Duncan RP, Bomford M, Moore G. 2004. Climatic suitability, life-history traits, introduction effort, and the establishment and spread of introduced mammals in Australia. *Conserv. Biol.* 18:557–69

Friedman CS. 1996. Haplosporidian infections of the Pacific oyster, *Crassostrea gigas* (Thunberg), in California and Japan. *J. Shellfish Res.* 15:597–600

Friedman CS, Beaman BL, Chun J, Goodfellow M, Gee A, Hedrick RP. 1998. *Nocardia crassostreae* sp. nov., the causal agent of nocardiosis in Pacific oysters. *Int. J. Syst. Bacteriol.* 48:237–46

Friedman CS, Perkins FO. 1994. Range extension of *Bonamia ostrae* to Maine, USA. *J. Invert. Path.* 64:179–81

Galil BS. 2000. A sea under siege—alien species in the Mediterranean. *Biol. Inv.* 2:177–86

Galil BS. 2003. Exotics in the Mediterranean: bioindicators for a sea change. *BIOMARE Online Newsl.* Vol. 1. www.biomareweb.org/1.1.html

Galtsoff PS. 1932. Introduction of Japanese oysters into the United States. *Fish. Circ.* 12:1–16

Garcia-Meunier P, Martel C, Pigeot J, Chevalier G, Blanchard G, et al. 2002. Recent invasion of the Japanese oyster drill along the French Atlantic coast: identification of specific molecular markers that differentiate Japanese, *Ocinebrellus inornatus*, and European, *Ocenebra erinacea*, oyster drills. *Aquat. Living Resour.* 15:67–71

Glude J. 1981. *The Feasibility of Aquaculture in the Bahamas.* Rome: Food and Agriculture Organization, United Nations. 65 pp.

Gollasch S, Carlberg S, Hansen MM, eds. 2003. *ICES Code of Practice on the Introductions and Transfers of Marine Organisms*, Copenhagen, Denmark: International Council for Exploration of the Sea. 28 pp.

Gottlieb SJ, Schweighofer ME. 1996. Oysters and the Chesapeake Bay ecosystem: A case for exotic species introduction to improve environmental quality? *Estuaries* 19:639–50

Goulletquer P, Heral M. 1991. Aquaculture of *Crassostrea gigas* in France. In *The Ecology of* Crassostrea gigas *in Australia, New Zealand, France and Washington State.* ed. MC Greer, JC Leffler, pp. 13–22. College Park, MD: Md. Sea Grant Coll. Prog.

Grabowski JH. 2004. Habitat complexity disrupts predator-prey interactions but not the trophic cascade on oyster reefs. *Ecology* 85:995–1004

Grabowski JH, Peterson CH, Powers SP, Gaskill D, Summerson HC. 2004. Growth and survivorship of non-native (*Crassostrea gigas* and *Crossostrea ariakensis*) versus native Eastern oysters (*Crassostrea virginica*). *J. Shellfish Res.* 23:781–93

Grabowski JH, Powers SP. 2004. Habitat complexity mitigates trophic transfer on oyster reefs. *Mar. Ecol. Prog. Ser.* 277:291–95

Grizel H. 1979. *Marteilia refringens* and oyster disease—recent observations. *Mar. Fish. Rev.* 41:38–39

Grizel H. 1996. Some examples of the introduction and transfer of mollusc populations *Rev. Sci. Tech. Office Int. Epizooties* 15:401–08

Grizel H, Heral M. 1991. Introduction into France of the Japanese oyster (*Crassostrea gigas*). *J. Conseil.* 47:399–403

Guo XM, Allen SK. 1994. Reproductive potential and genetics of triploid Pacific oysters, *Crassostrea gigas* (Thunberg). *Biol. Bull.* 187:309–18

Harding JM, Mann R. 2001a. Diet and habitat use by bluefish, *Pomotomus saltatrix*, in a Chesapeake Bay estuary. *Env. Biol. Fish.* 60:401–09

Harding JM, Mann R. 2001b. Oyster reefs as fish habitat: opportunistic use of restored reefs by transient fishes. *J. Shellfish Res.* 20:951–59

Harding JM, Mann R. 2003. Influence of habitat

on diet and distribution of striped bass (*Morone saxatilis*) in a temperate estuary. *Bull. Mar. Sci.* 72:841–51

Hardy JD Jr. 1978a. *Development of Fishes of the Mid-Atlantic Bight*. Washington, DC: US Department of the Interior Fish and Wildlife Service. 458 pp.

Hardy JD Jr. 1978b. *Development of Fishes of the Mid-Atlantic Bight*. Washington, DC: US Department of the Interior Fish and Wildlife Service. 394 pp.

Harvell CD, Kim K, Burkholder JM, Colwell RR, Epstein PR, et al. 1999. Emerging marine diseases—climate links and anthropogenic factors. *Science* 285:1505–10

Hedgpeth JW, Obrebski S. 1981. *Willapa Bay: A Historical Perspective and a Rationale for Research*. Washington, DC: Office of Biological Services, U.S. Fish and Wildlife Service (FWS/OBS-81/03)

Heral M, Deslous-Paoli JM. 1991. Oyster culture in European countries. See Menzel 1991, pp. 153–90

Hickey JM. 1979. Culture of the Pacific oyster, *Crassostrea gigas*, in Massachusetts waters. See Mann 1979, pp. 129–39

Hidu H, Lavoie RE. 1991. The European oyster, *Ostrea edulis* L., in Maine and Eastern Canada. See Menzel 1991, pp. 35–46

Hine PM. 1996. The ecology of *Bonamia* and decline of bivalve molluscs. *NZ J. Ecol.* 20: 109–16

Hine PM, Cochennec-Laureau N, Berthe FCJ. 2001. *Bonamia exitiosus* n. sp (Haplosporidia) infecting flat oysters *Ostrea chilensis* in New Zealand. *Dis. Aquat. Org.* 47:63–72

Honkoop PJC, Bayne BL, Drent J. 2003. Flexibility of size of gills and palps in the Sydney rock oyster *Saccostrea glomerata* (Gould, 1850) and the Pacific oyster *Crassostrea gigas* (Thunberg, 1793). *J. Exp. Mar. Biol. Ecol.* 282:113–33

Hopkins AE. 1937. Experimental observations on spawning, larval development, and setting in the Olympia oyster *Ostrea lurida*. *Bull. US Bur. Fish.* 48:438–503

Hosack G. 2003. *Effects of* Zostera marina *and* Crassostrea gigas *Culture on the Intertidal Communities of Willapa Bay, Washington*. MS Thesis. Seattle: Univ. Wash.

Hughes-Games WL. 1977. Growing the Japanese oyster (*Crassostrea gigas*) in subtropical fish ponds. I. Growth rate, survival and quality index. *Aquaculture* 11:217–29

Iribarne O, Armstrong D, Fernandez M. 1995. Environmental impact of intertidal juvenile Dungeness crab habitat enhancement: effects on bivalves and crab foraging rate. *J. Exp. Mar. Biol. Ecol.* 192:173–94

Irlandi EA, Crawford MK. 1997. Habitat linkages: the effect of intertidal saltmarshes and adjacent subtidal habitats on abundance, movement, and growth of an estuarine fish. *Oecologia* 110:222–30

Islas RO. 1975. El ostion japones *Crassostrea gigas* en Baja California. *Cien. Mar.* 2:50–59

Jackson JBC, Kirby MX, Berger WH, Bjorndal KA, Botsford LW, et al. 2001. Historical overfishing and the collapse of marine ecosystems. *Science* 293:629–38

Juliano RO, Guerrero R III, Ronquillo I. 1989. The introduction of exotic aquatic species in the Philippines. In *Exotic Aquatic Organisms in Asia, Proc. Workshop Intro. Exotic Aquat. Org. Asia*, ed. SS De Silva, 3:83–90. Manila, Philippines: Asian Fisheries Soc.

Kaufmann MJ, Seaman MNL, Andrade C, Buchholz F. 1994. Survival, growth, and glycogen content of Pacific oysters, *Crassostrea gigas* (Thunberg, 1793), at Madeira Island (Subtropical Atlantic). *J. Shellfish Res.* 13:503–05

Kennedy VS. 1996. The ecological roles of the eastern oyster, *Crassostrea virginica*, with remarks on disease. *J. Shellfish Res.* 15:177–83

Kincaid T. 1951. *The Oyster Industry of Willapa Bay, Washington*. Ilwaco, WA: Tribune. 45 pp.

Kincaid T. 1968. *The Ecology of Willapa Bay, Washington, in Relation to the Oyster Industry*. Seattle, WA: Self-published, 84 pp. +30 illus.

Kirby MX. 2004. Fishing down the coast: historical expansion and collapse of oyster fisheries along continental margins. *Proc. Natl. Acad. Sci.* 101:13096–99

Kolar CS, Lodge DM. 2001. Progress in invasion biology: predicting invaders. *Trends Ecol. Evol.* 16:199–204

Kuris AM, Culver CS. 1999. An introduced sabellid polychaete pest infesting cultured abalones and its potential spread to other California gastropods. *Invert. Biol.* 118:391–403

Kusuki Y. 1991. Oyster culture in Japan and adjacent countries: *Crassostrea gigas* (Thunberg). See Menzel 1991, pp. 227–43

Langdon CJ, Robinson AM. 1996. Aquaculture potential of the Suminoe oyster (*Crassostrea ariakensis* Fugita 1913). *Aquaculture* 144:321–38

Leguerrier D, Hiquil N, Petiau A, Bodoy A. 2004. Modeling the impact of oyster culture on a mudflat food web in Marennes-Oleron Bay (France). *Mar. Ecol. Prog. Ser.* 273:147–62

Lenihan HS. 1999. Physical-biological coupling on oyster reefs: how habitat form influences individual performance. *Ecol. Monogr.* 69:251–75

Lenihan HS, Micheli F, Shelton SW, Peterson CH. 1999. The influence of multiple environmental stressors on susceptibility to parasites: an experimental determination with oysters. *Limnol. Oceangr.* 44:910–24

Lenihan HS, Peterson CH. 1998. How habitat degredation through fishery disturbance enhances impacts of hypoxia on oyster reefs. *Ecol. Appl.* 11:128–40

Lenihan HS, Peterson CH, Allen JM. 1995. Does flow also have a direct effect on growth of active suspension feeders? An experimental test with oysters. *Limnol. Oceangr.* 41:1359–66

Lenihan HS, Peterson CH, Byers JE, Grabowski JH, Thayer GH, Colby DR. 2001. Cascading of habitat degradation: oyster reefs invaded by refugee fishes escaping stress. *Ecol. Appl.* 11:748–64

Luckenbach MW, Mann R, Wesson JA, eds. 1999. *Oyster Reef Habitat Restoration: A Synopsis and Synthesis of Approaches.* Gloucester Point, VA: Va. Acad. Mar. Sci.

Macdonald IAW, Reaser JK, Bright C, Neville LE, Howard GW, et al. 2003. Invasive alien species in Southern Africa. *National Reports and Directory of Resources.* Cape Town, South Africa: Global Invasive Species Prog.

Mann R, ed. 1979. *Exotic Species in Mariculture.* Cambridge, MA: MIT Press. 363 pp.

Mann R. 1983. The role of introduced bivalve mollusc species in mariculture. *J. World Maric. Soc.* 14:546–59

Mann R, Burreson E, Baker P. 1991. The decline of the Virginia oyster fishery in Chesapeake Bay: considerations for introduction of a non-endemic species, *Crassostrea gigas* (Thunberg, 1793). *J. Shellfish Res.* 10:379–88

Margalef R. 1968. *Perspectives in Ecological Theory.* Chicago: Univ. Chicago Press. 111 pp.

Martel C, Guarini JM, Blanchard G, Sauriau PG, Trichet C, et al. 2004. Invasion by the marine gastropod *Ocinebrellus inornatus* in France. III. Comparison of biological traits with the resident species *Ocenebra erinacea*. *Mar. Biol.* 146:93–102

Maurin C, Gras P. 1979. Experiments on the growth of the mangrove oyster, *Crassostrea rhizophorae*, in France. See Mann 1979, pp. 123–28

McMillin HC, Bonnot P. 1932. Oyster pests in California. *Calif. Fish Game* 18:147–48

MDSG. 1991. The ecology of *Crassostrea gigas* in Australia, New Zealand, France and Washington State. Synopsis of the Oyster Ecology Workshop: *Crassostrea gigas. Maryland Sea Grant Symp. Rep.* College Park, MD: Md. Sea Grant Coll. Prog.

Menzel RW, ed. 1991. *Estuarine and Marine Bivalve Mollusk Culture.* Boca Raton, FL: CRC Press. 376 pp.

Meyer DL, Townsend EC. 2000. Faunal utilization of created intertidal eastern oyster (*Crassostrea virginica*) reefs in the southeastern United States. *Estuaries* 23:35–45

Micheli F, Peterson CH. 1999. Estuarine vegetated habitats as corridors for predator movements. *Conserv. Biol.* 13:869–81

Miller W III, Morrison SD. 1988. Marginal marine Pleistocene fossils from near mouth of Mad River, northern California. *Proc. Calif. Acad. Sci.* 45:255–66

Mobius K. 1877. The oyster and oyster-culture. In *United States Commission of Fish and Fisheries Part VIII Report of the Commissioner for 1880*, ed. HJ Rice, pp. 683–751. Washington, DC: Gov. Print. Off.

Nagabhushanam R, Mane UH. 1991. Oysters in India. See Menzel 1991, pp. 201–9

Nascimento IA. 1991. *Crassostrea rhizophorae* (Guilding) and *C. brasiliana* (Lamarck) in South and Central America. See Menzel 1991, pp. 125–34

NRC. 2004. *Nonnative Oysters in the Chesapeake Bay.* Washington, DC: Natl. Acad. 343 pp.

Neira CL, Levin LA, Grosholz ED. 2005. Benthic macrofaunal communities of three sites in San Francisco Bay invaded by hybrid *Spartina*, with comparsion to uninvaded habitats. *Mar. Ecol. Prog. Ser.* 292:111–26

Nell J. 2002. The Australian oyster industry. *W. Aquacult.* 33:8–10

Nelson KA, Leonard LA, Posey MH, Alphin TD, Mallin MA. 2004. Using transplanted oyster (*Crassostrea virginica*) beds to improve water quality in small tidal creeks: a pilot study. *J. Exp. Mar. Bio. Ecol.* 298:347–68

Neto AI. 2000. Ecology and dynamics of two intertidal algal communities on the littoral of the island of Sao Miguel (Azores). *Hydrobiologia* 432:135–47

Newell RIE. 1988. Ecological changes in the Chesapeake Bay: Are they the result of overharvesting the American oyster? In *Understanding the Estuary: Advances in Chesapeake Bay Research*, ed. MP Lynch, EC Krome, pp. 536–46. Baltimore, MD: Chesapeake Bay Res. Consort.

Newell RIE. 2004. Ecosystem influences of natural and cultivated populations of suspension-feeding bivalve molluscs: a review. *J. Shellfish Res.* 23:51–61

Newell RIE, Koch EW. 2004. Modeling seagrass density and distribution in response to changes in turbidity stemming from bivalve filtration and seagrass sediment stabilization. *Estuaries* 27:793–806

Nie ZQ. 1991. The culture of marine bivalve mollusks in China. See Menzel 1991, pp. 261–76

Officer CB, Biggs RB, Taft JL, Cronin LE, Tyler MA, Boynton WR. 1984. Chesapeake Bay anoxia: origin, development, and significance. *Science* 223:22–25

Orensanz JM, Schwindt E, Pastorino G, Bortolus A, Casa G, et al. 2002. No longer the pristine confines of the world ocean: a survey of exotic marine species in the southwestern Atlantic. *Biol. Invasions* 4:115–43

Paillard C, Le Roux F, Borreg JJ. 2004. Bacterial disease in marine bivalves, a review of recent studies: trends and evolution. *Aquat. Living Resour.* 17:477–98

Paine RT. 1966. Food web complexity and species diversity. *Am. Nat.* 100:65–76

Park MS, Kang CK, Choi DL, Jee BY. 2003. Appearance and pathogenicity of ovarian parasite *Marteilioides chungmuensis* in the farmed Pacific oysters, *Crassostrea gigas*, in Korea. *J. Shellfish Res.* 22:475–79

Pascual M, Martin AG, Zampatti E, Coatanea D, Defossez J, Robert R. 1991. Testing of the Argentina oyster, Ostrea puelchana, in several French oyster farming sites. *Conserv. Int. Explor. Mer.* C.M.1991/K:30. Copenhagen, Denmark: International Council for Exploration of the Sea, 17 pp.

Perdue JA, Erickson G. 1984. A comparison of the gametogenic cycle between the Pacific oyster *Crassostrea gigas* and the Suminoe oyster *Crassostrea rivularis* in Washington state. *Aquaculture* 37:231–37

Peterson BJ, Heck KL Jr. 1999. The potential for suspension feeding bivalves to increase seagrass productivity. *J. Exp. Mar. Biol. Ecol.* 240:37–52

Peterson CH, Grabowski JH, Powers SP. 2003. Estimated enhancement of fish production resulting from restoring oyster reef habitat: quantitative valuation. *Mar. Ecol. Prog. Ser.* 264:249–64

Pollard D, Hutchings P. 1990. A review of exotic marine organisms introduced to the Australian region II. Invertebrates and algae. *Asian Fish. Sci.* 3:223–50

Posey MH, Alphin TD, Powell CM, Townsend E. 1999. Oyster reefs as habitat for fish and decapods. See Luckenbach et al. 1999, pp. 229–37

Powell EN, Hofmann EE, Klinck JM, Ray SM. 1992. Modeling oyster populations I. A commentary on filtration rate. Is faster always better? *J. Shellfish Res.* 11:387–98

Powers SP, Peterson CH, Grabowski JH, Lenihan HS. 2005. The realities of native oyster restoration and why the myth of failure intensifies a conservation crisis. *Rest. Ecol.* Submitted

Price P. 1980. *Evolutionary Biology of Parasites*. Princeton, NJ: Princeton Univ. Press. 412 pp.

Quayle DB. 1969. *Pacific Oyster Culture in British Columbia*, 169–92. Ottawa: Queen's Press Can.

Quek T. 2004. Local oysters, anyone? Ten professionals invest in their passion—island's first oyster farm produces 50,000 oysters a month. *The Straits Times*, Sept. 26: Singapore Press

Raghukumar C, Lande V. 1988. Shell disease of rock oyster *Crassostrea cucullata*. *Dis. Aquat. Org.* 4:77–81

Ranson G. 1926. L'Huitre Portugaise tend-elle à remplacer l'Huitre Francaise? *Off. Sci. Tech. Peches Marit., Notes Mem.* 47:2–9

Reece KS, Bushek D, Hudson KL, Graves JE. 2001. Geographic distribution of *Perkinsus marinus* genetic strains along the Atlantic and Gulf coasts of the USA. *Mar. Biol.* 139:1047–55

Reichard SH, Hamilton CW. 1997. Predicting invasions of woody plants introduced into North America. *Conserv. Biol.* 11:193–203

Reise K. 1985. Tidal flat ecology: an experimental approach to species interactions. Berlin: Springer-Verlag. 289 pp.

Reise K. 1998. Pacific oysters invade mussel beds in the European Wadden Sea. *Senckenb. Marit.* 28:167–75

Rejmanek M, Richardson DM. 1996. What attributes make some plant species more invasive? *Ecology* 77:1655–61

Renault T, Chollet B, Cochennec N, Gerard A. 2002. Shell disease in eastern oysters, *Crassostrea virginica*, reared in France. *J. Invert. Path.* 79:1–6

Renault T, Novoa B. 2004. Viruses infecting bivalve molluscs. *Aquat. Liv. Res.* 17:397–409

Ricciardi A. 2001. Facilitative interactions among aquatic invaders: Is an "invasional meltdown" occurring in the Great Lakes? *Can. J. Fish. Aquat. Sci.* 58:2513–25

Richardson CA, Seed R, Alroumaihi EMH, McDonald L. 1993. Distribution, shell growth and predation of the New Zealand oyster, *Tiostrea* (=*Ostrea*) *lutaria* Hutton, in the Menai Strait, North Wales. *J. Shellfish Res.* 12:207–14

Robinson TB, Griffiths CL, Tonin A, Bloomer P, Hare MP. 2005. Naturalized populations of *Crassostrea gigas* along the South African coast: distribution, abundance and population structure. *J. Shellfish Res.* In press

Rothschild BJ, Ault JS, Goulletquer P, Heral M. 1994. Decline of the Chesapeake Bay oyster population: a century of habitat destruction and overfishing. *Mar. Ecol. Prog. Ser.* 111:29–39

Ruesink JL. 2005. Global analysis of factors affecting the outcome of freshwater fish introductions. *Conserv. Biol.* In press

Ruesink JL, Feist BE, Harvey CJ, Hong JS, Trimble AC, Wisehart LM. 2005. Change. in productivity associated with four introduced species: ecosystem transformation of a "pristine" estuary. *Mar. Ecol. Prog. Ser.* In press

Ruesink JL, Parker IM, Groom MJ, Kareiva PM. 1995. Reducing the risks of nonindiginous species introductions—guilty until proven innocent. *Bioscience* 45:465–77

Ruesink JL, Roegner GC, Dumbauld BR, Newton JA, Armstrong DA. 2003. Contributions of coastal and watershed energy sources to secondary production in a northeastern Pacific estuary. *Estuaries* 26:1079–93

Shafee MS, Sabatie MR. 1986. Croissance et mortalite des huitres dans la lague Oualidia (Maroc). *Aquaculture* 53:201–14

Shatkin G, Shumway SE, Hawes R. 1997. Considerations regarding the possible introduction of the Pacific oyster (*Crassostrea gigas*) to the Gulf of Maine: a review of global experience. *J. Shellfish Res.* 16:463–77

Shea K, Chesson P. 2002. Community ecology as a framework for biological invasions. *Trends Ecol. Evol.* 17:170–76

Shpigel M. 1989. Gametogenesis of the European flat oyster (*Ostrea edulis*) and Pacific oyster (*Crassostrea gigas*) in warm water in Israel. *Aquaculture* 80:343–49

Simberloff D, Von Holle B. 1999. Positive interactions of nonindigenous species: invasional meltdown? *Biol. Inv.* 1:21–32

Simenstad CA, Fresh KL. 1995. Influence of intertidal aquaculture on benthic communities in Pacific Northwest estuaries—scales of disturbance. *Estuaries* 18:43–70

Spencer BE. 2002. *Molluscan Shellfish Farming*. Oxford: Blackwell Scientific. 272 pp.

Stafford J. 1913. *The Canadian Oyster: Its Development, Environment and Culture*. Ottawa, Ontario: The Mortimer Company. 159 pp.

Sunila I, Stokes NA, Smolowitz R, Karney RC, Burreson EM. 2002. *Haplosporidium costale* (seaside organism), a parasite of the eastern oyster, is present in Long Island Sound. *J. Shellfish Res.* 21:113–18

Tan Y, Tong H. 1989. The status of the exotic aquatic organisms in China. In *Exotic Aquatic Organisms in Asia, Proceedings of the Workshop on Introduction of Exotic Aquatic Organisms in Asia*, ed. SS De Silva, 3: 35–43. Manila, Philippines: Asian Fisheries

Tavares M. 2003. On *Halicarcinus planatus* (Fabricius) (Brachyura, Hymenosomatidae) transported from Chile to Brazil along with the exotic oyster *Crassostrea gigas* (Thunberg). *Nauplius* 11:45–50

Thomson J. 1952. The acclimatization and growth of the Pacific oyster *Cryphaea gigas* in Australia. *Aust. J. Mar. Fresh. Res.* 3:64–73

Thomson J. 1959. The naturalization of the Pacific oyster in Australia. *Aust. J. Mar. Fresh. Res.* 10:144–49

Townsend CH. 1896. *The Transplanting of Eastern Oysters to Willapa Bay, Washington, with Notes on the Native Oyster Industry*. Rep. US Comm. Fish Fish. 1895. Washington DC: Gov. Print. Off.

Tuttle JH, Jonas RB, Malone TC. 1987. Origin, development and significance of Chesapeake Bay anoxia. In *Containment Problems and Management of Living Chesapeake Bay Resources*, ed. SK Malumdar, LW Hall Jr, HM Austin, pp. 422–72. Philadelphia, PA: Penn. Acad. Sci.

Ulanowicz RE, Tuttle JH. 1992. The trophic consequences of oyster stock rehabilitation in Chesapeake Bay. *Estuaries* 15:298–306

Utting S, Spencer B. 1992. Introductions of marine bivalve molluscs into the United Kingdom for commercial culture: case histories. *Int. Council Expl. Sea Mar. Sci. Symp.* 194:84–91

Velez A. 1991. Biology and culture of the Caribbean or Mangrove oyster, *Crassostrea rhizophorae* Guilding, in the Caribbean and South America. See Menzel 1991, pp. 117–24

Veltman CJ, Nee S, Crawley MJ. 1996. Correlates of introduction success in exotic New Zealand birds. *Am. Nat.* 147:542–57

Viallanes H. 1892. Recherches sur la filtration de l'eau par les Mollusques et applications a l'Ostreiculture et a l'Oceanographie. *Comptes Rendus Hebd. Sceances Acad. Sci.* 114:1386–88

Villalba A, Reece KS, Ordas MC, Casas SM, Figueras A. 2004. Perkinsosis in molluscs: a review. *Aquat. Liv. Resour.* 17:411–32

Walne PR, Helm MM. 1979. Introduction of *Crassostrea gigas* into the United Kingdom. See Mann 1979, pp. 83–105

Walters KW, Prinslow TE. 1975. Culture of the

mangrove oyster *Crassostrea rhizophorae* Guilding, in Puerto Rico. *Proc. W. Maricul. Soc.* 6:221–36

Wasson K, Zabin CJ, Bediger L, Diaz CM, Pearse JS. 2001. Biological invasions of estuaries without international shipping: The importance of intraregional transport. *Biol. Conserv.* 102:143–53

Wells HW. 1961. Fauna of oyster beds, with special reference to salinity factor. *Ecol. Monogr.* 31:239–66

Went AEJ. 1962. Historical notes on the oyster fisheries of Ireland. *Proc. R. Irish Acad.* 62:195–223

White ME, Powell EN, Kitting CL. 1985. The ectoparasitic gastropod *Boonea* (=*Odostomia*) *impressa* (Say): Distribution, reproduction, and the influence of parasitism on oyster growth rates (Abstr.). *J. Shellfish Res.* 5:43–44

Wilson EA, Powell EN, White ME, Ray SM. 1988. The effect of the ectoparasitic snail, *Boonea impressa* on oyster growth and health in the field with comments on patch forma-

tion in snail populations (Abstr.). *J. Shellfish Res.* 7:137–38

Woelke CE. 1955. Introduction of the Kumamoto oyster *Ostrea* (*Crassostrea*) *gigas* to the Pacific coast. *Fish. Res. Pap., Wash. Dep. Fish.* 1:1–10

Wolff WJ, Reise K. 2002. Oyster imports as a vector for the introduction of alien species into Northern and Western European coastal waters. In *Invasive aquatic species of Europe. Distribution, impacts and management*, ed. E Lappakoski, S Gollasch, S Olenin, pp. 193–205. Dordrecht: Kluwer Acad.

Wood BP, Fraser DI. 1996. *Bonamiasis*. Aberdeen: SOAEFD. 5 pp.

Yamada SB. 1993. Predation by the crab, *Cancer oregonensis* Dana, inside oyster trays. *J. Shellfish Res.* 12:89–92

Yonge CM. 1960. *Oysters*. London, UK: Collins. 209 pp.

Zibrowius H. 1992. Ongoing modification of the Mediterranean marine fauna and flora by the establishment of exotic species. *Mésogée* 51:83–107

Subject Index

A

Aging
  mitochondrial DNA
    (mtDNA) and, 627
Agriculture
  coevolutionary approach
    to, 566
  defining, 564
  human
    crop ecology and
      evolution in, 570–73
    lessons from insects for,
      587–89
  insect, 563–89
    disease management
      strategies, 583–89
    evolutionary origins of,
      569–75
    microbial management
      strategies, 583–89
    role of disease in,
      582–83
  niche conservatism and,
    529–31, 533, 535
Akaike information criterion
  (AIC)
  phylogenetic models and,
    455–57, 460–61
Allopatric speciation
  niche conservatism and,
    521–23, 533, 535
Amazon molly, 401–12
Androdioecy
  pollen limitation and, 470
Androgenesis, 400
Ant(s)
  adaptations of, 352–53
  agricultural behaviors of,
    565
  agricultural evolution and,
    575–82

agriculture, 566–67, 575
  crop ecology and
    evolution in, 570–73
  disease ecology and
    evolution in, 576–77
ant-aphid interactions,
  349–51
  dynamic context,
    360–63
  ecological framework
    and, 353–56
  evolutionary constraints
    on, 357–60
  frontiers in, 363–65
  plants and, 359–63
  processes shaping,
    352–60
ant-aphid relationships,
  345–65
  features of partners,
    346–48
  fungiculture, 567
  fungus-growing, 563, 566
  mulattos and, 348–49
Antagonism
  ant-aphid interactions and,
    350–51
Antibiotic defenses
  insect agriculture and,
    585–87
Antigen recognition
  polymorphism
  immunopathology and,
    388–89
Aphid(s)
  adaptations of, 352–53
  ant-aphid interactions,
    349–51
    dynamic context,
      360–63
    ecological framework

and, 353–56
evolutionary constraints
  on, 357–60
frontiers in, 363–65
plants and, 359–63
processes shaping,
  352–60
ant-aphid relationships,
  345–65
features of partners,
  346–47
mutualism and, 348–49
Aquatic systems
  multitrophic interactions,
    99
Asexual reproduction
  pollen limitation and, 488
Asexuality
  advantages of, 399, 400
Asymmetry
  fluctuating, 1–15
ATP
  mitochondrial DNA
    (mtDNA) and, 622–24
Automatic selection
  plants and, 48–49
Avoidance
  plant secondary
    metabolites (PSMs) and,
    170–73, 180

B

Bateman's principle, 125,
  126, 140
Bayesian information
  criterion (BIC)
  phylogenetic models and,
    457–58
Bayesian model selection
  phylogenetic models and,
    457–58, 461

691

CUMULATIVE INDEXES

CONTRIBUTING AUTHORS, VOLUMES 32–36

CHAPTER TITLES, VOLUMES 32–36

Volume 32 (2001)

Volume 33 (2002)

Volume 34 (2003)